AF886134

Alfons Mersmann • Matthias Kind • Johann Stichlmair

Thermal Separation Technology

Alfons Mersmann • Matthias Kind • Johann Stichlmair

Thermal Separation Technology

Principles, Methods, Process Design

Springer

Prof. Dr.-Ing. Alfons Mersmann
Kolumbusstraße 5 b
81543 München
Germany
alfons.mersmann@mnet-online.de

Prof. Dr.-Ing. Matthias Kind
Karlsruher Institut für
Technologie (KIT)
Institut für Thermische
Verfahrenstechnik
Kaiserstraße 12
76131 Karlsruhe
Germany
matthias.kind@kit.edu

Prof. Dr.-Ing. Johann Stichlmair
TU München
Lehrstuhl für
Fluidverfahrenstechnik
Boltzmannstr. 15
85747 Garching
Germany
johann.stichlmair@fvt.mw.tum.de

ISBN 978-3-642-12524-9 e-ISBN 978-3-642-12525-6
DOI 10.1007/978-3-642-12525-6
Springer Heidelberg Dordrecht London New York

Library of Congress Control Number: 2011933560

© Springer-Verlag Berlin Heidelberg 2011
This work is subject to copyright. All rights are reserved, whether the whole or part of the material is concerned, specifically the rights of translation, reprinting, reuse of illustrations, recitation, broadcasting, reproduction on microfilm or in any other way, and storage in data banks. Duplication of this publication or parts thereof is permitted only under the provisions of the German Copyright Law of September 9, 1965, in its current version, and permission for use must always be obtained from Springer. Violations are liable to prosecution under the German Copyright Law.
The use of general descriptive names, registered names, trademarks, etc. in this publication does not imply, even in the absence of a specific statement, that such names are exempt from the relevant protective laws and regulations and therefore free for general use.

Cover design: WMXDesign GmbH, Heidelberg

Printed on acid-free paper

Springer is part of Springer Science+Business Media (www.springer.com)

Preface

Thirty years ago the first edition of this book was published in German. The concept of the work – at that time printed in leaden letters – was a cheap information source for students. Today our world is characterized by high-speed developments and more principal approaches in comparison to the past. Therefore, the authors put the question: Is it reasonable in the Internet age to write scientific books and – if yes – for whom and in which language?

This book is the answer to these questions. We hope that students as well as experienced engineers and scientists may find access to the representation of the topic and may gain from the lecture of the work. According to the concept of the book it is assumed that the book is read from the beginning to the end. However, we know very well that there are no such readers. The book may be a helpful tool for beginners and experts for consulting and for deepening their knowledge. In spite of the fact that the concept of a large number of unit operations is widely abandoned, the overall arrangement of the book corresponds to this structure because there are not only intelligent and abstractively thinking readers but also others with a more practical approach to problems of chemical engineering. The contents of this book are too extensive for students with time limitations for their studies but not sufficient for specialists.

The notation is a compromise between internationally and nationally applied symbols. With respect to the literature cited, only publications necessary for a deeper or more general study of the topic are mentioned.

Let us come back to the question: Is it reasonable to publish books in the modern world of communication and information? Since the times of Gutenberg, books have been printed and read. Two centuries ago Johann Wolfgang von Goethe wrote:

"*what you possess in black on white you can carry home.*"

However, Goethe could not have any idea of computers, copy machines, and Internet. If so, he probably would write:

"*Secure is what is stored in electronic brains as many dots,*
but often not in human brains with empty spots."

We think that not the persons believing in computers but thinking persons who are looking on black on white are in the position of getting access to problems of separation technology.

Munich

A. Mersmann
M. Kind
J. Stichlmair

Symbols

A		Constant
A	m^2	Area, exchange area, interfacial area
a	m^2/m^3	Volumetric interfacial area
a	m	Amplitude
a		Activity
B	kmol	Bottom fraction, bottom product
$\dot{B}$	kmol/s	Bottom product rate
b	m	Width
C_s	$W/(m^2 \times K^4)$	Radiation constant
c	kg/m^3	Mass concentration
$\tilde{c}$	$kmol/m^3$	Molar concentration
c_p	$kJ/(kg \times K)$	Specific heat capacity at constant pressure
$\tilde{c}_p$	$kJ/(kmol \times K)$	Molar heat capacity at constant pressure
c_v	$kJ/(kg \times K)$	Specific heat capacity at constant volume
$\tilde{c}_v$	$kJ/(kmol \times K)$	Molar heat capacity at constant volume
c_w		Friction factor
D	kmol	Distillate, overhead fraction
$\dot{D}$	kmol/s	Distillate rate
D	m	Diameter (apparatus)
D	m^2/s	Dispersion coefficient
D_{AB}, D_{ij}	m^2/s	Molecular diffusion coefficient
d	m	Diameter (small, sphere, particle, tube, stirrer)
d_h	m	Hydraulic diameter
d_p	m	Particle diameter
d_{32}	m	Sauter diameter ($d_{32} = \Sigma n \cdot d^3 / \Sigma n \cdot d^2$)
E	kJ	Energy
E		Efficiency, enhancement factor
E	kg	Extract
$\dot{E}$	kg/s	Extract rate
e	$A \times s$	Charge of an electron ($e = 1.06 \times 10^{-19} A \times s$)
F	N	Force
F	kmol	Feed
$\dot{F}$	kmol/s	Feed rate
F	kJ	Free inner energy
$\tilde{f}_i$	kJ/kmol	Partial molar free inner energy
f		Degree of freedom

f	Pa	Fugacity
f	m^2	Cross-sectional area
f	1/s	Frequency
G	kmol	Gas, G-phase
$\dot{G}$	kmol/s	Gas rate, G-phase rate
G	kJ	Free enthalpy, Gibbs enthalpy
g	kJ/kg	Specific free enthalpy
$\tilde{g}$	kJ/kmol	Molar free enthalpy
$\tilde{g}_i$	kJ/kmol	Partial molar free enthalpy
H	m	Total height, distance
h	m	Height
H	kJ	Enthalpy
h	kJ/kg	Specific enthalpy
$\tilde{h}$	kJ/kmol	Molar enthalpy
$\tilde{h}_i$	kJ/kmol	Partial molar enthalpy
$\Delta\tilde{h}_i$	kJ/kmol	Partial molar enthalpy of mixing
$\Delta\tilde{h}_{iB}$	kJ/kmol	Partial molar enthalpy of bonding
Δh	kJ/kg	Specific phase change enthalpy, latent heat of evaporation
$\Delta\tilde{h}$	kJ/kmol	Molar phase change enthalpy
I	kg×m/s	Momentum
K		Equilibrium ratio, equilibrium constant
k		Number of components
k	kJ/K	Boltzmann constant ($k = 1.38 \times 10^{-26}$ kJ/K)
k	$W/(m^2 \times K)$	Overall heat transfer coefficient
k	$kmol/(m^2 \times s)$	Molar mass transfer coefficient
k_d	m/s	Mass transfer coefficient
L	m	Total length
l	m	Length
L	kmol	Liquid (L-phase)
$\dot{L}$	kmol/s	Liquid rate (L-phase rate)
M	kg	Mass
$\dot{M}$	kg/s	Mass rate
$\tilde{M}$	kg/kmol	Molar mass
$\dot{m}$	$kg/(m^2 \times s)$	Mass flux
N		Number (transfer units, molecule layers)
N	$1/m^3$	Particles per volume
N_A	1/kmol	Avogadro constant or Loschmidt constant ($N_A = 6.022 \times 10^{26}$ 1/kmol)
N	kmol	Amount of substance
$\dot{N}$	kmol/s	Substance rate
$\dot{n}$	$kmol/(m^2 \times s)$	substance flux
n		Number (stages, particles)
n	1/s	Rate of revolutions

P	W	Power
p		Number of phases
p	Pa	Total pressure
p_c	Pa	Critical pressure
p_i	Pa	Partial pressure of component i
p_i^o	Pa	Vapor pressure of component i
p_r		Reduced pressure
Δp	Pa	Pressure difference, pressure loss
Q	kJ	Heat
$\dot{Q}$	kJ/s = kW	Heat rate
$\dot{q}$	W/m^2	Heat flux
R	m	Radius (tube, sphere, particle)
R		Reflux ratio
R	kg	Raffinate
$\dot{R}$	kg/s	Raffinate rate
$\tilde{R}$	kJ/(kmol×K)	General gas constant ($\tilde{R}$ = 8.314 kJ/(kmol×K))
R_i	kJ/(kg×K)	Specific gas constant of component i ($R_i = \tilde{R}/\tilde{M}_i$)
r	m	Radius
r	kJ/kg	Specific heat of evaporation
$\tilde{r}$	kJ/kmol	Molar heat of evaporation
S	kg	Solid
$\dot{S}$	kg/s	Solid rate
S	kJ/K	Entropy
s	kJ/(kg×K)	Specific entropy
$\tilde{s}$	kJ/(kmol×K)	Molar entropy
s_i	kJ/(kmol×K)	Partial molar entropy of component i
s	m	Thickness
T	$kg \times m^2/s^2$	Torque
T	kg	Carrier substance
$\dot{T}$	kg/s	Carrier substance rate
T	K	Absolute temperature
T_b	K	Boiling temperature at normal pressure
T_c	K	Critical temperature
T_r		Reduced temperature
T_s	K	Melting temperature
t	s	Time
U	kJ	Inner energy
u	kJ/kg	Specific inner energy
$\tilde{u}$	kJ/kmol	Molar inner energy
$\tilde{u}_i$	kJ/kmol	Partial molar inner energy of component i
u	m/s	Velocity (x-coordinate), superficial velocity
V	m^3	Volume
$\dot{V}$	m^3/s	Volumetric flow rate

$\dot{v}$	$m^3/(m^2 \times s)$	Volumetric flux
v	m/s	Velocity (y-coordinate)
v	m^3/kg	Specific volume
$\tilde{v}$	$m^3/kmol$	Molar volume
$\tilde{v}_i$	$m^3/kmol$	Partial molar volume
W	kJ	Work
$\dot{W}$	W	Power (=work per time)
w	kJ/kg	Specific work
w	m/s	Velocity (z-coordinate)
w_o	m/s	Velocity in an opening (hole, orifice, nozzle)
w_s	m/s	Terminal settling or rising velocity of a single particle
w_{ss}	m/s	Terminal settling or rising velocity of a swarm of particles
X_i	kg/kg	Mass loading of component i (adsorbate, liquid, raffinate)
$\tilde{X}_i$	kmol/kmol	Mole loading of component i (adsorbate, liquid, raffinate)
x_i	kg/kg	Mass fraction of component i (adsorbate, liquid, raffinate)
$\tilde{x}_i$	kmol/kmol	Mole fraction of component i (adsorbate, liquid, raffinate)
x	m	Rectangular coordinate
Y_i	kg/kg	Mass loading of component i (adsorptive, gas, extract)
$\tilde{Y}_i$	kmol/kmol	Mole loading of component i (adsorptive, gas, extract)
y_i	kg/kg	Mass fraction of component i (adsorptive, gas, extract)
$\tilde{y}_i$	kmol/kmol	Mole fraction of component i (adsorptive, gas, extract)
y	m	Rectangular coordinate
Z		Compressibility factor
z	m	Rectangular coordinate
z_i	kg/kg	Overall mass fraction in a multiphase system
$\tilde{z}_i$	kmol/kmol	Overall mole fraction in a multiphase system

Greek Symbols

α		Relative volatility
α		Degree of dissociation
α	$W/(m^2 \times K)$	Heat transfer coefficient
β	m/s	Mass transfer coefficient
β_h	m/s	Mass transfer coefficient (semipermeable interface)
β	1/K	Cubical expansion coefficient

γ		Activity coefficient
δ	m	Thickness, film thickness
ε	W/kg	Specific power input
ε		Voidage, porosity
ε_c		Volume fraction of continuous phase
ε_d		Volume fractioin of dispersed phase
η	Pa×s	Dynamic viscosity
ϑ	°C	Celsius temperature
Λ	m	Mean path length
λ	W/(m×K)	Heat conductivity
μ	kJ/kmol	Chemical potential
ρ	kg/m^3	Density
ρ	$kmol/m^3$	Molar density
σ	$J/m^2 = N/m$	Surface or interfacial tension
τ	Pa	Shear stress
τ	s	Residence time
φ		Fugacity coefficient
φ		Relative saturation
φ		Relative free area, volume fraction
ω	1/s	Angular velocity

Indices

A	Area
a	Activity
ads	Adsorption
agg	Agglomeration
anh	Anhydrate
AS	Avoidance of settling
at	Atom
ax	Axial
BCF	Burton–Cabrera–Frank
BL	Bottom lifting
B + S	Birth and spread
b	Boiling
c	Crystal, critical
circ	Circulation
col	Collision
dif	Diffusion
dis	Disruption, dispersion
eff	Effective
for	Foreign
G	Gas
g	Geometrical

h	Semipermeable
het	Heterogeneous
hom	Homogeneous
hyd	Hydrate
I	Interface
i	Component *i*
id	Ideal
L	Liquid
lam	Laminar
m	Molecule, molar
macro	Macro
max	Maximum value
micro	Micro
min	Minimum
opt	Optimum
p	Particle
PN	Poly nuclear
rel	Relative
S	Solid
s	Settling, seed
ss	Settling of a swarm
sus	Suspension
T	Total
tot	Total
turb	Turbulent
V	Vickers, volume
vdW	van der Waals
w	Velocity, superficial velocity
α	Start
ω	End

Dimensionless Numbers

Single and Two-Phase Flow

$$Ar = \frac{d^3 \cdot g \cdot \rho_c \cdot \Delta\rho}{\eta_c^2} = \frac{Re^2}{Fr^*}$$ Archimedes Number

$$Eu = \frac{\Delta p}{\rho \cdot w^2}$$ Euler number

$$Fo_E = \frac{D_E \cdot t}{L^2}$$ Dispersion or Fourier number

$$Fr = \frac{w^2}{d \cdot g}$$ Froude number

$Fr^* = \dfrac{w^2 \cdot \rho_c}{d \cdot g \cdot \Delta\rho}$	Modified Froude number in two-phase systems
$Ga = \dfrac{d^3 \cdot \rho^2 \cdot g}{\eta^2} = \dfrac{Re^2}{Fr}$	Galilei number
$Re = \dfrac{w \cdot d \cdot \rho}{\eta}$	Reynolds number
$We = \dfrac{w^2 \cdot d \cdot \rho}{\sigma}$	Weber number
$\dfrac{We}{Fr} = \dfrac{d^2 \cdot \rho \cdot g}{\sigma}$	Bond number (in US literature)
$\dfrac{We}{Fr^*} = \dfrac{d^2 \cdot \Delta\rho \cdot g}{\sigma}$	Modified number in two-phase systems
$Sr = \dfrac{L \cdot f}{w}$	Strouhal number

Heat and Mass Transfer

$Fo = \dfrac{\lambda \cdot t}{\rho \cdot c \cdot s^2}$	Fourier number (heat transfer)
$Fo^* = \dfrac{D \cdot t}{s^2}$	Fourier number (mass transfer)
$Gr = \dfrac{L^3 \cdot g \cdot \Delta\vartheta \cdot \rho^2 \cdot \beta}{\eta_c^2}$	Grashof number (heat transfer)
$Le = \dfrac{\lambda}{\rho \cdot D \cdot c}$	Lewis number
$Nu = \dfrac{\alpha \cdot L}{\lambda}$	Nusselt number
$Pe = \dfrac{w \cdot L \cdot \rho \cdot c}{\lambda}$	Peclet number
$Ph = \dfrac{c_L \cdot (\vartheta_G - \vartheta_O)}{\Delta h_{LG}}$	Phase change number
$Pr = \dfrac{\eta \cdot c}{\lambda}$	Prandtl number
$Sc = \dfrac{\eta}{\rho \cdot D}$	Schmidt number
$Sh = \dfrac{\beta \cdot L}{D}$	Sherwood number
$Bo_c = \dfrac{w_c \cdot H}{(1-\varepsilon_d) \cdot D_{ax,c}}$	Bodenstein number (continuous phase)

$Bo_{\mathrm{d}} = \dfrac{w_{\mathrm{d}} \cdot H}{\varepsilon_{\mathrm{d}} \cdot D_{\mathrm{ax,\,d}}}$	Bodenstein number (dispersed phase)
$B^* = \left(\dfrac{\eta_{\mathrm{L}}}{\rho_{\mathrm{L}} \cdot g^2}\right)^{1/3} \cdot \dfrac{\varepsilon_{\mathrm{d}} \cdot \dot{v}_{\mathrm{L}}}{\varepsilon_{\mathrm{c}} \cdot d_{\mathrm{p}}}$	Dimensionless irrigation rate
$d^*_{\mathrm{p,max}} = d_{\mathrm{p,\,max}} \cdot \left(\dfrac{\rho_{\mathrm{c}}^3 \cdot \varepsilon^2}{\sigma^3}\right)^{1/5}$	Maximum diameter of fluid particles in turbulent field
$Fl = \dfrac{\sigma^3 \cdot \rho_{\mathrm{c}}^2}{\eta_{\mathrm{c}}^4 \cdot \Delta\rho \cdot g}$	Fluid number of dispersed systems
$K_{\mathrm{F}} = \dfrac{\sigma^3 \cdot \rho^2}{\eta^4 \cdot g}$	Film number
$K_{\mathrm{w}} = \zeta \cdot \dfrac{\rho_{\mathrm{G}}}{\rho_{\mathrm{L}}} \cdot \dfrac{w_{\mathrm{G}}^2}{(g \cdot \eta_{\mathrm{L}}/\rho_{\mathrm{L}})^{2/3}}$	Gas-film number
$l_{\mathrm{s}}^* = l_{\mathrm{s}} \cdot \left(\dfrac{\rho_{\mathrm{c}}^3 \cdot \varepsilon}{\eta_{\mathrm{c}}^3}\right)^{1/4}$	Microscale of turbulence
$Ne = \dfrac{P}{\rho \cdot n^3 \cdot d^5}$	Newton number of stirrers
$N_{\mathrm{v}} = \dfrac{\dot{V}_{\mathrm{circ}}}{n \cdot d^3}$	Flow number of stirrers
$\mathrm{NTU} = \int \dfrac{\mathrm{d}y}{y^* - y}$	Number of transfer units
$n \cdot t$	Mixing time number
$t^*_{\mathrm{macro}} = t_{\mathrm{macro}} \cdot \left(\dfrac{\bar{\varepsilon}}{D^2}\right)^{1/3}$	Macromixing time
$t^*_{\mathrm{micro}} = t_{\mathrm{micro}} \cdot \left(\dfrac{\rho_{\mathrm{L}} \cdot \varepsilon}{\eta_{\mathrm{L}}}\right)^{1/2}$	Micromixing time
$\dot{v}_{\mathrm{d}}^* = \dot{v}_{\mathrm{d}} \cdot \left(\dfrac{\rho_{\mathrm{c}}^2}{\sigma \cdot \Delta\rho \cdot g}\right)^{1/4}$	Dimensionless flow density in bubble and drop columns
$\eta = \dfrac{X - X^*}{X_{\mathrm{Kn}} - X^*}$	Dimensionless humidity

Contents

1 Introduction

Starting with natural as well as chemically or biologically produced substances the separation of mixtures is an important task for chemical engineers. This discipline can be subdivided into thermal, mechanical, chemical, biological, and biomedical processes to effect changes of substances or to avoid such changes. Today, an additional objective of thermal separation technology is process engineering and product development.

A general problem of thermal separation technology is the identification of the most economical process for the separation of mixtures into pure components or specified fractions. The separation principles are based on:

- Differences of vapor pressures of the components (e.g., evaporation, condensation, distillation, rectification, drying)
- Differences of solubilities (e.g., extraction, reextraction, crystallization, absorption, desorption)
- Differences of sorption behavior (e.g., adsorption, desorption, chromatography, drying)
- Differences of forces (mechanical, capillary, electric, magnetic) in porous solids or membranes (e.g., dialysis, electrolysis, electrophoresis, osmosis, reverse osmosis, pervaporation, ultrafiltration)
- Differences in chemical equilibria (e.g., chemisorption, ion exchange)

As a rule, the addition of substances to a mixture is not helpful. However, there are exceptions, for instance, the addition of entrainers, desorbents, drowning-out agents, and solvents.

In industry, separations effected by porous solid materials (adsorbents, membranes, ion exchange resins, special kinds of solid matrices) are seldom applied in comparison to separations carried out in systems with fluid phases. The reason may be a poorer state of the art than in fluid systems. Furthermore, the handling of fluid systems is much easier than the handling of systems with solid phases. On the other

A. Mersmann et al., *Thermal Separation Technology: Principles, Methods, Process Design*, VDI-Buch, DOI 10.1007/978-3-642-12525-6_1,
© Springer-Verlag Berlin Heidelberg 2011

hand, there is a great potential of microporous material with respect to innovative and tailor-made membranes. The volumetric mass transfer coefficients are in the same order of magnitude in fluid/fluid and fluid/solid systems. Since, at the present state of the art, fluid/fluid systems dominate separation technology, they are preferentially presented in this book.

It is common practice to divide separation technologies into the unit operations absorption, adsorption, crystallization, distillation, rectification, extraction, evaporation, drying, etc. Such an approach is no longer recommended because the reader would have difficulties to recognize common and general items and laws of all these separation techniques. Our intention is to stress common features and methods and to demonstrate how to proceed in solving a concrete separation problem.

The following complexes are the basis of all models in chemical engineering:

- Chemical and physical equilibria (or more general, thermodynamic behavior of substances)
- Conservation laws of energy, mass, components, and numbers
- Single and multiphase flow (or, in general, the conservation law of momentum)
- Kinetics of changes in systems which are not at equilibrium (heat transfer, mass transfer, chemical reaction)

If the underlying laws of these complexes are well understood then separation processes can be developed and the equipment dimensioned. However, with respect to special features of separation technology, a gross subdivision in unit operations is indeed reasonable. Therefore, some parts of the book are structured in unit operations according to the principle "from the molecule to the plant."

Methods of conceptual process design are a prerequisite for the combination of process steps to an integrated process which is economically and ecologically efficient and meets modern safety standards. Special attention is paid to the question how high product qualities can be achieved. This is especially important for products produced in crystallization and drying steps. Furthermore, the purity of distillates or off-gases purified by absorption or desorption processes can be decisive in process development. The authors are aware that computer-aided process simulation is widely applied in industry. In this book tools for this approach are presented. The models are based on general laws of natural sciences and simplified but tolerable engineering methods.

It is important to note that in biological or chemical laboratories, where novel products are developed, the aspects of separation technology should be taken into consideration since in many cases waste streams of worthless side products, large recirculation streams, and mixture which are difficult to separate (water or azeotropic mixtures) can be reduced or avoided from the very beginning. Often, there is a good chance to simplify the process and to save energy and raw material.

Many examples of energy saving in thermal separation processes are presented in the last chapter of this book. In addition, energy saving is a general task for engineers with respect to the carbon dioxide problem. Therefore, the first pages of this book deal with process technologies for mitigation of carbon dioxide emissions into the atmosphere.

With respect to process optimization, at first high-efficient and selective catalysts in a general sense (e.g., chemical and biological catalysts, algae, bacteria, fungi, and mammal cells viruses, yeasts) are very important. An optimum process is characterized by minimum consumption of resources (energy, raw material) and minimum production of waste (heat, materials without any chance on the market). Safety aspects play a big role. Another goal of chemical engineering is the anticipation of possible improvements of the process during its production life. However, this is a difficult task in times of hectic simultaneous process development.

1.1 Contributions of Chemical Engineering to the Carbon Dioxide Problem

The main concern of the climate discussion in recent years was the anthropogenic increase of carbon dioxide in the atmosphere. Let us have a look on the global energy supply, see Fig. 1.1-1. The circles at the top and in the upper row mark primary renewable energies (i.e., sun, water, wind, biogas, and biomass) whereas primary energies from the earth crust (i.e., fossil combustibles, radioactive ores, geothermal heat, etc.) can be found in the circles of the lower row. Nearly all energies are consumed as thermal energy, mechanical energy, or electrical energy, finally dissipated as thermal energy in the environment. These secondary energies can be found in the circles of the intermediate row of Fig. 1.1-1.

Primary energy can be converted into secondary energy by energy converters like water or wind turbines, solar cells, sun collectors, furnaces, combustion engines, gas and vapor turbines, generators, fuel cells, nuclear reactors, and heat pumps. Arrows mark such converters. Their maximum efficiencies, e.g., how much of the

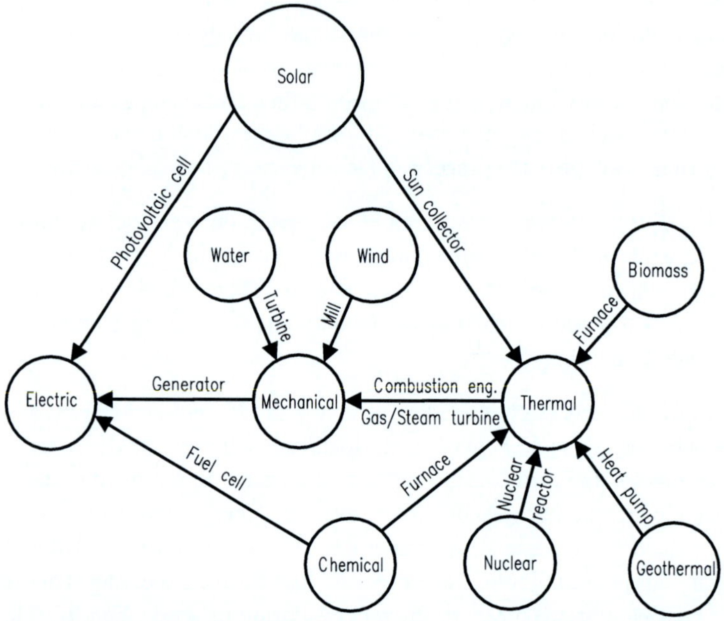

Fig. 1.1-1 Global energy supply

energy is converted from one state into the other, are between 0.95 and 0.99 valid for electrical generators and 0.05 up to 0.18 valid for solar cells.

There are many possibilities to reduce carbon dioxide emissions caused by fossil combustibles as carbon or hydrocarbons:

- Replacement of fossil combustibles by other energy sources, especially by renewable energies
- Recovery of waste heat (e.g., heat/power generation)
- Higher efficiencies of the energy converters mentioned in Fig. 1.1-1
- Lower energy losses of energy transportation systems
- Higher energy efficiency of equipment used by consumers (industry, traffic, household, etc.)

Since process industry is very energy consuming much information is given in this book about how to save or recover energy.

In addition, two further possibilities to avoid carbon dioxide emissions are discussed:

- Sequestration of carbon dioxide as compressed gas or liquid
- Cracking of methane or other hydrocarbons, combustion of hydrogen, and sequestration of carbon

Let us have a short look on the energy efficiencies of these processes.

In Fig. 1.1-2 simple schemes of feasible processes starting from methane (25 mass% hydrogen content) or carbon (no hydrogen content) as fossil combustibles are burnt with oxygen or air (the mass percentage of hydrogen of all hydrocarbons is between 0 and 25%). Note that the production of oxygen requires the separation of air. In Fig. 1.1-3 the cracking of methane, the sequestration of carbon, the compression of hydrogen (up to 70 MPa), and the liquefaction of hydrogen at a temperature of –254°C are considered in four processes. Hydrogen combustion may be promising for all kinds of vehicles. Chemical reactions and heats of reaction are

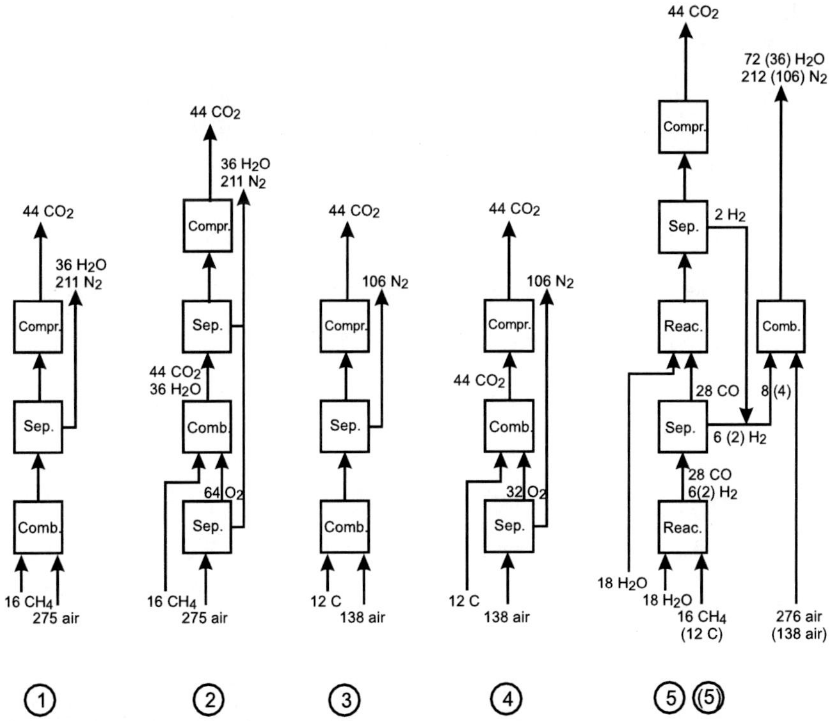

Fig. 1.1-2 Schemes of six combustion processes (separation of CO_2): Comb(ustion), Compr(ession), Reac(tion), Sep(aration)

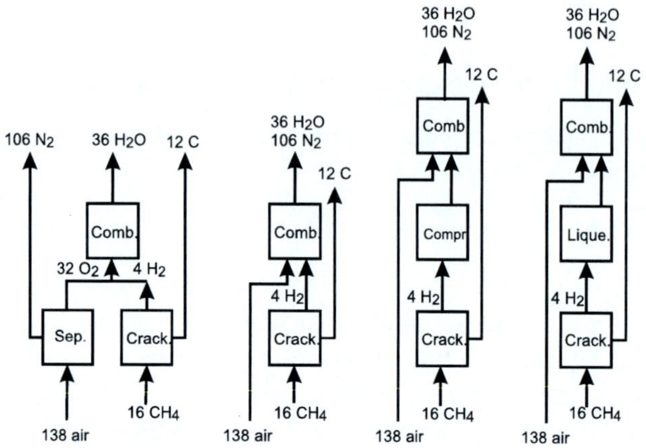

Fig. 1.1-3 Schemes of four combustion processes (sequestration of carbon): Crack(ing)

listed in Table 1.1-1. The vapor pressure of carbon dioxide vs. temperature is shown in Fig. 1.1-4. Carbon dioxide has to be separated from flue gases for instance by absorption or adsorption processes when the combustion is performed with air.

Calculations elucidate the minimum energy demand for the separation and compression of carbon dioxide based on the minimum separation energy of gaseous mixtures (flue gases) and on the adiabatic compression of carbon dioxide (up to 10 MPa). These minimum energy demands related to the heat of combustion of the combustibles (C or CH_4) are for

- The processes 1 to 4 approximately 7–8%
- The processes 5 and (5) approximately 21–23%

Note that, as a rule, compressors are driven by electrical engines, and efficiencies of power stations are below 0.5 (conversion of chemical into electrical energy). Using such electrical engines the above minimum percentages of the processes 1–4 have approximately to be doubled.

Let us now have a short look on the energetic efficiency of the processes depicted in Fig. 1.1-3 for the production of either compressed (70 MPa) or liquefied hydrogen produced by the cracking of methane. In the following, all energies are valid for 1 kmol = 16 kg methane or 2 kmol = 4 kg hydrogen (heat of combustion = –499,400 kJ) or 1 kmol = 12 kg carbon (heat of combustion = –395,800 kJ). The cracking of of 1 kmol methane (heat of combustion = –803,500 kJ) requires 91,700 kJ, and for

Table 1.1-1 Chemical reactions

	CH_4	$\underset{}{\overset{1{,}000°C}{\rightleftarrows}}$	$C + 2H_2$	$+91.7\frac{kJ}{mol\ CH_4}$
CH_4	$+2O_2$	$\overset{1{,}000°C}{\rightleftarrows}$	$CO_2 + 2H_2O$	$-803.5\frac{kJ}{mol\ CH_4}$
C	$+O_2$	$\overset{1{,}000°C}{\rightleftarrows}$	CO_2	$-395.8\frac{kJ}{mol\ C}$
H_2	$+\frac{1}{2}O_2$	$\overset{1{,}000°C}{\rightleftarrows}$	H_2O	$-249.7\frac{kJ}{mol\ H_2}$
C	$+H_2O$	$\rightleftarrows$	$CO + H_2$	$+131\frac{kJ}{mol\ C}$
CO	$+H_2O$	$\rightleftarrows$	$CO_2 + H_2$	$-41\frac{kJ}{mol\ CO}$
CH_4	$+H_2O$	$\rightleftarrows$	$CO + 3H_2$	$+206.3\frac{kJ}{mol\ CH_4}$
C	$+\frac{1}{2}O_2$	$\rightleftarrows$	CO	$-110.6\frac{kJ}{mol\ C}$
CH_4	$+\frac{1}{2}O_2$	$\rightleftarrows$	$CO + 2H_2$	$-35.7\frac{kJ}{mol\ CH_4}$
CO_2	$+C$	$\rightleftarrows$	2CO	$+173\frac{kJ}{mol\ CO_2}$

the compression of 4 kg hydrogen (up to 70 MPa) the energy demand is as high as 72,000 kJ (Romm, 2006; Kreysa, 2008). Since carbon is sequestered it is not considered here. Therefore, – 499,400 kJ + 91,700 kJ + 72,000 kJ = – 335,700 kJ are obtained by the combustion of 4 kg compressed hydrogen. This means that nearly 70% of the chemical energy of methane is either not utilized (sequestration of carbon) or necessary for cracking and compression.

If compression of hydrogen is replaced by liquefaction (+50,000 kJ/kg_{H_2}), approximately 85% of the heat of combustion of methane is lost as discarded carbon and required for cracking of methane and liquefaction of hydrogen.

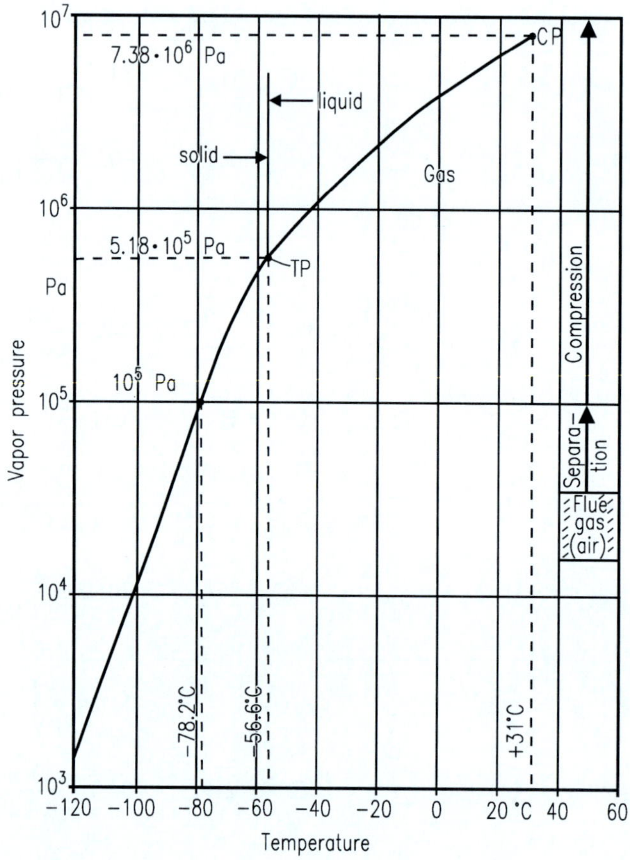

Fig. 1.1-4 Vapor pressure of CO_2 vs. temperature

After this disillusioning considerations let us have a short look on concerns and restrictions encountered in the area of primary energy sources:

- *Fossil combustibles.* Emissions of carbon dioxide or sequestration of carbon dioxide (availability of secure and tight caverns) or carbon.
- *Nuclear fuels.* Security and secure disposal of nuclear waste.
- *Geothermal heat.* Upper limit of operating temperature and large heat transfer areas because of heat flux densities $< 10^{-4}\ \mathrm{kW/m^2}$
- *Biogas and biomass.* In the case of energy plant production competition with human food supply.

- *Solar energy*. With respect to solar radiation of $<0.2\ \mathrm{kW/m^2}$, large absorption areas (photovoltaic) or large mirrors (solar power station) are necessary.
- *Water*. With the power $P \approx 0.9 \cdot \dot{V} \cdot H \cdot \rho_{\mathrm{L}} \cdot g$, high flow rates $\dot{V}$ or height H of water column are required. Availability of reservoirs and rivers.
- *Wind*. According to $P \approx 0.3 \cdot \rho_{\mathrm{G}} \cdot v^3 \cdot D^2$ and given wind velocities of $v \approx 10\ \mathrm{m/s}$, large diameters D (>100 m) are necessary. Great differences of local and momentary wind velocities.

Note that the storage of energy is difficult and the locations of energy offer and that of energy demand can be rather distant. This requires high efficiencies of transportation systems and much international cooperation in the future.

2 Thermodynamic Phase Equilibrium

In this chapter the thermodynamic behavior of single- and multiphase systems of pure substances and their mixtures are described in a general way. In the References section some general textbooks and data compilations are recommended.

Multiphase systems are often found in machinery and apparatuses of the processing industry because most thermal separation processes are based on the transfer of one or more components from one phase to another.

A phase is the entirety of regions, where material properties either do not change or only change continually, but never change abruptly. However, it makes no difference whether the regions are spatially coherent or not (continuous or dispersed phase). A phase can consist of one or more chemically uniform substances, which are called components. A system can contain one phase (gas, liquid, solid), two phases (e.g., liquid/gas, fluid/solid, fluid/fluid), or even more (in an evaporative crystallizer, e.g., there are a solid, a liquid and a gaseous phase) This chapter describes the thermodynamic equilibrium between phases.

Gibbs phase-rule states how many degrees of freedom f fully describe a multicomponent and multiphase system:

$$f = k - p + 2 . \tag{2.0-1}$$

Here k is the number of components and p is the number of phases. For liquid water, f equals 2, because the water's state is fully described by stating its pressure and temperature. For saturated steam, however, which is a liquid/gas system, statement of either pressure or temperature is sufficient. At the triple point there are three coexistent phases: one solid, one liquid, and one gaseous. The system is exactly defined at this point. When dealing with a two-component system or a binary mixture the degree of freedom increases by one. Then more information, e.g., the concentration, is necessary to fully describe the system.

The description of the thermodynamical equilibrium is based upon the laws of thermodynamics. The first law of thermodynamics is the law of conservation of energy. For a resting, closed, and nonreacting system, only the conservation of the internal energy u has to be considered. It can only be changed by transferring heat dq and work dw across the system boundaries:

A. Mersmann et al., *Thermal Separation Technology: Principles, Methods, Process Design*, VDI-Buch, DOI 10.1007/978-3-642-12525-6_2,
© Springer-Verlag Berlin Heidelberg 2011

$$\mathrm{d}u = \mathrm{d}q + \mathrm{d}w\,.$$ [1] (2.0-2)

Work dw is transferred across the system boundary by compressing the volume v of the system against its pressure p:

$$\mathrm{d}u = \mathrm{d}q - p \cdot \mathrm{d}v \tag{2.0-3}$$

From the definition of enthalpy,

$$h = u + p \cdot v \tag{2.0-4}$$

follows

$$\mathrm{d}h = \mathrm{d}u + p \cdot \mathrm{d}v + \mathrm{v} \cdot \mathrm{d}p\,. \tag{2.0-5}$$

The internal energy u of a system correlates to the translational, rotational, and vibrational energy of the molecules and only depends on the temperature of the system:

$$\mathrm{d}u = \left(\frac{\mathrm{d}u}{\mathrm{d}T}\right) \cdot \mathrm{d}T = c_v \cdot \mathrm{d}T\,. \tag{2.0-6}$$

Corresponding to the equations above, the heat capacity c_v has to be determined in a way that for a change in the system temperature in $\mathrm{d}T$ there is no transfer of work dw. Hence, the volume v has to be kept constant:

$$c_v = \left(\frac{\partial u}{\partial T}\right)_v\,. \tag{2.0-7}$$

And analogously

$$c_p = \left(\frac{\partial h}{\partial T}\right)_p \tag{2.0-8}$$

is to be measured at constant pressure.

The thermodynamic state of a system is defined not only by its internal energy u, but also by its entropy s, which is defined by

$$\mathrm{d}s = \frac{\mathrm{d}q}{T}\,. \tag{2.0-9}$$

[1] This equation and all of the following are valid for specific, i.e., for mass-based or molar-based quantities

The entropy s of a closed system, which is at temperature T, is changed by the heat $\mathrm{d}q$ transferred across its boundary. The second law of thermodynamics states that any process within the system enhances the entropy of the system, except if this process is reversible.

Furthermore, it is convenient to define the free enthalpy g as a state variable

$$g = h - T \cdot s\,, \tag{2.0-10}$$

or in its differential form:

$$\mathrm{d}g = \mathrm{d}h - T \cdot \mathrm{d}s - s \cdot \mathrm{d}T\,. \tag{2.0-11}$$

With $\mathrm{d}q = T \cdot \mathrm{d}s$, *Gibbs fundamental equation* follows for reversible processes within the system:

$$\mathrm{d}g = v \cdot \mathrm{d}p - s \cdot \mathrm{d}T\,. \tag{2.0-12}$$

This is an important relation for the description of thermodynamic equilibria as it contains the measurable dimensions $\mathrm{d}p$ and $\mathrm{d}T$ of the system.

2.1 Liquid/Gas Systems

In this chapter important thermodynamic fundamentals will be introduced for liquid/gas systems (L/G). In analogy, they also apply to the other two- (L/S, L/L, G/S) and multiphase systems.

2.1.1 Characteristics of Pure Substances

2.1.1.1 Vapor pressure

Vapor pressure p^0 is a function of temperature T only. The vapor pressure curve provides information about the behavior of pure substances. In Fig. 2.1-1, the vapor pressure curves of water, benzene, and naphthalene are depicted. Vapor pressure curves show a bend at the triple point TP and end at the critical point CP.

The pressure that is reached at a given temperature of a closed, equilibrated two-phase system (L/G) of a pure substance is called vapor pressure p^0. The irreversible exchange processes within the system come to rest if temperature and pressure of liquid and gas are in equilibrium. Then the entropy of the entire system reaches its maximum.

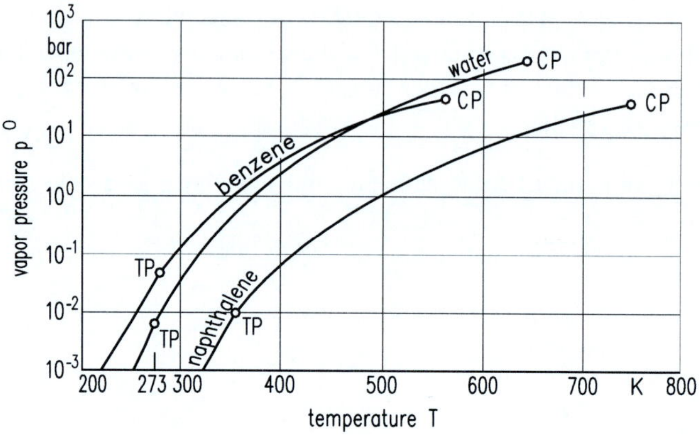

Fig. 2.1-1 Vapor pressure of some substances

$$T_L = T_G = T, \tag{2.1-1}$$

$$p_L = p_G = p^0. \tag{2.1-2}$$

At equilibrium, any small changes in the state variables T and p are reversible and therefore

$$\mathrm{d}g_L(T, p) = \mathrm{d}g_G(T, p) \tag{2.1-3}$$

holds.

The pressure reached in equilibrium is denoted vapor pressure p^0 (of the pure component). From this and with *Gibbs fundamental equation*, which is valid for each of the phases, it follows that

$$(v_G - v_L) \cdot \mathrm{d}p^0 = (s_G - s_L) \cdot \mathrm{d}T. \tag{2.1-4}$$

Sufficiently far from the critical point ($T \ll T_c$, and $p^0 \ll p_c^0$) it follows that $v_L \ll v_G$ and that the *law of ideal gases*

$$v_G = \frac{R \cdot T}{p^0} \tag{2.1-5}$$

is valid.

From this, the equation of *Clausius–Clapeyron* is derived:

$$R \cdot T \cdot \frac{\mathrm{d}p^0}{p^0} = (s_G - s_L) \cdot \mathrm{d}T. \tag{2.1-6}$$

To vaporize the substance in a system at $p = \text{const.}$, an amount of heat $\mathrm{d}q$, which is called specific vaporization enthalpy Δh_{LG}, has to be added. Therefore, $\Delta h_{\mathrm{LG}} = (s_{\mathrm{G}} - s_{\mathrm{L}}) \cdot T$, and

$$\frac{\mathrm{d}p^0}{p^0} = \frac{1}{R} \cdot \Delta h_{\mathrm{LG}} \cdot \left(\frac{\mathrm{d}T}{T^2}\right) = -\frac{1}{R} \cdot \Delta h_{\mathrm{LG}} \cdot \mathrm{d}\left(\frac{1}{T}\right) \tag{2.1-7}$$

or

$$\Delta h_{\mathrm{LG}} = -R \cdot \frac{\mathrm{d}(\ln p^0)}{\mathrm{d}(1/T)}, \tag{2.1-8}$$

or integrated

$$\ln p^0 = -\frac{\Delta h_{\mathrm{LG}}}{R \cdot T} + \text{const}. \tag{2.1-9}$$

Below the triple point

$$\ln p^0 = -\frac{\Delta h_{\mathrm{SG}}}{R \cdot T} + \text{const}'. \tag{2.1-10}$$

results, with Δh_{SG} being the sublimation enthalpy. The last equation provides simple vapor pressure relations for small temperature ranges in which Δh_{LG} and Δh_{SG} only change insignificantly:

$$\ln p^0 = \frac{A}{T} + B. \tag{2.1-11}$$

For greater temperature spans the *Antoine equation* provides better results:

$$\ln p^0 = \frac{A'}{T + C'} + B'. \tag{2.1-12}$$

Today, a modified form of an equation given by Wagner (1972) is considered to be state of the art

$$\ln\frac{p^0}{p_{\mathrm{c}}} = \frac{1}{T_{\mathrm{r}}}[A \cdot (1 - T_{\mathrm{r}}) + B \cdot (1 - T_{\mathrm{r}})^{1.5} + C \cdot (1 - T_{\mathrm{r}})^{2.5} + D \cdot (1 - T_{\mathrm{r}})^5]. \tag{2.1-13}$$

The VDI-Heat Atlas (2010) provides values for the coefficients of this equation for 275 substances.

A simple vapor pressure diagram with the possibility for extrapolation is given by *Hoffmann and Florin* and is shown in Fig. 2.1-2 . For many systems the modified temperature scale $1/\tau$ obeys the relation

$$\frac{1}{\tau} = \frac{1}{T} + 2.6726 \times 10^{-3} \log T - 0.8625 \times 10^{-6} T - 7.9151 \times 10^{-3}. \qquad (2.1\text{-}14)$$

Here T is the absolute temperature in Kelvin.

The vapor pressure curve of each component is rendered by two straight lines. The vapor pressure curve has a sharp bend at the triple point because of the difference between sublimation and vaporization enthalpy. The diagram offers the possibility to extraploate measured vapor pressures.

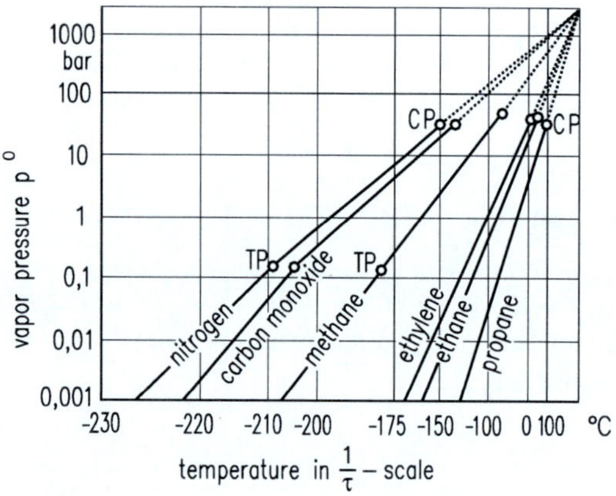

Fig. 2.1-2 Vapor pressure vs. modified temperature

Above, it has been shown that the enthalpy of vaporization can be obtained from the vapor pressure curve. However, an estimation of the enthalpy of vaporization can also be obtained with the help of *Trouton's rule*. Experience shows that for many substances and at standard conditions the molar entropy of vaporization $(\tilde{s}_G - \tilde{s}_L)$ is about 80–110 kJ/(kmol · K) . In Fig. 2.1-3, the entropy of vaporization of water and the homologues of different organic compounds is plotted vs. their respective molar mass. The stronger the dipole moment of the single molecules, the stronger the association of the molecules and the higher the entropy of vaporization. Water with its extremely high dipole moment has an entropy of vaporization of 109 kJ/(kmol · K) . The enthalpy of vaporization decreases when approaching the critical point and vanishes at the critical point. With good accuracy the enthalpy of vaporization can be calculated by the following equation for the entire temperature range between T_b and T_c (Watson 1943). T_b is the boiling temperature at standard pressure.

$$\Delta h_{LG} = (\tilde{s}_G - \tilde{s}_L) \cdot \frac{T_b}{\tilde{M}} \cdot \left(\frac{T_c - T}{T_c - T_b}\right)^{0.38} \tag{2.1-15}$$

The PPDS equation

$$\Delta h_{LG} = R \cdot T_c \cdot (A \cdot \tau^{1/3} + B \cdot \tau^{2/3} + C \cdot \tau + D \cdot \tau^2 + E \cdot \tau^6) \tag{2.1-16}$$

with $\tau = 1 - T/T_c$ is suitable for correlation of data of heat of vaporization. VDI-Heat Atlas (2010) provides values for the coefficients of this equation for 275 substances.

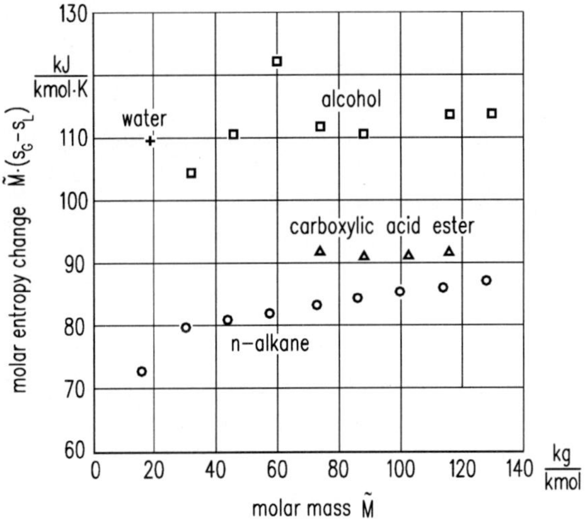

Fig. 2.1-3 Entropy of vaporization vs. the molar mass for different substances

After Thiesen (1923), it seems favorable to plot the enthalpy of vaporization vs. the difference of the critical temperature T_c and the temperature T in a double logarithmic diagram (see Fig. 2.1-4). A straight line results as long as the critical point is not very close. This diagram also shows that the enthalpy of phase transition changes at the triple point. At the triple point the enthalpy of vaporization and the enthalpy of melting add up to the enthalpy of sublimation:

$$\Delta h_{LG} + \Delta h_{SL} = \Delta h_{SG} . \tag{2.1-17}$$

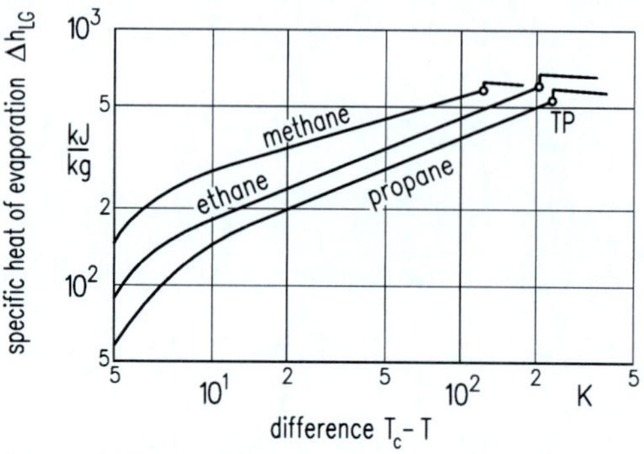

Fig. 2.1-4 Specific heat of vaporization vs. the difference $T_c - T$ for some substances

The enthalpy of melting Δh_{SL} can be calculated with the equation of *Clausius–Clapeyron* and with the change of specific volume from v_S to v_L :

$$\Delta h_{SL} = (v_L - v_S) \cdot T \cdot \frac{dp^0}{dT} . \tag{2.1-18}$$

The entropy of melting $(\tilde{s}_G - \tilde{s}_L) = \Delta\tilde{h}_{SL}/T$ after Green and Perry (2008) is about 9.2 kJ/(kmol K) for metals, 22–29 kJ/(kmol K) for inorganic compounds, and 38–58 kJ/(kmol K) for organic compounds.

2.1.1.2 Vapor Pressure at Strongly Curved Liquid Surfaces

The vapor pressure at liquid surfaces with pronounced concave curvature is smaller than that of flat surfaces. In case of convex curvature, the vapor pressure of the liquid is greater than that of flat surfaces. A concave liquid surface is found, e.g., for bubbles or in a capillary if the liquid wets the capillary's wall. Droplets on the contrary have a convexly bent surface. As prerequisite to vaporization, tiny steam bubbles have to be formed initially while for condensation small droplets have to form first.

The vapor pressure $(p^0)_r$ at strongly curved liquid surfaces with radius r can be calculated from the *Gibbs–Thomson* equation:

$$\ln\frac{(p^0)_r}{p^0} = \frac{2\sigma}{R \cdot T \cdot \rho_L \cdot r} , \tag{2.1-19}$$

with σ being the surface tension. This relation is obtained if the isothermal work of compression of a small amount of substance from vapor pressure $(p^0)_r$ of the bent surface to the vapor pressure p^0 is set equal to the increase of surface energy. According to this relation, the logarithm of the pressure ratio is directly proportional to the surface tension σ and inversely proportional to the radius. At convex liquid surfaces, the radius r is positive. In this case the pressure is larger than that of flat ones. In contrast, for concave surfaces with a negative radius of curvature a lowering of vapor pressure results. These effects are only significant, if the radius of curvature is very small.

In Fig. 2.1-5, the ratio $(p^0)_r/p^0$ is plotted vs. the radius of curvature r for methyl alcohol, water, and mercury, which shows surface tensions of 0.0226, 0.0727, and 0.435 N/m, respectively at 20°C. Figure 2.1-5 shows that the vapor pressure at bent surfaces deviates by more than 1% from that of flat liquid surfaces for many organic substances as well as for water, if the radii of bubbles and drops are smaller than 100 nm. It can be derived from this diagram that the effects are considerable for bubbles or drops made up of only a few hundred molecules. The *Gibbs–Thomson* equation is often used to explain retardation of boiling (superheating of the liquid) as well as condensation–inhibiting processes (supercooling of vapor). In addition it gives information about unusual sorption isotherms of adsorbent materials or about goods with very narrow capillary tubes which have to be dried, see Chap. 10.

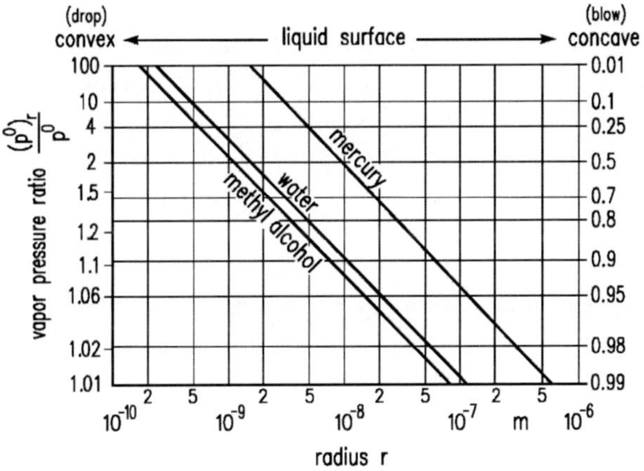

Fig. 2.1-5 Vapor pressure ratio vs. the radius of curvature

2.1.2 Behavior of Binary Mixtures

Process engineering often deals with multicomponent mixtures. The behavior of multicomponent mixtures in general does not differ from the behavior of two component mixtures, which are technically and practically easier to describe. Therefore, it is advantageous to acquire the basics for the behavior of binary systems. How the equations can be transferred to fit multicomponent mixtures is shown in Chap. 5.

A closed, binary two-phase system is in thermodynamic equilibrium if the entropy s of the whole system has reached its maximum. For two-phase systems of liquid (L) and vapor (G) the already mentioned conditions for equilibria are

$$T_L = T_G = T, \tag{2.1-20}$$

$$p_L = p_G = p^0, \tag{2.1-21}$$

$$dg_L(T, p) = dg_G(T, p). \tag{2.1-22}$$

Description of equilibria becomes particularly easy when dealing with dilute solutions. A solution is considered to be dilute if there is just one molecule of dissolved substance in about 100 or more molecules of solvent. The dissolved molecule is then only influenced by the interaction energy with the solvent molecules, but not by the interaction energy with molecules of its own kind. Therefore, the assumption of a dilute solution is only valid for a mole fraction of the dissolved substance smaller than 0.01.

Mole fraction is one possible measure for the composition of a mixture. It relates the amount of substance of one component to the total amount of substances in the system. In analogy, the mass fraction is defined. Other measures for the composition of a mixture (mass or molar loading) are given in Table 2.1-1. In addition, this table also gives the rules of conversion between these measures. The term concentration c or $\tilde{c}$ shall only be used if the amount of a substance is related to the volume.

2.1.2.1 *Vapor Pressure of Dilute Binary Solutions*

In a gas, every gas component aims to disperse into the whole space available, and dissolved components in a mixture of liquids aim to disperse all over the liquid.

When separating two liquid mixtures of different composition by a membrane which is only permeable for component a of the mixtures (e.g., the solvent) this component will flow through the membrane until its chemical potential μ_a is equal on both sides of the membrane. To avoid this flow, a pressure difference has to build up between both sides of the membrane. This phenomenon is called osmosis.

For pure solvent a on one side of the membrane and a dilute ideal solution of a substance b in solvent a on the other side, the osmotic pressure difference is calculated with *van't Hoff's law*:

$$p_{osm} = \tilde{c}_b \cdot \tilde{R} \cdot T. \tag{2.1-23}$$

Table 2.1-1 Definitions and conversion rules for measures of concentration

In terms of mass (kg) $M_i = \rho_i V_i$			In terms of amount of substance (kmol) $N_i = \frac{M_i}{\tilde{M}_i}$		
Components	2	k	Components	2	k
Total mass	$M_{total} = M_a + M_b$	$M_{total} = \sum_{j=1}^{k} M_j$	Total amount of Substance	$N_{total} = N_a + N_b$	$N_{total} = \sum_{j=1}^{k} N_j$
Mass fraction	$x_a = \frac{M_a}{M_{total}}$	$x_i = \frac{M_i}{M_{total}}$	Mole fraction	$\tilde{x}_a = \frac{N_a}{N_{total}}$	$\tilde{x}_i = \frac{N_i}{N_{total}}$
Mass loading	$X_a = \frac{M_a}{M_b}$	$X_i = \frac{M_i}{M_{Carrier}}$	Molar loading	$\tilde{X}_a = \frac{N_a}{N_b}$	$\tilde{X}_i = \frac{N_i}{N_{Carrier}}$
	Mass fraction x from mole fraction $\tilde{x}$ and vice versa				
	$x_a = \frac{1}{1 + \frac{\tilde{M}_b}{\tilde{M}_a} \cdot \frac{1 - \tilde{x}_a}{\tilde{x}_a}}$	$x_i = \frac{\tilde{x}_i \cdot \tilde{M}_i}{\sum_{j=1}^{k} (\tilde{x}_j \cdot \tilde{M}_j)}$		$\tilde{x}_a = \frac{1}{1 + \frac{\tilde{M}_a}{\tilde{M}_b} \cdot \frac{1 - x_a}{x_a}}$	$\tilde{x}_i = \frac{\frac{x_i}{\tilde{M}_i}}{\sum_{j=1}^{k} \left(\frac{x_j}{\tilde{M}_j}\right)}$
	Mass loading – mass fraction $X = \frac{x}{1 - x}$; $x = \frac{X}{1 + X}$		Molar loading – mole fraction $\tilde{X} = \frac{\tilde{x}}{1 - \tilde{x}}$; $\tilde{x} = \frac{\tilde{X}}{1 + \tilde{X}}$		

The osmotic pressure difference can be demonstrated with the help of a simple experiment. In Fig. 2.1-6, a dish is shown which contains pure solvent. A funnel, closed at its lower end by a membrane and immersed into the liquid, contains a solution which consists of solvent *a* and dissolved component *b*. In steady state, the hydrostatic pressure of the liquid column above the liquid's surface in the dish is identical to the osmotic pressure difference. If the imposed pressure difference is larger than the osmotic pressure difference, the solvent flows from the solution through the membrane. This is called reverse osmosis. Reverse osmosis is used, e.g., to gain drinking water from sea or wastewater. To do so, the employed pressure difference has to be larger than the osmotic pressure difference so that the water molecules from the solution permeate the membrane. For desalination of seawater with a salt concentration around 40,000 ppm a pressure difference of about 70 bar is necessary.

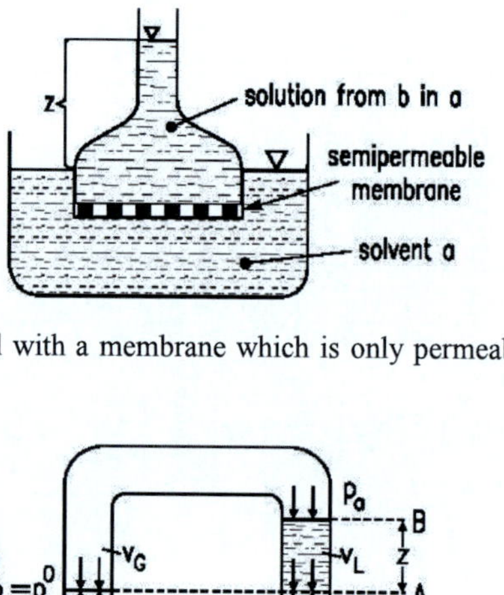

Fig. 2.1-6 Osmotic cell with a membrane which is only permeable for solvent *a* (semipermeable membrane)

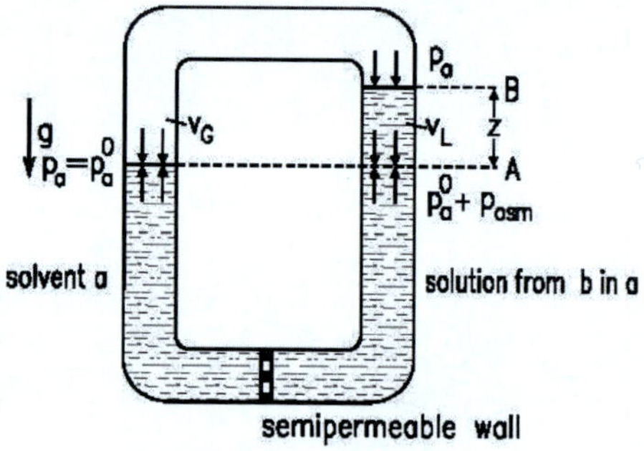

Fig. 2.1-7 Cell with semipermeable wall for the determination of osmotic pressure

Furthermore, the osmotic behavior of dilute binary solutions can be used to describe the partial pressure of its components. In the gas phase, which is in contact with a solution, each component of the solution shows a partial pressure, which is lower than the vapor pressure of this component. Therefore, in the case of a binary solution where solute *b* has a negligible vapor pressure, an elevation of the boiling point occurs in relation to that of the pure solvent *a*. This reduction compared to the vapor pressure can be modeled by describing the pressures of a hypothetical experimental setup as shown in Fig. 2.1-7. This closed and isothermal setup has a semi-permeable wall at the lower end which is permeable for the solvent (component *a*), but not for the dissolved component *b*. The left leg contains the solvent, the right leg the solution with the composition $\tilde{x}_a = 1 - \tilde{x}_b$. Due to the osmotic pressure difference, the level of the liquid in the right leg is higher than that in the left. (*Note*: In the case of marine water as solution, z would be about 700 m.) The same pressure p has to prevail at level B on both left and right branches. On the left branch it is

$$p = p_a^0 - \frac{z \cdot g}{v_G}. \tag{2.1-24}$$

On the right branch it is

$$p = p_a. \tag{2.1-25}$$

For equal densities of liquid solvent and solution and with the pressure equilibrium at the membrane (level A), the following relation results:

$$p_a^0 + p_{osm} = p_a + \frac{z \cdot g}{v_L}. \tag{2.1-26}$$

From these three equations, the term $(z \cdot g)$ can be eliminated and the following relation for vapor pressure reduction $(p_a^0 - p_a)$ is obtained:

$$p_{osm} \cdot v_L = (p_a^0 - p_a) \cdot (v_G - v_L). \tag{2.1-27}$$

Far away from the critical pressure the *law of ideal gases* is valid. Furthermore, the specific volume of the condensed phase v_L is much smaller than the volume of the gaseous one v_G. Therefore, the osmotic pressure difference is

$$p_{osm} = \tilde{\rho}_L \cdot \frac{\tilde{R} \cdot T}{p_a^0} \cdot (p_a^0 - p_a). \tag{2.1-28}$$

With *van't Hoff's law* (2.1-23), the following relation for a relative reduction of vapor pressure of binary, dilute solutions is (*Raoult–van't Hoff's law*)

$$\frac{p_a^0 - p_a}{p_a^0} = \frac{\tilde{c}_b}{\tilde{\rho}_L} = \tilde{x}_b = 1 - \tilde{x}_a, \tag{2.1-29}$$

with $\tilde{x}_a$ being the mole fraction of the solvent *a* and $(1 - \tilde{x}_a)$ being the mole fraction of the solute b. This law states that the relative reduction of vapor pressure of a solution is equal to the solute's mole fraction. This leads to the following path for calculating vapor pressure curves of solutions. In Fig. 2.1-8, a vapor pressure diagram is displayed which shows the vapor pressure p_a^0 of the pure solvent and the pressure p_a of a solution. The higher the vapor pressure reduction $(p_a^0 - p_a)$, the higher the elevation of boiling point ΔT_b. *Raoult–van't Hoff's law* is only valid, if

- The solute has a negligible vapor pressure, because otherwise the solute also adds to the vapor pressure
- The molecules are not dissociated into ions

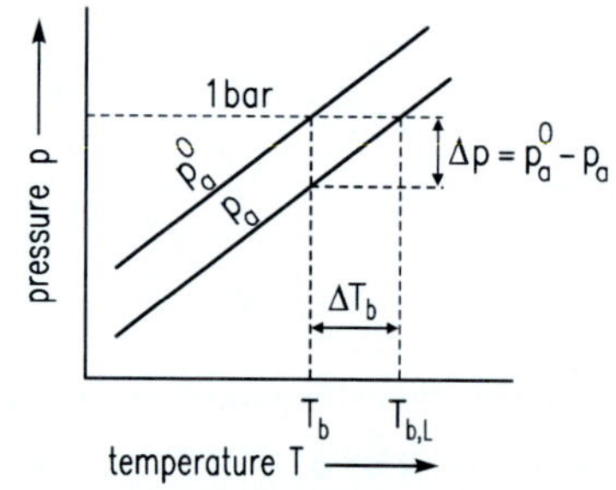

Fig. 2.1-8 Vapor pressure of a pure substance and a solution vs. the temperature (T_b, $T_{b,L}$ boiling temperature of solvent and solution at 1 bar)

Otherwise, the real mole fraction has to be used corresponding to the true number of particles. This is why in general the law is only valid for dilute solutions.

The elevation of boiling point ΔT_b can be calculated with the *relation of Clausius–Clapeyron*. Let $\tilde{q}$ be the heat of phase transition (i.e., vaporization), then with $T_b \approx T_{b,L} = T$ follows

$$\tilde{q} = \frac{\tilde{R} \cdot T^2}{p_a} \cdot \frac{dp_a^0}{dT}. \tag{2.1-30}$$

With substitution of the differential by the difference, it follows that

$$\Delta T_b = \frac{(p_a^0 - p_a)}{p_L} \cdot \frac{\tilde{R} \cdot T^2}{\tilde{q}}. \tag{2.1-31}$$

With consideration of *Raoult-van't Hoff's law* (2.1-29) ΔT_b finally results as

$$\Delta T_b = \frac{(1 - \tilde{x}_a)}{\tilde{x}_a} \cdot \frac{\tilde{R} \cdot T^2}{\tilde{q}}. \tag{2.1-32}$$

The elevation of boiling point then again depends on the concentration of solute.

If the solute dissociates the true mole fraction, that is the number of particles of the solute related to the total number of particles, has to be used. In this case, the elevation of boiling point is

$$\Delta T_{\mathrm{b}} = i \cdot \frac{(1-\tilde{x}_{\mathrm{a}})}{\tilde{x}_{\mathrm{a}}} \cdot \frac{\tilde{R} \cdot T^2}{\tilde{q}}, \tag{2.1-33}$$

with

$$i = 1 + (\gamma_{\mathrm{a}} \cdot m_{\mathrm{a}} - 1) \cdot \alpha \tag{2.1-34}$$

Herein m_{a} indicates the number of ions, which originate from one molecule a, and γ_{a} is the activity coefficient (see Sect. 2.1.4.1). The number α is the fraction of dissociated molecules.

For strong electrolytes, $\alpha \approx 1$, so that the equation simplifies to

$$i = m_{\mathrm{a}} \cdot \gamma_{\mathrm{a}}, \tag{2.1-35}$$

see Table 2.1-2. Values for activity coefficients can be found in many sources, e.g., in D'Ans-Lax Taschenbuch für Chemiker (1976).

Table 2.1-2 Data of m_{a} and i for some salts with a concentration of $0.1\ \mathrm{mol/kg_{H2O}}$ and $\vartheta = 25°\mathrm{C}$

	m_{a}	γ_{a}	i
NaCl	2	0.786	1.572
KCl	2	0.771	1.542
$BaCl_2$	3	0.499	1.497
KOH	2	0.754	1.508
NaOH	2	0.772	1.544

The vapor ascending from a solution is only made up of solvent, if the vapor pressure of the solute is negligible. With consideration of the elevation of boiling point, the vapor is superheated, so that the enthalpy of evaporation is greater than the enthalpy of vaporization at system pressure. The specific enthalpy of evaporation q therefore consists of the vaporization enthalpy at system pressure $p = p_{\mathrm{a}}^0$ and of the enthalpy of superheating $c_p \cdot \Delta T_{\mathrm{b}}$:

$$q = (\Delta h_{\mathrm{LG}})_{p_{\mathrm{a}}^0} + c_p \cdot \Delta T_{\mathrm{b}}. \tag{2.1-36}$$

Fig. 2.1-9 shows the vapor pressure vs. the mole fraction $\tilde{x}_a$ (upper left) and vs. the temperature T (upper right) and a temperature–entropy diagram for the pure solvent. The superheated vapor corresponds to point A with pressure p_a^0 and temperature $T_{b,L} = T_b + \Delta T_b$.

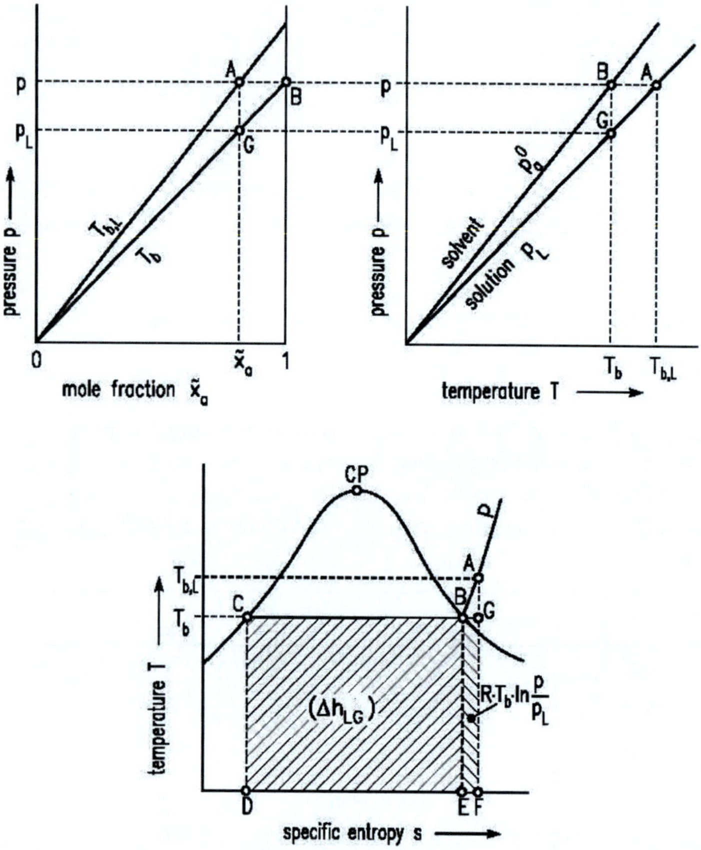

Fig. 2.1-9 Vapor pressure of solutions for temperatures T_b and $T_{b,L}$ vs. the mole fraction (*upper left*), vapor pressure of the solvent and the solution vs. the temperature (*upper right*) as well as a temperature–entropy diagram for the pure solvent (*below*)

If pure solvent is vaporized at system pressure p_a^0 (e.g., 1 bar), the vapor has to be supplied with additional enthalpy to reach the superheated point A. The required enthalpy for superheating is equal to the area $BEFG$, which is adequate to the heat quantity to be supplied for isothermal expansion from p to p_a. Therewith q is about

$$q \approx (\Delta h_{LG})_p + R \cdot T_b \cdot \ln \frac{p}{p_a}. \tag{2.1-37}$$

The derivation of this second equation is shown in the upper diagram of Fig. 2.1-9.

With *Dühring's rule*, it is possible to estimate the vapor pressure curve of a solution on the basis of the known vapor pressure curve of the pure solvent and of the known boiling temperature of the solution. The pure substance's vapor pressure curve can for instance be determined by the equation of *Clausius–Clapeyron* (T^0 is illustrated in Fig. 2.1-10):

$$\ln p_a^0 = -\frac{\Delta h_{LG}}{R \cdot T^0} + C_1. \tag{2.1-38}$$

At the boiling point, the vapor pressure and the boiling temperature are denoted by the index b:

$$\ln p_{b,a}^0 = -\frac{(\Delta h_{LG})_b}{R \cdot T_b} + C_1, \tag{2.1-39}$$

for small temperature ranges $\Delta h_{LG} \approx (\Delta h_{LG})_b$. Subtraction of these two equations results in

$$\ln p_{b,a}^0 - \ln p_a^0 = \frac{\Delta h_{LG}}{R} \cdot \frac{T_b - T^0}{T^0 \cdot T_b} \tag{2.1-40}$$

or finally

$$\ln p_{b,a}^0 - \ln p_a^0 = \frac{1}{\tilde{R}} \cdot \frac{\Delta \tilde{h}_{LG}}{T_b} \cdot \frac{T_b - T^0}{T^0}. \tag{2.1-41}$$

In this equation, the molar vaporization entropy appears, which is approximately constant for many substances according to Fig. 2.1-3. The following then results if the same derivation is done for the solution:

$$\ln p_{b,a} - \ln p_a = \frac{1}{\tilde{R}} \cdot \frac{\Delta \tilde{h}_{LG}}{T_b} \cdot \frac{T_{b,L} - T_L}{T_L}. \tag{2.1-42}$$

The molar vaporization entropies of both equations are equated. Then *Dühring's rule* follows

$$T_L \approx \frac{T_{b,L} \cdot T^0}{T_b}. \tag{2.1-43}$$

Often, this equation is fulfilled rather well, as long as the concerned temperature intervals are not too big. As an example, in Fig. 2.1-10 the vapor pressure of

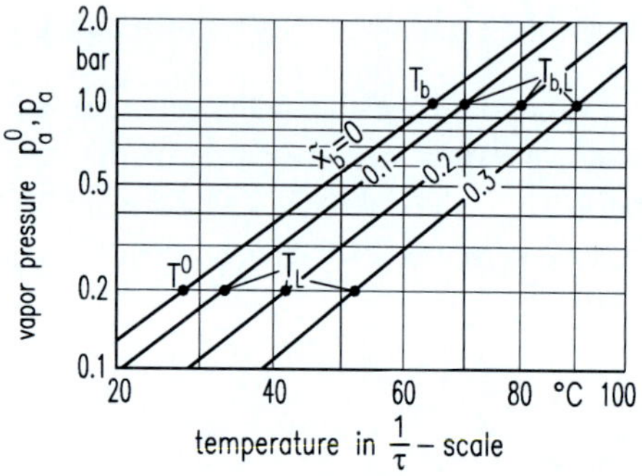

Fig. 2.1-10 Vapor pressure of methanol as well as of sodium methylate–methanol solutions vs. the temperature

sodium methylate–methanol solution is plotted logarithmically vs. the temperature in $1/\tau$-scaling. The different straight lines show experimentally determined data for pure methanol as well as for sodium methylate–methanol solution with different mass fractions of sodium methylate. Recalculation of the curves using *Dühring's rule* coming from the vapor pressure curve of pure methanol and the boiling temperatures of methanolic sodium methylate solution shows that calculation and experiment are quite in good accordance.

2.1.2.2 Freezing Point Depression

Solutions show a depression ΔT_s of their freezing temperature compared to pure solvents. In the following way, this freezing point depression can be calculated easily for a dilute solution of a nonvolatile component. In Fig. 2.1-11, the vapor pressure curve of solvent *a* is given in the vicinity of the triple point. At this point, the vapor pressure curve is bent.

With the quantities illustrated in Fig. 2.1-11, it follows that

$$\frac{p_a^0 - p_a}{\Delta T_s} = \frac{\Delta p_s^0 - \Delta p_a}{\Delta T_s} \approx \frac{dp_s^0}{dT} - \frac{dp_a}{dT}. \tag{2.1-44}$$

The vapor pressure curve of the liquid phase can be described as

$$\Delta h_{LG} = T_s \cdot (v_G - v_L) \cdot \frac{dp_a}{dT}. \tag{2.1-45}$$

For the solid phase accordingly

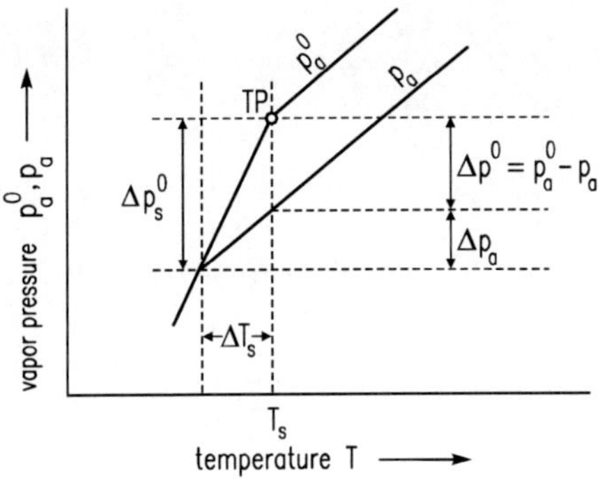

Fig. 2.1-11 Vapor pressure of a pure solvent as well as of a solution vs. the temperature in the environment of the triple point

$$\Delta h_{SG} = T_s \cdot (v_G - v_S) \cdot \frac{dp_s^0}{dT}. \tag{2.1-46}$$

Substituting the differentials in the equations above by differences results in

$$\frac{p_a^0 - p_a}{\Delta T_s} = \frac{dp_s^0 - dp_a}{dT} = \frac{\Delta h_{SL} \cdot p_a}{R \cdot T_s^2} \text{ with } \Delta h_{SL} = \Delta h_{SG} - h_{LG}. \tag{2.1-47}$$

In this equation, the vapor pressures of liquid and solid phases were set equal. In conclusion, the freezing-point depression is

$$\Delta T_s = \frac{p_a^0 - p_a}{p_a} \cdot \frac{R \cdot T_s^2}{\Delta h_{SL}} = \frac{1 - \tilde{x}_a}{\tilde{x}_a} \cdot \frac{R \cdot T_s^2}{\Delta h_{SL}} = \tilde{X}_b \cdot \frac{R \cdot T_s^2}{\Delta h_{SL}}. \tag{2.1-48}$$

According to this, the freezing-point depression is proportional to the relative reduction of vapor pressure or to the molar loading of the solute. For dissociation of the solute, the number of emerging ions has to be considered again. Also, here it is required to assume that the solute has a negligible vapor pressure and that the solution is dilute.

2.1.2.3 Raoult's Law

In many cases, the solute's vapor pressure has to be taken into account as well. This is clarified in Fig. 2.1-12. The left diagram describes the case where the solute has a much lower vapor pressure than the solvent. In the diagram on the right the partial pressures are plotted vs. the solvent's mole fraction. The equation

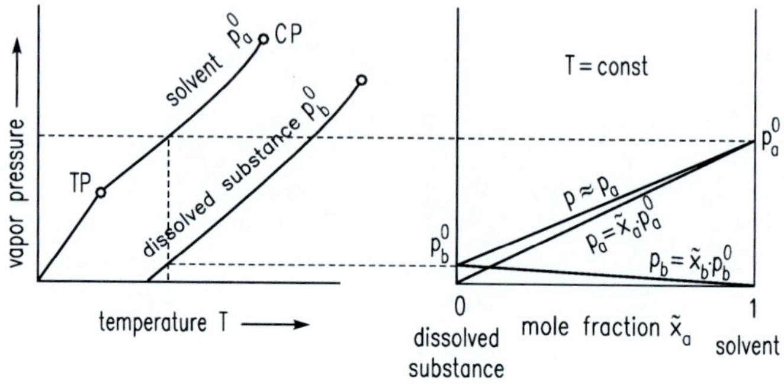

Fig. 2.1-12 Vapor pressure of solvent and of solute vs. the temperature (*left*) and the partial pressure p_a vs. the mole fraction at a constant temperature (*right, Raoult-van't Hoff's law* for $p_b^0 \ll p_a^0$)

$$p_a = \tilde{x}_a \cdot p_a^0 \tag{2.1-49}$$

is *Raoult's law* for the solvent. This law is also applicable for the solute:

$$p_b = \tilde{x}_b \cdot p_b^0 = (1 - \tilde{x}_a) \cdot p_b^0 . \tag{2.1-50}$$

The partial pressures p_a and p_b of the two components add up to the total pressure p:

$$p_a + p_b = p . \tag{2.1-51}$$

From the right diagram it follows that the partial vapor pressure of the solvent and the total pressure differ only insignificantly. If it is possible to neglect the vapor pressure and therefore also the partial vapor pressure of the solute this is the aforementioned *Raoult-van't Hoff's law.*

However, this simplification is no longer valid if solvent's and solute's vapor pressures are of the same magnitude. *Raoult's law* then has to be formulated for both components. This situation is illustrated in Fig. 2.1-13. The left diagram depicts vapor pressure curves of the two components, which are marked by the indices *a* and *b*. Now, it no longer makes sense to speak of solvent and solute, but rather in general of light end or low boiling component and heavy end or high boiling component. In the right diagram, the partial pressures p_a and p_b and the total pressure p are plotted vs. the mole fraction of the low boiling component. The straight lines added on the right of the diagram represent partial pressures and the total pressure and are valid for a defined temperature. For higher temperatures steeper straight lines result for the partial pressures.

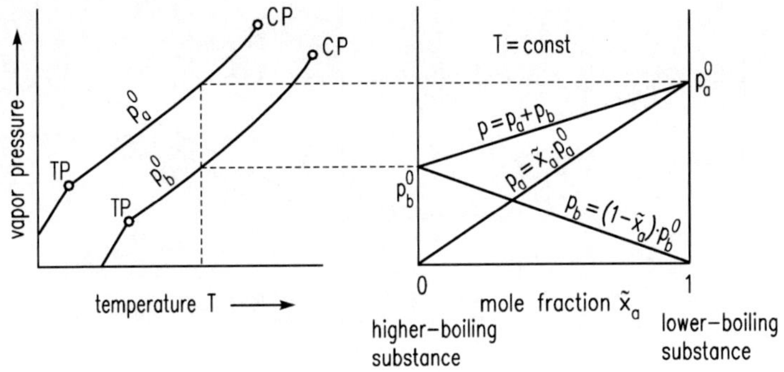

Fig. 2.1-13 Vapor pressure of two substances dependent of temperature (*left*) and of pressures p, p_a, and p_b vs. the mole fraction at constant temperature (*right*, Raoult's law)

Raoult's law is an important relation for boiling solutions, which are to be separated by distillation or rectification, see Chap. 5.

2.1.2.4 Henry's Law

Nevertheless, the case of a dissolved component, which has a much higher vapor pressure than the solvent, still has to be dealt with. This component, usually a gas, can even be existent in its supercritical state. This is shown in Fig. 2.1-14. At temperature T the gaseous component "a" is in its supercritical state so that no vapor pressure can be given. If the vapor pressure curve would be extended up to this temperature a reference vapor pressure $p_a^{0'}$ could be given instead. On the diagram's right side again the partial pressures are plotted vs. the mole fraction of the light ends component that is of the gaseous component "a". *Raoult's law* for the dissolved gas "a" therefore is

$$p_a = \tilde{x}_a \cdot p_a^{0'} . \tag{2.1-52}$$

Further on, it will be insinuated that the vapor pressure of the other component is much smaller than the fictional vapor pressure $p_a^{0'}$ of the supercritical gas: In this case, the partial pressures of both components again add up to the total pressure, which is about equal to the partial pressure of the light end component, however, because it may be acceptable to neglect the partial pressure of the high ends component. Since the extrapolation of the gaseous, light ends component's vapor pressure up to temperature T does not have a physical meaning the *Henry–pressure* $He_a = p_a^{0'}$ is a fictional pressure only. With this pressure *Henry's law* for component a is:

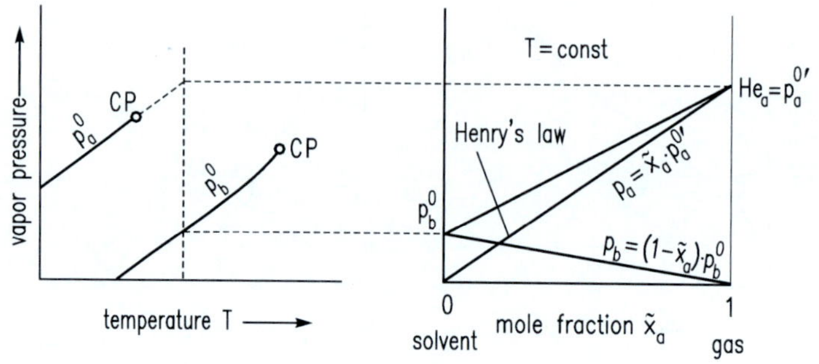

Fig. 2.1-14 Vapor pressure of a solute gas and solvent vs. the temperature (*left*), partial pressures vs. mole fraction at a certain temperature (*right*, Henry's law)

$$p_a = \tilde{x}_a \cdot He_a \,. \tag{2.1-53}$$

This law states that the partial pressure of a dissolved gaseous gas solute is proportional to its mole fraction in the liquid. It is in fairly good agreement for small mole fractions. The pressure He_a is strongly dependent on temperature.

Henry's law is significant for the description of processes of ab- and desorption of gases in liquids. Absorption takes place when the partial pressure of the gas in contact with the liquid is greater than the equilibrium pressure and therefore the gas is solved in the liquid. Desorption is the opposite process, thus the removal of gas out of liquids, see Chap. 5.

2.1.3 Behavior of Ideal Mixtures

The behavior of ideal binary mixtures can be described by *Raoult's law*:

$$p_a = \tilde{x}_a \cdot p_a^0 \text{ and } p_b = (1 - \tilde{x}_a) \cdot p_b^0 \,. \tag{2.1-54}$$

There is a number of mixtures which show this ideal behavior. Among these are especially mixtures with very small nonpolar molecules like nitrogen, oxygen, and argon as well as hydrogen and methane. Mixtures of nonpolar molecules of about the same size also show ideal behavior, e.g., benzene, toluol as well as *n*-butane and *i*-butane. Furthermore, mixtures of succeeding substances in a homologues series obey *Raoult's law* to a large extent, e.g., methanol, ethanol.

According to *Dalton's law of partial pressures* the partial pressure of a component is proportional to its mole fraction $\tilde{y}_a$ in the gas and to the total pressure p:

$$p_a = \tilde{y}_a \cdot p \text{ and } p_b = (1 - \tilde{y}_a) \cdot p \,. \tag{2.1-55}$$

Ideal behavior of gaseous mixtures requires low to moderate pressures ($p \leq 10$ bar). A combination of these equations provides a correlation between mole fraction $\tilde{y}_a^*$ in the gaseous phase and the total pressure p:

$$p = p_a + p_b = \tilde{x}_a p_a^0 + (1 - \tilde{x}_a) p_b^0 \,. \tag{2.1-56}$$

A combination of both equations leads to the relation for the thermodynamic equilibrium of a two-component mixture:

$$\tilde{y}_a^* = \frac{p_a}{p_a + p_b} = \frac{\tilde{x}_a \cdot p_a^0}{\tilde{x}_a \cdot p_a^0 + (1 - \tilde{x}_a) \cdot p_b^0} \,. \tag{2.1-57}$$

This equation is applicable for ideal liquids and vapor phases only. A liquid is ideal if *Raoult's law* can be applied for both components over the whole concentration range.

Figure 2.1-15 shows the partial pressure and the total pressure vs. the mole fraction $\tilde{x}_a$ for the system benzene/toluol at temperature $\vartheta = 100°C$. In addition, total pressure p is plotted vs. the mole fraction $\tilde{y}_a^*$.

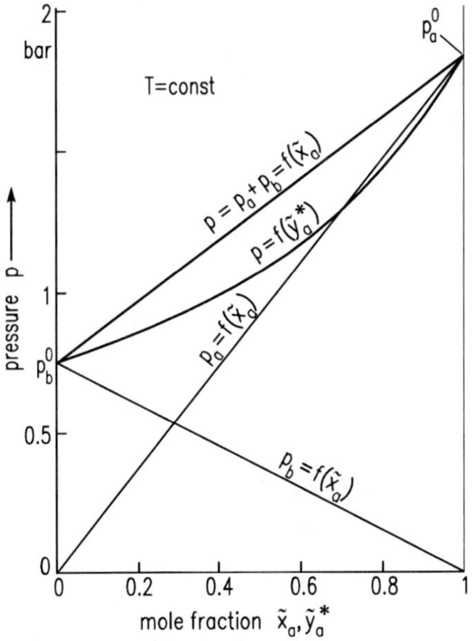

Fig. 2.1-15 Partial pressure and total pressure vs. the mole fraction of the liquid and vapor, respectively, at constant temperature

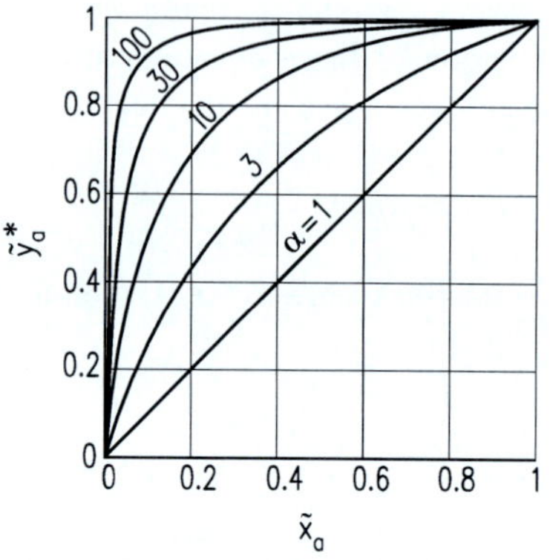

Fig. 2.1-16 Mole fraction of the vapor vs. the mole fraction of the liquid for different ideal liquids at constant temperature with relative volatility α as parameter

The equation above can be plotted as the selectivity or equilibrium diagram given in Fig. 2.1-16. This diagram shows that the gaseous phase is always richer in light ends component than the liquid phase. The ratio of partial pressures of both components is called relative volatility α

$$\alpha = \frac{p_a^0}{p_b^0}, \tag{2.1-58}$$

which is almost independent of temperature for many ideal two-component mixtures. Using this definition, (2.1-57) can be written as

$$\tilde{y}_a^* = \frac{\alpha \cdot \tilde{x}_a}{1 + (\alpha - 1) \cdot \tilde{x}_a}. \tag{2.1-59}$$

If the relative volatility is independent of temperature its geometrical mean is used for sectionwise calculation.

Figure 2.1-17 shows the evolution of the boiling temperature vs. the mole fraction $\tilde{x}_a$ in the liquid phase. This curve is called boiling curve. When cooling superheated steam of mole fraction $\tilde{y}_a$ it condensates at the so-called condensation temperature. The connecting line of all condensation temperatures is called saturated vapor line or dew-point curve. The diagram is valid for a constant pressure p. As an example, first there may only be liquid with a composition of $\tilde{x}_a = 0.4$. If this liquid is heated, it starts boiling at temperature T'' on the boiling curve. The evolving vapor has the composition $\tilde{y}_a^{*}{}''$. If more heat is supplied the boiling liquid heats up to the temperature T''' and the liquid attains composition $\tilde{x}_a'''$, while the vapor attains the mole fraction $\tilde{y}_a^{*}{}'''$. At temperature T'''' the liquid is almost vaporized, and the last

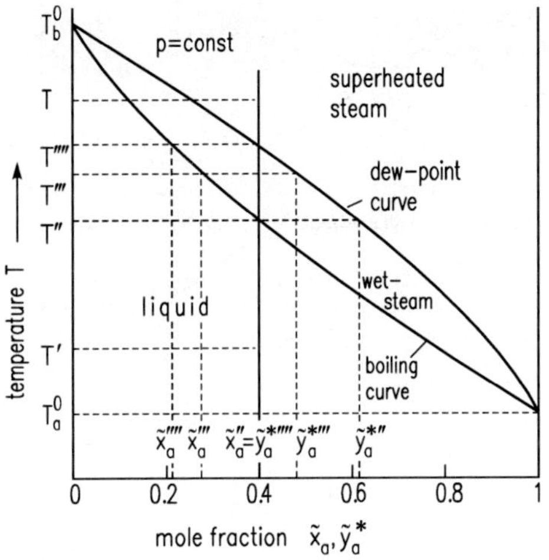

Fig. 2.1-17 Saturation temperature vs. the mole fraction in the liquid, respectively, vapor phase for constant pressure

remaining drop has a mole fraction $\tilde{x}_a''''$. Under further heat supply the vapor becomes superheated and has the same mole fraction as the primarily existing liquid. At thermodynamic equilibrium the temperatures in both phases have to be equal. Figure 2.1-18 contains a boiling diagram and a corresponding equilibrium diagram for a certain total pressure p.

Sometimes it is appropriate to define the mole fraction of a binary mixture although it is not known whether the mixture exists as liquid or vapor or in the twophase state. In this case the mole fraction is named $\tilde{z}_a$. This mole fraction states the number of moles of the light end in both phases based on the total number of moles. During vaporization the number of moles of the light ends component of the liquid and the vapor always has to add up to the total number of moles in the system. Thus

$$L \cdot \tilde{x}_a + G \cdot \tilde{y}_a^* = (L + G) \cdot \tilde{z}_a \tag{2.1-60}$$

or

$$G \cdot (\tilde{y}_a^* - \tilde{z}_a) = L \cdot (\tilde{z}_a - \tilde{x}_a) \text{ or } G \cdot b = L \cdot a . \tag{2.1-61}$$

This equation can be depicted in an easily interpretable boiling point diagram, see Fig. 2.1-19. For a known composition $\tilde{z}_a$ the ratio L/G of the amount of liquid and vapor can be given directly. If the amount of moles of gas and liquid perceived as weights in terms of mechanics this equation is the "lever rule." It can be used

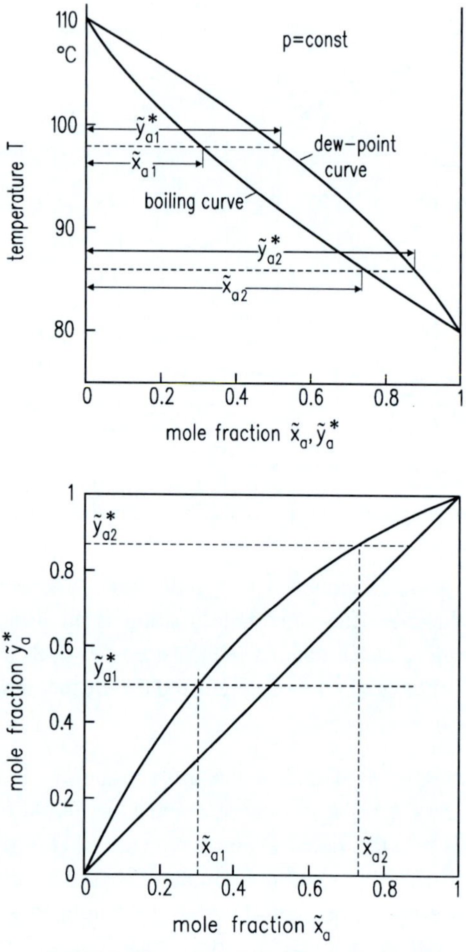

Fig. 2.1-18 Saturation temperature in dependence of the mole fraction at constant pressure (*above*) and mole fraction in the vapor phase in dependence of the mole fraction in the liquid phase for a constant pressure

effectively in many diagrams and is shown on the right in Fig. 2.1-19. After all it is nothing but the law of mass conservation of the light ends substance as well as the total mass $(L + G)$.

It is common practice to use the following equilibrium constant K_a:

$$K_a = \frac{\tilde{y}_a^*}{\tilde{x}_a} . \tag{2.1-62}$$

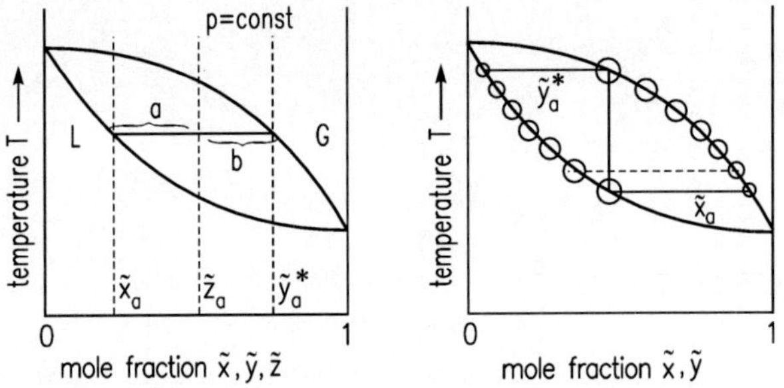

Fig. 2.1-19 Boiling curve and dew-point curve of a binary mixture (*left*) with iconographic representation of the lever rule (*right*)

If *Raoult's* and *Dalton's laws* are applicable, K_a is

$$K_a = \frac{p_a^0}{p}. \tag{2.1-63}$$

A combination of the last equations results in

$$\tilde{x}_a = \frac{\frac{L}{G}+1}{\frac{L}{G}+K_a}\tilde{z}_a. \tag{2.1-64}$$

For given values for $\tilde{z}_a$, K_a, and the ratio L/G, the concentrations $\tilde{x}_a$ or $\tilde{y}_a^*$ can be calculated. This is also true, if $\dot{L}/\dot{G}$ is given.

The laws of binary mixtures can also be applied to ideal multicomponent mixtures. The mole fraction in liquid and vapor have to add up to 1 and the sum of the partial pressures has to be equal to the total pressure. From this the following equations result:

$$\begin{aligned} &\tilde{x}_a + \tilde{x}_b + \tilde{x}_c + \ldots \tilde{x}_k = 1 \text{ or } \sum_1^k \tilde{x}_i = 1, \\ &\tilde{y}_a + \tilde{y}_b + \tilde{y}_c + \ldots \tilde{y}_k = 1 \text{ or } \sum_1^k \tilde{y}_i = 1, \\ &p_a + p_b + p_c + \ldots p_k = p \text{ or } \sum_1^k p_i = p. \end{aligned} \tag{2.1-65}$$

Raoult's and *Dalton's law* provide these equations

$$
\begin{aligned}
p_{\mathrm{a}} &= \tilde{x}_{\mathrm{a}} \cdot p_{\mathrm{a}}^{0} = \tilde{y}_{\mathrm{a}}^{*} \cdot p,\\
p_{\mathrm{b}} &= \tilde{x}_{\mathrm{b}} \cdot p_{\mathrm{b}}^{0} = \tilde{y}_{\mathrm{b}}^{*} \cdot p,\\
p_{\mathrm{c}} &= \tilde{x}_{\mathrm{c}} \cdot p_{\mathrm{c}}^{0} = \tilde{y}_{\mathrm{c}}^{*} \cdot p,\\
&\vdots\\
p_{k} &= \tilde{x}_{k} \cdot p_{k}^{0} = \tilde{y}_{k}^{*} \cdot p.
\end{aligned}
\tag{2.1-66}
$$

A liquid multicomponent mixture starts vaporizing when the following condition is true:

$$\frac{\tilde{x}_{\mathrm{a}} \cdot p_{\mathrm{a}}^{0}}{p} + \frac{\tilde{x}_{\mathrm{b}} \cdot p_{\mathrm{b}}^{0}}{p} + \frac{\tilde{x}_{\mathrm{c}} \cdot p_{\mathrm{c}}^{0}}{p} + \cdots + \frac{\tilde{x}_{k} \cdot p_{k}^{0}}{p} = 1 \text{ or} \tag{2.1-67}$$

$$\frac{1}{p} \cdot \sum(\tilde{x}_i \cdot p_i^{0}) \text{ and } \sum(K_i \cdot \tilde{x}_i) = 1\,. \tag{2.1-68}$$

Respectively, a vaporous multicomponent mixture starts condensing if the following relation is true:

$$\frac{\tilde{y}_{\mathrm{a}}^{*} \cdot p}{p_{\mathrm{a}}^{0}} + \frac{\tilde{y}_{\mathrm{b}}^{*} \cdot p}{p_{\mathrm{b}}^{0}} + \frac{\tilde{y}_{\mathrm{c}}^{*} \cdot p}{p_{\mathrm{c}}^{0}} + \cdots + \frac{\tilde{y}_{k}^{*} \cdot p}{p_{k}^{0}} = 1 \tag{2.1-69}$$

or

$$p \cdot \sum\left(\frac{\tilde{y}_i^{*}}{p_i^{0}}\right) = 1 \text{ and } \sum\left(\frac{\tilde{y}_i^{*}}{K_i}\right) = 1\text{, respectively.} \tag{2.1-70}$$

Determination of vaporization or condensation temperatures is done by an iterative procedure. This is due to the fact that the vapor pressure curve is specified as a transcendental equation. A system of transcendental equations cannot be solved analytically.

The vapor mole fraction $\tilde{y}_i^{*}$ of a two-phase mixture of a total composition $\tilde{z}_i$ depends on the relation L/G of the liquid and the vapor phase. For component i it is

$$\tilde{y}_i^{*} = \frac{\tilde{z}_i \cdot \left(1 + \frac{L}{G}\right)}{1 + \frac{L}{G \cdot K_i}}\,. \tag{2.1-71}$$

In Fig. 2.1-20 a vapor/liquid separator is shown with inlet flow $\dot{L}+\dot{G}$ and composition $\tilde{z}_i$. In this case x_i^* and y_i^* can be calculated for any flow ratio $\dot{L}/\dot{G}$ using the above equation.

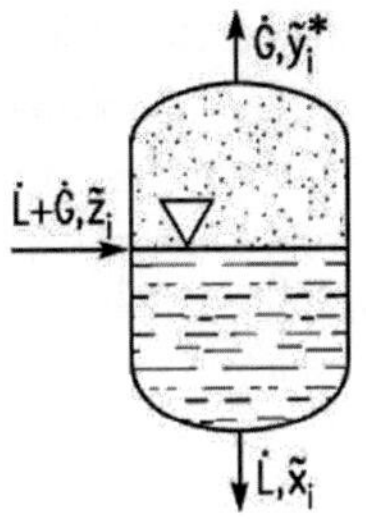

Fig. 2.1-20 Gas-liquid separator

2.1.4 Real Behavior of Liquid Mixtures

In the previous section some criteria were established which determine whether *Raoult's law* can be applied for a binary mixture over the whole range of concentration. The lesser the extent to which these criteria are true the greater the deviations of the partial pressures from their linear dependency on the mole fractions $\tilde{x}_i$. These deviations can be positive as well as negative. In the first case the partial pressure is higher and in the second case it is lower than the partial pressure calculated with *Raoult's law.* If the attractive interactions between the two mixture components are much smaller than the ones between identical molecules, positive deviations from *Raoult's law* result. The stronger the interactions between heterogeneous molecules the smaller the partial pressures compared to *Raoult's law.*

A summary of different binary mixtures is shown in Fig. 2.1-21. The upper row contains partial pressures and the total pressure of different mixtures plotted vs. the mole fraction of the liquid phase. Below, the boiling curve and the dew-point curve are shown for $p = 1$ bar. The lower row contains the respective selectivity diagrams.

The middle column depicts the mixture of benzene/toluene. The interaction energies between benzene and toluene molecules are about as big as the respective ones of benzene and toluene molecules among themselfes. Therefore, *Raoult's law* is about accurate over the whole concentration range. The partial pressures and the total pressure, according to the upper picture, and the boiling curve and the dew-point curve, according to the middle diagram, and the selectivity curve, according to the lower diagram, can be easily calculated with the help of *Raoult's* and *Dalton's law.*

For the mixture isopropyl ether/isopropyl alcohol, positive deviations from *Raoult's law* occur. They are so strong that the total pressure reaches a maximum for a certain concentration. The vapor pressure maximum corresponds to a boiling temperature minimum as it is depicted in the middle row. Boiling curve and dew-point curve contact each other in the so-called azeotropic point A. Here the mole fractions of vapor $\tilde{y}_a^*$ and liquid $\tilde{x}_a$ have the same value which also follows from the equilibrium diagram below. In the azeotropic point A the equilibrium curve intersects the diagonal $\tilde{y}_a = \tilde{x}_a$. To the left of point A the mole fraction of component a in the vapor is higher than that in the liquid. This is to be expected from *Raoult's law*. On the other hand, to the right of the azeotropic point the vapor mole fraction $\tilde{y}_a^*$ is smaller than that of the liquid $\tilde{x}_a$. Now, there is less light ends component in vapor than in liquid, and the equilibrium curve runs below the diagonal.

The interaction energies between the heterogeneous molecules can be so weak that both mixture components do no longer dissolve in each other and exhibit a miscibility gap. This is true for the system water/n-butanol, depicted in the left column. The miscibility gap stretches from $\tilde{x}_a = 0.6$ to $\tilde{x}_a = 0.97$. Liquid mixtures with a concentration in miscibility gap are always in equilibrium with vapor of azeotropic composition. The equilibrium curve runs above the diagonal in the region to the left of the azeotropic point; compared to the liquid the vapor is rich in light ends component. The opposite is the case to the right of the azeotropic point up to the value $\tilde{x}_a = 1$ for pure water. In this region $\tilde{y}_a^* < \tilde{x}_a$.

In the right columns, mixtures are displayed which show negative deviations from *Raoult's law*. For the mixture acetone/chloroform these deviations are so high that a vapor pressure minimum results which corresponds to a boiling point maximum. This can be seen in the boiling curve and the dew-point curve. In the azeotropic point both lines intersect. To the right of the azeotropic point the vapor is richer in light ends component which is to be expected after *Raoult's law*. This is why the equilibrium curve runs above the diagonal. At the azeotropic point it is again $\tilde{y}_a^* = \tilde{x}_a$. Here the equilibrium curve intersects the diagonal. To the left of the azeotropic point up to $\tilde{x}_a = 0$ the vapor has a smaller concentration of light ends component than the liquid for pure chloroform. That is why the equilibrium curve spans below the diagonal.

In the right column the mixture nitric acid/water is shown. For this the explanations given for the mixture acetone/chloroform hold as principal. Taking into account that there are extremely strong interaction energies between nitric acid and water molecules the described effects are very distinct in this case. There are, for example, very negative deviations from *Raoult's law* and the distance of the equilibrium curve to the diagonal is considerable. Solutions of small amounts of nitric acid in water show very small partial pressures of the nitric acid. This effect is even more

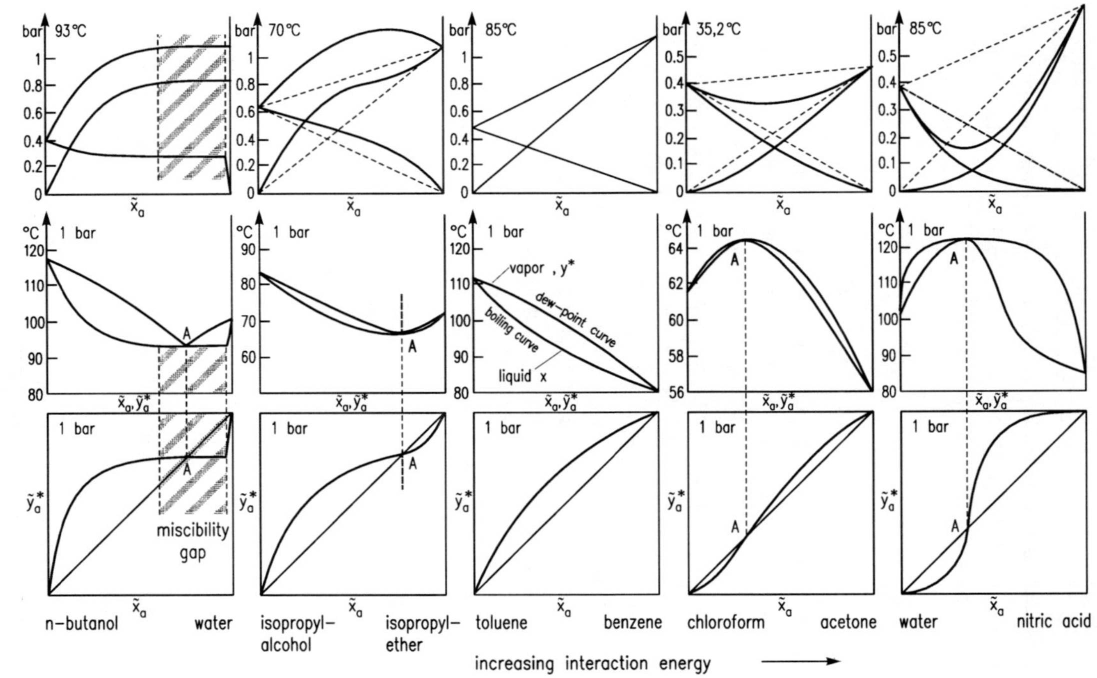

Fig. 2.1-21 *Top row:* partial pressure and total pressure vs. the mole fraction in the liquid phase for different binary mixtures. *Center row:* boiling temperature vs. the mole fraction in the liquid, respectively, in the vaporous phase for different mixtures. *Lower row:* equilibrium diagrams for different mixtures at 1 bar; in the center column a mixture is shown that behaves ideal. The mixtures to the left show positive, the mixtures to the right show negative deviations from Raoult's law

pronounced in case of aqueous hydrochloric acid. Low concentrated hydrochloric acids have a vanishingly small partial pressure of hydrogen chloride.

2.1.4.1 The Gibbs–Duhem Equation

Figure 2.1-21 shows that for many mixtures there are positive or negative deviations from *Raoult's law.* To still be able to use the introduced equations the activity coefficient γ_i is established for every component *i*. For nonideal liquid behavior and if the *law of ideal gases* holds, then

$$p_i = \gamma_i \cdot \tilde{x}_i \cdot p_i^0 = a_i \cdot p_i^0 . \qquad (2.1\text{-}72)$$

The product of activity coefficient γ_i and mole fraction $\tilde{x}_i$ is called activity a_i. Consequently, the partial pressure of component *i* is proportional to the activity and the saturation pressure of the component. The activity coefficient is 1 for ideal behavior, higher than 1 for positive deviations of the partial pressure, and smaller than 1 for negative deviations. Thus the mixture isopropyl ether/isopropyl alcohol has an activity coefficient higher than 1, whereas the mixture acetone/chloroform has values smaller than 1. This is shown in Fig. 2.1-22. In the upper diagrams the partial pressures and the total pressure are plotted vs. the mole fraction. The lower diagrams show the activity coefficient plotted vs. the mole fraction. It was pointed out before that in diluted solutions *Raoult's law* is true for the surplus component (as this is the component contained in excess). This also becomes clear in the upper diagrams where the partial pressure p_a for $\tilde{x}_a \rightarrow 1$ adapts to the straight line according to *Raoult's law.* The same is true for the partial pressure p_b for $\tilde{x}_a \rightarrow 0$. Therefore, for a binary mixture

$$p_a = \gamma_a \cdot \tilde{x}_a \cdot p_a^0; \quad \text{for} \quad \tilde{x}_a \rightarrow 1: \quad \gamma_a = 1 \quad \text{and} \quad \frac{d\gamma_a}{d\tilde{x}_a} = 0 \qquad (2.1\text{-}73)$$

$$p_b = \gamma_b \cdot (1 - \tilde{x}_a) \cdot p_b^0; \quad \text{for} \quad (1 - \tilde{x}_a) \rightarrow 1: \ \gamma_b = 1 \quad \text{and} \quad \frac{d\gamma_b}{d\tilde{x}_a} = 0 . \qquad (2.1\text{-}74)$$

More general statements can be obtained from mixed phase thermodynamics. The dimension ψ may be an extensive state variable, e.g., volume *V*, enthalpy *H*, free enthalpy $G = H - (T \cdot S)$, and entropy *S*. The value of the state variable dimension ψ is a function of pressure p, temperature *T*, and the amount of substances n_a, n_b, etc. of its components. The total differential $d\psi$ of the state variable ψ is

$$d\psi = \left(\frac{\partial \psi}{\partial T}\right)_{p,\, n_j} \cdot dT + \left(\frac{\partial \psi}{\partial p}\right)_{T,\, n_j} \cdot dp + \left(\frac{\partial \psi}{\partial n_a}\right)_{T,\, p,\, n_j \neq n_a} \cdot dn_a + \left(\frac{\partial \psi}{\partial n_b}\right)_{T,\, p,\, n_j \neq n_b} \cdot dn_b + \cdots \qquad (2.1\text{-}75)$$

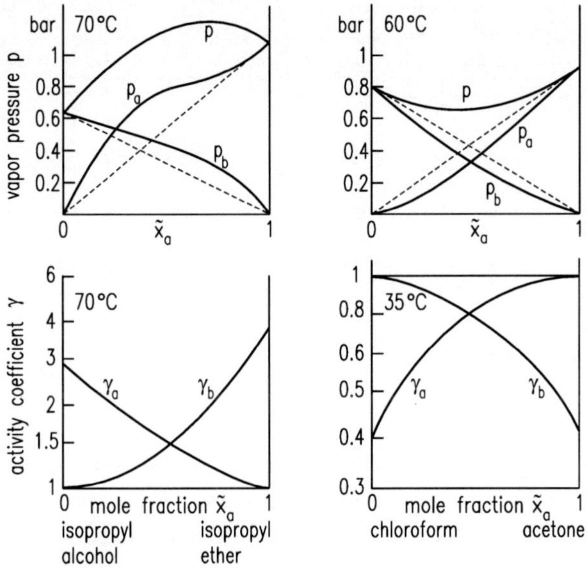

Fig. 2.1-22 Vapor pressure (*upper row*) and activity coefficients (*lower row*) vs. the mole fraction in the liquid phase. (*Left*) a mixture with positive and (*right*) a mixture with negative deviations from Raoult's law are plotted

For the index j of the amount of substance n all other amounts of substance $n_j (j = \mathrm{a, b}, \ldots)$ are to be kept constant for differentiation.

Partial molar state variables ψ_i are defined as follows:

$$\psi_i = \left(\frac{\partial \psi}{\partial n_i}\right)_{T, p, n_{j \neq i}} . \tag{2.1-76}$$

It was shown before that the phase equilibrium of a single component system can be described by the free enthalpy G. The partial molar free enthalpy G_i is

$$G_i = \left(\frac{\partial G}{\partial n_i}\right)_{T, p, n_{j \neq i}} \equiv \mu_i . \tag{2.1-77}$$

The variable μ_i is called the chemical potential of component i.

The total differential of the molar free enthalpy is then

$$\mathrm{d}G = \left(\frac{\partial G}{\partial T}\right)_{p, n_j} \cdot \mathrm{d}T + \left(\frac{\partial G}{\partial p}\right)_{T, n_j} \cdot \mathrm{d}p + \mu_\mathrm{a} \cdot \mathrm{d}n_\mathrm{a} + \mu_\mathrm{b} \cdot \mathrm{d}n_\mathrm{b} + \cdots . \tag{2.1-78}$$

A combination of the first and second law of thermodynamics supplies

$$(\mathrm{d}G)_{n_j} = V \cdot \mathrm{d}p - S \cdot \mathrm{d}T \tag{2.1-79}$$

or

$$\left(\frac{\partial G}{\partial T}\right)_{p,\, n_j} = -S \quad \text{and} \quad \left(\frac{\partial G}{\partial p}\right)_{T,\, n_j} = V\,. \tag{2.1-80}$$

Therefore it follows

$$\mathrm{d}G = -S \cdot \mathrm{d}T + V \cdot \mathrm{d}p + \mu_\mathrm{a} \cdot \mathrm{d}n_\mathrm{a} + \mu_\mathrm{b} \cdot \mathrm{d}n_\mathrm{b} + \cdots . \tag{2.1-81}$$

The following is true if a G phase and an L phase are in equilibrium:

$$\mathrm{d}G_\mathrm{G} = \mathrm{d}G_\mathrm{L} = 0\,. \tag{2.1-82}$$

In case of phase equilibrium of a mixture, the pressures (mechanical equilibrium), the temperatures (thermal equilibrium), and the chemical potential of all components (materials equilibrium) have to be equal in both phases:

$$p_\mathrm{G} = p_\mathrm{L} \qquad T_\mathrm{G} = T_\mathrm{L} \qquad (\mu_i)_\mathrm{G} = (\mu_i)_\mathrm{L}\,. \tag{2.1-83}$$

For constant pressure and constant temperature, it follows

$$(\mathrm{d}G)_{p,\,T} = \mu_\mathrm{a} \cdot \mathrm{d}n_\mathrm{a} + \mu_\mathrm{b} \cdot \mathrm{d}n_\mathrm{b} + \cdots = 0\,. \tag{2.1-84}$$

After Euler's homogeneous function theorem, see, e.g., Stephan et al. (2010), the extensive state variable G is

$$G = \mu_\mathrm{a} \cdot n_\mathrm{a} + \mu_\mathrm{b} \cdot n_\mathrm{b} + \cdots \tag{2.1-85}$$

or differentiated

$$(\mathrm{d}G)_{p,\,T} = (\mu_\mathrm{a} \cdot \mathrm{d}n_\mathrm{a} + \mu_\mathrm{b} \cdot \mathrm{d}n_\mathrm{b} + \cdots + n_\mathrm{a} \cdot \mathrm{d}\mu_\mathrm{a} + n_\mathrm{b} \cdot \mathrm{d}\mu_\mathrm{b} + \cdots)_{p,\,T}\,. \tag{2.1-86}$$

Subtraction of (2.1-84) from (2.1-86) leads to the *Gibbs–Duhem* equation

$$(n_\mathrm{a} \cdot \mathrm{d}\mu_\mathrm{a} + n_\mathrm{b} \cdot \mathrm{d}\mu_\mathrm{b} + \cdots)_{p,\,T} = 0\,. \tag{2.1-87}$$

For binary mixtures it follows

$$(\mathrm{d}\mu_\mathrm{a})_{p,\,T} = \left(\frac{\partial \mu_\mathrm{a}}{\partial \tilde{x}_\mathrm{a}}\right)_{p,\,T} \cdot \mathrm{d}\tilde{x}_\mathrm{a} \quad \text{and} \quad (\mathrm{d}\mu_\mathrm{b})_{p,\,T} = \left(\frac{\partial \mu_\mathrm{b}}{\partial \tilde{x}_\mathrm{a}}\right)_{p,\,T} \cdot \mathrm{d}\tilde{x}_\mathrm{a}\,, \tag{2.1-88}$$

and substituted in the next to last equation the *Gibbs–Duhem equation* for a binary mixture finally results

$$\tilde{x}_a \cdot \left(\frac{\partial \mu_a}{\partial \tilde{x}_a}\right)_{p,T} + (1-\tilde{x}_a) \cdot \left(\frac{\partial \mu_b}{\partial \tilde{x}_a}\right)_{p,T} = 0 . \tag{2.1-89}$$

In the next step the chemical potential of gases is supposed to be described by the state variables pressure and temperature. The chemical potential of a pure, ideal gas μ^{id} is

$$\mu^{id}(p,T) = \mu_0(p^+,T) + \int_{p^+}^{p} \tilde{v}^{id} dp = \mu_0(p^+,T) + \tilde{R} \cdot T \cdot \ln\left(\frac{p}{p^+}\right) \tag{2.1-90}$$

with the molar volume $\tilde{v}^{id} = \tilde{R} \cdot T/p$ of an ideal gas. The dimension p^+ is a reference pressure. A real pure gas has a molar volume $\tilde{v}$. The deviation of a real gas from the ideal gas behavior can either be described by the difference in volume

$$\Delta\tilde{v} = \tilde{v}^{id} - \tilde{v} = \frac{\tilde{R} \cdot T}{p} - \tilde{v} \tag{2.1-91}$$

or by the compressibility factor

$$Z = \frac{p \cdot \tilde{v}}{\tilde{R} \cdot T} = 1 - \frac{\Delta\tilde{v} \cdot p}{\tilde{R} \cdot T} . \tag{2.1-92}$$

By the introduction of fugacity f the chemical potential μ^{real} of a real gas can be expressed as

$$\mu^{real}(p,T) = \mu_0(p^+,T) + \tilde{R} \cdot T \cdot \ln\left(\frac{f}{p^+}\right) . \tag{2.1-93}$$

The fugacity coefficient φ is the ratio of fugacity f and pressure p

$$\varphi = \frac{f}{p} . \tag{2.1-94}$$

Conversions provide the following relation between fugacity and pressure:

$$\varphi = \frac{f}{p} = \exp\left[-\int_0^p \frac{\Delta\tilde{v}}{\tilde{R}T} dp\right] = \exp\left[\int_0^p (Z-1)\frac{dp}{p}\right] . \tag{2.1-95}$$

Hence, the fugacity can be calculated as a function of pressure and temperature by using the excess volume $\Delta\tilde{v}$ or the compressibility factor Z.

In the following, the aforementioned relations are applied to gas mixtures. The chemical potential of component i in a mixture of ideal gases is

$$\mu_i^{\text{id}}(p, T) = \mu_{0i}(p^+, T) + \tilde{R} \cdot T \cdot \ln\left(\frac{p_i}{p^+}\right), \tag{2.1-96}$$

with $\mu_{0i}(p^+, T)$ as the standard potential of the pure component. If the mixture consists of real gases the partial pressure p_i has to be substituted by the fugacity f_i:

$$\mu^{\text{real}}(p, T) = \mu_{0i}(p^+, T) + \tilde{R} \cdot T \cdot \ln\left(\frac{f_i}{p^+}\right). \tag{2.1-97}$$

Herein, f_i is the fugacity of component i which can be calculated with the help of the fugacity coefficient φ_i and the partial pressure p_i:

$$\varphi_i = \frac{f_i}{p_i}. \tag{2.1-98}$$

The differential of μ^{real} in (2.1-97) is

$$(\mathrm{d}\mu_i)_{p,T}^{\text{real}} = \tilde{R} \cdot T \cdot \mathrm{d}(\ln f_i), \tag{2.1-99}$$

and this leads to the following formulation of the *Gibbs–Duhem* equation:

$$\tilde{x}_{\text{a}} \cdot \left[\frac{\partial(\ln f_{\text{a}})}{\partial \tilde{x}_{\text{a}}}\right]_{p,T} + (1 - \tilde{x}_{\text{a}}) \cdot \left[\frac{\partial(\ln f_{\text{b}})}{\partial \tilde{x}_{\text{a}}}\right]_{p,T} = 0. \tag{2.1-100}$$

If the pressure is sufficiently small, fugacity fades to pressure:

$$\lim_{f_i \to 0} f_i = p_i. \tag{2.1-101}$$

Then the *Duhem–Margules* equation results

$$\tilde{x}_{\text{a}} \cdot \left[\frac{\partial(\ln p_{\text{a}})}{\partial \tilde{x}_{\text{a}}}\right]_{p,T} + (1 - \tilde{x}_{\text{a}}) \cdot \left[\frac{\partial(\ln p_{\text{b}})}{\partial \tilde{x}_{\text{a}}}\right]_{p,T} = 0. \tag{2.1-102}$$

This equation allows the calculation of the partial pressure of one component from the behavior of the partial pressure of the other component as a function of concentration. Apart from that, it is possible to test experimentally obtained equilibrium data for their thermodynamic consistency.

There are two limiting solutions to the *Duhem–Margules* equation:

- The first term of this equation is set to 1 and the second term is set to -1. The integration of both terms, in consideration of the boundary conditions, yields *Raoult's law* for both components. This means that the activity coefficients are equal to 1 over the whole range of concentrations.

- Both terms are set to 0. In this case the vapor pressures of both components add up to the total pressure. The partial pressures are independent of the mole fraction.

The second statement applies for binary mixtures, whose components are completely insoluble in one another. Such mixtures have a distinct vapor pressure maximum and therefore boil at temperatures lower than the boiling temperatures of each of the components. The boiling temperature of such a mixture can be easily determined, as it is shown in Fig. 2.1-23. In this graph the vapor pressures of water and some other organic substances that are almost completely insoluble in water are shown. Drawing the line, which depicts the total pressure minus the saturation pressure of water vs. the temperature (in the graph, 1 and 0.1 bar are chosen for example), the boiling temperature can be read off the intersect of this curve with the vapor pressure curve of each organic component. Pure toluene would boil at 115°C (at 1 bar); however the mixture water/toluene already boils at a temperature of about 85°C. This behavior is used for the so-called water vapor distillation to gently vaporize organic substances at the lowest possible temperatures.

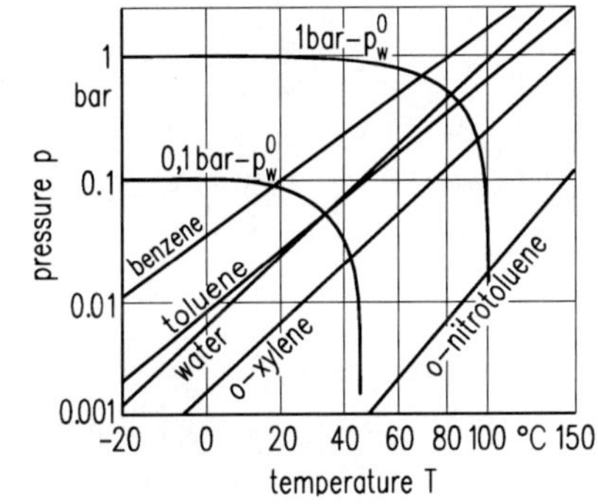

Fig. 2.1-23 Vapor pressure of some substances vs. the temperature. The diagram contains two curves for a constant difference between total pressure and the saturation pressure of water

Apart from these exceptions, the activity coefficient generally is a function of concentration, temperature, and pressure. Before discussing these correlations it is advisable to discuss the heat effects of solutions or binary systems, which consist of a vaporous and a condensed phase (liquid or solid).

2.1.4.2 *Heat of Phase Transition, Mixing, Chemical Bonding*

Assume two liquid components at equal temperature and pressure. When mixing one with the other at isobaric conditions, the temperature may rise, fall, or remain constant. In ideal mixtures the interaction potential between molecules of both constituents mimics those potentials between molecules within each of the starting liquids. Then the molar enthalpy $\bar{\tilde{h}}^{\mathrm{id}}$ of the mixture follows from the molar enthalpy $\tilde{h}_i^0$ of the respective constituents and their molar fractions $\tilde{x}_i$. For a binary mixture this gives

$$\bar{\tilde{h}}^{\mathrm{id}} = \tilde{x}_{\mathrm{a}} \cdot \tilde{h}_{\mathrm{a}}^0 + \tilde{x}_{\mathrm{b}} \cdot \tilde{h}_{\mathrm{b}}^0 \ . \tag{2.1-103}$$

The corresponding enthalpy vs. concentration plot exhibits an isotherm in the form of a straight line cf. Fig. 2.1-24. In case of real mixtures, $\overline{\Delta\tilde{h}} = \bar{\tilde{h}} - \bar{\tilde{h}}^{\mathrm{id}}$ deviates from zero. The expression $\overline{\Delta\tilde{h}}$ is known as molar enthalpy of mixing. It assumes negative values for exothermic mixing where heat has to be removed from the system to obtain isobaric and isothermal conditions. Supplied heat in case of endothermal mixing is denoted with positive values.

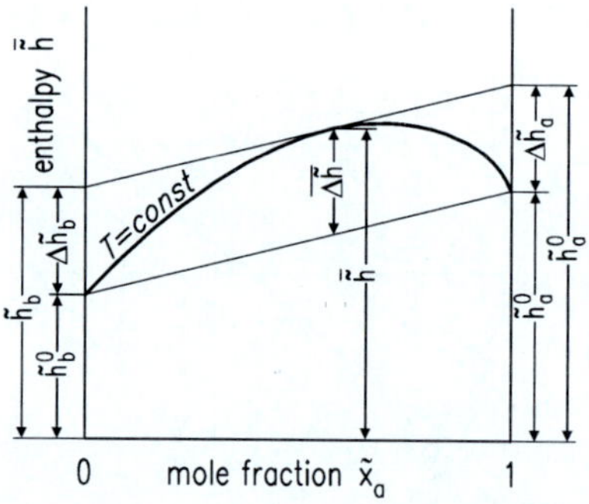

Fig. 2.1-24 Molar enthalpy of a real mixture dependent on the mole fraction of the liquid phase. The isotherm is valid for the case of endothermal mixing behavior

Figure 2.1-24 shows the characteristics of an endothermal mixing, i.e., the isotherms are running above the isotherm valid for an ideal system. The molar mixing enthalpy for k constituents is

$$\overline{\Delta\tilde{h}} = \bar{\tilde{h}} - \sum_1^k \tilde{h}_i^0 \cdot \tilde{x}_i = \sum_1^k (\tilde{h}_i - \tilde{h}_i^0) \cdot \tilde{x}_i . \tag{2.1-104}$$

The difference $\tilde{h}_i - \tilde{h}_i^0$ of the component i is the partial molar enthalpy of mixing

$$\tilde{h}_i - \tilde{h}_i^0 = \Delta\tilde{h}_i . \tag{2.1-105}$$

For a binary mixture it follows

$$\overline{\Delta\tilde{h}} = \tilde{h}_a \cdot \tilde{x}_a + \tilde{h}_b \cdot \tilde{x}_b - \tilde{h}_a^0 \cdot \tilde{x}_a - \tilde{h}_b^0 \cdot \tilde{x}_b = (\tilde{h}_a - \tilde{h}_a^0) \cdot \tilde{x}_a + (\tilde{h}_b - \tilde{h}_b^0) \cdot \tilde{x}_b. \tag{2.1-106}$$

Let b denote the solute, which in general may be present in the form of a gaseous phase, as in the case of absorption, or in the form of a solid phase, as in the case of solid solution. The enthalpy of the pure component b is referred to as $\tilde{h}_b^{0\prime}$. The expressions $\tilde{h}_b^{0\prime}$ and $\tilde{h}_b^0$ differ from each other in terms of the molar phase transition enthalpy λ_b of each pure component b:

$$\lambda_b = \tilde{h}_b^0 - \tilde{h}_b^{0\prime}. \tag{2.1-107}$$

This amount of heat is to be removed (negative sign) on condensation and needs to be supplied (positive sign) on melting if isobaric isothermal conditions are intended. The following expression is called differential solution enthalpy or differential heat of solution of the constituent b:

$$\Delta\tilde{h}_b' = (\tilde{h}_b - \tilde{h}_b^0) + \lambda_b = \Delta\tilde{h}_b + \lambda_b.$$

This term accounts for the difference between the partial molar enthalpy of a component b after mixing, as compared to the molar enthalpy of a gaseous or solid pure component b before mixing. For the special case of a liquid initial component the relation

$$\Delta\tilde{h}_b' = \Delta\tilde{h}_b \tag{2.1-108}$$

holds, i.e., the differential solution enthalpy equals the partial molar mixing enthalpy of this component.

Mixed phase thermodynamics provides the following important relations between the fugacity or activity coefficient on the one hand and the mixing enthalpy on the other hand:

$$\left(\frac{\partial \ln f_i}{\partial T}\right)_{p,\tilde{x}} = \left(\frac{\partial \ln(\gamma_i \cdot \tilde{x}_i \cdot f_i^0)}{\partial T}\right)_{p,\tilde{x}} = \frac{\tilde{h}_i^{0\prime} - \tilde{h}_i}{\tilde{R} \cdot T^2} = \frac{(\tilde{h}_i^{0\prime} - \tilde{h}_i^0) + \tilde{h}_i^0 - \tilde{h}_i}{\tilde{R} \cdot T^2} \tag{2.1-109}$$

and further

$$\left(\frac{\partial \ln \gamma_i}{\partial T}\right)_{p,\tilde{x}} = \frac{\tilde{h}_i^0 - \tilde{h}_i}{\tilde{R} \cdot T^2} = -\frac{\Delta\tilde{h}_i}{\tilde{R} \cdot T^2}. \tag{2.1-110}$$

These expressions can be used to calculate the differential solution enthalpy $\tilde{h}_i^{0\prime} - \tilde{h}_i = -\Delta\tilde{h}_i'$ for each constituent of the mixture as a function of the temperature dependence of the fugacity and to calculate the partial molar mixing enthalpy $\Delta\tilde{h}_i$ as a function of the temperature dependency of the activity coefficient.

For small pressures fugacity can be substituted by the pressure. Starting with

$$p_i = \gamma_i \cdot \tilde{x}_i \cdot p_i^0 = \tilde{x}_i \cdot He_i = a_i \cdot p_i^0 , \tag{2.1-111}$$

the following relations for the phase transition enthalpy in a system of gas and condensed phase (e.g., valid for desorption where $\Delta\tilde{h}_i'$ is supplied and therefore is denoted with a positive sign) can be obtained:

$$\Delta\tilde{h}_i' = -\tilde{R} \cdot \left[\frac{\partial \ln p_i}{\partial(1/T)}\right]_{p,\tilde{x}} = -\tilde{R} \cdot \left[\frac{\partial \ln He_i}{\partial(1/T)}\right]_{p,\tilde{x}} \tag{2.1-112}$$

or

$$\begin{aligned} \Delta\tilde{h}_i' &= -\tilde{R} \cdot \left[\frac{\partial \ln \gamma_i}{\partial(1/T)}\right]_{p,\tilde{x}} \\ &-\tilde{R} \cdot \left[\frac{\partial \ln p_i^0}{\partial(1/T)}\right]_{p,\tilde{x}} = \Delta\tilde{h}_i + \lambda_i \end{aligned} \tag{2.1-113}$$

In case of absorption, the condensed phase is the absorptive which accommodates the gaseous material to be absorbed. Whereas in case of the adsorbent the condensed phase is a solid, whose surface allows for accumulation of the gaseous or liquid material to be adsorbed. The absorption and adsorption enthalpies $\tilde{q}$, which are normalized to the amount of substance, read

$$\tilde{q} = \Delta\tilde{h}_i' . \tag{2.1-114}$$

The presented relations are applicable to the calculation of absorption as well as adsorption enthalpies $\Delta\tilde{h}_i'$ from temperature dependence of partial pressures and from temperature dependence of the activity coefficients.

In case of gaseous phase adsorption, the phase transition enthalpy $\Delta\tilde{h}_i'$ is released during adsorption, and can be discriminated into the phase transition enthalpy λ_i of the pure component i and the so-called bond enthalpy $\Delta\tilde{h}_{i\mathrm{B}}$:

$$\Delta\tilde{h}_i' = \lambda_i + \Delta\tilde{h}_{i\mathrm{B}} . \tag{2.1-115}$$

For the process of sorption it is common practice to normalize the partial pressure p_i by the saturation vapor pressure p_i^0. This quotient is denoted relative saturation φ_i:

$$\varphi_i = \frac{p_i}{p_i^0} \cong \gamma_i \cdot \tilde{x}_i . \tag{2.1-116}$$

The bond enthalpy $\Delta\tilde{h}_{iB}$ may be written in terms of the temperature dependence of the relative saturation φ_i at a given constant concentration $\tilde{x}_i$ in the adsorbed phase or at a given constant loading $\tilde{X}_i$:

$$\Delta\tilde{h}_{iB} = -\tilde{R} \cdot \left[\frac{\partial \ln \varphi_i}{\partial(1/T)}\right]_{p,\tilde{x}} . \tag{2.1-117}$$

The bond enthalpy may be of considerable magnitude in case of monolayer coverage on the adsorbing material. With increasing number of adsorbed molecule layers, i.e., with increased loading, the bond enthalpy decreases, cf. Chaps. 9 and 10.

As discussed earlier, experimentally determined activity coefficients may be extrapolated to other temperatures making use of the temperature dependence of the partial molar mixing enthalpy. This temperature dependence of the enthalpy shows small enthalpy changes for moderate temperature changes. Thus, if the expression of heat of mixing

$$\left[\frac{\partial \ln \gamma_i}{\partial(1/T)}\right]_{p,\tilde{x}} = \frac{\Delta\tilde{h}_i}{\tilde{R}} \tag{2.1-118}$$

is integrated and if the enthalpy is approximated as constant, the following expression is obtained. It is valid in a limited range of concentration, and $(\Delta\tilde{h}_i)_m$ is an averaged partial molar mixing enthalpy:

$$(\ln \gamma_i)_{T_1} - (\ln \gamma_i)_{T_2} = \frac{(\Delta\tilde{h}_i)_m}{\tilde{R}} \cdot \left(\frac{1}{T_1} - \frac{1}{T_2}\right). \tag{2.1-119}$$

Approximately straight lines are obtained in a plot of the logarithm of the activation coefficient vs. $1/T$ or, better, vs. $1/\tau$, cf. Fig. 2.1-25. Figure 2.1-25 shows the activity coefficient vs. the vapor pressure and the respective boiling temperature for the system water/acetic acid. Provided a small variation of the partial molar mixing enthalpy with temperature, the following simple relation is applicable to recalculate the activity coefficient at different temperatures:

$$T \cdot \ln \gamma_i = \text{const.} \tag{2.1-120}$$

The pressure dependence of the activity coefficient is usually small and is accessible through the following relation:

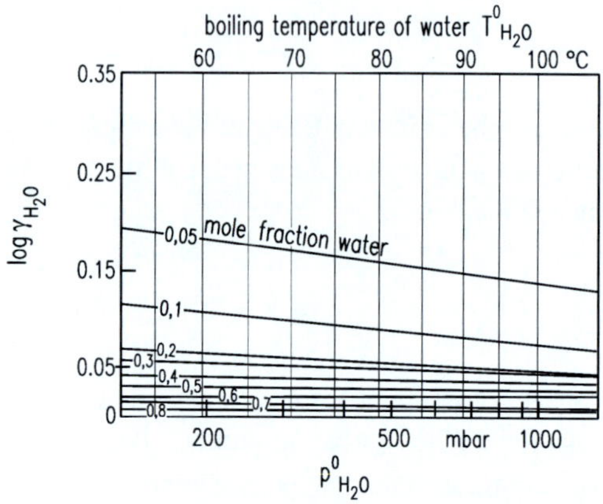

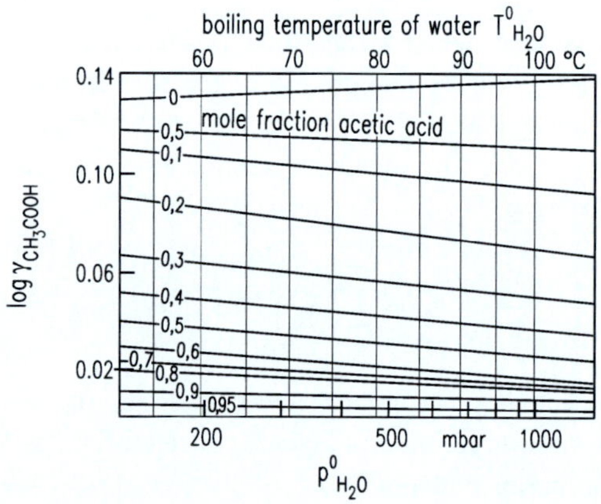

Fig. 2.1-25 Logarithm of the activity coefficient vs. temperature for water (*top*) and acetic acid (*bottom*)

$$\left(\frac{\partial \ln \gamma_i}{\partial p}\right)_{T,\tilde{x}} = \frac{\tilde{v}_i - \tilde{v}_i^0}{\tilde{R} \cdot T}, \tag{2.1-121}$$

where $\tilde{v}_i$ is the partial molar volume in the solution and $\tilde{v}_i^0$ is the molar volume of the pure component. As the volume contraction and expansion are small for the mixing of liquid components, the pressure dependence of the activation coefficient is frequently neglected.

Furthermore, the discussed methods allow the calculation of heat effects emerging for solution of solids in liquids. With the activity a_i it follows

$$\Delta\tilde{h}_i' = -\tilde{R} \cdot \left[\frac{\partial \ln a_i}{\partial(1/T)}\right]_p - \tilde{R} \cdot \left[\frac{\partial \ln p_i^0}{\partial(1/T)}\right]_p. \tag{2.1-122}$$

wherein the first term of the sum

$$\Delta\tilde{h}_i^* = -\tilde{R} \cdot \left[\frac{\partial \ln a_i^*}{\partial(1/T)}\right]_p \tag{2.1-123}$$

is the molar phase transition enthalpy, which accounts for heat transition during dissolving a solid in a nearly saturated solution. Such a saturation almost results in equilibrium between both phases. The symbol for the activity in this regime has a star as superscript. According to the previous equation the solution enthalpy from the temperature dependence of the saturation activity a_i^* can be obtained, if the solid dissolves in a liquid almost at saturation concentration $\tilde{y}_i^*$. In case of an ideal liquid behavior with an activity coefficient of 1, the equation reads

$$\Delta\tilde{h}_i^* = -\tilde{R} \cdot \left[\frac{\partial \ln \tilde{y}_i^*}{\partial(1/T)}\right]_p . \tag{2.1-124}$$

In this case the heat of solution equals the pure solid's heat of melting. Naphthalene and benzene molecules resemble one another to a large extent. Due to that, an almost ideal behavior of the liquid results and activity coefficients are close to 1. Thus, when dissolving solid naphthalene in liquid benzene, the released heat closely resembles the heat of melting of naphthalene. The heat effects of solidification and crystallization yield heat quantities identical to melting and dissolving, but with opposite sign.

2.1.4.3 Excess Quantities

The molar enthalpy of mixing $\overline{\Delta\tilde{h}}$,

$$\overline{\Delta\tilde{h}} = \bar{\tilde{h}} - \sum_1^k \tilde{h}_i^0 \cdot \tilde{x}_i , \tag{2.1-125}$$

describes the difference of enthalpies before and after mixing. Analogously for the entropy of mixing $\overline{\Delta\tilde{s}}$ this becomes

$$\overline{\Delta\tilde{s}} = \bar{\tilde{s}} - \sum_1^k \tilde{s}_i^0 \cdot \tilde{x}_i . \tag{2.1-126}$$

This equation can be rewritten in terms of two other expressions which distinguish ideal from nonideal mixture behavior, i.e., the entropy of mixing of ideal fluids and the so-called real part:

$$\overline{\Delta\tilde{s}} = -\tilde{R} \cdot \sum_1^k \tilde{x}_i \cdot \ln \tilde{x}_i + \bar{\tilde{s}}^{E} . \tag{2.1-127}$$

The real part $\bar{\tilde{s}}^{\mathrm{E}}$ is also called excess entropy. The free enthalpy $\bar{\tilde{g}}$ is defined as

$$\bar{\tilde{g}} = \sum_1^k \mu_i^0 \cdot \tilde{x}_i + \overline{\Delta\tilde{g}}\,, \tag{2.1-128}$$

where $\overline{\Delta\tilde{g}}$ is the free enthalpy of mixing. This expression can be split into two terms accounting for ideal and nonideal mixing properties, respectively. It follows

$$\overline{\Delta\tilde{g}} = \overline{\Delta\tilde{h}} - T \cdot \overline{\Delta\tilde{s}} = \overline{\Delta\tilde{h}} + \tilde{R} \cdot T \cdot \sum_1^k \tilde{x}_i \cdot \ln\tilde{x}_i - T \cdot \bar{\tilde{s}}^{\mathrm{E}} \quad , \tag{2.1-129}$$

and the excess enthalpy reads

$$\bar{\tilde{g}}^{\mathrm{E}} = \bar{\tilde{h}}^{\mathrm{E}} - T \cdot \bar{\tilde{s}}^{\mathrm{E}}\,. \tag{2.1-130}$$

These excess quantities can be closely connected to the activity coefficients. With the excess molar free enthalpy $\bar{\tilde{g}}^{\mathrm{E}} = \overline{\Delta\tilde{g}} - \overline{\Delta\tilde{g}}^{\mathrm{id}}$ this gives

$$\bar{\tilde{g}}^{E} = \tilde{R} \cdot T \cdot \sum_1^k \tilde{x}_i \cdot \ln \gamma_i. \tag{2.1-131}$$

This free excess enthalpy accounts for every nonideal thermodynamic behavior within the whole mixture and may be considered as the fingerprint of the system's nonideal mixing properties. The evolution of this quantity is accessible through measurement of parameters at phase equilibrium, e.g., isothermally recorded values of $p, \tilde{x}$, and $\tilde{y}$. In case of a binary mixture the graph of g^{E} vs. the concentration has the shape of an arch, starting at the abscissa $\tilde{x}_{\mathrm{a}} = 0$ with the value 0, rising to the maximum and declining again to 0 at the abscissa point $\tilde{x}_{\mathrm{a}} = 1$, cf. Fig. 2.1-26.

The evolution of the free excess enthalpy vs. the concentration may be analyzed with regard to the individual activity coefficients by means of the following relation:

$$\tilde{R} \cdot T \cdot \ln \gamma_k = \bar{\tilde{g}}^{\mathrm{E}} - \sum_1^{k-1} \tilde{x}_i \cdot \left(\frac{\partial \bar{\tilde{g}}^{\mathrm{E}}}{\partial \tilde{x}_i}\right)_{T,p,\tilde{x}_j \neq \tilde{x}_i} . \tag{2.1-132}$$

For binary mixtures this equation simplifies to

$$\tilde{R} \cdot T \cdot \ln \gamma_a = \bar{\tilde{g}}^{\mathrm{E}} + (1 - \tilde{x}_{\mathrm{a}}) \cdot \left(\frac{\partial \bar{\tilde{g}}^{\mathrm{E}}}{\partial \tilde{x}_{\mathrm{a}}}\right)_{T,p} \tag{2.1-133}$$

and

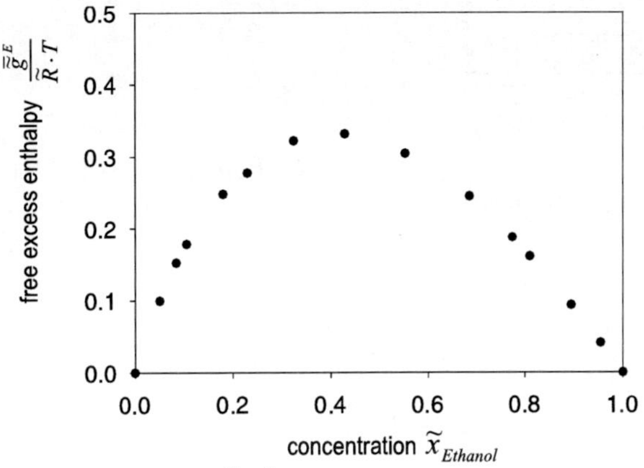

Fig. 2.1-26 Free excess enthalpy $\tilde{\bar{g}}^{E}/(\tilde{R}\cdot T)$ vs. the mole fraction of ethanol, for the mixture ethanol–water, at 55°C (From DECHEMA Chemistry Data Series.)

$$\tilde{R}\cdot T\cdot \ln\gamma_b = \tilde{\bar{g}}^{E} - \tilde{x}_a\cdot\left(\frac{\partial\tilde{\bar{g}}^{E}}{\partial\tilde{x}_a}\right)_{T,p} \quad . \tag{2.1-134}$$

Following this expression, further relations are derived Sect. 2.1.4.4

2.1.4.4 *Activity and Activity Coefficient*

Currently, there is no generalized method to predict activity coefficients for given concentration, temperature, and pressure. As previously discussed, the activity coefficient varies only slightly with pressure and a method has been introduced to extrapolate activity coefficients for given temperatures.

From parameters measured at the phase equilibrium, activity coefficients for each component can be determined and correlated empirically to the state variables. This implies the risk of thermodynamic inconsistency among several correlations.

Best practice is to assess one single correlation, i.e., between the free excess enthalpy of the whole system $\tilde{\bar{g}}^{E}/(\tilde{R}\cdot T)$ and the state variables, and then to derive the activity coefficients for each component consistently by means of (2.1-132). Thus the data drawn in Fig. 2.1-26 have to be approximated by an analytical function, which is differentiable with respect to the concentration and which complies with the boundary condition $\tilde{\bar{g}}^{E}/(\tilde{R}\cdot T) = 0$ for each of the pure substances. The literature provides a huge variety of such expressions, e.g., Margules (Margules 1895), van Laar (van Laar 1935), Wilson (Wilson 1964), NRTL (Renon and Prausnitz 1968), or UNIQUAC (Abrams and Prausnitz 1975). The procedure is elucidated following the method after Margules for a binary mixture:

$$\frac{\tilde{\tilde{g}}^{E}}{\tilde{R} \cdot T} = \tilde{x}_a \cdot \tilde{x}_b \cdot (A_{ab} \cdot \tilde{x}_a + A_{ba} \cdot \tilde{x}_b) . \qquad (2.1\text{-}135)$$

The values for the empirical constants A_{ab} and A_{ba} result after fitting the function to the measured data as shown in Fig. 2.1-26. Equation (2.1-132) then yields values for the activity coefficients

$$\ln \gamma_a = \tilde{x}_b^2 \cdot (A_{ba} + 2 \cdot \tilde{x}_a \cdot (A_{ab} - A_{ba})) \qquad (2.1\text{-}136)$$

and

$$\ln \gamma_b = \tilde{x}_a^2 \cdot (A_{ab} + 2 \cdot \tilde{x}_b \cdot (A_{ba} - A_{ab})) . \qquad (2.1\text{-}137)$$

Both equations yield a set of intrinsically consistent parameter values.

Another frequently used correlation is the *Wilson equation*:

$$\frac{\tilde{\tilde{g}}^{E}}{\tilde{R} \cdot T} = -\sum_{1}^{k} \tilde{x}_i \cdot \ln\left(\sum_{1}^{k} \Lambda_{ij} \cdot \tilde{x}_j\right) . \qquad (2.1\text{-}138)$$

From this the activity coefficients of the binary mixture can be calculated:

$$\ln \gamma_a = -\ln[\tilde{x}_a + \Lambda_{ab}(1-\tilde{x}_a)] + (1-\tilde{x}_a)\left[\frac{\Lambda_{ab}}{\tilde{x}_a + \Lambda_{ab}(1-\tilde{x}_a)} - \frac{\Lambda_{ba}}{\Lambda_{ba}\tilde{x}_a + (1-\tilde{x}_a)}\right] ,$$

$$\ln \gamma_b = -\ln[(1-\tilde{x}_a) + \Lambda_{ba}\tilde{x}_a] - \tilde{x}_a\left[\frac{\Lambda_{ab}}{\tilde{x}_a + \Lambda_{ab}(1-\tilde{x}_a)} - \frac{\Lambda_{ba}}{\Lambda_{ba}\tilde{x}_a + (1-\tilde{x}_a)}\right] .$$

(2.1-139)

The Wilson parameters Λ_{ab} and Λ_{ba} involve some physical meaning and can be derived from the molar volume of each constituent and the interaction force between molecules of same or different types:

$$\Lambda_{ab} = \frac{\tilde{v}_{bL}^{0}}{\tilde{v}_{aL}^{0}} \cdot \exp\left(-\frac{\lambda_{ab} - \lambda_{aa}}{\tilde{R} \cdot T}\right) , \qquad (2.1\text{-}140)$$

$$\Lambda_{ba} = \frac{\tilde{v}_{aL}^{0}}{\tilde{v}_{bL}^{0}} \cdot \exp\left(-\frac{\lambda_{ab} - \lambda_{bb}}{\tilde{R} \cdot T}\right) . \qquad (2.1\text{-}141)$$

To solve these equations two terms have to be provided from experimental data, i.e., the two differences involving the interaction energies λ_{aa}, λ_{bb}, λ_{ab}, which are generally not predictable. The variables $\tilde{v}_{aL}^{0}$ and $\tilde{v}_{bL}^{0}$ are the molar volumes of the pure components. The major advantage of the *Wilson equations* is that they render the temperature dependence of the activity coefficient reasonably well. As a drawback the *Wilson equation* does not include maxima or minima of activity coefficients. Such maxima and minima appear, e.g., in mixtures of chloroform with alcohol. Further difficulties arise for systems with a miscibility gap. In the latter case, the UNIFAC method (Fredenslund 1977) has proved successful, which allows for the calculation of activity coefficients on a semiempirical basis resolving influences from each molecular cluster. The approaches after Wilson, NRTL, UNIFAC allow the calculation of activity coefficients even in multicomponent ($k > 2$) systems.

The most common method for treatment of phase equilibria in nonideal mixtures is the previously discussed method to calculate the activity coefficients from the free excess enthalpy. It should be emphasized that this treatment is basically just an interpolation of measured data retaining the thermodynamic consistency. Only nonideal behavior of the liquid phase is comprehended within this approach. If nonideal behavior of the gaseous phase is encountered (e.g., at high pressures), the treatment by means of equations of state is recommended.

2.1.4.5 Fugacity and Fugacity Coefficient, Equilibrium Constant

Thermodynamic equilibria involving vapor mixtures are only predictable with known fugacity coefficients. According to the theory of corresponding states, their behavior depends on the reduced pressure p_r and on the reduced temperature T_r. Near the critical point (p_r > 0.6) the fugacity coefficient is not predictable straightforward for any kind of gas. Some properties of gases of nonassociated, loosely or tightly associated molecules are reasonably well modeled by the real gas factor Z_c at the critical point. For many substances this parameter exhibits a value of 0.27 (Hecht et al. 1966):

$$Z_c = \frac{p_c \cdot v_c}{R \cdot T_c} = 0.27\,. \tag{2.1-142}$$

Figure 2.1-27 shows the fugacity coefficient vs. the reduced pressure p_r, each line being parametrized by a value of the reduced temperature T_r. The line for the saturation limit (vapor pressure curve) ends at the critical point. If the critical real gas constant Z_c deviates from 0.27 then the fugacity coefficient complies with the following empirical law:

$$\left(\frac{f}{p}\right)' = \left(\frac{f}{p}\right) 10^{D(Z_c - (0.27))}\,. \tag{2.1-143}$$

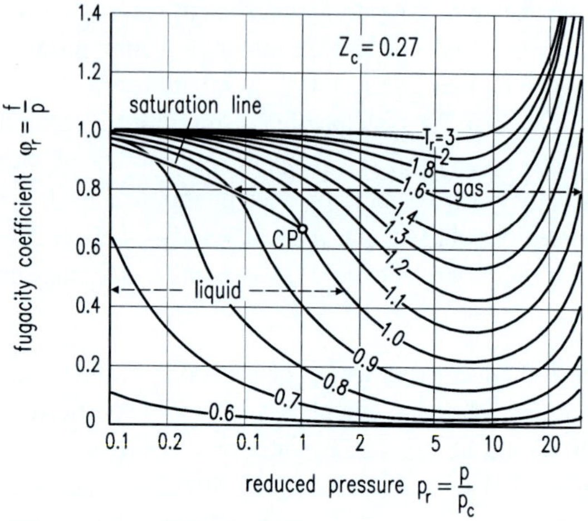

Fig. 2.1-27 Fugacity coefficient dependent on the reduced pressure, parametrized with the reduced temperature, at critical real gas factor $Z_c = 0.27$

Values for the parameter D are given in Hougen et al. (1959). Further reference regarding the behavior of gaseous phases can be found in e.g. Green and Perry (2008).

Far from the critical point ($p_r < 0.6$), the equilibrium constant $K_i = \tilde{y}_i^* / \tilde{x}_i$ can be calculated from the coefficients of fugacity and activity. With the fugacity coefficient

$$\varphi_i = \frac{f_i}{p_i} \text{ at system pressure } p \tag{2.1-144}$$

and with $\varphi_i^0 = f_i^0 / p_i^0$ at the saturation pressure p_i^0 of the pure component i this becomes

$$K_i \equiv \frac{\tilde{y}_i^*}{\tilde{x}_i} = \frac{\gamma_i \cdot \varphi_i^0 \cdot p_i^0}{\varphi_i \cdot p} \cdot \exp\left[\int_{p_i^0}^{p} \frac{\tilde{v}_{iL}^0}{\tilde{R} \cdot T} \mathrm{d}p\right] \tag{2.1-145}$$

with $\tilde{v}_{iL}^0$ the molar volume of the pure liquid component i. As this volume changes only slightly with the pressure, K_i can be rewritten in terms of the average value $(\tilde{v}_{iL}^0)_m$ (Stephan et al. 2010)

$$K_i \equiv \frac{\tilde{y}_i^*}{\tilde{x}_i} = \frac{\gamma_i \cdot \varphi_i^0 \cdot p_i^0}{\varphi_i \cdot p} \cdot \exp\left[\frac{(\tilde{v}_{iL}^0)_m \cdot (p - p_i^0)}{\tilde{R} \cdot T}\right]_T . \tag{2.1-146}$$

Based on this expression the equilibrium constant can be determined from given activity coefficients and material parameters of each component *i*.

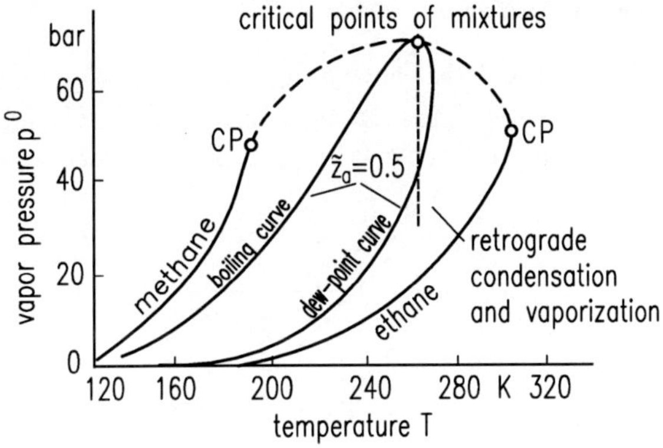

Fig. 2.1-28 Vapor pressure curve for methane and ethane, with boiling curve as well as condensation curve of this mixture at a mole fraction of $\tilde{z}_a = 0.5$ (Hougen et al. 1959)

The thermodynamic characteristics of gas mixtures near the critical point have to be evaluated experimentally. Usually, the critical pressure of a mixture exceeds that of the pure components, as will be elucidated with the example of a methane/ ethane mixture. Figure 2.1-28 shows the vapor pressures of methane and ethane as functions of the temperature. The boiling curve and dew-point curve are drawn for a mole fraction of $\tilde{z}_a = 0.5$. The critical point lies on a point locus from all vapor pressure graphs that emerge for different concentration ratios. This locus spreads between the two critical points of the pure components and features a maximum in between. The two-phase regime spreads between the boiling line and the saturated vapor line, where a liquid and a vapor phase are in equilibrium with one another. In this area isolines of vapor loading or of liquid/vapor ratios L/G can be drawn. The vertical dashed line in Figure 2.1-28 comprises the phenomena of retrograde condensation and evaporation: When reducing the pressure along this line, the mixture partially condenses at constant temperature, and in a subsequent stage evaporates again.

For any liquid mixture component that behaves almost ideally, it is convenient to write the equilibrium constant as a function of total pressure and to incorporate the temperature as a parameter. Figure 2.1-29 shows this type of function in a double logarithmic plot, again for the system methane/ethane. According to the relation

$$K_i^{\mathrm{id}} = \frac{p_i^0}{p} \text{ (with } \gamma_i = 1\text{)}, \tag{2.1-147}$$

the equilibrium constant is inversely proportional to the pressure p, presuming that the *ideal gas law* holds. It appears that deviations rise with increasing pressure. In this regime more reliable results are achieved using the fugacity instead of the pressure as an independent variable.

As pressure values approach the convergence pressure at $K = 1$ this method also suffers accuracy, leaving just the experiment to gain exact values. The fugacity of the ith component is higher than the vapor pressure at the same temperature.

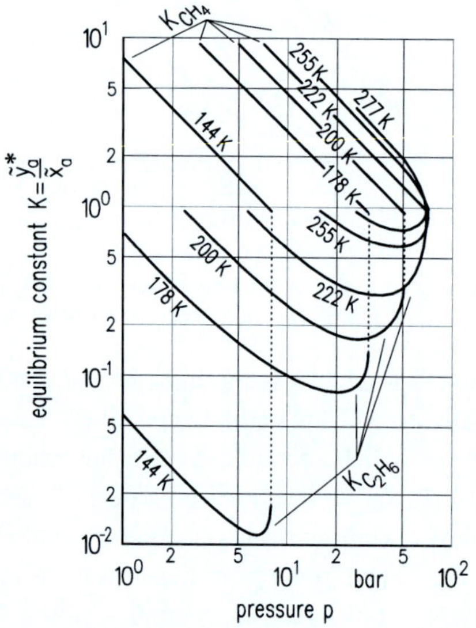

Fig. 2.1-29 Equilibrium constant vs. pressure, for binary mixture methane/ethane (Förg and Stichlmair 1969)

2.2 Liquid/Liquid Systems

Key characteristics in liquid/liquid systems are the mutual solubility and the concentration of both liquid phases in the equilibrium state. The previously discussed laws of mixed phase thermodynamics can still be applied. The feasibility of a liquid/liquid extraction depends on the occurrence of a miscibility gap and its width at different temperatures. Figure 2.2-1 shows three exemplary plots of the solubility temperature vs. the mass fraction. The letter z symbolizes mass fractions in any one- or two-phased mixture. The binary mixture formic acid/benzene has an upper critical solubility temperature whereas the mixture di-n-propylamine/water features a lower critical temperature. The third example system nicotine/water shows both a lower and an upper critical temperature. From these diagrams the concentrations of

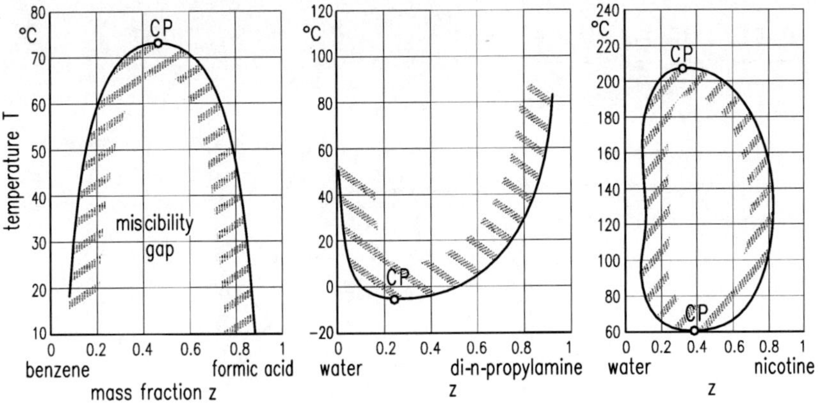

Fig. 2.2-1 Miscibility gaps of several binary mixtures with upper critical point (*lhs*), lower critical point (*center*), upper plus lower critical point (*rhs*)

both liquid phases and their mass fractions at given temperature and given overall phase concentration z can be obtained.

In the process of a liquid/liquid extraction a substance b is transferred from one liquid phase into the other. The donor is called raffinate phase and the recipient is called extract phase. The raffinate phase consists of the carrier substance a and the substance b to be released, whereas the extract phase consists of solvent s and substance b. For the raffinate phase, mass fractions and mass concentrations of the component i are denoted x_i and X_i, respectively, whereas in case of the extract phase, mass fractions and mass concentrations are denoted y_i and Y_i, respectively. Such systems are made up of at least three components. They can be represented in a triangular diagram as illustrated in Fig. 2.2-2 for the system water/benzene/acetic acid.

The three components are associated with the corners. In case of the three exemplary constituents, three binary mixtures are conceivable, i.e., the mixtures water/benzene, acetic acid/water, acetic acid/benzene. Each edge represents the respective binary mixture on a mass fraction scale from 0 to 1. The bottom edge in Fig. 2.2-2 represents the mixture water/benzene, which exhibits a widespread miscibility gap. The left edge represents the binary mixture acetic acid/water, whose components are miscible in any arbitrary relation at temperature 25°C. This holds likewise for the binary mixture acetic acid/benzene which is at the right edge of the triangle.

The mixture of all three components can either form a single-phase system or a two-phase system. Both regimes are delimited by the so-called binodal curve.

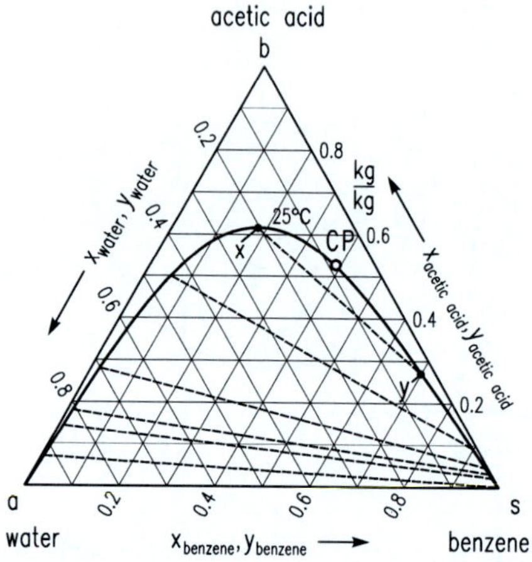

Fig. 2.2-2 Triangular diagram for the system water/benzene/acetic acid at 25°C

Exceeding this line in downward direction, a ternary mixture decomposes into two liquid phases. If for example acetic acid is extracted from water by means of benzene, then the water-rich phase is referred to as raffinate phase, and the benzene-rich phase is the extract phase. Every point with concentration z_i below the binodal curve decomposes into the benzene-rich extract phase with concentration y_i^* and the water-rich raffinate phase with concentration x_i. The dashed lines between the points x_i and y_i^* through some point z_i are called tie lines.

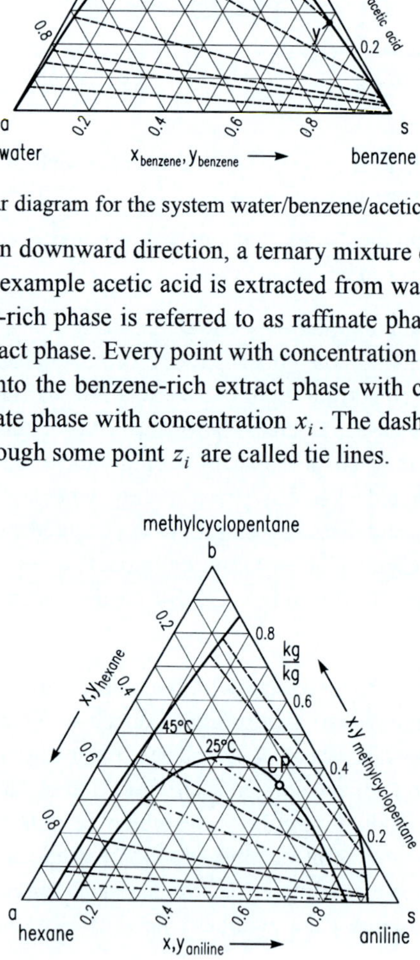

Fig. 2.2-3 Triangular diagram for the system hexane/aniline/methylcyclopentane

The evolution of the binodal curve is strongly coupled with temperature. In Fig. 2.2-3 the binodal curves are given for the system hexane/aniline/methylcyclopentane at temperatures 25 and 45°C, respectively. At 25°C the binary mixture aniline/methylcyclopentane exhibits a miscibility gap, which at 45°C completely vanishes. The tie lines steeply decay from the left to the right, which reflects that the aniline-rich phase is more depleted of methylcyclopentane than the hexane-rich phase. At higher fractions of methylcyclopentane the tie lines become shorter until both ends cojoin in the critical point.

For most mixtures the two-phase region fades away at higher temperatures. As an example the system phenol/water/acetone is shown in Fig. 2.2-4. The two-phase region is extended at 30°C, attenuated at 87°C, and vanishes at 92°C.

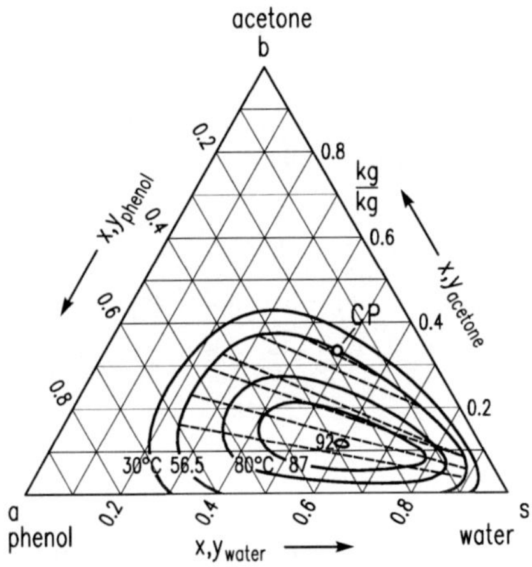

Fig. 2.2-4 Triangular diagram for the system phenol/water/acetone with binodal curves at various temperatures

There are mixtures featuring a miscibility gap for all three binary systems. This is, e.g., known for the system water/ether/succinyl-nitrile as well as for perfluortributylamine/nitroethane/trimethylpentane, which is illustrated in Fig. 2.2-5 for 25°C and 1 bar. Next to each corner three-phase regions with their characteristic total solubility of all three components can be found. Within the two-phase regions, tie lines are drawn. Tie lines provide a partition of the mixture, at a given point, into both coexisting phases. At last, a three-phase region spreads in the center, the corners of this region representing the respective three phases. If the overall composition of the three-phase system is known, then the material ratios of the three phases are found by twofold application of the *lever rule* .

There are some systems where the solvent hardly solves the carrier and vice versa. In such a system the miscibility gap spans almost the full range of concentration,

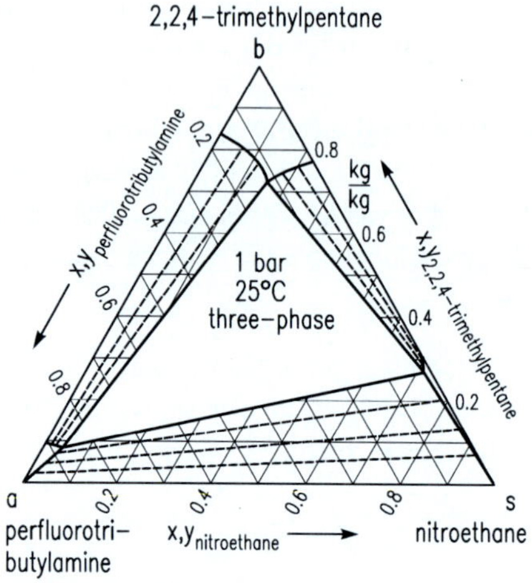

Fig. 2.2-5 Triangular diagram for the system perfluortributylamine/nitroethane/trimethylpentane (Landolt-Börnstein , 2. Band 2. Teil a)

e.g., water/benzene. If the material *b* to be extracted is present only in small concentrations, then the concentration x_b in the raffinate phase is proportional to the concentration y_b^* in the extract phase. In good approximation the *Nernst law* for diluted solutions can be applied. With the equilibrium constant, K_b for component *b* follows

$$K_b = y_b^*/x_b \, . \tag{2.2-1}$$

The equilibrium constant in the *Nernst law* can be approximated by calculation if the real behavior of component *b* in the raffinate phase and in the extract phase are known. According to the Wilson equation, the activity coefficient depends on concentration and on temperature. At a constant temperature – for the concentrations $x_b \to 0$ and $y_b \to 0$ – the logarithmized activity coefficient approaches a certain value called critical value:

$$\begin{aligned} \ln(\gamma_b)_x &= -\ln\Lambda_{ba,x} + 1 - \Lambda_{ab,x} \quad \text{(raffinate phase)} \\ \ln(\gamma_b)_y &= -\ln\Lambda_{ba,y} + 1 - \Lambda_{ab,y} \quad \text{(extract phase)} \end{aligned} \tag{2.2-2}$$

Thus, the activity coefficients can be cast into simplified expressions:

$$\begin{aligned} (\gamma_b)_x &= \exp(1 - \Lambda_{ab,x} - \ln\Lambda_{ba,x}) \\ (\gamma_b)_y &= \exp(1 - \Lambda_{ab,y} - \ln\Lambda_{ba,y}) \end{aligned} \tag{2.2-3}$$

For equal activities in both liquid phases, it follows that

$$(a_{\mathrm{b}})_x = (a_{\mathrm{b}})_y \quad \text{or} \quad (\gamma_{\mathrm{b}})_x \cdot x_{\mathrm{b}} = (\gamma_{\mathrm{b}})_y \cdot y_{\mathrm{b}}^* . \tag{2.2-4}$$

Finally

$$K_{\mathrm{b}} = \frac{y_{\mathrm{b}}^*}{x_{\mathrm{b}}} = \frac{\exp(1 - \Lambda_{\mathrm{ab},x} - \ln \Lambda_{\mathrm{ba},x})}{\exp(1 - \Lambda_{\mathrm{ab},y} - \ln \Lambda_{\mathrm{ba},y})} = \text{const.} \tag{2.2-5}$$

is obtained. This expression is equivalent to the *Nernst law.*

2.3 Solid/Liquid Systems

In general, thermodynamic equilibrium between a solid and a liquid phase (melt, solution) depends on temperature, pressure, and composition. However, in process technology, the pressure dependence of solid–liquid equilibrium may often be neglected, because of the virtual incompressibility of the solid and the liquid phases at moderate pressures.

At thermodynamic equilibrium the chemical potential of each component i in both liquid and solid phases has to be equal. For simple systems and certain simplifications, like pure crystalline solid phase of component b (see Walas 1985), thermodynamic considerations lead to the well-known Clausius–Clapeyron equation

$$\tilde{y}_{\mathrm{b}} = \frac{1}{\gamma_{\mathrm{b}}}\left[\frac{\Delta \tilde{h}_{\mathrm{b,m}}}{\tilde{R}}\left(\frac{1}{T_{\mathrm{b,m}}} - \frac{1}{T}\right)\right] . \tag{2.3-1}$$

(Note: Here, we follow the aforementioned rule and denote the molar composition of the liquid, the high energetic phase, as $\tilde{y}_i$. The composition of the solid, the low energetic phase, is denoted as $\tilde{x}_i$.)

At solid–liquid equilibrium, (2.3-1) relates the mole fraction $\tilde{y}_{\mathrm{b}}$ of component b in the liquid phase to the temperature T. Often, this relation is called solubility. The dependence of this solubility on the composition of the liquid phase is represented by the activity coefficient γ_{b} (see Sect. 2.1.4.4). To apply this equation, only the melting temperature $T_{\mathrm{b,m}}$ and the enthalpy of melting $\Delta h_{\mathrm{b,m}}$ of the pure component b have to be known. Unfortunately, most solid–liquid systems are complex and can therefore only be described piecewise by the above equation or not at all. Often, theoretical prediction of solid–liquid equilibria with the above formula is not successful, because either activity coefficients or the pure component's properties are not known with sufficient accuracy. Furthermore, the formation of associates in the liquid phase gives rise to uncertainties.

There is no clear distinction between solution and melt. Practically, liquid mixtures which are highly concentrated with the crystallizing substance (solute) are named melt. Liquid mixtures of low concentrated solute are named solution and their non-crystallizing components are called solvent.

Generally, the solubility composition of a solute in a solvent can be determined by two experimental methods. According to the first method, a suspension of crystalline solute and solution is kept at constant conditions until equilibrium is reached, then the composition of the clear solution is determined by a suitable method. According to the second method, the equilibrium temperature for solution of a given composition is determined. This temperature may be found by measuring the turbidity or the clear point of the solution.

Solubility data for some binary aqueous systems with crystalline anhydrate as solid are given in Fig. 2.3-1, and for some crystalline hydrates such data are given in Fig.

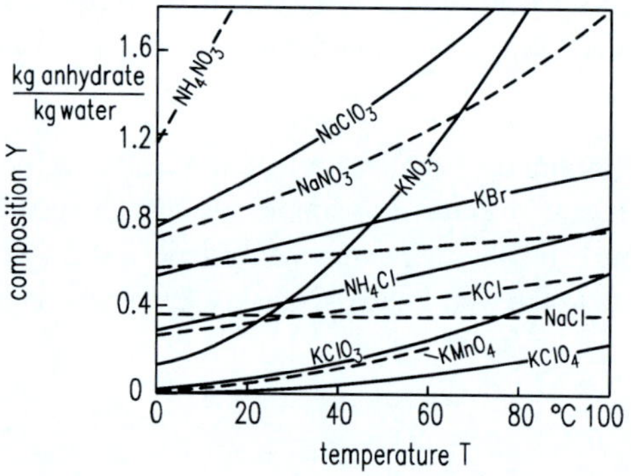

Fig. 2.3-1 Solubility curves of selected aqueous systems, which form crystalline anhydrates

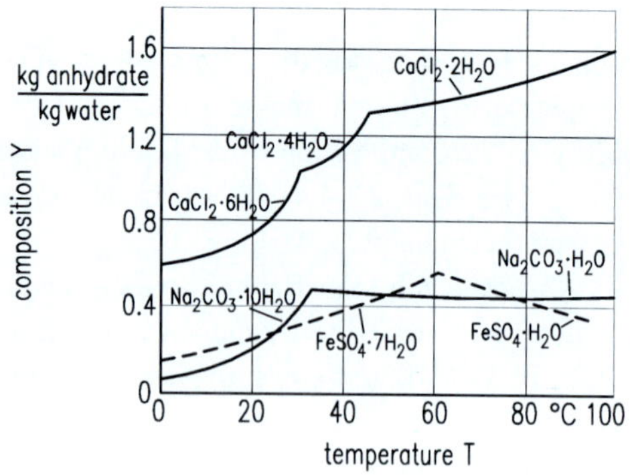

Fig. 2.3-2 Solubility curves of selected aqueous systems, which form crystalline hydrates

2.3-2. Usually, the solubility increases with temperature. However, there are also systems with inverse solubility in certain temperature ranges. In the case of hydrates, the solubility curves have a sharp bend at the points of conversion from one hydrate to another.

When two solutes are dissolved in one solvent, the solubility behavior can be conveniently depicted in a triangular diagram. Figure 2.3-3 shows such a diagram for the

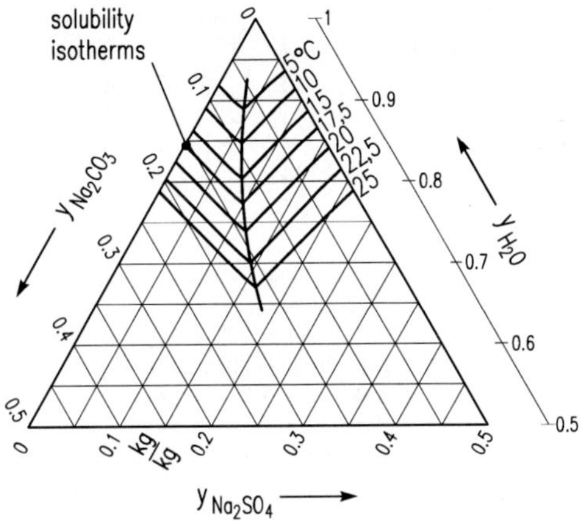

Fig. 2.3-3 Triangular solubility diagram of the system sodium–carbonate sodium sulphate–water

crystallizing system sodium carbonate, sodium sulphate, and water. The regions above the solubility isotherms demark undersaturated solution. Any compositions below the solubility isotherms are supersaturated and not stable. They decompose to either one of the two or both crystalline solids (precipitate) and to the depleted solution (mother liquor).

Crystallizing systems may be grouped into systems which show a eutectic point *E*, systems which show molecular miscibility of the solid phase, and systems which show both (Rittner and Steiner 1985). Respective phase diagrams of such systems are given in Fig. 2.3-4 for the binary case.

Roseboom (1982) classifies the latter systems into the following five types:

•Type I: Components *a* and *b* are miscible at any composition of the solid phase, i.e., anthracene/carbazole (Funakubo 1950)

•Type II: Components *a* and *b* are miscible at any composition of the solid phase, but show a maximum temperature, i.e., d-carvoxime/l-carvoxime (Reinhold and Kircheisen 1926)

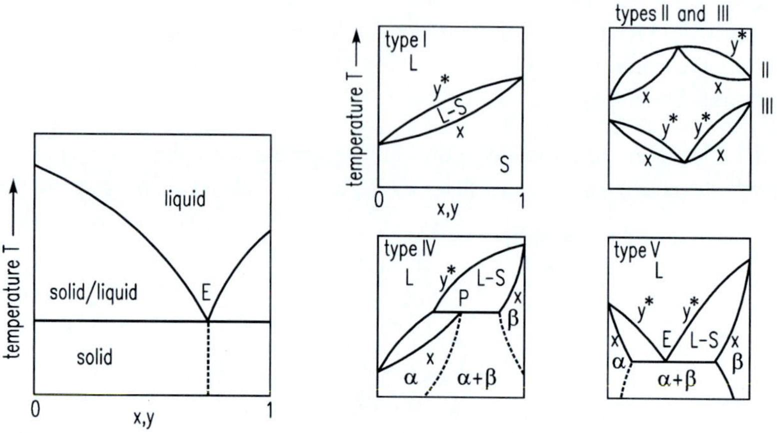

Fig. 2.3-4 Phase diagram of a eutectic binary system with total immiscibility in the two solid phases (*left*); phase diagrams of various types of binary systems with perfect (Type I) or partial (Type II –V) miscibility in the solid phase (*right*)

•Type III: Components *a* and *b* are miscible at any composition of the solid phase, but show a minimum temperature, i.e., *m*-chloronitrobenzene/*m*-flouronitrobenzene (Hasselblatt 1913)

•Type IV: Components *a* and *b* are partially miscible in the solid phases and show a peritectic point *P*, i.e., eicosanole/hexacosanole (Schildknecht 1964)

•Type V: Components *a* and *b* are partially miscible in the solid phases and show a eutectic point *E*, i.e., azobenzene/azoxybenzene (Polaczkowa et al. 1954)

Upon undercooling of a solution of eutectic composition below the eutectic temperature, both crystalline solids precipitate simultaneously and form a physical mixture of these crystals. However, upon undercooling a solution of peritectic composition below the peritectic temperature the second solid crystallizes "around" the first.

Most technically relevant systems are eutectic, and only very few show full miscibility. In the following, the behavior of these systems is explained in more detail for three real systems with the help of Fig. 2.3-5.

The left column of this figure shows the phase diagrams of the three systems. The right column shows the selectivity diagrams, as they have already been introduced in Sect. 2.1.3 for gas–liquid systems. x_i, $\tilde{x}_i$, y_i, and $\tilde{y}_i$ are the compositions of the solid and the liquid phases, either in mass or in mole fractions. The top row shows

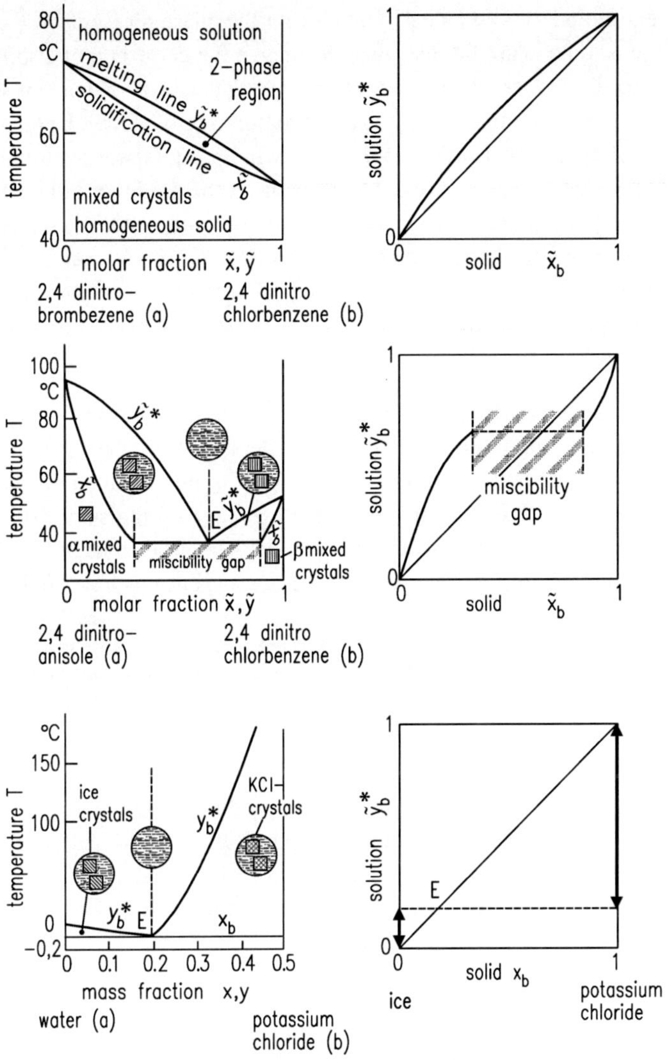

Fig. 2.3-5 Phase diagrams (*left*) and selectivity diagrams (*right*) of selected binary systems with perfect miscibility (*top row*), partial miscibility (*middle row*), and total immiscibility (*bottom row*) of two components in the two solid phases

the case of an almost ideal miscibility of both components in the solid phase. The middle row shows partial miscibility, and the bottom row shows a system where both components are immiscible in both solid phases.

Because an equilibrated system is isothermal, the compositions of the equilibrated solids x and liquids y can be easily read from the phase diagrams in the left column. These isotherms are tie lines, which connect the corresponding solid and the liquid phases. From these diagrams it can also be deduced, at which temperature T a solution of a given composition y is at saturation (melting line). Upon supersaturating this solution crystals of composition x will appear (solidification line). Corresponding x- and y-values are used to construct the selectivity diagrams of the right column. These selectivity diagrams for solid–liquid systems are analogous to the ones of vapor–liquid systems discussed in Sect. 2.1.3 and can be used in the same manner. The rule of lever can be used for instance to graphically represent mass and component balances for the mixing of two volumes of different solutions (rules of mixing).

For systems with a miscibility gap (middle row) melting and solidification lines are found. The melting lines of both components intersect in the eutectic point E. This point divides the phase diagram into two regions. In one region a certain component is depleted in the mother liquor; in the other region the same component is concentrated with respect to the starting solution. The little icons in the phase diagrams are meant to clarify which appearance the suspensions have in the different regions.

Finally, in the bottom row the system potassium chloride and water is given. In this case the miscibility gap of the solid phase is almost total. Potassium chloride crystals contain water only in traces and the same is true vice versa, i.e., ice crystals contain only traces of potassium chloride. If an aqueous potassium chloride solution with $x_{KCl} < 0.2$ is cooled below the melting line, then almost pure water ice will form. If $x_{KCl} > 0.2$, then almost pure potassium chloride crystals will form upon cooling below the melting line. The corresponding selectivity diagram shows that the miscibility gap spans over the entire concentration range $0 \leq x \leq 1$. This behavior of many crystallizing systems is favorable for crystallization processes in (ultra)purification applications.

Enthalpy–concentration diagrams are helpful engineering tools for the design of crystallization processes. Their advantage is that they allow easy formulation and graphical representation of mass and energy balances. The development and the use of such diagrams are explained in Sect. 2.5.

2.4 Sorption Equilibria

Adsorption is based on the energetic properties of solid surfaces. At the solid–fluid interface, attractive and repulsive forces are acting on the molecules of the adsorbate (adsorbed molecules). The most important forces are van der Waals or dispersion forces and electrostatic forces. It will be shown later that the Hamaker constant, the electrical charge, the polarizability of the adsorbent molecules, and the dipole and quadrupole moments as well as the polarizability of the adsorptive molecules are the decisive properties for gas–solid equilibria. These equilibria describe the relationship between the concentration of the adsorptive in the fluid phase and the loading of the adsorbent. Principally speaking, two or more components of a fluid can be adsorbed.

2.4.1 Single Component Sorption

Contrary to condensation or crystallization the process of accumulation of molecules is called adsorption when the fluid concentration $\tilde{c}_i$ is smaller than the saturation concentration $\tilde{c}_i^*$. In the case of gas adsorption the partial pressure p_i of the adsorptive is smaller than the vapor pressure p_i^0 at the temperature of the system. Dealing with the adsorption of a liquid component the concentration $\tilde{y}_i$ is smaller than the saturation concentration $\tilde{y}_i^*$. As a rule the relative saturation $\varphi_i = p_i/p_i^0$ or $\varphi_i = \tilde{y}_i/\tilde{y}_i^*$, respectively., is plotted against the loading $\tilde{X}_i$ of the solid adsorbent, see Fig. 2.4-1. The curve a is representative for an adsorption isotherm which

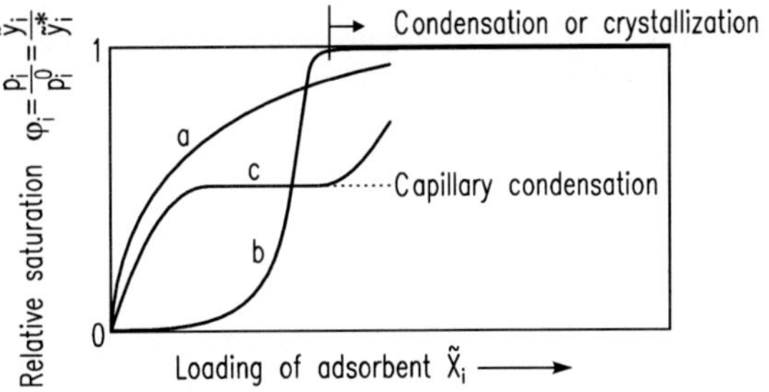

Fig. 2.4-1 Types of adsorption isotherms. Relative saturation vs. adsorbent loading

is unfavorable because the loading is small for a given relative saturation. The contrary is true for an isotherm according to the curve b.

The process of adsorption takes place when the concentration of the adsorptive is greater than the equilibrium value valid for the given temperature; however, desorption requires a fluid concentration of the adsorptive which is smaller than the equilibrium concentration. An adsorption isotherm favorable for adsorption is unfavorable for desorption and vice versa. Condensation of gases or vapors and solidification or crystallization will start when the relative supersaturation becomes $\varphi_i > 1$. In the case of adsorbents with capillary or very narrow pores, capillary condensation is observed for relative saturations $\varphi_i < 1$. Based on (2.1-28) the hysteresis of the adsorption isotherm valid for adsorption and desorption can sometimes be explained, see Fig. 2.4-2. Solid materials exposed to drying (see Chap. 10) often show such hysteresis behavior which can sometimes be explained by the curvature of the liquid surface in capillaries: The radius of this surface is greater in the case of adsorption in comparison to the radius valid for a desorption process, see Fig. 2.4-2.

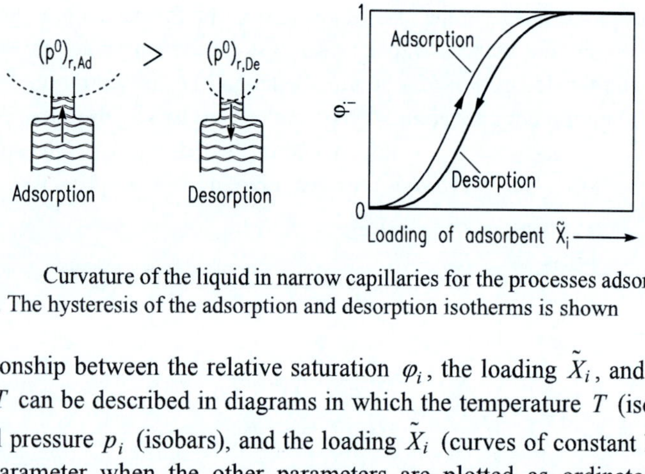

Fig. 2.4-2 Curvature of the liquid in narrow capillaries for the processes adsorption and desorption. The hysteresis of the adsorption and desorption isotherms is shown

The relationship between the relative saturation φ_i, the loading $\tilde{X}_i$, and the temperature T can be described in diagrams in which the temperature T (isotherms), the partial pressure p_i (isobars), and the loading $\tilde{X}_i$ (curves of constant loadings) are the parameter when the other parameters are plotted as ordinates against abscissa, see Fig. 2.4-3.

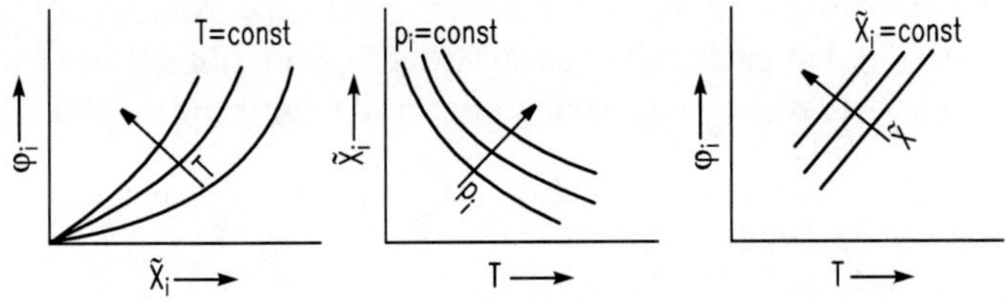

Fig. 2.4-3 Different diagrams for the presentation of adsorption equilibria. (*left*) adsorption isotherms; (*center*) adsorption isobars; (*right*) lines of constant loading

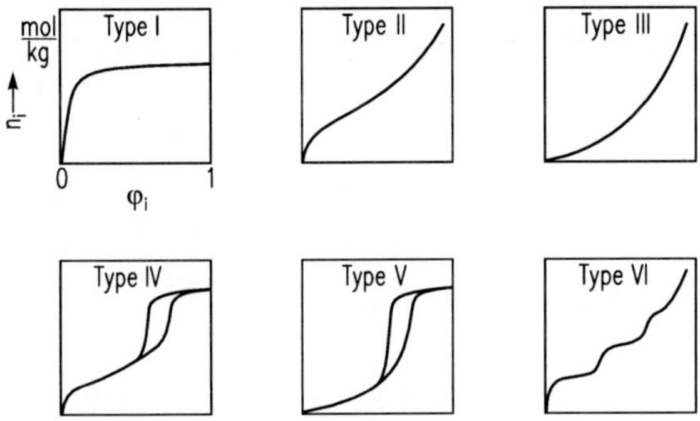

Fig. 2.4-4 IUPAC classification of adsorption isotherms

Energetic interrelationships between the molecules of the solid and the fluid phase in combination with different pore systems in an adsorbent can lead to a variety of adsorption isotherms, see Fig. 2.4-4. In the case of a given adsorbent–adsorptive system exposed to a given temperature, an adsorption isotherm is obtained for different partial pressures. The majority of isotherms can be classified according to the types I up to VI recommended by the IUPAC (International Union of Pure and Applied Chemistry), see Fig. 2.4-4. Types I up to V have already been proposed in the classification of Brunauer, Deming, Deming, and Teller (BDDT) in 1940. The loading n_i is plotted vs. the relative saturation φ_i according to

$$\varphi_i = \frac{p_i}{p_i^0} \tag{2.4-1}$$

for a given temperature T.

1. Type I isotherms are favorable for adsorption and the loading assumes a final value for $\varphi_i \to 1$. Inorganic and organic gases or vapors adsorbed on microporous adsorbents like active carbon or zeolites often lead to isotherms of this type.

2. Type II isotherms are typical for adsorbents with wide or no pores and applicable for one and multilayer adsorption.

3. Type III isotherms are unfavorable for adsorption and exhibit strong interactions between molecules of the adsorbate.

4. Type IV isotherms show a hysteresis caused in many cases by capillary condensation of mesoporous adsorbents. In the range of small relative saturations, the isotherm is (as for type II) favorable for adsorption; however, the loading assumes a final value for $\varphi_i \to 1$ as is the case of type I isotherms.

5. Type V isotherms are similar to type IV at medium and high partial pressure; however, in the range of small loadings weak energetic interrelationships between adsorbate–adsorbate molecules are important. The adsorption of water on activated carbon is an example (Brunauer et al. 1940).

6. Type VI isotherms are representative for a stepwise multilayer adsorption on homogeneous solid surfaces. Examples are the adsorption of argon and krypton on graphite at the temperature of liquid nitrogen.

The isotherms of the types II and V exhibit points of inflexion and the isotherm of type IV shows two such points. The equations of Table 2.4-1 can be used to express isotherms mathematically. Equations which are valid in the Henry range ($p \to 0$) and also in the saturation range ($p \to \infty$) are recommended. The last equation of the table according to Akgün and Mersmann is based on the critical data p_c and T_c of the adsorptive and the Hamaker constant Ha_j, the molecule diameter σ_j and the BET-surface of the adsorbents. v_{micro} denotes the micropore volume and $\beta(T)$ is the molar volume. An approach for polar adsorptives and electrically charged adsorbents is given later.

Dealing with adsorbents which can be characterized by a well-defined BET-surface (activated carbon) the coverage θ can be used. This quantity of loading is the substance n based on the substance n_{mon} which is adsorbed as a complete monolayer of adsorbate molecules. However, zeolites have a microporous structure with cages. In this case it is difficult to define a surface based on a defined geometry. Therefore, the pore filling v/v_{max} is used. The pore filling assumes the value 1 when the micropores are completely filled. Data of micropore volumes are presented in Chap. 9. Besides the substance n (expressed as mol/m^2, m_N^3/kg, or later as mol/kg) as the quantity adsorbed also mole or mass loadings are used.

Table 2.4-1 Models for adsorption isotherms

Equation for isotherm	Consistent in the Henry range?	Consistent in the saturation range?
Langmuir $\Theta \equiv \frac{n}{n_{\text{mon}}} = \frac{b \cdot p}{1 + b \cdot p}$ with $b = \frac{\alpha}{\sqrt{2\pi \tilde{M} \tilde{R} T} \cdot \dot{n}_{\text{des},\,\Theta = 1}}$	Yes	Yes
n-Layer BET (Brunauer, Emmet, and Teller) $\Theta \equiv \frac{n}{n_{\text{mon}}} = \frac{C \cdot \varphi_i \cdot [1 - (N+1) \cdot \varphi_i^N + N \cdot \varphi_i^{N+1}]}{(1 - \varphi_i) \cdot [1 + (C-1)\varphi_i - C \cdot \varphi_i^{N+1}]}$ with $\varphi_i = p_i / p_i^0$	Yes	Depends on N
Freundlich $n = K \cdot p^{1/m}$ with $K = b^* \cdot n_{\text{mon}}$	No (exception $m = 1$)	No
Sips $\Theta \equiv \frac{n}{n_{\text{mon}}} = \frac{b^* \cdot p^{1/m}}{1 + b^* \cdot p^{1/m}}$	No (exception $m = 1$)	Yes
Tóth $\Theta \equiv \frac{n}{n_{\text{mon}}} = \frac{p}{\left(\frac{1}{K_{\text{T}}} + p^t\right)^{1/t}}$ special case: $t = 1 \Rightarrow$ s. Langmuir equation	Yes	Yes
Dubinin–Astakov $\frac{v}{v_{\text{max}}} = \exp\left[-\left(\frac{\tilde{R} \cdot T \cdot \ln(p^0/p)}{\tilde{e}}\right)^{\nu}\right]$ $\tilde{e}$ is a characteristic energy $\nu = 2 \Rightarrow$ equation of Dubinin–Radushkevich	No	Yes
Akgün–Mersmann $n = He \cdot p(1 - n/n_\infty)^B$ with $n_\infty = v_{\text{micro}} / \beta(T)$; $B = 0.55 \times 10^{-3} \left(\frac{Ha_j}{\sigma_j^3 p_c}\right)^{3/2} \cdot \frac{T_c}{T_{\text{ads}}}$ He : see later	Yes	Yes

The quantity of adsorptive in the fluid phase can be expressed as partial pressure p_i, the relative saturation $\varphi_i = p_i / p_i^0$, or the adsorption potential $\varepsilon = -\tilde{R} \cdot T \cdot \ln(p_i / p_i^0) = \tilde{R} \cdot T \cdot \ln(1/\varphi_i)$

In the Henry range of an adsorption isotherm with p_i or $\varphi_i \to 0$ or $\varepsilon \to \infty$ the Henry coefficient $He \equiv n/p_i$ should be constant (with sufficient accuracy for chemical engineering). In Table 2.4-1 this requirement is fulfilled by the equations of Langmuir, Brunauer–Emmet–Teller, Tóth, and Akgün–Mersmann. In the saturation range the final loading is $n_\infty = v_{\text{micro}}/\beta(T)$ with v_{micro} as the micropore volume of microporous adsorbents (activated carbon, zeolites) and $\beta(T)$ as the molar volume of the adsorbate. This requirement is fulfilled by the equations of Langmuir, Tóth, Sips, Dubinin–Astakov, and Akgün–Mersmann.

Sometimes the adsorption isotherm has been experimentally determined only for a certain temperature, for instance for the room temperature. The equation of Dubinin–Astakov can be used as a basis to extrapolate loadings to other temperatures when the relative pore filling $v/v_{\max}$ is plotted against the adsorption potential $\varepsilon = \tilde{R}T\ln(1/\varphi_i)$ since the characteristic energy $\tilde{e}$ is constant for a certain adsorptive–adsorbent combination, see Fig. 2.4-5. The Henry coefficient of a certain component i depends on the temperature according to the equation of van't Hoff:

$$\left|\frac{\partial \ln p_i}{\partial(1/T)}\right|_{n_i = \text{const.}} = \left|\frac{\partial \ln He_i}{\partial(1/T)}\right|_{n_i = \text{const.}} = -\frac{\Delta\tilde{h}'_i}{\tilde{R}} \tag{2.4-2}$$

Here $\Delta\tilde{h}'_i$ is the difference of the enthalpy of the component i in the fluid and in the adsorbate phase and is called heat of adsorption for gases and heat of immersion for liquids. The factor b of the Langmuir equation depends on the temperature:

$$b = b_0 \cdot \exp\left(-\frac{\Delta\tilde{h}'_i}{\tilde{R}\cdot T}\right). \tag{2.4-3}$$

According to (1.4-1) for the relative saturation

$$\varphi_i = \frac{p_i}{p_i^0} = \gamma_i \cdot \tilde{x}_i = \gamma_i \cdot \frac{\tilde{X}_i}{1+\tilde{X}_i}, \tag{2.4-4}$$

the loading increases with the partial pressure for a given temperature. A comparison between the heat of bonding $\Delta\tilde{h}_{i,\text{B}}$ and the molar heat of mixing $\Delta\tilde{h}_i$ or

$$\Delta\tilde{h}_{i,\text{B}} = -\tilde{R}\cdot\left[\frac{\partial \ln\varphi_i}{\partial(1/T)}\right]_{p,\tilde{x}} \text{ and } \Delta\tilde{h}_i = -\tilde{R}\cdot\left[\frac{\partial \ln\gamma_i}{\partial(1/T)}\right]_{p,\tilde{x}} \tag{2.4-5}$$

leads to the result that the heat of bonding $\Delta\tilde{h}_{i,\text{B}}$ can also be interpreted as the "heat of mixing" of the adsorptive component i with the adsorbent. As a rule the

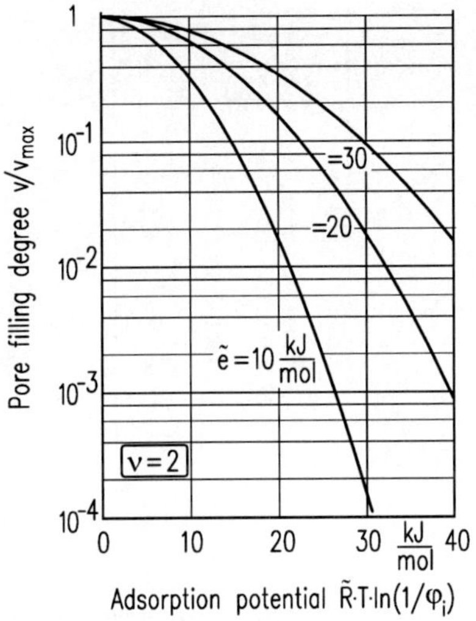

Fig. 2.4-5 Pore filling degree against the adsorption potential with $\nu = 2$ according to Dubinin–Astakov

heat of bonding is a function of the mole fraction $\tilde{x}_i$ or of the mole loading $\tilde{X}_i$. Therefore, the last equation cannot be simply integrated. In the special case of $\Delta h_{i,\mathrm{B}} \neq f(\tilde{x}_i)$ the result would be

$$\varphi_i = C_1 \cdot \exp\frac{\Delta\tilde{h}_{i,\mathrm{B}}}{\tilde{R} \cdot T}. \tag{2.4-6}$$

This means that the loading would increase with the relative supersaturation φ_i but would decrease with increasing temperature (the same behavior is valid for absorption equilibria). The logarithm of the partial pressure p_i plotted against the reciprocal of the absolute temperature leads approximately to straight lines, see Fig. 2.4-6 (as is also valid for $\ln p_i^0$ vs. $1/T$). This figure shows the adsorption equilibrium for the system propane–activated carbon (Szepesy and Illes 1963). The partial pressure of propane is plotted against the reciprocal of the temperature in the left diagram, whereas in the right diagram the temperature dependency of the relative saturation is shown.

2.4.2 Heat of Adsorption and Bonding

The differential heat of adsorption q and also the heat of bonding Δh_B can be derived from the dependency of the partial pressure p_i or the relative supersaturation φ_i, respectively, as a function of the temperature. The corresponding equations have already been presented in the Sect. 2.1.4:

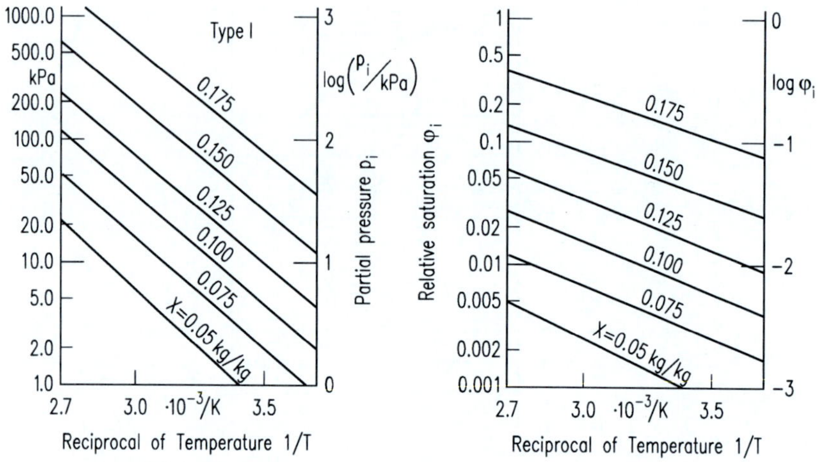

Fig. 2.4-6 Partial pressure against the reciprocal of the absolute temperature for propane on activated carbon, loading as parameter (*left*). Relative saturation against the reciprocal of the absolute temperature, loading as parameter (*right*)

$$q_i = -R_i \cdot \left[\frac{\partial \ln p_i}{\partial(1/T)}\right]_X = R_i \cdot T \cdot \left[\frac{\partial \ln p_i}{\partial \ln T}\right]_X .$$
$$\Delta h_{i,B} = -R_i \cdot \left[\frac{\partial \ln \varphi_i}{\partial(1/T)}\right]_X = R_i \cdot T \cdot \left[\frac{\partial \ln \varphi_i}{\partial \ln T}\right]_X \tag{2.4-7}$$

In Fig. 2.4-7 the heat of adsorption and the heat of bonding are plotted against the loading for the system propane–activated carbon. The difference of these two heats is equal to the heat of condensation Δh_{GL} which is identical with the heat of adsorption at high (multilayer) loadings. The heat of bonding can be derived from the slope of lines valid for constant loading in a $\ln \varphi$–$1/T$ diagram, see Fig. 2.4-6 right. The differential heat of adsorption q decreases in most cases with increasing loading for a constant temperature. This is valid for the system propane–activated carbon. The heat of bonding disappears at high loadings (approximately three up to five molecular layers) with the result $\Delta h_B = q - \Delta h_{GL} = 0$ or $q = \Delta h_{GL}$. In the case of capillary condensation, the heat of bonding can be calculated from the Thomson equation derived for the reduction of the vapor pressure of concave liquid surfaces:

$$\Delta h_B = \frac{2\sigma}{r \cdot \rho_L}, \tag{2.4-8}$$

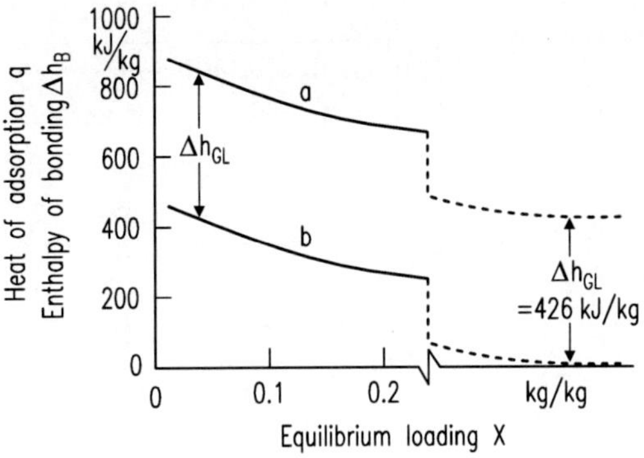

Fig. 2.4-7 Heat of adsorption (*curve a*) and enthalpy of bonding (*curve b*) against the loading of propane on activated carbon

with the pore radius r and the density ρ_L of the adsorbate. Very narrow pores lead to high heats of bonding. The exothermic heat of adsorption is equal to the endothermic heat of desorption and reflects the intermolecular energies of the molecules of the adsorptive and the adsorbate molecules, see Sect. 2.4.4

The integral heat of adsorption $\bar{q}$ is the amount of heat which is released from a mass unit of an adsorbent when the final loading X_1 is reached. Let us now assume an adsorbent which is already loaded according to the loading X_1. The differential heat of adsorption is equal to the heat released when the loading X_1 is increased by the small adsorbate mass dX based on the adsorbent mass. These considerations lead to the relationship

$$\bar{q} = \frac{1}{X_1} \cdot \int_0^{X_1} q \cdot \mathrm{d}X. \tag{2.4-9}$$

In Table 2.4-2 heats of adsorption of some adsorptives are listed.

2.4.3 Multicomponent Adsorption

Triangular diagrams can be used when two adsorptives are adsorbed. Fig. 2.4-8 shows the adsorption of nitrogen and oxygen on activated carbon. The conodes are straight lines between a concentration in the fluid phase (N_2–O_2 mixture) and the corresponding equilibrium concentration of the three-component (solid adsorbent + adsorbate) phase. In the case that the conodes are not passing through the adsorbent corner the binary concentrations or mole fractions $\tilde{y}'^{*}_{N_2}$ and the binary fraction x'_{N_2} are different. This can be seen more clearly when the binary mole fraction in

Table 2.4-2: Isosteric heats of adsorption

System	Polarity	Approximate isosteric heat of adsorption (kJ/mol)	Literature
O_2/NaX	Nonpolar	≈14	Münstermann (1984)
N_2/NaX		≈19	”
N_2/MS5A		≈20	Sievers (1993)
CH_4/MS5A		≈20	”
CH_4/AC		≈20	”
CHF_3/AC	Polar	≈25	Markmann (2000)
CO_2/AC		≈30 → ≈25	Sievers (1993)
CO/MS5A		≈35 → ≈30	”
CHF_3/MS5A		≈60 → ≈40	Markmann (2000)
CO_2/NaX		≈60 → ≈40	Münstermann (1984)
CO_2/MS5A		≈60 → ≈52	Sievers (1993)
CO_2/MS5A		≈70 → ≈45	Markmann (2000)
H_2O/MS5A		≈65 → ≈50	Münstermann (1984)

the fluid phase is plotted against the binary mole fraction of the solid phase, see Fig. 2.4-9. Such a diagram gives no information on the loading of the adsorbent. Therefore, an additional diagram above the equilibrium diagram is drawn where the reciprocal of the loading is plotted against the mole fractions. Adsorption equilibria are represented by conodes in the upper diagram and by the equilibrium curve in the diagram below. The reciprocal of the loading is zero in the fluid phase.

In Figs. 2.4-10–2.4-13 it is shown how binary equilibria can be depicted in diagrams. The system is methanol–water on silica gel at 50°C. In Fig. 2.4-10 the methanol loading (here in cm_N^3/g) is plotted against the partial pressure of methanol for different partial pressures of water. Figure 2.4-11 shows the water loading against the partial pressure of water for different partial pressures of methanol. In both diagrams the loading of one of the two components is reduced by an increase of the other component in the fluid phase. Figure 2.4-12 shows the loading of methanol and water as a function of the mole fraction $\tilde{y}_M$ of methanol (again for silica

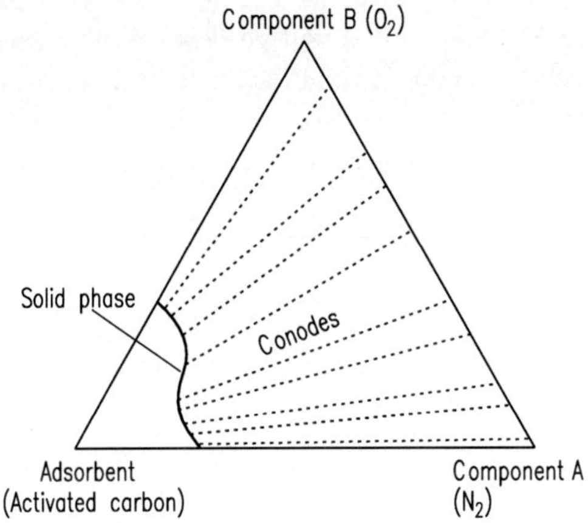

Fig. 2.4-8 Adsorption equilibrium of the gas mixture with the components nitrogen and oxygen on activated carbon

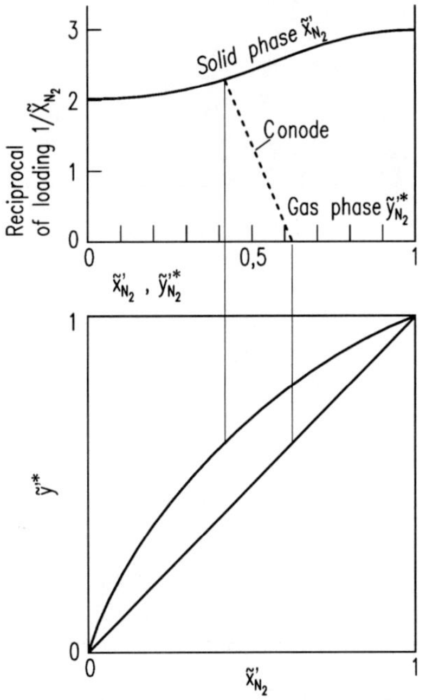

Fig. 2.4-9 Reciprocal of adsorbent loading against the mole fraction in the gas and the solid phase (*above*); equilibrium curve of the adsorbate (N_2-O_2–activated carbon)

gel at 50°C) for different active pressures which are the sum of the partial pressures of the sorptive–active components methanol and water. In the case of a lower temperature than 50°C these curves would move upward in such a diagram and for the temperature above 50°C the curves will go down. With respect to separation of mixtures by adsorption, equilibria curves are helpful. In Fig. 2.4-13 the mole fraction $\tilde{x}_M$ of the methanol in the adsorbate (here considered binary) is plotted against the mole fraction $\tilde{y}_M$ of the same component in the gas phase for different active pressures.

The selectivity S according to

$$S = \frac{\tilde{x}_M \cdot \tilde{y}_W}{\tilde{x}_W \cdot \tilde{y}_M} \tag{2.4-10}$$

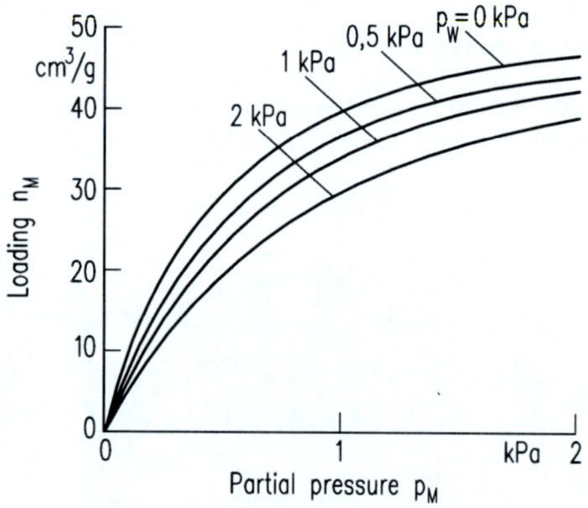

Fig. 2.4-10 Methanol loading in cm^3_N/g against the partial pressure of methanol p_M for different partial pressures of water p_W

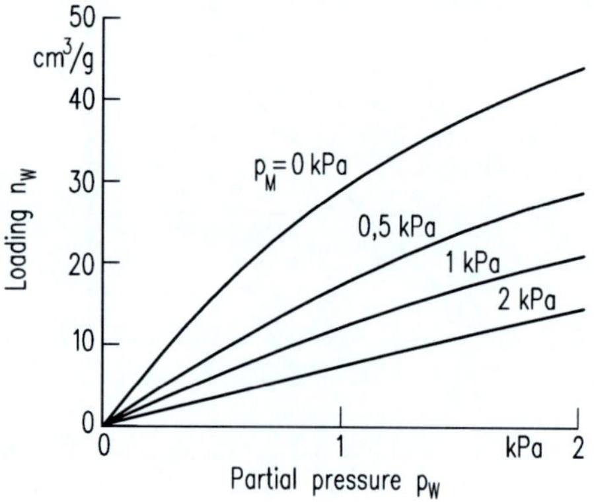

Fig. 2.4-11 Water loading in cm^3_N/g against the partial pressure of water p_W for different partial pressures of methanol p_M

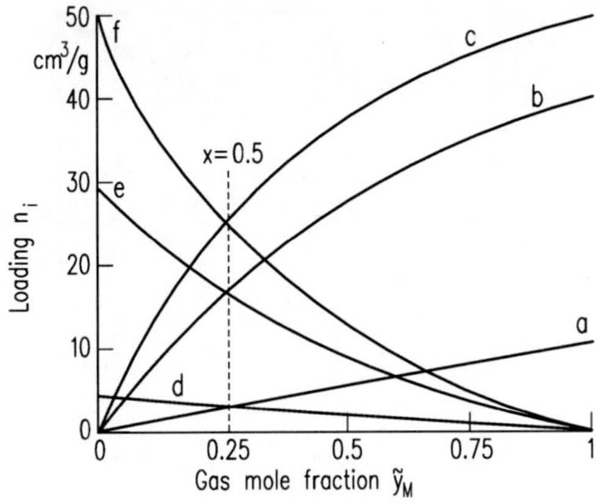

Fig. 2.4-12 Loading in cm_N^3/g against the mole fraction of methanol $\tilde{y}_M$ in the gas phase for different active pressures at 50°C. (**a**) Loading of methanol ($p_t = 0.1$ kPa). (**b**) Loading of methanol ($p_t = 1.0$ kPa). (**c**) Loading of methanol ($p_t = 3.0$ kPa). (**d**) Loading of water ($p_t = 0.1$ kPa). (**e**) Loading of water ($p_t = 1.0$ kPa). (**f**) Loading of water ($p_t = 3.0$ kPa)

decreases with increasing partial pressures; however, azeotropic behavior is not observed. Contrary to this behavior, azeotropic equilibria have often been found for adsorptives with polar components adsorbed on adsorbents with ions like zeolites. According to Fig. 2.4-14 the system toluene–1-propyl alcohol molecular sieve DAY13 shows an azeotropic point for the temperature 25°C and the pressure 1.05 kPa. In this diagram the mole fraction of toluene in the adsorbate phase is plotted against the mole fraction of this component in the gas phase. The lower diagram shows the total loading of both components in $mol_{adsorbate}/kg_{adsorbent}$. (The abbreviations MIAST and MSPDM will be explained in Sect. 2.4-5).

The system $C_2H_4-C_2H_6$ MS13X is not azeotropic at 423 K and 1.379 bar, see Fig. 2.4-14 right. However, azeotropic behavior has been observed for binary and ternary systems which contain CO_2 as one of the components. Note that the dipole moments of the molecules of C_2H_4, C_2H_6 and C_3H_8 are zero but the dipole moment of CO_2 is $\mu_{CO_2} = 0.7\text{ debye} = 2.34 \times 10^{-30}\text{Cm}$. Schweighart (Schweighart 1994) observed azeotropic behavior for the systems $C_2H_4-CO_2$ MS5A and $C_3H_8-CO_2$ MS5A. This is also true for the mixture $C_2H_4-C_2H_6-CO_2$ adsorbed on the zeolite MS5A, see Fig. 2.4-15.

The azeotropic lines separate the triangular diagram in certain ranges with a distinct selectivity. The selectivity is changed when a boundary of the ranges is crossed. Equilibria data of the quaternary system $C_2H_4-C_2H_6-C_3H_8-CO_2$ on MS5A can be found in Schweighart (1994).

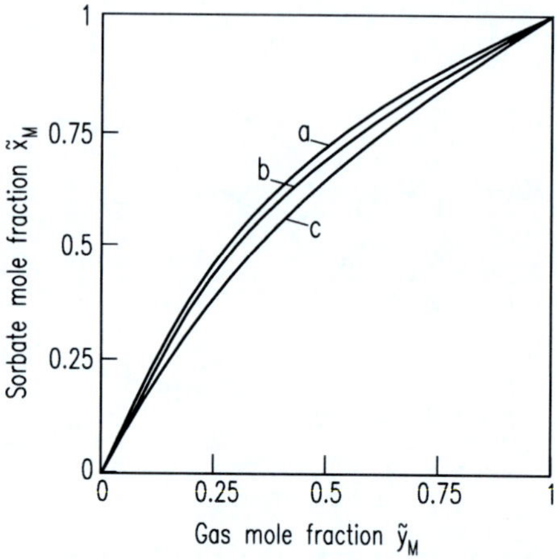

Fig. 2.4-13 Mole fraction $\tilde{x}_M$ of methanol in the adsorbate against the mole fraction $\tilde{y}_M$ in the gas for different pressures at 50°C. (**a**) $p_W + p_M = 0.1$ kPa. (**b**) $p_W + p_M = 1$ kPa. (**c**) $p_W + p_M = 3$ kPa

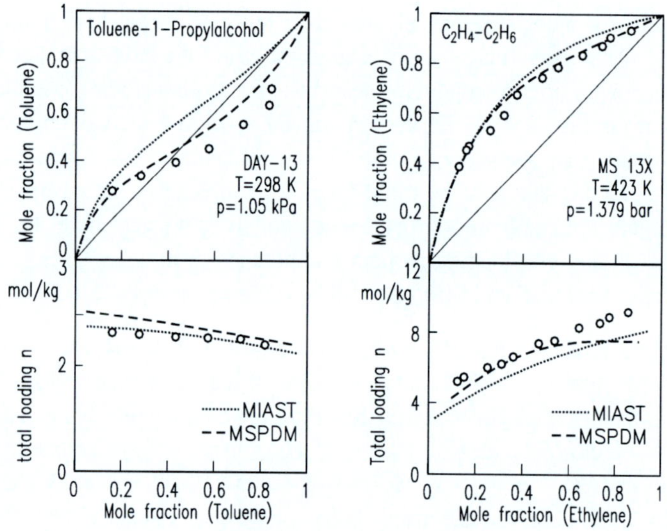

Fig. 2.4-14 Binary adsorption equilibria of toluene and 1-propyl alcohol on zeolite DAY13 and of C_2H_4 and C_2H_6 on molecular sieve 13X (Experimental data: Sakuth 1993, Kaul 1987.)

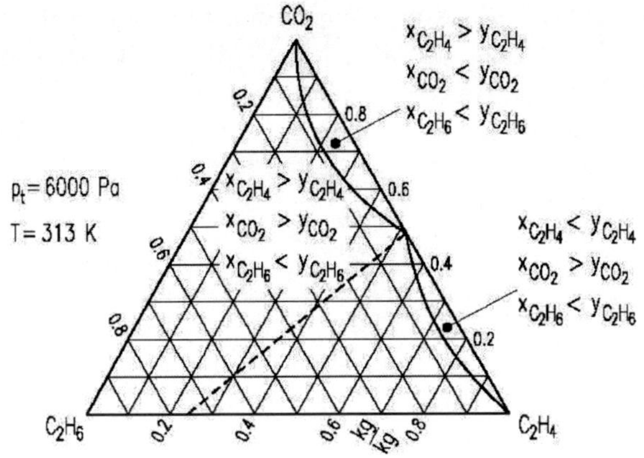

Fig. 2.4-15 System $C_2H_4-CO_2-C_2H_6$. Azeotropic lines and lines which indicate the change of selectivity

2.4.4 Calculation of Single Component Adsorption Equilibria

With respect to an adsorption isotherm starting at $p \to 0$ and ending at $p \to \infty$, two special ranges can be distinguished, the

- Henry range according to $n = He \cdot p$ with the Henry coefficient $He \neq f(p)$ based on the assumption that there are no interactions between adsorbate molecules
- Saturation range with the maximum loading

$$n_\infty = \frac{v_{\text{micro}}}{\beta(T)} \text{ for } p \to \infty$$

which results in the slope $\mathrm{d}n/\mathrm{d}p = \mathrm{d}(\ln n)/\mathrm{d}(\ln p) = 0$ when the micropore volume is completely filled with adsorbate (Mersmann and Akgün 2009).

The molar volume $\beta(T)$ is inter- and extrapolated between the molar volume $\tilde{v}_b$ of the adsorptive at the normal boiling point and the van der Waals molar volume $\tilde{v}_{vdW} = RT_c/(8p_c)$:

$$\beta(T_{ads}) = \tilde{v}_b + \frac{T_{ads} - T_b}{T_c - T_b}(\tilde{v}_{vdW} - \tilde{v}_b). \quad (2.4\text{-}11)$$

T_b and T_c are the absolute temperatures at the normal boiling point and at the critical point, respectively. Data of $\tilde{v}_b$ can be found in Reid et al. (1988).

It is assumed that in the saturation range the loading capacity of the macropores is small in comparison to that of micropores. The Henry coefficient of nonpolar adsorptives (H_2, N_2, O_2, He, Kr, CH_4, C_2H_6, C_3H_8, etc.) adsorbed on non-charged adsorbents (active carbon, silicalite) is mainly depending on the critical pressure p_c and the critical temperature T_c of the adsorptive and the Hamaker constant Ha of the adsorbent (Maurer 2000) with $Ha_{\text{active carbon}} = 6 \times 10^{-20}$ J and $Ha_{\text{silicate}} = 7.89 \times 10^{-20}$ J.

However, dealing with polar adsorptives (NH_3, CHF_3, CH_3OH, $C_2H_2F_4$, etc.) the electrical properties

- Dipole moment μ_i in C m
- Quadrupole moment Q_i in C m^2
- Polarizability α_i in C^2 m^2/J of the adsorptive i
- Polarizability α_j in C^2 m^2/J of the adsorbent j
- The number s_{cat} of cations per solid atom
- The electrical charge q_j in C

of the adsorbent j are additionally effective for the Henry coefficient, see Fig. 2.4-16.

With the abbreviations

indind for induced dipole–induced dipole interactions

indcha for induced dipole–electrical charge interactions

dipind for permanent dipole–induced dipole interactions

dipcha for permanent dipole–electrical charge interactions

quadind for quadrupole–induced dipole interactions

quadcha for quadrupole–electrical charge interactions

the Henry coefficient He defined by the equation

$$n = He \cdot p \qquad (2.4\text{-}12)$$

is a function of the temperature and the interaction energy $\phi(z)$ between an adsorbed molecule and the adsorbent molecules in the immediate neighborhood of

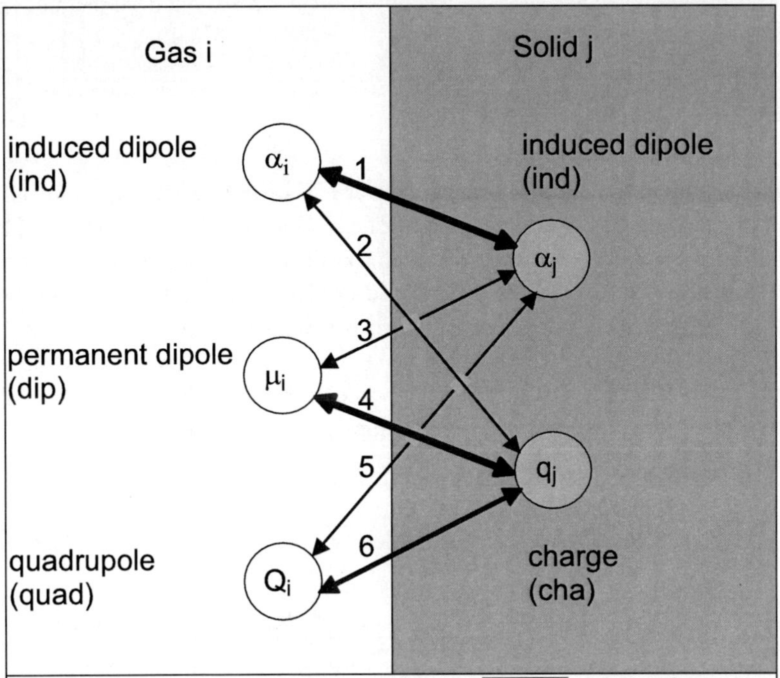

(1)	indind	$\frac{-\Phi_{indind}(z)}{kT_{ads}} = \frac{3}{4}\sqrt{\frac{2Ha_j}{\pi\sigma_j^3 p_c}}\,\frac{T_c}{T_{ads}}\,\frac{\sigma_{i,j}^3}{(\sigma_{i,j}+z)^3}$
(2)	indcha	$\frac{-\Phi_{indcha}(z)}{kT_{ads}} = \frac{\alpha_i s_{cat}^2 e^2 \rho_{j,at}}{8\pi\varepsilon_0^2\varepsilon_r^2 kT_{ads}}\,\frac{1}{(\sigma_{i,j}+z)}$
(3)	dipind	$\frac{-\Phi_{dipind}(z)}{kT_{ads}} = \frac{\mu_i^2\alpha_j\rho_{j,at}}{12\pi\varepsilon_0^2\varepsilon_r^2 kT_{ads}}\,\frac{1}{(\sigma_{i,j}+z)^3}$
(4)	dipcha	$\frac{-\Phi_{dipcha}(z)}{kT_{ads}} = \frac{\mu_i^2 s_{cat}^2 e^2\rho_{j,at}}{12\pi\varepsilon_0^2\varepsilon_r^2(kT_{ads})^2}\,\frac{1}{(\sigma_{i,j}+z)}$
(5)	quadind	$\frac{-\Phi_{quadind}(z)}{kT_{ads}} = \frac{3Q_i^2\alpha_j\rho_{j,at}}{40\pi\varepsilon_0^2\varepsilon_r^2 kT_{ads}}\,\frac{1}{(\sigma_{i,j}+z)^5}$
(6)	quadcha	$\frac{-\Phi_{quadcha}(z)}{kT_{ads}} = \frac{Q_i^2 s_{cat}^2 e^2\rho_{j,at}}{240\pi\varepsilon_0^2\varepsilon_r^2(kT_{ads})^2}\,\frac{1}{(\sigma_{i,j}+z)^3}$

Fig. 2.4-16 Interactions between adsorptive molecules i and adsorbent molecules j

the adsorbate molecule. The Henry coefficient can be derived from thermodynamics:

$$He = \frac{S_{\mathrm{BET}}}{\tilde{R} \cdot T} \cdot \int_0^{z_{\max}} \exp\left(-\frac{\phi(z)}{k \cdot T} - 1\right) \cdot \mathrm{d}z \tag{2.4-13}$$

with z as the distance from the solid surface. The BET surface S_{BET} in m^2/kg and the maximum distance $z_{\max}$ are constants which result from the integral to be solved for a certain adsorbent. The main problem is to find an equation for the energy $\phi(z)$ which is valid for arbitrary adsorptives and adsorbents. Maurer (Maurer 2000) introduced an expression which is similar to the three-dimensional van der Waals equation in combination with mixing rules; however, his model is restricted to uncharged adsorbents (activated carbon, graphite, silicalite). Taking into account the most important interaction energies the Henry coefficient He is given by Akgün (2007)

$$He = \frac{S_{\mathrm{BET}} \cdot \sigma_{i,j}}{\tilde{R} T_{\mathrm{ads}}} \int_0^{z_{\max}} \left[\exp\left\{ \frac{A_{\mathrm{indcha}} + A_{\mathrm{dipcha}}}{(1 + z/\sigma_{i,j})} + \right.\right.$$
$$\left.\left. + \frac{A_{\mathrm{indind}} + A_{\mathrm{dipind}} + A_{\mathrm{quadcha}}}{(1 + z/\sigma_{i,j})^3} + \frac{A_{\mathrm{quadind}}}{(1 + z/\sigma_{i,j})^5} \right\} - 1 \right] dz \tag{2.4-14}$$

with

$$A_{\mathrm{indind}} = \frac{3}{4} \cdot \sqrt{\frac{2 \cdot Ha_j}{\pi \cdot \sigma_j^3 \cdot p_{\mathrm{c}}}} \cdot \frac{T_{\mathrm{c}}}{T_{\mathrm{ads}}}, \tag{2.4-15}$$

$$A_{\mathrm{indcha}} = \frac{\alpha_i \cdot s_{\mathrm{cat}}^2 \cdot e^2 \cdot \rho_{j,\mathrm{at}}}{8 \cdot \pi \cdot (\varepsilon_0 \cdot \varepsilon_{\mathrm{r}})^2 \cdot (k \cdot T_{\mathrm{ads}}) \cdot \sigma_{i,j}}, \tag{2.4-16}$$

$$A_{\mathrm{dipind}} = \frac{\mu_i^2 \cdot \alpha_j \cdot \rho_{j,\mathrm{at}}}{12 \cdot \pi \cdot (\varepsilon_0 \cdot \varepsilon_{\mathrm{r}})^2 \cdot (k \cdot T_{\mathrm{ads}}) \cdot \sigma_{i,j}^3}, \tag{2.4-17}$$

$$A_{\mathrm{dipcha}} = \frac{\mu_i^2 \cdot s_{\mathrm{cat}}^2 \cdot e^2 \cdot \rho_{j,\mathrm{at}}}{12 \cdot \pi \cdot (\varepsilon_0 \cdot \varepsilon_{\mathrm{r}})^2 \cdot (k \cdot T_{\mathrm{ads}})^2 \cdot \sigma_{i,j}}, \tag{2.4-18}$$

$$A_{\mathrm{quadind}} = \frac{3 \cdot Q_i^2 \cdot \alpha_j \cdot \rho_{j,\mathrm{at}}}{40 \cdot \pi \cdot (\varepsilon_0 \cdot \varepsilon_{\mathrm{r}})^2 \cdot (k \cdot T_{\mathrm{ads}}) \cdot \sigma_{i,j}^5}, \tag{2.4-19}$$

$$A_{\text{quadcha}} = \frac{Q_i^2 \cdot s_{\text{cat}}^2 \cdot e^2 \cdot \rho_{j,\text{at}}}{240 \cdot \pi \cdot (\varepsilon_0 \cdot \varepsilon_r)^2 \cdot (k \cdot T_{\text{ads}})^2 \cdot \sigma_{i,j}^3}. \tag{2.4-20}$$

S_{BET} is the inner specific surface in m^2/kg according to Brunauer, Emmet, and Teller, and $\rho_{j,\text{at}}$ denotes the average number density of atoms per volume.

In the last equation z is the distance from the surface of a solid adsorbent and $\sigma_{i,j}$ is the mean diameter of the molecule diameter of the adsorptive

$$\sigma_i = \left(\frac{3 \cdot k \cdot T_c}{16 \cdot \pi \cdot p_c}\right)^{1/3} \tag{2.4-21}$$

and the "effective" diameter σ_j of an adsorbent molecule

$$\sigma_{i,j} = \frac{\sigma_i + \sigma_j}{2}. \tag{2.4-22}$$

The "effective" diameter σ_j according to

$$\sigma_j \approx \underbrace{\frac{1}{0.93 + 0.25 \cdot (n_{\text{Al}}/n_{\text{Si}})^3}}_{\text{correction factor}} \left(\frac{\tilde{M}_j}{n_{\text{Si}} \cdot \rho_s \cdot N_A}\right) \tag{2.4-23}$$

mainly depends on the molar mass $\tilde{M}_j$ of the adsorbent macromolecule based on the number n_{Si} of silicon atoms and the true density. The correction factor is 1.08 valid for silicalite (the number of aluminium atoms n_{Al} is zero), 0.847 valid for MS4A and MS5A and 1.06 for NaY with $n_{\text{Al}}/n_{\text{Si}} = 0.412$.

The dimensionless interaction energies $A_{\text{dipind}} \sim \sigma_{i,j}^{-3}$, $A_{\text{quadcha}} \sim \sigma_{i,j}^{-3}$, and especially $A_{\text{quadind}} \sim \sigma_{i,j}^{-5}$ are very sensitive to the mean diameter $\sigma_{i,j}$; however, as a rule, the contributions A_{dipind} and A_{quadind} are very small in comparison to the four other and can be neglected for nearly all adsorptives.

The relative permittivity ε_r can be calculated from the value $\varepsilon_r \approx 1$ valid for the gas and the contribution $\varepsilon_r \approx n^2$ with n as the refractive index of the adsorbent according to the equation $\varepsilon_r \approx \varphi + (1 - \varphi) \cdot n^2$. Here, φ is the porosity of an adsorbent pellet. With $k = 1.38 \times 10^{-26}$ kJ/K, $e = 1.06 \times 10^{-19}$ As and $\varepsilon_0 = 8.85 \times 10^{-12}$ A s/(V m) data of the Hamaker constant are necessary for the general prediction of Henry coefficients. Akgün has developed a model for the calculation of Hamaker constants based on the mean London constant $\bar{\beta}$, the mass

m_e, and the frequency ν_0 of an electron and the mean polarizability $\overline{\alpha}_j$. The Hamaker constant Ha_j can be calculated from

$$Ha_j = \pi^2 \cdot \rho_{j,\,\mathrm{at}}^2 \cdot \underbrace{\frac{3}{4}\sqrt{s} \cdot h \cdot \underbrace{\left(\frac{e}{2 \cdot \pi \cdot \sqrt{m_e \cdot \overline{\alpha}_j}}\right)}_{\text{frequency } \nu_0} \cdot \left(\frac{\overline{\alpha}_j}{4 \cdot \pi \cdot \varepsilon_0}\right)^2}_{\text{London constant } \bar{\beta}} \tag{2.4-24}$$

in combination with the refractive index according to

$$\frac{N_A \cdot \overline{\alpha}_j}{3\varepsilon_0} = \frac{\tilde{M}_j}{n_a \cdot \rho_{\mathrm{app}}} \cdot \frac{(n^2 - 1)}{(n^2 + 2)} \quad \text{and} \tag{2.4-25}$$

$$n = \sqrt{\frac{1 + 2\rho_{\mathrm{app}} \cdot R_m / \tilde{M}_j}{1 - \rho_{\mathrm{app}} \cdot R_m / \tilde{M}_j}}. \tag{2.4-26}$$

The ionic molar refractions $R_{m,i}$ of the ions Na^+, K^+, Ca^{2+}, Si^{4+}, Al^{3+}, and O_2^{2-} can be derived from the polarizability volumes tabulated in Lide (2004–2005). Data of the adsorbent properties

Number n_a of atoms in a macromolecule

Number n_{Si} of silicon atoms in a macromolecule

Number n_{Al} of aluminium atoms in a macromolecule

Average number s of valence electron bondings

Molar mass in kg/kmol

Solid density ρ_{app} in kg/m^3

Atomic density $\rho_{j,\,\mathrm{at}} = \dfrac{N_A \cdot n_a \cdot \rho_{\mathrm{app}}}{\tilde{M}_j}$ in $1/m^3$

Refractive index $n = \sqrt{\varepsilon_r}$

Mean polarizability $\overline{\alpha}_j$ in $(C^2 \cdot m^2)/J$

Hamaker constant Ha_j in J

of four different aluminosilicates and of silicalite-1 are compiled in Table 2.4-3.

Principally speaking, Henry coefficients can be calculated by means of the (2.4-14)–(2.4-20); however, the integral is difficult to solve. Some ways of solving

Table 2.4-3 Material properties of some adsorbents

		n_a	n_{Si}	n_{Al}	s	Molar mass $\tilde{M}_j$ (kg/kmol)	Solid density ρ_{app} in (kg/m^3)	Density $\rho_{j,at}$ (1/m^3)	Refractive index	Polarizability $\overline{\alpha}_j$ (C^{2}m^2/J)	Hamaker constant Ha_j (10^{-20} J)
MS4A	$Na_{12}((AlO_2)_{12}(SiO_2)_{12})$	84	12	12	2.30	1,700	1,680	5.00×10^{28}	1.44	1.40×10^{-40}	6.50
MS5A	$Ca_5Na_2((AlO_2)_{12}(SiO_2)_{12})$	79	12	12	2.43	1,680	1,150	3.27×10^{28}	1.57	2.67×10^{-40}	7.66
NaX	$Na_{56}((AlO_2)_{86}(SiO_2)_{106})$	662	106	86	2.32	12,000	1,100	3.64×10^{28}	1.62	2.55×10^{-40}	8.70
NaY	$Na_{56}((AlO_2)_{56}(SiO_2)_{136})$	632	136	56	2.43	12,760	1,100	3.28×10^{28}	1.61	2.82×10^{-40}	8.40
Silicalite-1	$Si_{96}O_{192}$	298	96	0	2.67	5,800	1,760	5.58×10^{28}	1.46	1.32×10^{-40}	7.89

are described in the literature (Akgün 2007). Here, only an empirical equation based on a dimensional analysis approach will be presented. In this model the dimensionless interaction energies A_{dipind} and A_{quadind} are omitted because their contribution is very small. It is assumed that the dimensionless Henry coefficient He^* according to

$$He^* = \frac{He \cdot RT}{S_{\text{BET}} \cdot \sigma_{i,j}} \tag{2.4-27}$$

depends on the following dimensionless numbers:

$$A_{\text{indind}} = \frac{3}{4} \cdot \left(\frac{2 \cdot Ha_j}{\pi \cdot \sigma_j^3 \cdot p_c}\right)^{1/2} \cdot \frac{T_c}{T_{\text{ads}}}, \tag{2.4-28}$$

$$A_{\text{indcha}} = \frac{\alpha_i \cdot s_{\text{cat}}^2 \cdot e^2 \cdot \rho_{j,\text{at}}}{8\pi(\varepsilon_0 \varepsilon_r)^2 \cdot (kT_{\text{ads}}) \cdot \sigma_{i,j}}, \tag{2.4-29}$$

$$\mu_i^* = \frac{\mu_i^2}{\alpha_i \cdot (kT_{\text{ads}})}, \tag{2.4-30}$$

$$Q_i^* = \left(\frac{Q_i}{\sigma_i}\right)^2 \cdot \frac{1}{\alpha_i (kT_{\text{ads}})}. \tag{2.4-31}$$

These four dimensionless expressions mainly depend on the gas properties α_i, μ_i, and Q_i and on the charge $q_j = s_{\text{cat}} \cdot e$ of the adsorbent molecules. The polarizability α_j is neglected because the dimensionless interactions A_{dipind} and A_{quadind} are omitted.

The following equation is based on many experimental results obtained for nonpolar and polar adsorptives (NH_3, CHF_3, and CH_3OH with strong dipole moments and CO, CO_2, and C_2H_4 with strong quadrupole moments):

$$\ln He^* \approx 0.465(A_{\text{indind}} + A_{\text{indcha}})^{1.18} + 0.025(\mu_i^* + Q_i^*)(A_{\text{indind}} + A_{\text{indcha}}). \tag{2.4-32}$$

The general equation can be simplified for nonpolar adsorptives ($\mu_i^* = Q_i^* = 0$) and the assumption that the interaction A_{indcha} can be neglected:

$$\ln He^* = 0.465 \cdot (A_{\text{indind}})^{1.18} = 0.465 \cdot \left(\frac{3}{4} \cdot \sqrt{\frac{2 \cdot Ha_j}{\pi \cdot \sigma_j^3 \cdot p_c}} \cdot \frac{T_c}{T_{\text{ads}}}\right)^{1.18}. \tag{2.4-33}$$

This equation delivers minimum Henry coefficients. Dealing with polar adsorptive molecules and adsorbents with electrical charges such as NH_3 ($\mu_i = 5 \times 10^{-30}$ Cm) and CHF_3 ($\mu_i = 5.34 \times 10^{-30}$ Cm), the Henry coefficient can be by a factor up to 2,000 higher than the minimum value based only on the Hamaker energy and critical data of adsorptive molecules.

The complete adsorption isotherm can be described by the differential equation (Mersmann and Akgün 2009)

$$\frac{d(\ln n)}{d(\ln p)} = \frac{1}{\times\ 1 + 0.55 \times 10^{-3} \left(\frac{Ha_j}{\sigma_j^3 \cdot p_c} \right)^{3/2} \cdot \frac{T_c}{T_{ads}} \cdot \left(\frac{n/n_\infty}{1 - n/n_\infty} \right)} . \qquad (2.4\text{-}34)$$

Integration leads to

$$n = He \cdot p \cdot (1 - n/n_\infty)^B \qquad (2.4\text{-}35)$$

with

$$B = 0.55 \times 10^{-3} \cdot \left(\frac{Ha_j}{\sigma_j^3 \cdot p_c} \right)^{3/2} \cdot \frac{T_c}{T_{ads}} . \qquad (2.4\text{-}36)$$

In Fig. 2.4-17 the dimensionless loading $n/(He \cdot p)$ is shown as a function of the residual pore filling $(1 - n/n_\infty)$ for different values of B.

2.4.5 Prediction of Multicomponent Adsorption Equilibria

The theory of the adsorbed solution is a very effective and widely used tool to predict multicomponent adsorption equilibria based on isotherms of single components in the state of gases or vapors. In Table 2.4-4 various submodels of this theory can be found characterizing different adsorbate properties and / or different surface properties of the adsorbent (homogeneous, heterogeneous).

The basis of all models is the ideal adsorption solution theory (IAST) published by Myers and Prausnitz (Myers and Prausnitz 1965). The basic assumption is the equality of the chemical potential of the component i in the gas phase

$$\mu_{i,G} = \mu_i^0(T) + \tilde{R} \cdot T \cdot \ln\left(\frac{\tilde{y}_i \cdot p}{p^*} \right) \qquad (2.4\text{-}37)$$

and the chemical potential of the same component in the adsorbed phase or adsorbate

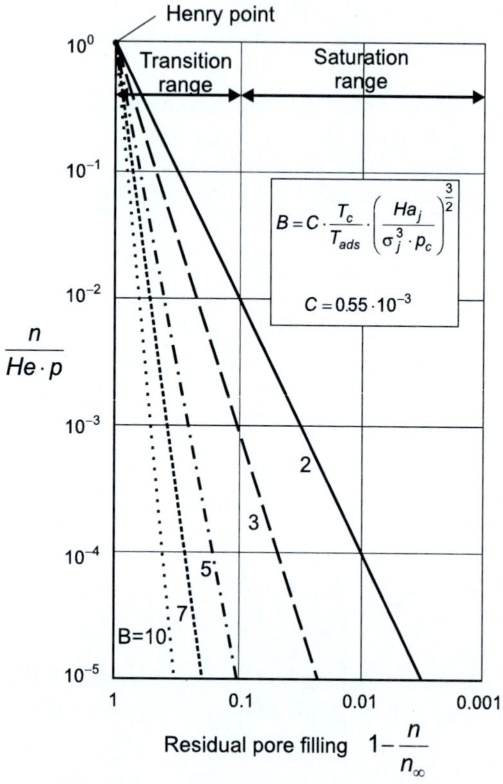

Fig. 2.4-17 Ratio $n/(He \cdot p)$ vs. residual pore filling

$$\mu_{i,\mathrm{S}} = \overset{0}{\mu_i}(T, \pi) + \tilde{R} \cdot T \cdot \ln(\gamma_i \cdot \tilde{x}_i) \tag{2.4-38}$$

with
$$\overset{0}{\mu_i}(T, \pi) = \overset{0}{\mu_i}(T) + \tilde{R} \cdot T \cdot \ln\left(\frac{\overset{0}{p_i}(\pi)}{p^*}\right). \tag{2.4-39}$$

In these equations, $\overset{0}{\mu_i}(T)$ is the chemical standard potential of the component i, p^* is a standard pressure , π is the spreading pressure of the adsorbed solution, and γ_i is an activity coefficient. The virtual pressure $\overset{0}{p_i}(\pi)$ (not the vapor pressure) depends on the spreading pressure π and is the pressure of the component i which has the same spreading pressure either in the pure state or in the gaseous mixture. The equality $\mu_{i,\mathrm{G}} = \mu_{i,\mathrm{S}}$ in the case of thermodynamic equilibrium leads to Raoult's law of adsorption

$$\tilde{y}_i \cdot p = \gamma_i \cdot \tilde{x}_i \cdot \overset{0}{p_i}(\pi), \tag{2.4-40}$$

Table 2.4-4 Adsorbed solution theories for the description or prediction of multicomponent adsorption equilibria. In the light gray area new theoretical models are listed. The theories in the double-framed area require experimental data of binary adsorptives. VLE denotes vapor liquid equilibrium. The meaning of VAE is vapor adsorbate equilibrium

Adsorbed solution theories		Energy distribution of the adsorbent surface	
		Homogeneous	Heterogeneous
Ideal adsorbed solution	Activity coefficient $\gamma = 1$	Ideal adsorbed solution theory (IAST)	Multiphase ideal adsorbed solution theory (MIAST) Heterogeneous ideal adsorbed solution theory (HIAST)
Real adsorbed solution	γ (VLE)	Real adsorbed solution theory (RAST)	Multiphase real adsorbed solution theory (MRAST)
	γ (VAE)	Predictive real adsorbed solution theory (PRAST) Spreading pressure-dependent model (SPDM)	Multiphase spreading pressure-dependent model (MSPDM) Multiphase predictive real adsorbed solution theory (MPRAST)

which is the basis of all variations of the theory of adsorbed solutions. An adsorbed solution can be ideal ($\gamma_i = 1$) or nonideal ($\gamma_i \neq 1$). In an ideal equilibrium the adsorbate–adsorbate interactions can be neglected (especially in the Henry range). The activity coefficient is unity, see Table 2.4-4; however, in the case that there are differences of the adsorbed molecules with respect to size, polarity, and polarizability, the activity coefficients are $\gamma_i \neq 1$ (especially at higher loading than the Henry loading). An activity coefficient can be drawn from VAE (vapor–adsorbate Equilibrium) data or from VLE (vapor–liquid Equilibrium) data (problematic) in the case of high loadings with two or more molecule monolayers. Besides the differences of the properties of the adsorbed molecules, different adsorbent surfaces can result in $\gamma_i \neq 1$. A surface is energetically homogeneous if all possible sites available for adsorptive molecules have the same potential. This requirement is not exactly fulfilled even for activated carbon (often said homogeneous) because industrially used adsorbents contain impurities. Dealing with microporous zeolites, a distribution of surface energies exists mainly due to their ionic character. Therefore, every site has a potential which is different from other sites and has its own selectivity $S = x_i(1 - y_i)/[(1 - x_i)y_i]$ for a given concentration in the gas phase. As a consequence the energy distribution of the adsorbent surface has to be taken into account in the models for the prediction of adsorption equilibria, see the last line in Table 2.4-4. In the case that a nonideal ($\gamma_i \neq 1$) adsorbed solution covers a heterogeneous adsorbent surface, both parameters have to be incorporated in the modeling.

2.4.5.1 Ideal Adsorbed Solution Theory

The ideal adsorbed solution theory (IAST) assumes a perfect mixture of the adsorbate. Therefore, the simple Raoult's law holds:

$$\tilde{y}_i \cdot p = \tilde{x}_i \cdot p_i^0(\pi) . \tag{2.4-41}$$

Myers and Prausnitz (1965) developed the basic model of a two-dimensional gas with the two-dimensional spreading pressure π.[1]

The spreading pressure π can be expressed as a reduced spreading pressure ψ in the case of a constant temperature T and a constant specific surface A in m^2/kg. Then ψ depends on the hypothetical vapor pressure $p_i^0(\pi)$:

$$\psi = \frac{\pi \cdot A}{\tilde{R} \cdot T} = \int_0^{p_i^0} \frac{n_i(p_i)}{p} \mathrm{d}p . \tag{2.4-42}$$

The reduced spreading pressure ψ can be derived from the Gibbs adsorption isotherm. With the condition

$$\sum_i \tilde{x}_i = 1 , \tag{2.4-43}$$

there are N equations for the calculation of N mole fractions $\tilde{x}_i$.

The total adsorbed material n_{tot} is

$$\frac{1}{n_{\mathrm{tot}}} = \sum_i \frac{\tilde{x}_i}{n_i^0} \tag{2.4-44}$$

with n_i^0 as the amount of substance which would be adsorbed in the case of a single component adsorption at the pressure p_i^0. The loading n_i of the component i is

$$n_i = \tilde{x}_i \cdot n_{\mathrm{tot}} . \tag{2.4-45}$$

---

1 In the case of negligible nondispersion forces or dominant dispersion forces (superscript d) the spreading pressure π is mainly dependent on the surface tension $\gamma_{\mathrm{LG}}^{\mathrm{d}}$ of the adsorptive, the Hamaker energy Ha_j of the adsorbent, the characteristic atom diameter σ_j of the adsorbent, and of the contact angle θ

$$\pi \approx \gamma_{\mathrm{LG}}^{\mathrm{d}} \left[\sqrt{\frac{Ha_j}{0.96 \cdot 3.14 \cdot \sigma_j^2 \cdot \gamma_{\mathrm{LG}}^{\mathrm{d}}}} - (1 + \cos\theta) \right] .$$

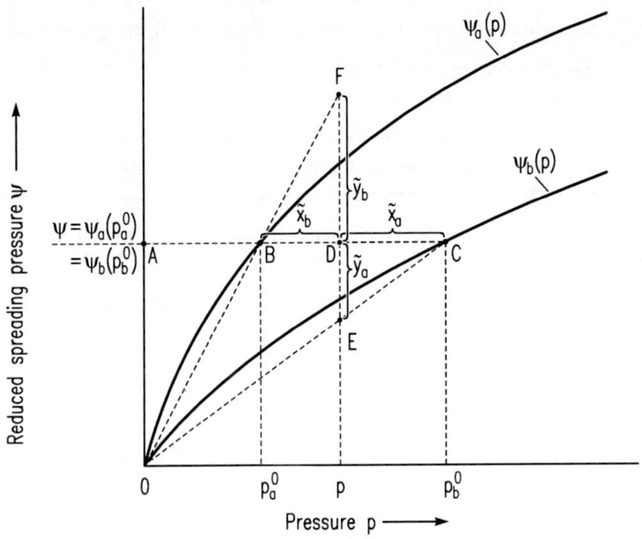

Fig. 2.4-18 Graphical solution according to the theory of adsorbed solution for a binary solution. $\tilde{x}_a = \overline{CD}/\overline{BC}$, $\tilde{x}_b = \overline{BD}/\overline{BC}$, $\tilde{y}_a = \overline{DE}/\overline{EF}$, and $\tilde{y}_b = \overline{DF}/\overline{EF}$

In Fig. 2.4-18 the basic equations of the IAST model are illustrated for a binary adsorbate mixture. The relationships ψ_a and ψ_b of the reduced spreading pressure are drawn against the total pressure p of the binary system. In the case of equilibrium both components have the same reduced spreading pressure $\psi = \psi_a = \psi_b$. This common reduced spreading pressure can be found between the two ψ-curves for a given pressure p. The exact position of ψ in the diagram is determined by the mole fractions $\tilde{y}_a$ and $\tilde{y}_b$ of the components a and b in the gas phase:

$$\tilde{y}_a = \overline{DE}/\overline{EF} \text{ and } \tilde{y}_b = \overline{DF}/\overline{EF} \text{ or } \frac{\tilde{y}_a}{\tilde{y}_b} = \frac{\overline{DE}}{\overline{DF}} \text{ with } \overline{EF} = 1\,. \tag{2.4-46}$$

The point of intersection of the $\psi = $ const. line with the curves $\psi_a(p)$ and $\psi_b(p)$ leads to the hypothetical vapor pressures p_a^0 and p_b^0 and also to the mole fractions $\tilde{x}_a$ and $\tilde{x}_b$ of the adsorbate mixture:

$$\tilde{x}_a = \overline{CD}/\overline{BC} \text{ and } \tilde{x}_b = \overline{BD}/\overline{BC} \text{ or } \frac{\tilde{x}_a}{\tilde{x}_b} = \frac{\overline{CD}}{\overline{BD}} \text{ with } \overline{BC} = 1\,. \tag{2.4-47}$$

2.4.5.2 Simplified Version of the Equations

Dealing with gases with more than two components, the solution of the system of equations requires much computation time.

The FAST–IAST theory introduced by O'Brian and Myers (1985) and Moon and Tien (1987) provides a much faster way to calculate multicomponent adsorption equilibria based on the IAST method. The application of the FAST–IAST theory requires the determination of the spreading pressure analytically. Next the equations of IAST are remodeled in such a way that a linear system of equations is obtained. A gas mixture of N components which is in equilibrium with the adsorbate on the adsorbent has $(N+1)$ degrees of freedom and can be described by $(N+1)$ independent variables. These variables are the temperature, the pressure, and $(N-1)$ mole fractions of the gas. The N equations for N unknowns can be solved for a given constant temperature. $(N-1)$ equations are obtained by setting equal all reduced spreading pressures:

$$\psi_i = \psi_N . \tag{2.4-48}$$

The missing last equation is the balance of substance in the adsorbate

$$\sum_{i=1}^{N} \tilde{x}_i = 1 \tag{2.4-49}$$

or the sum of mole fractions $\tilde{y}_i$ which is also unity. Applying this procedure, only the bottom row of the matrix has to be changed while all other rows remain constant. The set of equations can be solved numerically, for instance by means of a Gauß algorithm in the case of constant temperature. The problem to find appropriate initial values can be avoided in the Henry range of an adsorption isotherm with the linear relationship $n \sim p$. In this way, the first analytical solution is remodeled to a set of zeroes which is an appropriate initial value to solve the nonlinear equation system. In the case that the spreading pressure can be obtained analytically, the FAST–IAST is an efficient tool to calculate multicomponent adsorption equilibria. Appropriate adsorption isotherms are those of O'Brien and Myers (1984), Langmuir, Tóth, and equations according to statistical thermodynamics, however, not the equations of Langmuir–Freundlich and of Dubinin–Astakov.

2.4.5.3 Prediction of Binary Activity Coefficients

When binary activity coefficients can only be obtained from experimental equilibrium data, there is no way to predict multicomponent adsorption equilibria which are only based on single component isotherms; however, such a procedure would be desirable. The SPDM (spreading pressure dependent model) contains only predictive parameters with the exception of the binary parameter β_{ij} (Markmann 1999; Mersmann et al. 2002). Setting $\beta_{ij} = 0$, this method allows to calculate multicomponent adsorption equilibria without experimental data obtained for binary mixtures.

Sakuth, Meyer, and Gmehling (Sakuth 1993) have introduced another method which is based on activity coefficients of the single component adsorption isotherms at infinite dilution. They used VLE (Vapor–Liquid equilibria) expressions for the calculation of activity coefficients, for instance the Wilson equation (compare chapters 1 and 4).

This method is known as PRAST (predictive real adsorbed solution theory). The activity coefficients γ_a^∞ and γ_b^∞ of the two components a and b of a binary adsorbate solution, which is diluted, are given by the equations

$$\gamma_a^\infty = \frac{n_b}{He_a \cdot p_a^0} \text{ and } \gamma_b^\infty = \frac{n_a}{He_b \cdot p_b^0}. \tag{2.4-50}$$

Here He_a and He_b are the Henry coefficients of the corresponding components in their pure state.

2.4.5.4 Multiphase Theory of Ideal Adsorbed Solution

The multiphase ideal adsorbed solution theory (MIAST) is another model of the family of adsorbed solutions. Contrary to HIAST, an energy distribution function is assumed with differences of the local or molecular site energy. Therefore, every site has its own local adsorption isotherm and the adsorbate concentration differs from site to site. The adsorbate is not considered as a homogeneous phase but as a multiphase system.

The energy distribution according to MIAST is not based on a statistical model but on the single component adsorption isotherm. With respect to the condensation of adsorptive molecules, the following approximations are introduced (Cerofolini 1971, 1975; Rudzinsky and Everett 1992):

$$\theta(p, T, \varepsilon) = \begin{cases} 0 & \text{for} \quad p < p^*(T, \varepsilon^*) \\ 1 & \text{for} \quad p \geq p^*(T, \varepsilon^*) \end{cases}. \tag{2.4-51}$$

Here, $\theta = n/n_S$ is the coverage of the surface and p^* is the pressure as a function of the adsorption energy ε. If the pressure p is below the pressure p^*, the local coverage becomes zero. The coverage θ assumes the value 1 for $p \geq p^*(T, \varepsilon^*)$. During the early development of this model, the Dubinin–Astakov equation has been approximated this way. Principally speaking, this approximation of the condensation of adsorptive molecules can also be used in combination with other isotherm equations. The global Tóth isotherm of the component i which represents a number of local isotherms is applied when the total surface of the adsorbent is subdivided in N_S sections of equal area:

$$n(p, T) = \sum_{j=1}^{N_S} n(p, T, \varepsilon_j). \tag{2.4-52}$$

The loading n_{ij} of the component i in the section j is

$$\begin{aligned} n_{ij} &= \frac{n_{i,S}}{N_S} \quad \text{for} \quad p \geq p_j^* \\ n_{ij} &= 0 \quad \text{for} \quad p < p_j^*. \end{aligned} \tag{2.4-53}$$

In the case of thermodynamic equilibrium the spreading pressures of all components in the mixture are equal in this section. Thanks to a simple local isotherm equation, the integral of (2.4-51) becomes

$$\int_{p_{aj}^*}^{p_{aj}^0} \frac{n_{aj}}{p} dp = \int_{p_{ij}^*}^{p_{ij}^0} \frac{n_{ij}}{p} dp \quad \Rightarrow \quad n_{aj} \cdot \ln \frac{p_{aj}^0}{p_{aj}^*} = n_{ij} \cdot \ln \frac{p_{ij}^0}{p_{ij}^*}. \tag{2.4-54}$$

It can be solved because of the relationship

$$p_{ij} = \tilde{x}_{ij} \cdot p_{ij}^0 \text{ with } p_{ij} = p_i. \tag{2.4-55}$$

The loading n_i of the component i adsorbed on the total surface is equal to the sum of all loadings in the various sections:

$$n_i = \sum_{j=1}^{N_S} n_{ij}. \tag{2.4-56}$$

2.4.5.5 Theories of the Real Heterogeneous Adsorbed Solution

The theories of the heterogeneous adsorbed solution (MIAST, HIAST) and of the real adsorbed solution ($\gamma_i \neq 1$, RAST, PRAST, SPDM) can be combined to take into account the energetic heterogeneity of the adsorbent surface and also the adsorbate–adsorbate interactions which lead to $\gamma_i \neq 1$. Eiden (1989), Quesel (1995), and Markmann (1999) have published such combined-model theories. The MIAST based on Tóth equation is used on the one hand to incorporate the energetic heterogeneity of the adsorbent surface. On the other hand, the SPD model is an appropriate tool for activity coefficients depending on the spreading pressure. Raoult's law is valid for every section j and for every component i:

$$\tilde{y}_i \cdot p = \gamma_{ij}(T, \psi_i, x_{ij}) \cdot \tilde{x}_{ij} \cdot p_{ij}^0(\psi_j). \tag{2.4-57}$$

The total loading is (as has been shown for MIAST) the sum of all loadings in the sections j.

The following requirements must be fulfilled with respect to thermodynamically consistent activity coefficients:

- The activity coefficients of binary adsorbed mixtures must be computable from single component adsorption isotherms or the heat of adsorption of the pure components or the physical properties of the adsorbed molecules (polarity, polarizability, etc).
- The activity coefficient is depending on the section j and can differ from section j to section $j+1$ because the adsorbate concentration differs too.
- The parameter in the equations valid for activity coefficients should not depend on temperature.

The solution of such a combined model (MIAST and SPDM) will be called multiphase spreading pressure dependent model or MSPDM.

A wide variety of experimental and theoretical results have shown that the IAST is a very powerful tool for multicomponent adsorption equilibria of polar or non-polar adsorptives adsorbed on activated carbon; however, comprehensive physical models (for instance the MSPDM) are necessary in the case of polar (dipole and quadrupole moment, polarizability) adsorptives adsorbed on zeolites with free ions (Markmann 1999).

2.5 Enthalpy–Concentration Diagram

The conservation law of energy is part of the basis for the calculation of heat and mass transfer processes. According to the first law of thermodynamics, heat can be converted to enthalpy changes and mechanical work. If no work is added to an isobaric system and no work is removed from it (this is approximately true for many separation units of chemical engineering), heat transferred to the system corresponds to differences of enthalpies. Therefore, the calculation of enthalpies is an essential prerequisite for the design of separation equipment. The enthalpy of a fluid mixture depends on the pressure, the temperature, and the concentration, and can be calculated if

- The specific heat of every phase
- The heat caused by phase changes
- The heat of mixing of every phase

are known.

Dealing with ideal liquid systems (validity of Raoult's law in the entire range of concentration), the heat of mixing is zero. In most cases this is true for gas mixtures with exception of associating or dissociating molecules.

In the following, the calculation of an enthalpy–concentration diagram for a binary system will be demonstrated for a mixture of ethane and propane. These neighbored linear hydrocarbons are nonpolar without functional groups. There is neither association nor dissociation in the liquid or vapor phase. Therefore, heats of mixing are very small in comparison to the other contributions and can be neglected.

The molar enthalpy $\tilde{h}$ of a pure substance is given by the integral

$$\tilde{h} = \int_{T_0}^{T} \tilde{c}_p \cdot \mathrm{d}T . \qquad (2.5\text{-}1)$$

Here, the following question arises: What is a reasonable low integration limit T_0? Principally speaking, the absolute temperature $T = 0$ can be chosen; however, in most cases this is not practical with respect to the molar heat $\tilde{c}_p$ which is often only known in a limited temperature range. Any temperature can be chosen for T_0 because the calculation procedure is based on differences of enthalpies. Figure 2.5-1 is based on the agreement that the molar enthalpies of ethane and propane are zero at the absolute temperature $T_0 = 200\ \mathrm{K}$. At first, calculations are presented for the isotherm $T = 290\ \mathrm{K}$. The molar enthalpy of liquid propane is

$$(\tilde{h}_{\mathrm{C_3H_8}})_{\mathrm{L}} = (\tilde{c}_{p\ \mathrm{C_3H_8}})_{\mathrm{L}} \cdot (T - T_0) = 96.7 \frac{\mathrm{kJ}}{\mathrm{kmol \cdot K}} \cdot 90\ \mathrm{K} = 8{,}702 \frac{\mathrm{kJ}}{\mathrm{kmol}} \qquad (2.5\text{-}2)$$

and liquid ethane has the enthalpy

$$(\tilde{h}_{\mathrm{C_2H_6}})_{L} = (\tilde{c}_{p\ \mathrm{C_2H_6}})_{L} \cdot (T - T_0) = 72.8 \frac{\mathrm{kJ}}{\mathrm{kmol \cdot K}} \cdot 90\ \mathrm{K} = 6{,}536 \frac{\mathrm{kJ}}{\mathrm{kmol}} . \qquad (2.5\text{-}3)$$

Now the molar enthalpies of the pure components at the mole fractions $\tilde{x}_{\mathrm{a}} = 0$ and $\tilde{x}_{\mathrm{a}} = 1$, respectively, are fixed, see Fig. 2.5-1 where the enthalpy is plotted against the mole fraction. A straight line through these enthalpies at $\tilde{x}_{\mathrm{a}} = 0$ and $\tilde{x}_{\mathrm{a}} = 1$ is identical with the isotherm valid for 290 K if no heat of mixing exists in the entire concentration range. This is true for ideal mixtures for which the activity coefficient can be neglected ($\gamma_{\mathrm{a}} = \gamma_{\mathrm{b}} = 1$). Dealing with real mixtures, it is necessary to take into account negative heats of mixing (after mixing of the components with the same temperature, the temperature rises and the heat of mixing has to be removed to obtain the starting temperature) or positive heats of mixing (the mixture cools down and the heat of mixing has to be added). As has already been discussed, isotherms start upward-bent at $\tilde{x}_a = 0$ for endothermic mixtures but downward-curved in the case of an exothermic mixture. The behavior of the liquid mixture of ethane and propane is nearly ideal with the consequence that heats of mixing can be neglected.

Above the enthalpy–concentration diagram in Fig. 2.5-1 the dew-point temperature and the bubble-point temperature are shown as a function of the mole fraction and valid for the pressure of 14 bar. According to this diagram, a mixture with the mole fraction $\tilde{x}_\mathrm{a} = 0.26$, the temperature $T = 290\ \mathrm{K}$, and the pressure 14 bar starts boiling and is in equilibrium with a vapor which has the mole fraction $\tilde{y}_\mathrm{a} = 0.54$. Next, the enthalpy of the vapor (14 bar, 290 K) is calculated. This enthalpy is composed of three contributions the enthalpy of the liquid phase, the heat of evaporation, and the heat of superheating.

The molar enthalpy of the pure ethane vapor is given by

$$(\tilde{h}_{\mathrm{C_2H_6}})_\mathrm{G} = (\tilde{c}_{p\,\mathrm{C_2H_6}})_\mathrm{L} \cdot (T^0 - T_0) + (\Delta\tilde{h}_{\mathrm{C_2H_6}})_\mathrm{LG} + (\tilde{c}_{p\,\mathrm{C_2H_6}})_\mathrm{G} \cdot (T - T^0) \qquad (2.5\text{-}4)$$

$$= 72.6\,\frac{\mathrm{kJ}}{\mathrm{kmol\,K}} \cdot (253 - 200)\mathrm{K} + 10{,}967\frac{\mathrm{kJ}}{\mathrm{kmol}} + 46.9\,\frac{\mathrm{kJ}}{\mathrm{kmol\,K}} \cdot (290 - 253)\mathrm{K}$$

$$= 16{,}550\frac{\mathrm{kJ}}{\mathrm{kmol}}.$$

In a similar way, the enthalpy of the pure propane vapor is calculated:

$$(\tilde{h}_{\mathrm{C_3H_8}})_\mathrm{G} = (\tilde{c}_{p\,\mathrm{C_3H_8}})_\mathrm{L} \cdot (T^0 - T_0) + (\Delta\tilde{h}_{\mathrm{C_3H_8}})_\mathrm{LG} + (\tilde{c}_{p\,\mathrm{C_3H_8}})_\mathrm{G} \cdot (T - T^0) \qquad (2.5\text{-}5)$$

$$= 96.7\,\frac{\mathrm{kJ}}{\mathrm{kmol\,K}} \cdot (313 - 200)\mathrm{K} + 13{,}395\,\frac{\mathrm{kJ}}{\mathrm{kmol}} + 62.8\,\frac{\mathrm{kJ}}{\mathrm{kmol\,K}} \cdot (290 - 313)\mathrm{K}$$

$$= 22{,}877\,\frac{\mathrm{kJ}}{\mathrm{kmol}}.$$

Here T^0 is the boiling temperature for the component at the given pressure. The isotherm $T = 290\ \mathrm{K}$ for the vapor can be obtained by a straight line through the enthalpies calculated for the vapors of the two components. According to the dew-point temperature in the diagram above the system consists of only one phase, e.g., the vapor phase for the range $0.54 < \tilde{y}_\mathrm{a}^* < 1$ and the given temperature $T = 290\ \mathrm{K}$ and pressure 14 bar. Here the isotherm is represented by a straight line. The other isotherms can be obtained by the application of this calculation procedure.

Isothermal tie lines in the area of vapor–liquid equilibrium with the boundaries of the dew and the boiling line can be found by connecting the points on these lines, here the points $\tilde{x}_\mathrm{a} = 0.26$, $T = 290\ \mathrm{K}$ and $\tilde{y}_\mathrm{a}^* = 0.54$, $T = 290\ \mathrm{K}$. (Dealing with liquid–liquid equilibria such tie lines are also called tie lines.) Note that a

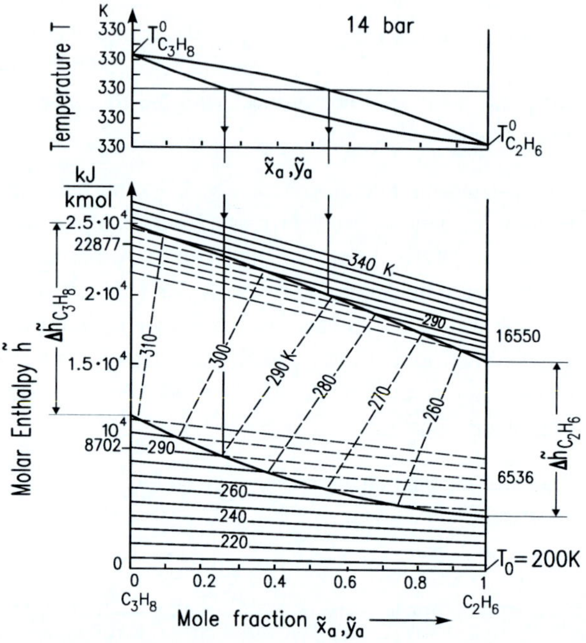

Fig. 2.5-1 Boiling and dew line (*above*) and enthalpy–concentration diagram of the binary mixture ethane/propane at 14 bar (Matschke 1962)

vapor–liquid equilibrium is represented by a point on the equilibrium curve shown in Fig. 2.1-18, by a horizontal line in a dew and boiling temperature diagram, see also Fig. 2.1-18 and by a tie line in an enthalpy–concentration diagram shown in Fig. 2.5-1.

In a similar way, enthalpy–concentration diagrams for liquid–solid systems can be calculated. Now the heat of melting and the heat of fusion are the decisive heats of phase change; however, in most cases the heats of mixing cannot be neglected. Strong heat effects can be expected for aqueous solutions of inorganic salts, and it is necessary to carry out numerous experiments to establish an enthalpy–concentration diagram. As an example, the enthalpy–concentration of the binary solution H_2O–$CaCl_2$ will be discussed.

In Fig. 2.5-2 the enthalpy of 1 kg mixture is shown as a function of the mass fraction x_b of calcium chloride for temperatures in the range between -100 and $300°C$. In general the vapor pressure of a mixture depends on temperature and concentration. With respect to the boiling rise of solutions, the vapor pressure curves have a higher enthalpy in comparison to the enthalpy of the solvent water. Besides isotherms, also curves of constant vapor pressure are drawn in Fig. 2.5-2.

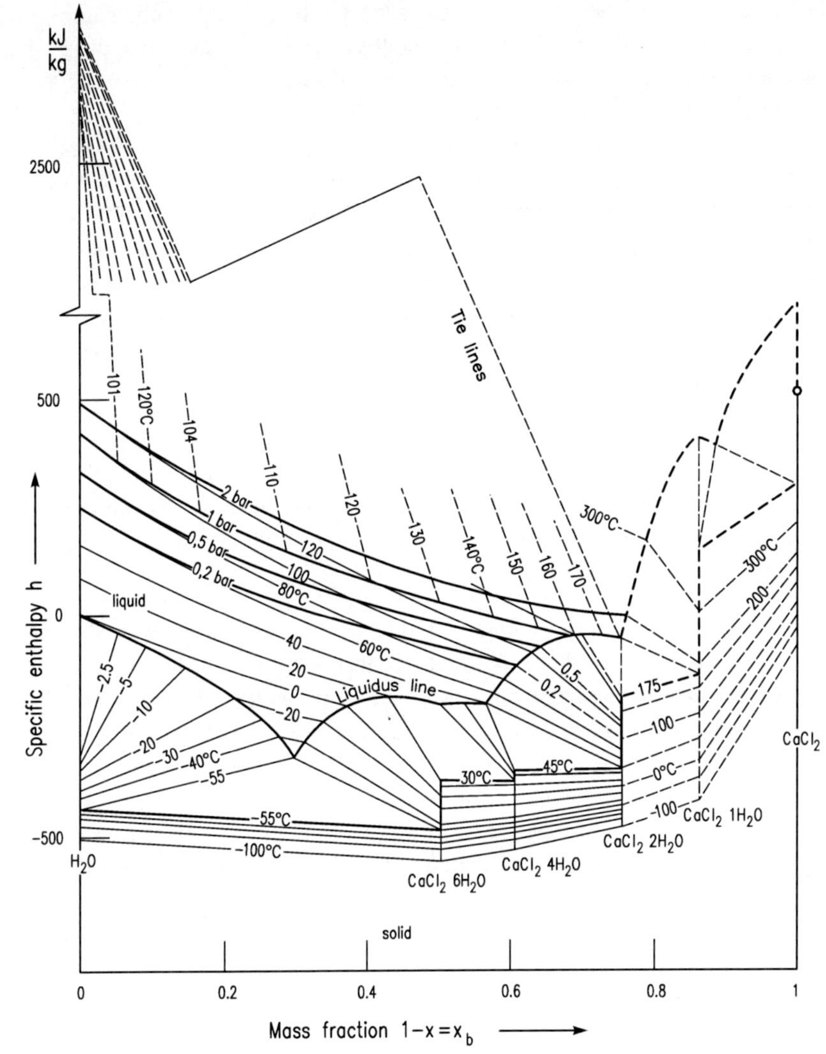

Fig. 2.5-2 Enthalpy–concentration diagram of aqueous calcium chloride solutions (Bosnjakovic 1965)

Figure 2.5-2 is rather complex because one calcium chloride molecule can be combined with one or two or four or six water molecules. Taking into account the molar masses of water and calcium chloride, the mass fractions of different hydrates can be calculated. There are areas in the diagram where only one phase (either solid, liquid, or gaseous) exists. The boundary between the liquid and the vapor phase is represented by the vapor pressure curves valid for different pressures. The liquidus line separates areas where a liquid phase is in equilibrium with a solid phase which can be ice for $x_b = 0$ or a hydrate. Below the liquidus line some two-phase areas (liquid and solid) can be found with some tie lines which indicate temperatures. A tie line gives the information which liquid solution is in equilibrium with a solid phase for a given temperature. As an example let us assume a temperature of $-10°C$ and a mass fraction $x_b = 0.1$ of a calcium chloride–water system. According to the enthalpy–concentration diagram, solid water or ice is in equilibrium with an aqueous calcium chloride solution of $x_b = 0.14$ (crossing point of the tie line and the liquidus line). A balance of a certain component (here $CaCl_2$) leads to the amounts of the two phases which are in equilibrium. The diagram shows a large three-phase area valid for the temperature $-55°C$. Here ice is in equilibrium with solid hexahydrate and an eutectic binary solution with $x_b \approx 0.3$.

A mass balance of a component present in one phase and present in the residual leads to the amounts of the phase and the residual. A further balance applied on the residual is established to obtain the amounts of the other phases present in the residual. Below a temperature of $-55°C$ the mixture with $x_b < 0.49$ is solid. Here, ice and $CaCl_2 \cdot 6H_2O$ are in equilibrium.

Evaporation of an aqueous calcium chloride solution results in the production of steam with $x_b = 0$. The enthalpy of this steam is composed of the contribution of liquid water and the heat of evaporation. Note that the vapor leaving a solution is superheated according to a boiling temperature rise. In the diagram tie lines between the enthalpies at $x_b = 0$ valid for steam and the enthalpies of aqueous solutions are drawn. These lines indicate that a vapor phase is in equilibrium with a boiling liquid solution. In chap. 7 and 8 it will be shown how problems encountered with the evaporation of a binary solution or the crystallization by cooling or evaporation can be solved.

As another example, Fig. 2.5-3 shows the temperature–concentration diagram and the enthalpy–concentration diagram of a magnesium sulfate–water system. Line EB is the melting-point line (or liquidus line) and the straight line EC is the solidification line (or solidus line) in the range up to the concentration of the eutectic point E. The melting temperature follows the line $EDAF$. In the EBC field, solid water (ice) and magnesium sulfate solutions are in equilibrium. The isothermal ECI triangle of $-3.89°C$ represents a three-phase system with a magnesium sulfate solution of composition E and of solid water (ice) and magnesium sulfate crystals that contain 12 molecules of water for every molecule of magnesium sulfate.

The triangle DHJ denotes also an isothermal three-phase system of $2.1°C$. Here, solid $MgSO_4 \cdot 12H_2O$ besides $MgSO_4 \cdot 7H_2O$ crystals are in equilibrium with

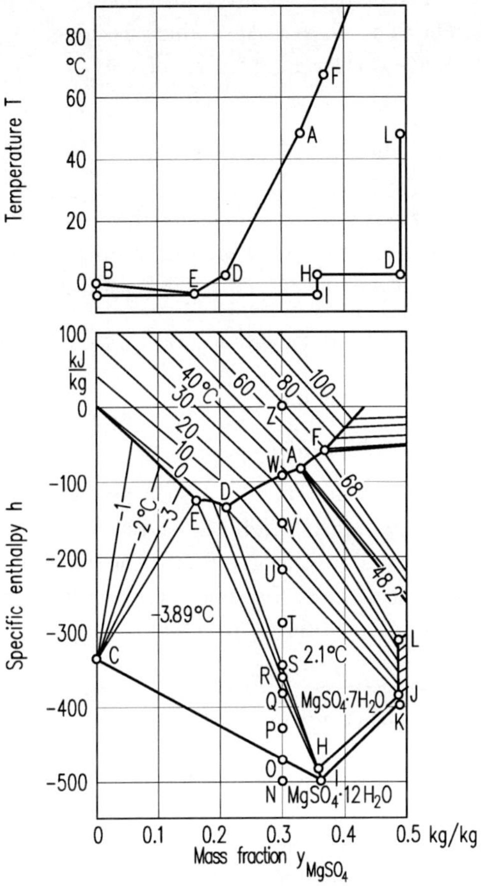

Fig. 2.5-3 Solubility (*above*) and enthalpy–concentration diagram of aqueous magnesium sulfate solutions

a saturated aqueous solution of magnesium sulfate represented by point D (International Critical Tables 1933).

The heat effect caused by the dissolution of a solid i in a liquid or the crystallization of a component i from a solution is a function of the interrelationship between the activity a_i or the activity coefficient $\gamma_i = a_i/\tilde{y}_i$ and the temperature T:

$$\Delta\tilde{h}_i^* = -\tilde{R} \cdot \left[\frac{\partial \ln a_i^*}{\partial(1/T)}\right]_p \tag{2.5-6}$$

Here, $\Delta\tilde{h}_i^*$ is the difference of the molar enthalpy of a component i which is subject to a phase change. Note that this heat is evolved from the final amount of solute which leads to the solubility for a given temperature. The heat of solution $\Delta\tilde{h}_i^*$ can be quite different from the heat $\Delta\tilde{h}_i^\infty$ which describes the heat effect when a small quantity of solute is dissolved in a pure solvent:

$$\Delta\tilde{h}_i^\infty = -\tilde{R} \cdot \left[\frac{\partial \ln a_i}{\partial(1/T)}\right]_p \quad \text{for } a_i \to 0. \tag{2.5-7}$$

The total heat of solution is the heat evolved by the system until a certain concentration is obtained. As a rule all these heats are based on 1 mol solute. Such heats have to be added or removed to carry out an isothermic–isobaric mixing process.

Dealing with hydrates, the heat of solution is composed of two contributions:

- The endothermic heat of melting
- The exothermic heat of hydration

and can be positive or negative (above zero or below zero). The sign depends on the ratio of the two contributions. In Table 2.5-1 the heats of solution of some inorganic compounds dissolved in water are listed. The dissolution of $Al_2(SO_4)_3$ in water leads to a strong increase of the temperature but the temperature is reduced after the addition of $Na_2SO_4 \cdot 10H_2O$ to water.

In ideal systems with the activity coefficient $\gamma_i = 1$ the activity a_i can be replaced by the mole fraction $\tilde{y}_i$ and (2.5-6) can be written as

$$\Delta\tilde{h}_i^* = -\tilde{R} \cdot \left[\frac{\partial \ln \tilde{y}_i^*}{\partial(1/T)}\right]_p . \tag{2.5-8}$$

In this special case the heat of solution is the same as the heat of melting of the pure solid solute.

Table 2.5-1 Heat of solution of some salts

Salt	$\Delta\tilde{h}_{SL}^\infty$ in (kJ/mol) (heat of solution)
$Al_2(SO_4)_3$	–500.0
Na_2SO_4	–1.18
NaCl	+5.0
$Na_2SO_4 \cdot 10H_2O$	+78.0

As has already been discussed the heat of mixing of gaseous components is zero in most cases. Therefore, enthalpy–concentration diagrams can be calculated on the basis of known or measured specific heats and heats of evaporation. Such a diagram for the system air / water is an excellent tool to solve problems encountered in drying, cooling towers, and air conditioning. In the following, the calculation of this diagram is discussed in more detail. Note that such diagrams can also be calculated for any organic vapor which is mixed with an inert gas like N_2.

Sometimes the specific enthalpy is based on a mass unit of the mixture; however, in this case it is practical to use loading in kg vapor per kg vapor-free inert gas. Note that in the following the loading Y is defined as Y kg water (solid, liquid, or gaseous) present in 1 kg dry air. With $\tilde{M}_{G_r}$ as the molar mass of dry air, p_i as the partial pressure of water, and p as the total pressure, the law of ideal gases for 1 kg dry air can be written as

$$(p - p_i) \cdot V = 1\ \text{kg} \cdot \frac{\tilde{R}}{\tilde{M}_{G_r}} \cdot T \tag{2.5-9}$$

Here V is the total volume and T denotes the absolute temperature. The same law applied on Y kg vapor reads

$$p_i \cdot V = Y \cdot \frac{\tilde{R}}{\tilde{M}_i} \cdot T. \tag{2.5-10}$$

With the relative saturation $\varphi_i = p_i / p_i^0$, the combination of the last two equations leads to

$$Y = \frac{\tilde{M}_i}{\tilde{M}_{G_r}} \cdot \frac{p_i^0}{p/\varphi_i - p_i^0} = 0.622 \cdot \frac{\varphi_i}{p/p_i^0 - \varphi_i}. \tag{2.5-11}$$

Note that there is no heat of mixing when dry air and vapor are blended. Therefore, the enthalpy h_{1+Y} of 1 kg dry air and Y kg vapor is the sum of the two contributions:

$$h_{1+Y} = 1 \cdot h_{G_r} + Y \cdot h_i = c_{pG_r} \cdot T + Y \cdot (r_0 + c_{pi} \cdot T) \tag{2.5-12}$$

or

$$h_{1+Y} = (c_{pG_r} + c_{pi} \cdot Y) \cdot T + r_0 \cdot Y. \tag{2.5-13}$$

Here c_{pG_r} denotes the specific heat of dry air and c_{pi} is the specific heat of vapor. The enthalpy h_{1+Y} is zero for air and liquid water at temperature 0°C. The heat of evaporation of water at 0°C is $r_0 = 2{,}500\ \mathrm{kJ/kg}$. The last equation can be transformed to

$$h_{1+Y} = (1 + 1.86 \cdot Y)\,\frac{\mathrm{kJ}}{\mathrm{kg \cdot K}} \cdot T + 2{,}500 \cdot Y\,\frac{\mathrm{kJ}}{\mathrm{kg}}. \tag{2.5-14}$$

A plot of the enthalpy h_{1+Y} as a function of the loading leads to a diagram which is not practical to solve problems encountered in drying and air conditioning. According to an early proposal, the product $r_0 \cdot Y$ of the heat of evaporation and the loading is subtracted from the horizontal axis for $Y = \text{const.}$ with the consequence that isenthalpes are not represented by horizontal lines as would be expected for other enthalpy–concentration diagrams but by steep straight lines, see Fig. 2.5-4. Besides advantages, such a diagram has the drawback that it is unusual and difficult to explain.

In Fig. 2.5-4 the enthalpy–concentration diagram for humid air according to Mollier is depicted. The enthalpy h_{1+Y} of 1 kg dry air and Y kg water vapor can be read from the falling straight lines. The relative humidity $\varphi_i = 1$ separates the entire diagram in the undersaturated region valid for $\varphi_i < 1$ and the range of saturated humid air below the curve $\varphi_i = 1$ as a function of the loading Y. The isotherms in the undersaturated region have a small positive slope. The addition of water with a definite temperature and saturation vapor pressure in the range $\varphi_i < 1$ leads to an increase of φ_i up to $\varphi_i = 1$. Additional water is not evaporated but is in thermodynamic equilibrium with the saturated humid air. The slopes of the tie lines in the saturated region are only a bit greater than the slopes of the isenthalpes according to the small enthalpy contribution of liquid water. Note that this liquid water can be in the system as small suspended droplets (fog) or as a liquid layer at the bottom. The loading Y' valid for the relative humidity $\varphi_i = 1$ or saturated air can be calculated by (2.5-11) by setting $\varphi_i = 1$ for a given temperature. In the range $Y > Y'$ the enthalpy h_{1+Y} is given by

$$h_{1+Y} = (c_{pG_r} + c_{pi} \cdot Y') \cdot T + r_0 \cdot Y' + (Y - Y') \cdot c_L \cdot T \tag{2.5-15}$$

valid for liquid water with $T > 0°\mathrm{C}$ and

$$h_{1+Y} = (c_{pG_r} + c_{pi} \cdot Y') \cdot T + r_0 \cdot Y' - (Y - Y') \cdot (\Delta h_{SL} - c_s \cdot T) \tag{2.5-16}$$

valid for ice. Here, c_L and c_s are the specific heats for water and ice, respectively, and Δh_{SL} denotes the heat of melting. In the following equations the index $1 + Y$ is omitted for reason of simplification:

$$h_{1+Y} = h. \tag{2.5-17}$$

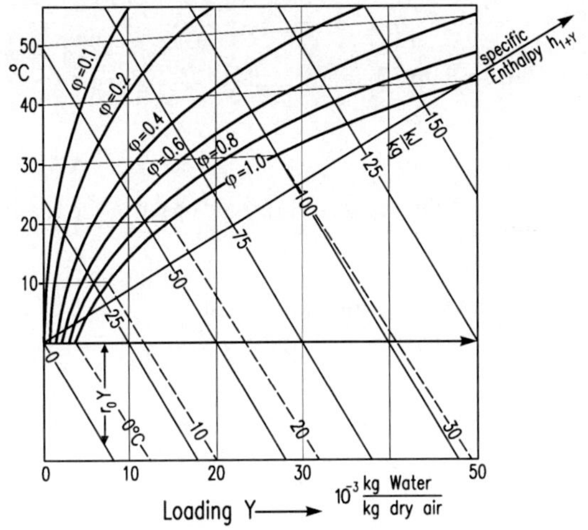

Fig. 2.5-4 Enthalpy–loading diagram of humid air at 1 bar according to Mollier

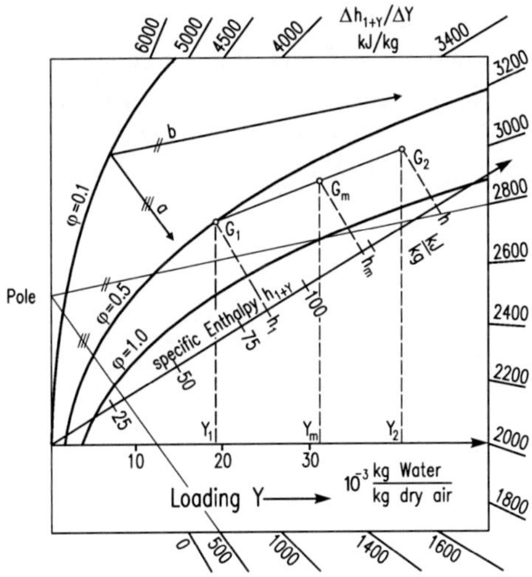

Fig. 2.5-5 Enthalpy–loading diagram for humid air with a scale at the border which gives the direction in the diagram for the addition of H_2O. Examples are the mixing of two humid air streams and the addition of water (liquid or vapor)

The enthalpy–concentration diagram for humid air can be easily used to solve problems encountered with the mixing of two moist air masses or with the addition or removal (drying) of humidity. A mass G_1 with the loading Y_1, the enthalpy h_1, and the temperature T_1 is mixed with the mass G_2 (with Y_2, h_2, and T_2). The loading Y_m and the enthalpy h_m of the mixture can be calculated by means of the balances of water and the enthalpies. The balance of the humidity is given by

$$G_{r1} \cdot Y_1 + G_{r2} \cdot Y_2 = (G_{r1} + G_{r2}) \cdot Y_m . \tag{2.5-18}$$

The formulation of the enthalpy balance leads to

$$G_{r1} \cdot h_1 + G_{r2} \cdot h_2 = (G_{r1} + G_{r2}) \cdot h_m . \tag{2.5-19}$$

Note that G_{r1} and G_{r2} are the masses of dry air because the loading is based on 1 kg dry air.

A combination of the last two equations results in

$$\frac{h_2 - h_m}{h_m - h_1} = \frac{Y_2 - Y_m}{Y_m - Y_1} . \tag{2.5-20}$$

and in the loading of the mixture

$$Y_m = \frac{Y_2 + (G_{r1}/G_{r2}) \cdot Y_1}{1 + G_{r1}/G_{r2}} . \tag{2.5-21}$$

This mixing process is described in Fig. 2.5-5. Both the loading Y_m and the enthalpy h_m of the mixture can be found on the straight line through the points Y_1, h_1, and Y_2, h_2. Let us now consider the addition of liquid water (spray) or water vapor to humid air (again G_{r1} denotes the mass of dry air). With L as the added water, the humidity balance is given by

$$G_{r1} \cdot Y_1 + L = G_{r1} \cdot Y_m \text{ or } Y_m = Y_1 + \frac{L}{G_{r1}} . \tag{2.5-22}$$

The enthalpy balance reads

$$G_{r1} \cdot h_1 + L \cdot h_L = G_{r1} \cdot h_m . \tag{2.5-23}$$

Here h_L is the enthalpy of the added humidity which can be liquid water or water vapor.

These equations lead to

$$\frac{h_m - h_1}{Y_m - Y_1} = \frac{\Delta h}{\Delta Y} = h_L . \tag{2.5-24}$$

Starting from the point Y_1, h_1 the direction of the straight line is given by these balances. The line describes the changes of the humidity and enthalpy caused by the addition of moisture. This direction can be found by the scale drawn at the border of the diagram. Note that all these addition lines are passing through the pole point on the ordinate. The addition of liquid water leads to saturated air with

$\varphi_i = 1$ and Y'. Additional water remains liquid. When humid air is mixed with water vapor with an enthalpy above a certain value, an undersaturated air–vapor mixture is obtained.

Drying means the removal of humidity. Such processes are discussed in Chap. 10.

Symbols

A	m^2/kg	Specific surface
a		Activity $a = \gamma \cdot x$
c	kg/m^3	Concentration
c	$kJ/(kg \cdot K)$	Specific heat capacity
$\tilde{e}$	$kJ/kmol$	Molar energy
f	Pa	Fugacity
G	kJ	Free enthalpy
g	kJ/kg	Specific free enthalpy
$\tilde{g}$	$kJ/kmol$	Molar free enthalpy
H	kJ	Enthalpy
h	kJ/kg	Specific enthalpy
$\tilde{h}$	$kJ/kmol$	Molar enthalpy
$\Delta\tilde{h}_B$	$kJ/kmol$	Molar heat of bonding
Ha	kJ	Hamaker energy
He	$mol/(kg \cdot Pa)$	Henry coefficient
K		Equilibrium constant
m		Number of ions
N_A	$1/kmol$	Avogadro constant
N_S		Number of sections
n	mol/kg	Loading
n		Refractive index
p	Pa	Pressure
Q	Cm^2	Quadrupole moment
q	C	Electric charge
q	kJ/kg	Specific heat, heat of adsorption
R	$kJ/(kg \cdot K)$	Specific gas constant
$\tilde{R}$	$kJ/(kmol \cdot K)$	Universal gas constant
r	m	Radius
S		Selectivity
S	kJ/K	Entropy
S_{BET}	m^2/kg	BET surface
s	$kJ/(kg \cdot K)$	Specific entropy
$\tilde{s}$	$kJ/(kmol \cdot K)$	Molar entropy
s_{cat}		Number of cations
T	K	Absolute temperature
u	kJ/kg	Specific inner energy

V	m^3	Volume
v	m^3/kg	Specific volume
$\tilde{v}$	$m^3/kmol$	Molar volume
w	kJ/kg	Specific work
x, y, z	kg/kg	Mass fraction
$\tilde{x}, \tilde{y}, \tilde{z}$	kmol/kmol	Mole fraction
Z		Compressibility factor
z	m	Height, distance

Greek symbols

α	$(Cm)^2/J$	Polarizability
$\beta(T)$	m^3/mol	Molar volume
γ		Activity coefficient
γ	J/m^2	Interfacial tension
ε	J	Adsorption potential
ε_0	C/(Vm)	Electrical permittivity ($\varepsilon_0 = 8.85 \times 10^{-12}$ C/(Vm))
ε_r		Relative permittivity
Λ		Parameter
Λ	kJ	Parameter
λ	kJ/mol	Phase change enthalpy
μ	Cm	Dipole moment
μ	kJ/mol	Chemical potential
ρ	$kmol/m^3$	Molar density
ρ	kg/m^3	Specific density
σ	m	Molecule diameter
φ		Relative saturation, voidage
φ		Fugacity coefficient
π	J/m^2	Spreading pressure
θ		Coverage
ψ	mol/kg	Reduced spreading pressure

Indices

$a, b, c, \ldots, i, j$	Components
ac	Active
b	Boiling
c	Critical
cat	Cations
E	Excess
G	Gas
i	Adsorptive
id	Ideal
j	Adsorbent

L	Liquid
osm	Osmotic
p	Constant pressure
r	Curved, pure
real	Real
S	Solid
s	Freezing, solid true density in (2.4-23)
t, tot	total
v	Constant volume

Dimensionless numbers

$$A_{\text{indind}} = \frac{3}{4} \cdot \sqrt{\frac{2 \cdot Ha_j}{\pi \cdot \sigma_j^3 \cdot p_c}} \cdot \frac{T_c}{T_{\text{ads}}}$$

$$A_{\text{indcha}} = \frac{\alpha_i \cdot s_{\text{cat}}^2 \cdot e^2 \cdot \rho_{j,\text{at}}}{8 \cdot \pi \cdot (\varepsilon_0 \cdot \varepsilon_r)^2 \cdot (k \cdot T_{\text{ads}}) \cdot \sigma_{i,j}}$$

$$A_{\text{dipind}} = \frac{\mu_i^2 \cdot \alpha_j \cdot \rho_{j,\text{at}}}{12 \cdot \pi \cdot (\varepsilon_0 \cdot \varepsilon_r)^2 \cdot (k \cdot T_{\text{ads}}) \cdot \sigma_{i,j}^3}$$

$$A_{\text{dipcha}} = \frac{\mu_i^2 \cdot s_{\text{cat}}^2 \cdot e^2 \cdot \rho_{j,\text{at}}}{12 \cdot \pi \cdot (\varepsilon_0 \cdot \varepsilon_r)^2 \cdot (k \cdot T_{\text{ads}})^2 \cdot \sigma_{i,j}}$$

$$A_{\text{quadind}} = \frac{3 \cdot Q_i^2 \cdot \alpha_j \cdot \rho_{j,\text{at}}}{40 \cdot \pi \cdot (\varepsilon_0 \cdot \varepsilon_r)^2 \cdot (k \cdot T_{\text{ads}}) \cdot \sigma_{i,j}^5}$$

$$A_{\text{quadcha}} = \frac{Q_i^2 \cdot s_{\text{cat}}^2 \cdot e^2 \cdot \rho_{j,\text{at}}}{240 \cdot \pi \cdot (\varepsilon_0 \cdot \varepsilon_r)^2 \cdot (k \cdot T_{\text{ads}})^2 \cdot \sigma_{i,j}^3}$$

$$He^* = \frac{He \cdot RT}{S_{\text{BET}} \cdot \sigma_{i,j}}$$

$$Q^* = \left(\frac{Q_i}{\sigma_i}\right)^2 \cdot \frac{1}{\alpha_i \cdot (k \cdot T_{\text{ads}})}$$

$$\mu^* = \frac{\mu_i^2}{\alpha_i \cdot (k \cdot T_{\text{ads}})}$$

3 Fundamentals of Single-Phase and Multiphase Flow

Single-phase or multiphase flow in chemical engineering apparatus and reactors such as evaporators, columns, fixed and fluidized beds, and stirred vessels is decisive for the efficiency and capacity of these equipment units. The mixing of two or more liquid components without the formation of a second liquid phase leads to a one-phase flow, for instance in a stirred vessel. In the case of an evaporative crystallizer a three-phase system is moved by the stirrer and the rising bubbles. In many cases two phases are flowing in countercurrent direction, e.g., in columns for absorption, rectification, and extraction. Two phases are also moving in bubble and drop columns, froth layers on plates, fluidized beds, and stirred vessels used for suspension flow or the breakup of gases and liquids. Dealing with packed or film columns, the liquid phase runs down in films and rivulets in countercurrent to the gas. In many apparatus of chemical engineering, internal pieces of equipment are mounted and the fluid flow is passing around or through these internals. Such internals are

- Perforated plates such as sieve trays used in absorption, distillation or extraction columns. The holes can be covered by caps or valves to avoid weeping in the range of low superficial gas or vapor velocities. The two phases are moving in a crossflow on a tray.
- Packings, often installed in columns used for separation processes. Films or rivulets are running down in countercurrent flow of the gas or vapor which are passing the voidage of the packing. Packings can consist of elements such as particles to be dried or extracted or mass separation elements such as adsorbents and ion exchange resins
- Solid or fluid particles (bubbles or drops) suspended in another fluid are moving in a large-scale flow. Such a flow can be present in fluidized beds with gas or liquid as continuous phase, for instance dryers, adsorbers, or crystallizers. Fluid particles are moving in bubble or drop columns or in froth (spray or bubble regime) present on column trays.

With respect to the design of such apparatus and the placement of internals or packings it is very important

A. Mersmann et al., *Thermal Separation Technology: Principles, Methods, Process Design*, VDI-Buch, DOI 10.1007/978-3-642-12525-6_3,
© Springer-Verlag Berlin Heidelberg 2011

- To obtain an even distribution of the phase flows in the entire cross-section area
- To perform nearly a plug flow of the phases and to reduce mostly their back-mixing
- To limit the phase flows in countercurrent direction to maintain the countercurrent flow (for example, the liquid moving in crossflow to the gas on a tray of a column should not be entrained by the gas or vapor)
- To perform a large interfacial area with respect to heat and mass transfer

The goal of this chapter is to describe the fluid motion laws in a very general way based on flow systems rather than valid for special apparatus in narrow ranges of velocities and fluid properties. At first, fundamental laws of one-phase flows are presented because they are the basis for two- and multiphase flow.

3.1 Basic Laws of Single-Phase Flow

All fluids with particles in the molecular range (atoms, ions, molecules) can be described by laws valid for one-phase flow.

3.1.1 Laws of Mass Conservation and Continuity

The equation of mass conservation will be written first in the general way, e.g., valid for the three-dimensional and instationary case. With the density ρ, the time t, and the velocity vector $\boldsymbol{v}$ with the velocities u, v, and w in the x-, y- and z-directions, respectively., the mass balance of a volume element $\mathrm{d}V = \mathrm{d}x \cdot \mathrm{d}y \cdot \mathrm{d}z$ is given by

$$\frac{\partial \rho}{\partial t} + \frac{\partial(\rho \cdot u)}{\partial x} + \frac{\partial(\rho \cdot v)}{\partial y} + \frac{\partial(\rho \cdot w)}{\partial z} = 0 \tag{3.1-1}$$

or

$$\frac{\partial \rho}{\partial t} + \mathrm{div}(\rho \cdot \boldsymbol{v}) = 0\,. \tag{3.1-2}$$

In the case of a stationary two-dimensional flow in an $x-y$ plane this equation can be reduced for incompressible fluids (approximately valid for liquids) to

$$\frac{\partial u}{\partial x} + \frac{\partial v}{\partial y} = 0\,. \tag{3.1-3}$$

The mass flow $\dot{M}$ in a uniformly diverging duct (see Fig. 3.1-1) can be written for the cross sections 1 and 2 in the z direction with $w_1 \neq f(t)$ as

$$\dot{M} = w_1 \cdot f_1 \cdot \rho_1 = w_2 \cdot f_2 \cdot \rho_2 \tag{3.1-4}$$

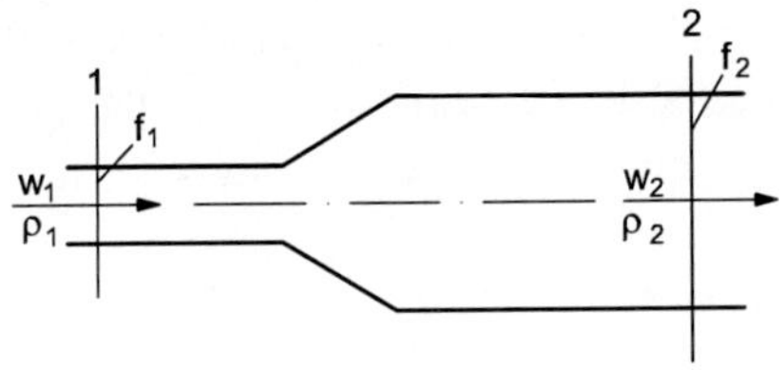

Fig. 3.1-1 Flow channel

This means that the mass flow in pipes with changing cross sections remains constant. In the special case $f_1 = f_2$ the mass flow density $\dot{m} = \rho \cdot w$ remains constant:

$$\dot{m} = \rho_1 \cdot w_1 = \rho_2 \cdot w_2 \text{ or } \rho \cdot w = \text{const.} \tag{3.1-5}$$

This equation can be further simplified for incompressible fluids:

$$w_1 = w_2 \text{ or } w = \text{const.} \tag{3.1-6}$$

3.1.2 Irrotational and Rotational Flow

Principally speaking, a flow can be an irrotational or potential flow or a rotational flow. If the velocity vector $\boldsymbol{v}$ can be described as the gradient of a velocity potential Φ according to

$$\boldsymbol{v} = \text{grad } \Phi = \boldsymbol{i} \cdot \frac{\partial \Phi}{\partial x} + \boldsymbol{j} \cdot \frac{\partial \Phi}{\partial y} + \boldsymbol{k} \cdot \frac{\partial \Phi}{\partial z}, \tag{3.1-7}$$

a potential flow exists. This can be written for incompressible fluids (div $\boldsymbol{v} = 0$) as

$$\Delta \Phi = \text{div}(\text{grad } \Phi) = \frac{\partial^2 \Phi}{\partial x^2} + \frac{\partial^2 \Phi}{\partial y^2} + \frac{\partial^2 \Phi}{\partial z^2} = 0. \tag{3.1-8}$$

The difference between a potential flow and a rotational flow becomes clear by the equations

$$\text{rot}(\text{grad } \Phi) = 0 \text{ valid for potential flow,}$$

$$\text{rot}(\text{grad } \Phi) \neq 0 \text{ in the case of rotational flow.}$$

The flow problems discussed in this book can be solved by the assumption of potential flow. This makes the treatment of fluid motion easy.

3.1.3 The Viscous Fluid

During the flow of a real Newtonian fluid, a certain quantity of the mechanical energy is converted or dissipated into heat or acoustic energy. The same is true when a flow around a solid resistance (tubes, spheres, packing elements) takes place. Consider a plate with the area A which is moving with the velocity w on the surface of a fluid, see Fig. 3.1-2. The force F based on the area A or the shear stress τ is proportional to the shear rate $\dot{\gamma} = \mathrm{d}w/\mathrm{d}y$ and the viscosity η:

$$\frac{F}{A} = \tau = -\eta \cdot \frac{\mathrm{d}w}{\mathrm{d}y} . \tag{3.1-9}$$

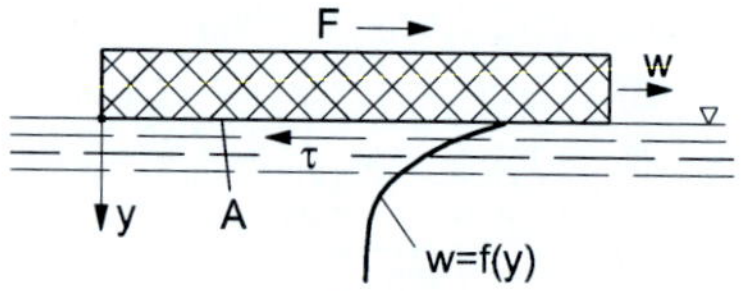

Fig. 3.1-2 Shear stress in a liquid induced by a moving plate

For Newtonian fluids $\eta \neq f(\dot{\gamma})$ but $\eta = f(T, p)$ is valid.

In the case of non-Newtonian liquids according to $\eta = f(\dot{\gamma})$ the shear stress is usually expressed by the relationship $\tau = k \cdot (\dot{\gamma})^n$ with $k = \eta$ and $n = 1$ valid for the special case of a Newtonian liquid. The symbol k is the consistency and η the fluidity ($n < 1$: pseudoplastic and $n > 1$: dilatant). Such fluids are not discussed here. The fluid property viscosity quantifies the inner friction within a fluid or the friction between molecules and is zero for ideal fluids. Increasing temperatures lead to an increase of viscosity in gases but to a reduction of this property in liquids.

3.1.4 Navier–Stokes, Euler, and Bernoulli Equations

The force balance written for a volume element $\mathrm{d}x \cdot \mathrm{d}y \cdot \mathrm{d}z$ leads to the Navier-Stokes equation of motion:

$$\frac{\mathrm{d}\boldsymbol{v}}{\mathrm{d}t} = \boldsymbol{g} - \frac{1}{\rho} \cdot \operatorname{grad} p + \frac{\eta}{\rho} \cdot \Delta \boldsymbol{v} . \tag{3.1-10}$$

Here $\mathrm{d}\boldsymbol{v}/\mathrm{d}t$ is the substantial derivative of the velocity or the sum of derivatives with respect to time and space. The symbols $\boldsymbol{g}$ and p denote the acceleration due to gravity and the pressure, respectively. Dealing with ideal fluids the viscosity is zero and the Navier-Stokes equation can be simplified to the Euler equation:

$$\frac{\partial u}{\partial t} + u \cdot \frac{\partial u}{\partial x} + v \cdot \frac{\partial u}{\partial y} + w \cdot \frac{\partial u}{\partial z} = g_x - \frac{1}{\rho} \cdot \frac{\partial p}{\partial x} \quad \text{for the } x\text{-direction,}$$

$$\frac{\partial v}{\partial t} + u \cdot \frac{\partial v}{\partial x} + v \cdot \frac{\partial v}{\partial y} + w \cdot \frac{\partial v}{\partial z} = g_y - \frac{1}{\rho} \cdot \frac{\partial p}{\partial y} \quad \text{for the } y\text{-direction,}$$

$$\frac{\partial w}{\partial t} + u \cdot \frac{\partial w}{\partial x} + v \cdot \frac{\partial w}{\partial y} + w \cdot \frac{\partial w}{\partial z} = g_z - \frac{1}{\rho} \cdot \frac{\partial p}{\partial z} \quad \text{for the } z\text{-direction.}$$

Or

$$\frac{\mathrm{d}\boldsymbol{v}}{\mathrm{d}t} = \boldsymbol{g} - \frac{1}{\rho} \cdot \mathrm{grad}\, p. \tag{3.1-11}$$

The integration of the Euler equations for the special case of a one-dimensional steady-state flow in z-direction leads to the Bernoulli equation:

$$\frac{w^2}{2 \cdot g} + z + \frac{p}{\rho \cdot g} = \text{const. in } [\text{m}, \text{pressure head}] \tag{3.1-12}$$

or

$$\frac{\rho \cdot w^2}{2} + z \cdot \rho \cdot g + p = \text{const}'. \text{ in } \left[\text{kg}/(\text{s}^2\text{m}), \text{N}/\text{m}^2, \text{J}/\text{m}^3, \text{pressure} = \frac{\text{energy}}{\text{volume}}\right] \tag{3.1-13}$$

or

$$\frac{w^2}{2} + z \cdot g + \frac{p}{\rho} = \text{const}''. \text{ in } \left[\text{m}^2/\text{s}^2, \text{ J/kg}, \frac{\text{energy}}{\text{mass unit}}\right]. \tag{3.1-14}$$

The first term is the kinetic energy, the second is the energy according to a height difference z and the third is the energy based on the unit volume of the fluid or – in general – the pressure. According to this energy balance, energy can be converted, for instance, from pressure to kinetic energy or vice versa. The last equation is the law of conservation of energy for an isothermal system without the addition or removal of work. The momentum $\boldsymbol{I}$ is the product of the mass M and the velocity $\boldsymbol{w}$:

$$\boldsymbol{I} = M \cdot \boldsymbol{w}. \tag{3.1-15}$$

The derivative of the momentum with respect to the time is equal to the sum of all forces $\Sigma \boldsymbol{F}$:

$$\frac{\mathrm{d}\boldsymbol{I}}{\mathrm{d}t} = \Sigma \boldsymbol{F} . \qquad (3.1\text{-}16)$$

Multiplying the mass flow $\dot{M} = \mathrm{d}M/\mathrm{d}t$ with the velocity $\boldsymbol{w}$ yields the momentum flux $\dot{M} \cdot \boldsymbol{w}$. The momentum flux density is the momentum flux based on the cross-sectional area f:

$$\frac{\dot{M}}{f} \cdot \boldsymbol{w} = \dot{\boldsymbol{m}} \cdot \boldsymbol{w} . \qquad (3.1\text{-}17)$$

The scalar product of the vectors $\dot{\boldsymbol{m}}$ and $\boldsymbol{w}$ yields a volumetric work or volumetric energy which has the dimension of a pressure. Let us assume a horizontal flow of an ideal ($\eta = 0$) fluid. According to the law of conservation of momentum, the differential of the momentum flux density is equal to the differential of the pressure:

$$\mathrm{d}(\dot{\boldsymbol{m}} \cdot \boldsymbol{w}) = \mathrm{d}p \qquad (3.1\text{-}18)$$

or for the special case of constant $\dot{m}$

$$\mathrm{d}(\dot{\boldsymbol{m}} \cdot \boldsymbol{w}) = \rho \cdot \boldsymbol{w} \cdot \mathrm{d}\boldsymbol{w} = \mathrm{d}p . \qquad (3.1\text{-}19)$$

The integration of this equation between the boundaries $w = 0$ and w and p_1 and p_2 leads to a special form of the Bernoulli equation:

$$\frac{\rho \cdot w^2}{2} = p_1 - p_2 . \qquad (3.1\text{-}20)$$

This equation describes the conversion of kinetic energy into potential energy and vice versa when an ideal fluid is flowing through a horizontal pipe with changing cross-sectional area.

In the case of very small inertia forces (very slow flow) of viscous fluids flowing through horizontal ducts the pressure change is given by

$$\text{grad}\, p = \eta \cdot \Delta v , \qquad (3.1\text{-}21)$$

which can be simplified for a one-dimensional flow and the equation valid for the shear stress:

$$\frac{\partial p}{\partial z} = \eta \cdot \frac{\partial^2 w}{\partial y^2} . \qquad (3.1\text{-}22)$$

3.1.5 Laminar and Turbulent Flow in Ducts

In laminar flow, all liquid fluid elements are moving in parallel paths without crossing each other.

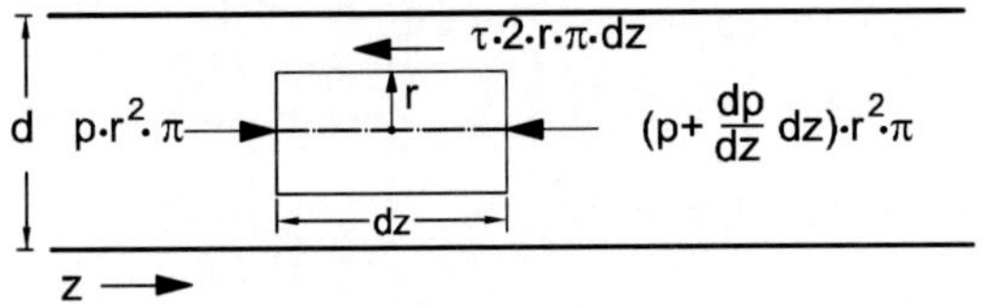

Fig. 3.1-3 Pressure and shear stress on a cylindrical liquid cylinder valid for laminar flow

The flow in a tube is an appropriate example to demonstrate the differences of laminar and turbulent flow. According to Fig. 3.1-3 the force balance for a cylindrical element with the radius r and the length $\mathrm{d}z$ can be written as

$$\tau \cdot 2 \cdot \pi \cdot r \cdot \mathrm{d}z = p \cdot r^2 \cdot \pi - \left(p + \frac{\mathrm{d}p}{\mathrm{d}z} \cdot \mathrm{d}z\right) \cdot r^2 \cdot \pi \tag{3.1-23}$$

or after introduction of the equation valid for the shear stress

$$-\eta \cdot \frac{\mathrm{d}w(r)}{\mathrm{d}r} = -\frac{r}{2} \cdot \frac{\mathrm{d}p}{\mathrm{d}z}. \tag{3.1-24}$$

Integration yields

$$w(r) = \frac{1}{4 \cdot \eta} \cdot \frac{\mathrm{d}p}{\mathrm{d}z} \cdot (r^2 - R^2). \tag{3.1-25}$$

According to the assumption that the velocity is zero at the pipe wall,

$$w(r = R) = 0. \tag{3.1-26}$$

The mean velocity w can be derived by the application of the law of continuity:

$$\dot{V} = R^2 \cdot \pi \cdot w. \tag{3.1-27}$$

The volumetric flow $\dot{V}$ can be expressed as a result of the integration of the local velocity $w(r)$:

$$\dot{V} = \int_0^R w(r) \cdot 2 \cdot \pi \cdot r \cdot \mathrm{d}r. \tag{3.1-28}$$

By this the velocity profile in a pipe valid for laminar flow is obtained:

$$w(r) = 2 \cdot w\left[1 - \left(\frac{2 \cdot r}{d}\right)^2\right]. \qquad (3.1\text{-}29)$$

The maximum velocity $w_{\max}$ at the center of the pipe is by a factor 2 greater than the mean velocity or

$$w_{max} = 2 \cdot w, \qquad (3.1\text{-}30)$$

where w is based on the equation of continuity

$$\dot{M} = w \cdot f \cdot \rho. \qquad (3.1\text{-}31)$$

The pressure drop $\mathrm{d}p$ of a fluid per unit length $\mathrm{d}z$ is

$$\frac{\mathrm{d}p}{\mathrm{d}z} = -\frac{8 \cdot \eta \cdot w}{R^2}. \qquad (3.1\text{-}32)$$

This Hagen–Poiseuille equation is a measure for the conversion of mechanical energy into heat or thermal energy.

The value of the Reynolds number,

$$Re = \frac{w \cdot d \cdot \rho}{\eta}, \qquad (3.1\text{-}33)$$

is decisive for the question whether a pipe flow is laminar or turbulent. If this Reynolds number with the pipe diameter d and the mean velocity w exceeds 2,300 the laminar flow is no longer stable and turbulent flow with the velocity profile

$$\frac{w(r)}{w_{\max}} = \left(1 - \frac{2 \cdot r}{d}\right)^n \qquad (3.1\text{-}34)$$

occurs. The exponent n is a function of the Reynolds number, see Fig. 3.1-4.

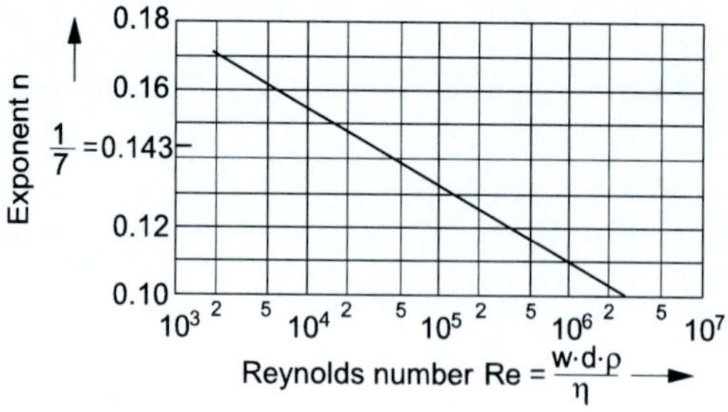

Fig. 3.1-4 Exponent n of the velocity profile equation as a function of the Reynolds number valid for turbulent flow

The pressure drop Δp of a fluid flowing through a pipe with the diameter d and the length L is given by

$$\Delta p = \lambda \cdot \frac{w^2 \cdot \rho}{2} \cdot \frac{L}{d}. \tag{3.1-35}$$

The friction factor λ can be derived theoretically (as already shown) according to the Hagen–Poiseuille equation:

$$\lambda = \frac{64}{Re} \text{ for } Re < 2{,}300. \tag{3.1-36}$$

In Fig. 3.1-5 the friction factor λ for pipes with a surface roughness k (k is the mean height of protuberances) is plotted against the Reynolds number. The equations presented here in combination with the diagram are general tools to calculate the pressure drop Δp in circular tubes with constant cross-sectional area.

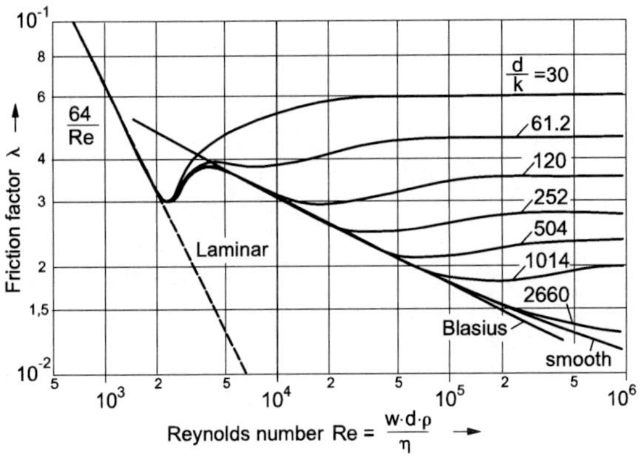

Fig. 3.1-5 Friction factor as a function of the Reynolds number with k as the roughness of the inside surface

An additional pressure change occurs in tubes with a sudden contraction or a sudden expansion with the cross-sectional area f_0 in the narrow and f in the wide tubes, see Fig. 3.1-6. These pressure changes are for a fully turbulent flow

sudden contraction sudden expansion

$$\Delta p = \left(\frac{f_0}{f_e} - 1\right)^2 \cdot \frac{\rho}{2} \cdot w_0^2 \qquad \Delta p = \left(1 - \frac{f_0}{f}\right)^2 \cdot \frac{\rho}{2} \cdot w_0^2. \tag{3.1-37}$$

The contraction coefficient $\mu = f_e/f_0 = w_0/w_e$ is the smallest cross-sectional area (within a vortex close to the wall in a certain distance from the sudden contraction) based on f_0. In the case of sharp-edged orifices it is

$$\mu = \frac{\pi}{\pi + 2} = 0.61. \tag{3.1-38}$$

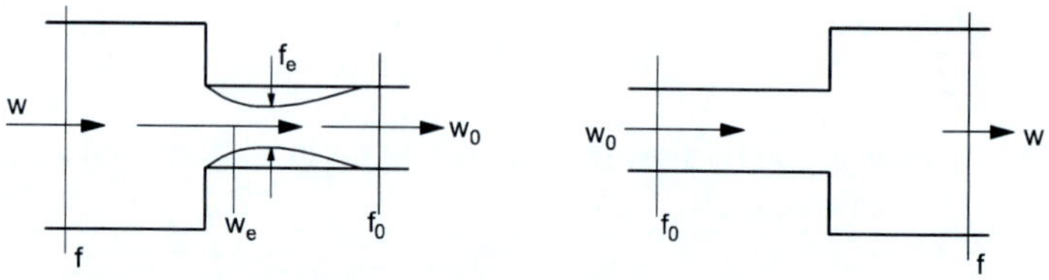

Fig. 3.1-6 Sudden contraction (*left*) and sudden enlargement (*right*)

Many apparatus are equipped with hole plates with the thickness L and d_0 as the hole diameter. Dealing with sharp-edged holes and small pressure drops Δp in comparison to the system pressure p the pressure drop Δp is given by

$$\Delta p = \zeta \cdot \frac{\varrho}{2} \cdot w_0^2 . \tag{3.1-39}$$

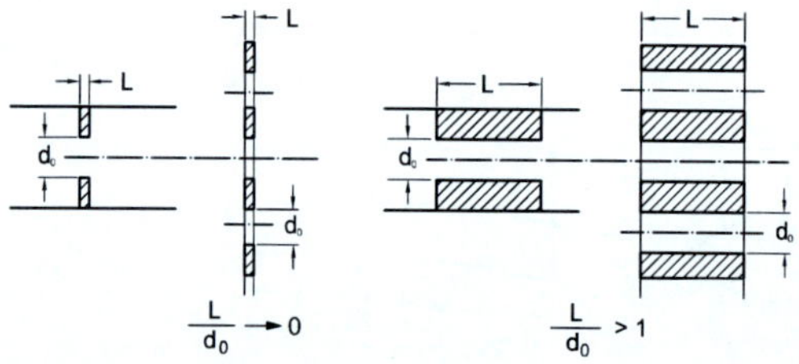

Fig. 3.1-7 Orifice or sieve plate with a small ratio L/d_0 (*left*) or a large ratio L/d_0 (*right*)

The friction factor ζ is a function of the ratio L/d_0 and the fraction φ of the total cross-section area of the perforated plate. Two different boundary cases ($L/d_0 \to 0$ and $L/d_0 \to \infty$) can be distinguished (see also Chap. 5):

(a) $\dfrac{L}{d_0} \to 0$.

$$\zeta = \left(\frac{1}{\mu} - \varphi\right)^2$$

or for $\varphi \to 0$

$$\zeta_0 = \left(\frac{1}{\mu}\right)^2 = 2.69 .$$

(b) $\dfrac{L}{d_0} \to \infty$.

$$\zeta = \left(\frac{1}{\mu} - 1\right)^2 + (1 - \varphi)^2$$

or for $\varphi \to 0$

$$\zeta_0 = \left(\frac{1}{\mu} - 1\right)^2 + 1 = 1.41 . \tag{3.1-40}$$

The friction factors for arbitrary Reynolds numbers and L/d_0 ratios can be taken from Fig. 3.1-8. The friction factor of a rounded flow nozzle without sharp edges is approximately $\zeta_0 \approx 1$.

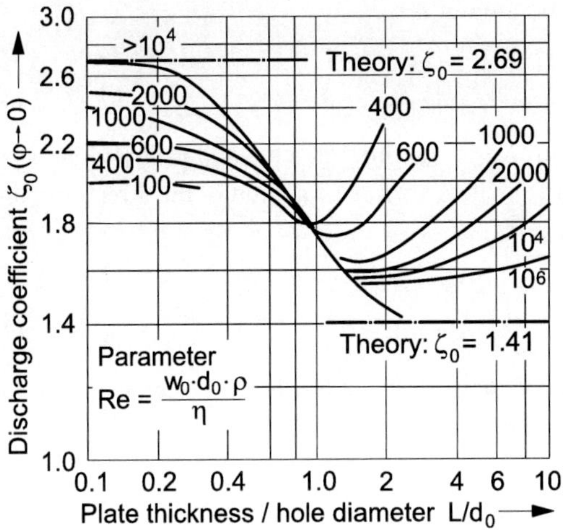

Fig. 3.1-8 Discharge coefficient as a function of the ratio L/d_0 for several Reynolds numbers

3.1.6 Turbulence

In industrial containments a laminar flow can only be expected if the viscosity of the fluid is very high. Dealing with gases at ambient temperature and low viscous liquids the flow will be turbulent. In a turbulent flow the mean velocity is time independent; however, local and momentary fluctuation velocities yield additional contributions. The instantaneous values of the velocity and pressure are fluctuating about their mean values. In rectangular coordinates the instantaneous velocities can be written as

$$u(t) = u + u'(t),$$

$$v(t) = v + v'(t),$$

$$w(t) = w + w'(t). \tag{3.1-41}$$

The time averaged local value in the z-direction is defined by an integral from $t = 0$ to $t = \tau$ according to

$$w = \frac{1}{\tau} \cdot \int_0^{\tau} w(t) \mathrm{d}t. \tag{3.1-42}$$

Besides this mean value (here w) the mean values of the fluctuating velocities are important. The mean values of these fluctuating velocities are zero according to their definitions

$$\bar{u}' = 0, \bar{v}' = 0, \bar{w}' = 0. \tag{3.1-43}$$

However, the quadratic mean values are not zero:

$$\overline{u'^2} \neq 0\,;\ \overline{v'^2} \neq 0\,;\ \overline{w'^2} \neq 0\,. \tag{3.1-44}$$

With the assumption of isotropic turbulence (approximately valid in stirred vessels) the mean values of the fluctuating velocities are not a function of the direction. The definition of the effective value w'_{eff} is

$$w'_{\text{eff}} = \sqrt{\overline{w'^2}}\,. \tag{3.1-45}$$

The ratio of an effective fluctuation velocity based on the mean velocity at a certain point is called the degree of turbulence Tu :

$$\text{Tu} = \frac{w'_{\text{eff}}}{w} = \frac{\sqrt{\overline{w'^2}}}{w}\,. \tag{3.1-46}$$

Additional information on turbulence is given in the chapter dealing with mixing.

3.1.7 Molecular Flow

All equations presented up to here are only valid if the mean free path lengths of the molecules are very short in comparison to the diameter d of pipes or pores. This condition is not fulfilled for gases flowing in pipes with a diameter in the millimeter range if the pressure is below 1 Pa. The pores in adsorbents or in porous materials to be dried have a width of some nanometers in most cases or even smaller than 1 nm with the result that the mean free path length is approximately the same or more than the pore diameter for gas pressures of 10^5 Pa. The ratio of the molecular mean free path Λ based on the tube or pore diameter d is the Knudsen number

$$Kn = \frac{\Lambda}{d}\,. \tag{3.1-47}$$

The flow is laminar for $Kn < 0.01$. The flow in the range $0.01 < Kn < 1$ is called slip flow and free molecule flow or Knudsen flow occurs for $Kn > 1$. The special characteristic of this flow is that the friction between molecules can be neglected and the pressure drop depends on the friction between the molecules and the wall of the tubes or pores.

In Fig. 3.1-9 on the left side a tube with diameter d and length L is shown where impinging molecules are reflected by the wall due to elastic impacts. On the right side a diffuse reflexion is illustrated as a result of rough walls and partially inelastic collisions. Single molecules can move in a direction opposite to the main Knudsen flow.

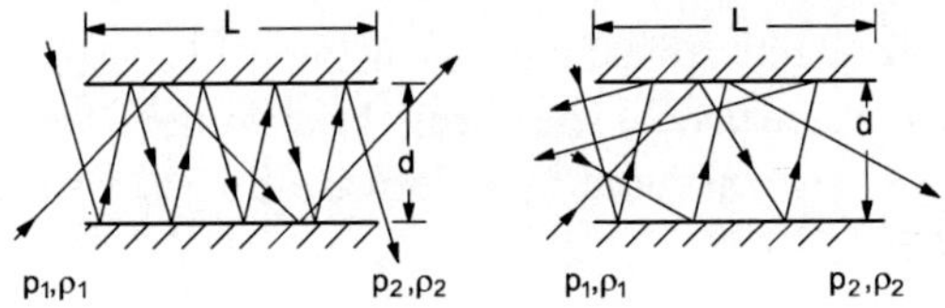

Fig. 3.1-9 Molecular flow in the case of completely elastic collisions (*left*) and diffusive reflection of the molecules (*right*)

The number of molecules $\dot{N}$ per unit time entering the tube is

$$\dot{N} = \frac{n_v}{4} \cdot \overline{w_{Mol}} \cdot \frac{d^2 \cdot \pi}{4}, \tag{3.1-48}$$

with n_v as the number of molecules per unit volume and $\overline{w_{Mol}}$ as the mean value of the molecule velocity according to

$$\overline{w_{Mol}} = \sqrt{\frac{8 \cdot R \cdot T}{\pi}}, \tag{3.1-49}$$

which can be derived from the kinetic theory of gases. The probability of an impact of a molecule on the inner tube wall is proportional to the ratio L/d. The number $\dot{N}_1$ of molecules flowing in one direction per unit time is

$$\dot{N}_1 = C_1 \cdot \frac{n_v}{4} \cdot \overline{w_{Mol}} \cdot \frac{d^2 \cdot \pi}{4} \cdot \frac{d}{L} = C \cdot n_v \cdot \frac{\overline{w_{Mol}} \cdot d^3}{L}, \tag{3.1-50}$$

with $C = \pi/12$ which can be derived from a bit lengthy relationships. When M_{Mol} denotes the mass of a molecule and ρ_1 the density of the entering gas the mass flow $\dot{M}_1$ at the entrance is given by

$$\dot{M}_1 = \dot{N}_1 \cdot M_{Mol} = \frac{\pi}{12} \cdot \overline{w_{Mol}} \cdot \frac{d^3}{L} \cdot \rho_1 . \tag{3.1-51}$$

The mass flow in the opposite direction can be described by an analogous equation. Let us consider a mean density ρ which is equal to the mean value between the entrance and the exit. The net mass flow rate $\dot{M} = \dot{M}_1 - \dot{M}_2$ is

$$\dot{M} = \dot{M}_1 - \dot{M}_2 = \frac{\pi}{12} \cdot \left(\rho + \frac{\Delta\rho}{2}\right) \cdot \overline{w_{Mol}} \cdot \frac{d^3}{L} - \frac{\pi}{12} \cdot \left(\rho - \frac{\Delta\rho}{2}\right) \cdot \overline{w_{Mol}} \cdot \frac{d^3}{L}. \tag{3.1-52}$$

With the equation for the mean molecule velocity presented earlier the mass flow rate $\dot{M}$ is

$$\dot{M} = \frac{\sqrt{\pi}}{3 \cdot \sqrt{2}} \cdot \frac{d^3 \cdot \Delta p}{\sqrt{R \cdot T} \cdot L}. \tag{3.1-53}$$

Let us consider the diffusion of an adsorptive in an inert gas through an adsorbent pore or the diffusion of vapor in a moist material to be dried. The mass flow due to molecular or Knudsen diffusion ($Kn > 1$) induced by a decrease $\mathrm{d}p_i/\mathrm{d}l$ of the partial pressure per unit length $\mathrm{d}l$ is

$$\dot{M} = -\frac{\sqrt{\pi}}{3 \cdot \sqrt{2}} \cdot \frac{d^3}{\sqrt{R \cdot T}} \cdot \frac{\mathrm{d}p_i}{\mathrm{d}l}. \tag{3.1-54}$$

When the pore is replaced by an orifice or a hole in a plate the mass flow density $\dot{m}$ for the pressure difference $p_1 - p_2$ before and behind the plate is given by

$$\dot{m} = \frac{p_1 - p_2}{\sqrt{2 \cdot \pi \cdot R \cdot T}}. \tag{3.1-55}$$

The common feature of all these equations is that the flow rate is inversely proportional to the root $\sqrt{R \cdot T}$ (or the molecule velocity) and the viscosity plays no role. The slip velocity in the range $0.01 < Kn < 1$ can be quantified by the equations valid for pure laminar and pure Knudsen flow by an appropriate superposition.

3.1.8 Falling Film on a Vertical Wall

In a variety of apparatus and chemical reactors, liquid films or rivulets are flowing around pipes, walls, or packing elements (packed columns, cooling towers, liquid film coolers). Things are very easy in the case of a laminar film running on a vertical wall, see Fig. 3.1-10.

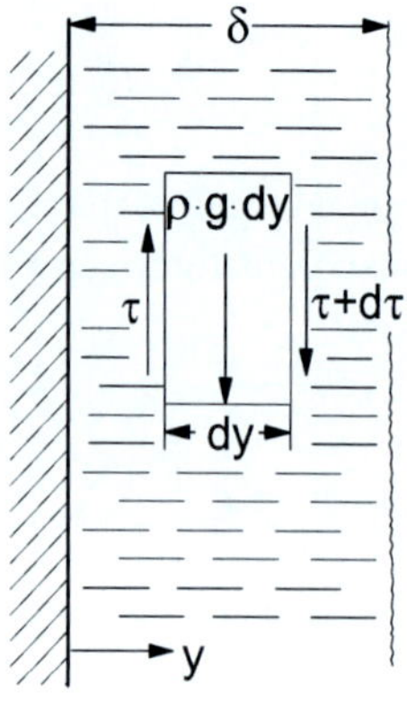

Fig. 3.1-10 Shear stresses on the outer area of a liquid element within a running film (laminar flow)

Only viscosity and gravity forces are acting on a film element with the thickness $\mathrm{d}y$. A force balance (Nusselt 1916) leads to the equation

$$\tau \cdot \mathrm{d}A = \left(\tau + \frac{\mathrm{d}\tau}{\mathrm{d}y} \cdot \mathrm{d}y\right) \cdot \mathrm{d}A + \mathrm{d}A \cdot \rho_\mathrm{L} \cdot g \cdot \mathrm{d}y\,. \tag{3.1-56}$$

With the already presented equation valid for the shear stress for Newtonian liquids, the last equation can be written as

$$\frac{\mathrm{d}\left(\eta_\mathrm{L} \cdot \frac{\mathrm{d}w(y)}{\mathrm{d}y}\right)}{\mathrm{d}y} + \rho_\mathrm{L} \cdot g = 0 \tag{3.1-57}$$

or

$$\eta_\mathrm{L} \cdot \frac{\mathrm{d}^2 w(y)}{\mathrm{d}y^2} + \rho_\mathrm{L} \cdot g = 0\,. \tag{3.1-58}$$

There are two boundary conditions. On the wall the velocity is zero:

$$w_{y=0} = 0\,. \tag{3.1-59}$$

If the liquid film is flowing between this wall and a stagnant gas without shear stress on the film, the slope of the velocity profile at the film–gas boundary is zero:

$$\left(\frac{\mathrm{d}w(y)}{\mathrm{d}y}\right)_{y=\delta} = 0\,. \tag{3.1-60}$$

Application of these boundary conditions for the integration of the above differential equations leads to the velocity profile

$$w(y) = \frac{g \cdot \rho_\mathrm{L}}{\eta_\mathrm{L}} \cdot \left(\delta \cdot y - \frac{y^2}{2}\right). \tag{3.1-61}$$

In some film apparatus, the volumetric flow $\dot{V}_\mathrm{L}$ is distributed on a wall with the total length or perimeter b. The liquid loading $\dot{V}_\mathrm{L}/b$ per unit irrigated length can be expressed as a function of the film thickness δ and the liquid properties ρ_L and η_L. Integrating the velocity profile differential equation

$$\frac{\dot{V}_\mathrm{L}}{b} = \int_0^\delta w(y) \cdot \mathrm{d}y = \frac{g \cdot \rho_\mathrm{L}}{\eta_\mathrm{L}} \cdot \int_0^\delta \left(\delta \cdot y - \frac{y^2}{2}\right) \cdot \mathrm{d}y \tag{3.1-62}$$

leads to

$$\frac{\dot{V}_L}{b} = \frac{\rho_L \cdot g}{\eta_L} \cdot \frac{\delta^3}{3} \,. \tag{3.1-63}$$

Let us introduce a Reynolds number Re_L with the mean film velocity w and the film thickness δ as the characteristic length. The film thickness can be expressed in the following way:

$$\delta = \left(\frac{3 \cdot \eta_L^2}{\rho_L^2 \cdot g}\right)^{1/3} \cdot {Re_L}^{1/3} = \left(\frac{3 \cdot \eta_L^2}{\rho_L^2 \cdot g}\right)^{1/3} \cdot \left(\frac{w \cdot \delta \cdot \rho_L}{\eta_L}\right)^{1/3} = \left(\frac{3 \cdot \dot{V}_L}{b} \cdot \frac{\eta_L}{\rho_L \cdot g}\right)^{1/3} . \tag{3.1-64}$$

This equation is valid only for the laminar film with a smooth surface. With an increasing film Reynolds number waves are formed on the film surface with first sinusoidal and then irregular plug shapes. Dealing with water as film liquid, the film is laminar and smooth for film Reynolds numbers up to 3.5. Sinusoidal waves have been registered in the range of Reynolds numbers between 3.8 and 8 and irregular waves for Reynolds numbers up to 400. The following equation allows the general prediction of the wavy film flow of arbitrary liquids flowing in a film when no drag forces are exerted on the surface and the density of the surrounding fluid is negligible ($\rho_G \ll \rho_L$):

$$Re_L = C_i \cdot \left(\frac{\rho_L \cdot \sigma^3}{\eta_L^4 \cdot g}\right)^{1/10} = C_i \cdot K_F^{1/10} . \tag{3.1-65}$$

The data of the factor C_i are presented in Fig. 3.1-11. K_F is a film number and important in film systems the flow of which is determined by forces due to gravity, viscosity, and surface tension. In pseudolaminar films the velocity of film elements on the liquid surface is greater than the mean velocity. The ratio of the surface velocity based on the mean velocity is between 1.5 and 2.15 (Brauer 1956). According to an empirical law based on experimental results the mean film thickness δ of a turbulent film is given by

$$\delta = 0.37 \cdot \left(\frac{3 \cdot \eta_L^2}{\rho_L^2 \cdot g}\right)^{1/3} \cdot Re_L^{1/2} = 0.285 \cdot \left(\frac{\eta_L}{\rho_L \cdot g^2}\right)^{1/3} \cdot w \,. \tag{3.1-66}$$

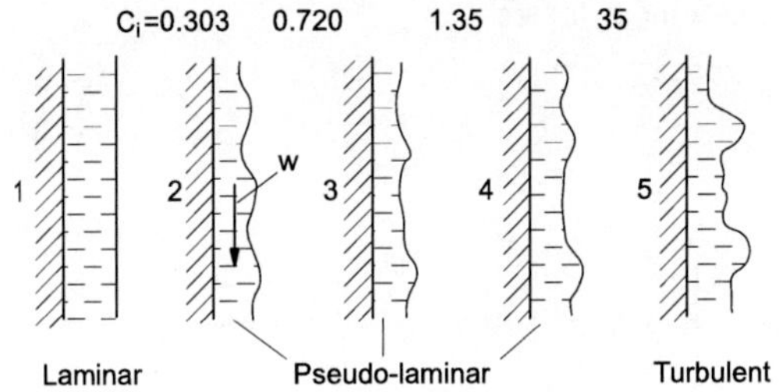

Fig. 3.1-11 Flow patterns of liquid films (stagnant gas)

3.2 Countercurrent Flow of a Gas and a Liquid in a Circular Vertical Tube

In a film evaporator the vapor is flowing in cocurrent or countercurrent flow with respect to the falling liquid. Here only the countercurrent flow will be discussed because flooding can limit the phase loadings. In a packed column the liquid is running around packing elements in countercurrent flow of the gas or vapor. In Fig. 3.2-1 some flow patterns are depicted which can be expected with increasing velocity of the gaseous phase (Feind 1960). At first waves and later slugs occur on the liquid surface. The gaseous flow causes a liquid entrainment due to breakup effects on the wave tops. Finally a spray flow followed by liquid bridging and bubble or froth flow is observed. All these phenomena lead to an increase of the liquid holdup in the tube. It is reasonable to introduce a Reynolds number of the gaseous phase with the mean gas velocity w_G based on the equation of continuity and the remaining width $d - 2\delta$ as the characteristic length:

$$Re_G = \frac{w_G \cdot (d - 2 \cdot \delta) \cdot \rho_G}{\eta_G}. \tag{3.2-1}$$

In the range of low gas velocities the film thickness δ remains constant and can be calculated as has been presented above. Starting at certain Reynolds numbers Re'_G the liquid holdup is increasing with rising gas velocities and the Reynolds number Re'_G becomes smaller with increasing liquid loading. This is shown in Fig. 3.2-2 in which the ratio δ'/δ is plotted against the Reynolds number Re_G of the gas with the Reynolds number Re_L as parameter. The film thickness $\delta' > \delta$ according to an increase of the liquid holdup is caused by high gas velocities. The curves in Fig. 3.2-2 are valid for the system air–water at 20°C. The loading of the tubes can be predicted for arbitrary systems by the equation

$$k \cdot \frac{Re_G}{Re_L^m} \cdot \left(\frac{\rho_L}{\rho_G}\right)^{2/5} \cdot \left(\frac{\eta_G}{\eta_L}\right)^{3/4} + 1.4 \times 10^4 = 1300 \cdot \left(\frac{d}{2 \cdot \delta}\right)^{5/4}. \tag{3.2-2}$$

The data of the constant k and the exponent m are

$Re_L < 400$; $k = 58.2$; $m = 1/3$,

$Re_L > 400$; $k = 158$; $m = 1/2$.

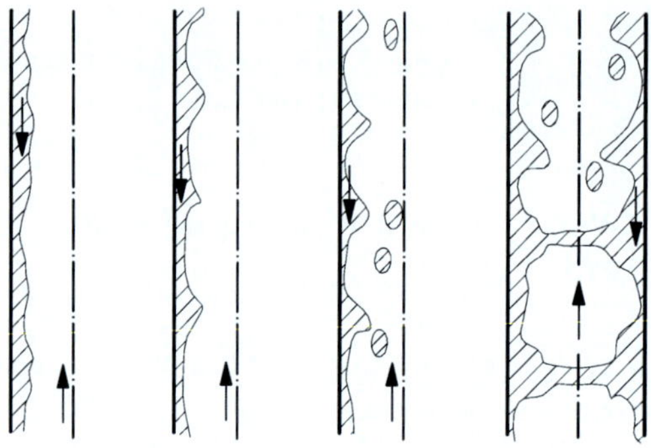

Fig. 3.2-1 Flow patterns of liquid films in vertical tubes for increasing volumetric flow densities of a gas

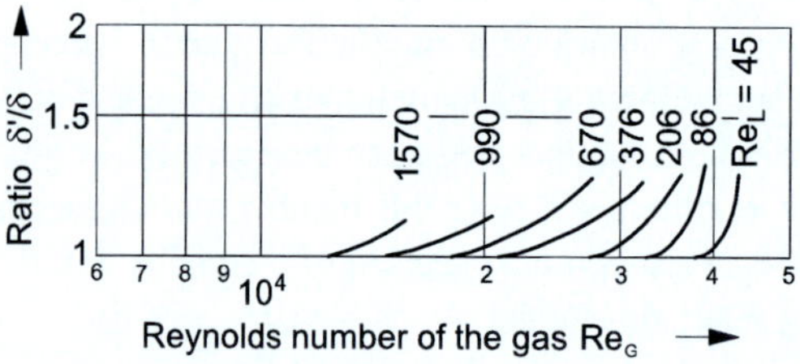

Fig. 3.2-2 Ratio of the film thicknesses as a function of the gas Reynolds number for some liquid Reynolds numbers

3.3 Similarity Hypothesis, Dimensional Analysis, and Dimensionless Numbers

Scale-up of apparatus and reactors is often carried out in such a way that the industrial unit and the model are geometrically similar. Geometrical similarity means that any length ratio is the same in both units. Similar pipes have the same ratio of length L based on the diameter d. Fluid motion is the result of forces acting on a fluid element. Remember that the Reynolds number is the ratio of inertia forces based on forces due to viscosity. The Navier–Stokes equations show that forces due to gravitation and pressure fields can be effective. The ratio of inertia forces based on gravitational forces is known as the Froude number:

$$Fr = \frac{w^2}{d \cdot g} . \qquad (3.3\text{-}1)$$

This number is mainly important for liquids because the density of gases at low pressure is small. The rise or fall of solid or fluid particles is the result of the difference of buoyancy and gravity forces and proportional to the density difference $|\rho_c - \rho_d| = \Delta\rho$. Therefore, an extended version of the Froude number is known:

$$Fr^* = \frac{w^2 \cdot \rho_c}{d \cdot \Delta\rho \cdot g}. \tag{3.3-2}$$

Since the velocity is present in both the Reynolds and the Froude numbers it is practicable to combine both numbers in such a way that the velocity is eliminated. By this, the Archimedes number Ar can be derived:

$$Ar \equiv \frac{Re^2}{Fr^*} \equiv \frac{d^3 \cdot g \cdot \rho_c \cdot \Delta\rho}{\eta_c^2}. \tag{3.3-3}$$

The elimination of the particle diameter d leads to

$$Re \cdot Fr^* \equiv \frac{w^3 \cdot \rho_c^2}{\eta_c \cdot \Delta\rho \cdot g}. \tag{3.3-4}$$

The surface or interfacial tension σ is an energy per unit interfacial area for neighboring fluid phases with different densities. The resulting pressure p_σ depends on the two main radii R_1 and R_2 of a convex or concave interface:

$$p_\sigma = \sigma \cdot \left(\frac{1}{R_1} + \frac{1}{R_2}\right) \tag{3.3-5}$$

or in the special case of a spherical fluid particle with $R_1 = R_2 = R = d/2$:

$$p_\sigma = \frac{2 \cdot \sigma}{R} = \frac{4 \cdot \sigma}{d}. \tag{3.3-6}$$

The ratio of inertia forces based on surface forces by virtue of an interfacial tension is called Weber number We:

$$We = \frac{w^2 \cdot d \cdot \rho}{\sigma}. \tag{3.3-7}$$

Dealing with the breakup of fluids resulting in the formation of bubbles or drops the Weber number is the decisive criterion. The stability of fluid particles can be described by a certain value of this number as will be shown later.

In the case of turbulent flow the pressure drop $\Delta p = p_1 - p_2$ of a fluid is proportional to the kinetic energy $\rho \cdot w^2/2$ per unit fluid volume. The ratio of the pressure difference $(p_1 - p_2)$ based on this energy is known as the Euler number Eu:

$$Eu = \frac{p_1 - p_2}{w^2 \cdot \rho}. \tag{3.3-8}$$

Let us consider a pipe flow. The pressure drop of a fluid in a pipe with diameter d, length L, and friction factor λ is given by

$$\Delta p = \lambda \cdot \frac{w^2 \cdot \rho}{2} \cdot \frac{L}{d} \quad \text{or} \quad Eu = \frac{\lambda}{2} \cdot \frac{L}{d}. \tag{3.3-9}$$

By combination of dimensionless numbers any other dimensionless group can be found, for instance, the already mentioned film number

$$K_\mathrm{F} = \frac{Re^4 \cdot Fr}{We^3} = \frac{\sigma^3 \cdot \rho}{\eta^4 \cdot g}, \tag{3.3-10}$$

or the fluid number which is very important in two-phase particle systems within a gravity field:

$$Fl = \frac{Re^4 \cdot Fr^*}{We^3} = \frac{\sigma^3 \cdot \rho_c^2}{\eta_c^4 \cdot \Delta\rho \cdot g}. \tag{3.3-11}$$

3.4 Particulate Systems

In the area of chemical engineering there is a big variety of apparatus and reactors in which a stagnant or moving dispersed phase (fluid or solid particles) is surrounded by a moving or nonmoving continuous fluid phase. In Table 3.4-1 some examples are presented.

All these two-phase systems can be characterized by the following parameter:

- Volume fraction of the continuous phase $\varepsilon_c = V_c/V_{tot}$ or volume fraction of the dispersed phase $\varepsilon_d = V_d/V_{tot}$. It is $V_c + V_d = V_{tot}$ and $\varepsilon_c + \varepsilon_d = 1$
- The mean particle diameter (given for solid particles but depending on flow conditions for fluid particles)
- Particle size distribution

Table 3.4-1 Examples of two-phase systems

Dispersed phase	Continuous phase	Examples
Solid, nonmoving	Gas	Fixed beds (adsorber, dryer, reactor)
Solid, nonmoving	Liquid	Fixed beds (adsorber, ion exchanger)
Solid, moving	Gas	Fluidized bed (dryer, reactor)
Solid, moving	Liquid	Fluidized bed (crystallizer, solid–liquid extractor, reactor)
Bubbles	Liquid	Bubble column, evaporator, tray column for absorption and distillation
Drops	Gas	Spray column, cooling tower, tray column
Drops	Liquid	Drop column, liquid–liquid extractor, decanter, emulsifier

The diameter of a spherical particle is d; however, nearly all particles are not spherical. In this case the following definition of irregular-shaped particles is helpful:

$$d_\mathrm{p} = \frac{6 \cdot V_\mathrm{p}}{A_\mathrm{p}} = \frac{6 \times \text{particle volume}}{\text{particle surface}} . \tag{3.4-1}$$

Let us consider the flow of a gas or a liquid around solid particles which are nonmoving (fixed bed) or moving in suspense (fluidized bed). The mean width of channels between the particles for the passage of the fluid is often characterized by a hydraulic diameter d_h which can be derived from the following assumption.

The real system and the idealized model system have the same voidage and the same interfacial area a, see Fig. 3.4-1, with

$$a = \frac{\text{interfacial area between the continuous and dispersed phase}}{\text{total volume of the system}} . \tag{3.4-2}$$

Based on the definition

$$d_\mathrm{h} = \frac{4 \cdot \varepsilon_\mathrm{c}}{a} \tag{3.4-3}$$

of the hydraulic diameter d_h, this parameter can be written as

$$d_\mathrm{h} = \frac{\varepsilon_\mathrm{c} \cdot d_\mathrm{p}}{1.5 \cdot \varepsilon_\mathrm{d} + d_\mathrm{p}/D} \tag{3.4-4}$$

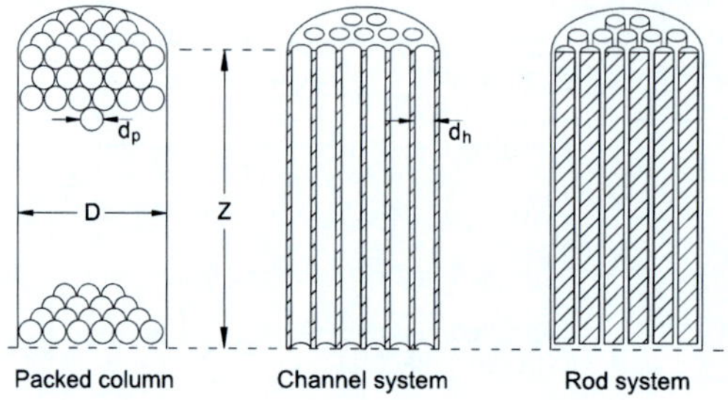

Fig. 3.4-1 Packed column (*left*) and model systems for the definition of a hydraulic diameter (*center, right*)

which takes into account the additional area provided by the wall of the vessel with diameter D. If the particle diameter d_p is very small in comparison to the column diameter D this equation can be simplified:

$$d_h = \frac{2 \cdot \varepsilon_c \cdot d_p}{3 \cdot \varepsilon_d} \tag{3.4-5}$$

and for $\varepsilon_c \to 1$ or $\varepsilon_d \to 0$ (entrainment zone above fluidized beds) it becomes

$$d_h = D. \tag{3.4-6}$$

In a system of spheres with the same diameter d the volumetric area a is

$$a = \frac{6 \cdot \varepsilon_d}{d}. \tag{3.4-7}$$

The definition of the so-called Sauter diameter d_{32} of a system with a particle size distribution with n_i particles of size d_i is given by

$$d_{32} = \frac{\sum n_i \cdot d_i^3}{\sum n_i \cdot d_i^2} \tag{3.4-8}$$

and the volumetric area becomes

$$a = \frac{6 \cdot \varepsilon_d}{d_{32}}. \tag{3.4-9}$$

In the literature several equations are known for the mathematical description of particle size distributions. With respect to problems in the area of chemical engineering there are two expressions often used:

- The logarithmic probability distribution
- The so-called RRSB (Rosin–Rammler–Sperling–Bennett) distribution

The logarithmic distribution has a theoretical background whereas the empirical RRSB distribution

$$R = \exp\left[-\left(\frac{d}{d'}\right)^n\right] \qquad (3.4\text{-}10)$$

can be treated mathematically in an easy way. In this equation R denotes the cumulative amount (mass or volume) smaller than the standard size in percent. The characteristic size d' is defined by the relationship $R = 1/e = 0.368$. The higher the dimensionless exponent n (or equalibility coefficient) the more narrow the particle size distribution. Differentiation of the cumulative amount R leads to the unit interval curve, and by integration of this curve the cumulative amount curve is obtained. In Fig. 3.4-2 two smoothed interval curves are shown valid for the exponent 4 for a broad and the exponent 10 for a narrow size distribution. A monodisperse distribution (all particles have the same size) would have $n = \infty$. In Fig. 3.4-3 cumulative amount curves (based on volume) valid for fluid particles produced in stirred vessels and columns are shown as a function of the size d based on the Sauter diameter d_{32}. Drop size distributions of drops resulting from breakup above holes of sieve trays are often narrow (Mersmann 1978).

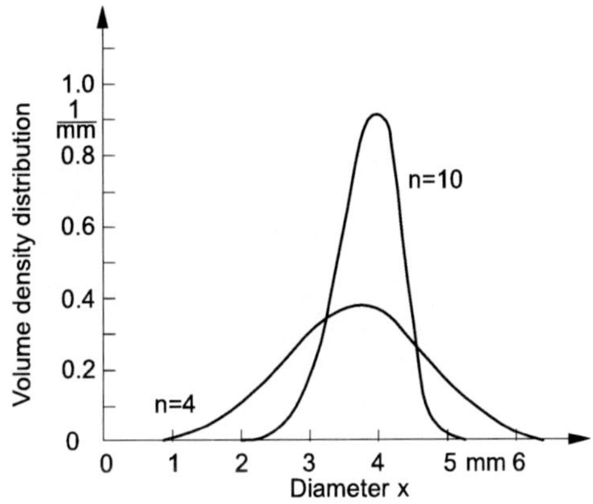

Fig. 3.4-2 Volume density distribution as a function of the particle diameter for two RRSB distributions

3.5 Flow in Fixed Beds

In Fig. 3.5-1 a fixed bed of solid particles supported on a grill gate or perforated plate is shown. In the case of a downward flow the fixed bed state is maintained. However, such a bed is converted into a fluidized bed for an upward flow beginning at the minimum fluidizing velocity w (mean superficial velocity based on the

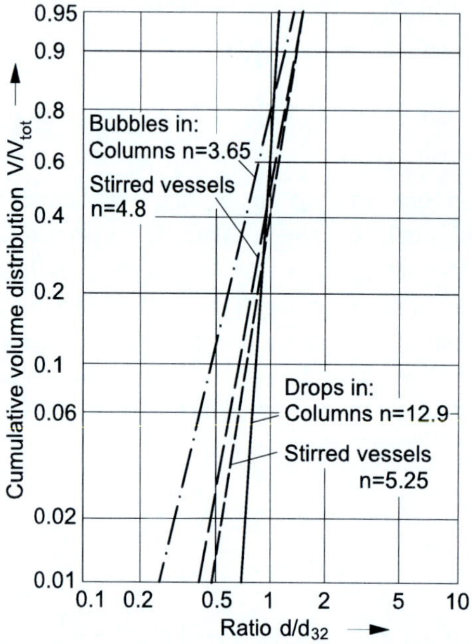

Fig. 3.4-3 Cumulative volume distribution of bubbles and drops in columns and stirred vessels

empty column). This happens when the pressure drop of the fluid becomes equal to the gravity force per unit cross-sectional column area. Fluidized beds are later discussed in more detail. With ε_c as the voidage of the fixed bed the volumetric particle holdup in the bed is $(1-\varepsilon_c)$. The effective or relative mean fluid velocity in the void spaces of the bed is

$$w_{eff} = \frac{w}{\varepsilon_c} = \frac{\dot{v}_c}{\varepsilon_c} = w_{rel}\,. \tag{3.5-1}$$

Here $\dot{v}_c$ is the volumetric flux density of the fluid in $m^3/(m^2\ s)$ which is equal to the superficial velocity w of the fluid above the bed in the particle-free column. The pressure drop of the fluid in the bed is given by

$$\Delta p = \mu_p \cdot \lambda \cdot \frac{L}{d_h} \cdot \frac{w^2 \cdot \rho}{\varepsilon_c^2 \cdot 2} = \mu_p \cdot \lambda \cdot \frac{3}{4} \cdot \frac{L \cdot \varepsilon_d \cdot w^2 \cdot \rho}{d_p \cdot \varepsilon_c^3}. \tag{3.5-2}$$

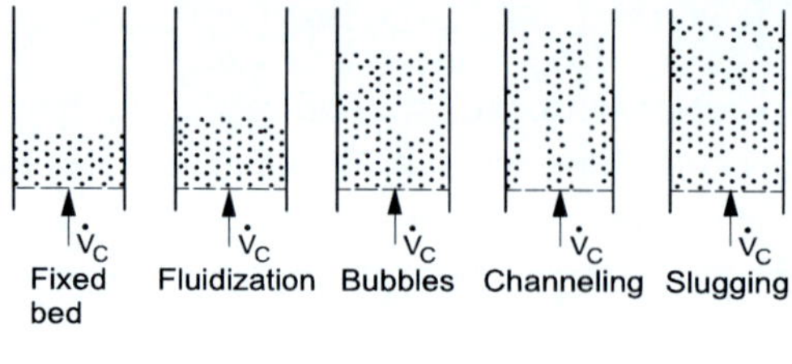

Fig. 3.5-1 Fixed bed (*left*) and different patterns of fluidized beds (compare Fig. 3.6-10)

Here d_h is the hydraulic diameter and μ_p is the tortuosity factor (compare later the section "Adsorption").

As a rule the last equation is simplified to

$$\Delta p = \xi \cdot \frac{\varepsilon_d}{\varepsilon_c^3} \cdot \frac{L}{d_p} \cdot \frac{w^2 \cdot \rho}{2} \quad \text{with} \quad \xi = f\left(Re = \frac{w \cdot d_p \cdot \rho_c}{\varepsilon_d \cdot \eta_c}\right). \tag{3.5-3}$$

Here the factor $2/3$ of the Reynolds number with the parameter d_h and w_{eff} is omitted (Empirical results show $\Delta p \sim \varepsilon_c^{-4.65}$, see later). In Fig. 3.5-2 the friction factor ξ is plotted against this Reynolds number for some solid particle materials according to this model, compare with (5.4-25).

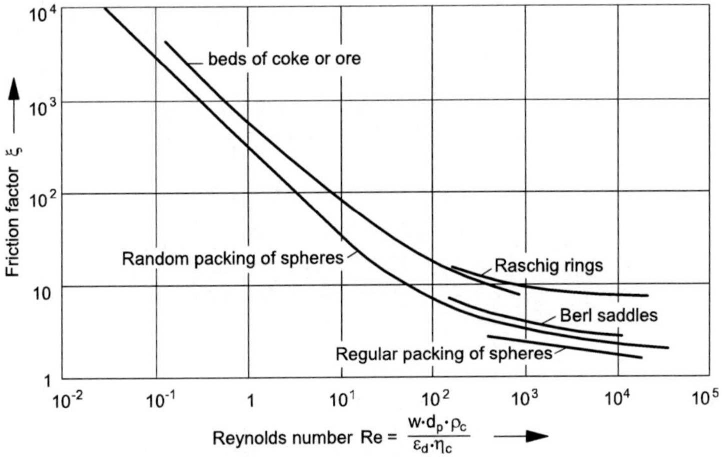

Fig. 3.5-2 Friction factor as a function of the Reynolds number for some packed beds

3.6 Disperse Systems in a Gravity Field

It is reasonable to present the flow in disperse systems in a general way to avoid repetitions of this topic. Such disperse systems are gas or liquid fluidized beds, bubble or drop columns, and spray columns. In all cases the solid or fluid particles are suspended or moving due to the density difference $\Delta\rho = (\rho_c - \rho_d)$ and the acceleration of gravity. In Fig. 3.5-1 a fixed bed on the left side and several fluidized beds with different flow patterns are depicted. The fluid flow density $\dot{v}_c$ in the fluidized beds is greater than the minimum flow density $\dot{v}_{cf}$ necessary to achieve fluidization. The volumetric holdup ε_c of the continuous phase increases for $\dot{v}_c > \dot{v}_{cf}$ with the fluid throughput. The relative velocity w_{rel} between the fluid and the suspended particles is inversely proportional to the volumetric holdup ε_c. With the rising or sinking velocity w_{ss} of many particles in a stagnant fluid the holdup of the continuous phase ε_c is

$$\varepsilon_c = \frac{\dot{v}_c}{w_{rel}} = \frac{w_{ss}}{w_{rel}}, \quad \text{therefore} \quad w_{rel} = \frac{w_{ss}}{\varepsilon_c}. \tag{3.6-1}$$

General information on the rising or sinking velocity w_{ss} of particles in a swarm ($\varepsilon_d > 0$) is given in the Sects. 3.6.1 and 3.6.2. Similar results can be expected when the continuous phase is not flowing but the particles of the dispersed phase are moving through a stagnant phase. This is the case in a bubble or drop column or in a spray column (drops are falling through a gas). Fluid particles can be produced at the orifices of a distributor plate with holes. Let us assume that the volumetric flow $\dot{V}_d$ of the dispersed phase is passing a circular hole with diameter d_0. Nearly equal-sized fluid particles are formed in the range of Weber numbers $We_0 = w_0^2 \cdot d_0 \cdot \rho_d / \sigma < 7$; however, in the case $We_0 > 7$, the entering jet breaks

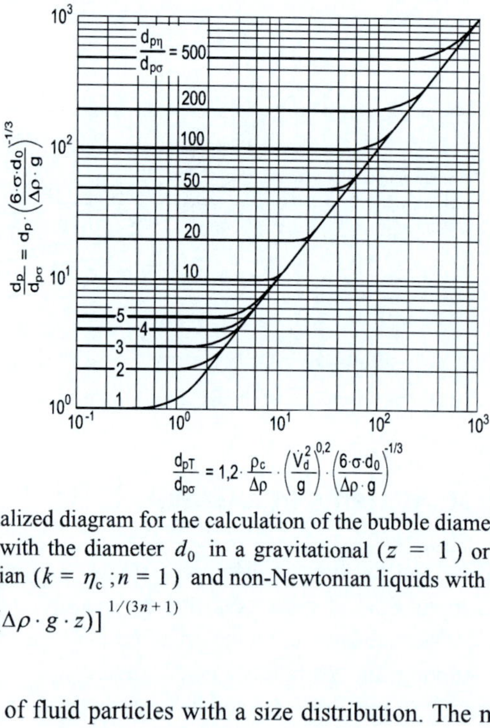

Fig. 3.6-1 Normalized diagram for the calculation of the bubble diameter d_p. Bubbles are produced at holes with the diameter d_0 in a gravitational ($z = 1$) or a centrifugal field ($z \cdot g$) for Newtonian ($k = \eta_c; n = 1$) and non-Newtonian liquids with

$d_{p\eta} = C \cdot [k \cdot \dot{V}_d^n / (\Delta\rho \cdot g \cdot z)]^{1/(3n+1)}$

up in a spectrum of fluid particles with a size distribution. The mean particle size d_p of bubbles can be calculated on the basis of Fig. 3.6-1. This diagram is valid for the formation of bubbles in Newtonian or non-Newtonian liquids with the consistency k and the fluidity n. The liquid may be existing in a gravitational field with the acceleration g due to gravity or in a centrifugal field with the acceleration $r \cdot \omega^2$, where r is the effective radius and ω is the angular velocity. The ratio z is defined as $z \equiv r \cdot \omega^2/g$. Dealing with Newtonian liquids it is $\eta_c = k$ and $n = 1$ (Voit et al. 1987). The factor C is $C = 2.32$ valid for $n = 1$ and $C = 4.48$ valid for $n = 0.1$. In the case $0.1 < n < 1$ the constant C is $4.48 > C > 2.32$. With increasing distance from the orifices or holes, fluid particles can break up or combine with the result that the bubble size distribution becomes wide, especially for coalescing liquids (liquids without surface active agents). In the upper part of the bubble column the bubble size distribution is mainly governed by the breakup of large bubbles and combination of smaller ones and the original distribution above the distributor is lost. At high superficial velocities of the gas huge bubbles are rising in the center of the column, see later.

In Fig. 3.6-2 a bubble or drop column is shown on the right-hand side (bubbles or drops are rising in an immiscible liquid) whereas a spray column (drops are falling in a gas) is depicted on the left-hand side. The fluid particles are formed at holes drilled or punched in a plate or tube distributor. Bubbles and drops are rising for $\rho_d < \rho_c$ or $\rho_c/\rho_d > 1$. Let us first assume that no continuous phase (gas in a spray column, liquid in a bubble or drop column) is entering the column.

The volumetric holdup ε_d of the dispersed phase increases with the volumetric flow density $\dot{v}_d$ (or superficial velocity) of the dispersed phase and is inversely proportional to the relative velocity w_{rel} or the (rising or falling) velocity w_{ss} of fluid particles in a swarm:

$$\varepsilon_d = \frac{\dot{v}_d}{w_{rel}} = \frac{\dot{v}_d}{w_{ss}} \cdot \varepsilon_c, \quad \text{or} \quad w_{rel} = \frac{w_{ss}}{\varepsilon_c}. \tag{3.6-2}$$

The relative velocity w_{rel} of particles in a swarm depends on the velocity w_s of a single rising or falling particle and of the volumetric holdup ε_c or ε_d with $\varepsilon_c + \varepsilon_d = 1$. The last equation can be rewritten as

$$\varepsilon_d = \frac{\dot{v}_d}{w_s} \cdot \frac{w_s}{w_{rel}} \tag{3.6-3}$$

and

$$\varepsilon_c = \frac{\dot{v}_c}{w_s} \cdot \frac{w_s}{w_{rel}}. \tag{3.6-4}$$

In the case of the same relative velocity in both equations it follows

$$\frac{\varepsilon_d}{\varepsilon_c} = \frac{\dot{v}_d}{\dot{v}_c}. \tag{3.6-5}$$

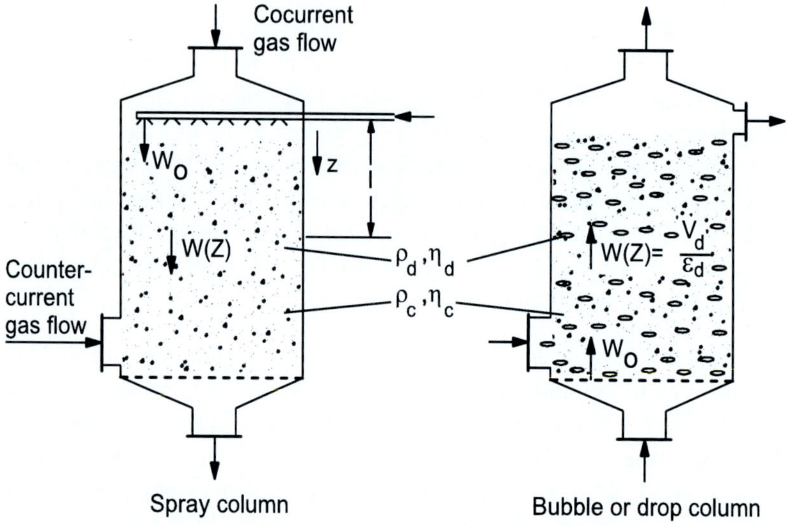

Fig. 3.6-2 Fluid dispersed systems. (*Left*) Spray column (drops moving in a gas); (*right*) bubble or drop column (fluid particles moving in a liquid)

Let us now assume that the continuous phase is moving ($\dot{v}_c \neq 0$) in countercurrent flow with respect to the rising or falling particles of the dispersed phase. In this case the relative velocity w_{rel} depends on $\dot{v}_c$, $\dot{v}_d$, ε_c and ε_d according to

$$w_{rel} = \frac{\dot{v}_d}{\varepsilon_d} + \frac{\dot{v}_c}{\varepsilon_c}. \tag{3.6-6}$$

All these equations make very clear: the volumetric holdup in such columns is a function of the velocity w_s of a single particle (in an extended continuous phase without wall effects) and on the ratio $w_s/w_{ss} = f(\varepsilon_d)$. Let us first consider the velocity w_s.

3.6.1 The Final Rising or Falling Velocity of Single Particles

After a distance governed by acceleration (falling particles) or retardation (rising fluid particles) the final falling or rising velocity of fluid or solid particles with the diameter d is the result of a force balance, e.g., the drag force with the drag coefficient c_W and the difference of the forces due to gravity and buoyancy:

$$c_W \cdot \frac{d^2 \cdot \pi}{4} \cdot \frac{w_s^2 \cdot \rho_c}{2} = \frac{d^3 \cdot \pi}{6} \cdot \Delta\rho \cdot g \tag{3.6-7}$$

or

$$c_W = \frac{4 \cdot d \cdot g \cdot \Delta\rho}{3 \cdot w_s^2 \cdot \rho_c} = \frac{4}{3 \cdot Fr^*}. \quad (3.6\text{-}8)$$

The drag coefficient is a function of the Reynolds number with the particle diameter d as the characteristic length:

$$c_W = f(Re_p) \quad \text{with} \quad Re_p = \frac{w_s \cdot d \cdot \rho_c}{\eta_c}. \quad (3.6\text{-}9)$$

With respect to practical applications it is reasonable to plot the dimensionless velocity according to

$$w_s \cdot \left(\frac{\rho_c^2}{\eta_c \cdot \Delta\rho \cdot g}\right)^{1/3} \quad (3.6\text{-}10)$$

against the dimensionless diameter (or $(Ar)^{1/3}$)

$$Ar^{1/3} = d \cdot \left(\frac{\rho_c \cdot \Delta\rho \cdot g}{\eta_c^2}\right)^{1/3}, \quad (3.6\text{-}11)$$

see Fig. 3.6-3. The straight line 1 represents Stokes' law

$$c_W = \frac{24}{Re_p}. \quad (3.6\text{-}12)$$

This law is valid for solid spherical particles up to $Re_p \approx 1$. The curve "rigid spheres" can be used to calculate the final rising or sinking velocity w_s of solid spheres for $Re_p > 1$.

Fluid particles are approximately rigid for $\eta_c / \eta_d \to \infty$; however, as a rule, an inner circulation takes place if no detergents are adsorbed at the interface. The law according to Hadamard and Rybczynski (Hadamard 1911; Rybczynski 1911) according to the straight line 2 in Fig. 3.6-3 describes the velocity w_s:

$$c_W = \frac{24}{Re_p} \cdot \frac{2/3 + \eta_d/\eta_c}{1 + \eta_d/\eta_c}. \quad (3.6\text{-}13)$$

This law is valid for $Re_p < 1.4$. With the assumption of spherical fluid particles and $\eta_c / \eta_d \to \infty$ the relationship according to Haas (Haas et al. 1972)

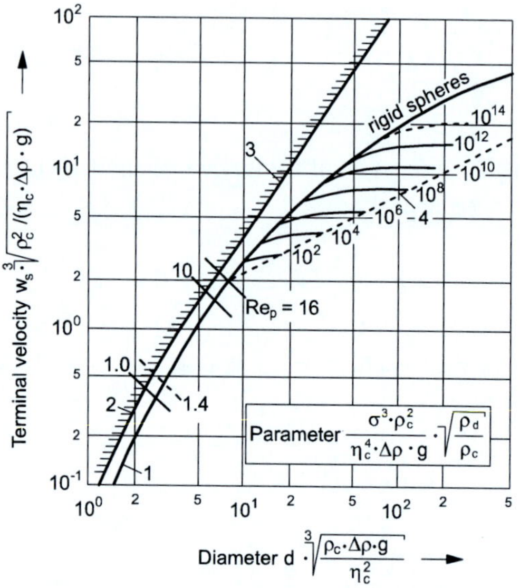

Fig. 3.6-3 Dimensionless terminal falling or rising velocity of solid and fluid particles as a function of the dimensionless diameter. The curves are valid for $\rho_c/\rho_d \ll 1$ and $\eta_c/\eta_d \ll 1$ (drops in gas). Line 4: $c_W = 2.69$

$$c_W = \frac{14.9}{Re_p^{0.78}}. \tag{3.6-14}$$

is valid for $Re_p > 1.4$, see the straight line 3 in Fig. 3.6-3. This law has been experimentally confirmed for bubbles in liquids and describes maximum velocities w_s for spherical bubbles.

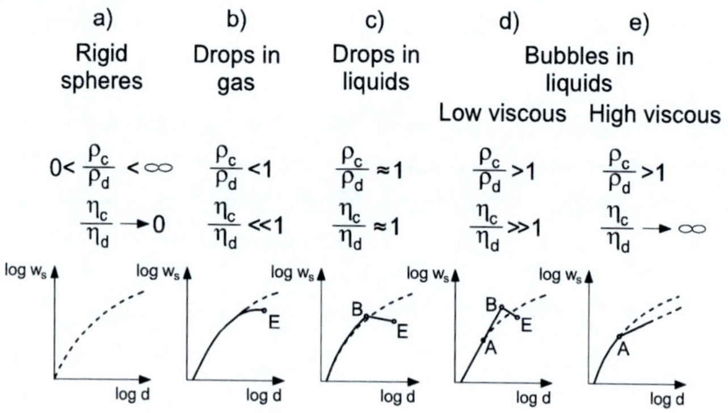

Fig. 3.6-4 Terminal rising or falling velocities of solid or fluid particles as a function of the particle diameter

Drops and especially bubbles show deformation and shape fluctuations with increasing size of the fluid particles, and these deviations from spheres become stronger with increasing size of the fluid particles and with increasing data of the parameter η_c / η_d and $(d^2 \cdot \Delta\rho \cdot g)/\sigma$. In Fig. 3.6-4 a qualitative survey of possible curves of the velocity w_s vs. the diameter d is given. The particle diameter d_E (E = end) denotes a particle size which cannot be exceeded with respect to the instability of fluid particles. This means that the probability of a particle breakup is high. Only by the use of special equipments (turnable spoon) bubbles with $d > d_E$ can be produced; however, their life time will be short (Mersmann 1978).

Figure 3.6-5 gives an answer to questions concerning the rising velocity w_s of fluid particles and their limits of deformation and particle stability. The particle Reynolds number

$$Re_p = \frac{w_s \cdot d_p \cdot \rho_c}{\eta_c} \tag{3.6-15}$$

is plotted against the ratio

$$\frac{We}{Fr^*} = \frac{d_p^2 \cdot \Delta\rho \cdot g}{\sigma} \tag{3.6-16}$$

with the fluid number

$$Fl = \frac{\sigma^3 \cdot \rho_c^2}{\eta_c^4 \cdot \Delta\rho \cdot g} \tag{3.6-17}$$

as parameter. This fluid number

$$Fl = \frac{Re_p^4}{We^3 \cdot Fr^*} \tag{3.6-18}$$

plays always an important role when fluid particles are rising or falling in a viscous fluid with the tendency of deformation due to weak surface forces in comparison to forces caused by gravity, buoyancy, or viscous friction. In Fig. 3.6-5 the boundaries of particle deformation (spherical fluid particles begin to deform) and stability (fluid particles start to break up) are drawn as solid or dashed curves, respectively. Note that fluid particles are always spherical for $Re_p < 1$, and the law according to Stokes

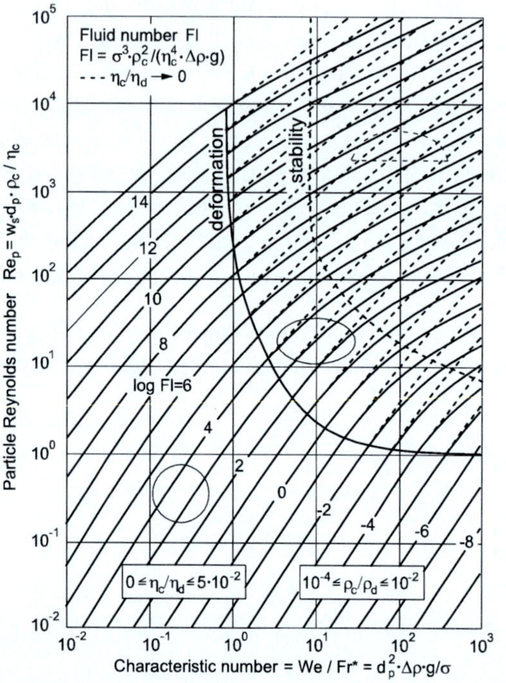

Fig. 3.6-5 Particle Reynolds number Re_{p} as a function of the number *We/Fr** (Mersmann 1986)

$$c_{\mathrm{W}} = \frac{24}{Re_{\mathrm{p}}} \quad \text{or} \quad w_{\mathrm{s}} = \frac{d_{\mathrm{p}}^2 \cdot \Delta\rho \cdot g}{18 \cdot \eta_{\mathrm{c}}} \qquad \text{(rigid spheres)} \tag{3.6-19}$$

or according to Hadamard and Rybczynski

$$c_{\mathrm{W}} = \frac{24 \cdot (2/3 + \eta_{\mathrm{d}}/\eta_{\mathrm{c}})}{Re_{\mathrm{p}} \cdot (1 + \eta_{\mathrm{d}}/\eta_{\mathrm{c}})} \quad \text{or} \quad w_{\mathrm{s}} = \frac{d_{\mathrm{p}}^2 \cdot \Delta\rho \cdot g}{12 \cdot \eta_{\mathrm{c}}} \quad \text{for } \eta_{\mathrm{d}}/\eta_{\mathrm{c}} \to 0 \tag{3.6-20}$$

is valid as has been shown earlier. When $Re_{\mathrm{p}} > 500$ the stability boundary of fluid particles with size d_{stab} can be approximately calculated from

$$\frac{d_{\mathrm{stab}}^2 \cdot \Delta\rho \cdot g}{\sigma} = C \tag{3.6-21}$$

with $C \approx 9$ valid for falling drops in a gas and $C \approx 12$ for bubbles rising in a liquid. Irregular-shaped bubbles which exhibit shape fluctuations can be characterized by the friction factor $c_{\mathrm{W,\,stab}} \approx 2.6$ (as also bubbles in a fluidized bed, see later), and their rising velocity $w_{\mathrm{s,\,stab}}$ (according to the point E in Fig. 3.6-4) is

$$w_{\mathrm{s,stab}} \approx 1.55 \cdot \left(\frac{\sigma \cdot \Delta\rho \cdot g}{\rho_{\mathrm{c}}^2}\right)^{1/4}. \tag{3.6-22}$$

This equation is also approximately valid for large drops at the stability curve in Fig. 3.6-5.

3.6.2 Volumetric Holdup (Fluidized Beds, Spray, Bubble and Drop Columns)

The objective of the following is to describe the volumetric holdup of two-phase systems in a general way. This is possible if

- The holdup is small ($\varepsilon_{\mathrm{d}} < 0.1$) and
- The ratio $\rho_{\mathrm{d}}/\rho_{\mathrm{c}}$ is in the range $1/3 < \rho_{\mathrm{d}}/\rho_{\mathrm{c}} < 3$ or
- The ratio $w_{\mathrm{ss}}/w_{\mathrm{s}}$ is close to unity

The knowledge of this ratio $w_{\mathrm{ss}}/w_{\mathrm{s}}$ is a prerequisite for the prediction of any holdup. It is difficult to describe the parameter $w_{\mathrm{ss}}/w_{\mathrm{s}}$ for solid and fluid particles by the same modelling because of the breakup and deformation of bubbles and drops. Let us assume a settling chamber filled with liquid in which solid spheres are sinking and the ratio $\rho_{\mathrm{d}}/\rho_{\mathrm{c}}$ is close to unity. In this case the volumetric holdup ε_{d} is only a function of the ratio $w_{\mathrm{ss}}/w_{\mathrm{s}}$ and the particle Archimedes number:

$$\frac{w_{\mathrm{ss}}}{w_{\mathrm{s}}} = f(\varepsilon_{\mathrm{d}} = 1 - \varepsilon_{\mathrm{c}}, Ar) \quad \text{with} \quad Ar \equiv \frac{d^3 \cdot g \cdot \Delta\rho \cdot \rho_{\mathrm{c}}}{\eta_{\mathrm{c}}^2}, \tag{3.6-23}$$

see Fig. 3.6-6.

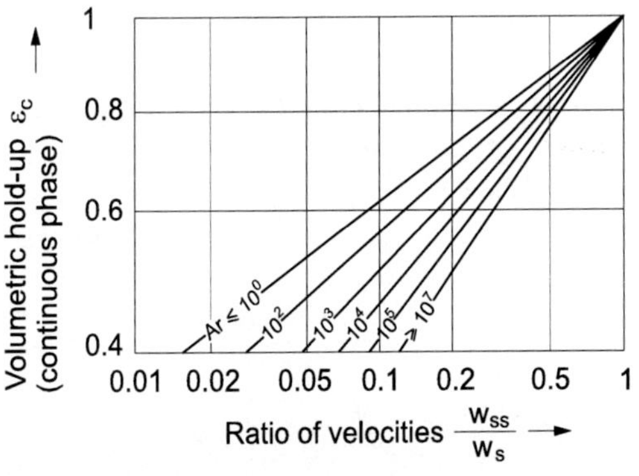

Fig. 3.6-6 Volumetric holdup of the continuous phase as a function of the ratio $w_{\mathrm{ss}}/w_{\mathrm{s}}$ of velocities with the Archimedes number as parameter

The relationship

$$\frac{w_{ss}}{w_s} = \varepsilon_c^{4.65} \tag{3.6-24}$$

is valid for $Ar < 1$. For $Ar > 1$ it can be written

$$\frac{w_{ss}}{w_s} = \varepsilon_c^{m}$$

where the exponent m is a function of the particle Reynolds number, see Fig. 3.6-7. The volumetric holdup ε_c of a fluidized bed can only be calculated with these

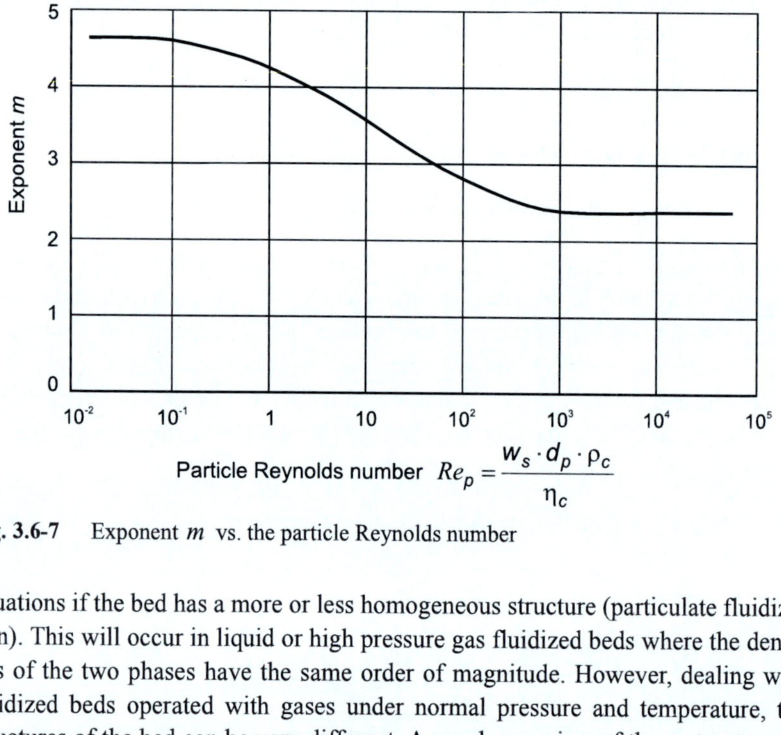

Fig. 3.6-7 Exponent m vs. the particle Reynolds number

equations if the bed has a more or less homogeneous structure (particulate fluidization). This will occur in liquid or high pressure gas fluidized beds where the densities of the two phases have the same order of magnitude. However, dealing with fluidized beds operated with gases under normal pressure and temperature, the structures of the bed can be very different. A rough overview of these structures is given in Fig. 3.6-8 where the ratio $\dot{v}_c/\dot{v}_{cf}$ is plotted against the particle Archimedes number. The minimum fluidization velocity $\dot{v}_{cf}$ and the volumetric holdup ε_c can be calculated by the use of a diagram where the dimensionless velocity

$$\dot{v}_c \cdot \left(\frac{\rho_c^2}{\eta_c \cdot \Delta\rho \cdot g} \right)^{1/3}$$

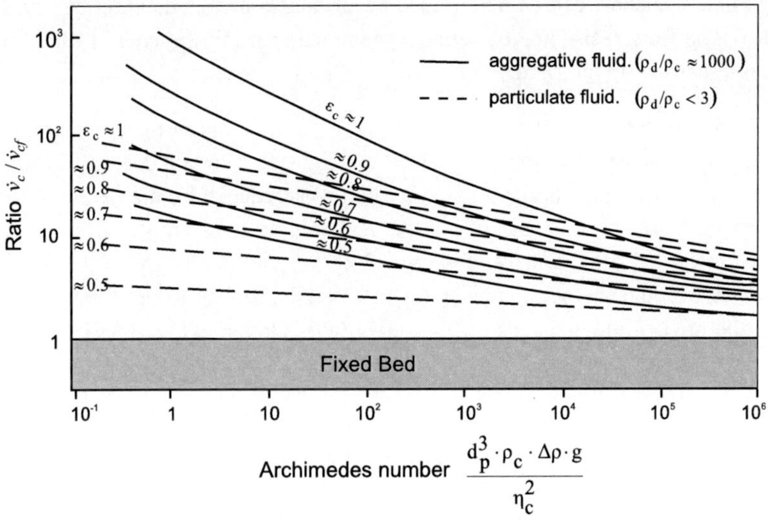

Fig. 3.6-8 Ratio $\dot{v}_c/\dot{v}_{cf}$ against the particle Archimedes number

is plotted against the dimensionless particle diameter

$$d_p \cdot \left(\frac{\rho_c \cdot \Delta\rho \cdot g}{\eta_c^2} \right)^{1/3}$$

with $\varepsilon_c = \text{const}$.-curves as parameter, see Fig. 3.6-9.

The curve ($\varepsilon_c = 0.4$) (or in general ε_c of the fixed bed) allows to determine the minimum fluidizing velocity $\dot{v}_{cf}$ from the expression

$$\dot{v}_{cf} \cdot \left(\frac{\rho_c^2}{\eta_c \cdot \Delta\rho \cdot g} \right)^{1/3}$$

for a given particle diameter d_p here expressed as

$$d_p \left(\frac{\rho_c \cdot \Delta\rho \cdot g}{\eta_c^2} \right)^{1/3} .$$

This is valid for beds operated with a gas or a liquid. The solid curves ($\varepsilon_c = \text{const.}$) for expanding homogeneous (liquid) fluidized beds are based on equations presented by Anderson (Anderson 1961) compared with Richardson and Zaki (Richardson and Zaki 1954). In the case of heterogeneous (low pressure gas) fluidized beds only the boundaries between the minimum fluidizing velocity ($\varepsilon_c \approx 0.4$) and the pneumatic transport ($\varepsilon_c \to 1$) are well known. The curve for pneumatic transport can be determined according to a consideration of Reh (Reh 1961): The flow resistance of isolated (demixed) particles is equal to the impact pressure caused by the fluid velocity.

The curves in the range $0.4 < \varepsilon_c < 1$ are mainly a function of the ratio ρ_d/ρ_c and also of the dimensions of the bed (diameter, height) to a certain degree. There is the tendency that the slope of the ε_c = const curves is reduced from 2 ($Ar < 10$) to 0.5 for $Ar > 10^6$ in the case $\rho_d/\rho_c \gg 1$. As an example lines are given in the figure valid for ρ_d = 2,500 kg/m³ and ρ_c = 0.3 kg/m³ and the height of the fixed bed H_0 = 0.2 m and the bed diameter 0.7 m (Wunder 1980). Rough information on the structure of fluidized beds for $\rho_c/\rho_d \geq 1{,}000$ is given in Fig. 3.6-10.

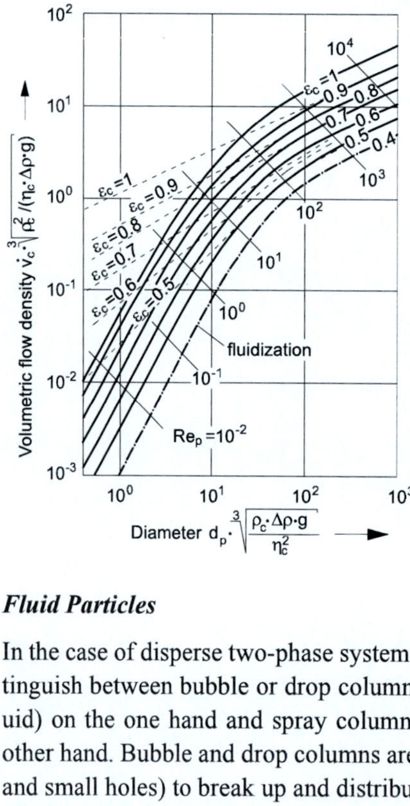

Fig. 3.6-9 Dimensionless volumetric flow density as a function of the dimensionless particle diameter for some volumetric holdups of the continuous phase
---------- homogeneous fluidization
- - - - - - heterogeneous fluidization
Height of fixed bed H_0 = 200 mm,
ρ_c = 0,3 kg/m³, ρ_d = 2500 kg/m³.

Fluid Particles

In the case of disperse two-phase systems with fluid particles it is reasonable to distinguish between bubble or drop columns (fluid particles rise in a nonmiscible liquid) on the one hand and spray columns (drops are falling through a gas) on the other hand. Bubble and drop columns are often equipped with sieve trays (only few and small holes) to break up and distribute the dispersed phase at the bottom.

The objective is that all the fluid passes through all holes without weeping of the continuous phase. With small holes according to the equation

$$d_0 \cdot \left(\frac{\Delta\rho \cdot g}{\sigma}\right)^{1/2} \cdot \left(\frac{\Delta\rho}{\rho_d}\right)^{1/8} < 2.32 \tag{3.6-25}$$

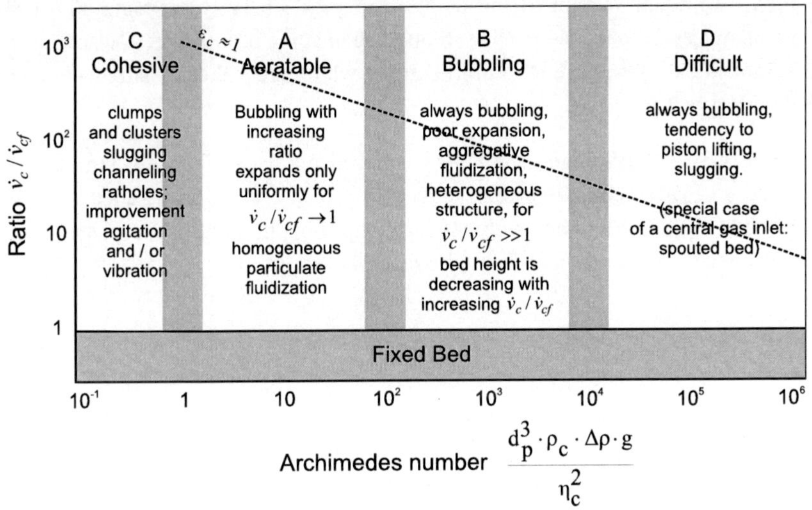

Fig. 3.6-10 Information on cohesive, aeratable, bubbling and difficult fluidized beds (Geldart 1973; Molerus 1982)

this can be realized for

$$We_0 \equiv \frac{w_0^2 \cdot d_0 \cdot \rho_d}{\sigma} \geq 2. \tag{3.6-26}$$

Note that this Weber number is formed with the density ρ_d of the dispersed phase, the velocity w_0 in the holes, and the surface tension σ. In the case of wide holes weeping can take place. This can be avoided when a certain Froude number is surpassed:

$$Fr_0^* \cdot \left(\frac{\rho_d}{\Delta\rho}\right)^{1/4} = \frac{w_0^2 \cdot \rho_d}{d_0 \cdot \Delta\rho \cdot g} \cdot \left(\frac{\rho_d}{\Delta\rho}\right)^{1/4} \geq 0.37. \tag{3.6-27}$$

The dispersed phase will be retarded (especially in bubble columns) and then broken up into fluid particles which at first rise with the velocity $w(z)$. However, after a short distance from the bottom plate fluid particles reach the relative velocity w_{rel}, see Fig. 3.6-2. In the range $0.3 < (d^2 \cdot \Delta\rho \cdot g)/\sigma < 9$, the approximation

$$w_{rel} \approx w_s \approx w_E \tag{3.6-28}$$

is valid for small volumetric holdups.

Dealing with spray columns (drops are falling in a gas) the velocity $w(z)$ is a function of the distance from the top distributor (tubes with holes) and can be determined by the application of the diagram in Fig. 3.6-11 (Mersmann 1978): The dimensionless distance $z \cdot g/w_E^2$ is plotted against the ratio $(w(z) - w_E)/w_E$. This diagram is based on certain simplifications which are valid for small volumetric holdups $\varepsilon_d < 0.1$.

In Fig. 3.6-12 the holdup ε_d of the dispersed phase is plotted against the ratio $\dot{v}_d/w(z)$ for a countercurrent flow $(\dot{v}_c/w(z) < 0)$ and a cocurrent flow $(\dot{v}_c/w(z) > 0)$. In a column without throughput of the continuous phase (neither countercurrent nor cocurrent flow) the result is simply

$$\varepsilon_d = \frac{\dot{v}_d}{w(z)} . \tag{3.6-29}$$

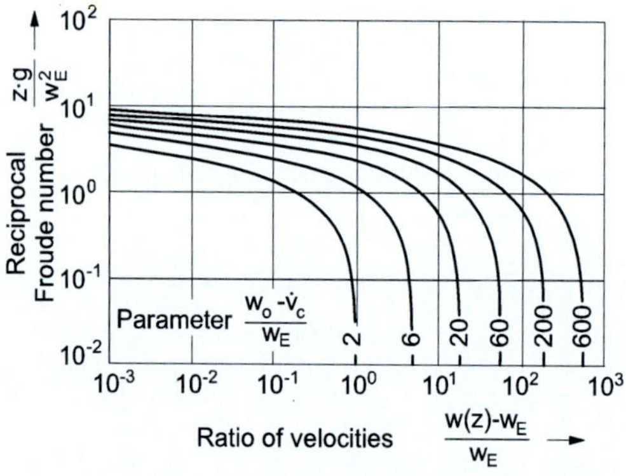

Fig. 3.6-11 Dimensionless rising (or falling) travelling distance of fluid particles decelerated after the formation at holes

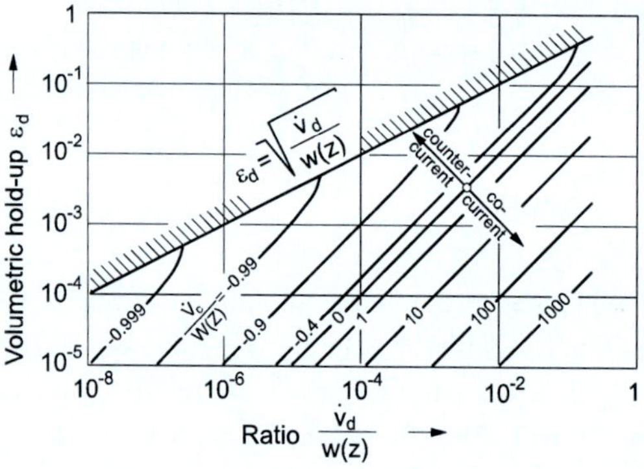

Fig. 3.6-12 Volumetric holdup of the dispersed phase as a function of the ratio $\dot{v}_d/w(z)$ of this phase for a cocurrent or countercurrent flow of the continuous phase

After a (rather long) distance, drops have obtained their final falling velocity $w_{s,stab}$ (drop stability) if there is no further breakup. Then the last equation reads

$$\varepsilon_d = \frac{\dot{v}_d}{w_{s,stab}} . \tag{3.6-30}$$

With increasing holdup ($\varepsilon_d > 0.1$) all these equations suffer a loss of validity. In the case of drop columns the reason is that the drop velocity is reduced in the presence of many drops in the neighborhood, compare Fig. 3.6-6. When the ratio ρ_c/ρ_d exceeds ≈ 300 ($\rho_c/\rho_d > 300$) the strong retardation of the dispersed phase after the passage of the distributor hole leads to a heterogeneous structure (bubbles with a wide particle size distribution, fast-rising large bubbles). In the center of the column large bubbles or bubble conglomerates can be observed with a rising velocity which can be a factor 5 greater than w_E or $w_{s,stab}$ of a single just stable bubble.

In Fig. 3.6-13 the volumetric holdup ε_d is plotted against the ratio $\dot{v}_d/w_E$ for drop columns ($\rho_c/\rho_d < 3$) and bubble columns $\rho_c/\rho_d > 300$ with the reciprocal of the fluid number as parameter. This holdup is only a function of the ratio $\dot{v}_d/w_E$ and the physical properties of the two phases. The figure is valid for $\dot{v}_c = 0$. The system water/mercury (water drops in mercury, $\rho_c/\rho_d = 13.6$) has the parameter $1/Fl = 0.26$. Measured data of ε_d follow approximately the diagonal (Mersmann 1977).

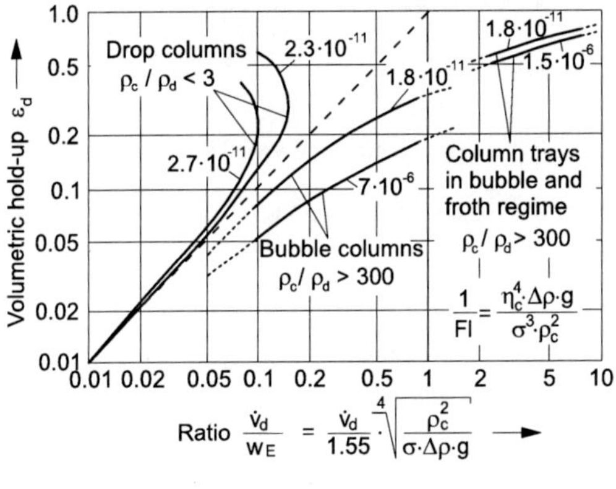

Fig. 3.6-13 Volumetric holdup of the dispersed phase as a function of the ratio $\dot{v}_d/w_E$ of this phase

3.7 Flow in Stirred Vessels

In the following the flow in large stirred vessels equipped with agitators of different shape and size will be discussed. It is assumed that no flow resistance internals are mounted in the vessel with the exception of four or more baffles (baffle width $\approx 0.1 \times$ vessel diameter) close to the wall to avoid a deep vortex.

This means that there are no draft tubes, tube bundles, or packings. A large-scale flow usually turbulent is maintained by the stirrer. In many cases solid or fluid particles (bubbles or drops) are more or less evenly distributed in a liquid continuous phase. In Fig. 3.7-1 often used stirred vessels are depicted:

- On the left side a vessel with a marine-type stirrer with $d/D = 1/2$ with a dominant axial flow
- In the center a vessel with a multiblade (six up to eighteen blades) with $d/D = 1/3$ with a dominant radial flow
- On the right side a vessel with a helical ribbon stirrer with $d/D = 0.9$ used for high viscous liquids

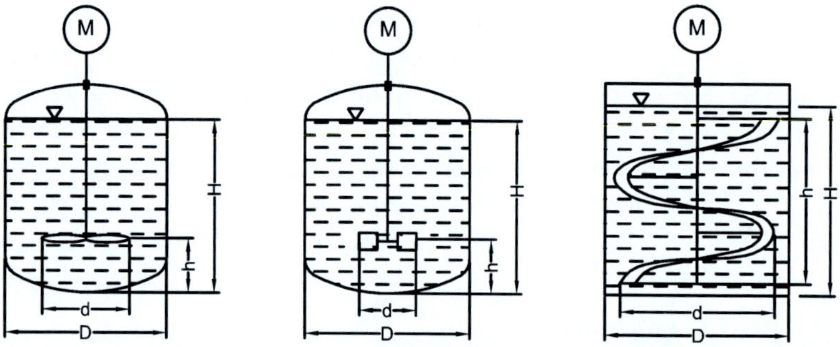

Fig. 3.7-1 Stirred vessel with a marine-type impeller (*left*), a multiblade impeller (*center*) for low viscous liquids and a helical ribbon stirrer (*right*) for high viscous liquids

As a rule in vessels with $H/D > 1$ multistage stirrers are installed with a distance $h \approx D$.

The large-scale flow within a vessel depends on

- The geometry of the system (d, D, h, H, etc.)
- The operating parameter as the stirrer speed and mass flow added or withdrawn
- Material properties (density, viscosity, consistence, fluidity, interfacial tension in liquids with fluid particles)

Agitated vessels are applied to carry out the following processes:

- Homogenization, e.g., reduction of differences of concentrations (micro-, meso-, macromixing, see later) and/or temperatures in liquids or multiphase systems, further the distribution of solid or fluid particles in solid/liquid systems (suspensions) or fluid/fluid systems (emulsions, gas in liquid dispersions)
- Suspending of particles, e.g., the off-bottom or off-top movement of particles and distribution of solid particles in a liquid (avoidance of settling)
- Breakup of fluids, e.g., the formation and distribution of fluid particles in a liquid

- Milling of solid particles in a liquid with the objective of particle size reduction

The Reynolds number according to

$$Re = \frac{n \cdot d^2 \cdot \rho}{\eta} \tag{3.7-1}$$

is an excellent tool to characterize the large-scale flow in a stirred vessel:

- Laminar range: $Re < 10$
- Intermediate range: $10 < Re < 10^4$
- Turbulent range: $Re > 10^4$

In vessels filled with liquids which contain solid or fluid particles the Froude number

$$Fr^* = \frac{n^2 \cdot d \cdot \rho_c}{\Delta\rho \cdot g} \tag{3.7-2}$$

is decisive for the tendency of particle settling. With the index c for the continuous phase and d for the dispersed phase the density difference $|\rho_c - \rho_d| = \Delta\rho$ is the driving force for particle settling or rising in a gravitational field due to the acceleration g.

The breakup of fluids depends on the surface or interfacial tension σ and is governed by the Weber number

$$We = \frac{n^2 \cdot d^3 \cdot \rho_c}{\sigma}. \tag{3.7-3}$$

Note that this number is written with the density ρ_c of the continuous phase.

The power consumption P of a stirrer is equal to

$$P = Ne \cdot \rho \cdot n^3 \cdot d^5 \tag{3.7-4}$$

(or analogous to centrifugal pumps with the volumetric flow rate $\dot{V}$ and the pressure head Δp

$$P \sim \underbrace{n \cdot d^3}_{\dot{V}} \cdot \underbrace{\frac{n^2 d^2 \rho}{2}}_{\Delta p} \sim \dot{V} \cdot \Delta p\,).$$

The so-called Newton (or power) number Ne depends on the Reynolds number and the special geometry of the system (vessel, stirrer, baffles, flow resistances). In Fig. 3.7-2 the Newton number is plotted against the Reynolds number for a multiblade stirrer in a vessel without and with baffles.

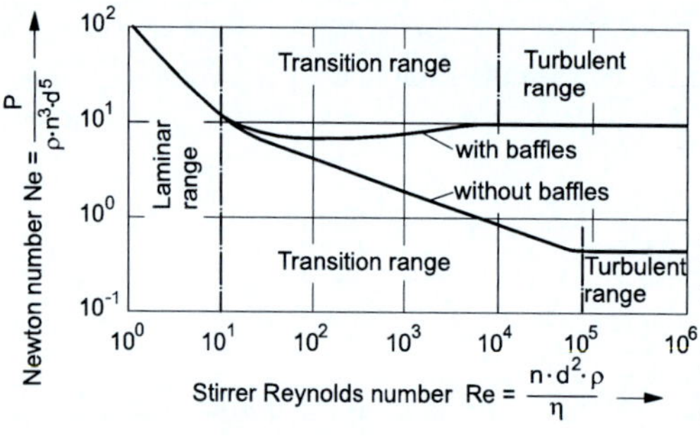

Fig. 3.7-2 Newton number of a multiblade impeller ($d/D = 0.3$) as a function of the Reynolds number with (*upper curve*) or without baffles (*lower curve*)

Three ranges can be distinguished:

- The laminar range with $Ne \cdot Re = \text{const.}$,
- The intermediate range with $Ne = f(Re, \text{baffles, etc.})$
- The turbulent range with $Ne_{\text{turb}} = \text{const.}$

It is approximately

$$Ne_{\text{turb}} \approx 1/3 \text{ for a well-designed marine-type impeller}$$

and

$$Ne_{\text{turb}} \approx 5 \text{ for multiblade stirrers.}$$

The Newton numbers for nearly all other stirrers in the completely turbulent flow range are between $1/3$ and 5 and depend on their flow resistances.

Many processes such as

- Breakup of gases or liquids introduced in a liquid phase immiscible with the other liquid
- Macro-, meso-, and micromixing
- Diffusion and chemical reaction rates

are strongly influenced by the intensity of turbulence. Therefore, important relationships of turbulent flow fields will be shortly introduced. They are mainly based on results published by Kolmogoroff (1958), Batchelor (1953), Hinze (1975), Brodkey (1975), and Corrsin (1964). The cascade of energy in eddies with different scales or the energy cascade is described by the energy spectrum $E(k)$ (here only one dimensional)

$$E(k) = \text{const.} \cdot \varepsilon^{2/3} \cdot k^{-5/3} \tag{3.7-5}$$

with k as the number of waves per unit length and ε as the local specific power input. Values for the const. in the last equation can be found in the literature in the range $0.42 < \text{const.} < 0.53$. In Fig. 3.7-3 the energy spectrum $E(k)$ is plotted against the wavenumber k with the ranges

$k < k_K$ called the inertial convective subrange

$k > k_K$ as the dissipation or viscous range in which the mechanical energy is dissipated in thermal energy

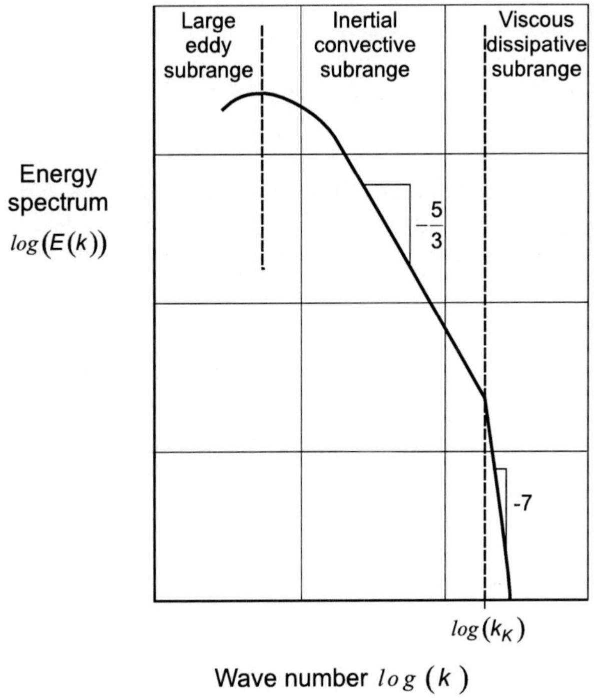

Fig. 3.7-3 Logarithm of the energy spectrum vs. the logarithm of the wavenumber

Kolmogoroff introduced the wavenumber $k_K = 1/\lambda_K$ with λ_K as the microscale of turbulence

$$\lambda_K = \left(\frac{\nu^3}{\varepsilon}\right)^{1/4}$$

valid for eddies with the scale $\lambda > 20\ \lambda_K$. Here ν is the viscosity of the fluid. The local specific energy k in J/kg depends on the fluctuating velocities u', v', and w' in the x-, y- and z-directions respectively and is given by

$$k = \frac{1}{2}[(u')^2 + (v')^2 + (w')^2]. \tag{3.7-6}$$

In the case of local isotropic turbulence (approximately valid in stirred vessels) with $u' = v' = w'$, the last equation becomes

$$k = \frac{3}{2}(u')^2. \tag{3.7-7}$$

The local specific power input ε is given by

$$\varepsilon = \frac{(u')^3}{\Lambda} = \left(\frac{2}{3}\right)^{3/2}\frac{k^{3/2}}{\Lambda}. \tag{3.7-8}$$

Here Λ is the macroscale of turbulence. In the literature some proposals for the prediction of this scale can be found. According to Brodkey, Λ is

$$\Lambda = \frac{d}{C} \tag{3.7-9}$$

with d as the stirrer diameter and $5 < C < 8$. The exact value depends on geometrical parameters, especially of the ratio d/D.

The local specific power input ε can also be derived from the longitudinal microscale λ_f for the corresponding lateral microscale λ_g.

According to Taylor ε is given by

$$\varepsilon = 30 \cdot \frac{\nu(u')^2}{\lambda_f^2} = 15 \cdot \frac{\nu(u')^2}{\lambda_g^2} \tag{3.7-10}$$

The longitudinal macroscale Λ of turbulence can be expressed by the longitudinal Taylor microscale:

$$\Lambda = \frac{(u')}{30 \cdot \nu} \cdot \lambda_f^2. \tag{3.7-11}$$

In low viscous liquids it is $\lambda_f \ll \Lambda$.

With respect to mixing (later discussed) further microscales have been introduced. The mixing-diffusion microscale λ_B according to Batchelor is given by

$$\lambda_B = \left(\frac{\nu \cdot D_{AB}^2}{\varepsilon}\right)^{1/4} = \lambda_K / \sqrt{Sc}. \tag{3.7-12}$$

In liquids with Schmidt numbers $Sc > 10^3$ the Batchelor scale λ_B is very small in comparison to the Kolmogoroff scale λ_K ($\lambda_B < 0.03\ \lambda_K$).

All these equations have a special feature in common: The dependence of microscales and fluctuating velocities on the local specific power input ε. Therefore, the prediction of specific power inputs (ε is the local, $\bar{\varepsilon}$ the mean, and ε_{max} the maximum value) is very important for many processes in the field of chemical engineering.

The maximum value ε_{max} in the immediate discharge region of a stirrer is given by

$$\frac{\varepsilon_{max}}{\bar{\varepsilon}} \approx 0.84 \cdot Ne_{turb}^{1/3} \cdot \left(\frac{D}{d}\right)^2 \cdot \frac{H}{D} \tag{3.7-13}$$

and the ratio $\varepsilon_{max}/\bar{\varepsilon}$ can be expected in the range $10 < \varepsilon_{max}/\bar{\varepsilon} < 30$.

The mean specific power input $\bar{\varepsilon}$ is the power P based on the mass of the contents in the vessel:

$$\bar{\varepsilon} = \frac{P}{V \cdot \bar{\rho}} = \frac{4 \cdot Ne}{\pi} \cdot n^3 \cdot d^2 \cdot \left(\frac{d}{D}\right)^2 \cdot \left(\frac{d}{H}\right). \tag{3.7-14}$$

The fluctuating velocities in the turbulent range ($Re > 10^4$) are proportional to the tip speed u_{tip} of the stirrer. This is true for the mean value $\overline{(u')}$ and also for the maximum value $(u')_{max}$:

$$\overline{(u')} \approx 0.18 \cdot Ne_{turb}^{4/9} \cdot u_{tip}, \tag{3.7-15}$$

$$(u')_{max} \approx 0.19 \cdot Ne_{turb}^{1/3} \cdot u_{tip}. \tag{3.7-16}$$

Dealing with chemical engineering processes carried out in stirred vessels the shear stress τ and the shear rate $\dot{\gamma}$ are important design parameters. The local shear rate $\dot{\gamma}$ is given by

$$\dot{\gamma} = \sqrt{\frac{\varepsilon}{\nu}}. \tag{3.7-17}$$

A characteristic shear rate $\dot{\gamma}_{\mathrm{cha}}$ depends on the mean specific power input $\bar{\varepsilon}$:

$$\dot{\gamma}_{\mathrm{cha}} = \mathrm{const.} \cdot \sqrt{\frac{\bar{\varepsilon}}{\nu}} \tag{3.7-18}$$

with const. depending on the type of stirrer.

Since the viscosity ν of a Newtonian liquid in an isothermal liquid is not dependent on the shear rate the relationship

$$\frac{\dot{\gamma}}{\gamma_{\mathrm{cha}}} = \frac{1}{\mathrm{const.}} \cdot \left(\frac{\varepsilon}{\bar{\varepsilon}}\right)^{1/2} \tag{3.7-19}$$

is valid. The mean shear stress $\bar{\tau}$ can be very important in stirred vessels applied for bioreactions, crystallization, or other processes which are very sensitive for local and mean shear stresses. The mean shear stress $\bar{\tau}$ is often described by the simple equation

$$\bar{\tau} = \eta \cdot \bar{\dot{\gamma}} = \eta \cdot k_{\mathrm{s}} \cdot n \tag{3.7-20}$$

with $10 < k_{\mathrm{s}} < 13$ for Newtonian liquids and $Re > 10^4$.

3.7.1 Macro-, Meso-, and Micromixing

In reaction engineering, mixing can be rate controlling in the case of very fast reactions. In general the reaction progress or the rate r_{A} of a chemical reaction (here for the component A) is described by the equation

$$r_{\mathrm{A}} = \frac{\mathrm{d}c_{\mathrm{A}}}{\mathrm{d}\tau_{\mathrm{r}}} = -k_{\mathrm{r}} \cdot (c_{\mathrm{A}})^n \tag{3.7-21}$$

with concentration c_{A}, reaction time τ_{r}, rate constant k_{r}, and the order of the reaction n. Integration of this equation for the special case $n = 1$ (first-order reaction) leads to

$$\ln\left(\frac{c_{\mathrm{A}\alpha}}{c_{\mathrm{A}}}\right) = k_{\mathrm{r}} \cdot \tau_{\mathrm{r}}, \tag{3.7-22}$$

where $c_{\mathrm{A}\alpha}$ is the concentration of the reactant A at the beginning. Let us assume that a reactant B is added to the reactant A already existing in the vessel. At first the two components are brought together by mesomixing in a zone very close to the feed point. Next the macromixing in the entire vessel volume is started. The mixing of the two components A and B on a molecular level is carried out in the smallest eddies of turbulence with extensions of the Batchelor microscale (micromixing). Corrsin has developed a general mixing model for the mixing time τ which is necessary to obtain a certain degree of mixing. In addition to the Schmidt number Sc, the macroscale L_c of concentration fluctuations and the fluid viscosity ν, the local specific power input ε is the most important parameter. The micromixing time τ_{micro} according to Corrsin is composed of two terms:

$$\tau_{micro} = \text{const.} \cdot \left(\frac{L_c^2}{\varepsilon}\right) + \text{const.}' \cdot \left(\frac{\nu}{\varepsilon}\right)^{1/2} \cdot \ln Sc. \qquad (3.7\text{-}23)$$

The first term describes the reduction of concentration fluctuations which are mostly reduced after approximately 10 up to 20 stirrer revolutions in the case of a fully turbulent flow with $Re > 10^4$. The second term depends on the limiting step in the microscale eddies and the degree of segregation. Many experimental results have shown that the minimum macromixing time (first term) τ_{macro} can be described by the simple equation

$$\tau_{macro} = C_{macro} \cdot \left(\frac{D^2}{\varepsilon}\right)^{1/3} \approx (5 \div 8.5) \cdot \left(\frac{D^2}{\varepsilon}\right)^{1/3} \qquad (3.7\text{-}24)$$

with D as the vessel diameter. In the case of a stirrer operated below the optimum speed or Reynolds number the macromixing time can be much longer. The geometry of the stirrer plays a minor role

According to Schäfer (Schäfer 2001) the macromixing time can be described by the last equation with $C_{macro} = 5.6$ for a special marine-type impeller, $C_{macro} = 6.5$ for an inclined blade impeller, and $C_{macro} = 8.5$ for a six-blade impeller. The number of revolutions or the dimensionless macromixing time is given by

$$n \cdot \tau_{macro} = 6.2\,(Ne_{turb})^{-1/3} \cdot \left(\frac{d}{D}\right)^{-5/3} \quad \text{(Mersmann et al. 1975),} \qquad (3.7\text{-}25)$$

$$n \cdot \tau_{macro} = 5.3 \cdot (Ne_{turb})^{-1/3} \cdot \left(\frac{d}{D}\right)^{-2} \quad \text{(Ruszkowski 1994).} \qquad (3.7\text{-}26)$$

Dealing with meso- and micromixing things are more complicated because there are several steps of turbulence which can be limiting for the progress of a chemical reaction characterized by the timescale $1/k_r$.

If any timescale of mixing is longer than $1/k_r$ the reaction time mixing is rate controlling. As will be shown later the root $\sqrt{\nu/\varepsilon}$ is the decisive parameter for meso- and micromixing with ν as the fluid viscosity and ε as the local specific power input. Therefore, the point and the kind of the addition of the reactant B play a big role. As a rule the reactant is fed at the point of the maximum specific power input ε_{max} in the immediate discharge region of the – preferable – blade impeller.

The entire mixing process can be split up in the following steps:

- In the large eddy subrange ($\lambda \gg \lambda_K$) different velocities of fluid elements and different fluctuation velocities result in a first equalization of large eddies with great concentration differences of the reactants.
- In the inertial convective subrange (sometimes called mesomixing) breakup of fluid elements into smaller ones takes place as a result of shear stress which is proportional to $\rho(u')^2$. The corresponding time constant is approximately

$$\tau_{meso} \approx \left(\frac{L_c^2}{\varepsilon}\right)^{1/3} \tag{3.7-27}$$

  with L_c as the macroscale of concentration fluctuations ($L_c > \lambda_K$).
- In the viscous convective subrange further mixing is achieved by stretching, thinning, and finally by the breakup of lamella in the smallest turbulent eddies. Their size is less than the Kolmogoroff microscale λ_K. Micromixing is started by this engulfment.
- In the viscous diffusive subrange with eddies of the Batchelor microscale size λ_B or even smaller the final complete concentration equalization takes place by molecular diffusion.

In general the micromixing time τ_{micro} depends on the root $\sqrt{\nu/\varepsilon}$, the Schmidt number Sc, and the degree I_s of segregation ($I_s = \overline{c^2}(t)/\overline{c_\alpha^2}$ with c as the concentration) and is given by the Corrsin model:

$$\tau_{micro} \approx \frac{1}{2} \cdot \sqrt{\frac{\nu}{\varepsilon}} \cdot (0.88 + \ln Sc) \cdot \left(\frac{1 - I_s}{I_s}\right). \tag{3.7-28}$$

With $I_s = 0.1$ and further simplifications valid for $Sc > 10^3$ the last equation can be reduced to

$$\tau_{micro} \approx 5 \cdot \sqrt{\frac{\nu}{\varepsilon}} \cdot \ln Sc. \tag{3.7-29}$$

It is important to note that the last equation describes the final step of viscous molecular diffusion; however, other homogenization processes such as gulf entrainment or engulfment and eddy breakup are not only rate controlling but also decisive for the importance of main and secondary reactions. A big variety of models for the prediction of time constants can be found in the literature. All these equations have in common that the times are always proportional to the root $\sqrt{\nu/\varepsilon}$. The time constant τ_{eng} for the engulfment step is

$$\tau_{eng} \approx 17 \cdot \sqrt{\frac{\nu}{\varepsilon}} \tag{3.7-30}$$

and is a bit longer than the mean life time

$$\tau_{diss} \approx 13 \cdot \sqrt{\frac{\nu}{\varepsilon}} \tag{3.7-31}$$

of eddies with energy dissipation valid for isotropic and homogeneous turbulence. The main results of this chapter are:

- Every mixing time is a function of the degree of mixing or segregation
- The macromixing time is mainly dependent on the mean specific power input $\bar{\varepsilon}$ and the vessel diameter D and increases with $\tau_{macro} \sim D^{2/3}$ for $\bar{\varepsilon} = $ const.
- The micromixing time is a function of the local specific power input ε, the viscosity ν, and the Schmidt number Sc.
- As a rule the micromixing time is short in comparison to the macromixing time; however, in fluid boundary layers close to solid surfaces (walls, heat exchanger, tube bundles, etc.), the local specific power input ε can be very low with the consequence of long micromixing times.
- The feed point of a reactant should be close to a spot of high $\varepsilon/\bar{\varepsilon}$-values when the progress of the chemical reaction is mixing-controlled.
- The lines $\varepsilon/\bar{\varepsilon} = $ const. are important parameters for the design and operation of stirred vessels applied as chemical reactors. In Fig. 3.7-4 such lines are shown for a six-blade impeller and a three-blade marine-type propeller.

3.7.2 Suspending, Tendency of Settling

Dealing with suspensions (solid particles in liquids) in stirred vessels, two important questions have to be answered:

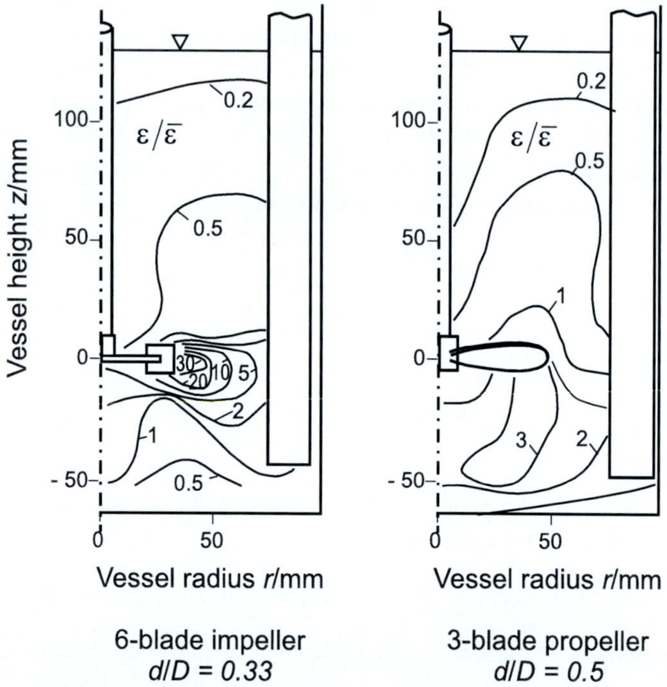

Fig. 3.7-4 Isoenergetic lines $\varepsilon/\bar{\varepsilon}$ valid for a turbulent flow ($Re > 10^4$)

- What is the minimum stirrer speed necessary for the suspending of particles or "Off-Bottom Lifting"? This process is difficult in small stirred vessels.
- What is the minimum mean specific power input to avoid settling of particles or to achieve "Avoidance of Settling"? This is decisive for large stirred vessels.

The mean specific power input $\bar{\varepsilon}$ is composed of the contributions

$\bar{\varepsilon}_{\mathrm{BL}}$ necessary for Off-Bottom Lifting and

$\bar{\varepsilon}_{\mathrm{AS}}$ to achieve Avoidance of Settling

according to (Mersmann et al. 1998)

$$\bar{\varepsilon} = \bar{\varepsilon}_{\mathrm{BL}} + \bar{\varepsilon}_{\mathrm{AS}} \tag{3.7-32}$$

with

$$\bar{\varepsilon}_{\mathrm{BL}} \approx 200 \cdot Ar^{1/2} \cdot [\varphi \cdot (1-\varphi)^m]^{3/4} \cdot \frac{\eta_{\mathrm{c}} \cdot \Delta\rho \cdot g}{\rho_{\mathrm{c}}^2 \cdot H} \cdot \left(\frac{D}{d}\right)^{5/2} \tag{3.7-33}$$

and

$$\bar{\varepsilon}_{\mathrm{AS}} \approx 0.4 \cdot Ar^{1/8} \cdot [\varphi \cdot (1-\varphi)^m] \cdot \left(d_\mathrm{p} \cdot \left(\frac{\Delta\rho \cdot g}{\rho_\mathrm{c}}\right)^3\right)^{1/2}. \tag{3.7-34}$$

The exponent m as a function of the Archimedes number $Ar = d_\mathrm{p}^3 \cdot \rho_\mathrm{c} \cdot \Delta\rho \cdot g / \eta_\mathrm{c}^2$ can be read from Fig. 3.6-7 with $m = 4.65$ for $Ar < 0.1$ and $m = 2.4$ for $Ar > 10^3$. The minimum mean specific power input $\bar{\varepsilon}_{\mathrm{AS}}$ is given by

$$\bar{\varepsilon}_{\mathrm{AS}} \approx [\varphi \cdot (1-\varphi)^m] \cdot \frac{w_\mathrm{S} \cdot \Delta\rho \cdot g}{\rho_\mathrm{c}}. \tag{3.7-35}$$

The settling of solid or fluid particles can be described in a general way. In Fig. 3.7-5 the dimensionless mean specific power input is plotted against the Archimedes number for two different particle holdups φ. The power provided by the stirrer can be enforced by the power $\dot{v}_\mathrm{G} \cdot g$ introduced by rising bubbles with $\dot{v}_\mathrm{G}$ as the superficial gas velocity. This leads to a three-phase system. The total specific mean power input $\bar{\varepsilon}_{\mathrm{tot}}$ is $\bar{\varepsilon}_{\mathrm{tot}} = \bar{\varepsilon} + \dot{v}_\mathrm{G} \cdot g$. The quality of the homogeneity of suspensions or three-phase systems depends on the parameter Ar and φ. Such systems are nearly homogeneous for small Archimedes numbers but high volumetric holdup. The greater the Archimedes number the more heterogeneous becomes the system. In the case $\bar{\varepsilon}/\sqrt{d_\mathrm{p}(\Delta\rho \cdot g/\rho_\mathrm{c})^3} < \approx 0.01$ a disengagement of the phases and a heterogeneous structure in combination with particle settling is observed.

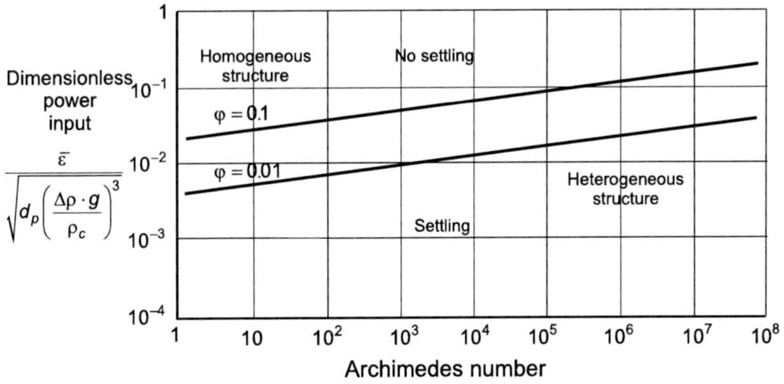

Fig. 3.7-5 Dimensionless mean specific power input vs. the Archimedes number

3.7.3 Breakup of Gases and Liquids (Bubbles and Drops)

A multiblade stirrer is a suitable tool to obtain small bubbles or drops because this stirrer produces high shear stresses in the discharge region close to the stirrer tip. In the case of a low stirrer intensity or low mean specific power input the diameter of the largest fluid particles can be calculated as has been already discussed. This size of fluid particles is reduced in the immediate vicinity of the stirrer due to high shear rates $r \cdot (\mathrm{d}\omega/\mathrm{d}r)$:

$$d_{\mathrm{p,\,max}} \leq d_{\mathrm{E}} . \tag{3.7-36}$$

The diameter of fluid particles is mainly a function of a Weber number which is composed of the local specific power input ε or the square of the fluctuating velocity $(u')^2$:

$$We' = \frac{(u')^2 \cdot \rho_{\mathrm{c}} \cdot d_{\mathrm{p}}}{\sigma} . \tag{3.7-37}$$

According to the Bohr–Wheeler criterion and an approach of Werner (1997) the mean (Sauter) size d_{32} is given by

$$d_{32}^3 \cdot \pi \cdot \rho_{\mathrm{c}} \cdot (u')^2 = f\left(\frac{\rho_{\mathrm{c}}}{\rho_{\mathrm{d}}}\right) \cdot d_{32}^2 \cdot \pi \cdot \sigma . \tag{3.7-38}$$

This equation can be transformed into

$$d_{32} \approx 0.267 \cdot \left(\frac{\rho_{\mathrm{c}}}{\rho_{\mathrm{d}}}\right)^{0.16} \cdot \frac{\sigma^{0.6}}{\rho_{\mathrm{c}}^{0.6} \cdot \varepsilon_{\mathrm{max}}^{0.4}} . \tag{3.7-39}$$

With the stirrer Weber number We according to

$$We = \frac{d^3 \cdot n^2 \cdot \rho_{\mathrm{c}}}{\sigma} \tag{3.7-40}$$

(with the speed n and the stirrer diameter d), the result is

$$\frac{d_{32}}{d} \approx 0.12 \cdot \left(\frac{\rho_{\mathrm{c}}}{\rho_{\mathrm{d}}}\right)^{0.16} \cdot \left(\frac{d^3 \cdot n^2 \cdot \rho_{\mathrm{c}}}{\sigma}\right)^{-0.6} . \tag{3.7-41}$$

In Fig. 3.7-6 the ratio d_{32}/d is plotted against the stirrer Weber number for air bubbles in low viscous liquids with $(\rho_{\mathrm{c}}/\rho_{\mathrm{d}})^{0.16} = (1{,}000/1.2)^{0.16} = 2.93$ and for drops valid for $\rho_{\mathrm{c}}/\rho_{\mathrm{d}} \approx 1$. Most of all, experimental results can be found in the shadowed area. It is important to note that the minimum drop sizes can only be obtained after a long breakup time (approximately 1 h) because all large drops have to be passed through the zone of the highest fluctuating velocities or local specific power input. The equations presented here are only valid for very small holdups of the dispersed phase ($\varphi < 0.01$). An increase of the holdup φ leads to an increase of the mean size which is described by

$$\frac{d_{32}}{d} = A \cdot (1 + B \cdot \varphi) \cdot We^{-0.6} \tag{3.7-42}$$

with $0.04 < A < 0.08$ and $2.7 < B < 9$ valid for drops.

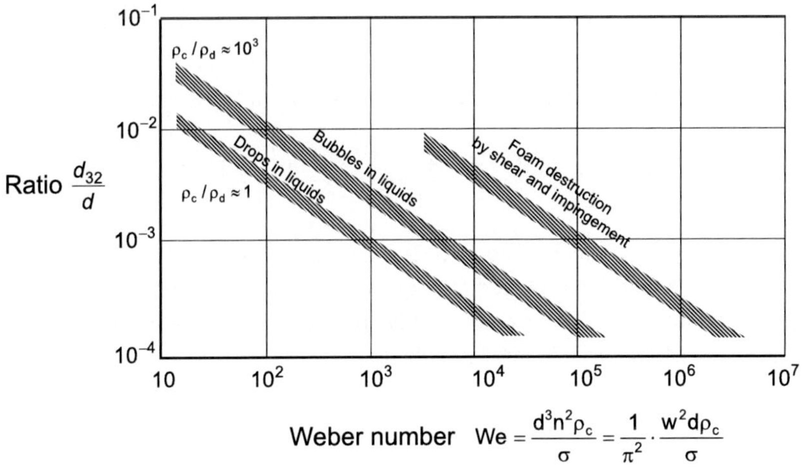

Fig. 3.7-6 Ratio d_{32}/d vs. the stirrer Weber number

3.7.4 Gas–Liquid Systems in Stirred Vessels

Many heterogeneous chemical reactions are carried out in stirred vessels often equipped with six- or eighteen-blade impellers. The gas can be added either at the top by insurgement of the gas in the headspace or by introduction of gas under pressure at the bottom. In the case of surface gas entrainment the minimum stirrer speed for the starting of gas uptake is (Zehner 1996)

$$n_{\min} = \text{const.} \cdot \frac{w_E \cdot D}{Ne_{\text{turb}}^{1/3} \cdot d^2} \tag{3.7-43}$$

with const. ≈ 1.6 for low viscous liquids and $d/D = 0.6$ and const. $= 2.8$ valid for a liquid viscosity of 50 mPa s and $d/D = 1/3$. As a rule the gas is introduced at the bottom of the vessel by a gas distributor (pipe or plate with holes). The gas holdup φ_G is mainly dependent on the ratio $\dot{v}_G/w_E$ (compare with bubble columns) and the ratio $\bar{\varepsilon}/(\dot{v}_G \cdot g)$ which is the mean specific power input $\bar{\varepsilon}$ based on the specific power input $\dot{v}_G \cdot g$ provided by the rising bubbles and is given by

$$\varphi_G = \frac{\dot{v}_G}{w_E} \cdot \left(1 + \frac{\bar{\varepsilon} \cdot d}{\dot{v}_G \cdot g \cdot D}\right)^{1/3} . \tag{3.7-44}$$

The maximum gas flow rate $\dot{V}_G$ which can be distributed in the entire vessel is sometimes described by the simple equation

$$\frac{\dot{V}_G}{n \cdot d^3} \approx 0.075 .$$

However, Zehner has shown that there is an additional influence of the diameter ratio d/D. The maximum gas flow rate density $\dot{v}_G = \dot{V}_G/(D^2 \pi/4)$ which can be distributed by a six- or eighteen-blade stirrer for a given stirrer speed n is given by

$$\frac{\dot{v}_G \cdot w_E}{(D-d) \cdot g} \cdot \left(\frac{D}{d}\right)^2 = \left[0.1 + \frac{n^2 \cdot d^3}{(D-d) \cdot g \cdot D}\right] \cdot \left(\frac{n^2 \cdot d^3}{(D-d) \cdot g \cdot D}\right) . \tag{3.7-45}$$

If the gas flow rate exceeds the value according to this equation the gas rises in form of large bubbles close to the center of the vessel with the result of a poor gas/liquid interfacial area. The same would happen in bubble columns when the ratio $\dot{v}_G/w_E$ exceeds values of approximately 0.1.

Symbols

A	m^2	Area
a	$1/m$	Volumetric surface
b	m	Width, circumference
C, C_1		Constants
c_w		Friction factor
D	m	Diameter (vessel)
d	m	Diameter (tube, pore, sphere, stirrer)
d_p	m	Particle diameter
$E(k)$	m^3/s^2	Energy spectrum
F	N	Force
f	m^2	Cross-sectional area
g	m/s^2	Acceleration due to gravity
H	m	Height
h	m	Height
I	$kg \cdot m/s$	Momentum

I_s		Degree of segregation
k		Constant
k	m	Roughness
k	1/m	Wavenumber
k	J/kg	Specific energy
k	$Pa \cdot s^m$	Consistency
L	m	Thickness, pore length, travelling distance, macroscale
l	m	Coordinate, characteristic extension
l_s	m	Microscale of turbulence
m		Exponent
$\dot{N}$	1/s	Rate of molecules
n		Number, exponent
n	1/s	Rate of rotation
n_v	$1/m^3$	Molecules per volume
P	W	Power
p	Pa	Pressure
Δp	Pa	Pressure drop
R		Fraction (mass, volume)
R	m	Tube radius
r	m	Radius
t	s	Time
u	m/s	Velocity (x-direction)
V	m^3	Volume
$\dot{V}$	m^3/s	Volumetric flow
$\dot{v}$	$m^3/(m^2 \cdot s)$	Volumetric flow density
v	m/s	Velocity (y-direction)
w	m/s	Velocity (z-direction)
w	$m^3/(m^2 \cdot s)$	Superficial velocity
x	m	x-coordinate
y	m	y-coordinate
z	m	z-coordinate
z		Multiple of the acceleration due to gravity

Greek symbols

$\dot{\gamma}$	1/s	Shear rate
δ	m	Film thickness
ζ		Discharge coefficient
η	$Pa \cdot s$	Dynamic viscosity
Λ	m	Mean free path length, macroscale
λ		Friction coefficient (tube)
λ	m	Microscale
μ		Contraction or tortuosity factor

ξ		Friction coefficient of a fixed bed
ρ	kg/m^3	Density
σ	N/m	Surface or interfacial tension
τ	s	Time
τ	Pa	Shear stress
ϕ	m^2/s	Velocity potential
φ		Holdup

Indices

c	Continuous phase
cf	Starting of fluidization
circ	Circulation
d	Dispersed phase
e	Minimum cross-sectional area
eff	Effective
F	Final value
G	Gas
h	Hydraulic
macro	Macromixing
max	Maximum
micro	Micromixing
Mol	Molecule
o	Opening
p	Particle
r	Radial, reaction
rel	Relative
s	Falling of a single particle
ss	Falling of particles in a swarm
stab	Stable
tip	Value at the tip of a stirrer
x	x -coordinate
y	y -coordinate
z	z -coordinate
σ	Related to surface tension
32	Sauter diameter

Dimensionless numbers

$Ar = \dfrac{d^3 \cdot \Delta\rho \cdot g \cdot \rho_c}{\eta_c^2}$ Archimedes number

$Eu = \dfrac{p_1 - p_2}{w^2 \cdot \rho}$ Euler number

$Fl = \dfrac{\sigma^3 \cdot \rho_c^2}{\eta_c^4 \cdot \Delta\rho \cdot g}$	Fluid number (two-phase systems)
$Fr = \dfrac{w^2}{d \cdot g}$	Froude number of pipe flow
$Fr^* = \dfrac{w^2 \cdot \rho_c}{d \cdot \Delta\rho \cdot g}$	Extended Froude number in a two-phase system
$Fr_0 = \dfrac{w_0^2 \cdot \rho_d}{d_0 \cdot \Delta\rho \cdot g}$	Froude number of an opening
$Fr^* = \dfrac{n^2 \cdot d \cdot \rho_c}{\Delta\rho \cdot g}$	Froude number of a stirrer in a two-phase system
$K_F = \dfrac{\sigma^3 \cdot \rho_L}{\eta_L^4 \cdot g}$	Film number
$Kn = \Lambda / d$	Knudsen number
$Ne = \dfrac{P}{\rho \cdot n^3 \cdot d^5}$	Newton number of a stirrer
$N_V = \dfrac{\dot{V}_{circ}}{n \cdot d^3}$	Volumetric flow number of a stirrer
$n \cdot t$	Mixing time number
$Re = \dfrac{w \cdot d \cdot \rho}{\eta}$	Reynolds number of a pipe flow
$Re_L = \dfrac{w \cdot \delta \cdot \rho_L}{\eta_L}$	Reynolds number of a film flow
$Re_G = \dfrac{w_G \cdot (d - 2 \cdot \delta) \cdot \rho_G}{\eta_G}$	Reynolds number of a gas/liquid film flow
$Re_p = \dfrac{w_s \cdot d \cdot \rho_c}{\eta_c}$	Reynolds number of a particle
$Re = \dfrac{w \cdot d_p \cdot \rho_c}{\varepsilon_d \cdot \eta_c}$	Reynolds number of a packed bed
$Re = \dfrac{n \cdot d^2 \cdot \rho}{\eta}$	Reynolds number of a stirrer
$Re \cdot Fr^* = \dfrac{w^3 \cdot \rho_c^2}{\eta_c \cdot \Delta\rho \cdot g}$	Dimensionless superficial velocity in a packed or fluidized bed

$Tu = \frac{w'_{\mathrm{eff}}}{w}$	Degree of turbulence
$We_0 = \frac{w_0^2 \cdot d_0 \cdot \rho_{\mathrm{d}}}{\sigma}$	Weber number of an opening
$We = \frac{w^2 \cdot d \cdot \rho}{\sigma}$	Weber number
$We = \frac{n^2 \cdot d^3 \cdot \rho_{\mathrm{c}}}{\sigma}$	Weber number of a stirrer
$We' = \frac{(\bar{u}')^2 \cdot \rho_{\mathrm{c}} \cdot d_{\mathrm{p,\,max}}}{\sigma}$	Weber number of a fluid particle
$\frac{We}{Fr^*} = \frac{d_{\mathrm{p}}^2 \cdot \Delta\rho \cdot g}{\sigma}$	Bond number (in US literature)

4 Balances, Kinetics of Heat and Mass Transfer

4.1 Introduction

Process engineering attends to yield products with a set of desired properties, given some raw materials. This aim may be achieved through several different process variants, which are found by *process design*. They usually differ with respect to feasibility, safety and particularly to cost effectiveness. It is thus necessary to appraise all process variants by means of *process analysis*.

Process variants are to be modeled mathematically to evaluate them quantitatively. For this purpose a process is modeled by process elements and by their interlinks as they appear due to flows of energy and mass. The mathematical evaluation of the process model for any given instance is denoted simulation of the process. *Process simulations* allow for assessment of the fixed and variable costs, the design of process equipment, of the necessary utilities, and thus for the assessment of the economic viability of the process variants.

The interlinked elements comprise functionalities regarding material conversion. These functionalities are described by physicochemical *models*. Depending on the level of itemization, the respective element is for example a unit operation "distillation" or a unit operation "chemical reactor." It may also be a plant component which in turn consists of several interlinked unit operations.

As mentioned before, the linkage of elements is rendered by flows of energy and mass. A flow may be caused by a pressure difference effected through a pump. This material flow carries a specific energy and transports it to the receiving element. Just as well the transport of energy can be maintained in form of electrical energy transformed to heat within the ohmic resistance of the target system, or in form of mechanical energy. A flow may also arise in the wake of leveling processes which may be driven by a temperature difference between neighboring elements or by concentration differences. In this case the kinetics of *heat and mass transfer* have to be taken into consideration.

A. Mersmann et al., *Thermal Separation Technology: Principles, Methods, Process Design*, VDI-Buch, DOI 10.1007/978-3-642-12525-6_4,
© Springer-Verlag Berlin Heidelberg 2011

The following sections cover methods to work with the equations of change from the first steps of raising a mathematical model up to the evaluation of a given technical case. This concept is extended later to the treatment of kinetic processes involving heat and mass transport. Worked examples illustrate the content.

4.2 Balances

4.2.1 Basics

Prior to putting up an equation of change for any property the system under consideration has to be defined. Thus the first step is to identify the system boundary, which allows to recognize interaction between system and environment. The boundary may be open, adiabatic, closed, or completely isolated. A closed boundary is impermeable to any mass flow, and an adiabatic boundary prevents any heat flow. An isolating boundary is impenetrable for heat and mass.

The state of a system is described by external and internal *properties of state*. The external properties characterize the system's arrangement and velocity relative to other systems, whereas the internal properties characterize the thermodynamical state of the system relative to a virtual reference state. The change of a system's state is termed a process. The only processes to occur within the boundaries of an isolated system are leveling processes. Leveling processes in isolated systems are decaying with time and finally come to rest.

If a system is split into subsystems the properties of state may either remain unchanged in every subsystem and then they are denoted intensive properties. The properties of state in the splitted subsystems may otherwise be altered with respect to the unsplit system, in which case they are classified as extensive properties. Pressure and temperature are intensive properties. Energy and mass are extensive properties.

A special case of an extensive property is the total energy of a system: It is independent of the choice of reference frame. This invariance leads to the known conservation laws of energy, linear momentum, and angular momentum.

Only those properties Z of a system, that are conserved properties, may enter into an equation of change. Energy, linear momentum, and angular momentum are conserved quantities and the respective equations of change are named *energy balance*, *linear momentum balance*, and *angular momentum balance*.

The basic assumption for any such balance is that the conserved physical quantity may only change due to flows of the quantity Z across the system's boundary. The change of a system's conserved physical quantity with time is equal to the sum of all inbound and outbound flows:

$$\frac{dZ}{dt} = \sum \dot{Z}_{in} - \sum \dot{Z}_{out}. \tag{4.2-1}$$

To balance any extensive property, additional source and drain terms which account for gain and loss of those properties have to be invoked:

$$\frac{dZ}{dt} = \sum \dot{Z}_{in} - \sum \dot{Z}_{out} + \sum \dot{Z}_{source} - \sum \dot{Z}_{drain}. \tag{4.2-2}$$

If the time resolved progression itself is not under consideration, then the above equation may be rewritten in terms of differences:

$$\Delta Z = \sum Z_{in} - \sum Z_{out}, \tag{4.2-3}$$

$$\Delta Z = \sum Z_{in} - \sum Z_{out} + \sum Z_{source} - \sum Z_{drain}. \tag{4.2-4}$$

The equations of change of linear and angular momentum are relevant to the mechanics of fluids and solids. The equation of change of energy and the subsequently derived equations are of fundamental importance to thermal process engineering and are discussed in the following.

Energy appears in various forms, which are mutually convertible and which are subject to certain constraints. These constraints are accounted for in form of the second law of thermodynamics.

The total energy of an isolated system is the sum of the several forms of energy contained in it:

$$E_{total} = E_0 + E_{external} + E_{internal} \tag{4.2-5}$$

with (4.2-6)

$$E_0 = M_0 c^2,$$

$$E_{external} = E_{kinetic} + E_{potential},$$

$$E_{internal} = E_{mechanical} + E_{thermal} + E_{chemical} + E_{nuclear},$$

where the greatest share falls upon the rest energy E_0. Any energy increase is equivalent to an increase of relativistic rest mass, as described by the relation $E_0 = M_0 c^2$. In the majority of process engineering tasks, relativistic effects are disregardable. If furthermore no external mass flow enters through the system's boundary, the relativistic rest mass M_0 is regarded constant. Further balance laws for matter result from this, as discussed below. The assumption of constant relativistic rest mass implies the constant of total energy E_0 which quantity is thus not further evaluated.

The kinetic energy of nonrelativistic systems that move with velocity v is described by

$$E_{\text{kinetic}} = \frac{1}{2} \cdot M_0 \cdot v^2. \tag{4.2-7}$$

The potential energy content of a system is regarded as a difference of two absolute potential energies. For example, in the earth's gravitational field, each mass gets assigned a potential energy defined by a height difference ΔH by which the mass M_0 is shifted while being retarded owing to the gravitational force which is assumed to be constant:

$$E_{\text{potential}} = M_0 \cdot g \cdot \Delta H. \tag{4.2-8}$$

The internal mechanical energy $E_{\text{mechanical}}$ is related to inner mechanical stresses which balance each other on a macroscopic scale, so-called residual stresses.

The internal thermal energy E_{thermal} is denoted by the letter U. This energy subsumes the kinetic and potential energy of the molecules which make up the system. The chemical energy is due to the configuration of the electron shell of the molecules. Finally, the nuclear energy is stored in the inneratomic bonds.

Energy can be transported across a system's boundary by being linked to a mass flow, or in form of radiative or conductive heat, or in form of mechanically transferred power. In case of an open system under isobaric conditions, the convective energy flow can be derived from the specific enthalpy. For example, for an inbound flow, it follows that

$$\dot{E}_{\text{in}} = \dot{M}_{\text{in}} \cdot h_{\text{in}} = \dot{M}_{\text{in}} \cdot c_{p,\text{in}} \cdot T_{\text{in}}. \tag{4.2-9}$$

In process engineering the external energy of an isolated system usually remains constant.

As explained before, the extensive state property mass is regarded a conserved quantity which can be utilized in equations of change. This is especially appropriate where no transformation of nuclear energy is involved. With the relative molar mass of molecules being constant, any *balance of mass* can be converted into a *balance of the amount of substance*.

If in addition the density $\rho = M/V$ (M mass, V volume) is constant, then also a *balance of volume* is applicable.

If no chemical reaction takes place, the balancing equations for mass and amount of substance may also be formulated for each constituent molecular species. This is referred to as a *component balance*. Chemical reactions are incorporated into the component balance by means of source and drain terms.

Some of the properties characterizing a system may be adherent to every individual of the constituting bulk of particles. If moreover such a property is a conserved quantity population balances may be raised, see Chap. 8. Examples for such properties are not only a particle's length, surface, volume, but also kinetic properties like reactivity, hardness, or growth rate.

The following relations (MESH-equations) are of prime importance for the mathematical description of the various process engineering phenomena:

M (Mass)	mass or material balance
E (Equilibrium)	$\tilde{z}''_i = f(\tilde{z}'_1, \tilde{z}'_2, \ldots, \tilde{z}'_i, \ldots \tilde{z}'_n)$
S (Summation)	$\sum_i z_i = 1$ or $\sum_i \tilde{z}_i = 1$
H (Heat)	energy balance

Furthermore, momentum balances, kinetic expressions and initial and boundary conditions may be required to completely describe a process engineering phenomenon.

Often description of a phenomenon requires a system of coupled equations that are hardly resolvable with analytical approaches.

When putting up a balance all properties of state are extensive properties. Therefore it is possible to split any system up or to add several of them together. Thus the same principles of balancing are applicable on all length scales, examples for the macroscale are: plant, process, subprocess (e.g., solvent recovery), a single device (e.g., rectification column), component parts (e.g., rectification tray); an example on the microscale is a differential height of a packing of a rectification column.

4.2.2 Balancing Exercises of Processes Without Kinetic Phenomena

Below are listed five consecutive balancing exercises, some of them on a very basic level. These examples illustrate the practical use of balancing. At first kinetic phenomena of heat and mass transfer (conduction, radiation, diffusion) as well as chemical reactions are not considered. These phenomena are addressed in other exercises in Sect. 4.3.3.

4.2.2.1 Exercise: Filling a Tank

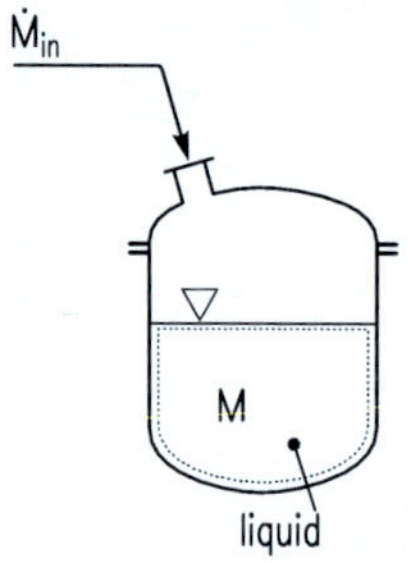

Fig. 4.2-1 Tank

Assume an empty tank, that is fed by a steady liquid mass flow $\dot{M}_{in}$ = const. (Fig. 4.2-1).

Question: How long does it take to reach M_{max}?

Material balance for the subsystem "liquid":

Application of (4.2-1) leads to

$$\frac{dM}{dt} = \dot{M}_{in}. \tag{4.2-10}$$

Separation of variables M and t and subsequent integration yield the filling time t_{fill}:

$$\frac{1}{\dot{M}_{in}} \cdot \int_{M=0}^{M_{max}} dM = \int_{t=0}^{t_{fill}} dt,$$

$$\frac{1}{\dot{M}_{in}} \cdot M\Big|_{M=0}^{M_{max}} = t\Big|_{t=0}^{t_{fill}},$$

$$t_{fill} = \frac{M_{max}}{\dot{M}_{in}}.$$

Dimensionless formulation of the filling process $M(t)$ can be achieved through expansion of both sides of (4.2-10) by $1/M_{max}$. After separation of variables and integration the following normalized expression is obtained:

$$\frac{M(t)}{M_{max}} = \frac{\dot{M}_{in}}{M_{max}} \cdot t. \tag{4.2-11}$$

The graph of $M(t)/M_{max}$ is shown in Fig. 4.2-2.

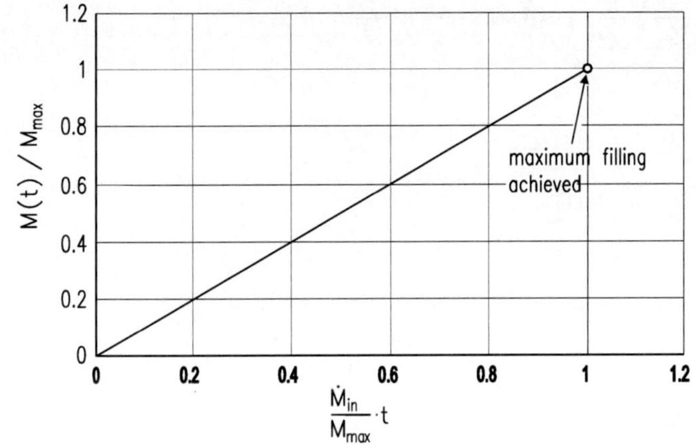

Fig. 4.2-2 Dimensionless representation of the liquid level in a tank with constant feed

4.2.2.2 *Exercise: Tank with Outlet*

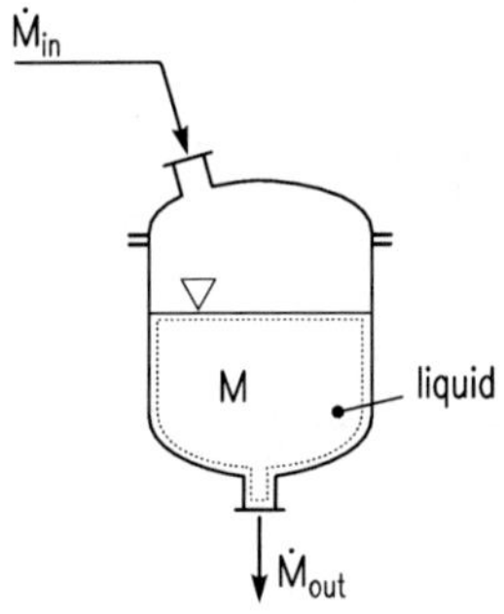

Fig. 4.2-3 Tank with outlet

Assume an initially empty tank which is fed by a constant liquid flow $\dot{M}_{in}$ = const. (cf. Sect. 4.2.2.1). A drainage flow leaves through the outlet. The outbound liquid flow $\dot{M}_{\text{out}}$ varies linearly with the filling quantity (Fig. 4.2-3):

$$\dot{M}_{\text{out}} = a \cdot M \text{ with } a = \text{const.} \tag{4.2-12}$$

Question: How does the filling quantity M change with time t?

Material balance for the subsystem "liquid"

Application of (4.2-1) yields

$$\frac{\mathrm{d}M}{\mathrm{d}t} = \dot{M}_{\text{in}} - \dot{M}_{\text{out}}$$
$$\frac{\mathrm{d}M}{\mathrm{d}t} = \dot{M}_{\text{in}} - a \cdot M \tag{4.2-13}$$

Separation of variables and integration yields

$$\int_{M=0}^{M(t)} \frac{\mathrm{d}M}{\dot{M}_{\mathrm{in}} - (a \cdot M)} = \int_{t=0}^{t} \mathrm{d}t\,,$$

$$-\frac{1}{a} \cdot \ln(\dot{M}_{\mathrm{in}} - a \cdot M)\Big|_{M=0}^{M(t)} = t\Big|_{t=0}^{t}\,,$$

$$-\frac{1}{a} \cdot \ln\frac{\dot{M}_{\mathrm{in}} - a \cdot M(t)}{\dot{M}_{\mathrm{in}}} = t,$$

$$M(t) = \frac{\dot{M}_{\mathrm{in}}}{a} \cdot (1 - \exp[-at])\,. \tag{4.2-14}$$

One recognizes that the filling level exponentially approaches its saturation value $\dot{M}_{\mathrm{in}}/a$ with time (Fig. 4.2-4).

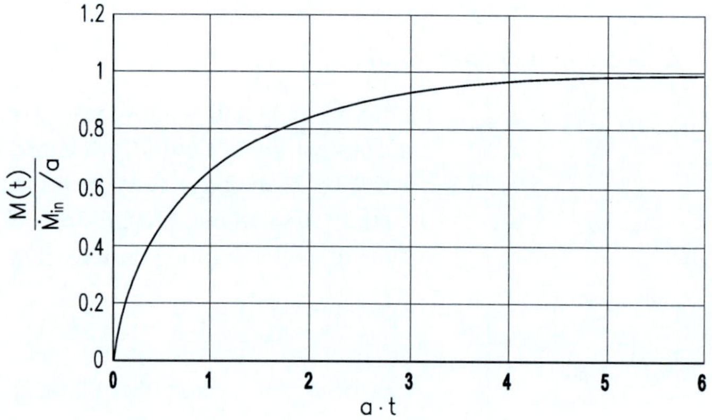

Fig. 4.2-4 Dimensionless representation of the liquid level in a tank with constant feed and variable outflow

4.2.2.3 Exercise: Temperature Evolution in an Agitated Tank

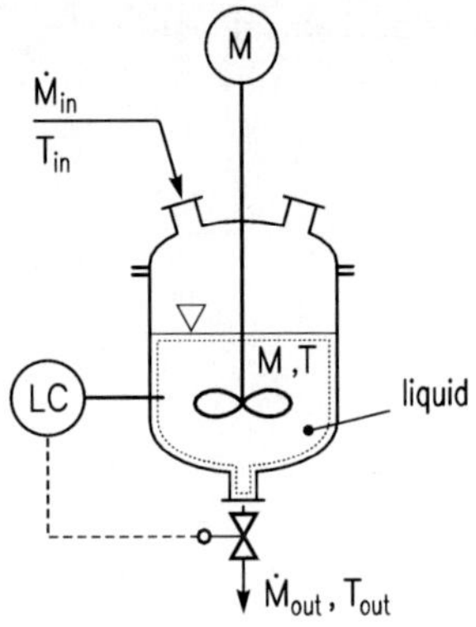

Fig. 4.2-5 Agitated tank

Assume that in a continuously operated stirred tank the liquid level M is kept constant, see Fig. 4.2-5. Furthermore, assume that the power $P_{agitator}$ is dissipated into heat. Let the tank content have an initial temperature T_0 and the inbound liquid flow $\dot{M}_{in}$ have temperature T_{in}. The specific heat capacity c_p is assumed to be independent of temperature.

Question: How does the temperature of the liquid in the tank change with time?

Material balance for subsystem "liquid"

(M = const.) Application of (4.2-1) yields

$$0 = \dot{M}_{in} - \dot{M}_{out}. \tag{4.2-15}$$

Energy balance for subsystem "liquid"

$$\frac{dE}{dt} = \dot{M}_{in} \cdot h_{in} + \dot{W}_{agitator} - \dot{M}_{out} \cdot h_{out} \tag{4.2-16}$$

with the reference temperature T_0 and the heat capacity c_p = const. the following holds:

$$E = M \cdot c_p \cdot (T - T_0),$$

$$\frac{dE}{dt} = M \cdot c_p \cdot \frac{dT}{dt}$$

and

$$\begin{aligned} h_{in} &= c_p \cdot (T_{in} - T_0), \\ h_{out} &= c_p \cdot (T_{out} - T_0), \end{aligned} \tag{4.2-17}$$

wherein $T_{out} = T$, as the liquid is regarded as ideally mixed.

Furthermore, it can be stated that

$$\dot{W}_{\text{mech}} = P_{\text{agitator}} = P_{\text{motor}} - P_{\text{bearing}} \tag{4.2-18}$$

The agitator power P_{agitator} is identically with the mechanical power $\dot{W}_{\text{mech}}$ which the agitator introduces into the subsystem "liquid." With (4.2-16) it follows that

$$M \cdot c_p \cdot \frac{\mathrm{d}T}{\mathrm{d}t} = \dot{M}_{\text{in}} \cdot c_p \cdot (T_{\text{ein}} - T) + P_{\text{agitator}}, \tag{4.2-19}$$

which can be normalized to

$$\frac{\mathrm{d}\Theta}{\mathrm{d}\tau} = \Pi - \Theta \tag{4.2-20}$$

by means of introducing a dimensionless temperature

$$\Theta = \frac{T - T_{\text{in}}}{T_0 - T_{\text{in}}} \tag{4.2-21}$$

as well as introducing a dimensionless time

$$\tau = \frac{t}{M/\dot{M}_{\text{in}}} \tag{4.2-22}$$

and a dimensionless agitator power

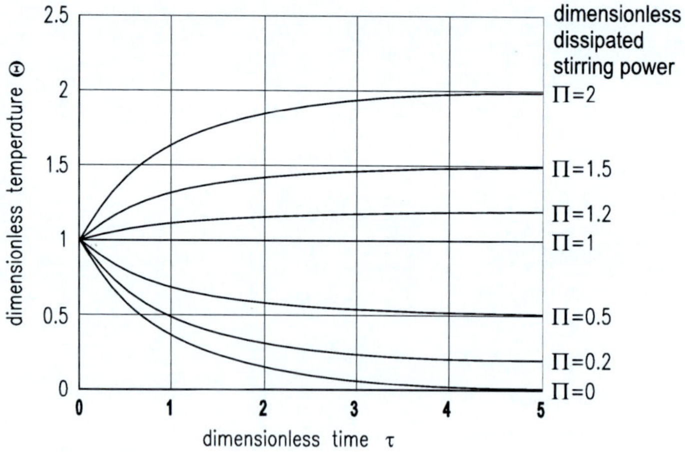

Fig. 4.2-6 Temperature evolution in a continuously operated tank with input of agitator power (Π = normalized dimensionless agitator power)

$$\Pi = \frac{P_{\text{agitator}}}{\dot{M} \cdot c_p \cdot (T_0 - T_{\text{in}})}. \tag{4.2-23}$$

With the initial condition $\Theta(\tau = 0) = 1$ the solution of (4.2-20) is the following relation between temperature Θ and time τ:

$$\Theta = \Pi - (\Pi - 1) \cdot \exp(-\tau). \tag{4.2-24}$$

This dependence is illustrated in Fig. 4.2-6 with several values of Π as a parameter. With the parameter value $\Pi = 0$ the result is an exponential decay of the temperature $\Theta = 0$.

4.2.2.4 Exercise: Isothermal Evaporation of Water

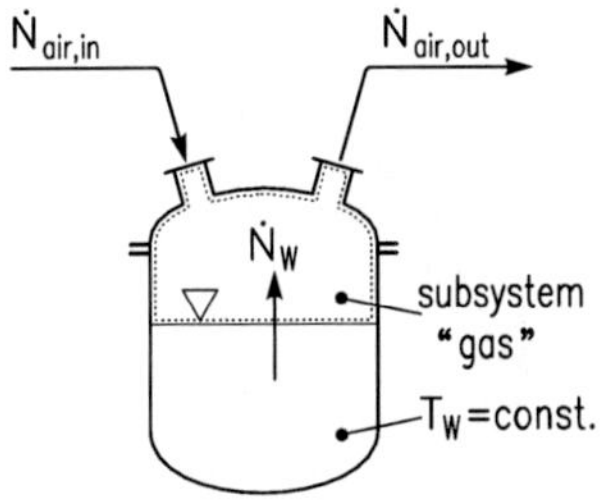

Fig. 4.2-7 Isothermal evaporation

Let the water surface in the tank be adjusted at temperature value T_W and assume a dry air flow ($\dot{N}_{\text{air,in}}; \tilde{Y}_{\text{in}} = 0$) circulating along the surface, slowly enough that the vapor flow leaving the tank is fully saturated with water. The pressure inside the tank is p. The vapor pressure of the water can be expressed by the *Antoine equation* (Fig. 4.2-7):

$$\log p_W^0 = A - \frac{B}{T + C}. \tag{4.2-25}$$

For water the Antoine-parameters occurring in this equation are tabulated as follows (p^0 in mbar and T in °C):

A	8.19625
B	1730.630
C	233.426

Question: What is the dependence of the evaporation rate $\dot{N}_W$ on the air flow rate $\dot{N}_{\text{air}}$ at temperature T_W?

Solution: Concerning the system "tank" only the subsystem "gas" needs to be evaluated. The gas shows steady-state behavior, i.e., its state does not change with time. Application of (4.2-1) yields the component balance for the air in the subsystem "gas":

$$\frac{dN_{\text{air}}}{dt} = 0 = \dot{N}_{\text{air,in}} - \dot{N}_{\text{air,out}}. \tag{4.2-26}$$

The component balance for the water in the gaseous phase reads

$$\frac{dN_{G,W}}{dt} = 0 = \dot{N}_{air,in} \cdot \underbrace{\tilde{Y}_{in}}_{=0} - \dot{N}_{air,out} \cdot \tilde{Y}_{out} + \dot{N}_W \tag{4.2-27}$$

$$\dot{N}_W = \dot{N}_{air,out} \cdot \tilde{Y}_{out} \, . \tag{4.2-28}$$

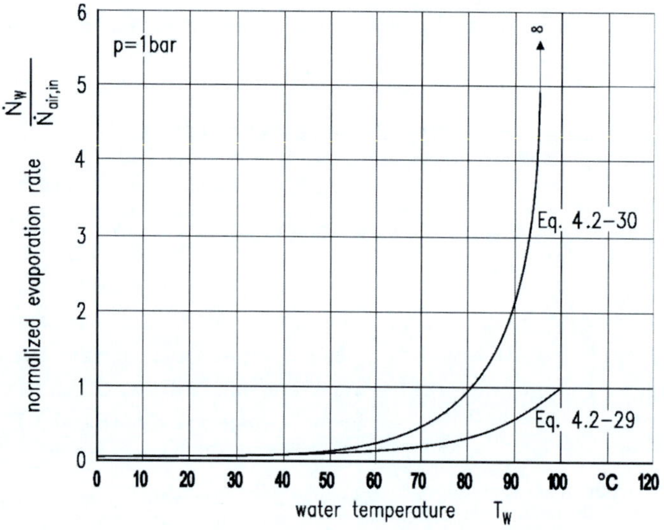

Fig. 4.2-8 Evaporation flow evolution vs. the liquid temperature under the conditions of isothermal evaporation

For temperatures notably below the boiling temperature of water ($T_W < 50°C$) the following relation holds:

$$\frac{\dot{N}_W}{\dot{N}_{air,in}} = \tilde{Y}_{out} \approx \tilde{y}_{out} = \frac{p_W}{p} \, . \tag{4.2-29}$$

Under the presumed condition of having reached a perfect equilibrium, and as the liquid is a pure substance, the partial pressure p_W appears to equal the vapor pressure $p_W^0(T = T_W)$. For temperatures near the boiling point of water, the simplified equation, (4.2-29) is no longer applicable and rather the following holds:

$$\frac{\dot{N}_W}{\dot{N}_{air,in}} = \tilde{Y}_{out} = \frac{\tilde{y}_{out}}{1-\tilde{y}_{out}} = \frac{p_W/p}{1-p_W/p} \, . \tag{4.2-30}$$

The resulting evolution of the normalized evaporation rate $\dot{N}_W/\dot{N}_{air,in}$ is illustrated in Fig. 4.2-8.

4.2.2.5 Exercise: Balancing a Crystallization Facility

This example elucidates how to systematically and swiftly perform balances of complex plants at steady-state conditions. Of key importance is the linearity of the governing equations describing the process. Due to linearity, which may be found as well for many similar tasks, the governing equations allow for a closed, analytical solution. An analogous problem is presented in Sect. 4.3.3.5 as regards the calculation of a heat exchanger.

Given is a typical connection of a mixing element M with an evaporative crystallizer C as illustrated in Fig. 4.2-9, showing also a filter F for liquid/solid separation and a splitter S. The splitter S is provided to limit the enrichment of contaminant in the system. In the mixer M the feed is mixed with the recycle from the splitter S. The flows are numbered consecutively by $j = 0–7$. The flows $j = 1–7$ are unknown and shall be resolved with respect to rate and composition.

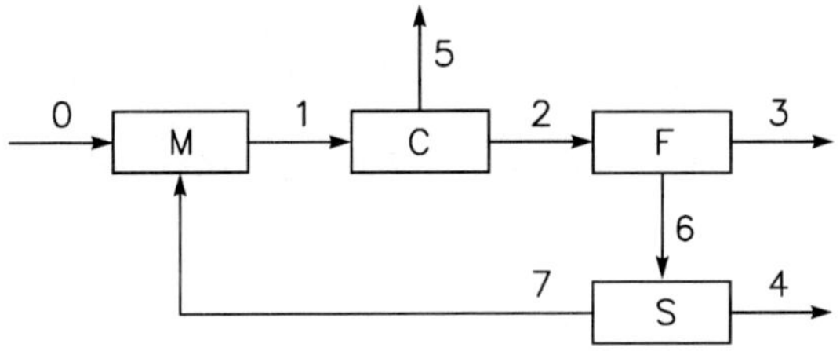

Fig. 4.2-9 Schematic of a crystallization plant consisting of : M mixer, C crystallizer, F filter, S splitter; in- and outbound flows are: 0 feed, 3 crystals, 4 bleed, 5 vapor

Let the feed $j = 0$ have the value $\dot{M}_0$ and assume it to be a mixture of three components i, i.e., salt (1), contaminant (2), and water (3). The composition of the inflow ($x_{i,0}$, for $i = 1,...,3$) shall be known.

For the solution, the following component flows shall be considered:

$$\dot{M}_{ij} = \dot{M}_j \cdot x_{ij}$$

There are 3 × 8 = 24 component flows in this example. The 3 component flows $\dot{M}_{i0}$ are given. Thus 21 component flows are unknown. The following 21 independent mathematical expressions can be raised:

- Twelve component balances (four elements with three components each; application of (4.2-1))

Mixer M: $-\dot{M}_{i0} = -\dot{M}_{i1} + \dot{M}_{i7}$ B1, B2, B3

Crystallizer C: $0 = \dot{M}_{i1} - \dot{M}_{i2} - \dot{M}_{i5}$ B4, B5, B6

Filter F: $0 = \dot{M}_{i2} - \dot{M}_{i3} - \dot{M}_{i6}$ B7, B8, B9

Splitter S: $0 = \dot{M}_{i6} - \dot{M}_{i4} - \dot{M}_{i7}$ B10, B11, B12

- Nine functionalities (operating conditions, boundary conditions, material properties, ...)

These functionalities result from the presumed material and plant behavior as discussed in the following.

First, assume that salt ($i = 1$) as well as contaminant ($i = 2$) are high boiling and do not evaporate:

Crystallizer C: $0 = \dot{M}_{15}$ F1

$0 = \dot{M}_{25}$ F2

Second, the filter should be regarded as flawless performing, i.e., it separates all emerged solid as a perfectly dry and pure filter cake:

Filter F: $0 = \dot{M}_{23}$ F3

$0 = \dot{M}_{33}$ F4

Furthermore, assume the splitter S to divert a fraction $\alpha = \dot{M}_4/\dot{M}_6$ as bleed. This fraction is an operating parameter and subject to choice. The inbound and outbound flows of the splitter have the same composition ($x_{i6} = x_{i4} = x_{i7}$), thus it follows:

Splitter S: $0 = \dot{M}_{i4} - \alpha \cdot \dot{M}_{i6}$ F5, F6, F7

Finally, the salt fractions in the liquid flows entering and leaving the filter have to be evaluated. Let both liquid flows be saturated with salt. Thus Stream 6 leaving the filter has the saturation concentration $x_{16} = x^*$. Usually this quantity depends on the solution temperature, composition, and contaminant. The crystallization temperature inside the evaporative crystallizer can be adjusted by the pressure in the crystallizer. The salt concentration x_{12} in Stream 2 leaving the crystallizer exceeds the saturation concentration due to the crystals contained in addition to the dissolved salt. The crystal fraction is an adjustable operation parameter. Thus, the value x_{12} is a parameter which describes the content of solid in the crystallizer. Thus, $x_{16} = x^*$ and x_{12} are the two functionalities that have been lacking so far:

Suspension after C: $x_{12} = \dot{M}_{12}/(\dot{M}_{12} + \dot{M}_{22} + \dot{M}_{32})$,

or $$0 = (1 - x_{12})\dot{M}_{12} - x_{12}\dot{M}_{22} - x_{12}\dot{M}_{32} \qquad \text{F8}$$

and saturated solution: $x_{16} = \dot{M}_{16}/(\dot{M}_{16} + \dot{M}_{26} + \dot{M}_{36}) = x^*$

or $$0 = (1 - x^*)\dot{M}_{16} - x^*\dot{M}_{26} - x^*\dot{M}_{36} \qquad \text{F9}$$

These 21 equations B1 to B12 and F1 to F9 are a system of linear equations, which can be posed in the form

$$\mathbf{A}\vec{\dot{M}} = \vec{\dot{F}} \qquad (4.2\text{-}31)$$

$\mathbf{A} =$

	11	12	13	14	15	16	17	21	22	23	24	25	26	27	31	32	33	34	35	36	37
	1	2	3	4	5	6	7	8	9	10	11	12	13	14	15	16	17	18	19	20	21
B1							1														
B2								−1						1							
B3															−1						1
B4	1	−1			1																
B5								1	−1			−1									
B6															1	−1			−1		
B7		1	−1			−1															
B8									1	−1			−1								
B9																1	−1			−1	
B10				−1		1	−1														
B11											−1		1	−1							
B12																		−1		1	−1
F1					1																
F2												1									
F3										1											
F4																	1				
F5				1		$-\alpha$															
F6											1		$-\alpha$								
F7																		1		$-\alpha$	
F8		$1-x_{12}$							$-x_{12}$							$-x_{12}$					
F9						$1-x^*$							$-x^*$					$-x^*$			

(note: all $\mathbf{A}_{ij} = 0$, except if value is given)

$$\vec{\dot{M}} = \begin{pmatrix} \dot{M}_{11} \\ \dot{M}_{12} \\ \dot{M}_{13} \\ \cdot \\ \cdot \\ \cdot \\ \cdot \\ \dot{M}_{17} \\ \dot{M}_{21} \\ \dot{M}_{21} \\ \dot{M}_{23} \\ \cdot \\ \cdot \\ \cdot \\ \cdot \\ \dot{M}_{27} \\ \dot{M}_{31} \\ \dot{M}_{32} \\ \dot{M}_{33} \\ \cdot \\ \cdot \\ \cdot \\ \cdot \\ \dot{M}_{37} \end{pmatrix} \quad \text{and} \quad \vec{\dot{F}} = \begin{pmatrix} -\dot{M}_{10} \\ 0 \\ 0 \\ \cdot \\ \cdot \\ \cdot \\ \cdot \\ 0 \\ -\dot{M}_{20} \\ 0 \\ 0 \\ \cdot \\ \cdot \\ \cdot \\ \cdot \\ 0 \\ -\dot{M}_{30} \\ 0 \\ 0 \\ \cdot \\ \cdot \\ \cdot \\ \cdot \\ 0 \end{pmatrix}$$

and is resolved for $\vec{\dot{M}}$ by expansion of both sides of the matrix equation by left sided multiplication with the inverse matrix $\mathbf{A}^{-1}$. With $\mathbf{A}^{-1}\mathbf{A} = \mathbf{E}$ being identical with the unit matrix the searched-for vector of component flows is directly obtained:

$$\vec{\dot{M}} = \mathbf{A}^{-1}\vec{\dot{F}}. \tag{4.2-32}$$

Both the inversion of a matrix and the multiplication of a vector by a matrix involve a great plenty of elementary calculations, for which reason they are conveniently achieved by a computer. Already general-use spreadsheet applications like Excel provide high-level commands *MINV* for matrix inversion and *MMULT* for matrix multiplication.

Table 4.2-1 shows a worked example for which it is assumed that the crystallization plant is supplied with a feed (1 t/h) of aqueous solution, which contains

50 mass percent of a salt and 1 mass percent of a contaminant. Let the solubility of the salt x^* be 0.5 ($\equiv x_{16}$). The salt fraction x_{12} after the crystallizer shall be 0.7. The resultant fraction of solid amounts to 20 mass percent and is extracted in the filter. The discharge ratio is given with $\alpha = 0.1$. For the current example, as apparent from Table 4.2-1 and Fig. 4.2-9, the feed flow has a contaminant fraction of of only 1 mass percent. Despite the designated discharge ratio with the α-value of only 0.1, the contamination fraction in the system is almost 15 mass percent. This may turn out to spoil the quality of the crystalline product.

Along the current example comprising a complex, interlinked assembly of process elements, a general method of proceeding is presented, which is applicable to a vast number of assignments. This procedure typically yields well-structured and robust application programmes, as no numerical iterations are involved and the solution is achieved in closed form. Not only the component flows as performed in this example are assessable, but also virtually any task involving large-scale calculations can be accomplished as long as the problem is traceable to a set of linear equations. Typical applications are complex heat exchangers or networks of heat exchangers. In case of nonlinear equation systems the presented program structure may be incorporated as substructure of an iterative solution strategy to the nonlinear problem. This is realized for example within the simulation of complex columns after, i.e., *Thiele and Gaddes*, which in addition account for the energy balance. As mentioned before, the complete set of equations is named MESH-equations (**M**ass, **E**quilibrium, **S**ummation, **H**eat).

Table 4.2-1 Crystallization plant with recycle. Results for $\alpha = 0.5$, $x_{12} = 0.7$ and $x_{16} = x^* = 0.5$. Salt: i=1; impurity: i=2; water: i=3

stream	component flow $\dot{M}_{ij}/t/h$			total flow $\dot{M}_j/t/h$	composition $x_{ij}/-$		
j	$\dot{M}_{1j}$	$\dot{M}_{2j}$	$\dot{M}_{3j}$	$\dot{M}_j$	x_{1j}	x_{2j}	x_{3j}
1	0.81	0.10	0.71	1.63	0.500	0.061	0.439
2	0.81	0.10	0.25	1.16	0.700	0.086	0.214
3	0.47	0.00	0.00	0.46	1	0	0
4	0.03	0.10	0.02	0.07	0.500	0.143	0.357
5	0.00	0.00	0.46	0.46	0	0	1
6	0.35	0.10	0.25	0.70	0.500	0.143	0.357
7	0.31	0.09	0.22	0.63	0.500	0.143	0.357

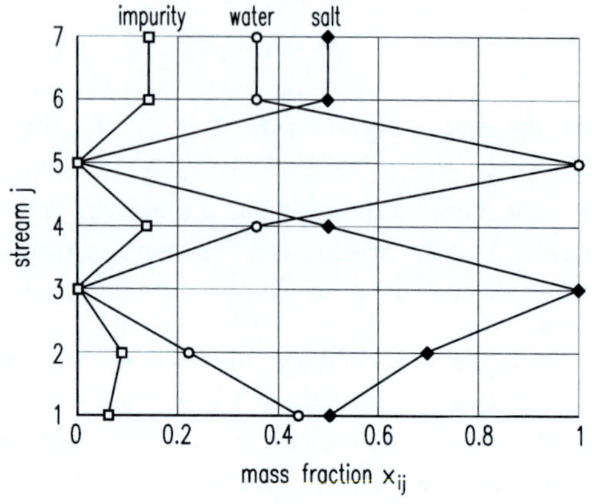

Fig. 4.2-9 Crystallization plant with recycle. Graphical representation of the data given in Table 4.2-1

4.3 Heat and Mass Transfer

4.3.1 Kinetics

The equation of change of a conserved quantity of a system, (4.2-1), comprises the flow terms $\sum \dot{Z}_{\text{in}}$ and $\sum \dot{Z}_{\text{out}}$. In the five aforementioned exercises, these were macroscopic flows (in- and outlet flows, mechanical power).

In the following, flows that are caused by molecular processes or by radiation shall be considered. The flows driven by molecular processes are heat conduction and molecular diffusion.

The fundamentals to describe the kinetics of molecular exchange processes come from Fourier (1768–1830) for the heat flux along a temperature gradient

$$\dot{q} = -\lambda \cdot \frac{\mathrm{d}T}{\mathrm{d}s} \tag{4.3-1}$$

and from Fick (1829–1901) for the component flux along a concentration gradient

$$j_i = -D_{ij} \cdot \frac{\mathrm{d}c_i}{\mathrm{d}s} \tag{4.3-2}$$

or

$$\tilde{j}_i = -D_{ij} \cdot \frac{d\tilde{c}_i}{ds} . \tag{4.3-3}$$

In these equations, λ is the thermal conductivity and D_{ij} is the binary diffusion coefficient of component i in medium j.

Because diffusional fluxes j_i can cause convective fluxes $\dot{k}_i$, both may have to be considered for the resulting mass flux $\dot{m}_i$

$$\dot{m}_i = j_i + \dot{k}_i = -D_{ij} \cdot \frac{dc_i}{ds} + \dot{m}_{tot} \cdot x_i . \tag{4.3-4}$$

The same applies for the molar flux $\dot{n}_i$

$$\dot{n}_i = \tilde{j}_i + \dot{\tilde{k}}_i = -D_{ij} \cdot \frac{d\tilde{c}_i}{ds} + \dot{n}_{tot} \cdot \tilde{x}_i . \tag{4.3-5}$$

Besides by conduction, heat can be exchanged between body 1 and 2 by radiation:

$$\dot{Q}_{12} = c_{12}(\sigma, \varepsilon_1, \varepsilon_2, \varphi_{12}) \cdot A_1 \cdot (T_1^4 - T_2^4) . \tag{4.3-6}$$

The radiative heat flux depends on the radiation exchange factor c12, which is a function of the Stefan-Boltzmann constant “sigma”, the emissivities “epsilon1” and “epsilon2” of the concernde bodies, and of the view factor “phi12”, and it depends on the radiation area A1 of body 1, and it depends on the difference of the 4th power of the absolute temperatures T1 and T2 of the involved bodies.

In the case of material exchange in the gas phase at low pressure or in small cavities and pores, the gas molecules collide more often with the enclosing walls than with other gas molecules, i.e., the mean free path length is longer than the distances to be travelled. In this case, the mass flux depends linearly on the mean velocity of the molecules w_i and on the concentration difference $\Delta\tilde{c}_i$, see Sect. 3.1.7:

$$\dot{n}_i = \frac{1}{4} \cdot w_i \cdot \Delta\tilde{c}_i . \tag{4.3-7}$$

The use of the above equations is often too complex for practical application. Instead heat transfer coefficients α and mass transfer coefficients β are introduced. Using these coefficients, heat and mass fluxes are to be calculated by the following equations:

$$\dot{q} = \alpha \cdot \Delta T \tag{4.3-8}$$

$$\dot{m}_i = \beta_i \cdot \Delta c_i \text{ respectively } \dot{n}_i = \beta_i \cdot \Delta \tilde{c}_i \,. \tag{4.3-9}$$

The Nusselt number is a dimensionless representation of the heat transfer coefficient:

$$Nu \equiv \frac{\alpha \cdot L}{\lambda}, \tag{4.3-10}$$

and the Sherwood number is a dimensionless representation of the mass transfer coefficient:

$$Sh \equiv \frac{\beta \cdot L}{D}. \tag{4.3-11}$$

Here, L is a length, which characterizes the geometry under consideration.

In the case of heat or material transfer in flowing media the dependency of the Nu-and the Sh-number on flow characteristics and material properties is often given in the following form:

$$Nu = C \cdot Re^m \cdot Pr^n \quad \text{respectively} \quad Sh = C \cdot Re^m \cdot Sc^n \tag{4.3-12}$$

with

$$Re = \frac{w \cdot L \cdot \rho}{\eta} \quad \text{Reynolds number,} \tag{4.3-13}$$

$$Pr = \frac{\eta \cdot c}{\lambda} \quad \text{Prandtl number,} \tag{4.3-14}$$

$$Sc = \frac{\eta}{D \cdot \rho} \quad \text{Schmidt number.} \tag{4.3-15}$$

Implementing (4.3-13) to (4.3-15) in (4.3-12) yields

$$Nu \equiv \frac{\alpha \cdot L}{\lambda} = C_1 \cdot \left(\frac{w \cdot L \cdot \rho}{\eta}\right)^m \cdot \left(\frac{\eta \cdot c}{\lambda}\right)^n \tag{4.3-16}$$

and

$$Sh \equiv \frac{\beta \cdot L}{D} = C_1 \cdot \left(\frac{w \cdot L \cdot \rho}{\eta}\right)^m \cdot \left(\frac{\eta}{\rho \cdot D}\right)^n. \tag{4.3-17}$$

The division of both (4.3-16) and (4.3-17) yields the following relation between the heat transfer coefficient α and the mass transfer coefficient β.

$$\frac{\alpha}{\beta} = \frac{\lambda}{D} \cdot \left(\frac{D \cdot \rho \cdot c}{\lambda}\right)^n . \tag{4.3-18}$$

This relation is very useful, because it allows to predict a mass transfer coefficient β if the heat transfer coefficient α is known and vice versa.

The characteristic $\lambda/(\rho \cdot c \cdot D)$ is indicated as Lewis number *Le*. The last equation can be written as follows:

$$\frac{\alpha}{\beta \cdot \rho \cdot c} = Le^{1-n} . \tag{4.3-19}$$

Table 4.3-1 contains Lewis numbers for several systems. The system water steam in air plays a big role in cooling tower, drying and air conditioning technologies, and its Lewis number is close to unity.

Table 4.3-1 Values of the Lewis number (0°C, 1 bar)

		Le
Water steam	in air	0.937
CO_2	in air	1.30
Benzene	in air	2.48
Water steam	in H_2	1.98
CO_2	in H_2	2.47
Benzene	in H_2	4.65

The dependence of exponent *n* on the type of flow is given in Table 4.3-2. A good example for the use of these data is the case of convective drying, where turbulent drying air flows past a wet surface and forms a laminar boundary layer. In this case $n = 1/3$ and

$$\alpha = \beta \cdot \rho \cdot c \cdot \left(\frac{\lambda}{\rho \cdot c \cdot D}\right)^{2/3} . \tag{4.3-20}$$

Table 4.3-2 Exponent n of the Lewis number depending on flow condition

Flow type	Exponent n	$\frac{\alpha}{\beta}$
Pure laminar	0	$c_p \cdot \rho \cdot Le$
Turbulent with laminar boundary layer	$1/3$	$c_p \cdot \rho \cdot Le^{2/3}$
Starting process in frictionless flow	$1/2$	$c_p \cdot \rho \cdot Le^{1/2}$
Turbulent flow	Experimental 0.42	$c_p \cdot \rho \cdot Le^{0,58}$
Frictionless flow ($Re \to \infty$)	1	$c_p \cdot \rho$

Using (4.3-20) it is possible to calculate the heat transfer coefficient α by using the mass transfer coefficient β which can be determined by gravimetric measurement.

The following section renders correlations, with which it is possible to determine heat and mass transfer coefficients for some cases significant to thermal process engineering.

A reliable literature source for heat and mass transfer coefficients is the VDI-Heat Atlas (2010).

4.3.2 Heat and Mass Transfer Coefficients

For better clarity, only correlations for α-values or Nu-numbers are given in the following sections. With (4.3-19) β-values can be derived and Sh-correlations can then be derived from Nu-correlations by exchanging the Prandtl number Pr with the Schmidt number Sc.

4.3.2.1 Heat and Mass Transfer at Forced Convection

If a fluid flows past a plate at bulk velocity w, then close to the surface a velocity, temperature and concentration profile develops. These profiles are approximated by laminar boundary layers, of characteristic thickness. In these boundary layers the resistances to momentum, heat and mass transfer are located. As long as the Reynolds number $Re = w \cdot L \cdot \rho / \eta$, calculated with plate length L, is smaller than 10^6, the flow is laminar.

Solving the system of equations for conservation of energy, momentum, and mass for the longitudinal flow past a plate at constant wall temperature results in the following equation for the heat transfer coefficient (Pohlhausen 1921):

$$Nu_L = \frac{\alpha \cdot L}{\lambda} = 0.664 \cdot \left(\frac{w \cdot L \cdot \rho}{\eta}\right)^{1/2} \cdot \left(\frac{\eta \cdot c}{\lambda}\right)^{1/3}. \tag{4.3-21}$$

In case of pipe flow, the pipe diameter d has to be chosen as characteristic length in the definition of the Nu - and the Re-numbers:

Nusselt number $Nu = \dfrac{\alpha \cdot d}{\lambda}$, (4.3-22)

Reynolds number $Re = \dfrac{w \cdot d \cdot \rho}{\eta}$. (4.3-23)

For pipe flow, which is neither developed concerning flow nor temperature profile, and which shows negligible viscosity differences near the wall, the following is valid (Stephan 1959):

$$Nu = 3.66 + \frac{0.664 \cdot (Re \cdot Pr \cdot d/L)^{1.33}}{1 + 0.1 \cdot Pr \cdot (Re \cdot d/L)^{0.83}}. \tag{4.3-24}$$

This equation shows simplified solutions for pipes which are either very long or very short. Regarding long pipes $(d/L \to 0)$, the profiles for velocity, temperature respectively concentration are fully developed and

$$Nu = 3.66 \tag{4.3-25}$$

is obtained. On the other hand, if the pipe is very short $(d/L \to \infty)$, the Nu-number is evaluated by

$$Nu = 0.664 \cdot Re^{1/2} \cdot Pr^{1/3}. \tag{4.3-26}$$

This correlation corresponds to the equation for the laminar flow past a plate.

But, if the pipe flow is developed concerning fluid dynamics but not concerning temperature or concentration profile, then the following correlation for short pipes holds (Stephan 1959; Hausen 1959):

$$Nu = 1.62 \cdot (Re \cdot Pr \cdot d/L)^{1/2}. \tag{4.3-27}$$

In this case, heat transfer depends on $Re \cdot Pr \cdot d/L = Pe \cdot d/L$, which dos not comprise the kinematic viscosity. The product $Re \cdot Pr$ is called Peclet number.

In Fig. 4.3-1 the Nusselt number is depicted as function of $Re \cdot Pr \cdot d/L$ respectively $Re \cdot Sc \cdot d/L$. The highest transfer coefficients are attained for frictionless fluids ($\nu = 0$ or $Pr = Sc = 0$). In case of developed laminar flow the transfer coefficients α and β only depend on the pipe diameter and the material properties. If the flow is not developed heat transfer increases with the flow velocity by d/L.

If a certain Reynolds number is exceeded, the flow is fully turbulent.

The average heat transfer coefficient of pipe flow at $Re = w \cdot d \cdot \rho/\eta \geq 10^4$ can be determined from the following equation, given by V. Gnielinski in VDI – Heat Atlas (2010)

$$Nu = \frac{(\xi/8) \cdot Re \cdot Pr}{1 + 12.7 \cdot \sqrt{\xi/8} \cdot \left(Pr^{2/3} - 1\right)} \left[1 + \left(d_i/L\right)^{2/3}\right], \tag{4.3-28}$$

with the friction factor $\xi = \left(1.8 \cdot \log_{10} Re - 1.5\right)^{-2}$ for turbulent flow in smooth pipes and L being the length of the pipe.

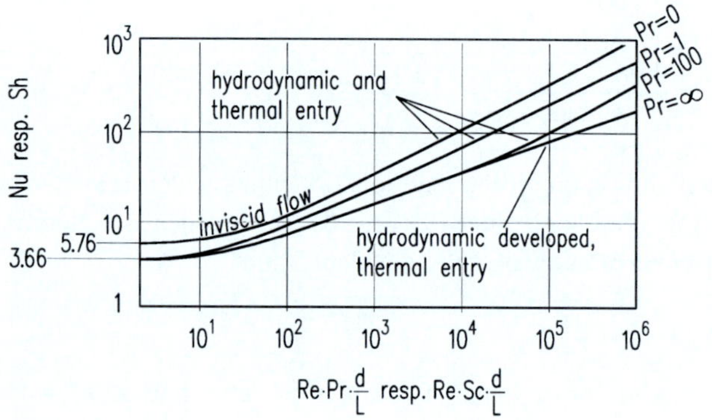

Fig. 4.3-1 Nusselt and Sherwood number depending on the inlet characteristic $Re \cdot Pr \cdot d/L$ respectively $Re \cdot Sc \cdot d/L$ for several flow regimes

The heat transfer coefficient for flow past a plane wall at $Re_L = w \cdot L \cdot \rho/\eta \geq 10^6$ with L being the length of the pane wall can be determined from the following equation, also given by V. Gnielinski in VDI-Heat Atlas (2010):

$$Nu = \frac{0.037 \cdot Re_L^{0.8} \cdot Pr}{1 + 2.443 \cdot Re_L^{-0.1} \cdot \left(Pr^{2/3} - 1\right)} \tag{4.3-29}$$

Krischer and Kast 1978 and Kast et al. 1974 found, that equations for the flow past a plane wall are well applicable for bodies of almost any form, as long as the average length of the stream line is chosen to be the characteristic length. For instance is d_p the characteristic length of a spherical particle.

4.3.2.2 Heat and Mass Transfer in Particulate Systems

Flow-through particulate systems occur in many apparatus of process engineering. During drying, heat is transferred towards the solid particles that have to be dried. In crystallizers the crystallizing substance migrates towards the particles, where the heat of crystallization has to be supplied or discharged. In many apparatus for distillation, absorption and fluid/fluid extraction there exist dispersed two-phase systems, in which fluid particles like bubbles and drops move through the continuous phase. In all these cases the transfer coefficients between the continuous and dispersed phase have to be known for apparatus design.

In the following it is demonstrated how heat and mass transfer correlations for pipe flow may be transferred to particulate systems. In Fig. 4.3-2 a particulate system with the particle diameter d_p and the overflown length $L' = d_\mathrm{p}$ is illustrated. The following is about a fixed or a fluidized bed, through which a fluid flows at volume flux $\dot{v}_\mathrm{c}$.

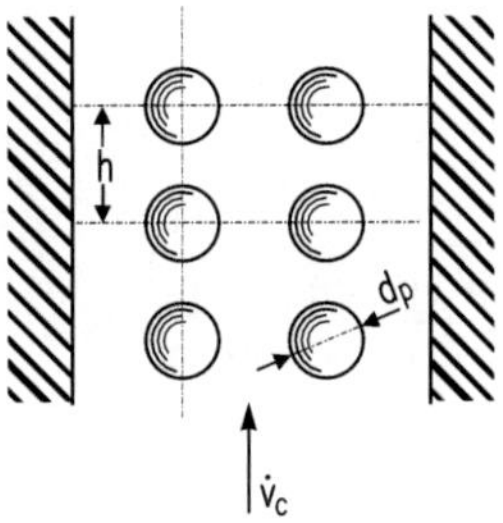

Fig. 4.3-2 Characterizing dimensions of a particulate system

In particular in drying technology, numerous experiments found that the use of a special hydraulic diameter is appropriate (Krischer and Kast 1978; Kast et al. 1974):

$$d_\mathrm{h}^* = d_\mathrm{h} \cdot \frac{L'}{h} . \qquad (4.3\text{-}30)$$

The hydraulic diameter d_h was defined in Chap. 3. If h denotes the mean distance of the particles in flow direction, then the following equation can be applied for randomly arranged systems of particle volume V_P:

$$h = \sqrt[3]{\frac{V_\mathrm{P}}{\varepsilon_\mathrm{d}}} = \sqrt[3]{\frac{\pi}{6 \cdot \varepsilon_\mathrm{d}}} \cdot d_\mathrm{p} . \qquad (4.3\text{-}31)$$

The hydraulic diameter related to the overflown length is then only a function of the volume fraction of the dispersed and the continuous phase:

$$\frac{d_h^*}{L'} = \sqrt[3]{\frac{16 \cdot \varepsilon_c^3}{9 \cdot \pi \cdot \varepsilon_d^2}} \quad \text{or} \quad d_h^* = \sqrt[3]{\frac{16 \cdot \varepsilon_c^3}{9 \cdot \pi \cdot \varepsilon_d^2}} \cdot d_p . \tag{4.3-32}$$

The relative velocity between the dispersed and the continuous phase decides upon the value of the transfer coefficients. This relative velocity is $w_{rel} = \dot{v}_c / \varepsilon_c$. Furthermore the hydraulic diameter is used as characterizing length. The following characteristics are introduced for particulate systems:

$$Nu_{d_h^*} = \frac{\alpha \cdot d_h^*}{\lambda}, \tag{4.3-33}$$

$$Pe_{d_h^*} = Re_{d_h^*} \cdot Pr = \frac{\dot{v}_c \cdot d_h^* \cdot \rho \cdot c}{\varepsilon_c \cdot \lambda}, \tag{4.3-34}$$

$$Re_{d_h^*} = \frac{\dot{v}_c \cdot d_h^* \cdot \rho}{\varepsilon_c \cdot \eta}. \tag{4.3-35}$$

The heat and mass transfer correlations for pipe flow can be applied for particulate systems, if the ratio $d_h^*/(L')$ and a corresponding parameter for pipe flow are known (Schlünder 1972; Krischer and Kast 1978; Kast et al. 1974).

Figure 4.3-3 illustrates this diagram for parameter allocation, which was created on the basis of numerous drying experiments. A particle's ratio $d_h^*/(L')$ is illustrated depending on the ratio $(d/L)_R$ of the diameter by the length of the pipe. Therewith the equations for heat and mass transfer for pipe flow can also be used for particulate systems. If the volume fractions for the continuous and the dispersed phase are known, the ratio $d_h^*/(L')$ can be calculated and in addition the ratio diameter by length can be determined using Fig. 4.3-3.

This calculation method was mainly developed for the heat transfer from air ($Pr \approx 0.7$) to goods that have to be dried. It is applicable for fixed and fluidized beds.

In the case of fluid dispersed two-phase systems the relations are often very different in comparison to systems with solid particles, see Wesselingh and Bollen (1999). Fluid particles can deform, divide, and aggregate. Sometimes there is an inner circulation.

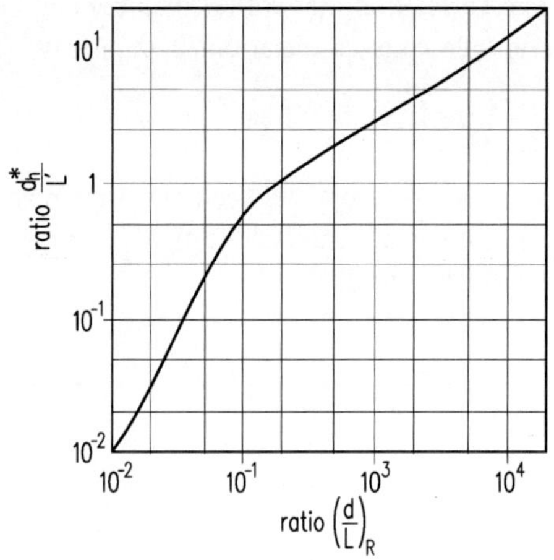

Fig. 4.3-3 Parameter allocation diagram. This diagram allows to use results of the heat and mass transfer at pipe flow for particle systems

4.3.2.3 Heat and Mass Transfer at Natural Convection

Natural convection is caused by density differences $\Delta\rho$, which result from temperature differences or from concentration differences. The characteristic length is the contact length, i.e., the height of a vertical wall, half of the circumference of a vertical cylinder or half of the circumference of a sphere. The volume related buoyancy energy, that is caused by the density difference $\Delta\rho$ in the gravity field of the earth, is transformed into kinetic energy:

$$L \cdot \Delta\rho \cdot g \sim \frac{w^2 \cdot \rho_c}{2} . \qquad (4.3\text{-}36)$$

By extension of both terms with $\rho_c \cdot L^2/\eta_c^2$, the Grashof number is obtained:

$$Gr = \frac{L^3 \cdot \rho_c \cdot \Delta\rho \cdot g}{\eta_c^2} \sim \frac{w^2 \cdot L^2 \cdot \rho_c^2}{2 \cdot \eta_c^2} \sim Re^2 ,$$

$$Gr = \frac{L^3 \cdot \rho_c \cdot \beta \cdot \Delta T \cdot g}{\eta_c^2} . \qquad (4.3\text{-}37)$$

Herein β is the thermal coefficient of volumetric expansion of the fluid.

Referring to the above correlation the Grashof number can be converted into the Reynolds number. The following was determined empirically (Schlünder 1970):

$$Gr = 2.5 \cdot Re^2 . \tag{4.3-38}$$

This conversion is necessary if natural convection is superposed by forced convection. It is recommended to use a combined Reynolds number according to the following equation (VDI-Heat Atlas 2010):

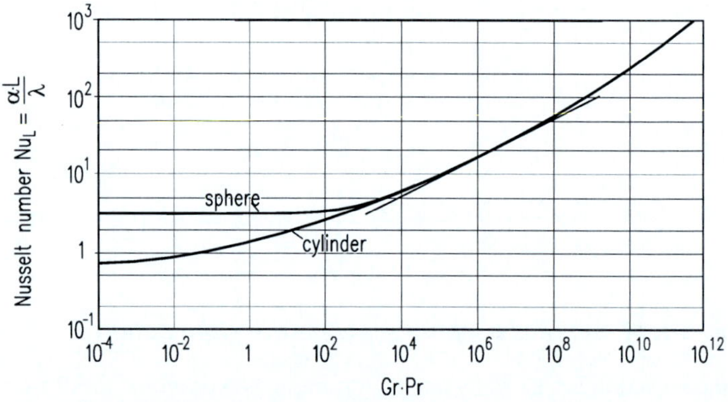

Fig. 4.3-4 Nusselt number depending on the product $Gr \cdot Pr$ for the sphere and the cylinder at natural convection. The illustrated straight line gives a theoretical solution (by Schmidt and Beckmann 1930) for a Prandtl number of 0.73

$$Re = \sqrt{Re_{\text{forced}}^2 + 0.4 \cdot Gr} . \tag{4.3-39}$$

In Fig. 4.3-4 the Nusselt number is given as a function of the product of Grashof and Prandtl number. The diagram comprises averaged curves for the sphere and the cylinder, as they result for small Prandtl numbers ($Pr < 1$). Note, that this diagram does not reproduce the dependency of the Nu-number on the Pr-number. The straight line renders a correlation that was set up on the basis of conservation laws for the case of the laminar boundary layer (Schmidt and Beckmann 1930). If the product of the Grashof and Prandtl number is higher than 2×10^9, then the flow becomes turbulent (Saunders 1936, 1939).

4.3.2.4 Heat Transfer in Fluidized Systems

As already discussed in Chap. 3, fluid flow in gas and liquid fluidized beds, as well as in bubble and drop columns, is governed by the product $\Delta\rho \cdot g$. In this case, the buoyancy energy is converted into kinetic energy. As shown in Chap. 3 the Archimedes number characterizes the flow. In contrast to the Grashof number, density differences do not result from differences in temperature or concentration, but stem from different phase densities.

The heat transfer coefficient between vertical heating and cooling surfaces and a dispersed system (gas and liquid fluidized beds, bubble and drop columns) has been determined experimentally. Often, a maximum heat transfer coefficient α_{max} is found at a certain volume flux $\dot{v}_c$ or $\dot{v}_d$ of the continuous respectively the dispersed phase (Mersmann and Wunder 1978). Here, only information about this maximum value is provided.

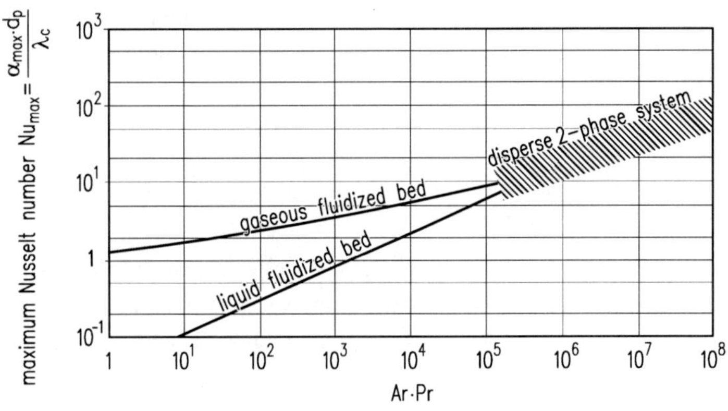

Fig. 4.3-5 Maximum Nusselt number depending on the product $Ar \cdot Pr$ for the gaseous fluidized bed, the liquid fluidized bed as well as for disperse two-phase systems (fluidized beds plus bubble and drop columns)

Figure 4.3-5 gives the Nusselt number as a function of the product of the Archimedes and the Prandtl number. In this case, the particle diameter is the characteristic length. The shaded area depicts experimental results from bubble and drop columns. The Prandtl number of the continuous phase has been varied over a wide range. The straight line represents experimental results that have been found for homogeneous liquid fluidized beds, in fact for Prandtl numbers between 7 and 14,000. Finally, the upper curve depicts test results for gaseous fluidized beds. In all cases only the maximum heat transfer coefficients have been considered. Basically, the transfer coefficients depend on the fluxes of the dispersed and continuous phase. In liquid dispersed two-phase systems a final value is attained, while in fluidized beds a maximum value is found.

4.3.2.5 Unsteady Heat and Mass Transfer

There are numerous processes during which the heat and mass flux changes with time, i.e., the drying rate decreases with time in the second period of drying, or, adsorption and desorption of an adsorbent grain is unsteady.

At steady state, the fluxes of heat and mass do not change with time. However, this does not mean that all the leveling processes of all material elements are steady-state processes. If, e.g., a newly formed bubble or drop may at first show high potential differences with respect to its environment then these differences decay while the bubble or drop moves in the spouted or sprayed bed.

Simple solutions can be given for two limiting cases. These limiting cases are distinguished either by very low or very high values of the Fourier number. The Fourier number is a nondimensional time. It is defined as follows:

$$Fo = \frac{\lambda \cdot t}{\rho \cdot c \cdot s^2} . \tag{4.3-40}$$

s is a characteristic length scale.

The Fourier number takes low values for short contact times. If the Fourier number is smaller than 0.1 the following solution for the heat transfer coefficient α is a good approximation for the heat exchange into a plate of semi-infinite thickness:

$$Nu_s = \frac{2}{\sqrt{\pi}} \cdot \sqrt{\frac{1}{Fo}} \quad \text{or} \quad \alpha = \frac{2}{\sqrt{\pi}} \cdot \sqrt{\frac{\lambda \cdot \rho \cdot c}{\tau}} . \tag{4.3-41}$$

This transfer coefficient α is the mean value for the time span $t = 0$ to τ, and is calculated from

$$\alpha = \frac{1}{\tau} \cdot \int_0^{\tau} \alpha_t \, dt . \tag{4.3-42}$$

The transfer coefficient is inversely proportional to the square root of the contact time τ.

For many unsteady processes, the contact time τ can be estimated with good accuracy. Often it is equal to the retention time of a fluid element in the contact region. In other cases it can be estimated from a characteristic velocity and length.

For long contact times and Fourier number values higher than 0.3 the following equation holds for the heat flux at the surface, if a plate of finite thickness s is considered which at $t = 0$ is at temperature T_α. At $t > 0$, the surface temperature is set to T_S:

$$\dot{q} = 4 \cdot \frac{\lambda}{s} \cdot (T_\alpha - T_S) \cdot \exp\left(-\frac{\pi^2 \cdot \lambda \cdot t}{\rho \cdot c \cdot s^2}\right) . \tag{4.3-43}$$

Here, the heat flux is proportional to the initial temperature difference and depends on the Fourier number.

The average temperature $\bar{T}$ can be calculated as follows:

$$\bar{T} - T_O = (T_\alpha - T_S) \cdot \frac{8}{\pi^2} \exp\left(-\frac{\pi^2 \cdot \lambda \cdot t}{\rho \cdot c \cdot s^2}\right) . \tag{4.3-44}$$

Using

$$\dot{q} = \alpha \cdot (\bar{T} - T_S), \tag{4.3-45}$$

the dimensionless transfer coefficient for long contact times can be deduced:

$$Nu_s = \frac{\pi^2}{2} \tag{4.3-46}$$

Summarizing, the transient heat transfer into or out off a plate can be quantified with the Nusselt and the Fourier number. Figure 4.3-6 illustrates the Nusselt number depending on the reciprocal of the Fourier number. The diagram comprises the discussed limiting cases and furthermore allows the calculation of data, which are not covered by the limiting cases.

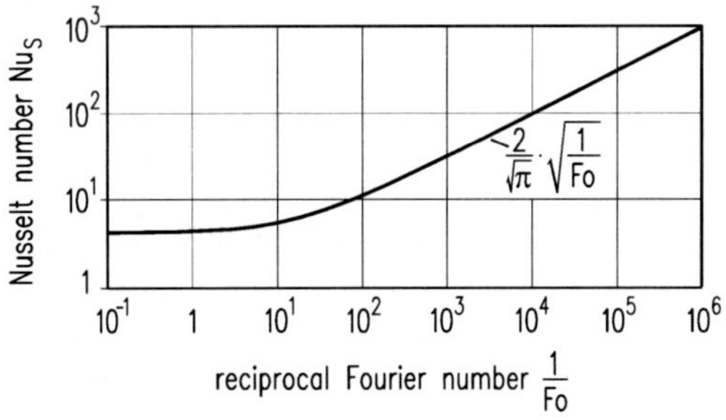

Fig. 4.3-6 Nusselt number vs. the reciprocal of the Fourier number for the transient heat transfer of a plate

4.3.2.6 Heat Transfer at Condensing Steams

Condensing steam on a cold surface either develops a closed film or drops. Although dropwise condensation delivers far higher heat transfer coefficients than film condensation, it cannot be maintained for long, because the drops coalesce into trickles.

At film condensation, the heat has to be transported through the film. If the film is laminar and the film surface is at thermodynamic equilibrium, then the heat is transferred by conduction. The transfer coefficient can be calculated, if the film thickness is known depending on the film length. In case of a laminar film the velocity profile is defined by an equilibrium between viscous and gravitational forces, see Chap. 3. Considering the conservation laws for mass and energy allows to derive the heat transfer coefficient on a theoretical basis.

At turbulent film flow it is recommended to use empirically determined correlations.

Circumstances are more complicated, if the steam has a noticeable velocity. Then a shear stress takes effect on the film surface, which thins the film and therewith enhances the heat transfer coefficient.

At film condensation, the heat transfer rate depends on the Prandtl number of the liquid film:

$$Pr_L = \frac{\eta_L \cdot c_{pL}}{\lambda_L}. \tag{4.3-47}$$

In the definition of the Nu -number, the characteristic length is chosen to be a function of the dynamic viscosity η_L and the gravitational acceleration g, because the thickness of the film depends on these quantities:

$$Nu = \frac{\alpha}{\lambda_L} \cdot \sqrt[3]{\frac{\eta_L^2}{\rho_L^2 \cdot g}}. \tag{4.3-48}$$

The flow regime of the film can be described with the aid of the Galilei number Ga, which is the same as the square of the Reynolds number, divided by the Froude number. Here, the characteristic length is the length L of the film, and the Galilei number is

$$Ga = \frac{L^3 \cdot g \cdot \rho_L^2}{\eta_L^2}. \tag{4.3-49}$$

After all, the transfer coefficient depends on the Phase Change number *Ph* which constitutes the ratio of sensible heat to latent heat:

$$Ph = \frac{c_{pL} \cdot (T_G - T_O)}{\Delta h_{LG}} . \tag{4.3-50}$$

If noticeable steam velocities w_G dominate, another parameter has to be considered, which constitutes the ratio of shear force of the steam to its gravitational force:

$$K_W = \xi \cdot \frac{\rho_G}{\rho_L} \cdot \frac{w_G^2}{\sqrt[3]{(\eta_L^2/\rho_L^2) \cdot g^2}} . \tag{4.3-51}$$

ξ is a drag coefficient of the steam flow and has a value of about 0.02 at turbulent flow.

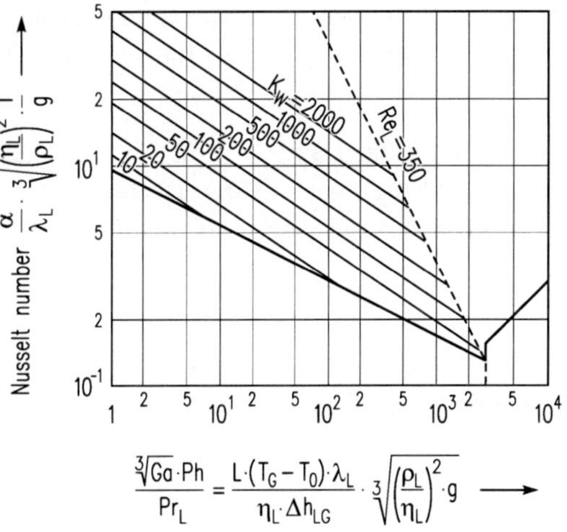

Fig. 4.3-7 Heat transfer at the condensation of pure steam. The lower straight line is valid for static steam

Figure 4.3-7 shows the Nusselt number depending on a combination of the Galilei, the Phase Change, and the Prandtl number with the flow parameter K_W as a parameter. It shows that the heat transfer coefficient increases with the steam velocity.

The equation for laminar film flow at vanishing steam velocity ($K_W = 0$), theoretically derived by Nusselt (1916), is

$$Nu \equiv \frac{\alpha}{\lambda_L} \cdot \sqrt[3]{\frac{\eta_L^2}{\rho_L^2 \cdot g}} = \frac{2\sqrt{2}}{3} \cdot \frac{\sqrt[4]{Pr_L}}{\sqrt[4]{Ga^{1/3} \cdot Ph}} \tag{4.3-52}$$

and is inscribed as limiting case in Fig. 4.3-7.

The proposed calculation method is only valid for the case of condensing steam without inert gas. If steams contain inert gases, an additional mass transfer resistance between the gas and the fluid phase appears, and the transfer coefficients degrade, see Chap. 7.

4.3.2.7 Heat Transfer at Evaporation of Pure Fluids

Heat transfer from a heating wall to a boiling liquid has to be divided into several boiling regimes. The first is convective boiling and takes place at low temperature differences between wall and liquid. Above a certain temperature difference bubbling appears at the heated wall and enhances the heat transfer coefficient (nucleate boiling). At even higher temperature differences, bubbles form so close to each other that they coalesce and build an insulating film of steam. This regime is called film boiling.

The three ranges can be distinguished in a diagram in which the heat flux is depicted depending on the temperature difference between the wall and the boiling temperature, see Fig. 4.3-8 top. This diagram is valid for water at 1 bar.

In the bottom of Fig. 4.3-8 the heat transfer coefficient is indicated depending on the temperature difference. In the range of convective boiling the heat flux and the heat transfer coefficient rise with the temperature difference. The heat transfer coefficient can be calculated with the correlations for natural or forced convection.

In range of nucleate boiling the heat flux $\dot{q}$ as well as the heat transfer coefficient α rise more sharply with the temperature difference as at convection boiling. In the maximum of the curve the maximum heat flux of $9 \times 10^5\ \mathrm{W/m^2}$ is reached at a temperature difference of about 30 K. Further increase of the temperature difference gives rise to the unstable region of film boiling. Here, the aforementioned steam film develops and constitutes a large heat transfer resistance. For this reason the driving temperature difference has to be raised significantly (in this case up to 800 K) to keep raising the heat flux. Prevalently, superheating destroys the heating surface (burn out).

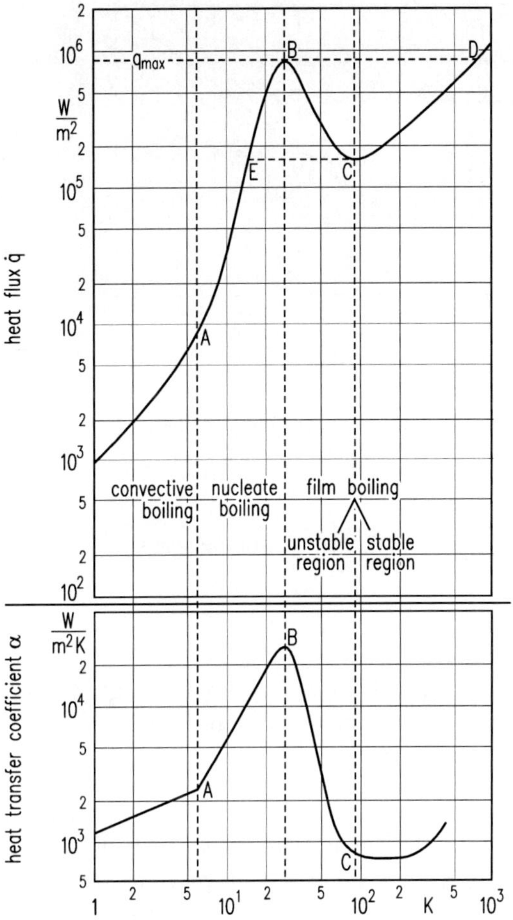

Fig. 4.3-8 Heat flux depending on temperature difference $T_O - T_L$ at evaporating water at 1 bar (above) and heat transfer coefficient depending on temperature difference $T_O - T_L$ at evaporating water at 1 bar (below)

Figure 4.3-9 illustrates the heat transfer coefficient depending on the heat flux for several organic fluids. This shows that the maximum heat flux has a value of about 3×10^5 W/m^2 and that for any pressures and temperatures the temperature difference $T_O - T_L$ at nucleate boiling is about three to four times bigger than the value at the beginning of nucleate boiling.

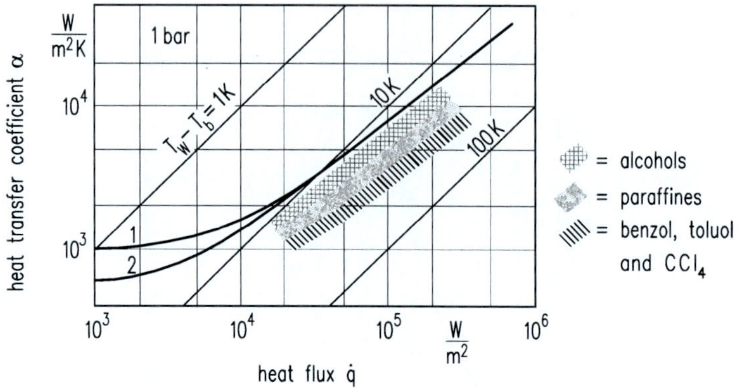

Fig. 4.3-9 Heat transfer coefficient depending on the heat flux for water as well as for several organic liquids. Curve 1: horizontal heating area. Curve 2: vertical heating area, both cases for water

The heat transfer coefficient at position z of the heat transfer surface is defined by

$$\alpha(z) = \frac{\dot{q}(z)}{T_W(z) - T_{sat}(z)} \tag{4.3-53}$$

This definition is also applicable for pure substances and for mixtures. In the case of mixtures, the local composition of the boiling liquid gives rise to the saturation temperature at the local pressure.

For quantitative calculation purposes, see VDI-Heat Atlas (2010), it is of utmost importance to determine the boiling regime. For pool boiling, this is done by a boiling regime map. For flow boiling the flow regime must be determined first. Then the heat transfer coefficient α has to be calculated according to various instructions.

Whereas the heat transfer coefficient α for convective pool boiling is calculated according to Sect. 4.3.2.3, the prediction of the heat transfer coefficient α for convective flow boiling is more complicated. Here the vapor fraction plays a significant role, because the flowing gas influences the flow pattern of the liquid.

The heat transfer coefficient α of nucleate pool boiling depends on the properties of the liquid, the operating parameters (heat flux $\dot{q}$, pressure p), and on the nature of the heated surfaces. This dependence is cast into the following equation:

$$\frac{\alpha}{\alpha_0} = F_q \cdot F_{p*} \cdot F_w, \tag{4.3-54}$$

where α_0 is the heat transfer coefficient for the specific fluid of interest at a defined reference state, which is set to: heat flux $q_0 = 20\ \mathrm{kW/m^2}$, reduced pressure $p_0* = (p/p_c)_0 = 0.1$ and mean roughness of the surface of the heater $R_{a0} = 0.4\ \mu\mathrm{m}$. This reference state is the same for all fluids. The functions F are nondimensional functions, with which the various influences are covered. For numerous systems α_0-values are tabulated. F_q, F_{p*}, and F_w are calculated by

$$F_q = \left(\frac{q}{\dot{q}_0}\right)^n \quad \text{with } n = 0.95 - 0.3 \cdot p*^3. \tag{4.3-55}$$

$$F_{p*} = 0.7 \cdot p*^{0.2} + 4 \cdot p* + \frac{1.4 \cdot p*}{1 - p*} \tag{4.3-56}$$

$$F_w = \left(\frac{R_a}{R_{a0}}\right)^{\frac{2}{15}} \cdot \left(\frac{\lambda \cdot \rho \cdot c}{(\lambda \cdot \rho \cdot c)_{\mathrm{Cu}}}\right)^{0.25} \tag{4.3-57}$$

In the case of flow boiling, this heat transfer coefficient depends on the above parameters and in addition on the following parameters:

- Mass flow rate of the liquid
- Vapor mass fraction
- State of the liquid in the pipe (subcooled or saturated)
- Wetting of the surface (wetted or dry out)

VDI-Heat Atlas (2010) provides detailed information about the calculation of the heat transfer coefficient taking these parameters into consideration.

4.3.3 Balancing Exercises of Processes with Kinetic Phenomena

The following exercises are the continuation of the exercises given in Sect. 4.2.2. They address balancing problems with kinetic processes of the heat and mass transfer.

4.3.3.1 Exercise: Stirred Tank Heated with Condensing Steam

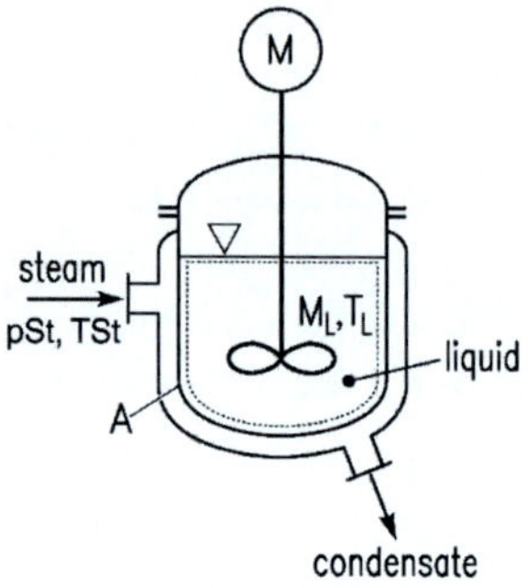

Fig. 4.3-10 Steam tank heated with condensing steam

A stirred tank shall be heated with condensing steam at pressure p_{St}. It is filled with mass M_{L} of a liquid, which is at starting temperature T_{L0}. The heat transfer area is A. The tank wall thickness is s. The tank wall has a heat conductivity of λ_{W}; its heat capacity is insignificant (Fig. 4.3-10).

Question: How does the tank temperature change T_{L} with time?

Applying (4.2-1) results in the

Energy balance for the subsystem "liquid"

$$\frac{\mathrm{d}E_{\mathrm{L}}}{\mathrm{d}t} = \dot{Q}, \tag{4.3-58}$$

$$E_{\mathrm{L}} = M_{\mathrm{L}} \cdot c_{p\mathrm{L}} \cdot T_{\mathrm{L}}. \tag{4.3-59}$$

The heat flow $\dot{Q}$ is calculated as follows by a series of heat transfer resistances and the existing temperature difference between the heating steam St and the liquid L (Fig. 4.3-11):

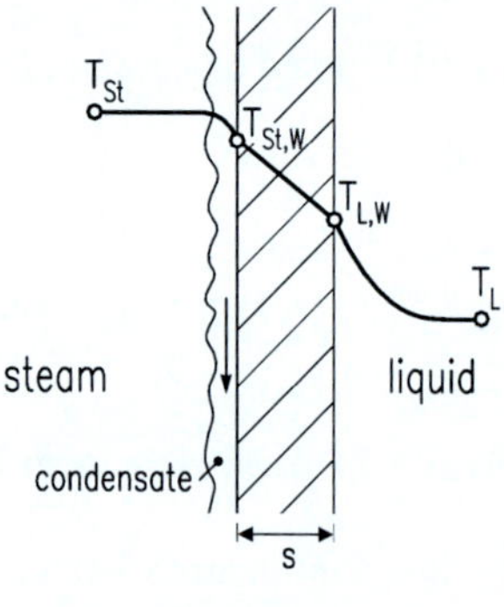

Fig. 4.3-11 Temperature profile across the wall of the tank

$$\dot{Q} = A \cdot \alpha_{\mathrm{St}} \cdot (T_{\mathrm{St}} - T_{\mathrm{St,W}}) \quad \text{or} \quad \frac{\dot{Q}}{A} \cdot \frac{1}{\alpha_{\mathrm{St}}} = T_{\mathrm{St}} - T_{\mathrm{St,W}},$$

$$\dot{Q} = A \cdot \frac{\lambda_W}{s} \cdot (T_{St,W} - T_{L,W}) \text{ or } \frac{\dot{Q}}{A} \cdot \frac{1}{\lambda_W/s} = T_{St,W} - T_{L,W}$$

$$\dot{Q} = A \cdot \alpha_L \cdot (T_{L,W} - T_L) \text{ or } \frac{\dot{Q}}{A} \cdot \frac{1}{\alpha_L} = T_{L,W} - T_L .$$

Addition of all three equations results in

$$\frac{\dot{Q}}{A} \cdot \left(\frac{1}{\alpha_{St}} + \frac{1}{\lambda_W/s} + \frac{1}{\alpha_L}\right) = T_{St} - T_L \text{ or } \dot{Q} = A \cdot k \cdot (T_{St} - T_L), \tag{4.3-60}$$

with the heat transfer coefficient

$$k = \frac{1}{\frac{1}{\alpha_{St}} + \frac{1}{\lambda_W/s} + \frac{1}{\alpha_L}} . \tag{4.3-61}$$

The steam temperature T_{St} results from the pressure p_{St} of the available saturated steam, see Chap. 2.

Applying (4.3-60) and (4.3-59) on (4.3-58) leads to

$$M_L \cdot c_{pL} \cdot \frac{dT_L}{dt} = A \cdot k \cdot (T_{St} - T_L) \tag{4.3-62}$$

with the initial condition $T_L(t = 0) = T_{L0}$.

The dimensionless form of (4.3-62) is

$$\frac{d\Theta}{d\tau} = -\Theta; \quad \Theta(\tau = 0) = 1 \tag{4.3-63}$$

with the dimensionless temperature

$$\Theta = \frac{T_{St} - T_L}{T_{St} - T_{L0}} \tag{4.3-64}$$

and with the dimensionless time τ

$$\tau = \frac{A \cdot k}{M_L \cdot c_{pL}} \cdot t . \tag{4.3-65}$$

Separating the variables in (4.3-63) and integration result in the temperature profile $\Theta(\tau)$

$$\ln\Theta\big|_1^{\Theta} = -\tau\big|_0^{\tau} \text{ and } \ln\Theta = -\tau \text{ or } \Theta = \exp(-\tau) \tag{4.3-66}$$

or

$$T_L(t) = T_{St} - (T_{St} - T_{L0}) \cdot \exp\left(-\frac{A \cdot k}{M_L \cdot c_{pL}} \cdot t\right). \tag{4.3-67}$$

As expected the temperature of the fluid T_L exponentially approaches the temperature of the steam T_{St}. Figure 4.3-12 is a graphical illustration of this profile.

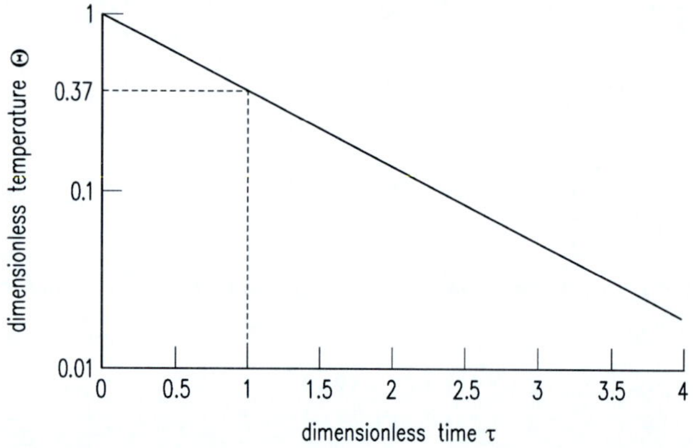

Fig. 4.3-12 Temperature profile in a steam heated stirred tank at semi logarithmic illustration

The dimensionless temperature Θ has decreased to the value $1/e = 0.37$ after the characteristical time $t = M_L \cdot c_{pL}/(A \cdot k)$.

4.3.3.2 Exercise: Cooling of a Stirred Tank with Cooling Water

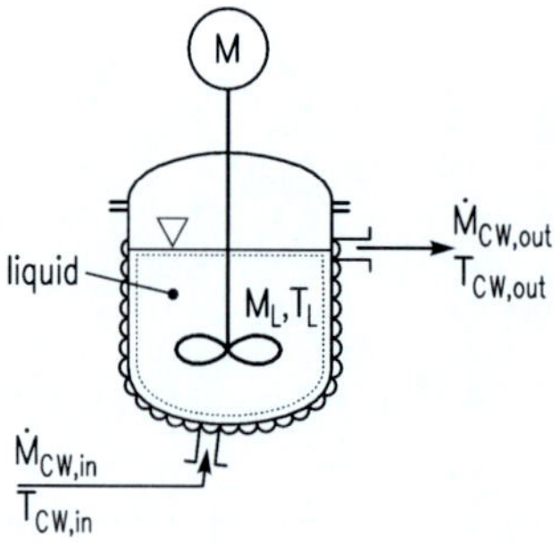

Fig. 4.3-13 Liquid cooled stirrer tank with pipe

As in the previous chapter, a stirred tank is filled with fluid mass M_L, which initially is at temperature T_{L0}. The stirred tank is cooled with cooling water $\dot{M}_{CW}$ of temperature $T_{CW,\,in}$, which is directed through a coiled pipe attached to the outside of the stirred tank (Fig.4.3-13).

Questions: (a) Calculate the temperature $T_{CW,\,out}$ of the cooling water leaving the apparatus.

(b) How does the temperature of the fluid L change with time?

About (a): Outlet Temperature of the Cooling Water

At first T_L is assumed to be constant. The time dependence of T_L is determined later on as part of the solution to question (b).

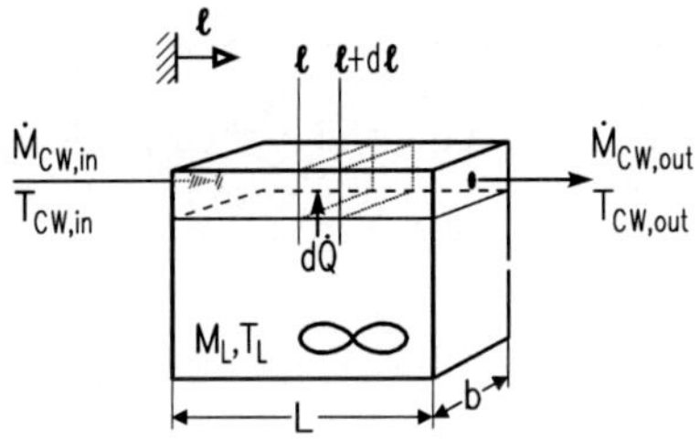

Fig. 4.3-14 Simplified illustration of Fig. 4.3-13

For a better clarity it is recommended to simplify the illustration above according to Fig. 4.3-14:

Obviously the described problem deals with the case that an ideally stirred subsystem "fluid" exchanges heat with a subsystem "cooling water" which itself has to be divided into differential subsystems "cooling water elements." That is the cooling water flow is assumed to be plug flow.

Applying (4.2-1) results in the steady state energy balance for the differential subsystem "cooling water element"

$$0 = \dot{M}_{CW} \cdot c_{pCW} \cdot (T_{CW,l} - T_{CW,l+dl}) + d\dot{Q}. \tag{4.3-68}$$

Taylor series development and derogation after the first element result in

$$0 = \dot{M}_{CW} \cdot c_{pCW} \cdot \left(-\frac{\partial T_{CW}}{\partial l} \cdot dl\right) + k \cdot b \cdot (T_L - T_{CW}) \cdot dl \tag{4.3-69}$$

and lead to

$$\frac{\partial T_{CW}}{\partial l} = \frac{k \cdot b}{\dot{M}_{CW} \cdot c_p} \cdot (T_L - T_{CW}) \tag{4.3-70}$$

with boundary condition $T_{CW}(l = 0) = T_{CW,in}$.

The dimensionless description is

$$\frac{\partial \Theta_{CW}}{\partial \xi} = -\Theta_{CW} \tag{4.3-71}$$

with the boundary condition $\Theta_{CW}(\xi = 0) = 1$ and the dimensionless temperature and length

$$\Theta_{\rm CW} = \frac{T_{\rm L} - T_{\rm CW}}{T_{\rm L} - T_{\rm CW,in}} \quad \text{and} \quad \xi = \frac{k \cdot b}{\dot{M}_{\rm CW} \cdot c_{p\rm CW}} \cdot l\,. \tag{4.3-72}$$

Integrating (4.3-71) after separating the variables results in

$$\ln \Theta_{CW}\Big|_{\Theta_{\rm CW}=1}^{\Theta_{\rm CW}} = -\xi\big|_0^{\xi} \quad \text{and} \quad \ln \Theta_{\rm CW} = -\xi\,. \tag{4.3-73}$$

With NTU: Number of transfer units the temperature at the end of the cooling coil is

$$\ln \Theta_{\rm CW,out} = -NTU \quad \text{and} \quad \Theta_{\rm CW,out} = \exp(-NTU) \tag{4.3-74}$$

with

$$\Theta_{\rm CW,out} = \frac{T_L - T_{\rm CW,out}}{T_L - T_{\rm CW,in}} \quad \text{and}\ NTU = \frac{k \cdot D \cdot L}{\dot{M}_{\rm CW} \cdot c_{p\rm CW}} = \frac{k \cdot A}{\dot{M}_{\rm CW} \cdot c_{p\rm CW}}.$$

The NTU number expresses to which extent the cooling water temperature $T_{\rm CW,out}$ approaches the temperature of the fluid $T_{\rm L}$. For $NTU \to 0$, i.e., bad heat transfer $k \cdot A$ or high capacity flow $\dot{M}_{\rm CW} \cdot c_{p\rm CW}$ of the temperature of the cooling water, $\Theta_{\rm CW,out}$, approaches 1; thus the cooling water temperature hardly changes in the coiled pipe. But for $NTU \to \infty$ it follows that $\Theta_{\rm CW,out} \to 0$ and $T_{\rm CW,out} \to T_{\rm L}$. Even at $NTU = 5$ less than 1% of the starting temperature difference $T_{\rm L} - T_{\rm CW,\,in}$ is available as a driving temperature difference $T_{\rm L} - T_{\rm CW,out}$ at the outlet of the cooling pipe (i.e., $\Theta_{\rm L} < 0.01$). In practice, increasing NTU above a value of 5 (in practical cases 2–3) is inefficient.

About (b) Liquid Temperature Profile $T_{\rm L}$

The transferred heat flow from the subsystem fluid to a differential subsystem "cooling water element" is calculated by

$$\mathrm{d}\dot{Q} = k \cdot b \cdot (T_{\rm L} - T_{\rm CW}(l))\ \mathrm{d}l\,. \tag{4.3-75}$$

The total heat flow $\dot{Q}_{\rm tot}$ transferred into the subsystem "cooling water" results from integration

$$\int_0^{\dot{Q}_{\text{tot}}} \mathrm{d}\dot{Q} = k \cdot b \cdot \int_0^L (T_\mathrm{L} - T_\mathrm{CW}(l))\, \mathrm{d}l$$

$$= k \cdot b \cdot (T_\mathrm{L} - T_\mathrm{CW,in}) \cdot \frac{\dot{M}_\mathrm{CW} \cdot c_{p\mathrm{CW}}}{k \cdot b} \cdot \int_0^{NTU} \Theta_\mathrm{CW}(\xi)\, \mathrm{d}\xi$$

$$\dot{Q}_\text{tot} = (T_\mathrm{L} - T_\mathrm{CW,\,in}) \cdot \dot{M}_\mathrm{CW} \cdot c_{p\mathrm{CW}} \cdot \int_0^{NTU} \exp(-\xi)\, \mathrm{d}\xi$$

$$\dot{Q}_\text{tot} = \dot{M}_\mathrm{CW} \cdot c_{p\mathrm{CW}} \cdot (T_\mathrm{L} - T_\mathrm{CW,in}) \cdot (1 - \exp[-NTU])\,. \tag{4.3-76}$$

Considering the definition of *NTU* and (4.3-74), the total transferred heat flow can be written as

$$\dot{Q}_\text{tot} = k \cdot A \cdot \frac{T_\mathrm{CW,out} - T_\mathrm{CW,\,in}}{\ln \dfrac{(T_\mathrm{L} - T_\mathrm{CW,\,in})}{(T_\mathrm{L} - T_\mathrm{CW,out})}} = k \cdot A \cdot \Delta T_\text{log}\,. \tag{4.3-77}$$

Hence the transferred heat flow $\dot{Q}_\text{tot}$ can be determined using the so-called logarithmical temperature difference ΔT_log. For determining the progress of the fluid temperature T_L, the energy balance for the subsystem “fluid” has to be analyzed.

$$\frac{\mathrm{d}E_\mathrm{L}}{\mathrm{d}t} = -\dot{Q}_\text{tot} \tag{4.3-78}$$

Use (4.3-76) for $\dot{Q}_\text{tot}$

$$M_\mathrm{L} \cdot c_{p\mathrm{L}} \cdot \frac{\mathrm{d}T_L}{\mathrm{d}t} = -\dot{M}_\mathrm{CW} \cdot c_{p\mathrm{CW}} \cdot (T_\mathrm{L} - T_\mathrm{CW,\,in}) \cdot (1 - \exp[-NTU])\,.$$

In dimensionless form

$$\frac{\mathrm{d}\Theta_\mathrm{L}}{\mathrm{d}\tau_\mathrm{L}} = -\Theta_\mathrm{L} \cdot (1 - \exp[-NTU]) \tag{4.3-79}$$

with the initial condition $\Theta_\mathrm{L}(\tau_\mathrm{L} = 0) = \Theta_\mathrm{L0} = 1$ and with

$$\Theta_\mathrm{L} = \frac{T_\mathrm{L} - T_\mathrm{CW,\,in}}{T_\mathrm{L0} - T_\mathrm{CW,\,in}}\,; \text{ and } \tau_L = \frac{\dot{M}_\mathrm{CW} \cdot c_{p\mathrm{CW}}}{M_\mathrm{L} \cdot c_{p\mathrm{L}}} \cdot t\,. \tag{4.3-80}$$

Separating the variables and integration result the time dependence of the temperature of the fluid:

$$\begin{aligned} \mathrm{d}\ln\Theta_{\mathrm{L}} &= -(1-\exp[-NTU])\cdot \mathrm{d}\tau_{\mathrm{L}} \\ \Theta_{\mathrm{L}} &= \Theta_{L0}\cdot \exp(-\{1-\exp[-NTU]\}\cdot \tau_{\mathrm{L}}). \end{aligned} \tag{4.3-81}$$

This correlation is illustrated in Fig. 4.3-15. In accordance with the above discussion of (4.3-74) an increase from $NTU = 1$ to $NTU = 2$ is more efficient in terms of Θ_{L} than its increase from $NTU = 2$ to $NTU = 5$. However, considering the definition of NTU and τ_{L}, it is clear that at constant NTU the transferred heat flow can still be increased more and more by mutually increasing $k \cdot A$ and $\dot{M}_{\mathrm{CW}} \cdot c_{p\mathrm{CW}}$.

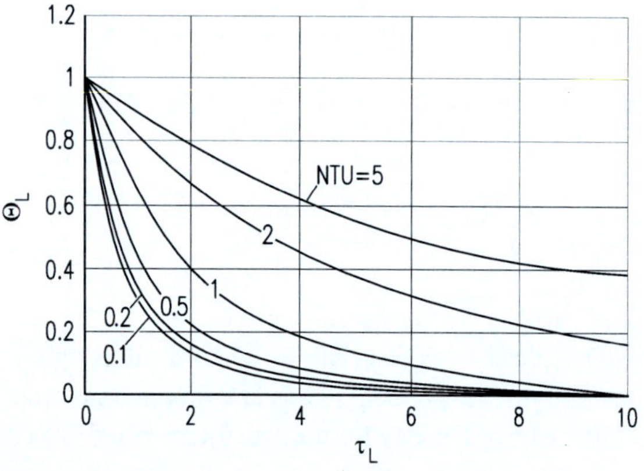

Fig. 4.3-15 Evolution of the dimensionless fluid temperature Θ_{L} of a cooled stirred tank vs. dimensionless time τ_{L}.

4.3.3.3 Exercise: Transient Mass Transport in Spheres

The transient mass transport in spheres plays an important role in many process engineering applications. At adsorption the adsorptive moves through porosities and is accumulated on the internal surface of the spherical adsorbent. At regeneration of such adsorbents, as well as at drying of capillary active solids the opposite process occurs. Mass transport caused by transient diffusion is also existent in fluid particles, as long as no convection occurs in the particle. This applies for small viscous droplets.

In the following the diffusion of component i in a sphere is discussed. The component balance for a sphere element leads to the following partial differential equation:

$$\frac{\partial c_a}{\partial t} = D_{ab} \cdot \left(\frac{\partial^2 c_a}{\partial r^2} + \frac{2}{r} \cdot \frac{\partial c_a}{\partial r} \right). \tag{4.3-82}$$

Herein r is the radius; D_{ab} is the (constant) diffusion coefficient of the component a in its environment b.

For solving the differential equation one initial and two boundary conditions have to be defined. If initially the sphere is free of substance a, then

$$c_a(r) = 0 \text{ at } t = 0 \text{ and } 0 \le r \le R.$$

At $t > 0$ the concentration at $r = R$ shall be set to a constant value (first boundary condition)

$$c_a(R) = c_{a\omega} \text{ at } t > 0.$$

Furthermore it is assumed that the concentration profile is symmetrical with respect to the center of the sphere at any time of the diffusion process. Hence, the concentration gradient is zero at the center of the sphere:

$$\left. \frac{\partial c_a}{\partial r} \right|_{r=0} = 0 \text{ at } t > 0.$$

Solution of (4.3-82) with these initial and boundary conditions is the following equation for the transient concentration profile:

$$\frac{c_a}{c_{a\omega}} = 1 + \frac{2 \cdot R}{\pi \cdot r} \sum_{n=1}^{\infty} (-1)^n \cdot \frac{1}{n} \exp\left(-\frac{n^2 \cdot \pi^2 \cdot D_{ab} \cdot t}{R^2} \right) \sin \frac{\pi \cdot n \cdot r}{R}. \tag{4.3-83}$$

The mean concentration $\bar{c}_a$ of the diffusing substance is obtained by integration:

$$\bar{c}_a = \frac{3}{R^3} \cdot \int_0^R r^2 \cdot c_a(r)\, \mathrm{d}r \tag{4.3-84}$$

which results in

$$\frac{\bar{c}_a}{c_\omega} = 1 - \frac{6}{\pi^2} \sum_{n=1}^{\infty} \frac{1}{n^2} \exp\left(-\frac{n^2 \cdot \pi^2 \cdot D_{ab} \cdot t}{R^2} \right). \tag{4.3-85}$$

Herein the value $D_{ab} \cdot t / R^2$ is the Fourier number of mass transfer. Equivalent to this equation is

$$\frac{\bar{c}_a}{c_{a\omega}} = \frac{6}{R} \cdot \sqrt{D_{ab} \cdot t} \cdot \left[\frac{1}{\sqrt{\pi}} + 2 \cdot \sum_{n=1}^{\infty} ierf\left(\frac{n \cdot R}{\sqrt{D_{ab} \cdot t}}\right)\right] - \frac{3 \cdot D_{ab} \cdot t}{R^2} \qquad (4.3\text{-}86)$$

with $ierfc(z) = \int_z^{\infty} erfc(\xi) d\xi$

Here the Fourier number of the dispersed phase

$$Fo_{\mathrm{d}} = \frac{D_{ab} \cdot t}{R^2}$$

appears as an important dimensionless characteristic. If $\bar{c}_a / c_{a\omega}$ is smaller than 0.95, then the second term of the sum in the angular bracket can be disregarded compared to the value $1/\sqrt{\pi}$, and

$$\frac{\bar{c}_a}{c_{a\omega}} = \frac{6}{\sqrt{\pi}} \cdot \sqrt{\frac{D_{ab} \cdot t}{R^2}} - \frac{3 \cdot D_{ab} \cdot t}{R^2}. \qquad (4.3\text{-}87)$$

The deviations between (4.3-86) and (4.3-87) are smaller than 1% if the concentration ratio $\bar{c}_a / c_{a\omega}$ is smaller than 0.95. If at $t = 0$ the concentration of the sphere is at $c_{a\alpha}$ then

$$\frac{\bar{c}_a - c_{a\alpha}}{c_{a\omega} - c_{a\alpha}} = \frac{6}{\sqrt{\pi}} \cdot \sqrt{\frac{D_{ab} \cdot t}{R^2}} - \frac{3 \cdot D_{ab} \cdot t}{R^2}. \qquad (4.3\text{-}88)$$

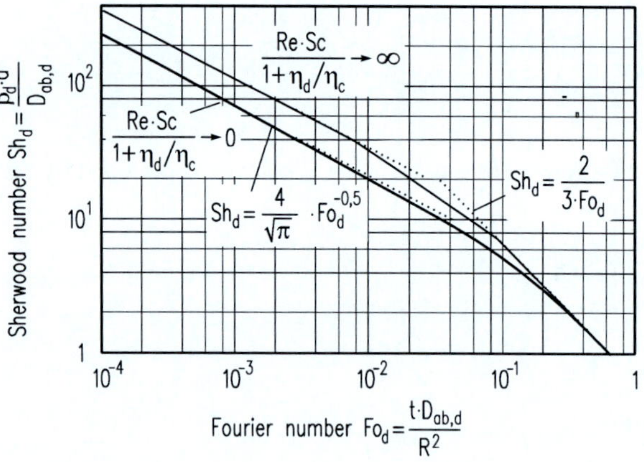

Fig. 4.3-16 Time averaged Sherwood number of fluid particles depending on the Fourier number with limiting laws

In the case of fluid particles (bubbles or drops), mass transfer coefficient β_{d} of the particle may be governed by internal flow. Then mass transfer is best described by an internal mass transfer coefficient, which is also a function of $Re = w_s \cdot d_P \cdot \rho_c/\eta_c$, of $Sc = \eta_c/(\rho_c \cdot D_{ab})$, and of the ratio η_d/η_c of the viscosities of both phases.

Figure 4.3-16 illustrates the time averaged Sherwood number Sh_d depending on the Fourier number of the dispersed phase $Fo_d = t \cdot D_{ab,d}/R^2$.

4.3.3.4 Exercise: Isothermal Evaporation of a Binary Mixture

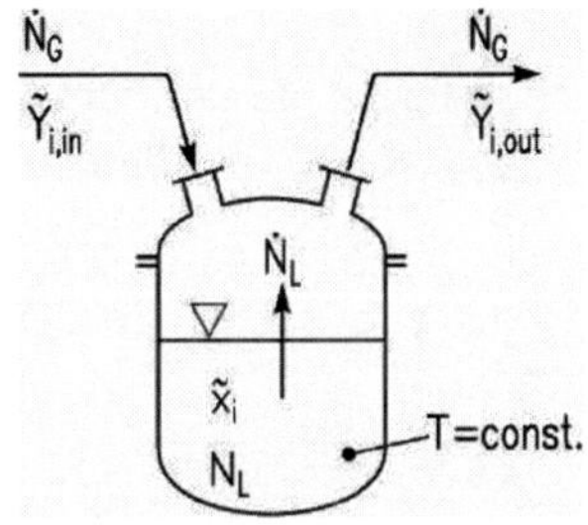

Fig. 4.3-17 Vessel filled with a binary liquid, which is overflown by a gas

A binary fluid mixture $i=a,b$, which is kept at constant temperature, is overflown by a dry carrier gas. How does the composition of the fluid $\tilde{x}_i$ change with advancing evaporation (Fig. 4.3-17) (Note: This example is taken from Schlünder (1996))

Combining the total balance for the fluid

$$\frac{dN_L}{dt} = -\dot{N} \tag{4.3-89}$$

and the component balance for the fluid

$$\frac{d(N_L \cdot \tilde{x}_i)}{dt} = -\dot{N}_i \tag{4.3-90}$$

results in

$$N_L \cdot \frac{d\tilde{x}_i}{dt} + \tilde{x}_i \cdot \frac{dN_L}{dt} = -\dot{N}_i, \tag{4.3-91}$$

$$\dot{r}_i = \frac{\dot{N}_i}{\dot{N}} = \frac{N_L}{dN_L} \cdot d\tilde{x}_i + \tilde{x}_i \tag{4.3-92}$$

with $\dot{r}_i$ as the relative flux of i. Separating the variables leads to

$$\frac{dN_L}{N_L} = \frac{d\tilde{x}_i}{\dot{r}_i - \tilde{x}_i}. \tag{4.3-93}$$

Integrating this differential equation from the beginning of evaporation $N_L = N_{L,0}$ to some later degree of evaporation N_L yields

$$\frac{N_L}{N_{L,0}} = \exp \int_{\tilde{x}_{i,0}}^{\tilde{x}_i} \frac{d\tilde{x}_i}{\dot{r}_i - \tilde{x}_i}. \tag{4.3-94}$$

For the determination of the composition of the liquid residue x_i as a function of the degree of evaporation, the relative mass flow $\dot{r}_i$ has to be known as a function of the fluid composition $\tilde{x}_i$.

If the evaporation process is controlled by thermodynamics and not by kinetics, then for a binary liquid mixture and for low molar fractions of i in the gas phase, i.e., $\tilde{y}_i \approx \tilde{Y}_i$ follows that

$$\frac{\dot{r}_a}{\dot{r}_b} = \frac{\tilde{y}_a}{\tilde{y}_b}\left(\approx \frac{\tilde{Y}_a}{\tilde{Y}_b}\right) = \frac{K_a \cdot \tilde{x}_a}{K_b \cdot \tilde{x}_b}. \tag{4.3-95}$$

Herein the K-values K_1 and K_2 specify the tendencies of components 1 and 2 to vaporize. With the definition of the relative molality $\alpha_{ab} = K_a/K_b$ it follows that

$$\frac{\dot{r}_a}{1-\dot{r}_b} = \alpha_{ab} \cdot \frac{\tilde{x}_a}{1-\tilde{x}_a}, \tag{4.3-96}$$

$$\dot{r}_a = \frac{1}{1 - \dfrac{1}{\alpha_{ab}} + \dfrac{1}{\alpha_{ab} \cdot \tilde{x}_a}}. \tag{4.3-97}$$

Inserting this equation in (4.3-94) results in the so-called residue curve, i.e., the dependence of the composition $\tilde{x}_a$, $\tilde{x}_b$ of the liquid residue on the degree of evaporation $N_L/N_{L,0}$.

In the case that the evaporation process is controlled by thermodynamics and by kinetics of mass transport in the gas phase, the concentration in the bulk of the (well-mixed) gas phase $\tilde{Y}_i = \tilde{Y}_{i,\text{out}}$ is different from that at the gas–liquid interphase, $\tilde{Y}_{i,Ph}$. The component balance for the gas phase at steady-state condition is as follows:

$$0 = \dot{N}_G \cdot (\underbrace{\tilde{Y}_{i,\text{in}}}_{=0} - \tilde{Y}_{i,\text{out}}) + \dot{N}_i, \tag{4.3-98}$$

$$\tilde{Y}_{i,\text{out}} = \frac{\dot{N}_i}{\dot{N}_G}. \tag{4.3-99}$$

At low gas concentrations, i.e., $\tilde{y}_i \approx \tilde{Y}_i$, the kinetics of mass transfer is described by

$$\dot{N}_i = \beta_{i,G} \cdot A_{Ph} \cdot \tilde{\rho}_G \cdot (\tilde{Y}_{i,Ph} - \tilde{Y}_{i,\text{out}}). \tag{4.3-100}$$

From (4.3-98) and (4.3-100) it follows that

$$\tilde{Y}_{i,\text{out}} = \frac{NTU_{\text{G},i}}{1+NTU_{\text{G},i}} \cdot \tilde{Y}_{i,\text{Ph}} \tag{4.3-101}$$

with $NTU_{\text{G},i} = \dfrac{\beta_{i,\text{G}} \cdot A_{\text{Ph}} \cdot \tilde{\rho}_{\text{G}}}{\dot{N}_{\text{G}}}$ (number of transfer units).

Following the procedure shown in (4.3-95) – (4.3-97) results in

$$\frac{\dot{r}_a}{\dot{r}_b} = \frac{\tilde{Y}_{a,\text{out}}}{\tilde{Y}_{b,\text{out}}} = K_G \cdot \frac{\tilde{Y}_{a,\text{Ph}}}{\tilde{Y}_{b,\text{Ph}}} = \frac{K_\text{G}}{\alpha_{ab}} \cdot \frac{\tilde{x}_{a,\text{Ph}}}{\tilde{x}_{b,\text{Ph}}} \tag{4.3-102}$$

with

$$K_\text{G} = \frac{NTU_{\text{G},a}}{1+NTU_{\text{G},a}} \cdot \frac{1+NTU_{\text{G},b}}{NTU_{\text{G},b}}.$$

Thus, in the case of mass transfer limitation in the gas phase ($\equiv$ no concentration profile in the liquid phase, $\tilde{x}_i = \tilde{x}_{i,\text{Ph}}$), the relative flux is

$$\dot{r}_a = \frac{1}{1-\dfrac{K_\text{G}}{\alpha_{ab}}+\dfrac{K_\text{G}}{\alpha_{ab}} \cdot \dfrac{1}{\tilde{x}_{a,\text{Ph}}}}, \tag{4.3-103}$$

which again has to be inserted in (4.3-94) for calculation of the residual curve.

In the case that the evaporation process is controlled by thermodynamics and by kinetics as well as in the gas phase and in the liquid phase, appreciable concentration profiles can exist in the liquid phase ($\tilde{x}_i \neq \tilde{x}_{i,\text{Ph}}$).

Mass transport in the binary liquid phase cannot be estimated with an equation analogous to (4.3-100) because of convective effects. Thus, (4.3-5) has to be used in this case. Integration of (4.3-5) in the limits of bulk and interphase of the liquid yields

$$\dot{N}_\text{L} = A_{\text{Ph}} \cdot \tilde{\rho}_\text{L} \cdot \beta_\text{L} \cdot \dot{r}_a \cdot \ln\left(\frac{\dot{r}_a - \tilde{x}_{a,\text{Ph}}}{\dot{r}_a - \tilde{x}_a}\right), \tag{4.3-104}$$

and further on

$$\dot{r}_a - \tilde{x}_{a,\text{Ph}} = \left[\exp\left(\frac{\dot{N}}{A_{\text{Ph}} \cdot \tilde{\rho}_\text{L} \cdot \beta_\text{L}}\right)\right] \cdot (\dot{r}_a - \tilde{x}_a). \tag{4.3-105}$$

Due to mass transport $\dot{N}/A_{\text{Ph}}$ the liquid interface recedes with velocity

$$v_L = \frac{\dot{N}}{A_{\text{Ph}} \cdot \tilde{\rho}_L} .$$

Furthermore, (4.3-105) can be converted into

$$\tilde{x}_{a,\text{Ph}} = \dot{r}_a - \frac{\dot{r}_a - \tilde{x}_a}{K_L} \tag{4.3-106}$$

with

$$K_L = \exp\left(-\frac{v_L}{\beta_L}\right) .$$

Insertion of (4.3-106) into (4.3-103) leads to the following quadratic expression for $\dot{r}_i$:

$$\dot{r}_a^2 + p \cdot (\tilde{x}_a) \cdot \dot{r}_a + q(\tilde{x}_a) = 0 \tag{4.3-107}$$

or

$$\dot{r}_{a,1/2} = \frac{p(\tilde{x}_a)}{2} \pm \sqrt{\left(\frac{p(\tilde{x}_a)}{2}\right)^2 - q(\tilde{x}_a)} \tag{4.3-108}$$

with

$$p(\tilde{x}_a) = -\frac{1 + \left(1 - \frac{K_G}{\alpha_{ab}}\right) \cdot (\tilde{x}_a - K_L)}{\left(1 - \frac{K_G}{\alpha_{ab}}\right) \cdot (1 - K_L)} , \tag{4.3-109}$$

$$q(\tilde{x}_a) = \frac{\tilde{x}_a}{\left(1 - \frac{K_G}{\alpha_{ab}}\right) \cdot (1 - K_L)} . \tag{4.3-110}$$

For a complete solution, (4.3-108) has to be inserted into (4.3-94). The following limiting cases can be distinguished:

Case I

The evaporation process is not determined by the mass transport processes, but only by the thermodynamic equilibrium, thus:

$$NTU_{G,i} \rightarrow \infty \text{ or } K_G = 1,$$

$$v_L/\beta_L \rightarrow 0 \quad \text{or } K_L = 1.$$

For this case (4.3-107) results in (4.3-97), derived above:

$$\dot{r}_a = \frac{1}{1 + \frac{1}{\alpha_{ab}} \cdot \frac{1-\tilde{x}_a}{\tilde{x}_a}} \tag{4.3-111}$$

Case II

The evaporation process is determined by the thermodynamic equilibrium and the mass transport in the gas phase, thus:

$$NTU_{G,i} \rightarrow 0 \text{ or } K_G = \frac{\beta_{Ga}}{\beta_{Gb}},$$

$$v_L/\beta_L \rightarrow 0 \quad \text{or } K_L = 1.$$

For this case (4.3-107) results in (4.3-103), derived above:

$$\dot{r}_a = \frac{1}{1 - \frac{K_G}{\alpha_{ab}} \cdot \left(\frac{1-\tilde{x}_{a,Ph}}{\tilde{x}_{a,Ph}}\right)}. \tag{4.3-112}$$

Case III

The evaporation process is controlled by the mass transport in the liquid phase, i.e.,

$$NTU_{G,i} \rightarrow \infty \text{ or } K_G = 1$$

$$v_L/\beta_L \rightarrow \infty \quad \text{or } K_L = 0.$$

In this case (4.3-105) results in

$$K_L = \exp\left(-\frac{v_L}{\beta_L}\right) = \frac{\dot{r}_a - \tilde{x}_a}{\dot{r}_a - \tilde{x}_{a,Ph}} = 0. \tag{4.3-113}$$

which can only be valid, if

$$\dot{r}_a = \tilde{x}_a. \tag{4.3-114}$$

Thus, in this case, the selectivity of the evaporation process disappears. Even the thermodynamic equilibrium is without importance. For the three limit cases the concentration profiles in the gaseous and liquid phase are illustrated in Fig. 4.3-18.

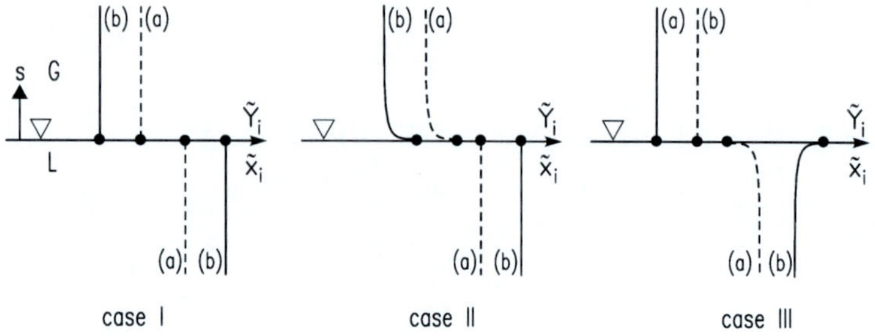

Fig. 4.3-18 Concentration profiles in the gaseous and liquid phase at the isothermal evaporation of a binary fluid compound into a carrier gas

4.3.3.5 Example: Balancing a Shell and Tube Heat exchanger

This example is in analogy to the already discussed balancing of a crystallization facility, see Sect. 4.2.2.5. It shows the mathematical modeling of a network of heat exchanging elements with the help of balancing equations and kinetic correlations, as well as their rather simple solution by a standard software. As an extension to the balancing example in Sect. 4.2.2.5, kinetic phenomena have to be taken into account. The baffled 2-pass shell and tube heat exchanger shown in Fig. 4.3-19 shall be considered.

Question: How do the outlet temperature of the tubes $T_{\mathrm{T,out}}$ and the outlet temperature of the shell $T_{\mathrm{S,out}}$ depend on the geometry and the operational mode of the apparatus?

To answer this question the apparatus is subdivided into six heat exchanging elements, each of which consists of two cells, the shell- and the tube-cell. The tube-side flow $\dot{M}_{\mathrm{T}}$ passes through these elements in the sequence $1\rightarrow 2\rightarrow 3\rightarrow 4\rightarrow 5\rightarrow 6$. The shell-side flow $\dot{M}_{\mathrm{S}}$ passes through these elements in the sequence $4\rightarrow 3\rightarrow 2\rightarrow 5\rightarrow 6\rightarrow 1$. As well the tube-side fluid as the shell-side fluid in each of the $i = 1,...,6$ cells of these elements shall be ideally mixed and shall attain the unknown temperatures $T_{\mathrm{T},i}$ and $T_{\mathrm{S},i}$. The shell- and the tube-side cells in each element i exchange heat according to

$$\dot{Q}_i = k \cdot A \cdot (T_{\mathrm{S},i} - T_{\mathrm{T},i}). \qquad (4.3\text{-}115)$$

The energy balance of a tube-side cell i $(i \neq 1)$, into which flows $\dot{M}_{\mathrm{T}}$ from cell k, is

$$0 = \dot{M}_{\mathrm{T}} \cdot c_{p\mathrm{T}} \cdot T_{\mathrm{T}k} - \dot{M}_{\mathrm{T}} \cdot c_{p\mathrm{T}} \cdot T_{\mathrm{T}i} + \dot{Q}_i, \qquad (4.3\text{-}116)$$

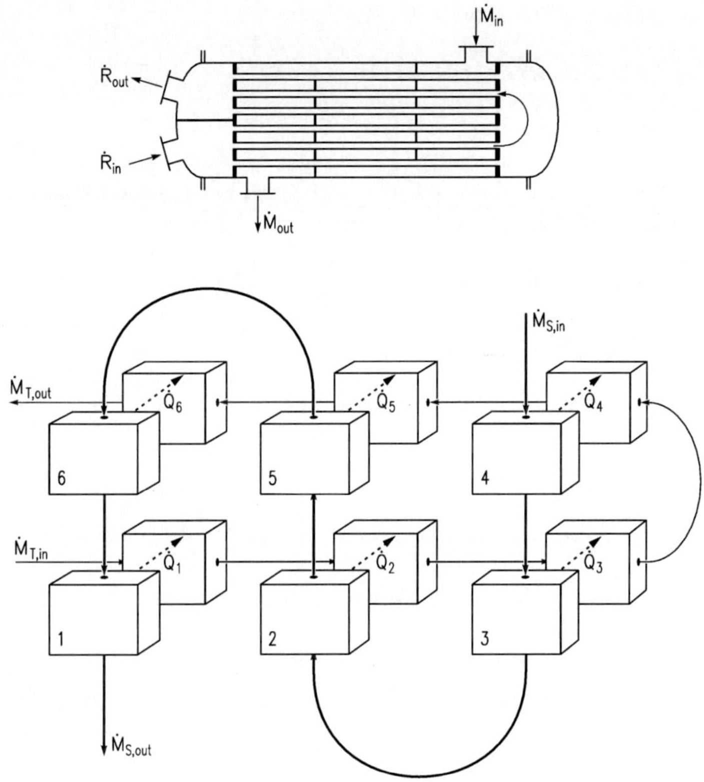

Fig. 4.3-19 Two-flow shell and tube heat exchanger with two baffles as well as an equivalent circuit consisting of six pairs of cells, in which heat is exchanged between the shell-side guided fluid (cells M_i) and the tube-side guided fluid (cells R_i)

or

$$0 = T_{Tk} - T_{Ti} + \frac{k \cdot A}{\dot{M}_T \cdot c_{pT}} \cdot (T_{Si} - T_{Ti}). \tag{4.3-117}$$

This equation can be made nondimensional by using the following definitions:

$$\Theta_{Si} = \frac{T_{Si} - T_{T,in}}{T_{S,in} - T_{T,in}} \; ; \Theta_{Ti} = \frac{T_{Ti} - T_{T,in}}{T_{S,in} - T_{T,in}} \tag{4.3-118}$$

$$NTU_S = \frac{k \cdot A}{\dot{M}_S \cdot c_{pS}} ; \; NTU_T = \frac{k \cdot A}{\dot{M}_T \cdot c_{pT}} . \tag{4.3-119}$$

It follows:

$$0 = \Theta_{Tk} - \Theta_{Ti} + NTU_T \cdot (\Theta_{Si} - \Theta_{Ti}). \tag{4.3-120}$$

In analogy, it follows for a shell-side cell i, $i \neq 4$:

$$0 = \Theta_{Sk} - \Theta_{Si} - NTU_S \cdot (\Theta_{Si} - \Theta_{Ti}). \tag{4.3-121}$$

The energy balance for tube-side cell $i = 1$ (inlet of $\dot{M}_T$) and for shell-side cell $i = 4$ (inlet of $\dot{M}_S$) is

$$-\Theta_{T,in} = -\Theta_{T1} + NTU_T \cdot (\Theta_{S1} - \Theta_{T1}), \tag{4.3-122}$$

$$-\Theta_{S,in} = -\Theta_{S4} - NTU_S \cdot (\Theta_{S4} - \Theta_{T4}). \tag{4.3-123}$$

The above set of linear equations can be written in matrix form (Table 4.3-3):

$$\mathbf{A} \cdot \vec{\Theta} = \vec{F} \tag{4.3-124}$$

with $\mathbf{A}$ and $\vec{F}$ given in the following table (all non-defined values are 0!)::

Table 4.3-3 Matrix **A** of coefficients and vector $\vec{F}$ of feed streams

	Θ												
Cell	T_1	T_2	T_3	T_4	T_5	T_6	S_1	S_2	S_3	S_4	S_5	S_6	$\vec{F}$
T_1	$-(1+NTU_T)$						NTU_T						$-\Theta_{T,in}$
T_2	1	$-(1+NTU_T)$						NTU_T					
T_3		1	$-(1+NTU_T)$						NTU_T				
T_4			1	$-(1+NTU_T)$						NTU_T			
T_5				1	$-(1+NTU_T)$						NTU_T		
T_6					1	$-(1+NTU_T)$						NTU_T	
S_1	NTU_S						$-(1+NTU_S)$					1	
S_2		NTU_S						$-(1+NTU_S)$	1				
S_3			NTU_S						$-(1+NTU_S)$	1			
S_4				NTU_S						$-(1+NTU_S)$			$-\Theta_{S,in}$
S_5					NTU_S			1			$-(1+NTU_S)$		
S_6						NTU_S					1	$-(1+NTU_S)$	

As already shown in Sect. 4.2.2.5 the solution to the above equation can be obtained by $\Theta = \mathbf{A}^{-1} \cdot \vec{F}$, whereas the inversion of **A** and its multiplication with $\vec{F}$ may be performed with the help of standard PC-software. In EXCEL the respective matrix-functions to be used are MINV and MMULT.

In Fig. 4.3-20, the result of such a calculation is shown for $\Theta_{T,\,in} = 1$, $\Theta_{S,\,in} = 1$, $NTU_T = 1$, and $NTU_S = 1.5$. It is interesting to see that the heat flux in cells 5 and 6 is inverted. The tube-side stream, which is to be cooled by the shell-side stream, is actually heated up in these two cells. Thus, the proposed heat exchanger is not behaving ideally.

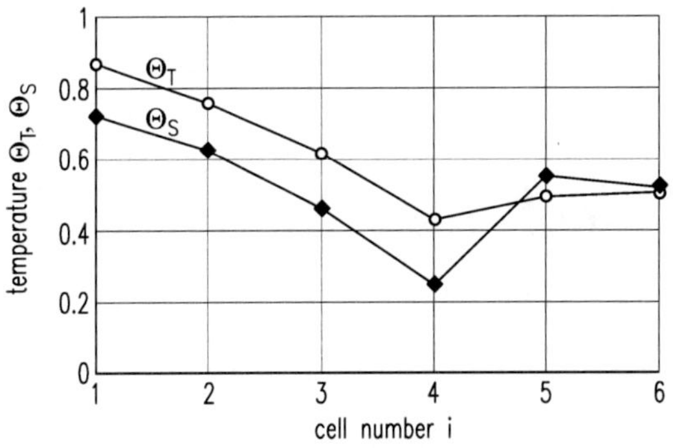

Fig. 4.3-20 Temperature profile for the shell and tube heat exchanger

5 Distillation, Rectification, and Absorption

The unit operations distillation, rectification, and absorption are by far the most important technologies for fractionating fluid mixtures. This great technical importance is founded on the fact that only fluid phases, which can be handled very easily, are involved. A further advantage is a high density difference between the coexisting phases. High density differences enable high velocities in the equipment and make the separation of the phases easier.

All three unit operations use a very simple separation principle that consists of three steps:

- Generation of a two-phase system
- Interfacial mass transfer
- Separation of the phases

In distillation and rectification the second phase (vapor) is generated by supply of heat. In absorption, however, an external liquid (solvent, absorbent) is added to the system at hand.

Of special importance is the unit operation rectification since it is the only technology which is capable to separate fluid mixtures into all pure substances. Disadvantages of distillation and rectification are a risk of thermal degradation of substances and, even more importantly, a high energy demand. This high energy demand, however, can be drastically reduced by special measures, e.g., by material and thermal coupling of the equipment, by implementing heat pumps or by complete heat integration of the process. Rectification is a highly developed technology with respect to thermodynamic fundamentals (e.g., vapor–liquid equilibrium, thermodynamics) as well as design and construction of the equipment. Rectification columns can be safely constructed and operated up to diameters of 10 m and heights of 100 m. Column internals (e.g., trays, packings) are very effective in enhancing the interfacial mass transfer (Kirschbaum 1969). All these advantages often make rectification the separation technology of choice. Whenever a fluid mixture can be fractionated by rectification then rectification is in most cases the winner of a comparison of several separation techniques (Fair 1990).

A. Mersmann et al., *Thermal Separation Technology: Principles, Methods, Process Design*, VDI-Buch, DOI 10.1007/978-3-642-12525-6_5,
© Springer-Verlag Berlin Heidelberg 2011

In the thermodynamic description of distillation, rectification, and absorption processes the concentrations of the mixture are advantageously expressed by molar fractions x (liquid phase) and y (gas or vapor phase) since molar fractions enable a more straightforward formulation of phase equilibrium. In turn, the amounts of the phases (G or L) have to be expressed in moles, too.

5.1 Distillation

Distillation denotes a process of partial evaporation (with succeeding condensation) of a close boiling liquid mixture. The name comes from the latin word *destillare* that describes the dripping down of the liquid product after condensation. Distillation splits a liquid mixture into two fractions having different concentrations. However, distillation is in most cases not able to produce pure products.

5.1.1 Fundamentals

This section deals with the thermodynamic fundamentals of distillation processes.

5.1.1.1 Modes of Operation

In principle, three different modes of operation are feasible for distillation processes:

- Closed distillation
- Open distillation
- Countercurrent distillation

Schemes of these operation modes are shown in Fig. 5.1-1. In closed distillation gas and liquid are kept in intimate contact within the evaporator. Consequently, the gas and liquid products are in equilibrium state.

In open distillation the vapor generated by a supply of heat is immediately removed from the liquid. Therefore, the content of the low boiler in the vapor product is higher than in closed distillation.

In countercurrent distillation the vapor product has the highest content of low-boiling constituents. The vapor is, in a first approximation, in equilibrium state with the liquid feed. In industrial practice only the operation modes closed distillation and open distillation are used. All operation modes can be performed continuously or

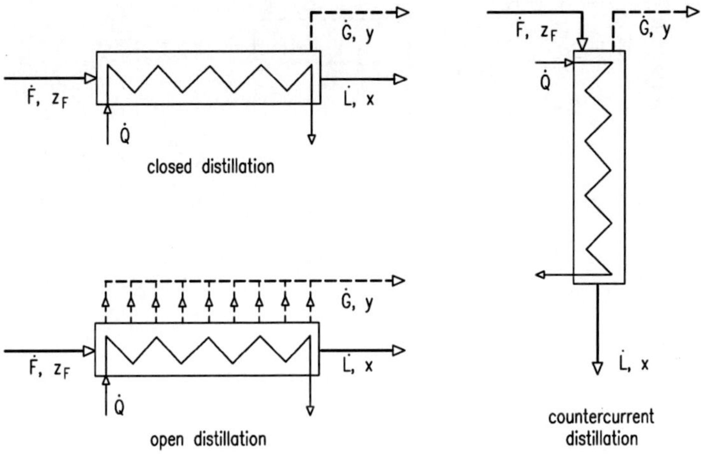

Fig. 5.1-1 Operation modes of single stage distillation: —— liquid; ----- vapor

discontinuously. However, discontinuous operation is preferred in open distillation (batch distillation, see Sect. 5.1.3).

A process analogous to distillation is partial condensation. Here, a vapor feed is fractionated in a condenser. In principle, the operation modes shown in Fig. 5.1-1 can also be applied to partial condensation by replacing the evaporator with a condenser.

5.1.1.2 Phase Equilibrium

According to the separation principle described before, vapor–liquid (or gas–liquid) equilibrium is the basis of any separation technology. No distinction is made between vapor and gas in this chapter.

Binary Mixtures

First of all, three special cases of vapor–liquid equilibrium of binary mixtures are presented qualitatively in Fig. 5.1-2. Considered are ideal mixtures (case A), mixtures with total miscibility gap in the liquid phase (case B), and mixtures with irreversible chemical reaction in the liquid phase (case C). By convention, the symbols x and y denote the molar fraction of the low-boiling component a in the liquid and the gas phase, respectively.

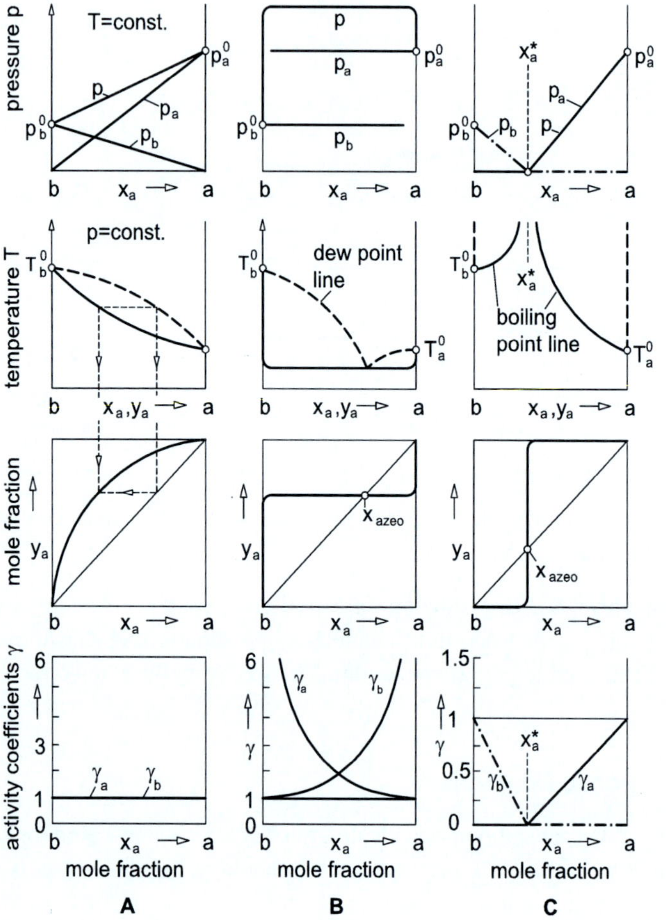

Fig. 5.1-2 Vapor–liquid equilibrium of three binary mixtures: (**A**) ideal system, (**B**) system with a total miscibility gap in the liquid phase, and (**C**) system with irreversible chemical reaction in the liquid phase

Ideal Mixtures (Case A)

Mixtures have a thermodynamic ideal behavior when the intermolecular forces are the same for all kinds of molecules present in the system. The partial pressure in the vapor phase is described by Dalton's law:

$$p_a = y_a \cdot p_t \quad \text{or} \quad y_a = \frac{p_a}{p_a + p_b}. \tag{5.1-1}$$

In the liquid phase, Raoult's law holds

$$p_a = x_a \cdot p_a^0 \quad \text{and} \quad p_b = x_b \cdot p_b^0 = (1 - x_a) \cdot p_b^0 . \tag{5.1-2}$$

From (5.1-1) and (5.1-2) follows:

$$y_a^* = \frac{p_a^0 \cdot x_a}{p_a^0 \cdot x_a + p_b^0 \cdot (1 - x_a)} \quad \text{with} \quad p_a^0 \text{ and } p_b^0 = f(T) . \tag{5.1-3}$$

As the vapor pressures p_a^o and p_b^o strongly depend on temperature, evaluation of (5.1-3) requires a knowledge of temperature. A simpler formulation of gas–liquid equilibrium is possible by using relative volatilities α according to the following definition:

$$\alpha_{ab} \equiv p_a^0 / p_b^0 . \tag{5.1-4}$$

Combining (5.1-3) and (5.1-4) yields

$$y_a^* = \frac{\alpha_{ab} \cdot x_a}{1 + (\alpha_{ab} - 1) \cdot x_a} \quad \text{or} \quad x_a^* = \frac{\alpha_{ab}^{-1} \cdot y_a}{1 + (\alpha_{ab}^{-1} - 1) \cdot y_a} . \tag{5.1-5}$$

The relative volatility α_{ab} is often constant, i.e., it does not depend on temperature. Therefore, application of (5.1-5) enables the calculation of gas–liquid equilibrium without knowledge of temperature.

Mixtures with Total Miscibility Gap in the Liquid Phase (Case B)

In mixtures with total miscibility gap each substance has its full vapor pressure. The individual vapor pressures are not weakened by the presence of the other component like in homogeneous mixtures (Fig. 5.1-3):

$$p_a = p_a^0 \quad \text{and} \quad p_b = p_b^0 \tag{5.1-6}$$

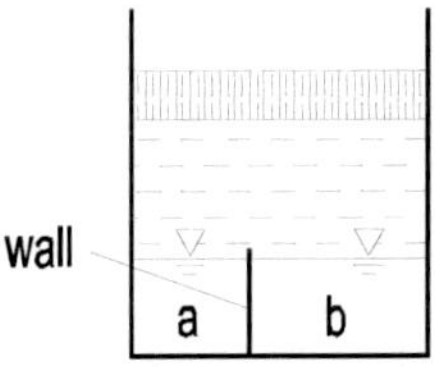

Fig. 5.1-3 System with a total miscibility gap in the liquid phase. The vapor pressures of components *a* and *b* are not weakened in the mixture

Within the miscibility gap the vapor has a constant concentration $y_a = p_a^0 / (p_a^0 + p_b^0)$. Therefore, the equilibrium line intersects the diagonal in the y/x diagram of Fig. 5.1-2B. At the point of intersection vapor and liquid have the same concentration. This feature is called *azeotrope* since it is very important to distillation.

Deviations from Raoult's law are accounted for by activity coefficients γ_i in (5.1-2), which transforms into

$$p_a = \gamma_a \cdot x_a \cdot p_a^0 = p_a^0 \quad \text{and} \quad p_b = \gamma_b \cdot x_b \cdot p_b^0 = p_b^0 . \tag{5.1-7}$$

In mixtures with total miscibility gap the following holds:

$$\gamma_a = 1/x_a \quad \text{and} \quad \gamma_b = 1/x_b . \tag{5.1-8}$$

Both activity coefficients are larger than 1. In real systems, however, there always exists some mutual miscibility of the substances and, in turn, the activity coefficients of diluted systems ($x \rightarrow 0$) are always smaller than infinite. These activity coefficients, called border activity coefficient γ_i^∞, very clearly reflect the intermolecular forces between the species involved.

Boiling temperature within the miscibility gap is easily determined from a p^o/T diagram (Fig. 5.1-4). The vapor pressure of one component, e.g., water, is plotted as difference $(p_t - p_i^0)$ vs. temperature. The point of intersection gives the boiling temperature (abscissa) and the gas concentrations p_i (ordinate) of the mixture. In the special case of total immiscibility, bubbles can only be generated at the interface between the coexisting phases.

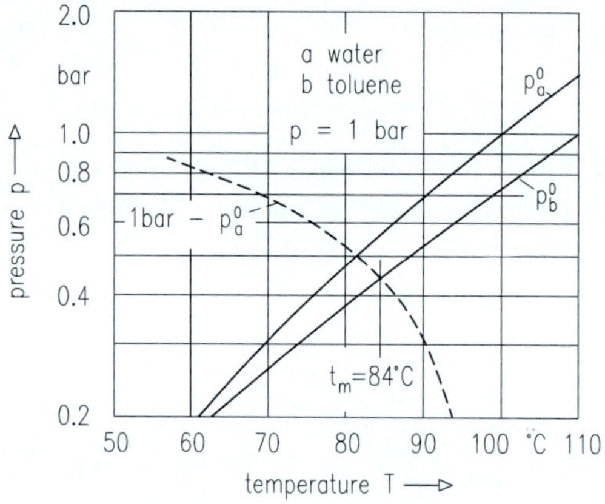

Fig. 5.1-4 Diagram for the determination of boiling temperatures of mixtures with total miscibility gap

Mixtures with Irreversible Chemical Reaction in the Liquid Phase (Case C)

In the third special case, shown in Fig. 5.1-2C, it is supposed that a fast and irreversible reaction takes place in the liquid phase:

$$a + \nu \cdot b \rightarrow z . \tag{5.1-9}$$

The reaction product z, being a very large molecule, is supposed to have a very low vapor pressure ($p_z^o \to 0$). Hence, the boiling temperature is very high at the stoichiometric concentration $x_a{}^* = 1/(\nu+1)$. There exists a high–boiling azeotrope at this concentration. Both activity coefficients are smaller than 1.

Most real systems show a behavior between the three special cases presented in Fig. 5.1-2. Some examples are shown in Fig. 5.1-5.

Azeotropes

Azeotropes are of great importance to distillation and rectification. At the azeotrope gas and liquid have the same concentration ($y = x$) and, in turn, no driving force for interfacial mass transfer exists. Azeotropic mixtures behave in some respects like pure substances. They cannot be fractionated by simple distillation. Azeotropes can exhibit a boiling point minimum (minimum azeotropes) or a boiling point maximum (maximum azeotropes). In multicomponent mixtures saddle point azeotropes with intermediate boiling temperature can also exist.

Ternary Mixtures

In ternary systems, the phase equilibrium can also be calculated from Dalton's and Raoult's laws. The following holds for ideal mixtures:

$$y_a^* = \frac{p_a^0 \cdot x_a}{p_a^0 \cdot x_a + p_b^0 \cdot x_b + p_c^0 \cdot x_c}. \tag{5.1-10}$$

The relative volatilities are defined as

$$\alpha_{ac} \equiv p_a^0/p_c^0, \quad \alpha_{bc} \equiv p_b^0/p_c^0, \quad \alpha_{cc} \equiv p_c^0/p_c^0 = 1. \tag{5.1-11}$$

With $x_c = 1 - x_a - x_b$, it follows

$$y_a^* = \frac{\alpha_{ac} \cdot x_a}{1 + (\alpha_{ac} - 1) \cdot x_a + (\alpha_{bc} - 1) \cdot x_b} = \frac{\alpha_{ac} \cdot x_a}{N_3}. \tag{5.1-12}$$

N_3 denotes the denominator of (5.1-12). The vapor concentrations of components *b* and *c* are

$$y_b^* = \frac{\alpha_{bc} \cdot x_b}{N_3} \quad \text{and} \quad y_c^* = \frac{x_c}{N_3} \quad \text{or} \quad y_c^* = 1 - y_a^* - y_b^*. \tag{5.1-13}$$

Some examples of vapor–liquid equilibria of ternary mixtures (zeotropic as well as azeotropic) are presented in Sect. 5.2.2.2.

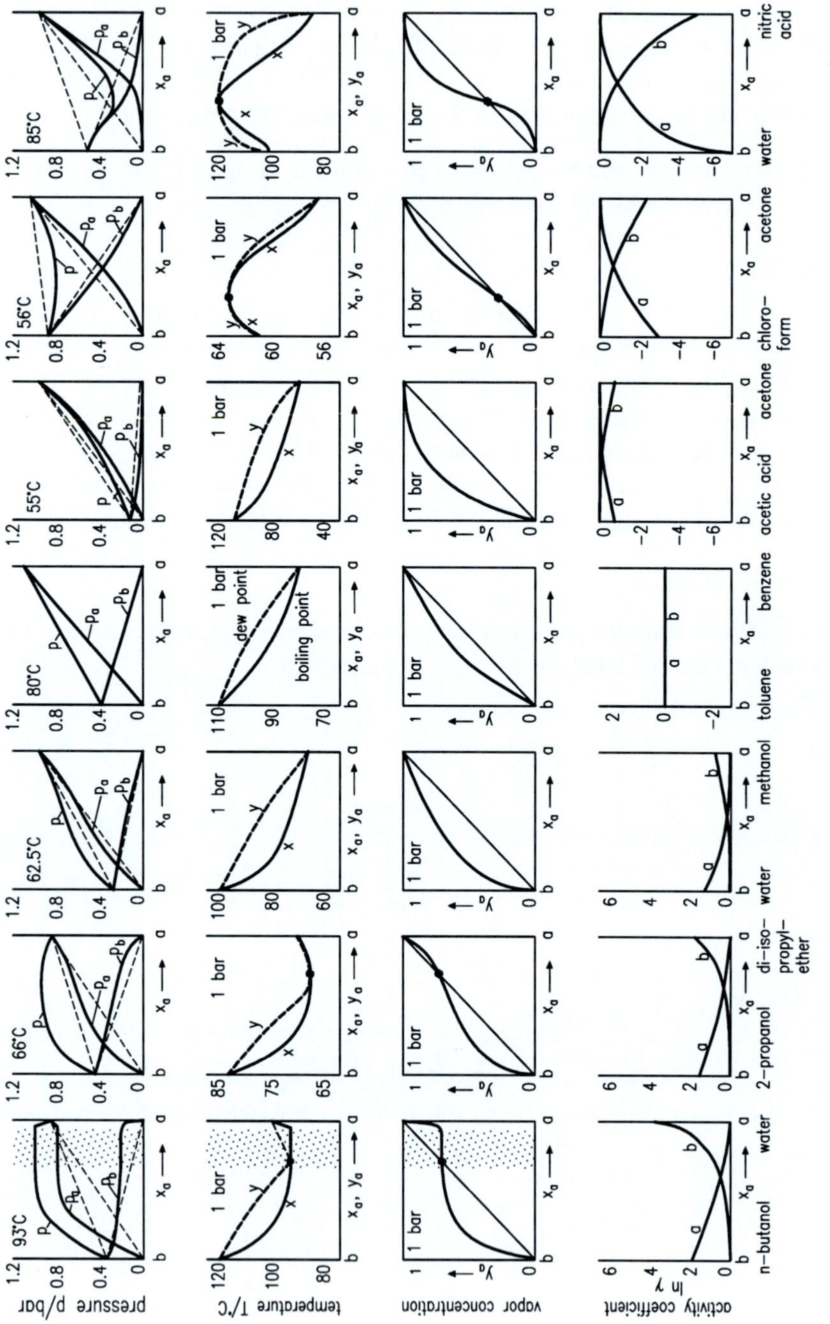

Fig. 5.1-5 Vapor–liquid equilibrium of selected real mixtures (Mersmann 1980)

Multicomponent Mixtures

Analogously, the vapor concentrations of ideal multicomponent mixtures follow from

$$y_i^* = \frac{\alpha_{ik} \cdot x_i}{1 + \sum_1^{k-1} (\alpha_{jk} - 1) \cdot x_j} . \qquad (5.1\text{-}14)$$

In the relative volatilities α_{ik} all individual vapor pressures p_i^0 are referred to the vapor pressure p_k^0 of the highest boiling component k.

Sometimes, the equilibrium ratio (or K-value) K_i is used for the formulation of the phase equilibrium of multicomponent mixtures. The definition of the equilibrium ratio K_i is

$$K_i \equiv y_i^* / x_i . \qquad (5.1\text{-}15)$$

Using equilibrium ratios K_i instead of relative volatilities α_{ik} often enables a simpler formulation of thermodynamic relationships. However, the numerical evaluation of such equations is more complex since the equilibrium ratios K_i strongly depend on temperature. The evaluation of the K-functions always requires an additional calculation of dew and boiling temperatures.

Practical Determination of Phase Equilibria

Many or even most liquid mixtures encountered in industrial practice exhibit a nonideal equilibrium behavior. In high-pressure systems both the vapor phase and the liquid phase can deviate from Dalton's and Raoult's laws (e.g., Prausnitz et al. 1998). In low-pressure systems deviations from Raoult's law prevail. The fundamental principles for formulating the thermodynamics of nonideal systems are presented in Chap. 2.

Very important for the practical determination of vapor–liquid equilibria are collections of experimental vapor–liquid-equilibrium (VLE) data. The collection of (Gmehling and Onken 1977ff) consists of about 20 volumes containing vapor–liquid equilibria of nearly all binary, ternary, and quaternary mixtures published so far. A collection of azeotropic data (Gmehling et al. 2004) lists data of 18,000 systems involving approximately 1,700 compounds.

5.1.1.3 Boiling Point, Dew Point

In distillation processes the liquid phase is at its boiling point and the gas phase at its dew point. As the phase equilibrium strongly depends on temperature, boiling and dew temperatures of mixtures have to be determined in process simulation.

Boiling Point

A liquid, consisting of a single substance only, is at its boiling point when the vapor pressure is equal to the total pressure p_t:

$$p^0 = p_t . \tag{5.1-16}$$

Analogously, a multicomponent liquid is at its boiling point when the sum of all partial pressures p_i equals the total pressure p_t.

$$\sum p_i = p_t \quad \text{or} \quad \sum p_i/p_t = 1 . \tag{5.1-17}$$

With $p_i/p_t = y_i$ follows:

$$\sum y_i = 1 . \tag{5.1-18}$$

Equation (5.1-18) demands that the first bubble meets the condition $\sum y_i = 1$. However, the vapor concentration y_i is a priori not known. It has to be determined from the liquid concentration via the equilibrium ratio K_i:

$$y_i = K_i \cdot x_i . \tag{5.1-19}$$

Combining (5.1-18) and (5.1-19) yields the boiling condition:

$$\sum K_i \cdot x_i = 1 . \tag{5.1-20}$$

In ideal systems the following relation holds:

$$K_i = p_i^0(T)/p_t = f(T, p_t) . \tag{5.1-21}$$

Therefore, the boiling condition of ideal mixtures is

$$\sum \frac{p_i^0(T)}{p_t} \cdot x_i = 1 . \tag{5.1-22}$$

From (5.1-22) follows that a subcooled liquid approaches boiling temperature either by increasing the temperature T or by decreasing the total pressure p_t.

The procedure for boiling point calculation is as follows. Firstly, a temperature T^1 is estimated. Secondly, the vapor pressure p^1 of the mixture is calculated from

$$p^1 = \sum x_i \cdot p_i^0(T^1) . \tag{5.1-23}$$

If the pressure p^1 is not equal to the system pressure p_t the calculation has to be repeated. A better estimation of the temperature is found by plotting the values T^1 and p^1 in a vapor pressure/temperature diagram and drawing a line parallel to the vapor pressure curves of the pure substances of the mixture at hand. After a few iterations the following condition will be met:

$$\sum x_i \cdot p_i^0(T)/p_t = 1 . \tag{5.1-24}$$

The values of the additive terms of (5.1-24) are the concentrations y_i of the gas phase.

Dew Point

In analogue to (5.1-18) the dew point condition of a gas mixture is

$$\sum x_i = 1 . \tag{5.1-25}$$

Hence, the first droplet generated has to meet the condition $\sum x_i = 1$. However, the concentration of the first droplet is unknown. It has to be calculated via the equilibrium ratio K_i as follows:

$$y_i = K_i \cdot x_i \rightarrow x_i = y_i/K_i \tag{5.1-26}$$

The dew point condition is

$$\sum y_i/K_i = 1 . \tag{5.1-27}$$

For ideal mixtures with $K_i = p_i^0(T)/p_t$:

$$\sum y_i \cdot p_t/p_i^0(T) = 1 . \tag{5.1-28}$$

The dew point of a superheated vapor is reached either by decreasing temperature or by increasing total pressure. The additive terms of (5.1-28) are the concentrations x_i of the first droplet.

In single component systems bubble point and dew point are represented by the same line, called vapor pressure curve (Fig. 5.1-6). In multicomponent systems

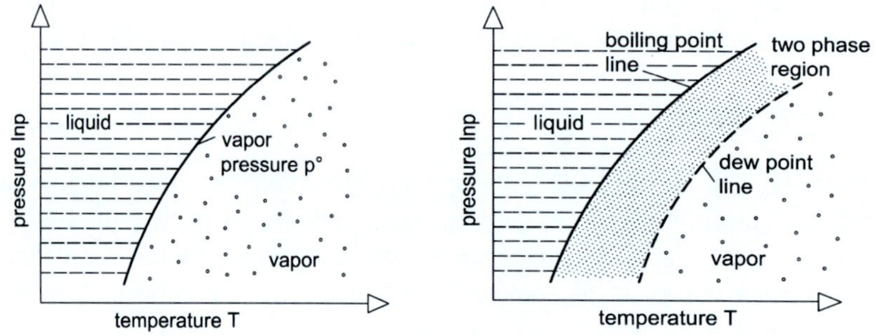

Fig. 5.1-6 Boiling point and dew point lines of coexisting vapor and liquid phases: (left) single component system; (right) multi component system

(mixtures), however, bubble point and dew point constitute separate curves with a region of two coexistent phases in between.

5.1.2 Continuous Closed Distillation

Closed distillation is performed as a continuous process in most applications. This process enables the separation of a liquid mixture into two fractions having different concentrations.

5.1.2.1 Binary mixtures

Thermodynamic calculation of continuous distillation is based on material balances:

$$\dot{F} = \dot{G} + \dot{L} \quad \text{and} \quad \dot{F} \cdot z_F = \dot{G} \cdot y + \dot{L} \cdot x . \tag{5.1-29}$$

Transformation yields the equation of the operating line:

$$y = -\dot{L}/\dot{G} \cdot x + (1 + \dot{L}/\dot{G}) \cdot z_F . \tag{5.1-30}$$

This equation formulates a linear relationship between gas and liquid concentration. As the coexisting phases have equilibrium state, the following condition applies:

$$y^* = \frac{\alpha \cdot x}{1 + (\alpha - 1) \cdot x} . \tag{5.1-31}$$

Combining (5.1-30) and (5.1-31) yields

$$x = \frac{-B + \sqrt{B^2 + 4 \cdot A \cdot z_F}}{2 \cdot A} \quad \text{with } A = (\alpha - 1) \cdot \left(1 - \frac{\dot{G}}{\dot{F}}\right) \text{ and}$$

$$B = 1 + (\alpha - 1) \cdot \left(\frac{\dot{G}}{\dot{F}} - z_F\right). \tag{5.1-32}$$

Often, (5.1-30) and (5.1-31) are solved graphically as demonstrated in Fig. 5.1-7. The point of intersection of operating and equilibrium lines gives the concentrations x_D and x_B of the two product fractions.

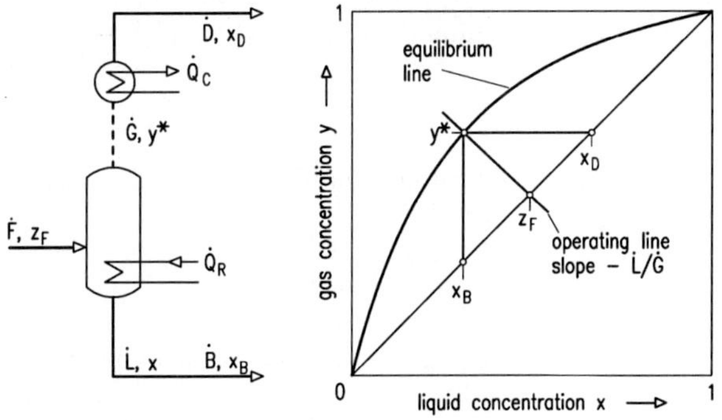

Fig. 5.1-7 Scheme of continuous distillation and graphical determination of product concentrations

5.1.2.2 Multicomponent Mixtures

Continuous distillation of multicomponent mixtures is also described by mass balances and phase equilibria:

$$\dot{F} = \dot{L} + \dot{G} \quad \text{and} \quad \dot{F} \cdot z_i = \dot{L} \cdot x_i + \dot{G} \cdot y_i. \tag{5.1-33}$$

After transformation:

$$y_i = z_i \cdot \left(1 + \frac{\dot{L}}{\dot{G}}\right) - x_i \cdot \frac{\dot{L}}{\dot{G}}. \tag{5.1-34}$$

Phase equilibrium is advantageously written with equilibrium ratios K_i (K-value):

$$y_i = K_i \cdot x_i \quad \rightarrow \quad x_i = y_i / K_i. \tag{5.1-35}$$

Combining (5.1-34) and (5.1-35) delivers

$$y_i = \frac{z_i \cdot (1 + \dot{L}/\dot{G})}{1 + (\dot{L}/\dot{G}) \cdot 1/K_i} \quad \text{for each component } i. \tag{5.1-36}$$

With the condition $\sum y_i = 1$ follows:

$$\sum \frac{z_i \cdot (1 + \dot{L}/\dot{G})}{1 + 1/K_i \cdot \dot{L}/\dot{G}} = 1 \quad \text{or} \quad \sum \frac{z_i \cdot K_i}{1 + \dot{G}/\dot{F} \cdot (K_i - 1)} = 1. \tag{5.1-37}$$

From a knowledge of temperature T, system pressure p_t, and overall system concentration z_i the ratio $\dot{L}/\dot{G}$ (or $\dot{G}/\dot{F}$) can be evaluated. The additive terms are the individual gas concentrations y_i. Liquid concentrations follow from

$$x_i = y_i/K_i. \tag{5.1-38}$$

Evaluation of (5.1-37) requires several iterations. Helpful is, for instance, the Newton Algorithm, see Stichlmair and Fair (1998).

Energy Demand

After evaluation of (5.1-37) the energy demand of multicomponent distillation is calculated from

$$\frac{\dot{Q}}{\dot{F}} = c_L \cdot (T - T_F) + \frac{\dot{G}}{\dot{F}} \cdot \sum (y_i \cdot r_i). \tag{5.1-39}$$

Here, r_i denotes the latent heat of evaporation whose values depend on temperature and species.

5.1.2.3 Flash Distillation

Distillation is a partial evaporation of a liquid mixture as gas and liquid have different concentrations. In most cases partial evaporation is performed by supply of heat. However, distillation can also be performed by decrease of pressure in a noz-

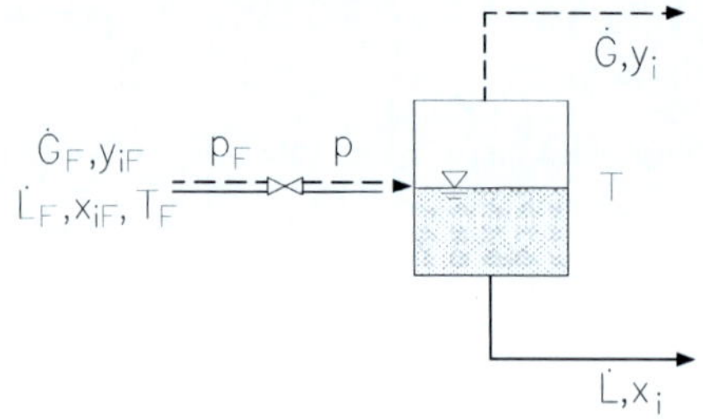

Fig. 5.1-8 Scheme of flash distillation by pressure reduction

zle (Fig. 5.1-8). The basis of thermodynamic calculation are material balances and phase equilibria:

$$\sum \frac{z_i \cdot (1 + \dot{L}/\dot{G})}{1 + 1/K_i \cdot \dot{L}/\dot{G}} = 1 \quad \text{or} \quad \sum \frac{z_i \cdot K_i}{1 + \dot{G}/\dot{F} \cdot (K_i - 1)} = 1\,. \tag{5.1-40}$$

Additionally, the energy balance has to be formulated. As the enthalpy of a fluid is constant at a nozzle, the following holds:

$$\dot{L}_F \cdot c_L \cdot T_F + \sum \dot{G}_F \cdot y_{Fi} \cdot (c_L \cdot T_F + r_i(T_F)) = \dot{L} \cdot c_L \cdot T + \sum \dot{G} \cdot y_i \cdot (c_L \cdot T + r_i(T)) \tag{5.1-41}$$

The vapor generated by flash distillation has a higher enthalpy than the liquid. The energy demand of the vapor formation is covered by a temperature decrease of the whole system. From (5.1-41) follows:

$$T_F - T = \frac{\dot{G} \cdot \sum y_i \cdot r_i(T) - \dot{G}_F \cdot \sum y_{Fi} \cdot r_i(T_F)}{\dot{F} \cdot c_L}\,. \tag{5.1-42}$$

In case of a liquid feed, (5.1-42) transforms into

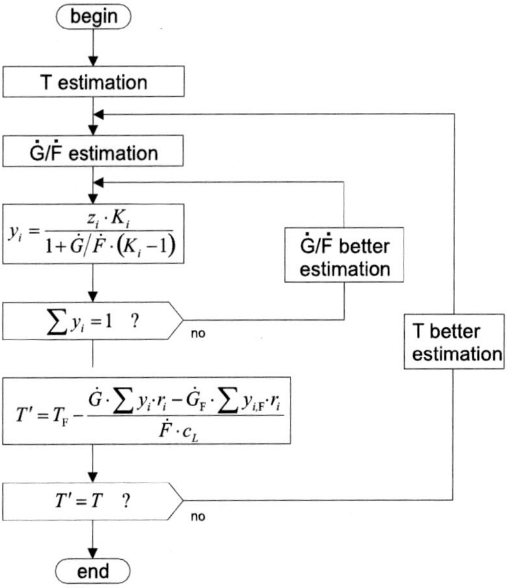

Fig. 5.1-9 Procedure for calculating flash distillation

$$T_F - T = \frac{\dot{G} \cdot \sum y_i \cdot r_i(T)}{\dot{F} \cdot c_L}. \tag{5.1-43}$$

Evaluation of (5.1-40) and (5.1-42) is very complex since two interlocking iterations have to be performed (see Fig. 5.1-9).

Industrial Example of Partial Condensation

A process for separating nitrogen from natural gas is shown in Fig. 5.1-10. The separation is performed by partial condensation and partial evaporation. The feed is cooled down and partially liquefied against the cold products. The two-phase system is flashed in a nozzle and, in turn, further cooled down. The two fractions after the nozzle are split in a vessel and heated up in separate product lines. The whole process can be simulated with the equations presented before. Especially complex is the evaluation of the cooling down and heating up curves in the enthalpy/temperature diagram of Fig. 5.1-10.

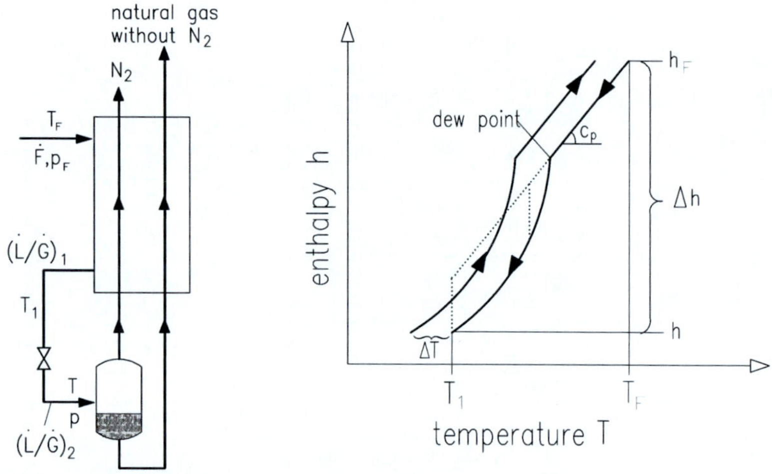

Fig. 5.1-10 Process for separating nitrogen from natural gas

5.1.3 Discontinuous Open Distillation (Batch Distillation)

In open distillation the vapor generated in the reboiler is steadily removed from the residual liquid. In practice this process is preferably performed as a discontinuous process, called batch distillation.

5.1.3.1 Binary Mixtures

The scheme of a batch distillation unit is shown in Fig. 5.1-11. The feed is charged into the vessel that is continuously heated by steam. The vapor is steadily removed, condensed, and collected in receivers that are changed periodically. During operation, amount and concentrations of the liquid in the vessel change. The concentration of the distillate D also varies over time.

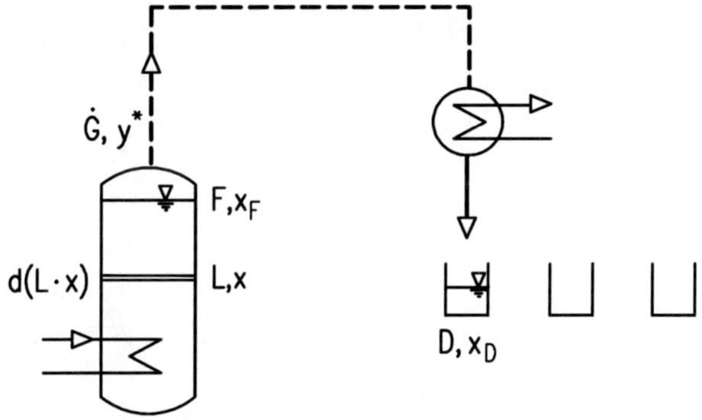

Fig. 5.1-11 Scheme of batch distillation

The material balance in a differential time $\mathrm{d}t$ yields

$$\dot{G} \cdot \mathrm{d}t + \mathrm{d}L = 0 \quad \text{or} \quad \frac{\mathrm{d}G}{\mathrm{d}t} \cdot \mathrm{d}t + \mathrm{d}L = 0\,. \tag{5.1-44}$$

After transformation:

$$\mathrm{d}G + \mathrm{d}L = 0 \quad \Rightarrow \quad \mathrm{d}G = -\mathrm{d}L\,. \tag{5.1-45}$$

Material balance of component i:

$$\dot{G} \cdot y_i^* \cdot \mathrm{d}t + \mathrm{d}(L \cdot x_i) = 0 \tag{5.1-46}$$

With $\dot{G} = \mathrm{d}G/\mathrm{d}t$ follows:

$$\frac{\mathrm{d}G}{\mathrm{d}t} \cdot \mathrm{d}t \cdot y_i^* + \mathrm{d}L \cdot x_i + L \cdot \mathrm{d}x_i = 0\,. \tag{5.1-47}$$

After some transformations:

$$\frac{\mathrm{d}x_i}{y_i^* - x_i} = \frac{\mathrm{d}L}{L}\,. \tag{5.1-48}$$

(5.1-48) is the well-known *Rayleigh equation*. It is important to note that the rising vapor and the residual liquid are in equilibrium state at each time.

Solution of the differential equation (5.1-48) requires knowledge of phase equilibrium. In ideal systems, the following simple equation holds:

$$y^* = \frac{\alpha \cdot x}{1 + (\alpha - 1) \cdot x}. \tag{5.1-49}$$

Combining (5.1-48) and (5.1-49) yields

$$\frac{dL}{L} = \frac{dx}{\dfrac{x \cdot \alpha}{1 + x \cdot (\alpha - 1)} - x}. \tag{5.1-50}$$

After integration:

$$\frac{L}{F} = \left(\frac{x}{x_F}\right)^{1/(\alpha-1)} \cdot \left(\frac{1 - x_F}{1 - x}\right)^{\alpha/(\alpha-1)} \quad \text{with} \quad D/F = 1 - L/F \tag{5.1-51}$$

A graphical plot of (5.1-51) is depicted in Fig. 5.1-12 for different values of the relative volatility α.

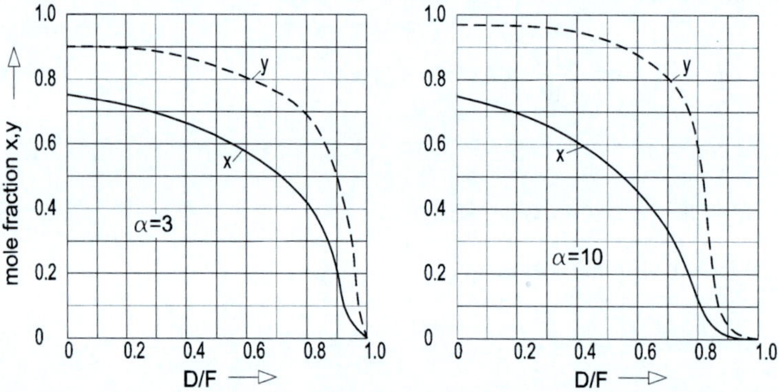

Fig. 5.1-12 Plot of gas and liquid concentrations vs. relative amount of distillate. Plots with different relative volatilities

The mean concentration of the distillate accumulated in the receiver follows from material balances:

$$D = F - L \quad \text{and} \quad D \cdot y_m = F \cdot x_F - L \cdot x. \tag{5.1-52}$$

The result is

$$y_m = x + \frac{x_F - x}{D/F}. \tag{5.1-53}$$

Figure 5.1-13 demonstrates the graphical determination of the mean concentration of the distillate in the receiver. Advantageously, the receivers are changed periodically (i.e., at constant values of D/F).

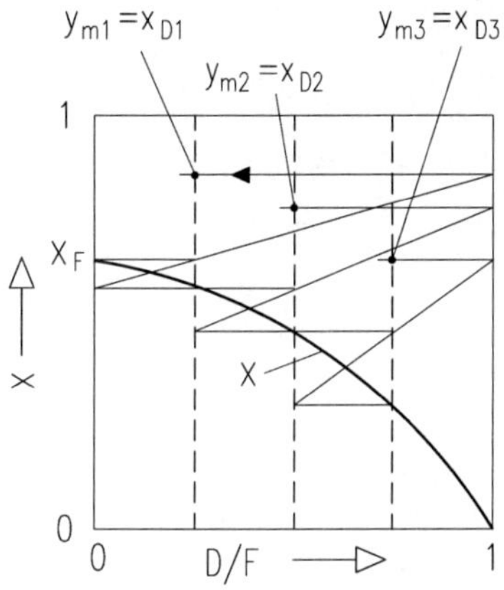

Fig. 5.1-13 Graphical determination of product concentrations of binary batch distillation

5.1.3.2 Batch Distillation of Ternary Mixtures

The Rayleigh equation (5.1-48) holds for each component i in a mixture. Application to ternary mixtures yields

$$\frac{dL}{L} = \frac{dx_a}{\overset{*}{y}_a - x_a} \quad \text{and} \quad \frac{dL}{L} = \frac{dx_b}{\overset{*}{y}_b - x_b}. \tag{5.1-54}$$

After elimination of dL/L:

$$\frac{dx_a}{dx_b} = \frac{\overset{*}{y}_a - x_a}{\overset{*}{y}_b - x_b}. \tag{5.1-55}$$

Equation (5.1-55) formulates the so-called *residuum line* of a liquid. This line describes the change of the state of the liquid phase during batch distillation as shown in the triangular concentration diagram in Fig. 5.1-14. The states of the coexistent vapor phase lie on a tangent to the residuum line.

In practice, the process of batch distillation is represented in diagrams shown in Fig. 5.1-15. Here, the concentrations in the gas and liquid phase are plotted vs. the relative amount of distillate. It is important to note that the intermediate boiling

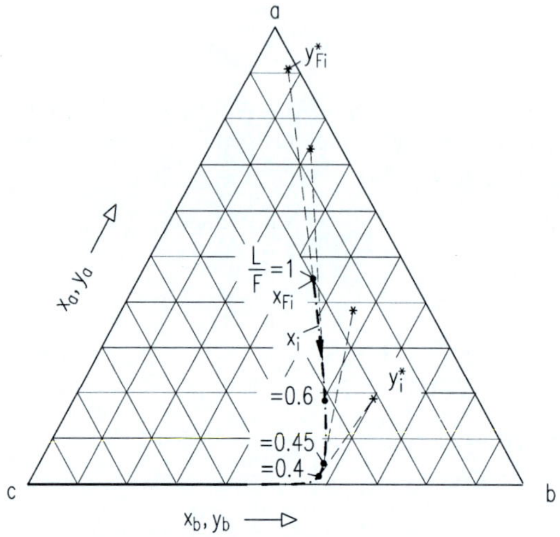

Fig. 5.1-14 Residuum line of a zeotropic ternary mixture

substances show concentration maxima. This is a very important feature also encountered in rectification processes (see Sect. 5.2.3).

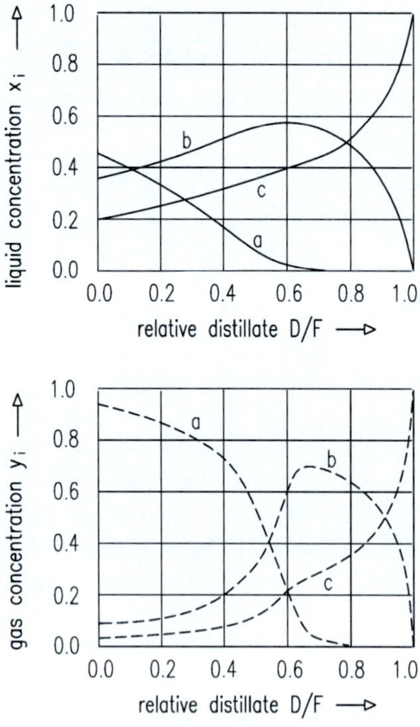

Fig. 5.1-15 Course of distillate and bottom fractions during batch distillation

5.2 Rectification

The technical term *rectification* comes from the latin words *recte facere* (= improve). Rectification denotes a change of concentrations beyond that of simple distillation. Rectification is a very effective separation technology that has the capability to fractionate a fluid mixture into all pure constituents.

5.2.1 Fundamentals

In principle, rectification is a multiple distillation as shown in Fig. 5.2-1. The concentrations of gas and liquid in all stages can be seen in the y/x diagram. Multiple distillation has the potential of producing two highly concentrated fractions. However, the yield of the process is rather poor since only a small fraction of the feed is concentrated up. Multiple distillation produces many side streams that do not meet product specifications.

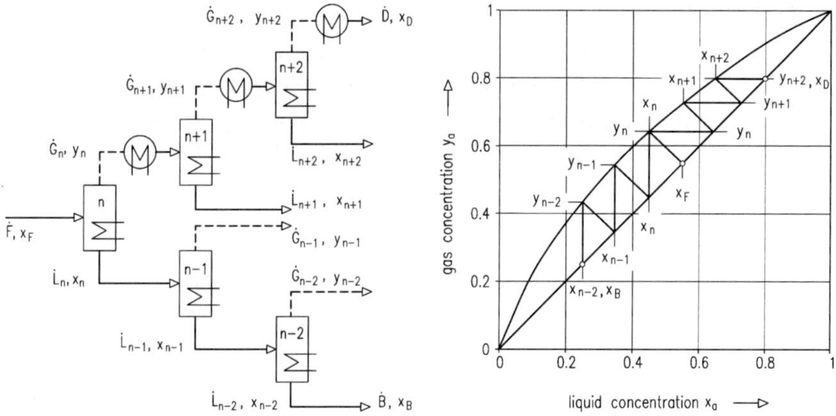

Fig. 5.2-1 Basic scheme of multiple distillation

The process of multiple distillation is significantly improved by two modifications (Fig. 5.2-2). The first modification is the recycling of the side streams ($\dot{L}$ in the upper stages, $\dot{G}$ in the lower stages) within the process. This measure increases the yield of the process drastically. The second modification is the removal of internal heat exchangers. For instance, the vapor from stage n does not need to be condensed before stage $(n+1)$ and reevaporated in stage $(n+1)$. Both gas and liquid can be directly brought into intimate contact.

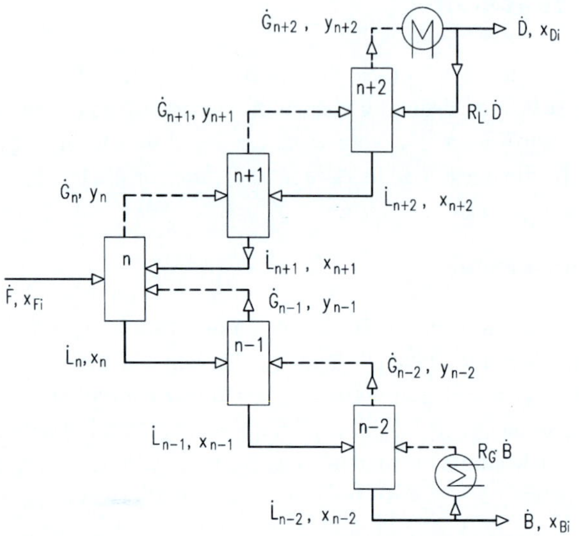

Fig. 5.2-2 Modified scheme of multiple distillation (rectification)

The process scheme of Fig. 5.2-2 is, in principle, identical with the scheme shown in Fig. 5.2-3. Here, the distillation stages are arranged in a vertical cascade with countercurrent flow of gas and liquid. From a material balance follows:

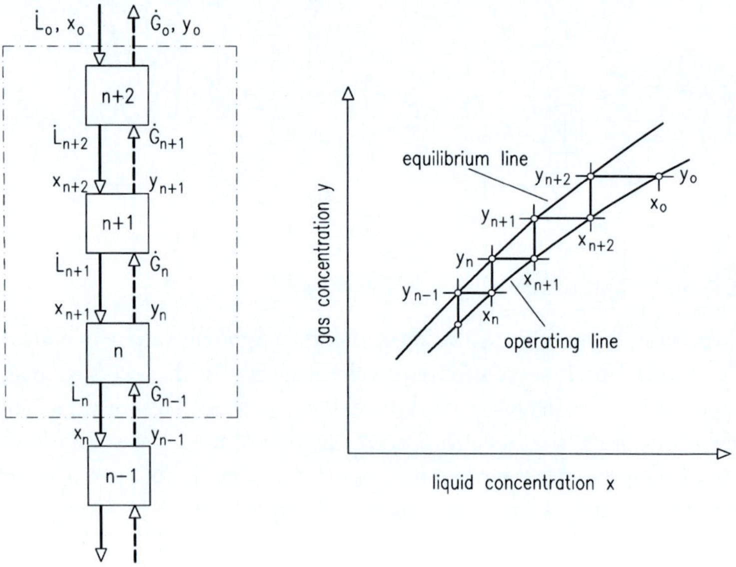

Fig. 5.2-3 Cascade of multiple distillation

$$y_{n-1} = \frac{\dot{L}_n}{\dot{G}_{n-1}} \cdot x_n + \frac{\dot{G}_0}{\dot{G}_{n-1}} \cdot y_0 - \frac{\dot{L}_0}{\dot{G}_{n-1}} \cdot x_0 . \qquad (5.2\text{-}1)$$

This relationship describes the operating line of multistage distillation. It formulates a relation between liquid and gas concentration within the cascade. In Fig. 5.2-3, the operating line and the equilibrium line are plotted in an y/x diagram. The distance between these two lines represents the driving force of the interfacial mass transfer. Distance and length of these lines indicate the difficulty of the separation performed in the cascade. It can be expressed either by the number of equilibrium stages or by the number of transfer units.

5.2.1.1 Concept of Equilibrium Stages

The concept of equilibrium stages is based on the idea of multiple distillation. The number of equilibrium stages required for a specified separation determines the height of a column. In the special case of linear operating and equilibrium lines the number n of equilibrium stages follows from:

$$n = \frac{1}{\ln J} \cdot \ln[(1 - 1/J) \cdot Q + 1] . \qquad (5.2\text{-}2)$$

J denotes the ratio of the slopes of operating and equilibrium lines and Q the ratio of the concentration change performed to the concentration difference at the end of the cascade.

5.2.1.2 Concept of Transfer Units

The concept of transfer units is based on the fundamental law of mass transfer that formulates a linear relationship between concentration change $\mathrm{d}y$ and driving force $(y^* - y)$. The number of transfer units N_{OG} is defined as ratio of these two quantities:

$$\mathrm{d}N_{\mathrm{OG}} = \frac{\mathrm{d}y}{y^* - y} \qquad (5.2\text{-}3)$$

The driving force $(y^* - y)$ is the vertical distance between equilibrium and operating line. In case of linear equilibrium and operating lines the following holds:

$$N_{\mathrm{OG}} = \frac{1}{1 - 1/J} \cdot \ln[(1 - 1/J) \cdot Q + 1] . \qquad (5.2\text{-}4)$$

5.2.1.3 Comparison of Both Concepts

In principle, the concept of equilibrium stages and the concept of transfer units are equivalent. However, the concept of equilibrium stages is preferred in practice because it is easier to understand and simpler to apply. The following relationship holds for linear equilibrium and operating lines:

$$N_{OG} = n \cdot \frac{\ln(1/J)}{1/J - 1} \quad \text{with} \quad n = N_{OG} \text{ for } J = 1 \,. \tag{5.2-5}$$

If both lines are parallel (i.e., $J = 1$) then the number N_{OG} of transfer units is equal to the number n of equilibrium stages.

5.2.2 Continuous Rectification

Continuous operation is the preferred mode of rectification in particular when large amounts of fluids have to be processed. The feed $\dot{F}$ is fractionated into an overhead fraction $\dot{D}$ and a bottom fraction $\dot{B}$ with significantly different concentrations.

5.2.2.1 *Binary mixtures*

The thermodynamic fundamentals of rectification are best explained by considering binary mixtures first.

Column Simulation with Material Balances

If the heats of vaporization of all components of the mixture do not differ very much, the process of binary rectification can be simulated on the basis of material balances alone. In Fig. 5.2-4 several balance envelopes that enable the derivation of fundamental relationships are marked.

Balance I

A material balance over balance envelope I delivers

$$\dot{F} = \dot{D} + \dot{B} \quad \text{and} \quad \dot{F} \cdot z_F = \dot{D} \cdot x_D + \dot{B} \cdot x_B \,. \tag{5.2-6}$$

From both equations follows:

$$\dot{D} = \dot{F} \cdot \frac{z_F - x_B}{x_D - x_B} \quad \text{or} \quad \dot{B} = \dot{F} \cdot \frac{x_D - z_F}{x_D - x_B} \,. \tag{5.2-7}$$

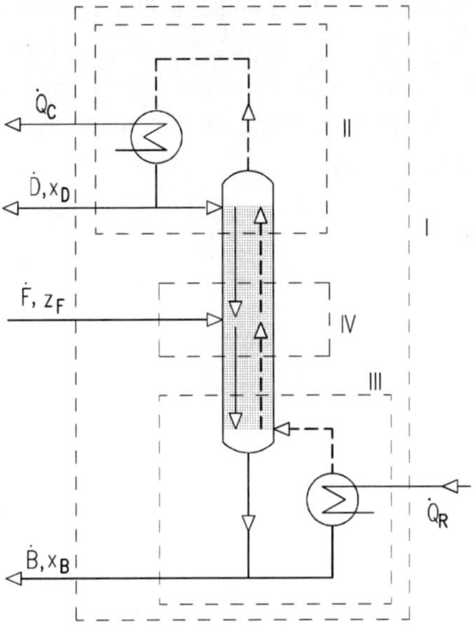

Fig. 5.2-4 Flow sheet of a rectification column with several material balance envelopes

After having specified the product qualities, the amount of overhead $\dot{D}$ and bottom product $\dot{B}$ can be calculated from (5.2-7).

Material Balance II

The material balance II yields

$$\dot{G} = \dot{L} + \dot{D} \quad \text{and} \quad \dot{G} \cdot y = \dot{L} \cdot x + \dot{D} \cdot x_D . \tag{5.2-8}$$

Rewriting gives the equation of the operating line in the upper section of the column, called rectifying section:

$$y = \frac{\dot{L}}{\dot{G}} \cdot x + \left(1 - \frac{\dot{L}}{\dot{G}}\right) \cdot x_D . \tag{5.2-9}$$

The term $\dot{L}/\dot{G}$ is the internal reflux ratio. In practice, the external reflux ratio R_L is preferred:

$$R_L \equiv \dot{L}/\dot{D} . \tag{5.2-10}$$

With $\dot{G} = \dot{L} + \dot{D}$ follows:

$$y = \frac{R_L}{R_L + 1} \cdot x + \frac{1}{R_L + 1} \cdot x_D . \tag{5.2-11}$$

Equation (5.2-11) formulates the operating line in the rectifying section.

Balance III

From balance III follows the operating line in the lower section of the column, which is called stripping section:

$$y = \frac{\dot{L}}{\dot{G}} \cdot x + \left(1 - \frac{\dot{L}}{\dot{G}}\right) \cdot x_B . \tag{5.2-12}$$

The definition of the external reboil ratio is

$$R_G \equiv \dot{G}/\dot{B} . \tag{5.2-13}$$

With $\dot{L} = \dot{G} + \dot{B}$ follows:

$$y = \frac{R_G + 1}{R_G} \cdot x - \frac{1}{R_G} \cdot x_B . \tag{5.2-14}$$

Equation (5.2-14) is the operating line in the stripping section formulated with the external reboil ratio.

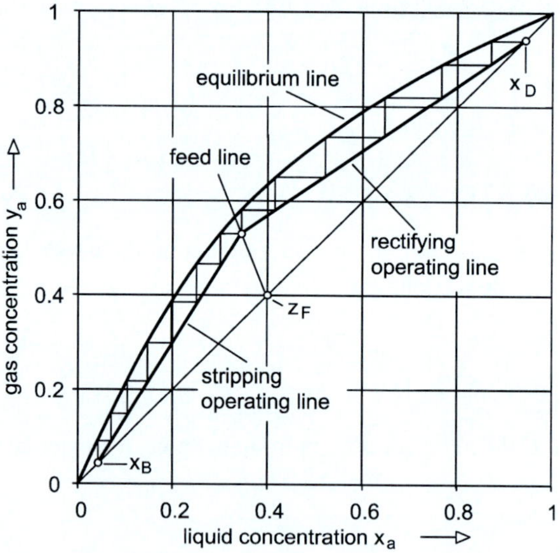

Fig. 5.2-5 McCabe–Thiele diagram with operating lines, feed line, and equilibrium line. The stages drawn between operating and equilibrium lines are a measure for the difficulty of the separation

The operating lines in the rectifying and the stripping section are depicted in Fig. 5.2-5. Additionally, the equilibrium line is shown. This diagram, called McCabe–Thiele diagram, represents the thermodynamics of binary distillation. The operating lines intersect the diagonal at the product concentrations x_D and x_B, respectively. The operating lines must not touch or intersect the equilibrium lines since the distance between these two lines is the driving force for mass transfer.

This condition can always be met by variation of the reflux R_L or the reboil ratio R_G, which are correlated by

$$R_G = \frac{(\dot{D}/\dot{F}) \cdot (R_L + 1) - (1 - q_F)}{1 - (\dot{D}/\dot{F})} . \tag{5.2-15}$$

The term q_F denotes the caloric state of the feed according to the definition

$$q_F \equiv \dot{L}_F / \dot{F} . \tag{5.2-16}$$

In a two-phase system the value of q_F represents the fraction of liquid in the feed, i.e., boiling liquid $q_F = 1$ and saturated vapor $q_F = 0$. The caloric state of the feed can also be defined by enthalpies:

$$q_F = \frac{\text{enthalpy required for feed vaporization}}{\text{latent heat of vaporization}} . \tag{5.2-17}$$

A subcooled liquid is characterized by $q_F > 1$, a superheated vapor by $q_F < 0$.

Feed Line

As shown in Fig. 5.2-5, the operating lines intersect each other at the so-called feed line, which passes through the feed concentration at the diagonal. Material balance IV of Fig. 5.2-4 delivers

$$\dot{F} \cdot z_F + \dot{L} \cdot x + \dot{G}' \cdot y = \dot{G} \cdot y + \dot{L}' \cdot x . \tag{5.2-18}$$

$\dot{G}'$ and $\dot{L}'$ represent the flow rate of gas and liquid below the feed point. With $\dot{L}_F = \dot{F} \cdot q_F$ and $\dot{G}_F = \dot{F} \cdot (1 - q_F)$ the equation of the feed line can be derived:

$$y = -\frac{q_F}{1 - q_F} \cdot x + \frac{z_F}{1 - q_F} . \tag{5.2-19}$$

Equation (5.2-19) is graphically represented in Fig. 5.2-6. All feed lines emerge from the feed concentration z_F at the diagonal of the McCabe–Thiele diagram.

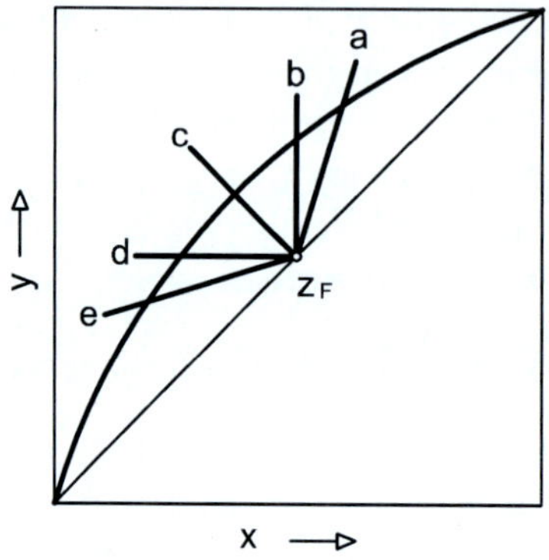

a: subcooled liquid $q_F > 1$

b: boiling liquid $q_F = 1$

c: vapor–liquid mixture $0 < q_F < 1$

d: saturated vapor $q_F = 0$

e: superheated vapor $q_F < 0$

Fig. 5.2-6 Positions of the feed line for different caloric states q_F of the feed

Multiple Feed

Rectification columns are often operated with several feed streams. At each feed point the internal flow rate of liquid $\dot{L}$ or vapor $\dot{G}$ changes. As the slope of the operating line is $\dot{L}/\dot{G}$, the operating line bends sharply at each feed point, see Fig. 5.2-7.

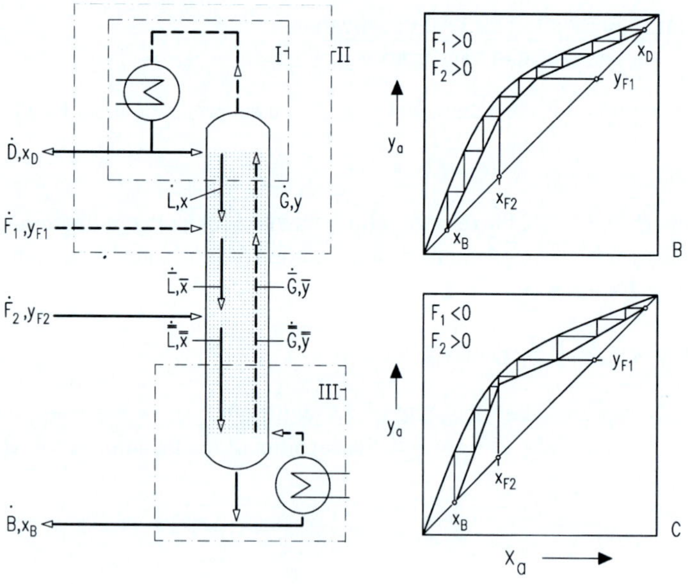

Fig. 5.2-7 Operating lines of multiple feed columns

Reflux and Reboil Ratios

Reflux and reboil ratios are very important parameters for column operation. They must not fall below critical values, called minimum reflux and minimum reboil, respectively. Two special cases have to be considered: total reflux (reboil) and minimum reflux (reboil).

Operation with Total Reflux

At operation with total reflux, no overhead product is withdrawn from the column since all overheads are recycled into the column. In this mode of operation the operating line becomes identical with the diagonal of the McCabe–Thiele diagram (Fig. 5.2-8). This special case, requiring the smallest number of equilibrium stages, can be easily treated for ideal systems.

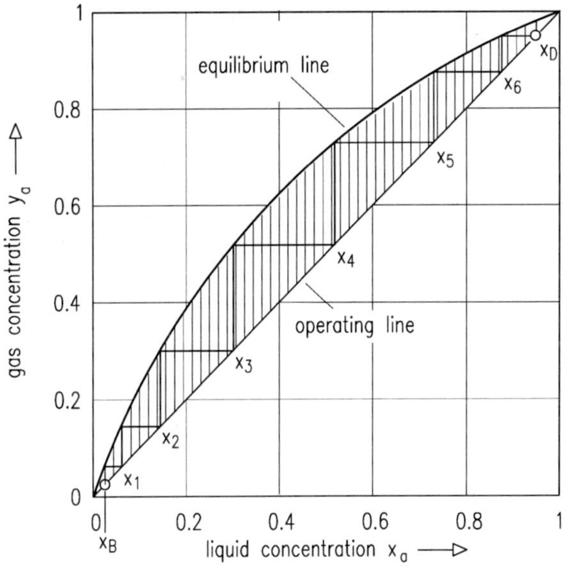

Fig. 5.2-8 McCabe–Thiele diagram at total liquid reflux and reboil, respectively

Operating line:

$$y = x . \tag{5.2-20}$$

Equilibrium line:

$$y = \frac{\alpha \cdot x}{1 + (\alpha - 1) \cdot x} \quad \text{or} \quad \frac{y}{1-y} = \alpha \cdot \frac{x}{1-x} . \tag{5.2-21}$$

Concentration changes within the column:

$$\text{Stage 1: } \frac{x_1}{1-x_1} = \alpha \cdot \frac{x_0}{1-x_0}, \tag{5.2-22}$$

$$\text{Stage 2: } \frac{x_2}{1-x_2} = \alpha \cdot \frac{x_1}{1-x_1} = \alpha^2 \cdot \frac{x_0}{1-x_0}, \tag{5.2-23}$$

$$\text{Stage 3: } \frac{x_3}{1-x_3} = \alpha \cdot \frac{x_2}{1-x_2} = \alpha^3 \cdot \frac{x_0}{1-x_0}, \tag{5.2-24}$$

$$\text{Stage } n\text{: } \frac{x_n}{1-x_n} = \alpha^n \cdot \frac{x_0}{1-x_0}, \quad \text{or} \quad x_n = \frac{\alpha^n \cdot x_0}{1+(\alpha^n - 1)\cdot x_0} \tag{5.2-25}$$

Rewriting yields the number of equilibrium stages required for a separation between x_D and x_B:

$$n_{\min} = \frac{1}{\ln\alpha} \cdot \ln\left(\frac{x_D}{1-x_D} \cdot \frac{1-x_B}{x_B}\right). \tag{5.2-26}$$

This equation enables a rough estimation of the number of equilibrium stages required for any specified product qualities.

Operation with Minimum Reflux

As is seen from Fig. 5.2-9, at operation with minimum reflux, the operating line intersects the equilibrium line at the feed concentration x_F (for $q_F = 1$). This point of intersection is called pinch.

The slope of the operating line is

$$\left(\frac{\dot{L}}{\dot{G}}\right)_{\min} = \frac{x_D - y_F^*}{x_D - x_F} \quad \text{internal reflux ratio} \tag{5.2-27}$$

Replacing the internal reflux ratio $\dot{L}/\dot{G}$ by the external one yields

$$R_{L\min} = \frac{x_D - y_F^*}{y_F^* - x_F}. \tag{5.2-28}$$

Equation (5.2-28) holds for all components in the mixture. For sharp separations of ideal systems:

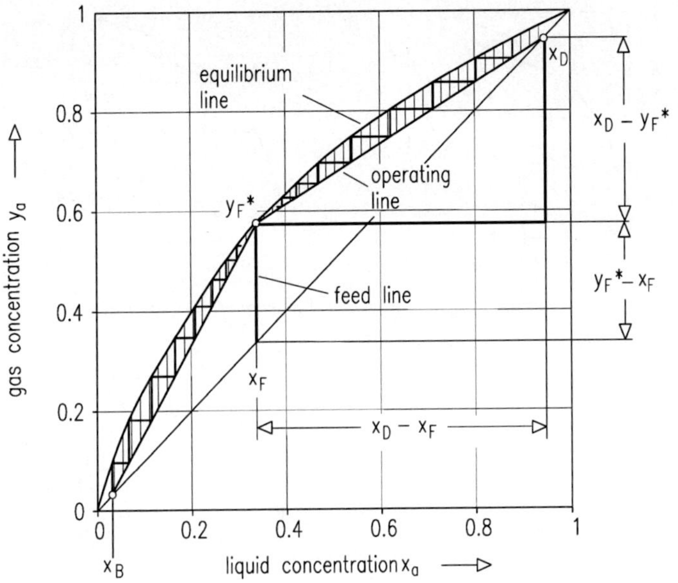

Fig. 5.2-9 McCabe–Thiele diagram at minimum reflux and reboil, respectively

$$R_{\text{Lmin}} = \frac{0 - y^*_{\text{F}b}}{y^*_{\text{F}b} - x_{\text{F}b}} \quad \text{and} \quad y^*_{\text{F}b} = \frac{\alpha_{bb} \cdot x_{\text{F}b}}{1 + (\alpha_{ab} - 1) \cdot x_{\text{F}a}}. \tag{5.2-29}$$

With $\alpha_{bb} = 1$ follows:

$$R_{\text{Lmin}} = \frac{1}{(\alpha_{ab} - 1) \cdot x_{\text{F}a}} \quad \text{for } q_{\text{F}} = 1 \quad \text{and} \quad x_{\text{D}b} = 0. \tag{5.2-30}$$

Equation (5.2-30) enables a very simple calculation of the minimum energy demand of sharp separations (see the next section). For nonsharp separations, the respective relation is

$$R_{\text{Lmin}} = \frac{1}{(\alpha_{ab} - 1)} \cdot \left(\frac{x_{\text{D}a}}{x_{\text{F}a}} - \alpha_{ab} \cdot \frac{1 - x_{\text{D}a}}{1 - x_{\text{F}a}} \right). \tag{5.2-31}$$

Some relations for minimum reflux and minimum reboil for liquid feed ($q_{\text{F}} = 1$) and vapor feed ($q_{\text{F}} = 0$) are presented in Fig. 5.2-10. All these relations are valid for sharp separations (i.e., pure products) only.

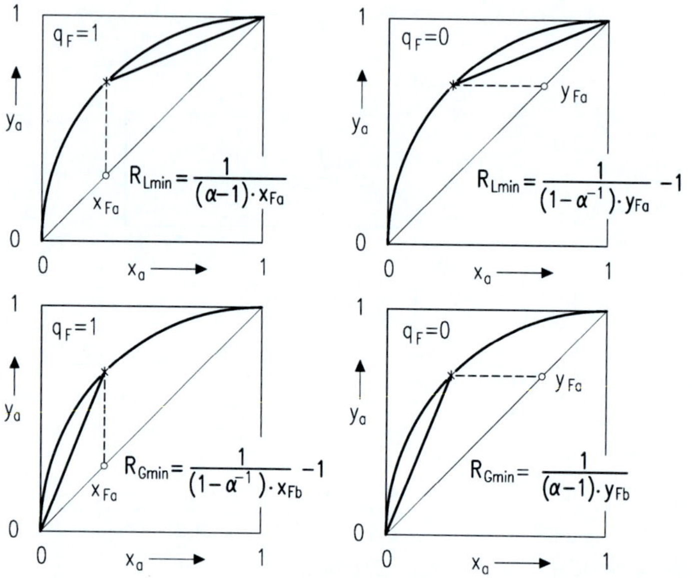

Fig. 5.2-10 Minimum reflux and minimum reboil ratios at different caloric states q_F of the feed

Energy Demand

The most important characteristic of rectification processes is the energy demand of the column. It can be determined without rigorous column simulation. The basic relation is

$$\dot{Q}_R = \dot{G}_{bottom} \cdot r \text{ and } \dot{Q}_c = -\dot{G}_{head} \cdot r \quad . \tag{5.2-32}$$

At the bottom:

$$\dot{G}_{bottom} = R_G \cdot \dot{B}. \tag{5.2-33}$$

From both equations follows:

$$\dot{Q}_R = R_G \cdot \dot{B} \cdot r \, . \tag{5.2-34}$$

An analogous relation holds at the top of the column with $R_L = \dot{L}/\dot{D}$:

$$\dot{L} = R_L \cdot \dot{D} \, . \tag{5.2-35}$$

After transformation:

$$\dot{Q}_\mathrm{C} = -r \cdot (R_\mathrm{L} + 1) \cdot \dot{D} . \tag{5.2-36}$$

Implementing (5.2-7) yields

$$\frac{\dot{Q}_\mathrm{R}}{\dot{F} \cdot r} = R_\mathrm{G} \cdot \frac{x_\mathrm{D} - z_\mathrm{F}}{x_\mathrm{D} - x_\mathrm{B}} \quad \text{or} \quad \frac{\dot{Q}_\mathrm{C}}{\dot{F} \cdot r} = -(R_\mathrm{L} + 1) \cdot \frac{z_\mathrm{F} - x_\mathrm{B}}{x_\mathrm{D} - x_\mathrm{B}} . \tag{5.2-37}$$

The energy demand depends on reflux ratio R_L and reboil ratio R_G, respectively. Both quantities cannot fall below minimum values. Hence, rectification columns have a minimum energy demand. In practice rectification columns are operated with an energy surplus of 10%–20%.

Minimum Energy Demand

For sharp separations ($x_\mathrm{D} = 1$) and boiling liquid feed ($q_\mathrm{F} = 1$) the minimum energy demand is

$$\frac{\dot{Q}_\mathrm{min}}{\dot{F} \cdot r} = (R_\mathrm{Lmin} + 1) \cdot x_\mathrm{F} . \tag{5.2-38}$$

Implementing (5.2-30) delivers

$$\frac{\dot{Q}_\mathrm{min}}{\dot{F} \cdot r} = \frac{1}{\alpha - 1} + x_\mathrm{F} \quad \text{for} \quad q_\mathrm{F} = 1 . \tag{5.2-39}$$

For small values of the relative volatility α the energy demand is nearly independent of feed composition.

For saturated vapor feed ($q_\mathrm{F} = 0$):

$$\frac{\dot{Q}_\mathrm{min}}{\dot{F} \cdot r} = R_\mathrm{Gmin} \cdot \frac{1 - y_{\mathrm{F}a}}{1 - 0} . \tag{5.2-40}$$

Replacing R_Gmin by the respective relation in Fig. 5.2-10 yields

$$\frac{\dot{Q}_\mathrm{min}}{\dot{F} \cdot r} = \frac{1}{\alpha - 1} \quad \text{for} \quad q_\mathrm{F} = 0 . \tag{5.2-41}$$

When operated with a saturated vapor feed the energy demand of a rectification column does not depend on feed concentration at all. This is a very important fact.

A comparison of minimum energy demands of different caloric states of the feed is shown in Fig. 5.2-11. A vaporous feed requires less energy than a liquid feed. However, the latent heat of the vapor is only utilized in the rectifying section of the

$\frac{\dot{Q}_C}{\dot{F}\cdot r} = -(x_{F\alpha} + \frac{1}{\alpha-1})$	$= -(1 + \frac{1}{\alpha-1})$	$= -(y_{F\alpha} + \frac{1}{\alpha-1})$	$= -(1 + \frac{1}{\alpha-1})$

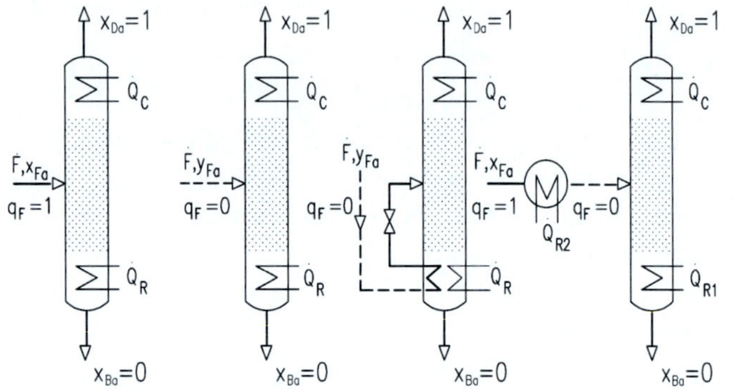

$\frac{\dot{Q}_R}{\dot{F}\cdot r} = x_{F\alpha} + \frac{1}{\alpha-1}$	$= \frac{1}{\alpha-1}$	$= y_{F\alpha} + \frac{1}{\alpha-1} - 1$	$= 1 + \frac{1}{\alpha-1}$

Fig. 5.2-11 Minimum energy demand of binary distillation for different caloric states q_F of the feed

column. A better mode of operation is heating the reboiler by the vapor feed before feeding it into the column. This mode of operation is, however, only possible if a sufficient pressure difference between feed and column exists. A prevaporization of a liquid feed is not recommended since the energy required for prevaporization is only utilized in the rectifying section of the column.

Column Simulation with Material and Enthalpy Balances

A rigorous column simulation requires, besides material balances, the implementation of enthalpy balances. In binary mixtures the enthalpies of gas and liquid streams can be taken from enthalpy/concentration diagrams.

Enthalpy/Concentration Diagram

The enthalpy/concentration diagram of a binary mixture is qualitatively shown in Fig. 5.2-12. For developing a quantitative h/x diagram the following data are required:

- ***Boiling lens***. For ideal systems the boiling lens can easily be calculated with Raoul's and Dalton's laws.
- Heat capacities of gas and liquid as well as latent heats of vaporization of the pure constituents.

- ***Heats of mixing***. In the vapor phase the heats of mixing can be neglected. The same holds for the liquid phase of ideal mixtures.

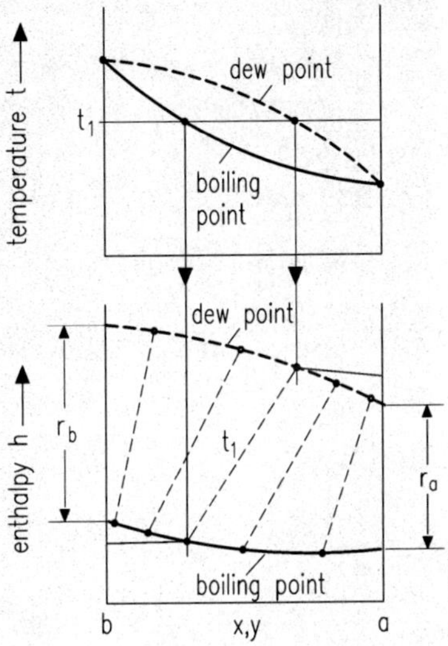

Fig. 5.2-12 Relationship between temperature/concentration diagram and enthalpy/concentration diagram

The relationship between boiling lens and h/x diagram are graphically presented in Fig. 5.2-12. In principle, the intensive quantity temperature is replaced by the extensive quantity enthalpy. The isotherms drawn between dew point and boiling point lines represent the phase equilibrium.

Representation of Rectification in Enthalpy/Concentration Diagrams

An enthalpy balance over the whole column (balance envelope I in Fig. 5.2-4) delivers the relationship

$$\dot{F} \cdot h_{\text{F}} + \dot{Q}_{\text{R}} = \dot{D} \cdot h_{\text{D}} + \dot{B} \cdot h_{\text{B}} + \dot{Q}_{\text{C}} . \tag{5.2-42}$$

After rewriting:

$$\dot{F} \cdot h_{\text{F}} = \dot{D} \cdot h_{\text{D}} + \dot{Q}_{\text{C}} + \dot{B} \cdot h_{\text{B}} - \dot{Q}_{\text{R}} . \tag{5.2-43}$$

By convention, the heats $\dot{Q}_{\text{C}}$ and $\dot{Q}_{\text{R}}$ are referred to their respective streams $\dot{D}$ and $\dot{B}$, respectively:

$$\dot{F} \cdot h_{\text{F}} = \dot{D} \cdot \left(h_D + \frac{\dot{Q}_{\text{C}}}{\dot{D}} \right) + \dot{B} \cdot \left(h_{\text{B}} - \frac{\dot{Q}_{\text{R}}}{\dot{B}} \right) . \tag{5.2-44}$$

With the definition $\pi_{\text{D}} = h_{\text{D}} + \dot{Q}_{\text{C}}/\dot{D}$ and $\pi_{\text{B}} = h_{\text{B}} - \dot{Q}_{\text{R}}/\dot{B}$ follows:

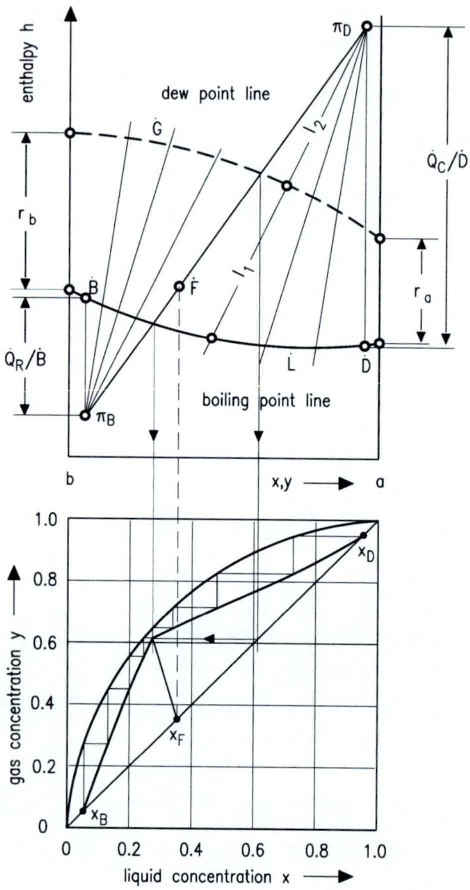

Fig. 5.2-13 Calculation of binary rectification by use of an enthalpy/concentration diagram

$$\dot{F} \cdot h_F = \dot{D} \cdot \pi_D + \dot{B} \cdot \pi_B . \tag{5.2-45}$$

This linear relationship demands that the states of the feed $\dot{F}$, the poles π_D, and π_B are collinear in the h/x diagram (Fig. 5.2-13).

Analogously, a balance over the upper column section (balance II in Fig. 5.2-4) gives

$$\dot{G} \cdot h_G = \dot{L} \cdot h_L + \dot{D} \cdot \pi_D . \tag{5.2-46}$$

Equation (5.2-46) represents a group of straight lines, all emerging from the pole π_D of the distillate. Their points of intersection with dew point and boiling point lines constitute the operating line in the McCabe–Thiele diagram. Generally, the operating lines are not straight but curved lines. Drawing steps between operating and equilibrium lines delivers the number of equilibrium stages required for the specified separation.

Conditions for Straight Operating Lines

The operating line in the McCabe–Thiele diagram (Fig. 5.2-5) is a straight line if the flow rates of gas $\dot{G}$ and liquid $\dot{L}$ are constant within the column (constant molar overflow).

$$\dot{L}/\dot{G} = \text{const.} \tag{5.2-47}$$

With the lever rule, valid in h/x diagrams, follows:

$$l_2 \cdot \dot{G} = (l_1 + l_2) \cdot \dot{L} \quad \text{or} \quad \frac{\dot{L}}{\dot{G}} = \frac{l_2}{l_1 + l_2}. \tag{5.2-48}$$

According to basic laws of geometry, the ratio $l_2/(l_1 + l_2)$ is constant if dew point and boiling point lines are parallel straight lines. Consequently, the conditions for straight operating lines are:

- Equal latent heats of vaporization, $r_a = r_b$
- Close boiling system
- No heats of mixing

The most important condition is the equality of the latent heats of vaporization. This condition is often met by using molar heats of vaporization instead of specific (referred to mass) ones. The use of molar heats of vaporization demands, in turn, the use of molar flow rates and molar concentrations. This is one of the reasons why the thermodynamics of rectification processes are preferably expressed with molar quantities.

5.2.2.2 Rectification of Ternary Mixtures

Binary mixtures are a very special case seldom encountered in separation processes. In practice, multicomponent mixtures have to be processed. A very interesting example are ternary mixtures because they show all the characteristic features of multicomponent distillation.

Presentation of Phase Equilibrium

Gas–liquid equilibrium of ternary mixtures is graphically presented by distillation lines. Distillation lines constitute a sequence of equilibrium stages. Beginning with the initial liquid concentration x_0 the concentration y_0* of the equilibrium vapor is determined. This vapor is totally condensed to generate a liquid with the same concentration, $x_1 = y_0*$. A sequence of such process steps constitutes the distillation line according to the recursion formula

$$x_0 \rightarrow y_0^* = x_1 \rightarrow y_1^* = x_2 \rightarrow y_2^* \cdots \tag{5.2-49}$$

In ideal mixtures, phase equilibrium is formulated with relative volatilities $\alpha_{ac} \equiv p_a^0/p_c^0$ and $\alpha_{bc} \equiv p_b^0/p_c^0$. The equation of a distillation line is

$$x_{an} = \frac{\alpha_{ac}^n \cdot x_{a0}}{1 + (\alpha_{ac}^n - 1) \cdot x_{a0} + (\alpha_{bc}^n - 1) \cdot x_{b0}} \tag{5.2-50}$$

and

$$x_{bn} = \frac{\alpha_{bc}^n \cdot x_{b0}}{1 + (\alpha_{ac}^n - 1) \cdot x_{a0} + (\alpha_{bc}^n - 1) \cdot x_{b0}}. \tag{5.2-51}$$

Here, n denotes the number of equilibrium stages. For $n = 1$ the above equations transform into (5.1-12) which are valid for the phase equilibrium.

Equation (5.2-50) describes the concentration profile within a column in case of a very high (infinite) reflux ratio. Furthermore, the product qualities can easily be determined from knowledge of distillation lines (see the next section). This feature of distillation lines is very important to process design.

Distillation lines follow, in a first approximation, the course of a ball rolling downward on a boiling point surface, see Fig. 5.2-14. Therefore, distillation lines (Fig. 5.2-15) always begin at the highest point of the boiling point surface and end at the lowest point. The distances between the dots marked on the distillation lines characterize the length of one equilibrium stage. Large distances between the dots characterize wide boiling mixtures. Hence, the difficulty of a separation can be easily estimated from a graphical plot of distillation lines (Fig. 5.2-15).

In systems with multiple peaks and valleys in the boiling point surface, i.e., mixtures with azeotropes, so-called boundary distillation lines can exist. Boundary distillation lines divide regions with different origins or termini of distillation lines, see Figs. 5.2-17 and 5.2-19. Boundary distillation lines always run between local temperature extremata of the same type (minima or maxima). They follow either a ridge or a valley in the boiling point surface.

A very interesting mixture is presented in Figs. 5.2-18 and 5.2-19. The mixture acetone/chloroform/methanol exhibits two binary minimum azeotropes, one binary

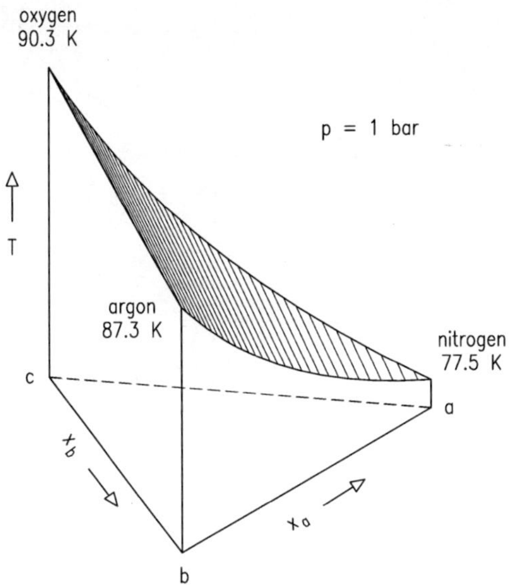

Fig. 5.2-14 Boiling point surface of the system nitrogen/argon/oxygen at a pressure of 1 bar

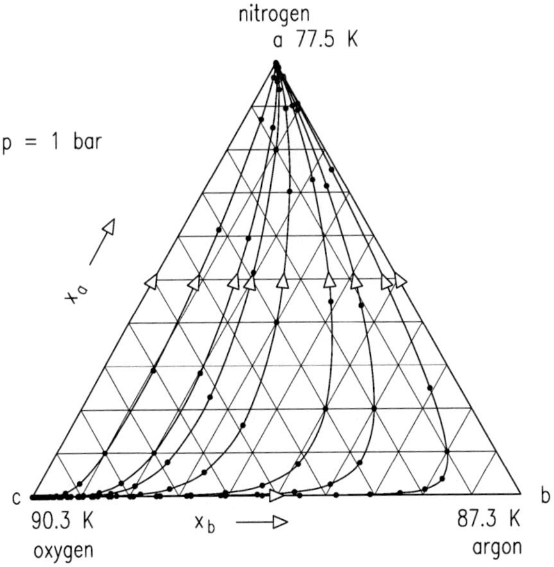

Fig. 5.2-15 Distillation lines of the system nitrogen/argon/oxygen a a pressure of 1 bar

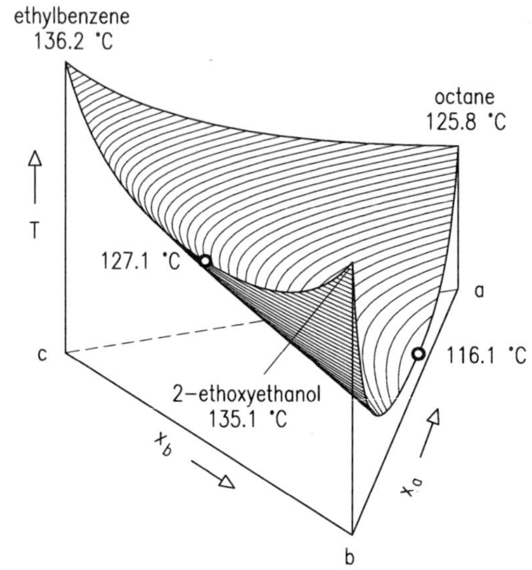

Fig. 5.2-16 Boiling point surface of the system octane/ethoxy-ethanol/ethylbenzene at a pressure of 1 bar

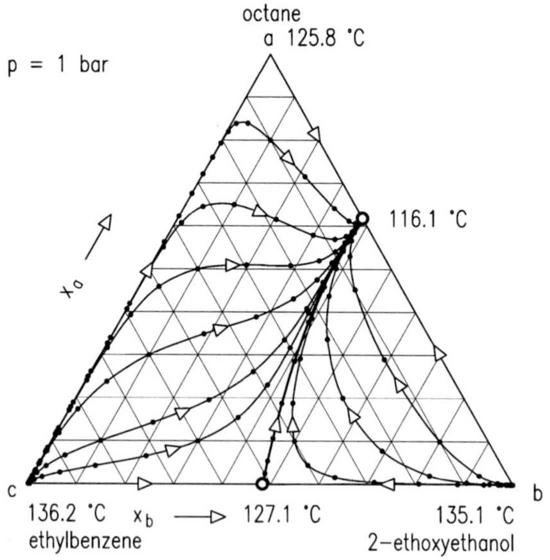

Fig. 5.2-17 Distillation lines of the system octane/ethoxy-ethanol/ethylbenzene at a pressure of 1 bar. A boundary distillation line runs between the two minimum azeotropes

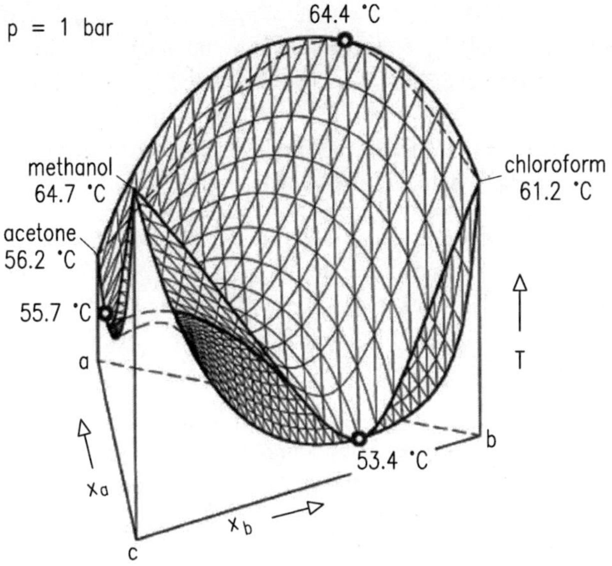

Fig. 5.2-18 Boiling point and dew point surfaces of the system acetone/chloroform/methanol at a pressure of 1 bar

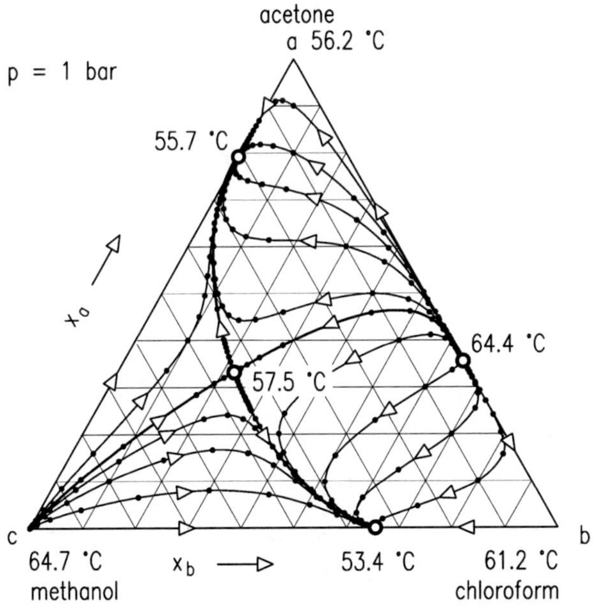

Fig. 5.2-19 Distillation lines of the system acetone/chloroform/methanol at a pressure of 1 bar. The boundary distillation lines, which run between the minimum and the maximum azeotropes, respectively, divide the mixture into four fields with different starting and end-points of distillation lines

maximum azeotrope and a ternary saddle point azeotrope, which is characterized by an intermediate boiling temperature. One boundary distillation line runs between the two minimum azeotropes, the other one between the maximum azeotrope and the high boiler (methanol). The boundary distillation lines intersect at the ternary azeotrope. In Fig. 5.2-18 the dew point surface is depicted covering most of the boiling point surface, which is also shown.

Boiling point and dew point surfaces touch each other only at the pure constituents and at the azeotropes. They do not touch at the ridges and the valleys, i.e., at the boundary distillation lines. Hence, boundary distillation lines do not have exactly the same features as azeotropes have with respect to rectification.

Separation Regions

Knowledge of the feasible products of a single distillation column is an important prerequisite for process design. Firstly, all substances entering the column with the feed $\dot{F}$ have to leave the column in the product fractions $\dot{D}$ and $\dot{B}$:

$$\dot{F} = \dot{D} + \dot{B} \quad \text{and} \quad \dot{F} \cdot z_i = \dot{D} \cdot x_{\mathrm{D}i} + \dot{B} \cdot x_{\mathrm{B}i} . \tag{5.2-52}$$

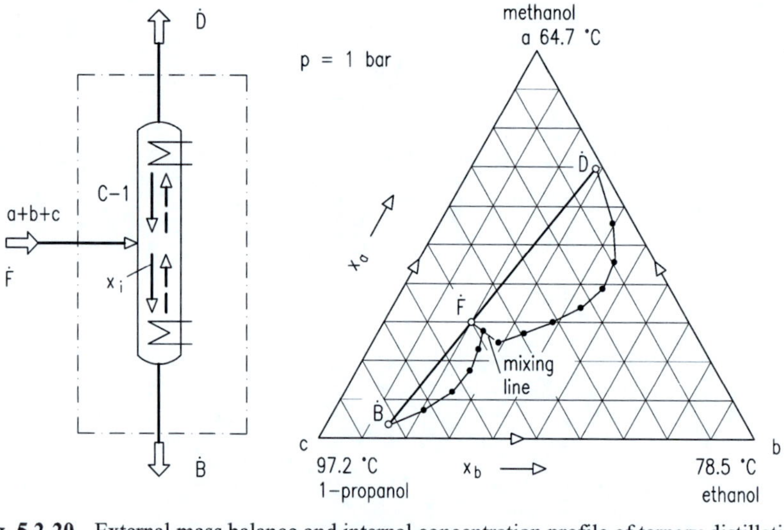

Fig. 5.2-20 External mass balance and internal concentration profile of ternary distillation

Equation (5.2-52) is represented by a straight line connecting the states of the feed $\dot{F}$ and the products $\dot{D}$ and $\dot{B}$. Secondly, the products $\dot{D}$ and $\dot{B}$ have to be end-points of the internal liquid concentration profile also shown in Fig. 5.2-20.

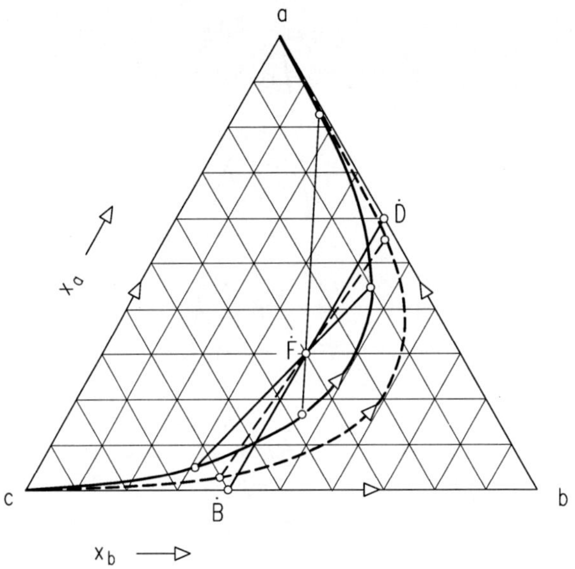

Fig. 5.2-21 Determination of the feasible products $\dot{D}$ and $\dot{B}$ of a zeotropic ternary mixture. The straight lines passing though the feed $\dot{F}$ constitute a chord to a distillation line

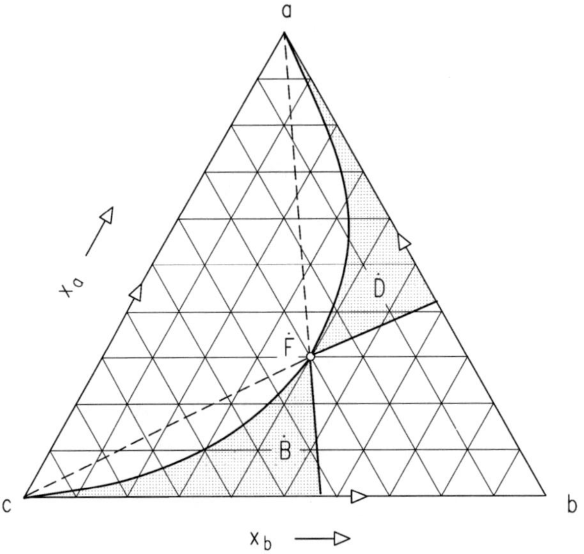

Fig. 5.2-22 Regions of feasible products of a zeotropic mixture at very high reflux ratio. The regions of feasible products are limited by a distillation line passing through the feed $\dot{F}$ and by straight lines emerging from the endpoints of this distillation line

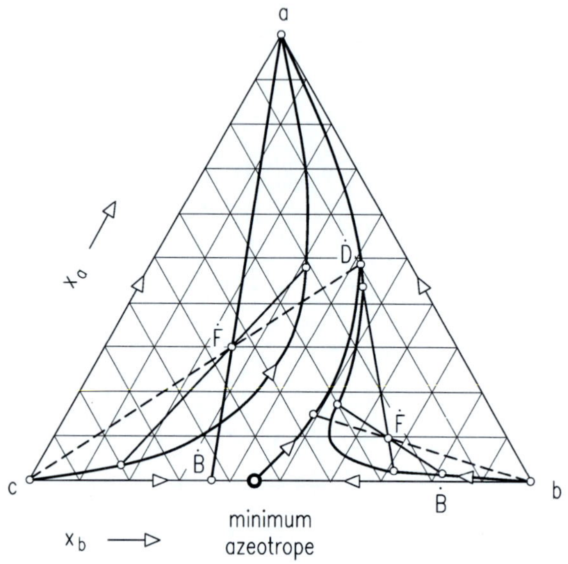

Fig. 5.2-23 Determination of the feasible products $\dot{D}$ and $\dot{B}$ of an azeotropic ternary mixture with a border distillation line. The straight lines passing though the feed $\dot{F}$ constitute a chord to a distillation line

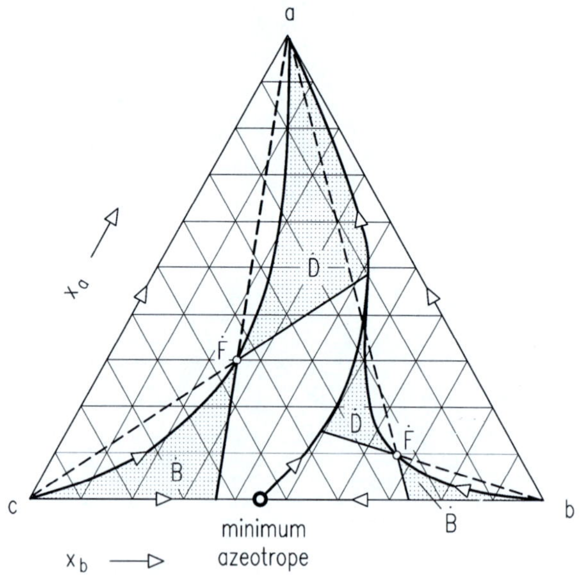

Fig. 5.2-24 Regions of feasible products of an azeotropic mixture at a very high reflux ratio. The regions of feasible products are limited by a distillation line passing through the feed $\dot{F}$ and by straight lines between the feed and the endpoints of this distillation line

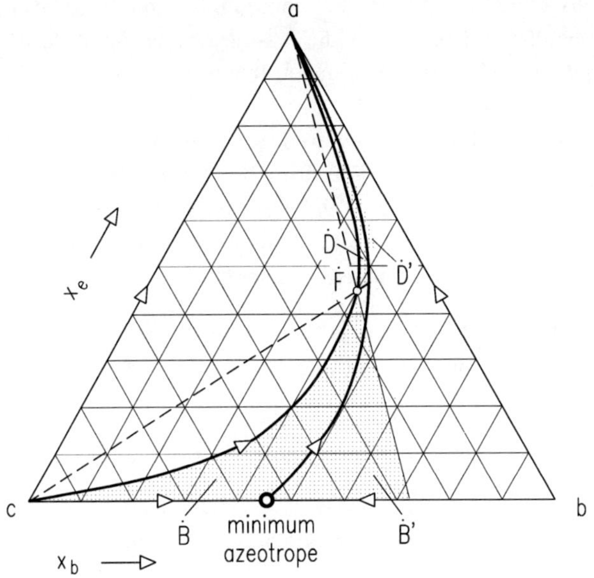

Fig. 5.2-25 Regions of feasible products when the feed is at the concave side of a border distillation line. The boundary distillation line is not a barrier for the bottom fraction if the overhead fraction can be gained in both distillation fields (and vice versa)

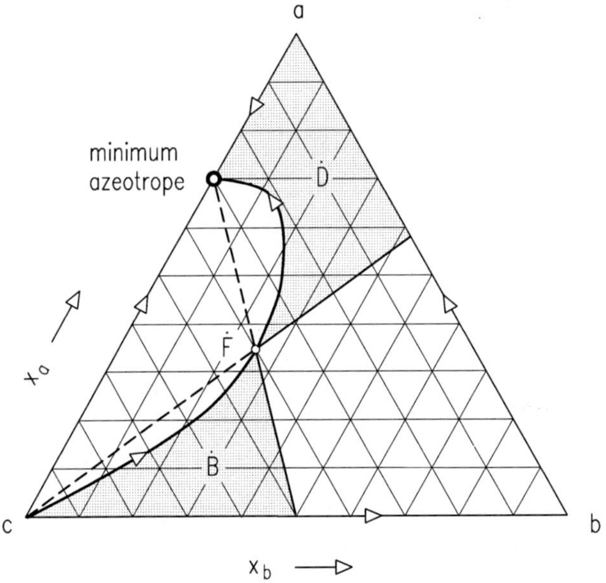

Fig. 5.2-26 Regions of feasible products when the minimum azeotrope lies between the low and the high boiler of a ternary mixture. The low boiler is a feasible overhead product, even though it is not an endpoint of distillation lines

At high reflux operation, the internal concentration profile is identical with the course of a distillation line. Hence, the feasible products have to lie on a straight line through $\dot{F}$ as well as on a distillation line, see Fig. 5.2-21. The straight lines through $\dot{F}$ constitute chords to a distillation line. From the multiplicity of feasible chords to different distillation lines follows the region of feasible products. The region of feasible products is restricted by a distillation line through the feed $\dot{F}$ and by straight lines through the feed $\dot{F}$ and through the origin and the terminus of that distillation line, respectively. Figure 5.2-22 makes it clear that an azeotropic mixture can be fractionated in a very restricted manner only.

The region of feasible products is further narrowed down in azeotropic mixtures with boundary distillation lines. From Figs. 5.2-23 and 5.2-24 the following rules can be formulated:

Rule 1: Only substances that are origins or termini of distillation lines can be recovered as pure products.

Rule 2: Boundary distillation lines are barriers to distillation.

However, there are some exemptions from these rules as demonstrated in Figs. 5.2-25 and 5.2-26. Very important to process synthesis is the insight that a boundary distillation line can be overcome if the specific product can be gained in both distillation fields (Fig. 5.2-25).

Energy Demand

The energy demand of a rectification column is described by (5.2-37):

$$\frac{\dot{Q}_{\mathrm{C}}}{\dot{F}\cdot r} = (R_{\mathrm{L}}+1)\cdot\frac{z_{\mathrm{F}i}-x_{\mathrm{B}i}}{x_{\mathrm{D}i}-x_{\mathrm{B}i}} \quad \text{or} \quad \frac{\dot{Q}_{\mathrm{R}}}{\dot{F}\cdot r} = R_{\mathrm{G}}\cdot\frac{x_{\mathrm{D}i}-z_{\mathrm{F}i}}{x_{\mathrm{D}i}-x_{\mathrm{B}i}}. \tag{5.2-53}$$

The energy demand depends on reflux and reboil ratio, respectively, that must not fall below a limiting value, called minimum reflux or minimum reboil ratio. For boiling liquid feed (5.2-28) holds

$$R_{\mathrm{Lmin}} = \frac{x_{\mathrm{D}i}-\overset{*}{y}_{\mathrm{F}i}}{\overset{*}{y}_{\mathrm{F}i}-x_{\mathrm{F}i}} \quad \text{for} \quad q_{\mathrm{F}}=1. \tag{5.2-54}$$

This equation is valid for each component i of a ternary mixture. It constitutes a system of three equations that is firstly evaluated for the high boiler c:

$$R_{\mathrm{Lmin}} = \frac{x_{\mathrm{D}c}-\overset{*}{y}_{\mathrm{F}c}}{\overset{*}{y}_{\mathrm{F}c}-x_{\mathrm{F}c}}. \tag{5.2-55}$$

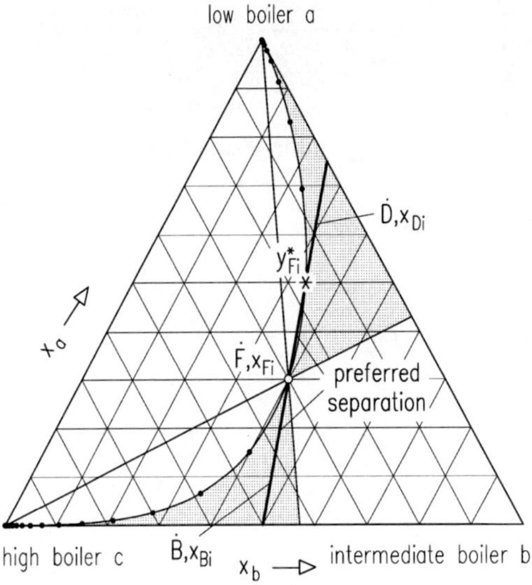

Fig. 5.2-27 Preferred separation of a ternary mixture. The external mass balance lies on a straight line passing through x_{Fi} and y^*_{Fi}

From the knowledge of R_{Lmin} the product compositions $x_{\mathrm{D}a}$ and $x_{\mathrm{D}b}$ can be calculated from (5.2-54):

$$x_{Da} = R_{\mathrm{Lmin}} \cdot (y^*_{\mathrm{F}a} - x_{\mathrm{F}a}) + y^*_{\mathrm{F}a} \quad \text{and} \quad x_{\mathrm{D}b} = R_{\mathrm{Lmin}} \cdot (y^*_{\mathrm{F}b} - x_{\mathrm{F}b}) + y^*_{\mathrm{F}b} . \tag{5.2-56}$$

Equation (5.2-56) constitutes a system of linear equations, which are graphically represented by a straight line in Fig. 5.2-27. The distillate $\dot{D}$ lies on a straight line through the points $x_{\mathrm{F}i}$ and $y_{\mathrm{F}i}{}^*$. This very special separation is called *preferred separation*. For sharp preferred separations ($x_{Dc} = 0$) the following holds:

$$R_{\mathrm{Lmin}} = \frac{-y^*_{\mathrm{F}c}}{y^*_{\mathrm{F}c} - x_{\mathrm{F}c}} . \tag{5.2-57}$$

For ideal mixtures, the phase equilibrium is

$$y^*_{\mathrm{F}c} = \frac{x_{\mathrm{F}c}}{N_3} \quad \text{with} \quad N_3 = 1 + (\alpha_{ac} - 1) \cdot x_{\mathrm{F}a} + (\alpha_{bc} - 1) \cdot x_{\mathrm{F}b} . \tag{5.2-58}$$

Combining both equations yields

$$R_{\mathrm{Lmin}} = \frac{-x_{\mathrm{F}c}/N_3}{x_{\mathrm{F}c}/N_3 - x_{\mathrm{F}c}} = \frac{1}{N_3 - 1} \tag{5.2-59}$$

or

$$R_{\text{Lmin}} = \frac{1}{(\alpha_{ac}-1)\cdot x_{\text{F}a}+(\alpha_{bc}-1)\cdot x_{\text{F}b}} \quad \text{for} \quad q_{\text{F}} = 1 . \tag{5.2-60}$$

The above equation transforms for $x_{\text{F}b} = 0$ into (5.2-30) which is valid for binary mixtures. The concentration $x_{\text{D}a}$ of the distillate is

$$x_{\text{D}a} = R_{Lmin}\cdot(\alpha_{ac}-1)\cdot x_{\text{F}a} \tag{5.2-61}$$

or

$$x_{\text{D}a} = \frac{(\alpha_{ac}-1)\cdot x_{\text{F}a}}{(\alpha_{ac}-1)\cdot x_{\text{F}a}+(\alpha_{bc}-1)\cdot x_{\text{F}b}} . \tag{5.2-62}$$

The minimum energy demand of a sharp separation (i.e., $x_{\text{B}a} = 0$) is

$$\frac{\dot{Q}_{\text{min}}}{\dot{F}\cdot r} = (R_{\text{Lmin}}+1)\cdot\frac{x_{\text{F}a}-x_{\text{B}a}}{x_{\text{D}a}-x_{\text{B}a}} = (R_{\text{Lmin}}+1)\cdot\frac{x_{\text{F}a}}{x_{\text{D}a}} . \tag{5.2-63}$$

With (5.2-60) for R_{Lmin} and (5.2-62) for $x_{\text{D}a}$ follows:

$$\frac{\dot{Q}_{\text{min}}}{\dot{F}\cdot r} = \frac{1+(\alpha_{ac}-1)\cdot x_{\text{F}a}+(\alpha_{bc}-1)\cdot x_{\text{F}b}}{\alpha_{ac}-1} \quad \text{for} \quad q_{\text{F}} = 1 . \tag{5.2-64}$$

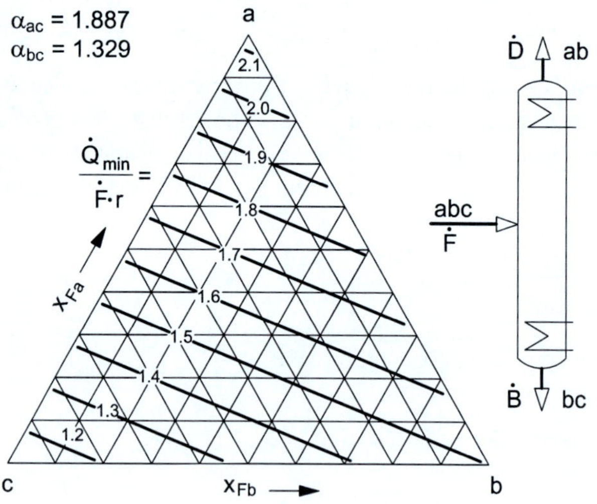

Fig. 5.2-28 Minimum energy demand of preferred separation of an ideal ternary mixture

For $x_{\text{F}b} = 0$ this equation also transforms into (5.2-39) derived for binary mixtures. Equation (5.2-64) is evaluated for a close boiling system in Fig. 5.2-28. The parameter lines are parallel and equidistant straight lines. Hence, the energy demand of a preferred separation is a linear function of feed concentration $x_{\text{F}i}$.

The preferred separation has the lowest energy demand of all sharp separations; however, it does not provide pure products. Distillate as well as bottom fractions are binary fractions. In practice, however, either a pure low boiler or a pure high boiler has to be gained. Such separations are feasible but they generally have a higher energy demand.

In Fig. 5.2-29 the internal concentration profiles of preferred separation, low-boiler separation, and high-boiler separation are qualitatively shown. Limiting for the energy demand is the pinch (i.e., point of intersection of operating and equilibrium lines). At preferred separations a double pinch exists whose concentration is equal to the feed concentration (like in binary distillation).

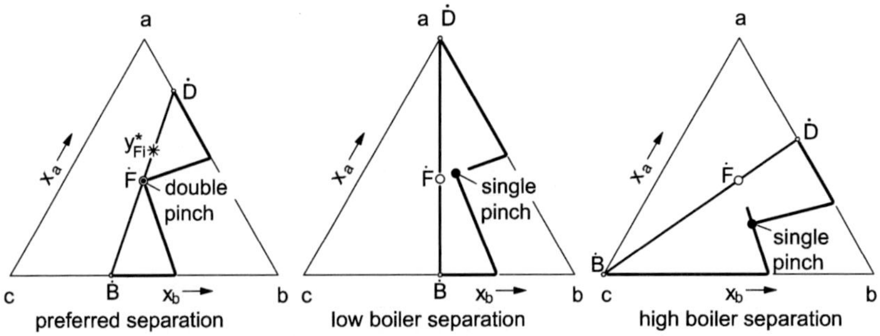

Fig. 5.2-29 Internal concentration profiles and pinch points of preferred separation, low boiler separation, and high boiler separation

At low-boiler separations the rectifying section of the column is operated with increased reflux to gain a pure overhead product. Just the stripping section is operated with minimum reboil. Hence, just a single pinch point exists whose concentration is different from the feed concentration.

Analogously, at high-boiler separations the stripping section is operated with a surplus of reboil and, in turn, no pinch exists in this section. The rectifying section, however, is operated with minimum reflux characterized by a pinch immediately above the feed point. Again, pinch concentration differs from feed concentration.

The calculation of the minimum reflux and reboil ratios of nonpreferred separations is based on the fact that, in ideal mixtures, the states of constant reflux ratio R_L = const. constitute a straight line in the triangular concentration space of Fig. 5.2-30. Their endpoints on the side lines of the triangle can easily be determined from the McCabe–Thiele diagram showing the equilibrium curves of the binary mixtures *a–b* and *a–c*. From a first estimation of the reflux ratio the operating line is drawn. Its points of intersection with the equilibrium lines deliver the endpoints of the line R_L = const. in the triangular diagram. The solution is found if this line passes through the feed concentration.

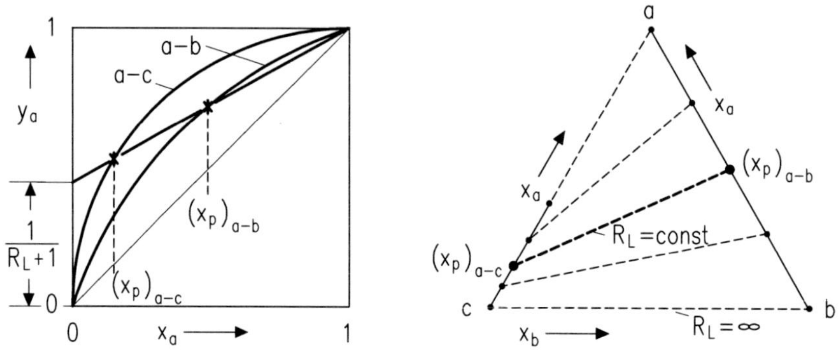

Fig. 5.2-30 Graphical determination of the minimum reflux ratio of the low-boiler separation from an ideal ternary mixture

An algebraic solution of the problem is also possible. For boiling liquid feed (q_F = 1) the following holds for low-boiler separations:

$$(R_{\text{Lmin}})_a = \overset{*}{R}_{\text{Lmin}} \cdot \frac{x_{\text{D}a} - \overset{*}{y}_a}{\overset{*}{x}_{\text{D}a} - \overset{*}{y}_a} \tag{5.2-65}$$

with $\overset{*}{R}_{\text{Lmin}} = \dfrac{1}{(\alpha_{ac}-1)\cdot x_{\text{F}a} + (\alpha_{bc}-1)\cdot x_{\text{F}b}}$,

$\overset{*}{x}_{\text{D}a} = \overset{*}{R}_{\text{Lmin}} \cdot (\alpha_{ac}-1)\cdot x_{\text{F}a}$, and

$\overset{*}{y}_a = A - \sqrt{A^2 - \dfrac{\overset{*}{x}_{\text{D}a}\cdot \alpha_{ab}}{(\alpha_{ab}-1)\cdot(\overset{*}{R}_{\text{Lmin}}+1)}}$,

with $A = \dfrac{\alpha_{ab} + (\alpha_{ab}-1)\cdot(\overset{*}{R}_{\text{Lmin}} + \overset{*}{x}_{\text{D}a})}{2\cdot(\alpha_{ab}-1)\cdot(\overset{*}{R}_{\text{Lmin}}+1)}$.

For high-boiler separations the minimum reboil ratio is

$$(R_{\text{Gmin}})_c = \overset{*}{R}_{\text{Gmin}} \cdot \frac{\overset{*}{x}_b - x_{\text{B}b}}{\overset{*}{x}_b - \overset{*}{x}_{\text{B}b}} \tag{5.2-66}$$

with $\overset{*}{R}_{\text{Gmin}} = \dfrac{1}{(1-\alpha_{ab}^{-1})\cdot x_{\text{F}b} + (1-\alpha_{ac}^{-1})\cdot x_{\text{F}c}} - 1$,

$$\overset{*}{x}_{Bb} = (\overset{*}{R}_{\text{Gmin}} + 1) \cdot (1 - \alpha_{ab}^{-1}) \cdot x_{Fb}, \quad \text{and}$$

$$\overset{*}{x}_b = -B + \sqrt{B^2 + \frac{\overset{*}{x}_{Bb}}{(\alpha_{bc} - 1) \cdot (\overset{*}{R}_{\text{Gmin}} + 1)}}$$

with $$B = \frac{1 - (\alpha_{bc} - 1) \cdot (\overset{*}{R}_{\text{Gmin}} + \overset{*}{x}_{Bb})}{2 \cdot (\alpha_{bc} - 1) \cdot (\overset{*}{R}_{\text{Gmin}} + 1)}.$$

Both equations formulate a nonlinear relationship between feed concentration and reflux and reboil ratio, respectively.

From the knowledge of the minimum reflux and reboil ratio the minimum energy demands can be calculated with (5.2-53). The results are plotted in Figs. 5.2-31 and 5.2-32. The parameter lines are curved in contrast to Fig. 5.2-28 that is valid for preferred separations.

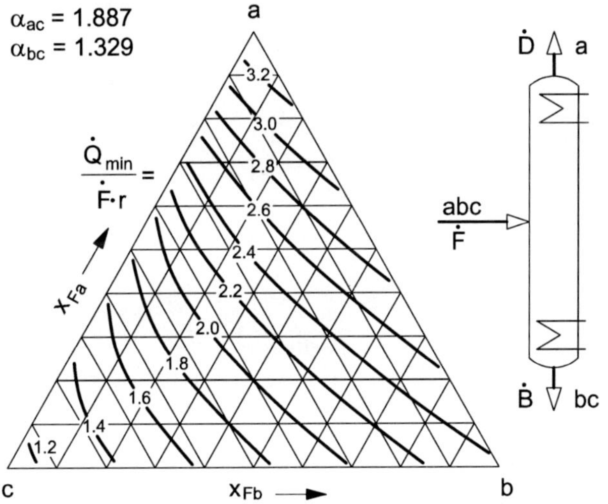

Fig. 5.2-31 Minimum energy demand of the low-boiler separation from an ideal ternary mixture

5.2.2.3 *Rectification of Multicomponent Mixtures*

Simulation of multicomponent rectification is based on the so-called MESH equations (**M** = material balances, **E** = equilibrium, **S** = summation of mole fractions, **H** = heat balance) formulated for a single equilibrium stage (Fig. 5.2-33) as follows:

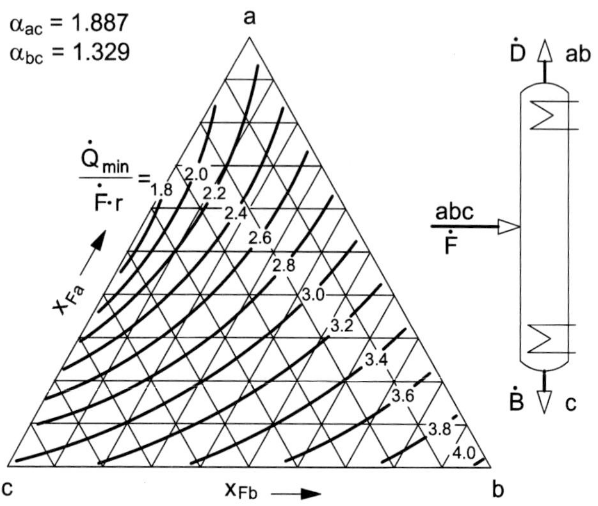

Fig. 5.2-32 Minimum energy demand of the high-boiler separation from an ideal ternary mixture

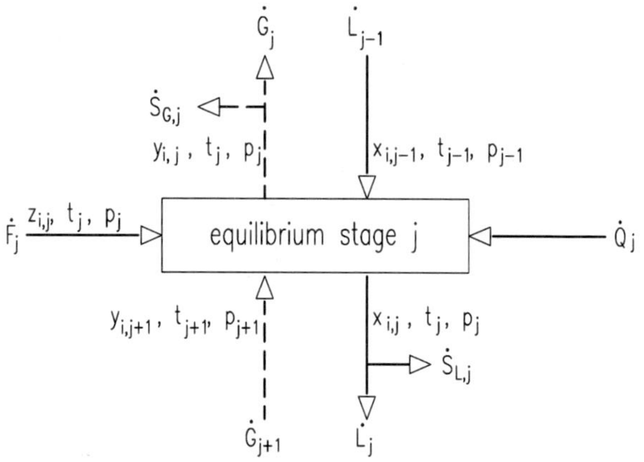

Fig. 5.2-33 Scheme of an equilibrium stage. A mass balance yields the MESH equations

M-equations:

$$\dot{L}_{j-1} \cdot x_{i,j-1} + \dot{G}_{j+1} \cdot y_{i,j+1} + F_j \cdot z_{i,j} - (\dot{L}_j + \dot{S}_{L,j}) \cdot x_{i,j} - (\dot{G}_j + \dot{S}_{G,j}) \cdot y_{i,j} = 0 \,. \tag{5.2-67}$$

E-equations:

$$y_{i,j} - K_{i,j} \cdot x_{i,j} = 0 \,. \tag{5.2-68}$$

S-equations:

$$\sum x_{i,j} - 1 = 0 \quad \text{and} \quad \sum y_{i,j} - 1 = 0\,. \tag{5.2-69}$$

H-equations:

$$\dot{L}_{j-1} \cdot h_{\mathrm{L},j-1} + \dot{G}_{j+1} \cdot h_{\mathrm{G},j+1} + F_j \cdot h_{\mathrm{F},j} - (\dot{L}_j + \dot{S}_{\mathrm{L},j}) \cdot h_{\mathrm{L},j} - (\dot{G}_j + \dot{S}_{\mathrm{G},j}) \cdot h_{\mathrm{G},j} = 0 \tag{5.2-70}$$

The MESH equations constitute a system of $n \cdot (2 \cdot k + 3)$ equations. A mixture with 5 components fractionated in a column with 50 equilibrium stages ($n = 50$) is modeled by 650 algebraic equations. Furthermore, there is a large number of additional equations for modeling the phase equilibrium and the enthalpies.

The MESH equations form a tridiagonal matrix since the states on stage j depend only on the states of stage $j - 1$ (due to the liquid flow) and that of stage $j + 1$ (due to gas flow).

Contrary to appearance, the MESH equations are not a system of linear equations. In particular, the equilibrium ratio K is a strong nonlinear function of temperature. In nonideal mixtures, K also depends on the concentration of all components in the mixture. The same holds for the enthalpies. Hence, the MESH equations constitute a strongly coupled system of nonlinear algebraic equations.

The MESH equations are solved by using the Newton–Raphson algorithm (e.g., Naphtali and Sandholm 1971). The equations are linearized by the Taylor algorithm in the vicinity of first estimations $\boldsymbol{x}_0$ of all variables:

$$\boldsymbol{F}(\boldsymbol{x}_0) + \boldsymbol{J}(\boldsymbol{x}_0) \cdot (\boldsymbol{x}_1 - \boldsymbol{x}_0) = 0\,. \tag{5.2-71}$$

$\boldsymbol{F}$ denotes the vector of all functions and $\boldsymbol{x}$ the vector of all variables. $\boldsymbol{J}$ stands for the Jacobi matrix which has a tridiagonal structure. Hence, all elements of the matrix consist of matrices with partial derivatives of all variables. The linear system (5.2-71) is directly solved:

$$\boldsymbol{x}_1 = \boldsymbol{x}_0 - \boldsymbol{F}(\boldsymbol{x}_0) \cdot \boldsymbol{J}^{-1}(\boldsymbol{x}_0)\,. \tag{5.2-72}$$

Here, $\boldsymbol{x}_1$ denotes the rigorous solution of the linear system (5.2-71). This solution, however, is just a better approximation of the nonlinear MESH equations. Hence, the calculation has to be repeated several times.

Calculation of the inverse Jacobi matrix $\boldsymbol{J}^{-1}$ is a very complex mathematical operation facilitated in this case by a tridiagonal structure. However, due to the large number of variables and equations, the procedure is very difficult. Furthermore, some of the partial derivatives in the Jacobi matrix have to be performed numerically.

The MESH equations constitute a nonlinear and strongly coupled system of algebraic equations since the equilibrium ratios $K_{i,j}$ and the enthalpies h_L and h_G are complex functions of temperature and concentrations. The system (5.2-71) is numerically solved by the iterative Newton–Raphson algorithm. Commercial software packages (e.g., ASPEN, HYSYS, CHEMCAD) contain both the mathematical solver and the required system properties, such as vapor liquid equilibria and enthalpies.

Typically, the column is completely specified first, including the definition of the number of equilibrium stages, the feed stage, the reflux ratio, and the heat supply. By solving the MESH equations, the internal temperature and concentration profiles and, in turn, the product concentrations are found. If the calculated product qualities do not meet the specifications, then a new column configuration has to be made and the calculation repeated.

A typical result of such a computer simulation is shown in Fig. 5.2-34. It is important to note that intermediate boiling components often exhibit maxima in the concentration profile within the column.

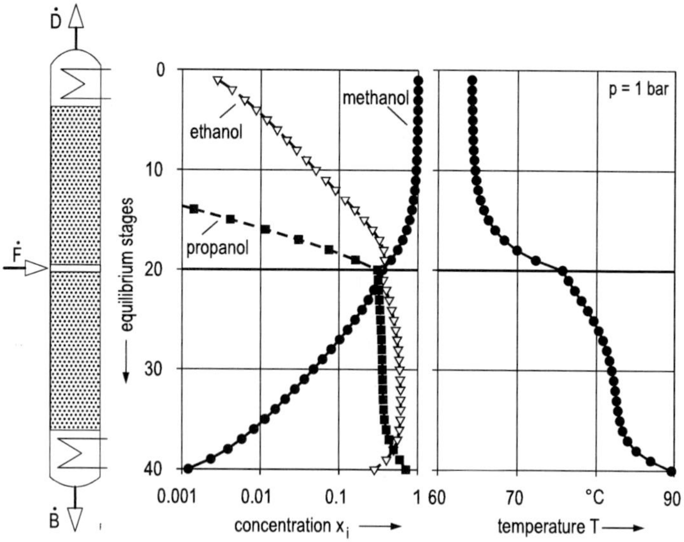

Fig. 5.2-34 Internal concentration and temperature profiles in a rectification column for the fractionation of the system methanol/ethanol/propanol

The equilibrium-stage model can be made more realistic by including tray efficiencies (see Sect. 5.4.1.4) in the E-equations (5.2-68). Another possibility for improving column simulation is the use of the concept of transfer units (see Sect. 5.2.1.1) instead of the concept of equilibrium stages. This concept is well suited for packed columns. Such models, called *rate-based models*, simultaneously solve the relevant thermodynamic and mass transfer equations describing the complex mechanisms in a column (e.g., Taylor et al. 1994; Kloecker et al. 2005). Thus, rate-based models are much more complex since, for instance, the mass transfer coefficients and the interfacial area (see Sect. 5.4.1.4) must a priori be known.

The concepts of equilibrium stages and of transfer units are, in principle, equivalent in case of parallel operating and equilibrium lines. The more the slopes of these lines differ the more superior are rate-based models to equilibrium-stage models. Further advantages of rate-based models are a better simulation of reactive distillation and absorption processes. At the present state of the art, however, equilibrium stage models are the standard tool for the simulation of distillation columns.

5.2.2.4 Reactive Distillation

In recent years reactive (or catalytic) distillation (or rectification) has gained some importance in the process industry. Reactive rectification denotes the simultaneous performance of chemical reaction and physical separation within a countercurrently operated column. The integration of these two unit operations in one column offers advantages for reversible liquid-phase reactions where the reaction products hinder the progress of the reaction (e.g., Sundmacher and Kienle 2003; Frey et al. 2003; Frey and Stichlmair 1998).

The principles of reactive rectification are explained for the following liquid-phase reaction:

$$\nu_a \cdot a + \nu_b \cdot b \leftrightarrow \nu_c \cdot c\,. \tag{5.2-73}$$

The symbols a, b, and c denote the substances, ordered in rising boiling temperatures, and ν_i the stoichiometric coefficients of the reaction. The chemical equilibrium is

$$K_R = \frac{x_c^{\nu_c}}{x_a^{\nu_a} \cdot x_b^{\nu_b}}\,. \tag{5.2-74}$$

Equation (5.2-74) is graphically represented in Fig. 5.2-35 by a curved line running between the pure reactants (feed components) a and b of the reaction. The larger the value of the equilibrium constant K, the more convex is this curve.

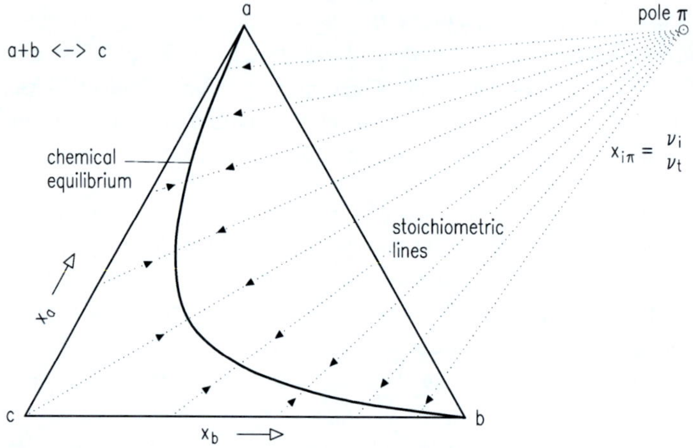

Fig. 5.2-35 Chemical equilibrium of the reaction $a + b \leftrightarrow c$ with $K_R = 8$. All stoichiometric lines emerge from the pole π

The concentration change effected by the progressing reaction is described via the stoichiometry (e.g., Stichlmair and Frey 1998):

$$x_i = \frac{x_{i0} \cdot (\nu_k - \nu_t \cdot x_k) + \nu_i \cdot (x_k - x_{k0})}{\nu_k - \nu_t \cdot x_{k0}} \quad \text{with} \quad \nu_t = \sum \nu_i . \tag{5.2-75}$$

This linear relation is graphically represented by dotted lines in Fig. 5.2-35. All those lines emerge from a single point, called pole π, whose coordinates are:

$$x_{\pi i} = \nu_i / \sum \nu_i . \tag{5.2-76}$$

The values of the stoichiometric coefficients are negative for reactants and positive for products.

In fast reactions, the states of the liquid within the column always lie on the curve of the reaction equilibrium. The superposition of reaction and rectification can easily be traced in Fig. 5.2-36.

Starting from any liquid state 1 on the chemical equilibrium curve the concentration of the equilibrium vapor is determined, point 1*. By total condensation a new liquid is generated having the same concentration. As this liquid is off the chemical

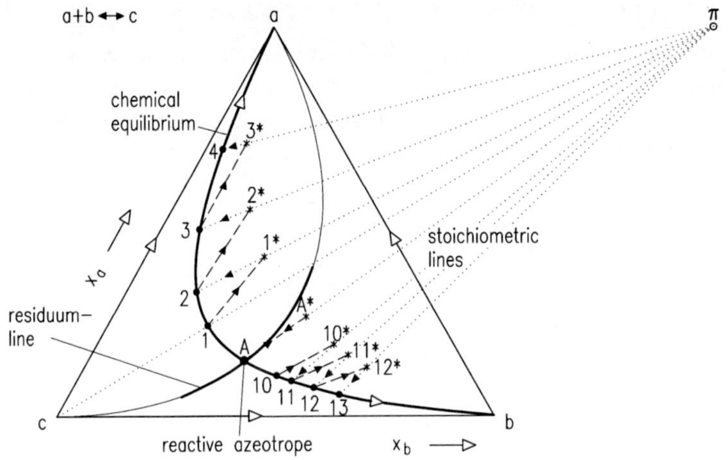

Fig. 5.2-36 Design of the reactive distillation line of a ternary mixture: - - - - vapor–liquid equilibrium; . . . . . stoichiometric lines

equilibrium, the reaction starts effecting a concentration change along the stoichiometric line through point 1* until chemical equilibrium is reached in point 2 .

In the succeeding distillation step the equilibrium vapor, point 2* , is generated. This vapor is transferred into a liquid by total condensation and, in turn, the reaction starts again. A sequence of such reaction and distillation steps constitutes the points 1 , 2 , 3 ,... on the chemical equilibrium curve. This curve is identical with the reactive distillation line, which is running towards the low boiler a .

If this procedure starts at point 10 , for instance, the states 11 , 12 , 13 ,... are found that run towards the intermediate boiler b . A very special situation comes up when the sequence of distillation and reaction steps starts at point A on the chemical equilibrium curve. Here, the equilibrium vapor has the state A^* . After total condensation the liquid-phase reaction effects a concentration change from A^* back to A . Hence, the concentration change of distillation, $A \rightarrow A^*$, is fully compensated for by the reaction step $A^* \rightarrow A$. In this case, a sequence of distillation and reaction does not effect any change of concentration. In analogy to distillation, point A is called *reactive azeotrope*. Reactive azeotropes constitute barriers to reactive distillation like azeotropes do to physical distillation.

At the reactive azeotrope A , the stoichiometric line forms a tangent to the residuum line. Points of tangential contact constitute a line whose point of intersection with the chemical equilibrium line is the reactive azeotrope. With this condition existence and location of reactive azeotropes can be identified (Frey and Stichlmair 1998).

Processes of Reactive Distillation

For instantaneous chemical reactions a reactive distillation line represents the concentration profile of the liquid within a column. From its knowledge processes of reactive distillation can be designed (Stichlmair and Frey 1998) as is demonstrated at the example of the reversible instantaneous reaction $a + b \leftrightarrow c$.

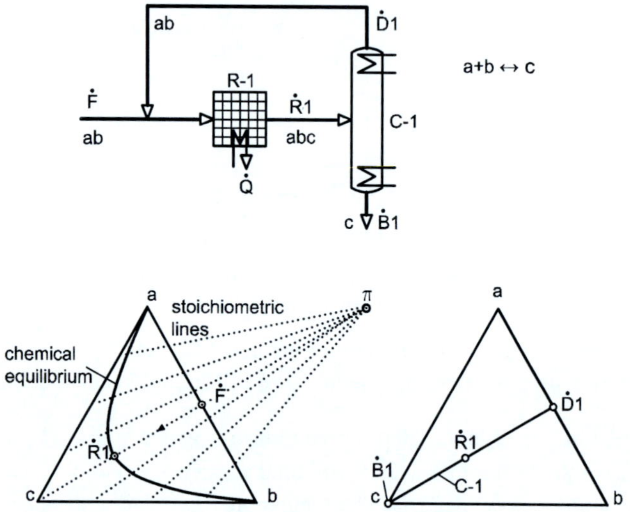

Fig. 5.2-37 Conventional process of the reaction $a + b \leftrightarrow c$ and sequential product recovery

A conventional process, i.e., sequential reaction and distillation, is presented in Fig. 5.2-37. The chemical reaction is established in the reactor R-1. The reaction product $\dot{R}1$ contains some unreacted feed components a and b due to the reversible nature of the reaction. In the succeeding distillation column C-1 the high-boiling product c is recovered as bottoms. The unconverted components a and b, being the overhead product of the column, are recycled to the reactor R-1. Depending on the reaction equilibrium this recycle stream can be very large. Generally, processes with large internal recycles are very disadvantageous.

The corresponding process with simultaneous reaction and distillation is depicted in Fig. 5.2-38. The countercurrently operated column consists of a reaction zone (above the feed point) and a distillation zone, i.e., stripping section of a nonreactive distillation column.

The concentration profile within the column is shown in the triangular concentration space of Fig. 5.2-38. In the upper section, the reaction section, the internal profile lies on the curve of the chemical equilibrium (for instantaneous reactions). In the lower section the internal concentration profile is approximately identical with a distillation line beginning at the high boiler c. A total conversion of the feed components a and b is accomplished.

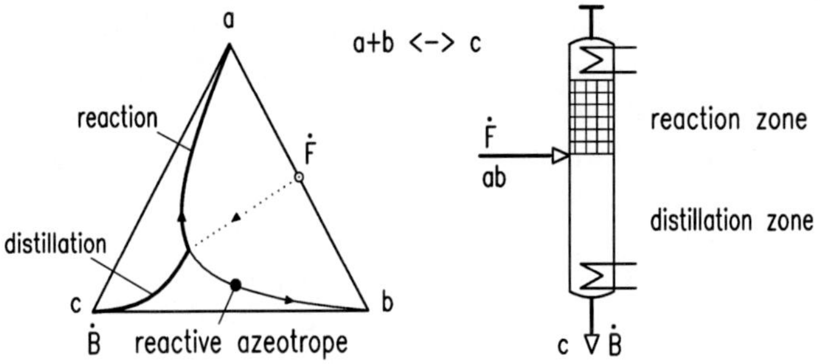

Fig. 5.2-38 Reactive distillation of the process in Fig. 5.2-37

Reactive distillation has a high potential for process intensification. Mostly, only one section of the column works as a reactor. The transition between reactive and distillative sections is easily established in heterogeneously catalyzed reactions where the solid catalyst is integrated in the column packings (Krishna 2003).

5.2.3 Batch Distillation (Multistage)

Batch distillation is a discontinuous form of a distillation process. It is characterized by the coupling of a liquid vessel, into which the mixture is initially charged, with a distillation column, from which the products are continuously withdrawn. Three different configurations of batch distillation processes are shown in Fig. 5.2-39.

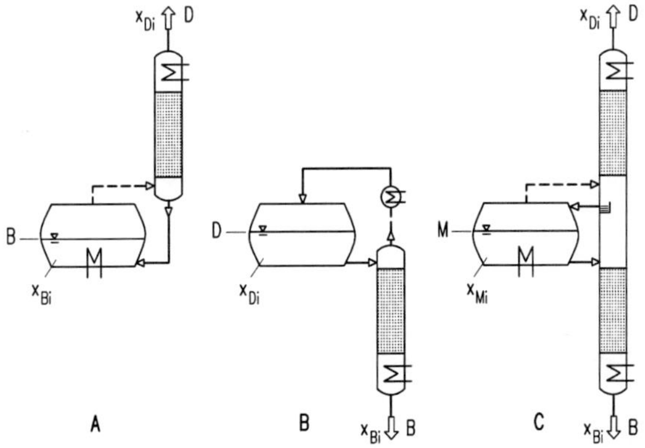

Fig. 5.2-39 Batch distillation configurations: (**A**) regular batch distillation, (**B**) inverse batch distillation, and (**C**) middle vessel batch distillation

The configuration A is the standard batch distillation process. It consists of a liquid vessel at the bottom of a rectifying distillation column. The liquid in the vessel is continuously heated and the emerging vapor is fed into the bottom of the column. Due to the countercurrent liquid flow established by the reflux in the column, the high-boiling components are removed from the rising vapor. Thus, the overhead fraction is rich in low-boiling constituents. By continuous removal of the overhead product, both the amount B and the concentration x_{Ba} of the liquid in the vessel decrease. In a sufficiently effective column, the low boiler a is recovered as first overhead product. After depletion of substance a in the vessel, substance b is the lowest boiling component and, in turn, the next overhead product. In this way, a zeotropic multicomponent mixture can be fractionated into all constituents. The products are recovered at the top of the column in a sequence of rising boiling temperatures (a, b, c, ...).

The process configuration B in Fig. 5.2-39 allows the removal of the pure constituents as bottom fractions in a sequence of decreasing boiling temperatures (…, c, b, a). This process is less important in industry.

The process configuration C enables the simultaneous removal of the lowest and the highest boiling constituents from the mixture in the vessel. The intermediate boiling components enrich in the middle vessel. Therefore, this process is advantageous when small amounts of low- and high-boiling impurities have to be removed from a mixture rich in intermediate boiling species.

5.2.3.1 Binary Mixtures

Multistage batch distillation (i.e., batch rectification) is governed by the Rayleigh equation

$$\frac{dB}{B} = \frac{dx_B}{x_D - x_B} \tag{5.2-77}$$

Contrary to single stage batch distillation, the product concentration x_D is not in equilibrium state with the bottom concentration x_B. Besides the phase equilibrium, the relationship between product and vessel concentrations (x_D and x_B) depends on the number of stages as well as on the reflux ratio R_L of the column. Two modes of operation have to be considered (Fig. 5.2-40).

- Operation with constant reflux ratio R_L and, in turn, decreasing concentration x_D of the distillate

- Operation with constant concentration x_D of the distiller and, in turn, increasing reflux ratio R_L

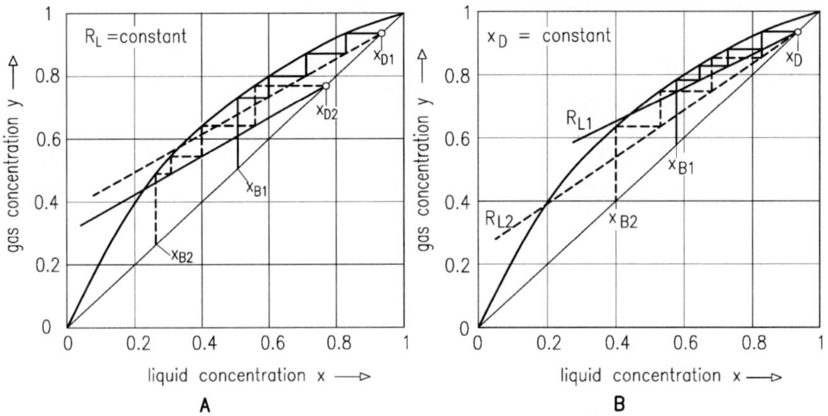

Fig. 5.2-40 Batch distillation with constant reflux (**A**) and constant product concentration (**B**)

The energy demand Q of multistage batch distillation can be found from the following equation:

$$dQ = r \cdot (R_L + 1) \cdot dB. \tag{5.2-78}$$

The term $(R_L + 1)$ accounts for the fact that some of the vapor generated in the reboiler is recycled as liquid within the column.

In the operation mode R_L = const. the energy equation (5.2-78) is directly solved:

$$Q = r \cdot (R_L + 1) \cdot B_0 \cdot (1 - B/B_0). \tag{5.2-79}$$

The term B/B_0 is determined by integrating the Rayleigh equation (5.2-77) resulting in

$$\frac{Q}{B_o \cdot r} = (R_L + 1) \cdot (1 - \int \frac{1}{x_D - x_B} \cdot dx_B). \tag{5.2-80}$$

The relationship between x_D and x_B can be pointwise determined from the McCabe–Thiele diagram shown in Fig. 5.2-40A.

In the operation mode x_D = const. (5.2-77) is solved first resulting in

$$\frac{B}{B_o} = \frac{x_D - x_{Bo}}{x_D - x_B} \quad \text{or} \quad dB = B_o \cdot \frac{x_D - x_{Bo}}{(x_D - x_B)^2} \cdot dx_B. \tag{5.2-81}$$

Inserting into the energy equation (5.2-78) and integrating yields

$$\frac{Q}{B_o \cdot r} = (x_D - x_{Bo}) \cdot \int \frac{R_L + 1}{(x_D - x_B)^2} \cdot dx_B . \tag{5.2-82}$$

The term $(R_L + 1)/(x_D - x_B)^2$ is pointwise determined from the McCabe–Thiele diagram in Fig. 5.2-40B.

In the special case of a high column with many equilibrium stages, the liquid concentration x_B in the vessel is approximately equal to the point of intersection of the operating and the equilibrium lines. Thus the relationship between x_D , x_B and R_L is given by (5.2-28) which is valid for minimum reflux ratio. After inserting this relationship (5.2-80) and (5.2-82) can be analytically solved for ideal systems. The results are shown in Fig. 5.2-41 for both modes of batch distillation and, additionally, for continuous distillation.

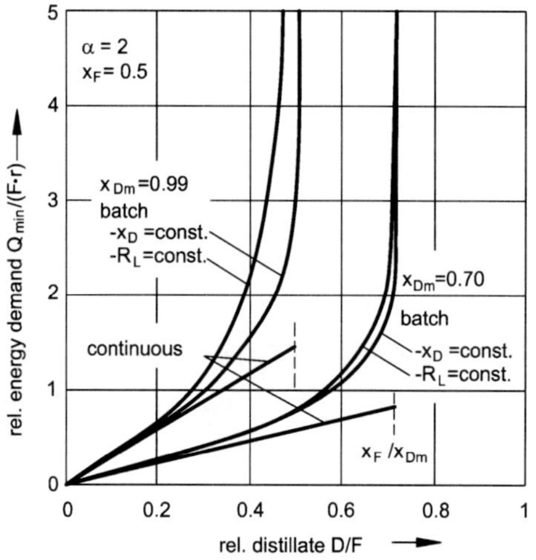

Fig. 5.2-41 Energy demand of both modes of batch distillation and continuous distillation

The energy demand of the operation mode x_D = const. is always lower than that of the operation mode R_L = const. However, the energy demand of both modes of batch distillation is always higher than the energy demand of an equivalent continuous distillation. Especially, in sharp separations, where the low-boiling component is nearly completely separated from the mixture, continuous distillation is greatly superior to batch distillation with respect to energy demand.

5.2.3.2 Ternary Mixtures

The Rayleigh equation is valid for all components in the mixture. Application of (5.2-77) to ternary mixtures yields

$$\frac{\mathrm{d}B}{B} = \frac{\mathrm{d}x_{\mathrm{B}a}}{x_{\mathrm{D}a} - x_{\mathrm{B}a}} \quad \text{and} \quad \frac{\mathrm{d}B}{B} = \frac{\mathrm{d}x_{\mathrm{B}b}}{x_{\mathrm{D}b} - x_{\mathrm{B}b}} \tag{5.2-83}$$

After elimination of $\mathrm{d}B/B$:

$$\frac{\mathrm{d}x_{\mathrm{B}a}}{\mathrm{d}x_{\mathrm{B}b}} = \frac{x_{\mathrm{D}a} - x_{\mathrm{B}a}}{x_{\mathrm{D}b} - x_{\mathrm{B}b}} \tag{5.2-84}$$

Equation (5.2-84) describes the change of the concentration $x_{\mathrm{B}i}$ of the liquid in the vessel during batch distillation (i.e., *residuum line*). The state of the distillate $x_{\mathrm{D}i}$ lies on a tangent to the actual liquid state in the vessel. Thus, as long as a pure product is withdrawn from the column, the state in the vessel moves along a straight line in the triangular concentration diagram. A curve in the residuum line indicates a change of distillate concentration. Figure 5.2-42A shows a typical residuum line of multistage batch distillation of a zeotropic ternary mixture. The graph B presents the standard plot of product concentration $x_{\mathrm{D}i}$ vs. the amount of distillate D/F. In mixtures with azeotropes and boundary distillation lines the process of batch distillation is more complex. From the system acetone/chloroform/benzene, shown in Figs. 5.2-43 and 5.2-44, it can be seen that a boundary distillation line cannot be crossed by batch distillation. Starting from any feed point F the residuum line first approaches the boundary distillation line and then follows the course of this distillation line. The concentration of the distillate plotted vs. the relative product ratio D/F shows a very untypical characteristic (maxima, minima) which can only be understood from a plot in the triangular diagram.

Batch distillation is of great importance in industry, especially for processing small quantities of liquid mixtures. The main advantages of batch distillation are the capability of fractionating even multicomponent mixtures into all constituents and of processing different liquid mixtures in the same unit (multiple purpose unit). Disadvantages of batch distillation are a high energy demand, a risk of thermal degradation of the substances and a rather complex process control.

5.2.3.3 Reactive Systems

Important applications of batch distillation are systems with reversible liquid-phase reactions. Assuming that the reaction takes place in the vessel only, such processes

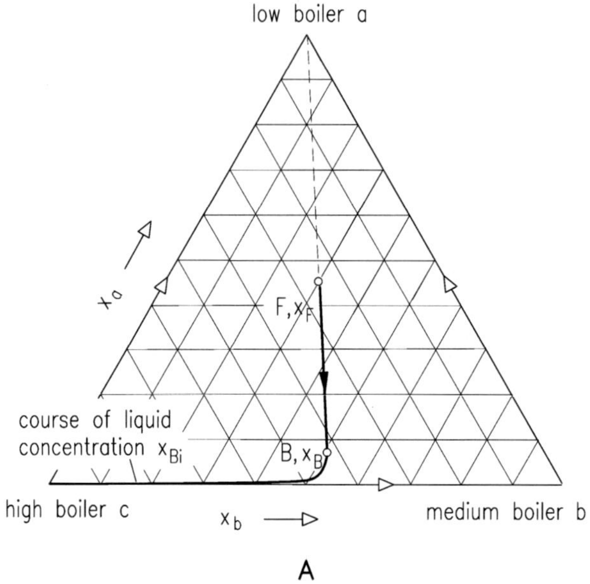

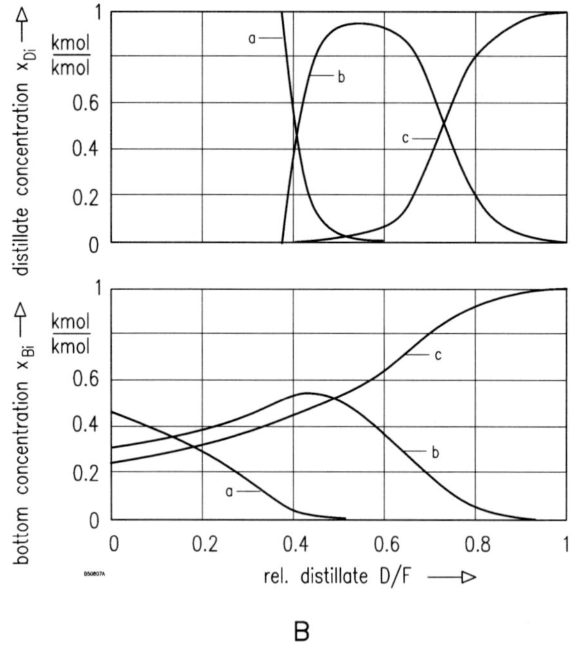

Fig. 5.2-42 Batch distillation of a zeotropic ternary mixture: (**A**) residuum line; (**B**) standard plot of product concentrations

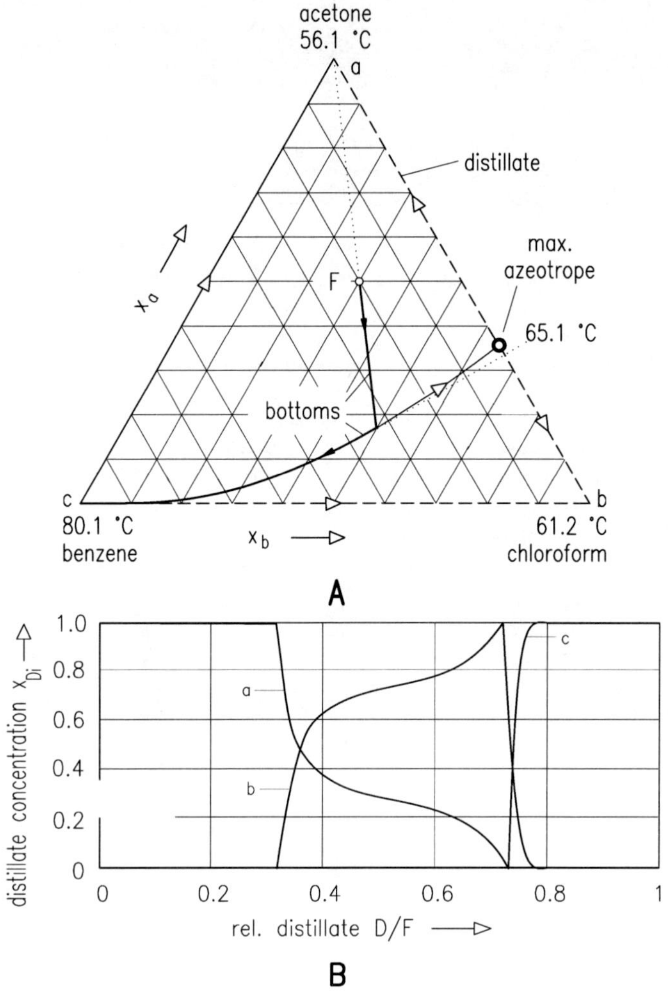

Fig. 5.2-43 Batch distillation of a ternary mixture with maximum azeotrope and boundary distillation line: **(A)** residuum line; **(B)** standard plot of product concentrations

can be treated with the methods outlined in the previous section. The process configurations have to be adjusted to the reaction at hand. The reactants have to remain in the vessel and only the products should be removed via the column, see Fig. 5.2-45. Following this principle the regular configuration (Fig. 5.2-39A) is suited for reactions where high-boiling reactants react to low-boiling products (e.g., $b + c \leftrightarrow v \cdot a$). Inverse batch distillation (Fig. 5.2-39B) is suited best for systems where low-boiling reactants react to high-boiling products (e.g., $a + b \rightarrow c$). A mid-

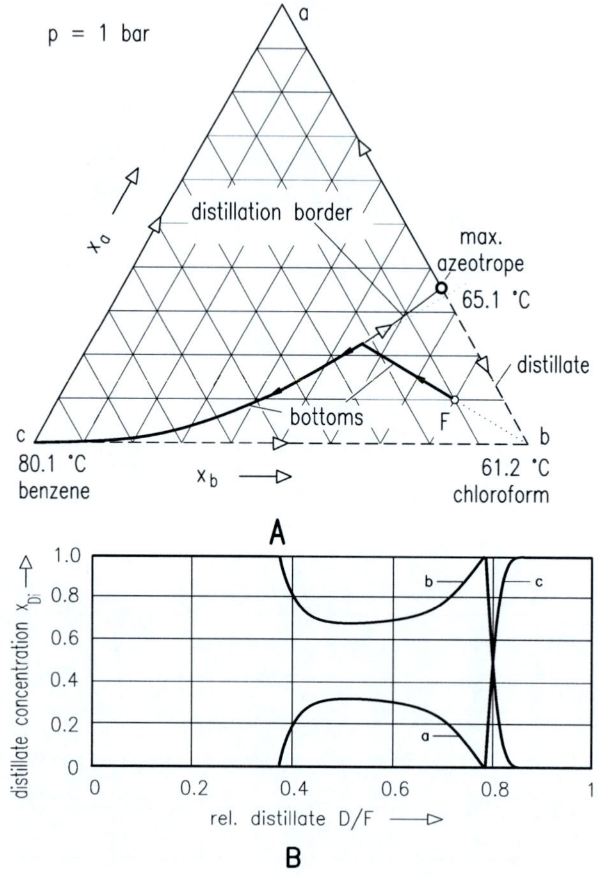

Fig. 5.2-44 Batch distillation of a ternary mixture with maximum azeotrope and boundary distillation line: (**A**) residuum lines; (**B**) standard plot of product concentrations

dle vessel configuration (Fig. 5.2-39C) has to be chosen for reactions of intermediate boilers to low- and high-boiling products (e.g., $b \rightarrow a + c$). By the removal of the products from the reaction zone in the vessel a complete conversion into the products can be established even in reversible reactions.

5.3 Absorption and Desorption

Absorption denotes the dissolution of gaseous substances (absorptives) in a liquid (absorbent, solvent, washing agent). The reversed process is called desorption.

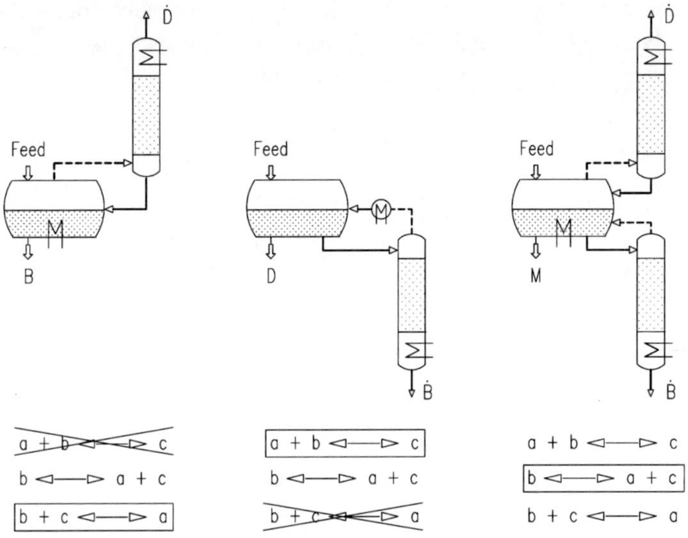

Fig. 5.2-45 Process configurations of reactive batch distillation for different types of chemical ractions

Process Scheme

In most cases absorption units consist of an absorption step and a regeneration step. Regeneration enables the internal recycling of the absorption agent as shown in Fig. 5.3-1.

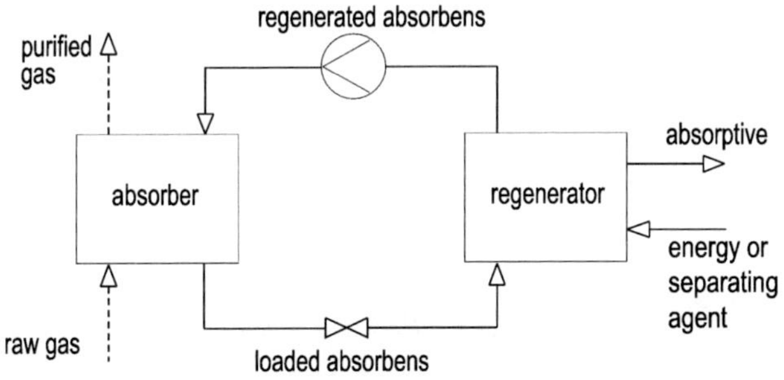

Fig. 5.3-1 Simplified scheme of an absorption process

In principle, all separation techniques can be applied to the regeneration of the loaded absorbents. Often used are flash, stripping with inert gases or steam and distillation. Very effective are combinations of different regeneration processes (Fig. 5.3-2), e.g., flash and distillation.

Absorption processes are often technically realized in packed columns or, more scarcely, in tray columns. For chemical absorption, however, many other types of equipment are used in industry, e.g., spray tower, venturi scrubber, bubble columns, etc.

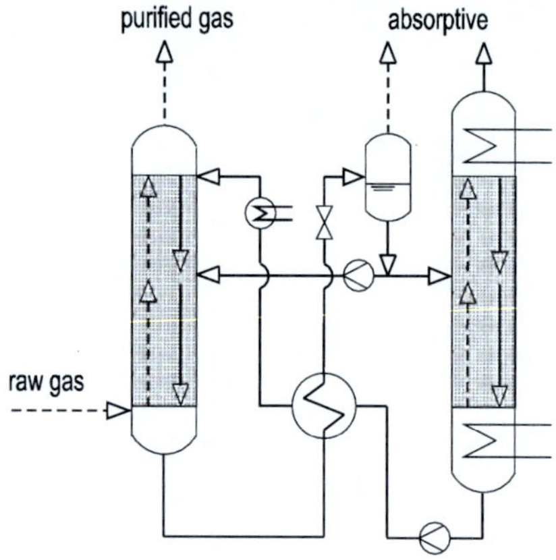

Fig. 5.3-2 Absorption process with absorbents regeneration by flash and distillation

5.3.1 Phase Equilibrium

Absorption equilibrium is preferably described by Henry's law since often, but not always, the absorptives are above their critical temperatures:

$$p_i = He_{ij} \cdot x_i . \tag{5.3-1}$$

The Henry coefficient He_{ij} depends on both the nature of the absorptive i and the absorbent j (washing agent). The values of the Henry coefficients significantly increase with rising temperatures. In practice, the Bunsen's absorption coefficients α are very often used. Their definition is

$$\alpha_i = \frac{m_N^3 \text{ dissolved absorptive } i}{m^3 \text{ pure absorbent}} \quad \text{at } p_i = 1 \text{ bar} . \tag{5.3-2}$$

In essence, α_i formulates the concentration of the liquid phase alone. The concentration of the coexisting gas phase is expressed by its partial pressure, mostly $p_i = 1$ bar . The following holds:

$$x_i = \frac{n_i}{n_i + n_j} = \frac{1}{1 + n_j/n_i} . \tag{5.3-3}$$

The symbol n denotes the moles of the absorptive i and of the absorbent j. Their values are:

$$n_i = \frac{\alpha_i\,[\mathrm{m}_N^3/\mathrm{m}^3]\cdot V_j\,[\mathrm{m}^3]}{\tilde{V}_N\,[\mathrm{m}_N^3/\mathrm{kmol}]} \quad \text{and} \quad n_j = \frac{V_j\,[\mathrm{m}^3]\cdot \rho_j\,[\mathrm{kg/m}^3]}{\tilde{M}_j\,[\mathrm{kg/kmol}]}. \tag{5.3-4}$$

$\tilde{V}_N$ denotes the standard molar volume with $\tilde{V}_N = 22.413\ \mathrm{m}^3/\mathrm{kmol}$.

Combining (4.3-3) and (4.3-4) yields

$$x_i = \frac{\alpha_i\cdot V_j/\tilde{V}_N}{\alpha_i\cdot V_j/\tilde{V}_N + V_j\cdot \rho_j/\tilde{M}_j} = \frac{1}{1+\tilde{V}_N\cdot \rho_j/(\alpha_i\cdot \tilde{M}_j)}. \tag{5.3-5}$$

After implementing into (5.3-1):

$$1\ \mathrm{bar} = He_{ij}\cdot \frac{1}{1+\tilde{V}_N\cdot \rho_j/(\alpha_i\cdot \tilde{M}_j)}. \tag{5.3-6}$$

Rewriting delivers

$$He_{ij} = \left(1\,bar + \frac{\tilde{V}_N\cdot \rho_j}{\alpha_i\cdot \tilde{M}_j}\right). \tag{5.3-7}$$

Selected values of Bunsen's absorption coefficient α are plotted vs. temperature in Fig. 5.3-3. Often used is also the technical absorption coefficient λ_i. Here, the amount of dissolved gases is referred to the mass of pure solvents.

5.3.2 Physical Absorption

The thermodynamics of physical absorption are formulated in analogue to distillation. Important quantities are the minimum demand of solvent (analogue: minimum reflux) and the minimum demand of stripping gas (minimum reboil). Both quantities can be formulated via material balances.

5.3.2.1 Minimum Demand of Solvent

A material balance around the column (Fig. 5.3-4) delivers

$$\dot{G}_u\cdot y_{iu} = \dot{L}_u\cdot x_{iu} \quad \Rightarrow \quad \dot{L}_u = \dot{G}_u\cdot \frac{y_{iu}}{x_{iu}}. \tag{5.3-8}$$

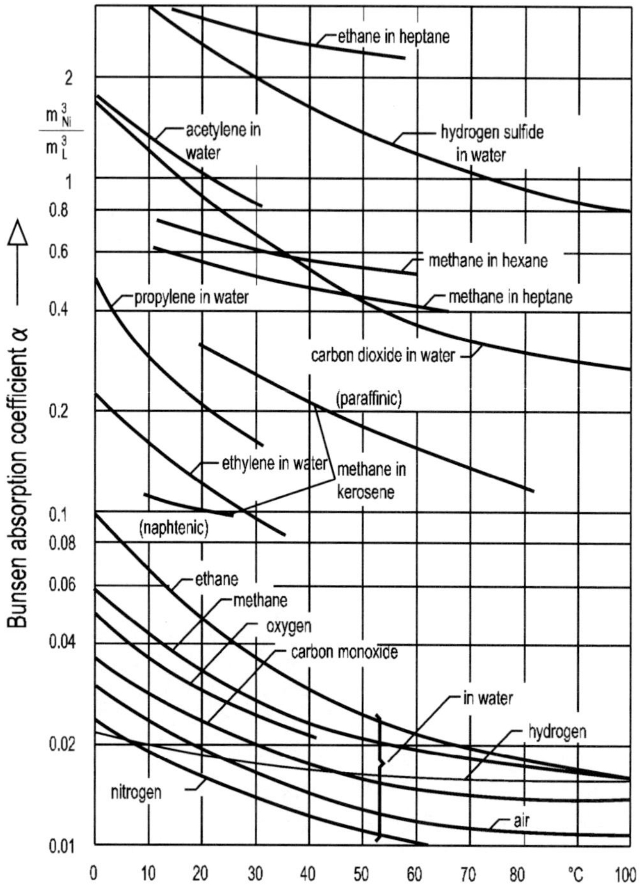

Fig. 5.3-3 Values of Bunsen absorption coefficient α of selected compounds vs. temperature taken from Landolt-Börnstein (1980)

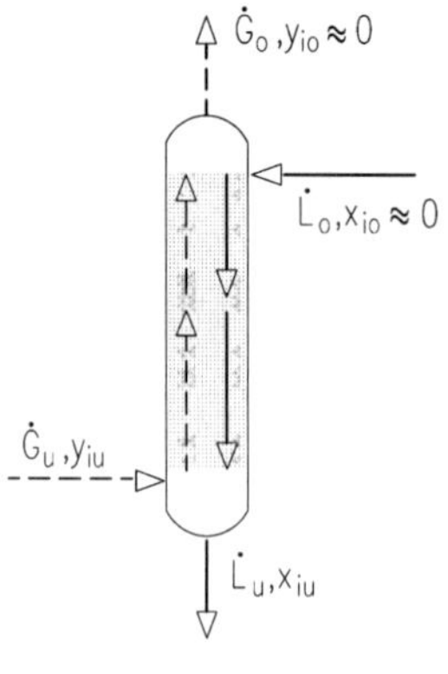

Fig. 5.3-4 Material balance of an absorption column for calculating the minimum demand of solvent

In the special case of a very high column, phase equilibrium is reached at the bottom of the column:

$$\left(\frac{y_{iu}}{x_{iu}}\right)^* = \frac{He_{ij}}{p}. \tag{5.3-9}$$

From the two equations above follows:

$$\dot{L}_{\mathrm{umin}} = \dot{G}_u \cdot \frac{He_{ij}}{p}. \tag{5.3-10}$$

According to (5.3-10) the demand of solvent is directly proportional to the amount of raw gas to be treated and inversely proportional to the system pressure. As the Henry coefficient *He* increases with temperature, low system temperatures are advantageous for absorption. It is important to note that the minimum demand of solvent does not depend on raw gas concentration. In technical processes, absorption is effected with a surplus of the solvent:

$$\dot{L} \approx 1.3 \cdot \dot{L}_{\mathrm{min}}. \tag{5.3-11}$$

5.3.2.2 Minimum Demand of Stripping Gas

The minimum amount of stripping gas can be found analogously. A material balance delivers (Fig. 5.3-5):

$$\dot{L}_o \cdot x_{io} = \dot{G}_o \cdot y_{io} \quad \Rightarrow \quad \dot{G}_o = \dot{L}_o \cdot \frac{x_{io}}{y_{io}}. \tag{5.3-12}$$

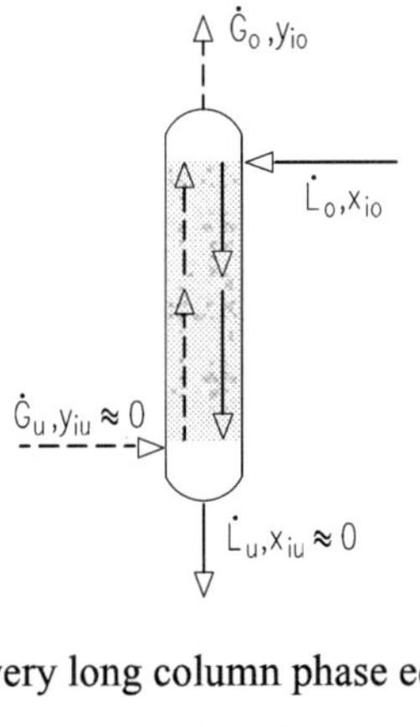

Fig. 5.3-5 Material balance of a desorption column for calculating the minimum demand of stripping gas

In a very long column phase equilibrium is reached at the top of the column:

$$\dot{G}_{\text{omin}} = \dot{L}_o \cdot \frac{p}{He_{ij}}. \tag{5.3-13}$$

Hence, desorption is enhanced by low system pressures and high temperatures. The minimum demand of stripping gas is directly proportional to the amount of liquid $\dot{L}_0$ to be processed. It does not depend on feed concentration of the liquid.

5.3.2.3 Number of Equilibrium Stages

In analogue to distillation, the number of equilibrium stages required for a specified separation is graphically determined in a y/x diagram (analogue to McCabe–Thiele diagram). However, modified concentrations X, Y have to be used (instead of x, y) for the following reasons:

The raw gas consists of an inert gas $\dot{G}_t$ and of the absorptives. On the way up through the column, the absorptives are removed from the inert gas and, in turn, the amount of the gas phase decreases. Analogously, the amount of the liquid phase is increased during the absorption process. Therefore, the ratio of liquid to gas $\dot{L}/\dot{G}$ changes and, in turn, the operating line becomes curved. This problem can be solved by using a concentration where the amounts of absorptives are referred to the amount of pure inert gas and pure solvent, respectively:

$$Y_i \equiv \frac{n_i}{n_t} \quad \text{and} \quad X_i \equiv \frac{n_i}{n_t}. \tag{5.3-14}$$

At very low concentrations often encountered in absorption processes the values of X, Y and x, y become equal.

The material balance in Fig. 5.3-6 delivers

$$\dot{G}_t \cdot Y_u + \dot{L}_t \cdot X = \dot{G}_t \cdot Y + \dot{L}_t \cdot X_u. \tag{5.3-15}$$

Rewriting yields

$$Y = \frac{\dot{L}_t}{\dot{G}_t} \cdot X + \left(Y_u - \frac{\dot{L}_t}{\dot{G}_t} \cdot X_u\right) \quad \text{or} \quad Y = \frac{\dot{L}_t}{\dot{G}_t} \cdot X + \left(Y_o - \frac{\dot{L}_t}{\dot{G}_t} \cdot X_o\right). \tag{5.3-16}$$

Obviously, the operating line is linear between Y and X (Fig. 5.3-7).

Isothermal Absorption

In the case of isothermal absorption, the equilibrium line is easily written with Henry's law, which formulates a linear relationship between y and x. The molar fractions have to be replaced by the new molar loadings according to

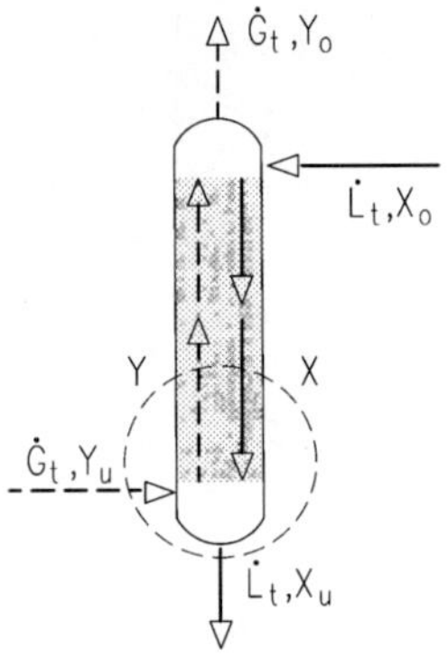

Fig. 5.3-6 Material balance for calculating the operating line of absorption processes

$$y = \frac{Y}{1+Y} = \frac{1}{1+1/Y} \quad \text{and} \quad x = \frac{X}{1+X} = \frac{1}{1+1/X}. \tag{5.3-17}$$

Implementation into Henry's law gives

$$\frac{1}{1+1/Y} = \frac{He_{ij}}{p} \cdot \frac{1}{1+1/X}. \tag{5.3-18}$$

After rewriting:

$$Y = \frac{He_{ij} \cdot X}{p + X \cdot (p - H_{ij})}. \tag{5.3-19}$$

Operating line and equilibrium line are qualitatively depicted in Fig. 5.3-7. In analogue to distillation, the difficulty of separation is expressed by the number of equilibrium stages that are determined by drawing stages between operating and equilibrium lines. The same procedure is applied to desorption, also shown in Fig. 5.3-7. It is important to note that in absorption processes the operating line lies above the equilibrium line since the light constituent has to be removed from the gas phase. In desorption processes, however, the operating line always lies below the equilibrium line since, like in distillation, the light constituents have to be removed from the liquid phase.

Nonisothermal Absorption

Absorption is always accompanied by a temperature increase since the heat of absorption q is set free by the dissolution of the absorptive. In particular at high concentrations of the absorptives in the raw gas the temperature increase has to be accounted for as the Henry coefficient He significantly depends on temperature. In a first approximation, the heat of absorption effects an increase of liquid tem-

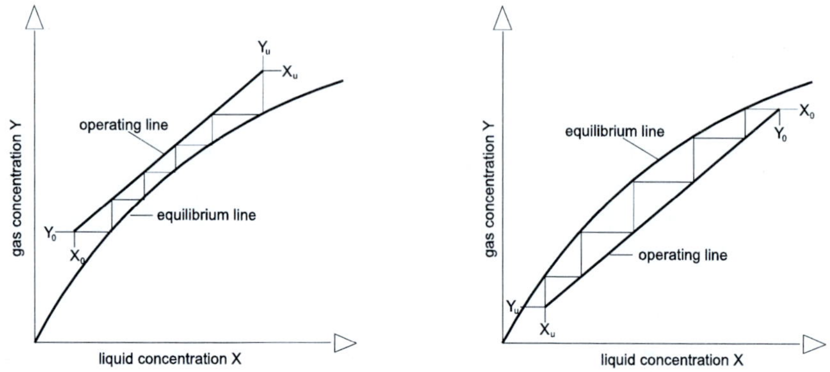

Fig. 5.3-7 Operating and equilibrium lines for the determination of the number of equilibrium stages. *Left*: absorption; *right*: desorption (stripping)

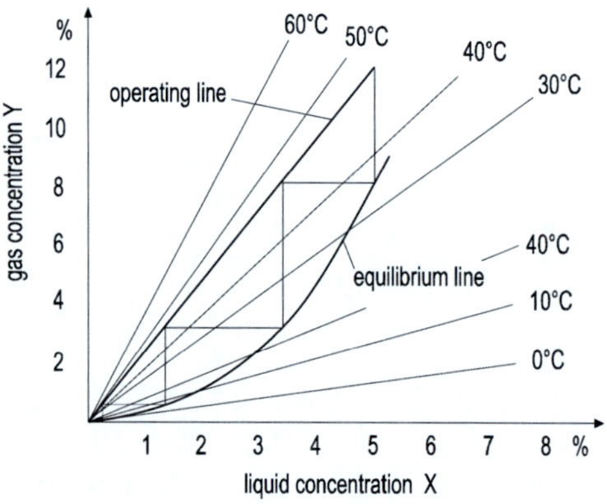

Fig. 5.3-8 Operating and equilibrium lines for nonisothermal (adiabatic) absorption

perature only (as the heat capacity of the gas is very low). An enthalpy balance delivers

$$\dot{L} \cdot c_L \cdot \Delta T = \dot{L} \cdot \Delta X \cdot q \quad \text{or} \quad \Delta T = \Delta X \cdot q / c_L . \tag{5.3-20}$$

The change of liquid temperature ΔT is directly proportional to the concentration change ΔX of the liquid. Starting from a plot of several equilibrium lines at constant temperature, the equilibrium line relevant to nonisothermal absorption can be drawn step by step (Fig. 5.3-8). Typical profiles of temperature and concentration within the column are shown in Fig. 5.3-9.

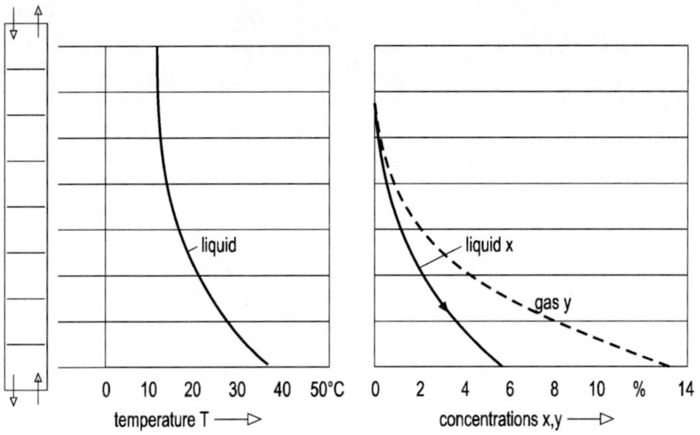

Fig. 5.3-9 Internal temperature and concentration profiles of nonisothermal absorption

5.3.2.4 Comparison Between Distillation and Absorption

The processes of absorption and distillation are similar in a high extent. In principle, the same processes are effected within the column. A countercurrent flow of gas and liquid is superimposed by a mass transfer between the phases. Both phases are saturated, i.e., the gas is at its dew point and the liquid at its boiling point.[1]

The principle difference between distillation and absorption is the condition at the column ends. As shown in Fig. 5.3-10, a liquid enters and a gas leaves at the top the column. At absorption, the entering liquid and the leaving gas are independent from each other. At distillation, however, the entering liquid is a part of the leaving gas. Therefore, gas and liquid are interrelated phases at the top of a distillation column. The same holds at the bottom of the column. This interrelation is twofold. It refers to the substances as well as to the amounts of the phases. This coupling leads to the fact that in distillation processes the operating line can only lie between equilibrium line and diagonal of the McCabe–Thiele diagram. These restrictions are not encountered in absorption and desorption, respectively. In turn, absorption and desorption are the more general case of mass transfer columns.

1. One should be aware that the terms "dew point" and "boiling point" lose their meaning in wide boiling systems encountered in absorption and desorption. For instance, water saturated with air at 20°C contains so little dissolved air that no bubbling can be effected by a small increase of temperature. Nevertheless, from a thermodynamic point of view, the water/air system is at its boiling point.

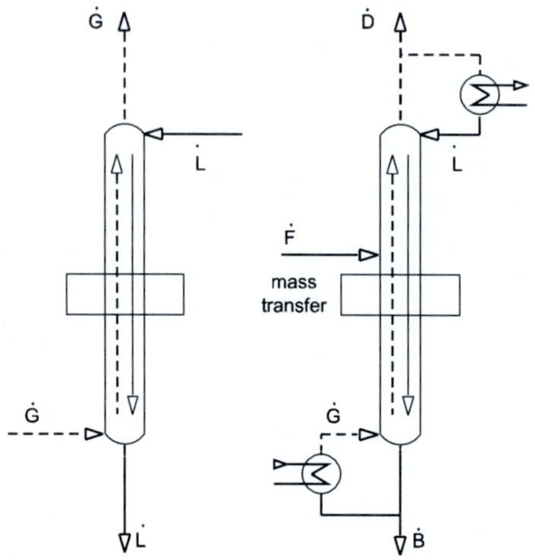

Fig. 5.3-10 Principle schemes of absorption (*left*) and rectification (*right*)

5.3.3 Chemical Absorption

In industrial absorption processes the specifications of gas purity are often very high. These requirements can only hardly be met by physical absorption. A very effective measure to enhance interfacial mass transfer is the addition of substances to the solvent that chemically react with the absorptives. By chemical reaction the capacity as well as the selectivity of absorption can be drastically improved. In a first step the gaseous substance a is physically dissolved in the liquid. In a second step the dissolved substance a reacts with the active compound b in the liquid to the product z according to

$$a + \nu \cdot b \rightarrow z \quad \text{(irreversible) or} \quad a + \nu \cdot b \leftrightarrow z \quad \text{(reversible).} \tag{5.3-21}$$

The symbol ν denotes the stoichiometric coefficient of the reaction. For technical absorption processes only very fast reactions which are reversible at higher temperatures are suited.

During absorption and succeeding reaction the heat of reaction q is set free. It can be determined from listed values of the "heat of formation", see, for instance, *Perry's Chemical Engineer's Handbook* (Perry et al. 1997). Its use is explained at the absorption of SO_2 in aqueous $NaOH$ solutions at a temperature of 25°C. The reaction is

$$(2NaOH)_{aq} + SO_2 \rightarrow (Na_2SO_3)_{aq} + H_2O + q\,. \tag{5.3-22}$$

Table 5.3-1 Heats of formation of selected compounds (taken from Perry et al. 1997), aq, 400 = 1 mol dissolved in 400 mol of water

Compound	State	Heat of formation at 25°C in kcal/mol	State	Heat of formation at 25°C in kcal/mol
$CaCO_3$	c, calcite	– 289.5		
$Ca(OH)_2$	c	– 235.58	aq, 800	– 239.2
$CaSO_4$	c, insoluble form	– 338.73	c, soluble form	– 309.8
CO_2	g	– 94.052		
HNO_3	g	– 31.99	l	– 41.35
H_2O	g	– 57.7979	l	– 68.3174
H_2S	g	– 4.77	aq, 2000	– 9.38
Na_2CO_3	c	– 269.46	aq, 100	– 275.13
$NaHCO_3$	c	– 226.0	aq	– 222.1
$NaOH$	c	– 101.96	aq, 400	– 112.193
Na_2S	c	– 89.8	aq, 400	– 105.17
Na_2SO_3	c	– 261.2	aq, 800	– 264.1
Na_2SO_4	c	– 330.50	aq, 1100	– 330.82
NH_3	g	– 10.96	aq, 200	– 19.27
$(NH_4)_2CO_3$	aq	– 223.4		
NH_4NO_3	c	– 87.40	aq, 500	– 80.89
$(NH_4)_2SO_4$	c	– 281.74	aq, 400	– 279.33
O_2	g	0.0		
K_2CO_3	c	– 274.01	aq, 400	– 280.90
$KHCO_3$	c	– 229.8	aq, 2000	– 224.85
KOH	c	– 102.02	aq, 400	– 114.96
SO_2	g	– 70.94		
SO_3	g	– 94.39	l	– 103.03

This equation is written with the heats of formation taken from Table 5.3-1

$$2 \cdot (-112.19) + (-70.94) = (-264.1) + (-68.32) + q\,. \tag{5.3-23}$$

The result is

$$q = 37.1\,\frac{\text{kcal}}{\text{mol}} = 155.34\,\frac{\text{kJ}}{\text{mol}}\,. \tag{5.3-24}$$

The reaction considered here is strongly exothermal. In nonideal systems the heat of reaction also depends on concentration.

A chemical reaction taking place in the liquid phase significantly influences the phase equilibrium of the absorptives. Typically, a small part of the absorptive remains physically dissolved in the liquid, the greater part has been converted to the product z. Hence, the total concentration of the absorptive, c_{at}, is

$$c_{at} = c_a + c_z \quad \text{or} \quad X_{at} = X_a + X_z\,. \tag{5.3-25}$$

In irreversible reactions, the equilibrium line of component a is shifted by the reactant concentration c_b/ν. The effective equilibrium lines are – in a first approximation – parallel to the line of physical absorption (Fig. 5.3-11). The absorption capacity is significantly increased by the presence of the reactive substance in the solvent.

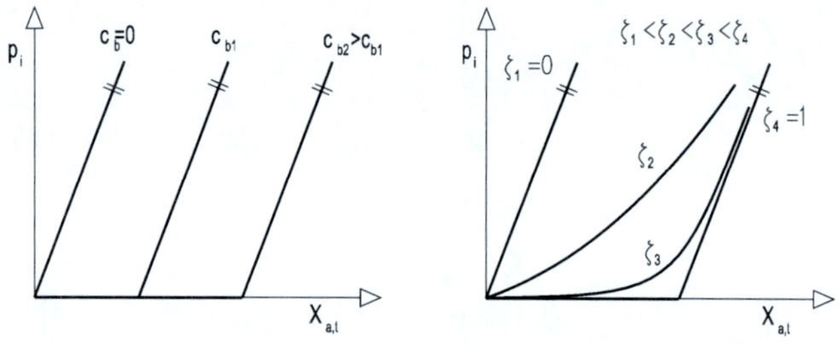

Fig. 5.3-11 Phase equilibrium of chemical absorption at irreversible (*left*) and reversible (*right*) chemical reactions

In reversible reactions, the partial pressure p_i of an absorptive slightly increases with increasing concentration in the liquid. Hence, the chemical reaction significantly changes the phase equilibrium in such systems. It is quantitatively described by the extent of reaction ζ:

$$|dn_a| = \left|\frac{dn_b}{\nu}\right| = d\zeta, \quad 0 \le \zeta \le 1. \tag{5.3-26}$$

At $\zeta = 0$, no reaction takes place and physical absorption alone determines the phase equilibrium. A value of $\zeta = 1$ characterizes an irreversible reaction.

In the design and dimensioning of absorption equipment the reaction kinetics is of great importance besides the phase equilibrium. If the reaction is very slow, the physical absorption is the dominant mechanism since the residence time of the liquid in the absorber is too low. In fast reactions, the equilibrium line of the chemical system is the relevant quantity. In very fast (instantaneous) reactions, the kinetics of the mass transfer is enhanced, too (see Chap. 4). In industrial practice, only fast reactions are applied to chemical absorption (Table 5.3-3).

5.3.3.1 Minimum Demand of Solvent of Chemical Absorption

The minimum demand of solvent $\dot{L}_{\min}$ of physical absorption is directly proportional to the raw gas $\dot{G}_u$ and inversely proportional to the system pressure p. In chemical absorption, however, the minimum demand of solvent is proportional to the amount of absorptives $\dot{G}_u \cdot y_u$ in the raw gas and inversely proportional to the concentration of the reactant x_b in the solvent. In many applications the amount of

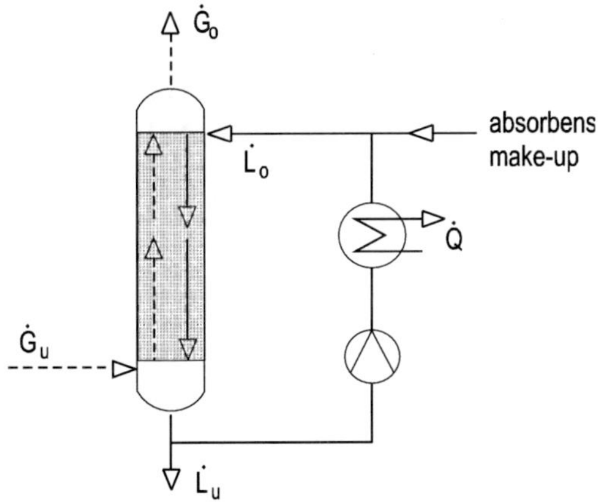

Fig. 5.3-12 Chemical absorption with partial washing agent recycling

Table 5.3-2 Comparison of physical and chemical absorption

	Physical absorption		Chemical absorption
Amount of liquid $\dot{L}$	$\dot{L} \sim \dot{G}_u$ $\dot{L} \sim \frac{1}{p}$	 $\dot{L} \neq f(p_i)$	$\dot{L}_{\min} = \dot{G}_u \cdot y_u \cdot \frac{1}{x_u + x_b/\nu}$
Application	Large amount of raw gas. Intermediate gas purity		Small amount of raw gas. Very high gas purity

solvent required is very small. In this case, the amount of solvent has to be determined via a heat balance instead of a material balance. The temperature increase caused by the heat of reaction has to be restricted to, for instance, 10°C. The resulting amount of solvent is in most cases much larger than the minimum amount of solvent determined from the stoichiometry of the reaction (Fig. 5.3-12).

Consequently, some of the reactive solvent is recycled to the top of the absorption column.

In Table 5.3-3 some examples of industrial processes of chemical absorption are listed. These processes are often used for the separation of acidic compounds from off gases and from synthesis gases.

5.4 Dimensioning of Mass Transfer Columns

The demands on effective mass transfer equipments can be understood from the basic equation of mass transfer (Taylor and Krishna 1993):

$$\dot{N} = k_{\mathrm{OG}} \cdot A \cdot (y^* - y) \tag{5.4-1}$$

To achieve a high mass transfer rate $\dot{N}$ each factor in (5.4-1) should be large. The driving force $(y^* - y)$ is high at a countercurrent flow of gas and liquid. A large interfacial area A is provided by column internals like trays or packings. The mass transfer coefficient k_{OG} has high values if the interfacial area is steadily renewed. All these demands are well met by tray columns and by packed columns.

Table 5.3-3 Industrial processes of chemical absorption

Process	Washing agent	Absorptive	T in °C	Chemical reaction	Regeneration
Cold caustic soda process	8 wt% sodium hydroxide	CO2, H2S	10–30	$2NaOH + CO_2 \rightarrow Na_2CO_3 + H_2O$ $2NaOH + H_2S \rightarrow Na_2S + 2H_2O$	60 kg CaO, 300 kg steam, 7 m^3 cooling water, 15 $kWh/m^3_{N,\,lye}$
Hot caustic soda process	(2–4) wt% sodium hydroxide	COS	≥60	$4NaOH + COS \rightarrow Na_2CO_3 + Na_2S + 2H_2O$	60 *kg* CaO, 300 *kg* steam, 7 m^3 cooling water, 15 $kWh/m^3_{N,\,lye}$
Cold potash lye process	(10–12) wt% K2CO3 in H2O	CO2	≈40	$K_2CO_3 + CO_2 + H_2O \rightarrow 2KHCO_3$	6.5 kg $steam/m^3_{N,\,CO_2}$
Hot potash lye process	(15–30) wt% K2CO3 in H2O	CO2, H2S	110–116	$K_2CO_3 + CO_2 + H_2O \rightarrow 2KHCO_3$ $K_2CO_3 + H_2S \rightarrow KHCO_3 + KHS$	2.1–4.2 kg $steam/m^3_{N,\,CO_2}$
Monoethanol amine MEA	(10–20) wt% MEA in H2O	CO2	<55	$2HOC_2H_4NH_2 + CO_2 + H_2O \rightarrow (HOC_2H_4NH_3)_2CO_3$	3.1–6.8 kg $steam/m^3_{N,\,CO_2}$
Aiethanol amine DEA	(10–25) wt% DEA in H2O	H2S	≈20	$(HOC_2H_4)_2NH + H_2S \rightarrow (HOC_2H_4)_2NH_3S$	6.0 kg $steam/m^3_{N,\,H_2S}$
Ammonium-water process	5 wt% NH3-water	CO2	≈20	$2NH_3 + CO_2 + H_2O \rightarrow (NH_4)_2CO_3$	6.0 kg $steam/m^3_{N,\,CO_2}$
Alkazid process	Alkazid	CO2, H2S	≈20	–	7.0 kg $steam/m^3_{N,\,CO_2,\,H_2S}$

At the present state of the art dimensioning of mass transfer columns is performed in two steps. The first step is the thermodynamic simulation of the processes taking place in the column (see Sect. 5.2). The results of this rigorous thermodynamic simulation are the flow rates of gas and liquid within the column, the energy demand, the number of equilibrium stages, the internal profiles of concentrations and temperature, and the product qualities. These data are the basis for the second step, the sizing of the equipment. In this step all geometrical structures and dimensions of the equipment (type and dimensions of columns and internals) have to be determined. This includes the calculation of the diameter and the height of the column (i.e., the number of actual trays and the height of the packing, respectively).

Two-phase flow and mass transfer in seperation columns are rather complex mechanisms that cannot be modeled from first principles alone. At the present state of the art, models for column dimensioning are – to a large extent – based on empirical know-how developed over decades of industrial application. Most know-how is well described in the open literature (e.g., Perry and Green 2008; Kister 1989, 1992; Stichlmair and Fair 1998; Lockett 1986; Mersmann et al. 2005; Kolev 2006; Strigle 1987). Additionally, there is a huge number of research reports from universities. Important contributions to the state of the art stem from industry funded research institutes as Fractionation Research Inc. (FRI) in Stillwater, OK, and Separation Research Program (SRP) of the University of Texas at Austin, TX. In recent years some computer software for column dimensioning is also available (e.g., KG-Tower from Koch-Glitsch, SulCol from Sulzer, TrayHeart from Welchem).

5.4.1 Tray Columns

A perspective view of a tray column is shown in Fig. 5.4-1. The gas flows upward and the liquid downward in the column. On the trays itself there exists a cross flow of the two phases. The horizontal trays have openings (holes, caps, valves) for enabling the gas flow through the plates.

5.4.1.1 Design Principles

Three principles for the design of the openings in the plates can be distinguished (Fig. 5.4-2):

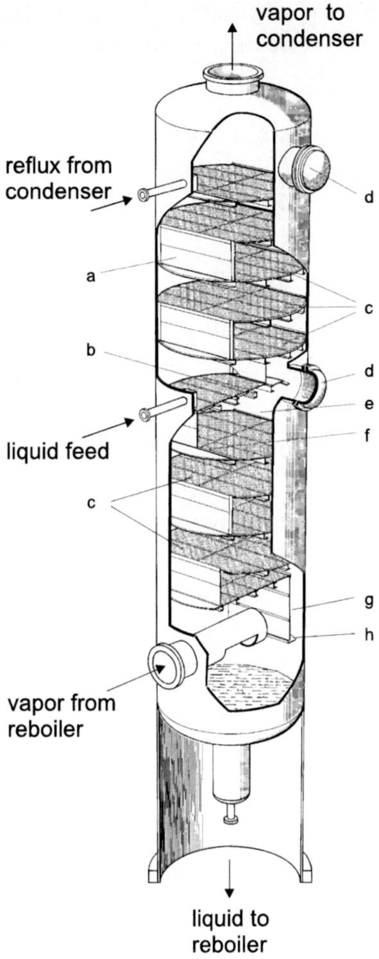

Fig. 5.4-1 Perspective view of a tray column: (a) downcomer, (b) tray support, (c) sieve trays, (d) man way, (e) outlet weir, (f) inlet weir, (g) side wall of downcomer, and (h) liquid seal

- Sieve trays have a large number of round holes in the plate (6–13 mm in diameter) with an open area of about 8–15% of the active area A_{ac}. The weeping of liquid is prevented by the upward flow of gas through the holes.
- In bubble cap trays weeping of liquid is prevented by design. A chimney with a cap covers the openings in the plate. Bubble caps typically have a diameter of 50–80 mm. The relative free area is about 10–20%. The pressure drop of bubble cap trays is rather high.

- Valve trays have movable elements in the tray openings for adjusting the free area to the actual gas load. This widens the operating range and reduces the pressure drop at high gas loads.

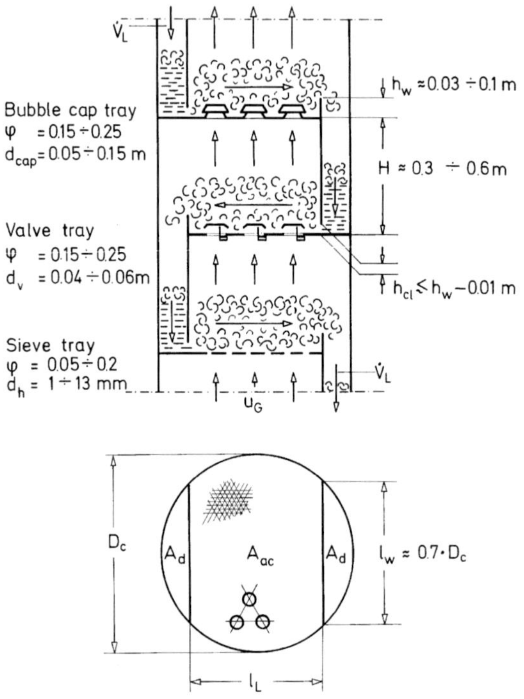

Fig. 5.4-2 Scheme of a tray column showing the most important design parameters of some standard tray designs

The two-phase layer on the trays is approximately 10–30 cm high and has a large volumetric interface of up to 1000 m^2/m^3. The liquid crosses the tray on its way to the outlet weir (3–8 cm high) and the downcomer. Through a side opening at the lower end of the downcomer the liquid is fed to the tray beneath. The downcomers are alternately arranged at the left-hand and right-hand side of the trays. Typically, tray spacing is in the range from 0.3 to 0.6 m. The active area on the trays is about 80% of column cross section. In large diameter columns special trays are often used with several downcomers evenly distributed over the active area (multiple downcomer trays). In systems with high fouling risk trays without downcomers (dual flow trays) are recommended. Table 5.4-1 lists some characteristic design parameters of trays.

Table 5.4-1 Characteristic dimensions of industrial tray designs

	Vacuum	Atmospheric pressure	High pressure
Tray spacing (m)	0.4–0.6	0.4–0.6	0.3–0.4
Weir length	$(0.5–0.6) \cdot D_c$	$(0.6–0.7) \cdot D_c$	$(0.6–0.8) \cdot D_c$
Weir height (m)	0.02–0.03	0.03–0.07	0.04–0.1
Skirt clearance	$0.7 \cdot h_w$	$0.8 \cdot h_w$	$0.8 \cdot h_w$
Bubble cap diameter (m)	0.08 – 0.15	0.08 – 0.15	0.08 – 0.15
Bubble cap spacing	$1.25 \cdot d_{cap}$	$(1.25 - 1.4) \cdot d_{cap}$	$1.5 \cdot d_{cap}$
Valve diameter (m)	0.04 – 0.05	0.04 – 0.05	0.04 – 0.05
Valve spacing	$1.5 \cdot d_V$	$1.5 \cdot d_V$	$1.5 \cdot d_V$
Hole diameter (m)	0.004 – 0.013	0.004 – 0.013	0.004 – 0.013
Hole spacing (triang.)	$(2.5 - 3) \cdot d_h$	$(3 - 4) \cdot d_h$	$(3.5 - 4.5) \cdot d_h$
Relative free area (%)	10–15	6–10	4–7.5

The gas load, expressed as F-factor, is typically in the range from 1 to 2 $Pa^{1/2}$. The liquid load $\dot{V}_L$ referred to weir length l_w can be varied from approximately 2 to 100 $m^3/(m \cdot h)$. The pressure drop of tray columns is rather high, typically 7 mbar per equilibrium stage. Thus, tray columns are not very well suited for vacuum services.

5.4.1.2 Operation Region of Tray Columns

Tray columns can only be operated within certain limits of gas and liquid flow as is shown in Fig. 5.4-3. The upper limits (bold lines) are strong limits which can never be crossed without causing mechanical damages. The lower limits (dashed lines) are soft limits which can be crossed if some losses of mass transfer efficiency are accepted. The operation point should be chosen so that a sufficient safety margin to the operation limits remains.

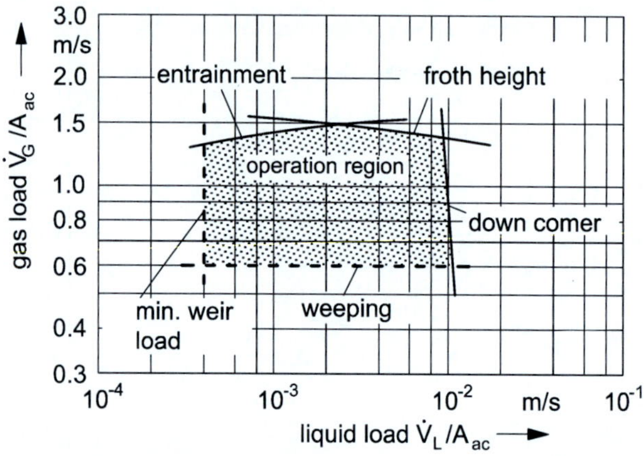

Fig. 5.4-3 Operation region of tray columns

Maximum Gas Load

The gas load in the column is a very important quantity since it governs the diameter of the column. By convention, the gas load in columns is expressed by the so-called F-factor according to the definition

$$F \equiv u_G \cdot \sqrt{\rho_G}. \tag{5.4-2}$$

The term u_G denotes the superficial gas velocity referred to the active area A_{ac} of the tray, i.e., the cross-sectional area of the column minus the area of two downcomers (Fig. 5.4-2). The gas load has to be kept between a maximum and a minimum value.

The maximum gas load is often determined from the equation of Souders and Brown (1934):

$$F \le C_G \cdot \sqrt{\rho_L - \rho_G} \qquad \text{or} \qquad u_G \le C_G \cdot \sqrt{(\rho_L - \rho_G)/\rho_G}. \tag{5.4-3}$$

The gas load factor C_G, defined by (5.4-3), has to be determined from experimental data or from empirical correlations of such data (e.g., Perry and Green 2008; Kister 1992; Lockett 1986). Well known and often used is the correlation of Fair (1961) originally developed for sieve trays (Fig. 5.4-4).

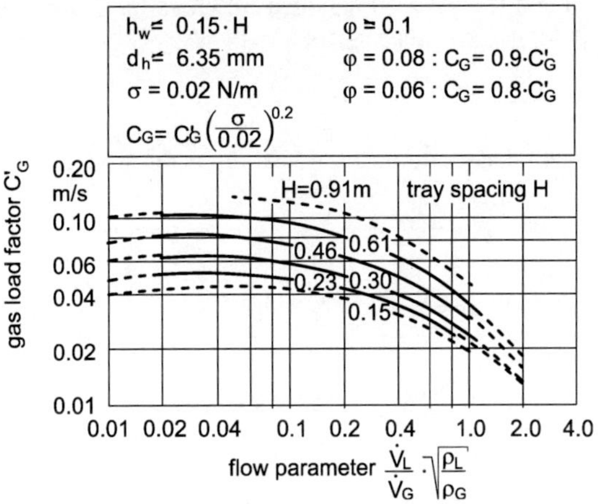

Fig. 5.4-4 Correlation of the maximum gas load of sieve trays according to Fair (1961)

A theoretically better founded approach has been developed by Stichlmair (1978; Stichlmair and Fair 1998). A balance of weight, buoyancy, and drag forces on a single droplet whose size is determined via a critical Weber number yields

$$F_{\max} = 2.5 \cdot (\varphi^2 \cdot \sigma \cdot (\rho_L - \rho_G) \cdot g)^{1/4} \quad \text{or}$$

$$C_{G\max} = 2.5 \cdot (\varphi^2 \cdot \sigma \cdot g / (\rho_L - \rho_G))^{1/4} \tag{5.4-4}$$

According to these equations the maximum feasible gas load depends on system properties (density ρ of gas and liquid, surface tension σ) as well as on tray design (relative free area φ). Equations (5.4-4) describe the so-called entrainment flooding of trays that is decisive at very large tray spacings. For smaller tray spacings the froth height on the tray sets a lower limitation (see Sect. 5.4.1.3).

Minimum Gas Load

A limitation of the minimum gas load is set either by weeping of liquid through the openings in the tray or by nonuniform gas flow through these openings. The condition for uniform gas flow through the holes of a sieve tray is according to (Mersmann 1963):

$$F_{\min} \geq \varphi \cdot \sqrt{2 \cdot \sigma / d_h} \tag{5.4-5}$$

According to Ruff et al. (1976), weeping of liquid through the holes of a sieve tray is prevented if the gas load in the hole is higher than

$$F_{\min} \geq \varphi \cdot \sqrt{0.37 \cdot d_h \cdot g \cdot (\rho_L - \rho_G)^{5/4} / \rho_G^{1/4}}\,. \tag{5.4-6}$$

Equations (5.4-5) and (5.4-6) are evaluated for sieve trays operated with three different systems in Fig. 5.4-5. At low hole diameters the minimum gas load in the holes, $F_{\mathrm{hmin}} = F_{\mathrm{min}}/\varphi$, decreases with increasing hole diameter d_{h}. At large hole diameters, however, the minimum gas load F_{min}/φ increases with increasing hole diameter since the condition of no-weeping is decisive in this case.

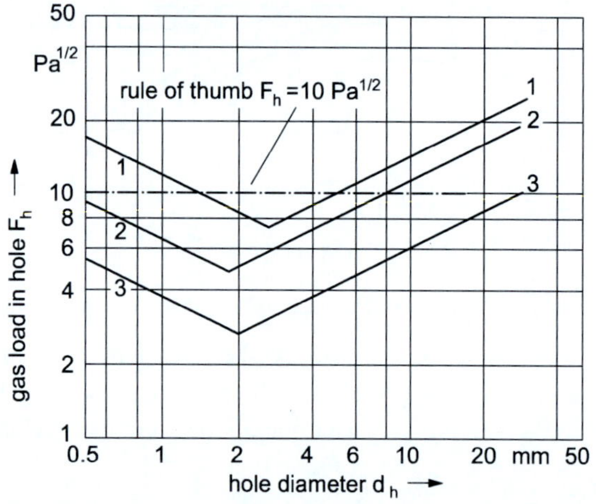

Fig. 5.4-5 Minimum gas load of sieve trays for three different systems: 1 air/water, 2 benzene/toluene, and 3 i-butane/n-butane

Maximum Liquid Load

The liquid flow downward through the downcomers is enforced by gravity forces which results in a limitation of the maximum liquid load. In practice, four empirical rules of thumb are recommended for the determination of maximum liquid load:

- The relative weir load $\dot{V}_{\mathrm{L}}/l_{\mathrm{w}}$ should be lower than $60\ \mathrm{m}^3/(\mathrm{m}\cdot\mathrm{h})$.
- The liquid velocity in the downcomer should not exceed the value of 0.1–0.2 $\mathrm{m/s}$.
- The volume of the downcomer should permit a residence time of the liquid of more than 5 s.
- The height of the clear liquid in the downcomer should not exceed half of the tray spacing H.

These rules have been developed by long-term practical experiences. However, they do not permit any insight into the limiting physical phenomena. In essence, liquid flow through a downcomer is comparable with the flow of liquid out of a vessel, which is described by Torricelli's equation:

$$u_{\mathrm{Ld}} = \alpha \cdot \sqrt{2 \cdot g \cdot H} \tag{5.4-7}$$

This basic equation has to be adapted to the conditions present at the outlet area of a downcomer. According to Stichlmair (1978; Stichlmair and Fair 1998) the following relationship holds for a standard tray design:

$$\frac{\dot{V}_{L,\max}}{l_{\mathrm{w}}} = \alpha \cdot \varepsilon_{\mathrm{Ld}} \cdot h_{\mathrm{cl}} \cdot \sqrt{2 \cdot g \cdot H \cdot \frac{\rho_{\mathrm{L}} - \rho_{\mathrm{G}}}{\rho_{\mathrm{L}}} \cdot \left(1 - \frac{h_{\Delta p} + h_{\mathrm{L}}}{\varepsilon_{\mathrm{Ld}} \cdot H}\right)} \tag{5.4-8}$$

Here, $\varepsilon_{\mathrm{Ld}}$ denotes the relative liquid holdup in the downcomer, h_{cl} the height of downcomer clearance, $h_{\Delta p}$ the pressure loss of a tray expressed in clear liquid height, and h_{L} the clear liquid height on the tray (see Sect. 5.4.1.3). The orifice discharge coefficient α has, in most cases, a constant value of 0.61. For nonfoaming systems, a value of 0.4 is recommended for the relative liquid holdup $\varepsilon_{\mathrm{Ld}}$ in the downcomer.

Minimum Liquid Load

In principle, a column tray can be operated even with very small liquid loads because the necessary height of the two-phase layer (froth) on the tray is provided by the exit weir. At extremely low liquid loads, however, the liquid will flow in an uneven pattern across the tray resulting in some degree of maldistribution of liquid. Accordingly, it is recommended to ensure a minimum liquid flow rate over the exit weir larger than $\dot{V}_{\mathrm{L}}/l_{\mathrm{w}} \approx 2\ \mathrm{m^3/(m \cdot h)}$. In small diameter columns, however, the liquid load can be considerably lower.

5.4.1.3 Two-Phase Flow on Trays

Decisive for the performance of tray columns is the behavior of the two-phase layer on the trays. Three different structures of the two-phase layer can be distinguished:

- *Bubble regime*: The liquid constitutes the continuous phase. The gas rises in the liquid in form of discrete bubbles.
- *Drop regime*: The gas forms the continuous phase and the liquid is dispersed into fine droplets.
- *Froth regime*: This regime represents the intermediate state between bubble and drop regime. The two-phase layer is intensively agitated and no definitely dispersed phase exists.

The froth regime is the dominant regime at normal column operation. Only in vacuum services drop regime may sometimes exist on trays.

Relative Liquid Holdup

Relative liquid holdup ε_L is a very important parameter of the two-phase layer on a tray. It is defined as the ratio of clear liquid height h_L to froth height h_f:

$$\varepsilon_L \equiv h_L/h_f. \tag{5.4-9}$$

There exists a large database of relative liquid holdup data and many empirical correlations in literature. Most published data are well summarized in the following correlation (Stichlmair 1978; Stichlmair and Fair 1998):

$$\varepsilon_L = 1-(F/F_{max})^{0.28}. \tag{5.4-10}$$

The term F_{max} (see (5.4-4)) accounts for gas load, several properties of the system and tray design. Liquid holdup decreases significantly with increasing gas load. Its dependency on liquid load is very small. In most cases the relative liquid holdup ε_L on a tray is as low as 5%–10%.

Froth Height

Knowledge of the height h_f of the two-phase layer on a tray is very important because it has to be significantly lower than tray spacing to avoid excessive liquid entrainment. The decisive mechanism for froth height is the liquid flow over the weir at the end of the tray. Thus, froth height cannot fall below weir height h_w. The flow of the two-phase mixture ($\dot{V}_L/\varepsilon_L$) over the outlet weir (length l_w) additionally contributes to froth height as well as the lifting effect ($F-0.2\cdot\sqrt{\rho_G}$) of the gas moving upward through the two-phase layer. From a large database the following correlation has been developed by (Stichlmair 1978; Stichlmair and Fair 1998):

$$h_f = h_w + \frac{1.45}{g^{1/3}}\cdot\left(\frac{\dot{V}_L/\varepsilon_L}{l_w}\right)^{\frac{2}{3}} + \frac{125}{(\rho_L-\rho_G)\cdot g}\cdot\left(\frac{F-0.2\cdot\sqrt{\rho_G}}{1-\varepsilon_L}\right)^2 \quad \text{for } F \geq 0.2\cdot\sqrt{\rho_G}. \tag{5.4-11}$$

An increasing liquid load increases the froth height h_f according to the second term in the above equation (i.e., Francis formula applied to a two-phase system). An increasing gas load F results in an increase of froth height according to both a decreasing relative liquid holdup ε_L (in the second term) and an increasing third term of (5.4-11).

At normal operation the froth height has values of approximately 0.1–0.3 m. This requires a tray spacing of 0.3–0.5 m for safe operation.

Entrainment of Liquid

Gas flowing through the two-phase layer on a tray always entrains some liquid in form of small droplets. This liquid entrainment is unfavorable because it affects the countercurrent flow of gas and liquid within the column and, in turn, the separation efficiency of the column. Figure 5.4-6 presents an empirical correlation that summarizes nearly all data published in literature so far (Stichlmair 1978; Stichlmair and Fair 1998).

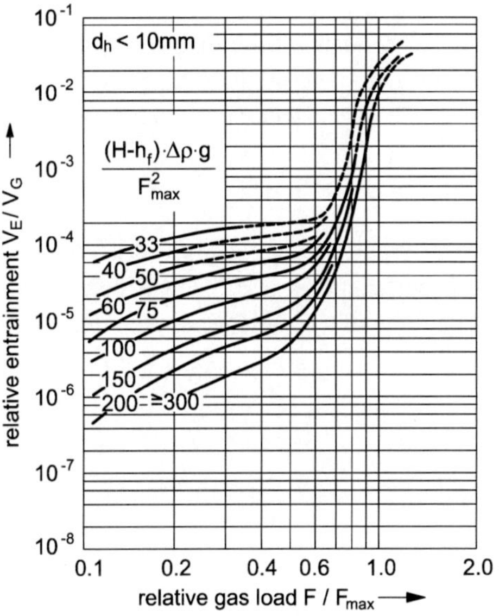

Fig. 5.4-6 Correlation of liquid entrainment vs. gas load. The parameter lines stand for tray spacing

The volume of the entrained liquid very strongly increases with increasing gas load, $\dot{V}_E \propto F^5$. Thus, the entrainment data spread over 5 decades. The sharp increase of the entrainment at a relative gas load of $F/F_{max} \approx 0.65$ is obviously caused by the transition from froth regime to drop regime (*phase inversion*).

In practice, the entrainment of approximately 10% of the liquid fed to a tray is tolerated. Even at higher entrainment rates two-phase flow within the column is not disturbed severely but mass transfer efficiency will be reduced to some degree. At very high entrainment rates, however, the pressure drop rises significantly and, eventually, the countercurrent flow in the column breaks down (*entrainment flooding*).

Liquid Mixing

The two-phase layer on a tray is heavily moved and agitated by the gas emerging from the tray openings. This intensive motion effects a partial mixing of the liquid on the tray resulting in leveling out concentration gradients. This mechanism affects mass transfer efficiency of the trays (see Sect. 5.4.1.4).

The degree of liquid mixing on a tray is expressed by the dimensionless Peclet number whose definition is

$$Pe \equiv \frac{l_L^2}{D_E \cdot \tau_L} . \tag{5.4-12}$$

The term τ_L denotes the residence time of the liquid in the two-phase layer. A value $Pe = 0$ stands for complete liquid mixing on a tray, i.e., no concentration gradients exist. A value $Pe \to \infty$ stands for no back mixing and, in turn, for plug flow of the liquid across the tray. Here, significant concentration gradients exist in the liquid on the tray due to cross flow of gas and liquid.

Predicting the degree of liquid mixing, expressed in *Pe* number, requires knowledge of the dispersion coefficient D_E. Only few experimental data have been published in literature. The following correlation is recommended:

$$D_E = 1.06 \cdot h_f \cdot F \cdot \left(\frac{\varepsilon_L}{\rho_L - \rho_G}\right)^{1/2} . \tag{5.4-13}$$

Maldistribution of Liquid

Undisturbed plug flow of liquid across the tray provides the highest mass transfer rate attainable in a tray column. In small diameter columns the plug flow is disturbed by back mixing of some liquid. In large diameter columns, however, back mixing of liquid is of minor importance. Here maldistribution, i.e., nonuniform liquid flow across the tray, is the dominant mechanism. Some characteristic flow patterns are shown in Fig. 5.4-7 (Stichlmair 1978; Stichlmair and Fair 1998). The lines plotted in the active area of the trays are liquid isotherms experimentally determined in the system air/water (hot). In such a system the mass transfer from liquid into gas causes a temperature decrease of liquid (like in a cooling tower). In a first approximation, the isotherms are lines of constant residence times of the liquid in the froth. In case of plug flow the liquid isotherms are supposed to be parallel straight lines. Any deviations from straight lines are caused by local variations of liquid velocities in the froth. The shape of the isotherms in Fig. 5.4-7 proves increased liquid velocities near the walls and in the center of the tray.

Liquid maldistribution affects mass transfer efficiency of columns. However, in tray columns maldistribution is restricted to a single tray since the liquid is well mixed in the subsequent downcomer and evenly redistributed to the next tray. In this respect, tray columns are better suited for large diameter columns than packed

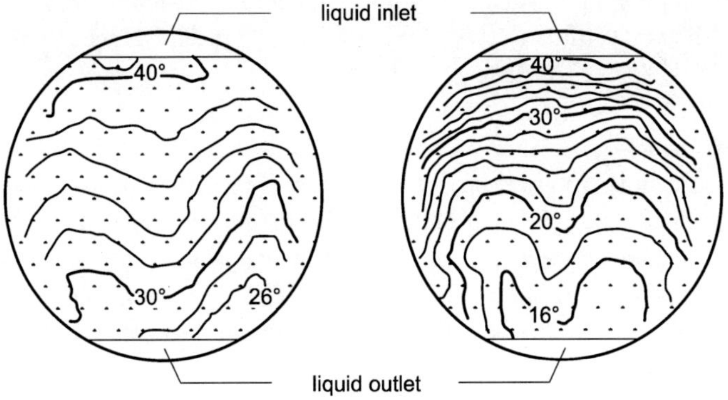

Fig. 5.4-7 Liquid isotherms in the two-phase layer of a bubble cap tray (2.3 m diameter) operated with hot water and air. Deviations from horizontal lines are caused by maldistribution (left) small gas load (right) large gas load

columns where the detrimental effects of maldistribution accumulate over the packing height, see Sect. 5.4.2.4.

Interfacial Area

One of the most important quantities of the two-phase layer on a tray is the size of the interfacial area contributing to mass transfer. Present knowledge on interfacial area on trays is rather poor and contradicting. Figure 5.4-8 shows the result of a theoretical estimation (Stichlmair 1978; Stichlmair and Fair 1998) of the relative interfacial area a (i.e., interfacial area referred to the volume of the two-phase layer) for three systems with different surface tensions. This rough estimation shows that in aqueous sys-

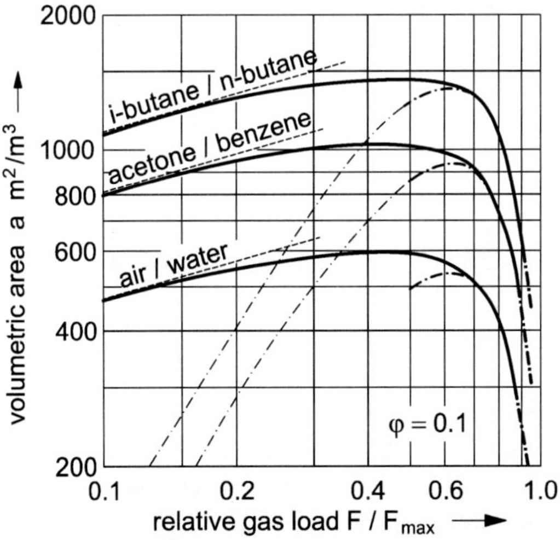

Fig. 5.4-8 Rough estimation of the relative interfacial area in the froth of sieve trays

tems the relative interfacial area is in the range of 500–600 m^2/m^3. In organic systems, however, the relative interfacial area typically reaches values up to 1000 m^2/m^3 or even larger.

Pressure Drop

The pressure drop of a gas flowing through a tray is a very important quantity. Generally, the pressure drop (better pressure loss) should be as low as possible, in particular in vacuum services. Pressure loss of the gas significantly depends on both gas and liquid load as shown in Fig. 5.4-9. During operation the pressure drop exerts an upward force on the mechanical structure of the tray.

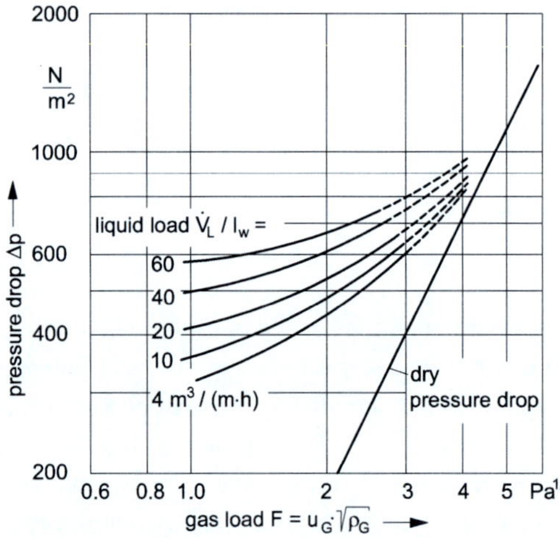

Fig. 5.4-9 Plot of pressure drop of a sieve tray vs. gas load. The parameter lines denote constant liquid flow rates across the outlet weir

The pressure loss is caused by the gas flow through the openings in the tray and by the liquid holdup h_L on the tray:

$$\Delta p = \xi \cdot \frac{\rho_G}{2} \cdot u_{Gh}^2 + h_f \cdot \varepsilon_L \cdot \rho_L \cdot g\,. \tag{5.4-14}$$

The first term in (5.4-14) refers to the dry pressure loss Δp_{dry} of the plate. It depends on the gas velocity u_{Gh} in the tray openings. The second term accounts for the liquid head on the tray.

Dry Pressure Loss of Sieve Trays

In sieve trays, the orifice coefficient ξ of the holes can be taken from Fig. 5.4-10 (Stichlmair 1978; Stichlmair and Fair 1998). The reading yields the orifice coefficient

ξ_0 of a plate with a single hole, i.e., for a relative open area $\varphi \rightarrow 0$. This value is transformed into the orifice coefficient of a tray with $\varphi > 0$ by

$$\xi = \xi_0 + \varphi^2 - 2 \cdot \varphi \cdot \sqrt{\xi_0} \quad \text{for} \quad s/d_h \rightarrow 0 \quad \text{or}$$

$$\xi = \xi_0 + \varphi^2 - 2 \cdot \varphi \quad \text{for} \quad s/d_h > 0 \tag{5.4-15}$$

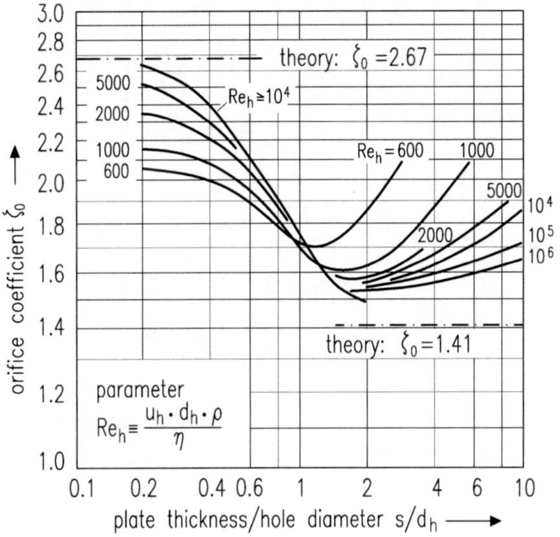

Fig. 5.4-10 Orifice coefficient ξ_0 for a very small relative free area $\varphi \rightarrow 0$ and for sharp edged holes of a sieve tray $\varphi \rightarrow 0$

Dry Pressure Loss of Bubble Cap and Valve Trays

Bubble cap and valve trays have very complex geometrical structures that make the rigorous prediction of the orifice coefficients rather difficult. As such elements are built in great numbers, experimental data of orifice coefficients are often provided by the vendors.

For the calculation of the dry pressure loss of bubble cap and valve trays a slightly different approach is recommended. The orifice coefficients are not related to the velocity in the smallest geometrical open area but to the whole active area of the tray since the quality of tray design depends not only on the pressure loss of a single element but also on the number of elements arranged on the tray. The following definition implies both quantities:

$$\Delta p_{dry} = \xi' \cdot \frac{\rho}{2} \cdot u^2 \tag{5.4-16}$$

A compilation of published tray orifice coefficients ξ' is presented in Fig. 5.4-11 (Stichlmair 1978; Stichlmair and Fair 1998). The symbols are plotted for the recommended number of elements per tray. The parameter lines show the orifice coefficients when less elements are installed. Such a modification is often necessary for geometrical reasons. The abscissa in Fig. 5.4-11 is the smallest relative free area φ' which can be, for instance, either the slots or the chimney of a bubble cap. For comparison's sake, the orifice coefficient ξ' of a sieve tray with $\xi_0 = 1.7$ is also plotted in Fig. 5.4-

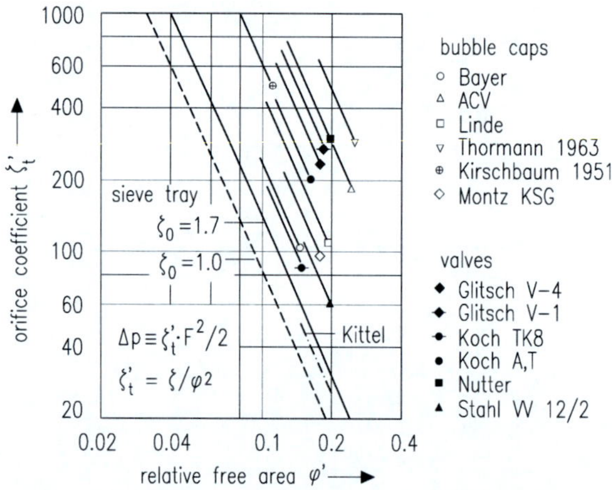

Fig. 5.4-11 Correlation of the orifice coefficients ξ' of bubble cap and valve trays

11. The quality of a tray design is determined by the distance of the relevant parameter line from the line of a sieve tray with $\xi_0 = 1.0$, which is the theoretical minimum of a sieve tray with rounded edges of the holes.

5.4.1.4 Mass Transfer in the Two-Phase Layer

At the present state of the art the mass transfer on a tray is expressed by tray efficiencies. Two definitions of tray efficiencies have to be distinguished, i.e., point efficiency E_{OG} and tray efficiency E_{OGM}.

The point efficiency is formulated with the concentrations along an individual stream line of the gas moving vertically through the froth

$$E_{OG} \equiv \frac{y - y_{n-1}}{y^*(x) - y_{n-1}} \tag{5.4-17}$$

The subscript OG indicates that the overall mass transfer resistance is formally placed into the gas phase. The point efficiency can reach a maximal value of 1 provided the liquid is thoroughly mixed over the height of the froth, an assumption that is well met. Hence, point efficiency is a very important theoretical quantity.

In practice, tray efficiencies (Murphree Efficiency) are used that are defined by

$$E_{\mathrm{OGM}} \equiv \frac{y_n - y_{n-1}}{y^*(x_n) - y_{n-1}}. \tag{5.4-18}$$

The total gas flow through the tray is considered here. The actual concentration change $(y_n - y_{n-1})$ is related to the maximal feasible concentration change $(y^*(x_n) - y_{n-1})$. The gas concentration $y^*(x_n)$ is in equilibrium state with the liquid leaving the tray. This definition is arbitrary since the liquid concentration on the tray varies due to the cross flow of the phases. Hence, values of tray efficiencies higher than 1.0 are feasible.

The above equations are the definitions of gas side efficiencies. Sometimes liquid side efficiencies are used to describe the interfacial mass transfer. However, the gas side definitions are preferred since gas-phase resistance is the dominant mechanism in distillation processes.

Relationship Between Point and Tray Efficiency

In case of plug flow of the liquid the following relationship between point and tray efficiency holds (Lewis 1936):

$$\frac{E_{\mathrm{OGM}}}{E_{\mathrm{OG}}} = \frac{\dot{L}/\dot{G}}{m} \cdot \frac{1}{E_{\mathrm{OG}}} \cdot \left[\exp\left(\left(\frac{m}{\dot{L}/\dot{G}} \cdot E_{\mathrm{OG}}\right) - 1\right)\right] \tag{5.4-19}$$

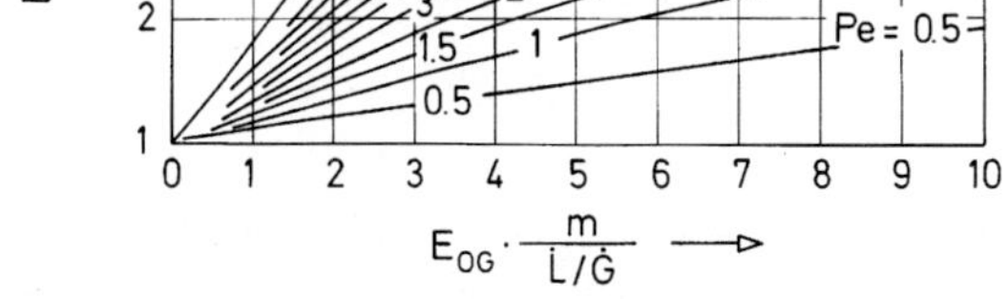

Fig. 5.4-12 Relationship between tray and point efficiency under consideration of liquid back mixing

Plug flow prevails on very large trays only. In smaller trays there is always some degree of back mixing of liquid, which is quantitatively expressed by the Peclet number (see Sect. 5.4.1.3). Figure 5.4-12 presents the relationship between tray and point efficiency under consideration of back mixing of the liquid. At total back mixing, tray efficiency E_{OGM} is equal to point efficiency E_{OG}. At very low back mixing (5.4-19) holds.

For large diameter trays very large values of tray efficiency are found from Fig. 5.4-12 since large trays have a very small degree of liquid back mixing. However, large trays normally have some degree of maldistribution that also exerts a detrimental effect on tray efficiency. In practice, tray efficiencies are only slightly higher than point efficiencies.

Point Efficiency

Point efficiency E_{OG} is calculated from fundamental laws of mass transfer. Adjusting these laws to tray columns yields

$$E_{OG} = 1 - \exp(-N_{OG}) \text{ with } N_{OG} = \frac{a \cdot \beta_G \cdot h_f / u_G}{1 + m \cdot \beta_G / \beta_L \cdot \tilde{M}_L / \tilde{M}_G \cdot \rho_G / \rho_L} \,. \quad (5.4\text{-}20)$$

The most crucial quantities in this equation are the interfacial area a (see Sect. 5.4.1.3), the gas side mass transfer coefficient β_G, and the liquid side mass transfer coefficient β_L.

The mass transfer coefficients can be estimated by the assumption of an unsteady mass transfer (Stichlmair 1978, 1998). Decisive is the residence time τ_G of the gas in the froth since it is the gas present in the liquid that generates the interfacial area and, in turn, enables the interfacial mass transfer:

$$\beta_G = \sqrt{\frac{4 \cdot D_G}{\pi \cdot \tau_G}} \quad \text{and} \quad \beta_L = \sqrt{\frac{4 \cdot D_L}{\pi \cdot \tau_G}} \quad \text{with} \quad \tau_G = h_f \cdot \varepsilon_L / u_G \quad (5.4\text{-}21)$$

In organic systems with low values of interfacial tension, (5.4-21) sometimes gives too low values due to a steady renewal of the interfacial area.

The weak point of mass transfer prediction, however, is poor knowledge of the interfacial area a that contributes to mass transfer (see Sect. 5.4.1.3). Therefore, empirical values of tray efficiencies are normally used in practice. In distillation systems values of tray efficiencies of 70% are often recommended. However, if a significant liquid side mass transfer resistance exists, tray efficiencies can be much lower. Some empirical values are presented in Fig. 5.4-13.

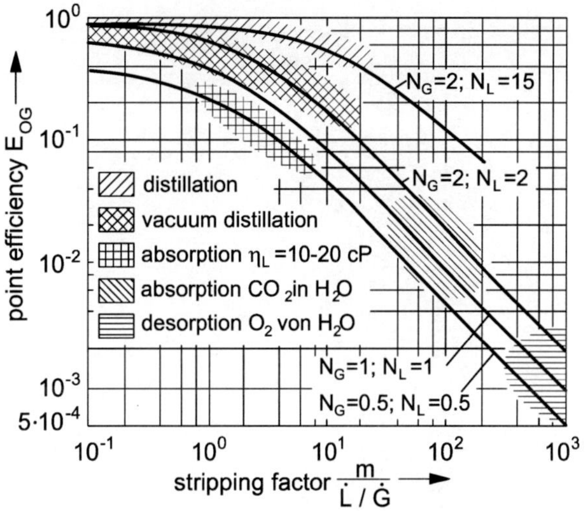

Fig. 5.4-13 Typical experimental values of gas side point efficiencies

5.4.2 Packed Columns

Packed columns are as important as tray columns in the process industry. Due to novel developments of packing elements the industrial use of packed columns is steadily increasing. In packed columns there exists a genuine countercurrent flow of gas and liquid as is shown in Fig. 5.4-14. An intimate contact between gas and liquid phases is established by packings that represent a solid structure with high porosity and large internal surface. The liquid proceeds downward in form of thin films or rivulets. Decisive for a good performance are a low pressure drop of the gas and a liquid flow that is uniform over the cross section of the column.

5.4.2.1 Design Principles

Two different design principles of packings can be distinguished

- *Random packings* consist of dumped beds of particles made of ceramic, metal, or plastic (Fig. 5.4-15). The shape of the particles should ensure a homogeneous structure of the dumped bed and a low pressure drop of gas flow. Standard particles (Raschig rings, Berl saddles) have a closed surface. Modern particles are characterized by net (Pall rings) or grid structures (Super rings). The size of the particles must not exceed 5% or 10% of column diameter. The porosity of a bed is as high as 70% (ceramic particles) or more than 90% (metal particles). The volumetric surface of a bed depends on particle size (and, in turn, on column diameter). Typical values are 50–300 m^2 per cubic meter packing vol-

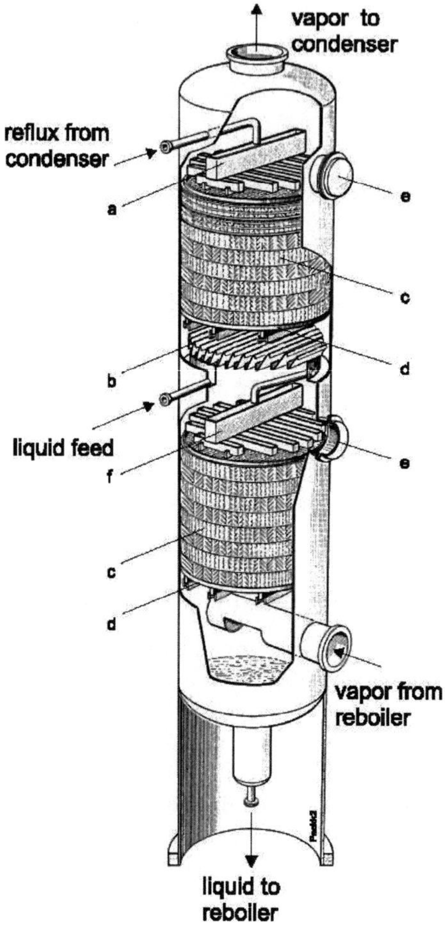

Fig. 5.4-14 Perspective view of a packed column (Sulzer): (a) liquid distributor, (b) liquid collector, (c) structured packing, (d) support grid, (e) man way, and (f) liquid re-distributor

ume (Table 5.4-2). The separation efficiency of random packings strongly depends on column diameter. Typical values of industrial columns are in the range of 1–2 equilibrium stages per meter packing height.

- *Structured packings* have been increasingly used since the mid-60s because they better meet the demand in homogeneous bed structures. The standard design consists of vertically arranged corrugated metal sheets, approximately 20 cm in height, with alternating orientation of subsequent layers (Fig. 5.4-16). The porosity of structured beds is significantly higher than 95%. The specific surface is typically 250 m^2/m^3 or higher (Table 5.4-3). Structured packings have a lower pressure drop and a better mass transfer efficiency than

Table 5.4-2 Characteristic data of random packings

Type	dN (mm)	ε	a (m^2/m^3)
Raschig rings (ceramic)	10	0.65	440
	25	0.726	192
	38	0.76	140
	50	0.77	98
Raschig rings (metal)	10	0.744	520
Raschig super rings (metal)	20	0.970	250
	25	0.98	180
	30	0.98	150
	38	0.98	120
	50	0.98	100
Pall rings (metal)	25	0.94	210
	35	0.945	141
	50	0.96	102
	90	0.97	65
Pall ring (ceramic)	25	0.73	220
	35	0.76	165
	50	0.78	120
Raflux rings (metal)	25	0.94	215
	35	0.95	145
	50	0.97	120
	90	0.97	65
Hiflow (metal)	25	0.97	200
	28	0.97	200
	50	0.98	99
	58	0.987	92
Hiflow (ceramic)	20	0.72	280
	35	0.76	128
	50	0.78	102
	75	0.76	70
IMTP (metal)	15	0.964	282
	25	0.972	225
	40	0.981	150
	50	0.979	100
Torus saddles (ceramic)	25	0.724	185
	38	0.76	128
	50	0.76	142
	75	0.78	92
	90	0.78	68

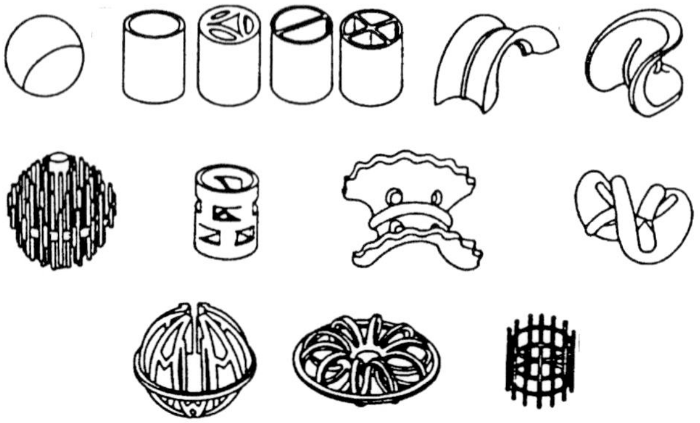

Fig. 5.4-15 Examples of random packing elements. *Upper row*: ceramic *middle row*: metal, and *lower row*: plastic

random packings. Mass transfer efficiency is as high as 2–4 equilibrium stages per meter. The pressure drop per equilibrium stage is typically as low as 0.5 mbar.

Fig. 5.4-16 Example of a structured packing of corrugated metal sheets (Sulzer 1982)

Very important for all types of packings is a uniform liquid distribution at the top of the bed and a limitation of bed height to 6 or 8 m. Beneath each bed, the liquid has to be collected, mixed, and redistributed. These measures intend to suppress the so-called maldistribution of liquid because it strongly affects mass transfer rates. The design of liquid distributors, liquid collectors, support grids, etc., should provide a large open area not to hinder the countercurrent flow of gas and liquid.

Table 5.4-3 Characteristic data of some structured packings

Type	ε	a (m^2/m^3)
Flexipac (metal)	0.91 9.94 0.96 0.98	558 223 135 70
Gempak (metal)	0.947 0.96 0.961 0.974	525 131 394 262
Mellapak (metal)	0.975 0.975	250 500
Montzpac (metal)	0.972 0.98 0.99	300 200 100

5.4.2.2 Operation Region of Packed Columns

Packed columns can be operated within certain limits of gas and liquid loads only. A typical operation region of packed columns is shown in Fig. 5.4-17. The mechanisms that set limitations to gas and liquid flow rates are flooding and poor surface wetting. Flooding is a strong limitation that cannot be surpassed. Nonsufficient surface wetting is a soft limitation that can be crossed at the expenses of poorer mass transfer efficiency. It is important to note that the operation range of the liquid is

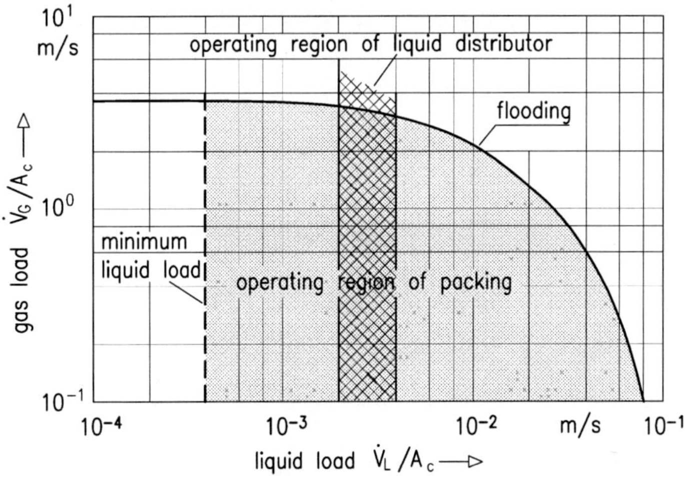

Fig. 5.4-17 Operation region of a packed column

further narrowed down by the liquid distributor. The gas load can be very low in packed columns.

The operation point has to be chosen so that a sufficient safety margin to the operation limits exists. In practice, an operating pressure loss of approx. 3 mbar/m is often recommended for column dimensioning.

Maximum Gas and Liquid Load

The upper operation limit of packed columns, called flooding, is caused by a too high pressure drop that blocks off the countercurrent flow of gas and liquid. There exist numerous empirical correlations for predicting flooding of packed columns (e.g., in Kister 1992). The correlation of Mersmann (1965; Mersmann and Bornhüb 2002) is illustrated in Fig. 5.4-18. In principle, the gas load is plotted vs. the liquid load in this disgram. However, the gas load is substituted by the dry pressure drop of the gas flowing through the packing. The abscissa is a dimensionless liquid load. The upper curve describes the condition of flooding. The parameter lines denote constant values of the dimensionless pressure drop of an irrigated packing.

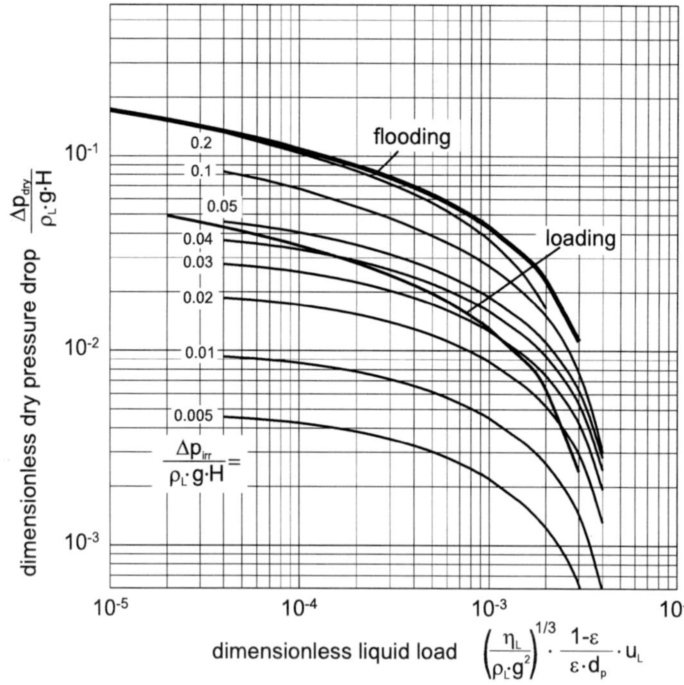

Fig. 5.4-18 Correlation of flooding of packed columns according to Mersmann (1965, 2002)

A more rigorous approach to flooding is based on pressure drop models of irrigated packed beds, see Sect. 5.4.2.3.

Minimum Liquid Load

The minimum liquid load of random packings can be estimated by the following equation (Schmidt 1979):

$$u_{\mathrm{L,\,min}} = 7.7 \cdot 10^{-6} \cdot \left(\frac{\rho_{\mathrm{L}} \cdot \sigma^3}{\eta_{\mathrm{L}}^4 \cdot g} \right)^{2/9} \cdot \left(\frac{g}{a} \right)^{1/2} \tag{5.4-22}$$

For industrial practice the following values are recommended:

$u_{\mathrm{L,\,min}} \geq 10\ \mathrm{m}^3/(\mathrm{m}^2 \cdot \mathrm{h})$ for aqueous systems,

$u_{\mathrm{L,\,min}} \geq 5\ \mathrm{m}^3/(\mathrm{m}^2 \cdot \mathrm{h})$ for organic systems.

Structured wire mesh packings can be operated with organic systems down to $0.2\ \mathrm{m}^3/(\mathrm{m}^2 \cdot \mathrm{h})$. This value is in the order of magnitude of a heavy tropical rainfall.

An additional limitation to liquid load is set by the liquid distributor as shown in Fig. 5.4-17. In most cases the operation of the distributor is less flexible than that of the packing itself.

5.4.2.3 *Two-Phase Flow in Packed Columns*

The geometrical structures of packed beds are too complex for simple mathematical description. Therefore, the real bed structures are replaced by model structures, for instance by a system of parallel channels or a system of dispersed particles. The model structures should have the same interfacial area a and the same porosity ε as the packed bed. From both conditions follows the hydraulic diameter d_h of the channel model structure:

$$d_{\mathrm{h}} = 4 \cdot \varepsilon / a \tag{5.4-23}$$

The same conditions give the particle diameter d_{p} of the particle model structure:

$$d_{\mathrm{p}} = 6 \cdot \frac{1 - \varepsilon}{a} \tag{5.4-24}$$

It must be noted that d_{h} characterizes the open structure and d_{p} the solid structure of the packed bed, respectively. The round channels intersect at large porosities of the bed. The particles (spheres) of the particle structure do not touch each other like in a fluidized bed. Thus, the particle model is more realistic and, in turn, preferred in this section.

Pressure Drop

The pressure drop of a gas flowing through a packed bed is one of the key quantities in the design of packed columns. It is dominated by the gas and liquid load as well as by the structure of the bed, see Fig. 5.4-20.

Pressure Drop of Dry Beds

The pressure drop of a gas flowing through a dry bed of dispersed particles is (Stichlmair 1998):

$$\frac{\Delta p_{\mathrm{dry}}}{H} = \frac{3}{4} \cdot \xi_o \cdot \frac{1-\varepsilon}{\varepsilon^{4.65}} \cdot \rho_{\mathrm{G}} \cdot \frac{u_{\mathrm{G}}^2}{d_{\mathrm{p}}}. \tag{5.4-25}$$

Equation (5.4-25) differs from the well-known Ergun equation (see Kister 1992) by the exponent 4.65 on the porosity ε. The improved porosity term allows the prediction of the pressure loss of a bed of dispersed particles from knowledge of the friction factor ξ_0 of a single particle. Hence, (5.4-25) holds for the whole range of bed porosity.

Figure 5.4-19 presents a graphical correlation of the friction factors ξ_0 valid for several dumped and structured packings (Stichlmair 1998).

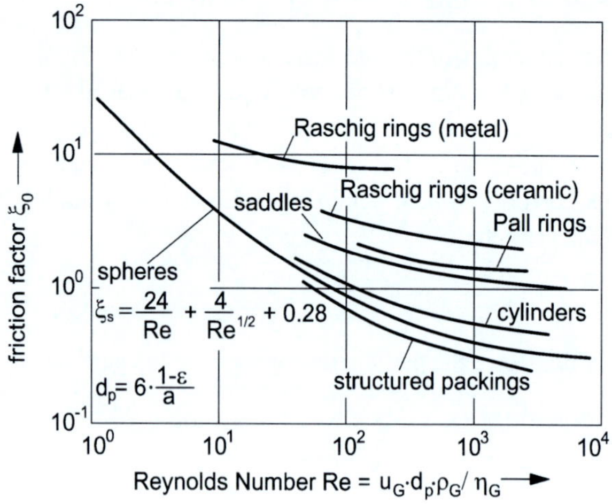

Fig. 5.4-19 Correlation of the friction factor ξ_0 according to the particle structure model

In a limited range of Reynolds numbers, the values presented in Fig. 5.4-19 are correlated by

$$\xi_0 = b \cdot Re_0^{(2+c)} \qquad \text{with} \qquad Re_0 = \frac{u_G \cdot d_p \cdot \rho_G}{\eta_G} \tag{5.4-26}$$

Some values of the factor b and the exponent c are listed in Table 5.4-4.

Table 5.4-4 Values of the factor b and the exponent c in (5.4-26)

Packing	Factor b	Exponent c	Reynolds numbers
Raschig rings (ceramic)	7.4	– 0.17	$6 \cdot 10^1 - 3 \cdot 10^3$
Raschig rings (metal)	16	– 0.13	$3 \cdot 10^1 - 2 \cdot 10^2$
Pall rings (metal)	4.8	– 0.15	$1 \cdot 10^2 - 3 \cdot 10^3$
Saddles	5.5	– 0.20	$7 \cdot 10^1 - 6 \cdot 10^3$
Spheres	1.7	– 0.20	$2 \cdot 10^2 - 5 \cdot 10^3$
Cylinders	3.0	– 0.23	$6 \cdot 10^1 - 2 \cdot 10^3$
Structured packings	4.8	– 0.38	$4 \cdot 10^1 - 3 \cdot 10^3$

Pressure Drop of Irrigated Beds

Typical values of the pressure drop of irrigated beds are presented in Fig. 5.4-20. The pressure loss of irrigated beds is always significantly higher than that of dry beds since the characteristic quantities of the bed are changed by the liquid accumulated within the bed (*liquid holdup* h_L).

The liquid holdup h_L changes the bed characteristics in several ways. Of great importance are the reduction of the bed porosity ε, the enlargement of the particle diameter d_p and, in turn, the reduction of the friction coefficient ξ_0. All three effects can be modeled giving (Stichlmair 1998):

$$\frac{\Delta p_{irr}}{\Delta p_{dry}} = \left[\frac{1 - \varepsilon \cdot (1 - h_L/\varepsilon)}{1 - \varepsilon}\right]^{(2+c)/3} \cdot \left(1 - \frac{h_L}{\varepsilon}\right)^{-4.65} \tag{5.4-27}$$

Evaluation of this equation demands knowledge of the liquid holdup h_L in the bed.

Liquid Holdup in Packed Beds

Liquid holdup is a decisive parameter of two-phase flow in packed beds. Typical values of the holdup h_L, defined as ratio of liquid volume to packing volume, are presented in Fig. 5.4-21 (Billet and Mackowiak 1984). Here, the experimental val-

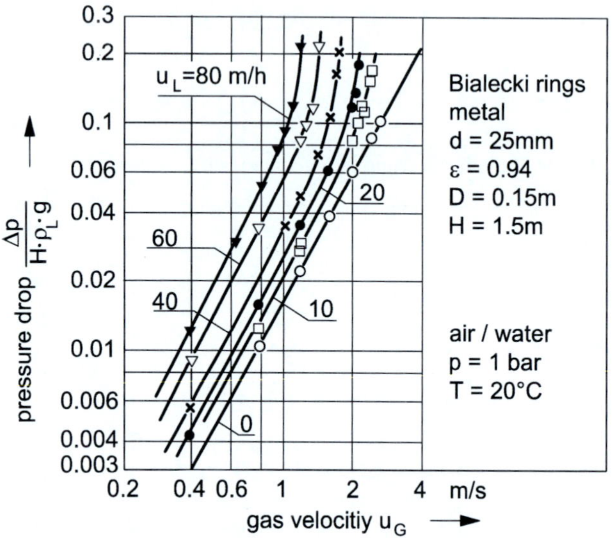

Fig. 5.4-20 Pressure drop of a packed column with 25 mm Bialecki rings (metal). The parameter lines denote constant liquid load (Billet and Mackowiak 1984)

ues of the holdup h_L are plotted vs. the superficial gas velocity. More advantageous is a plot of the holdup h_L vs. the pressure drop of the irrigated bed $\Delta p_{irr}/(H \cdot \rho_L \cdot g)$ as shown in Fig. 5.4-22.

Two regions can be distinguished in this diagram:

- A region of low pressure drop (low gas load) where the holdup depends on liquid load only (horizontal parameter lines)
- A region of high pressure drop (high gas load) where the holdup is further increased by pressure drop

The transition between these two regions is called loading point. The parameter lines in Fig. 5.4-22 end at the flooding point.

In the region below the loading point the liquid holdup h_{Lo} depends on liquid load, liquid properties, and packing structure only. The following correlation holds for random packings (Engel 1999):

$$h_{Lo} = 3.6 \cdot \left(\frac{u_L^2 \cdot a}{g}\right)^{0.33} \cdot \left(\frac{\eta_L^2 \cdot a^3}{\rho_L^2 \cdot g}\right)^{0.125} \cdot \left(\frac{\sigma \cdot a^2}{\rho_L \cdot g}\right)^{0.1} . \tag{5.4-28}$$

At column operation above the loading point, the liquid holdup is further influenced by the pressure drop of the column. This influence is described by

$$h_L = h_{Lo} \cdot \left[1 + 20 \cdot \left(\frac{\Delta p_{irr}}{H \cdot \rho_L \cdot g}\right)^2\right] \qquad (5.4\text{-}29)$$

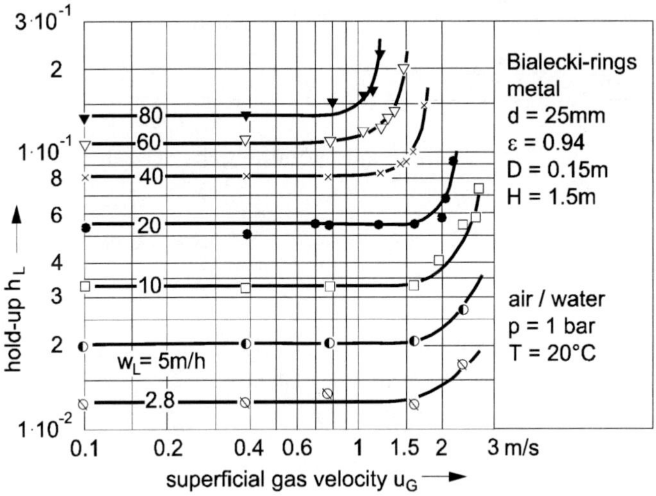

Fig. 5.4-21 Liquid holdup h_L of a dumped packing of 25 mm Bialecki rings (metal) plotted vs. superficial gas velocity (Billet and Mackowiak 1984)

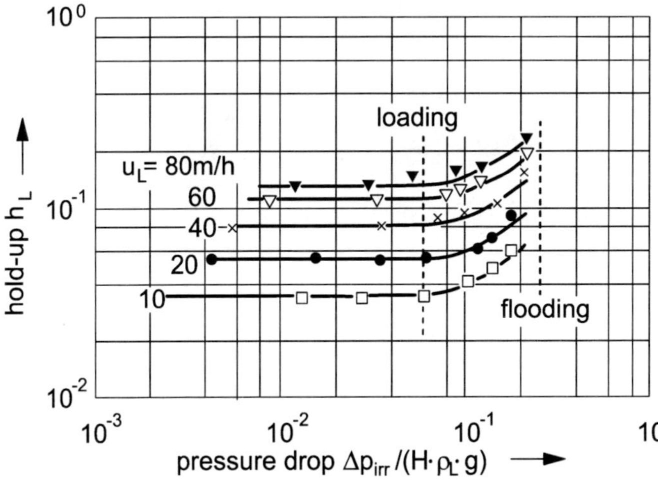

Fig. 5.4-22 Liquid holdup h_L of a dumped packing of 25 mm Bialecki rings (metal) plotted vs. pressure drop of the irrigated bed. The pressure drop is taken from Fig. 5.4-20

Flooding of Packed Columns

As is seen from Fig. 5.4-20, the condition for flooding is an infinite increase of pressure drop with increasing gas load. Taking the dry pressure loss as measure for the gas load gives the following flooding condition:

$$\frac{\partial \Delta p_{\text{irr}}}{\partial \Delta p_{\text{dry}}} = \infty \quad \text{or} \quad \frac{\partial \Delta p_{\text{dry}}}{\partial \Delta p_{\text{irr}}} = 0. \tag{5.4-30}$$

Substituting (5.4-27) into (5.4-29) and performing the derivation of (5.4-30) give (Stichlmair 1998):

$$\frac{1}{\Delta^2} = \frac{40 \cdot (2+c)/3 \cdot h_{\text{Lo}}}{1-\varepsilon + h_{\text{Lo}} \cdot (1+20 \cdot \Delta^2)} + \frac{186 \cdot h_{\text{Lo}}}{\varepsilon - h_{\text{Lo}} \cdot (1+20 \cdot \Delta^2)} \quad \text{with} \quad \Delta \equiv \frac{\Delta p_{\text{irr}}}{H \cdot \rho_{\text{L}} \cdot g} \tag{5.4-31}$$

Equation (5.4-31) can be solved by iteration only. As flooding of random packings typically occurs at values of the dimensionless pressure drop of $\Delta \approx 0.1$–0.3, a value of $\Delta = 0.1$ is a good first estimation.

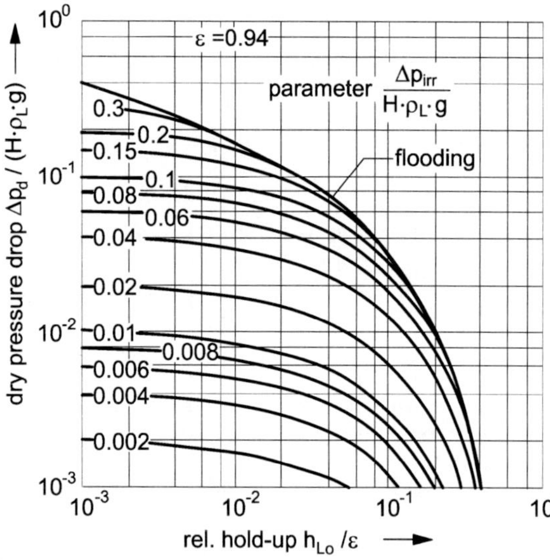

Fig. 5.4-23 Pressure drop and flooding of a packed column. The value of the abscissa is determined from (5.4-28)

Figure 5.4-23 presents the result of the evaluation of (5.4-27) together with (5.4-28) and (5.4-29) for a random packing of Bialecki rings. This plot is almost identical with the empirical flooding correlation of Mersmann (Fig. 5.4-18). Just the dimensionless liquid load has been replaced by the liquid holdup h_{Lo} below loading.

5.4.2.4 Mass Transfer in Packed Columns

By convention, mass transfer in packed columns is described by the model of transfer units since a countercurrent flow of gas and liquid prevails in packed beds. According to this model the required height H of the packing within the column is

$$H = H_{\mathrm{OG}} \cdot N_{\mathrm{OG}}, \tag{5.4-32}$$

The term N_{OG} denotes the number of transfer units, which is a measure of the difficulty of the separation (see Sect. 5.2.1). The mass transfer in a packed bed is expressed by the height of a transfer unit H_{OG}. In practice empirical values of the height of a transfer unit are often used for column dimensioning. Sometimes, empirical values of the height equivalent of an equilibrium stage (*HETP*) are also determined and applied to column dimensioning. A more rigorous approach is based on the following equation that is valid for undisturbed plug flow of gas and liquid within the packing:

$$H_{OG} = \frac{u_{\mathrm{G}}}{\beta_{\mathrm{G}} \cdot a_{\mathrm{eff}}} \cdot \left(1 + m \cdot \frac{\beta_{\mathrm{G}}}{\beta_{\mathrm{L}}} \cdot \frac{\tilde{M}_{\mathrm{L}}}{\tilde{M}_{\mathrm{G}}} \cdot \frac{\rho_{\mathrm{G}}}{\rho_{\mathrm{L}}}\right) \tag{5.4-33}$$

This rather simple equation contains three crucial quantities that depend on design and operation of the packed column. These quantities are the effective interfacial area a_{eff}, the gas side mass transfer coefficient β_{G}, and the liquid side mass transfer coefficient β_{L}. According to the present state of the art, no models based on first principles are available for the prediction of these rather crucial quantities. Only some empirical correlations have been published so far (e.g., in Kister 1992).

Empirical Correlations for Mass Transfer Prediction

Often recommended is the empirical model developed by Onda et al. (1968). His correlation of the effective interfacial area is

$$\frac{a_{\mathrm{eff}}}{a} = 1 - \exp\left(-1.45 \cdot \left(\frac{\sigma_{\mathrm{crit}}}{\sigma}\right)^{0.75} \cdot \left(\frac{u_L \cdot \rho_{\mathrm{L}}}{a \cdot \eta_{\mathrm{L}}}\right)^{0.1} \cdot \left(\frac{u_{\mathrm{L}}^2 \cdot a}{g}\right)^{-0.05} \cdot \left(\frac{u_{\mathrm{L}}^2 \cdot \rho_{\mathrm{L}}}{\sigma \cdot a}\right)^{0.2}\right) \tag{5.4-34}$$

Some values of the critical surface tension σ_{crit} are listed in Table 5.4-5. Typically, the effective interfacial area a_{eff} is much smaller than the geometrical area a.

Table 5.4-5 Values of critical surface tension in (5.4-34) (Onda et al. 1968)

Packing material	Polyethylene	Polyvinyl chloride	Ceramic	Glass	Stainless steel	Steel
σ_{crit} (N/m)	0.033	0.040	0.061	0.073	0.071	0.075

The corresponding equations for the mass transfer coefficients are

$$\beta_G = C \cdot \left(\frac{\rho_G \cdot u_G}{a \cdot \eta_G}\right)^{0.7} \cdot \left(\frac{\eta_G}{\rho_G \cdot D_G}\right)^{1/3} \cdot (a \cdot d_N)^{-2} \cdot (a \cdot D_G) \tag{5.4-35}$$

and

$$\beta_L = 0.0051 \cdot \left(\frac{g \cdot \eta_L}{\rho_L}\right)^{1/3} \cdot \left(\frac{\rho_L \cdot u_L}{a_{eff} \cdot \eta_L}\right)^{2/3} \cdot \left(\frac{\rho_L \cdot D_L}{\eta_L}\right)^{1/2} \cdot (a \cdot d_N)^{2/5} . \tag{5.4-36}$$

The factor C is 2 for small ($d_N < 15$ mm) and 5.23 for large packing elements.

Bravo and Fair (1982) published a model that combines the correlations of Onda for the mass transfer coefficients with a new correlation for the effective interfacial area:

$$\frac{a_{eff}}{a} = 19.76 \cdot \sigma^{0.5} \cdot H^{-0.4} \cdot \left(\frac{\eta_L \cdot u_L}{\sigma} \cdot \frac{u_G \cdot \rho_G}{a \cdot \eta_G}\right)^{0.392} . \tag{5.4-37}$$

In this correlation, the term $H^{-0.4}$ accounts for the influence of maldistribution on mass transfer. Equation (5.4-37) is valid for the units m and kg/s^2 for the height H and the surface tension σ, respectively.

Billet and Schultes (1995, 1999) developed the following correlations for all three unknowns:

$$\frac{a_{eff}}{a} = 1.5 \cdot (a \cdot d_h)^{-0.5} \cdot \left(\frac{u_L \cdot d_h \cdot \rho_L}{\eta_L}\right)^{-0.2} \cdot \left(\frac{u_L^2 \cdot \rho_L \cdot d_h}{\sigma}\right)^{0.75} \cdot \left(\frac{u_L^2}{g \cdot d_h}\right)^{-0.45} , \tag{5.4-38}$$

$$\beta_G = C_V \cdot (\varepsilon - h_L)^{-1/2} \cdot \left(\frac{a}{d_h}\right)^{1/2} \cdot D_G \cdot \left(\frac{u_G \cdot \rho_G}{a \cdot \eta_G}\right)^{3/4} \cdot \left(\frac{\eta_G}{\rho_G \cdot D_G}\right)^{1/3} , \tag{5.4-39}$$

$$\beta_L = C_L \cdot 12^{1/6} \cdot \left(\frac{u_L \cdot D_L}{h_L \cdot d_h}\right)^{1/2} . \tag{5.4-40}$$

Table 5.4-6 lists some values of the factors C_V and C_L .

Problems of Mass Transfer Prediction in Packed Columns

At the present state of the art all models for the prediction of mass transfer in packed columns are insufficient to some degree. One problem is a poor knowledge of the interfacial area effective for mass transfer. Published studies differ significantly and their results are often contradicting. Furthermore, the existing models for predicting mass transfer coefficients are not sufficiently reliable.

Table 5.4-6 Selected values of the factors C_V and C_L in (5.4-39) and (5.4-40) according to Billet and Schultes (1999)

Packing	C_V	C_L
Raschig rings (ceramic)		
50	0.210	1.416
25	0.412	1.361
Raschig super rings (metal)		
1	0.440	1.290
2	0.400	1.323
3	0.300	0.850
Bialecki rings (metal)		
50	0.302	1.721
35	0.390	1.412
25	0.331	1.461
Pall rings (metal)		
50	0.410	1.192
35	0.341	1.012
25	0.336	1.440
Pall rings (ceramic)		
50	0.333	1.278
Intalox (ceramic)	0.488	1.677
Ralu pak (metal)		
YC-250	0.385	1.334
Impulse packing		
(metal)	0.270	0.983
(ceramic)	0.327	1.317
Montz packing (metal)		
B1-200	0.390	0.971
B2-300	0.422	1.165

The main reason for the deficiencies of the present state of the art is, however, the maldistribution of gas and liquid in packed beds. Many studies reveal that there exist large deviations from plug flow of gas and in particular of liquid within the bed. (Hoek et al. 1986; Kammermaier 2008). The degree of maldistribution as well as its effect on mass transfer are unknown and, in turn, not accounted for in existing mass transfer models. Thus the published data for interfacial area and mass transfer coefficients comprise the maldistribution in an undefined manner. The data are not true but pseudo values which are not predictable within the plug flow model.

Figure 5.4-24 presents some experimental results of a packed column operated with hot water and air like a cooling tower (Stichlmair and Fair 1998). In this system the mass transfer of water from liquid into gas phase causes a decrease of liquid temperature, which can be easily measured. The lines drawn within the packed bed are

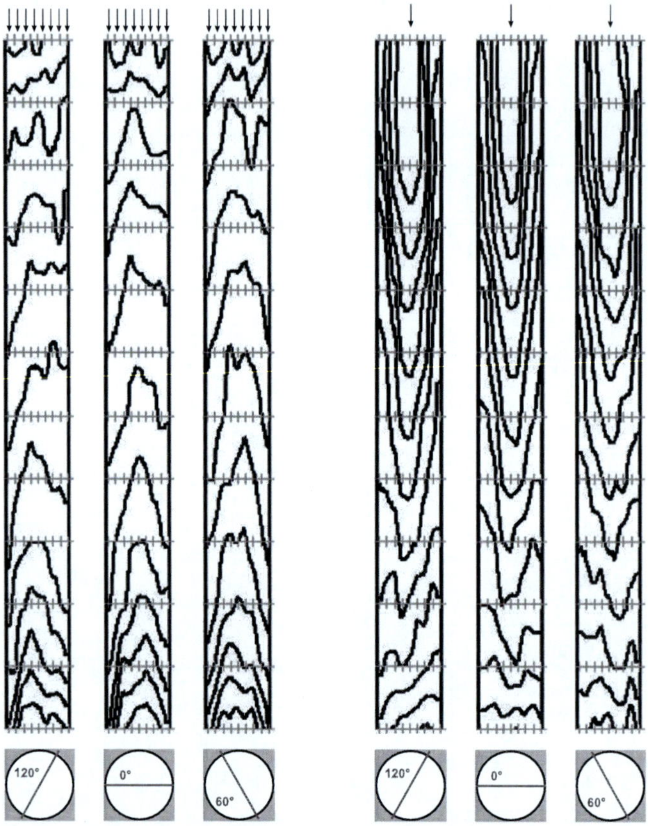

Fig. 5.4-24 Maldistribution in a packed column (metal Pall rings 35 mm). Column diameter $D = 0.63$ m, packing height $H = 6.8$ m. Gas load $F = 1.1\ Pa^{1/2}$, liquid load $B = 5.2\ m^3/(m^2 \cdot h)$. *Left*: Uniform liquid feed. *Right*: Single point liquid feed

liquid isotherms that are supposed to be horizontal and straight lines in case of plug flow. Any deviations from straight lines are caused by a nonuniform gas and, in particular, liquid flow within the packed bed.

In the experiments performed with uniform liquid distribution at the top (Fig. 5.4-24, left), the degree of maldistribution increases on the way of the liquid downward through the bed. At nonuniform initial liquid distribution (Fig. 5.4-24, right) however, the degree of maldistribution decreases. Thus, the degree of maldistribution depends on both the liquid distributor and the packed bed itself. At the present state of the art, it is not possible to predict either the degree of maldistribution or its effect on mass transfer. As the effects of maldistribution accumulate over packing height, it is important to suppress the maldistribution, for instance, by dividing the packing into several short sections. This measure for suppressing the maldistribution requires the installation of multiple support grids, liquid collectors, liquid mixers, and liquid distributors.

Symbols

A	m^2	Area
A_c	m^2	Cross section of column
a	$1/\mathrm{m}$	Volumetric interfacial area
a		Species (low boiler)
B	kmol	Amount of bottoms
$\dot{B}$	kmol/s	Bottoms (stream)
b		Species (intermediate boiler)
$C-1$		Column 1
C_G	m/s	Gas load factor
c		Species (high boiler)
c	$\mathrm{kJ/(kmol \cdot K)}$	Molar heat capacity
c	$\mathrm{kmol/m^3}$	Concentration
D	kmol	Amount of distillate
$\dot{D}$	kmol/s	Distillate (stream)
D	$\mathrm{m^2/s}$	Diffusion coefficient
D_c	m	Column diameter
D_E	$\mathrm{m^2/s}$	Dispersion coefficient
d	m	Diameter
d_N	m	Nominal diameter of packings
E_OG		Gas side point efficiency
E_OGM		Gas side tray efficiency
F	$\mathrm{Pa}^{1/2}$	F-factor $F \equiv u_\mathrm{G} \cdot \sqrt{\rho_\mathrm{G}}$
F	kmol	Amount of feed
$\dot{F}$	kmol/s	Feed (stream)
G	kmol	Amount of gas (vapor)
$\dot{G}$	kmol/s	Gas, vapor (stream)
g	$\mathrm{m/s^2}$	Acceleration of gravity
H	m	Height (of packing)
He_ij	bar	Henry coefficient
H_OG	m	Height of a transfer unit
h_L		Holdup (in packing)
h	kJ/kmol	Molar enthalpy
h_f	m	Height of two-phase layer (froth height)
h_w	m	Weir height
J		Stripping factor
K		Equilibrium ratio $K \equiv (y/x)^*$

K_R		Equilibrium constant of chemical reaction
k_{OG}	kmol/s	Gas side mass transfer coefficient (over all)
L	kmol	Amount of liquid
$\dot{L}$	kmol/s	Liquid (stream)
l_w	m	Weir length
l	m	Distance
$\tilde{M}$	kg/kmol	Molar mass
m		Slope of equilibrium curve
N_2, N_3		Denominator in equilibrium equation
$\dot{N}$	kmol/s	Molar flow
N_{OG}		Number of gas side transfer units
n		Number of equilibrium stages
n_i	kmol	Number of moles of i
Pe		Peclet number $Pe \equiv l^2/(D \cdot \tau)$
p	bar	Pressure, partial pressure
p^0	bar	Vapor pressure
Q		Dimensionless concentration
Q	kJ	Amount of heat
$\dot{Q}$	kJ/s	Heat (stream)
q	kJ/mol	Heat of absorption
q_F		Caloric factor of feed
R_L		Reflux ration
R_G		Reboil ratio
r	kJ/kmol	Latent heat of vaporization
T	°C, K	Temperature
t	s	Time
t	°C, K	Temperature
V	m^3	Volume
$\dot{V}$	m^3/s	Volume (stream)
w	m/s	Superficial velocity
w_{Bl}	m/s	Terminal velocity of bubbles
X		Molar loading of liquid
x		Molar fraction of liquid
Y		Molar loading of gas
y		Molar fraction of gas
z		Molar fraction

Greek Symbols

α		Relative volatility
α_i	m^3_{Ni}/m^3	Bunsen absorption coefficient
β	m/s	Mass transfer coefficient
γ		Activity coefficient
Δp	N/m^2	Pressure drop

ε		Porosity
ε_L		Volumetric fraction of liquid
ζ		Extent of reaction
η	kg/(m/s)	Viscosity
ν		Stoichiometric factor of chemical reaction
ξ		Friction factor
π		Pole
ρ	kg/m^3	Density
σ	kg/s^2	Surface tension
φ		Relative free area of a tray
τ	s	Residence time

6 Extraction

Extraction denotes the removal of some constituents from liquid or solid mixtures via a liquid solvent. The former process is referred to as solvent extraction, the latter as leaching. A precondition of all extraction processes is that the solvent is not (or not completely) miscible with the feed to be treated. Hence, there has to exist a large miscibility gap that creates a two-phase system with large interfacial area, which is effective for interfacial mass transfer. The treated feed is called raffinate, the loaded solvent extract.

The principal scheme of an extraction process is depicted in Fig. 6.0-1. The feed is brought into intimate contact with the solvent to enhance mass transfer between the two phases. Then the phases are separated. The raffinate is withdrawn from the unit. The loaded solvent (extract) is fed in to a regenerator for further processing. The regenerated solvent is recycled to the extractor.

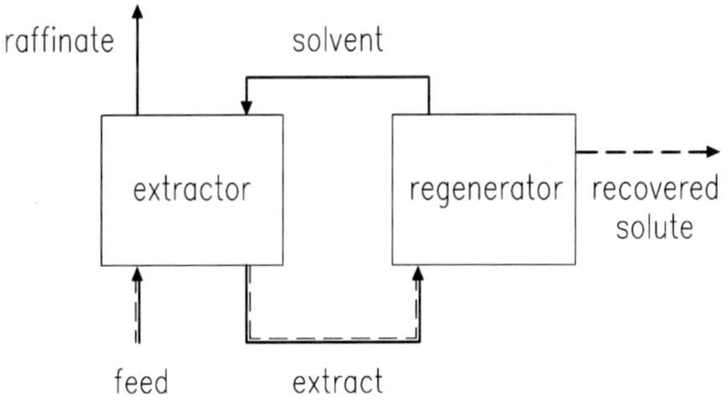

Fig. 6.0-1 Principal scheme of extraction processes

During extraction the feed is polluted with traces of the solvent since there always exists some mutual solubility of the two phases (i.e., raffinate and extract). This fact requires a further separation process for solvent recovery, making extraction processes more complex than, for instance, distillation processes.

In most cases, solvent extraction is operated as continuous process since the two fluid phases can easily be handled in the unit. Leaching, however, is more often performed as a batch process with the raffinate phase fixed in a vessel. Further-

A. Mersmann et al., *Thermal Separation Technology: Principles, Methods, Process Design*, VDI-Buch, DOI 10.1007/978-3-642-12525-6_6,
© Springer-Verlag Berlin Heidelberg 2011

more, the process of leaching typically requires rather large residence times and, in turn, large equipment dimensions that cannot be realized in continuous operation.

6.1 Phase Equilibrium

The prerequisite of all types of extraction processes is the existence of a large miscibility gap between raffinate and extract. The thermodynamic principles of phase equilibrium are dealt with in Chap. 2. An extensive collection of liquid–liquid equilibria is given in the Dechema Data Collection (Sorensen and Arlt 1980ff). Volume 1 contains data of miscibility gaps of binary systems. Phase equilibrium data (miscibility gaps and distribution equilibrium) of ternary and quaternary mixtures are listed in volumes 2–7.

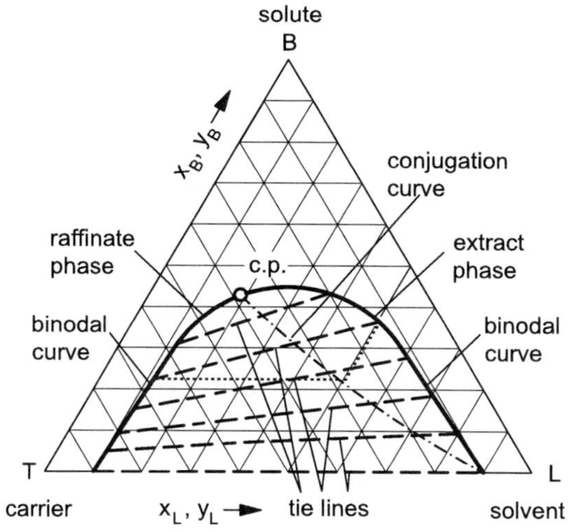

Fig. 6.1-1 Typical system for solvent extraction

A ternary system well suited for solvent extraction is shown in Fig. 6.1-1. The feed consists of substance T (carrier) and a solute B, which has to be removed or recovered in pure form. Both substances are mutually miscible. The solute B is also miscible with the solvent L which is often an organic compound. Substances T and L, however, must have a large miscibility gap that narrows down with increasing content of solute and, eventually, diminishes at the critical point (or plait point). The boundary of the two-phase region is called binodal curve. The dashed lines within the two-phase region, called tie lines (or conodes), describe the distribution equilibrium of solute B. The dashed-dotted line (conjugation curve) allows the interpolation between given tie lines.

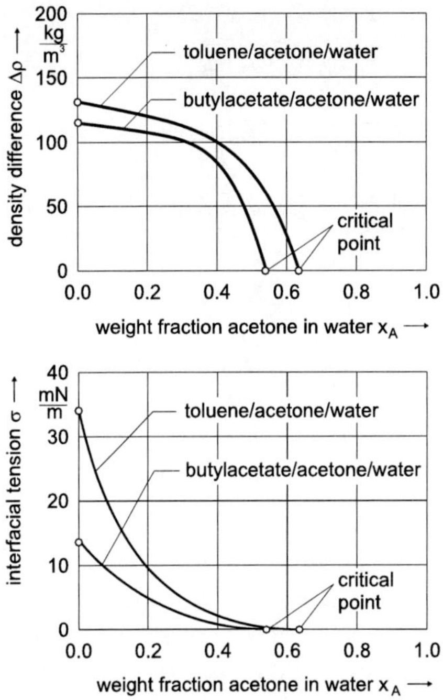

Fig. 6.1-2 Values of density differences and interfacial tensions vs. solute concentration at a temperature of 20°C. Both values approach zero at the critical point

With increasing content of solute B, the states of raffinate and extract approach each other and, finally, coincide at the critical point. In consequence, density difference $\Delta\rho$ and interfacial tension σ approach zero as shown in Fig. 6.1-2. Low values of $\Delta\rho$ and σ make the phase splitting more difficult or even impossible. Therefore, the region near the critical point cannot be utilized in solvent extraction processes.

A typical system suited for leaching is depicted in Fig. 6.1-3. The phase equilibrium of such a system can be understood from the following model.

The solid matrix consists of macropores (voids between particles of the bed) and micropores (voids within the particles). The transfer compound B (solute) is typically concentrated in the micropores. During leaching process the solvent penetrates into all pores and mixes with substance B in the micropores. At equilibrium state the liquid in the macropores has the same concentration as the liquid in the micropores. Therefore, the ratio of solute to solvent (B/L) has the same value in the raffinate and the extract. In this case, all tie lines, which represent the distribution equilibrium, emerge from the S-corner of the triangular diagram in Fig. 6.1-3. Supposed that only the liquid in the macropores can be removed from the matrix by

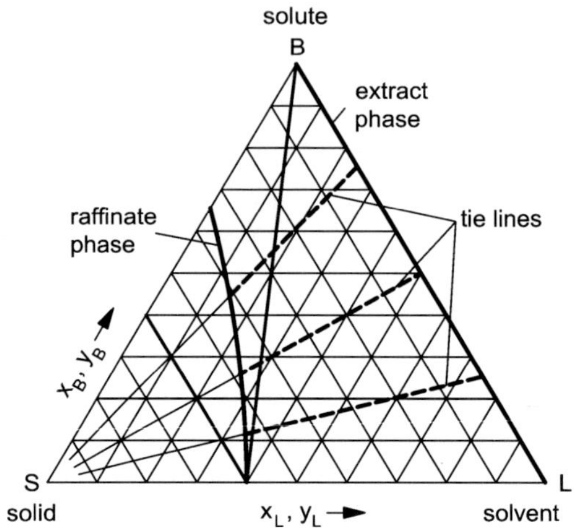

Fig. 6.1-3 Typical system for leaching

phase splitting, the residuum curve is a straight line running parallel to the right hand side of the concentration triangle. The liquid residuum in the solid matrix (raffinate) does not depend on liquid concentration in this case. In the other special case, the solvent L in the solid matrix is constant. Here, the ratio of solvent to solid (L/S) is constant and the raffinate is a straight line through the B-corner of the triangle. In reality, the raffinate is described by a curve running between the two special cases considered before. Generally, the amount of liquid in the raffinate depends on liquid concentration.

This model oversimplifies the real conditions. Often, the transfer species are solid compounds enclosed in the matrix (e.g., in cells of plants) that have to be dissolved in the solvent before they can be transported through the matrix.

6.1.1 Selection of Solvent

Essential for efficient extraction processes is the selection of an appropriate solvent. The solvent should have a very large miscibility gap with the raffinate and a high capacity for the transfer component (solute) B.

The capacity of a solvent is described by a solution parameter, which represents the molar cohesion energy δ^2. The energy δ^2 consists of the dispersion energy δ_d^2, the dipole energy δ_p^2, and the hydrogen bonding energy δ_h^2. For many substances the dispersion energies δ_d^2 have similar values. The energies δ_p^2 and δ_h^2, however, show large differences. The latter is very large in particular for water, alcohols, and amines. A solvent has a high capacity if its solution parameter δ^2 is similar to the corresponding parameter of the transfer component. Values of solution parameters δ^2 can be found in Hildebrand and Scott (1962), Hansen (1969), and Hampe (1978).

The decisive quantity for phase splitting is the density difference $\Delta\rho$ between the phases. As the density of the raffinate is given by the system at hand the solvent has to be selected with respect to density. The density difference must not fall below values of approximately 30–50 kg/m^3.

Table 6.1-1 Solvents for extracting different solutes from water (Stichlmair and Steude 1990)

Solute	Solvents
Organic aliphatic acids	Methyl isobutyl ketone (MIBK) Ethyl acetate/cyclohexane Butyl acetate Diisopropyl ether 2-Ethyl hexanol
Sulfonic acids	*n*-Butanol Butanol/butyl acetate Chelating agents in dodecane
Phenols	MIBK Ether Triamyl methyl ether (TAME) Butyl acetate Toluene, xylene
Amines	Toluene, xylene Cumene
Aliphatic amines	Ion exchangers in dodecane
Metal salts Inorganic acids	Chelates Ion exchangers in dodecane
Pharmaceuticals (of pesticides)	Octanol 2-Ethyl hexanol MIBK/butyl acetate

In industrial extraction processes the regeneration of the solvent is often established by distillation. Consequently, a further aspect for solvent selection is an easy separability of transfer component and solvent by distillation. Therefore, the transfer component B should form either an origin or a terminus of distillation lines in the ternary extract phase (see Sect. 5.2.2.2). If the transfer component B is the low boiler in the raffinate, then the boiling point of the solvent has to be higher than that of the transfer component. In case of a high boiling transfer component (higher than T) the boiling point of the solvent L has to be lower than that of the transfer component B. Furthermore, the solvent must not form azeotropes with the transfer component. Important aspects are also costs, availability, chemical stability, and nontoxicity of the solvent.

Of great industrial importance is the separation of substances from water by solvent extraction. Some solvents technically used for these processes are listed in Table 6.1-1. Obviously, the selection of the solvent strongly depends on the transfer component.

6.2 Thermodynamic Description of Extraction

The thermodynamics of extraction processes are advantageously described in triangular concentration diagrams since both phases, i.e., raffinate and extract, are ternary mixtures. Mass fractions or molar fractions can be used as a measure for the composition of the phases. Here, x denotes the concentration of the raffinate and y the concentration of the extract. The symbol z stands for the overall concentration of a two-phase system.

6.2.1 Single Stage Extraction

The scheme of continuous single stage extraction is shown in Fig. 6.2-1. The feed $\dot{F} = \dot{R}_0$ and the solvent $\dot{E}_0$ are mixed (state $\dot{M}$) in the separation stage to create a two-phase system with large interfacial area that enhances the interfacial mass transfer. After having reached phase equilibrium, the phases are split into the raffinate $\dot{R}_1$ and extract $\dot{E}_1$ that lie on a tie line passing through the mixing state $\dot{M}$. A mass balance yields

$$\dot{F} + \dot{E}_0 = \dot{M}_1 = \dot{R}_1 + \dot{E}_1 \,. \tag{6.2-1}$$

The state of the mixture $\dot{M}_1$ results from a mass balance of the transfer component:

$$\dot{F} \cdot x_F + \dot{E}_0 \cdot y_0 = \dot{M}_1 \cdot z_1 = \dot{R}_1 \cdot x_1 + \dot{E}_1 \cdot y_1 \tag{6.2-2}$$

or

$$z_1 = \frac{\dot{F} \cdot x_F + \dot{E}_0 \cdot y_0}{\dot{F} + \dot{E}_0} \,. \tag{6.2-3}$$

Hence, the amount of solvent $\dot{E}_0$ determines the position of the mixing point in the concentration space. The point $\dot{M}_1$ represents the overall concentration of the two phases. It has to lie within the two-phase region (miscibility gap). Points $\dot{R}_1$ and $\dot{E}_1$ define the concentrations of the raffinate and the extract, respectively. Their amounts are determined via mass balances or, graphically, via lever rule.

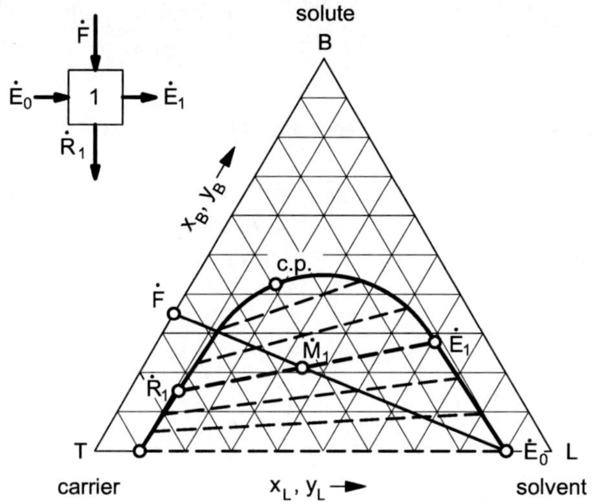

Fig. 6.2-1 Graphical presentation of single stage solvent extraction

The operation parameter of solvent extraction is the amount of the solvent. The larger the E_0 the lower the residual concentration in the raffinate. However, the concentration of transfer component in the extract becomes very low, too.

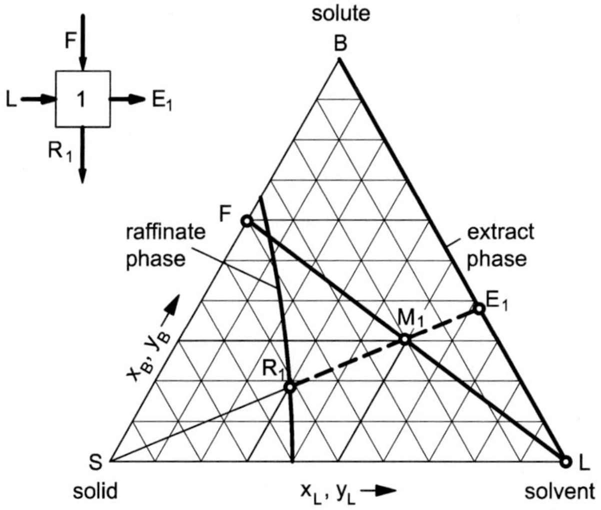

Fig. 6.2-2 Graphical presentation of single stage leaching

The process of leaching is shown in Fig. 6.2-2. The thermodynamic description is fully equivalent to solvent extraction. The raffinate phase lies on the residuum line which is a ternary mixture. The extract phase, however, is binary when no solid material is dissolved in the solvent.

6.2.2 Multistage Crossflow Extraction

Extraction processes can be improved by multiple treating the raffinate product with fresh solvent. In Fig. 6.2-3 a four stage crossflow process is depicted. The raffinate from each stage is mixed with fresh solvent.

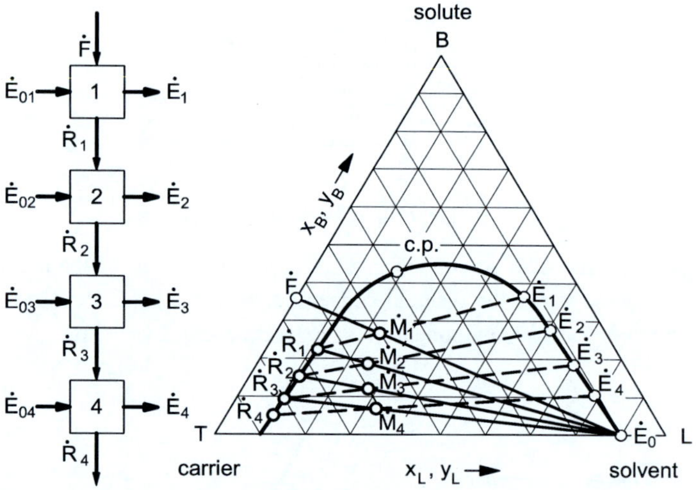

Fig. 6.2-3 Scheme and graphical presentation of four stage cross flow solvent extraction

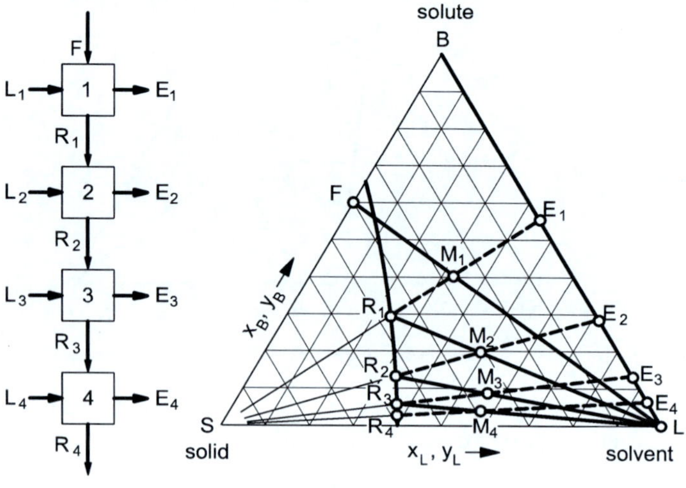

Fig. 6.2-4 Scheme and graphical presentation of four stage cross flow leaching

As can be seen from the graphical construction the concentration of the transfer component B in the raffinate is reduced step by step to very low values. However, the concentration in the extract also falls down making the recovery of a pure fraction of B by further separation processes more difficult. The analogous process of multiple crossflow leaching is shown in Fig. 6.2-4. The thermodynamics of both types of extraction are completely equivalent.

6.2.3 Multiple Stage Countercurrent Extraction

The most effective mode of operation of extraction processes is depicted in Fig. 6.2-5. The raffinate $\dot{R}$ and the extract $\dot{E}$ are in countercurrent contact in a cascade of n (here 4) equilibrium stages. The concentration of the feed decreases from x_F to x_n $(n = 4)$. The transfer component B enriches from y_0 to y_1. An overall mass balance delivers the mixing point of the system:

$$\dot{F} + \dot{E}_0 = \dot{R}_4 + \dot{E}_1 = \dot{M}. \tag{6.2-4}$$

A mass balance of the transfer component gives

$$\dot{F} \cdot x_F + \dot{E}_0 \cdot y_0 = \dot{R}_4 \cdot x_4 + \dot{E}_1 \cdot y_1 = \dot{M} \cdot z. \tag{6.2-5}$$

From a balance around the lower section of the cascade follows

$$\dot{R}_3 - \dot{E}_4 = \dot{R}_2 - \dot{E}_3 = \dot{R}_1 - \dot{E}_2 = \cdots\cdots = \text{const.} \tag{6.2-6}$$

Hence, the difference of the states of the streams between neighboring stages is constant. This difference is called pole π. With $\dot{F} = \dot{R}_0$,

$$\pi = (\dot{R}_0 - \dot{E}_1) = (\dot{R}_1 - \dot{E}_2) = (\dot{R}_2 - \dot{E}_3) = (\dot{R}_3 - \dot{E}_4) = (\dot{R}_4 - \dot{E}_0) \tag{6.2-7}$$

or

$$\pi + \dot{E}_1 = \dot{F};\quad \pi + \dot{E}_2 = \dot{R}_1;\quad \pi + \dot{E}_3 = \dot{R}_2;\quad \pi + \dot{E}_4 = \dot{R}_3. \tag{6.2-8}$$

This equations state that the pole π is collinear with the states of raffinate $\dot{R}$ and extract $\dot{E}$ between any neighboring stages of the separation cascade.

Typically, the pole is located outside the triangular concentration space at the left-hand or the right-hand side. All states of the coexisting $\dot{R}$ and $\dot{E}$ phases can be determined graphically:

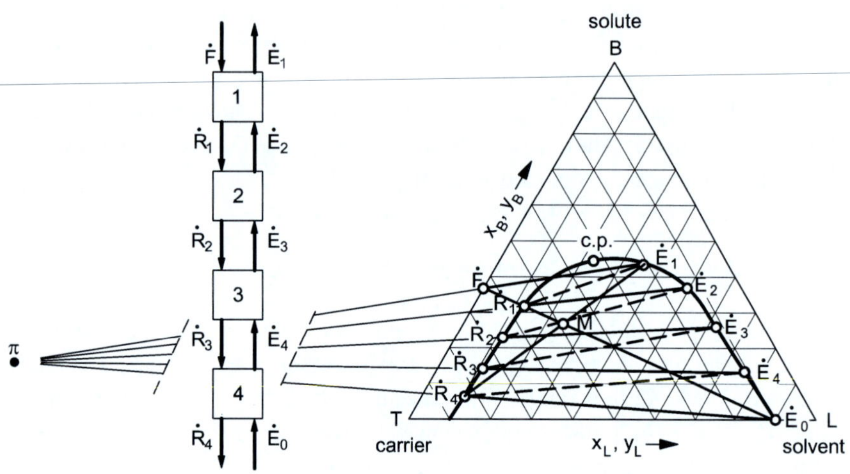

Fig. 6.2-5 Scheme and graphical presentation of four stage countercurrent solvent extraction

- Marking the states of the specified streams $\dot{F}$, $\dot{R}_n$ and $\dot{E}_0$. A mass balance yields the states of $\dot{M}$ and $\dot{E}_n$.
- Extension of the straight lines $\overline{FE_1}$ and $\overline{R_nE_0}$. The point of intersection is the pole π.
- Drawing the tie line through R_n (here R_4) yields point $\dot{E}_n$.
- Drawing a straight line (operating line) through points E_n and π. The result is point $\dot{R}_{n-1}$, etc.

The most important parameter in the graphical construction of the equilibrium stages is the amount of solvent E_0. The larger the amount of solvent, the greater the distance between pole π and feed F. In case of parallel lines $\overline{\pi E_0}$ and $\overline{\pi E_n}$, the pole π approaches infinity and, eventually, switches to the right-hand side of the triangle in Fig. 6.2-5.

The minimum solvent demand is reached if the straight lines through the pole (the operating line) coincides with a tie line between raffinate and extract.

The graphical construction of the equilibrium stages of solvent extraction can analogously be applied to the process of leaching, as demonstrated in Fig. 6.2-6.

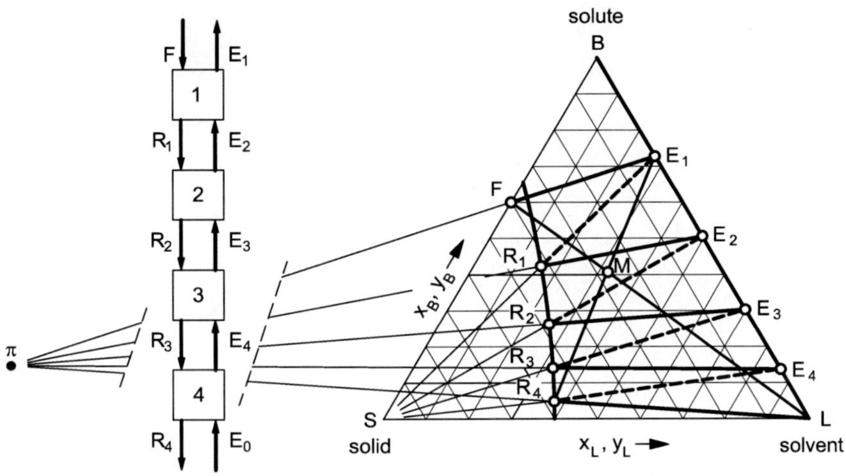

Fig. 6.2-6 Scheme and graphical presentation of four stage countercurrent leaching

For large numbers of equilibrium stages the graphical construction becomes unprecise since the angles between intersecting straight lines are very small. In such situations the following procedure is recommended

The phase equilibrium, i.e., the tie lines, is projected into a y_B/x_B diagram as shown in Fig. 6.2-7. The result is a curved equilibrium line that originates from zero and terminates on the diagonal at the critical point. Analogously, the states of the operating line are plotted in the y_B/x_B diagram (Fig. 6.2-8). Pairs of concentrations x_B and y_B are determined by the points of intersection of a straight line, which emerges from the pole π, with the binodal curve. The x values have to be reflected at diagonal of the y_B/x_B diagram. Typically, the operating line is slightly curved. The number of equilibrium stages required for the separation is determined by drawing steps between operating and equilibrium line in the y_B/x_B diagram (Fig. 6.2-9).

Processes of solvent extraction are mostly performed in countercurrent operation mode. Due to restrictions set by the equipment the number of equilibrium stages should be lower than 10. If more equilibrium stages are required to meet the product specifications, extraction is unfavorable and has to be replaced, if possible, by any other separation process.

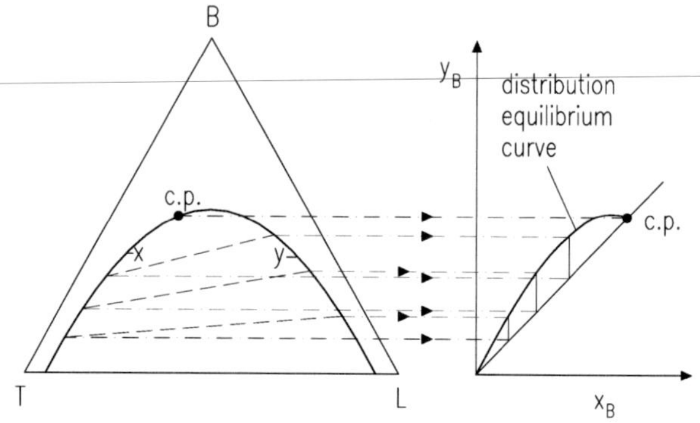

Fig. 6.2-7 Graphical determination of an equilibrium line in the y/x diagram

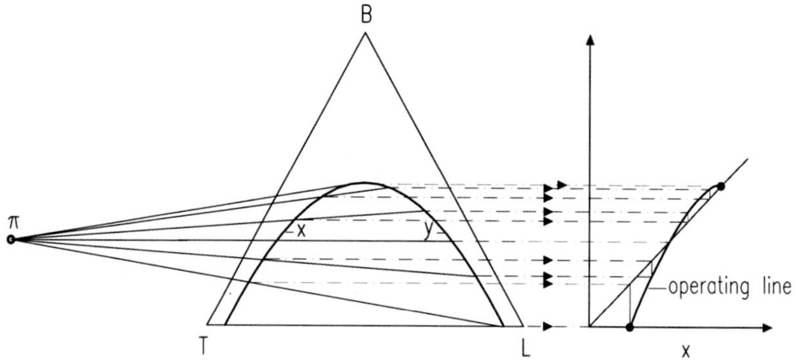

Fig. 6.2-8 Graphical determination of an operating line in the y/x diagram

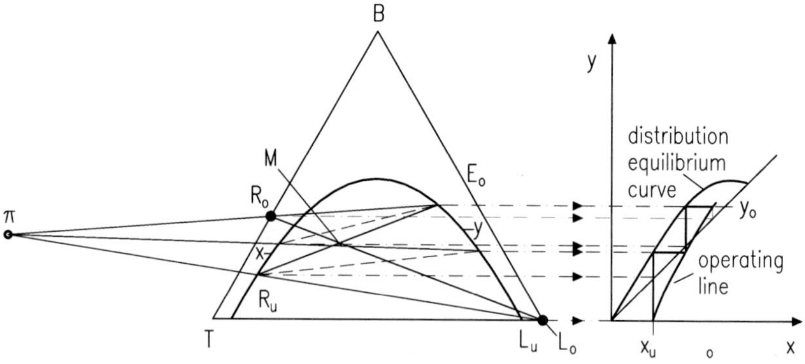

Fig. 6.2-9 Graphical determination of the number of equilibrium stages in the y/x diagram

6.3 Equipment

In the process industry a great variety of different equipment designs is used in extraction processes. This is due to the fact that the density difference between two liquid phases is very small, typically smaller than $\Delta\rho < 100\ \mathrm{kg/m^3}$. This small density difference restricts the velocities of the phases to very small values and, in turn, reduces the interfacial mass transfer rates. This problem is solved by putting external motions into the system, for instance, by pulsation or agitation. These externally induced motions are meant to enhance the interfacial mass transfer rates. Three types of solvent extractors can be distinguished: static devices, pulsed devices, and agitated devices (Blaß 1994; Godfrey and Slater 1994; Rydberg et al. 1992).

6.3.1 Equipment for Solvent Extraction

6.3.1.1 Static Columns

Three examples of static columns operated countercurrently are depicted in Fig. 6.3-1. The spray column does not have any internals. The dispersed phase (droplet phase) is moving countercurrently to the continuous phase. Spray columns can only be used if the density difference between the two phases is higher than $150\ \mathrm{kg/m^3}$ and if the required separation efficiency is low. Hence, spray columns are seldomly used in extraction.

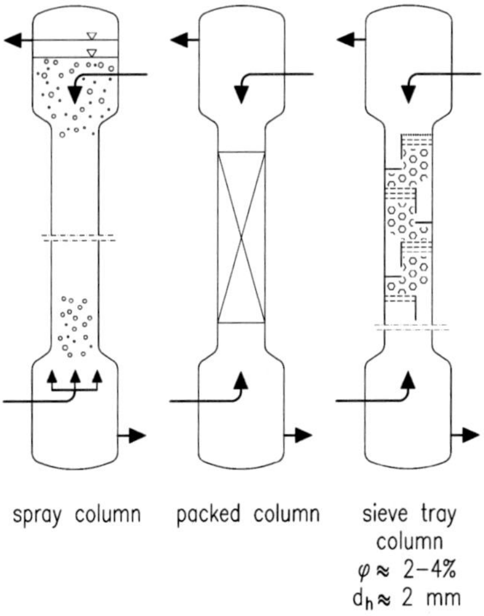

Fig. 6.3-1 Scheme of some static extraction columns

The static packed column, shown in Fig. 6.3-1, is more efficient with respect to mass transfer. Its design is similar to that of gas/liquid systems (e.g., absorption and distillation). In contrast, the design of a static tray column, used for solvent extraction, is completely different from the corresponding column for gas/liquid service. The sieve trays have very small hole diameters (2–4 mm) and a very small free area ($\varphi \approx 2-4\%$) (Fig. 6.3-1).

6.3.1.2 Pulsed Columns

Pulsed columns, depicted in Fig. 6.3-2, are frequently used in solvent extraction. The design of pulsed packed columns is identical with the design of static packed columns. Just the two-phase liquid hold-up is periodically vertically moved with the frequency f. Typically, the pulsation height is about 0.8–1.2 cm. Very common is a pulsation intensity of $a \cdot f \approx 1\text{–}2.5\ \mathrm{cm/s}$. The pulsation effects a decrease of droplet size and, in turn, an increase of interfacial mass transfer rates.

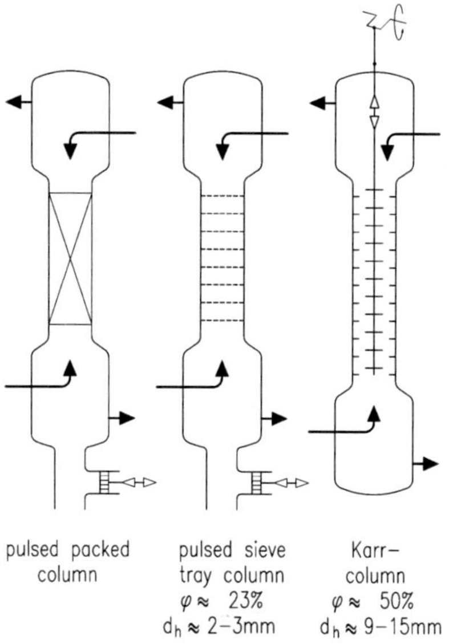

Fig. 6.3-2 Scheme of some pulsed extraction columns

Pulsed sieve tray columns differ significantly from static sieve tray columns as they do not have any down comers. The light and the heavy phase pass periodically through the same holes. Furthermore, pulsed tray columns have a much larger relative free area of up to $\varphi \approx 0.25$ with hole diameters of 2–4 mm.

A modification of a pulsed tray column is the Karr column shown at the right-hand side of Fig. 6.3-2. By an eccentric gear at the top of the column all trays (arranged as package) are periodically moved up and down (reciprocating trays). The relative free area φ is very high, typically $\varphi = 0.5$, with large hole diameters of 9–15 mm. The Karr column is often used for solvent extraction as well as for leaching services.

6.3.1.3 Agitated Extractors

Some examples of agitated extractors are depicted in Fig. 6.3-3. The rotating disc contactor (RDC) uses flat disc agitators. A modification of this design is the asymmetrical rotating disc contactor (ARD). Conventional agitator elements are used in the Kühni and the RZE extractors. All agitated extraction columns are equipped with stator rings, arranged between the agitators, to suppress the vertical back mixing of the phases.

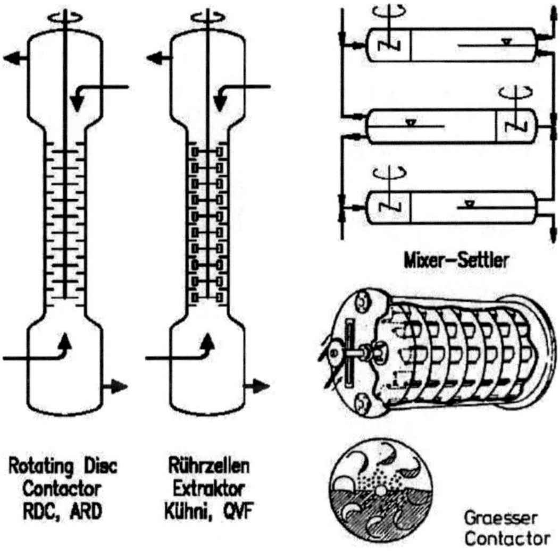

Fig. 6.3-3 Scheme of some agitated extractors

The oldest and most important agitated extractor is, however, the mixer settler. Its design is very simple (agitation and decantation zones). Mixer settler can be built in large units and with many stages. In each stage, the two phases are firstly mixed to form a two-phase system with large interfacial area for good mass transfer. The attached decantation zone is very large and provides a very effective phase splitting. In each stage the two phases move cocurrently. In a cascade (vertically or horizontally arranged) the two phases move countercurrently.

The Graesser contactor, also shown in Fig. 6.3-3, has some industrial importance for processing systems with very low interfacial tensions. The agitating elements (half pipes) rotate around a horizontal axis. During upward motion the elements transport some heavy liquid upward and disperse it into the light phase. After being filled with light phase the elements move downward and disperse the light phase into the heavy phase. To reduce the axial back mixing of the phases the device consists of a great number of parallel cells.

6.3.1.4 Comparison of the Performance

In Fig. 6.3-4 typical performance data of ten different extractor designs are presented (Stichlmair 1980). All extractors have been operated with the system toluene/acetone/water with toluene as dispersed phase and acetone as transfer component. The initial concentration of acetone in water was approximately 5 wt.%. The mass transfer took place from the continuous water phase into the dispersed toluene phase. The ratio of toluene to water has been chosen so that operating and equilibrium lines are parallel in the y/x diagram. Such operation conditions avoid pinches and, in turn, facilitate the calculation of the equilibrium stages from measured concentrations. The data presented in Fig. 6.3-4 are valid for small units only, e.g., with diameters from 50 to 100 mm.

Pseudovalues of the separation efficiencies (see Sect. 6.4.2) referred to the active length of the extractor are plotted on the ordinate. The abscissa is the total liquid load, i.e., the sum of dispersed and continuous phase.

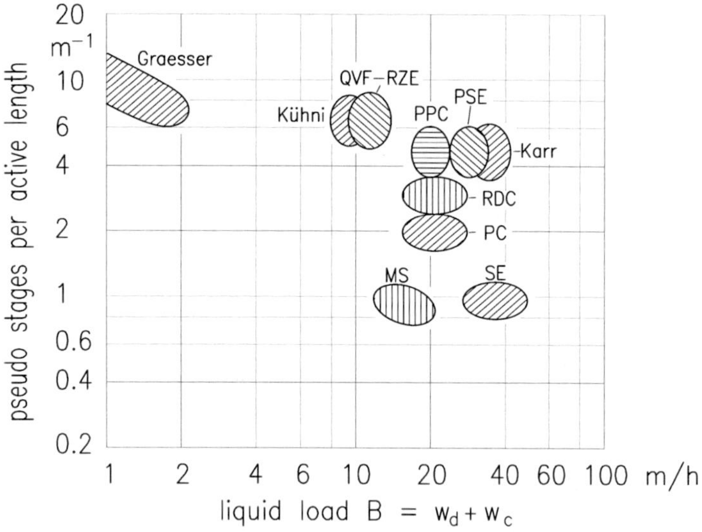

Fig. 6.3-4 Typical performance data of ten different extractor designs for the system tolulene/acetone/water. MS mixer-settler, SE static sieve tray column, PC static packed packed column, RDC rotating disc contactor, PSE pulsed sieve tray column, PPC pulsed packed column, RZE agitated cell extractors

The Graesser contactor takes an extreme position in the diagram. It has the highest separation efficiency (10 stages per meter) but the lowest capacity (1–2 m/h). The other extremum takes the static sieve tray extractor with only one equilibrium stage per meter and up to 50 m/h capacity. The capacity of a pulsed sieve tray column is as high as 30 m/h with a separation efficiency of 5 to 6 stages per meter.

The Karr column approximately shows the same separation efficiency but a slightly higher capacity. Packed columns have a capacity of 20 m/h. The separation efficiency of a static column is 2 stages per meter and that of a pulsed column is 4 to 6 stages per meter. Columns with structured packings have the same efficiency but an 80% higher capacity. The performance data of RDC columns are between that of static and pulsed packed columns. Agitated columns (Kühni and RZE) have a very good separation efficiency (6–8 stages per meter) with rather low capacity (10–15 m/h). As the design of the mixer settler is completely different from that of columns, the comparison is rather difficult. In a first approximation, the capacity is up to 20 m/h and the separation efficiency is as low as 1 stage per meter.

With respect to separation efficiency the different extractor designs differ up to a factor of 10, with respect to capacity up to a factor of 50. However, one should be aware that the performance of each design can be changed to some degree by variations of the geometrical data. But the principal performance characteristics remain.

6.3.2 Selection of the Dispersed Phase

Most contactors, in particular columns, can be operated in different modes. The heavy phase is fed into the column at the top and, in turn, the light phase at the bottom. The two-phase mixture is, in most cases, an univocal drop regime, i.e., one phase is dispersed into the other phase in the form of small droplets that move countercurrently against the continuous phase. The light phase is withdrawn at the top, the heavy phase at the bottom of the column (Fig. 6.3-5).

Within the extractor there exists the so-called principal interface whose location depends on the density of the dispersed phase. In case that the lighter liquid is the dispersed phase the principal interface is at the top of the column just before the withdrawal of the lighter liquid. When the heavier liquid is dispersed into droplets the principal interface is located at the bottom of the column. The dispersed phase can be selected and steered by the mode of start-up. That liquid, which is filled in first, constitutes the continuous phase. The other liquid fed into the full column will

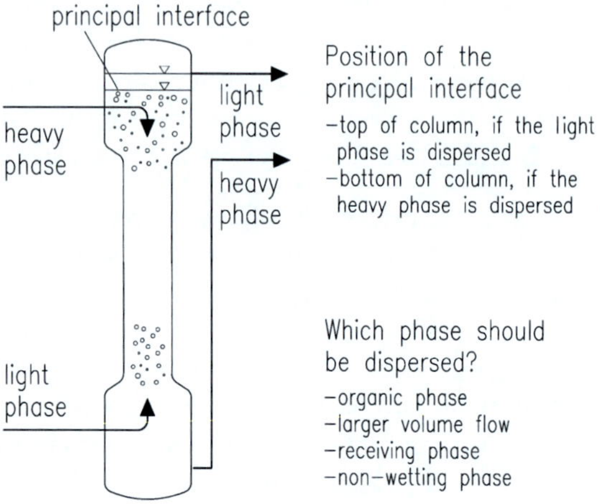

Fig. 6.3-5 Feasible positions of the principal interface and criteria for choosing the dispersed phase

be dispersed into droplets. This mode of operation remains even for long times of continuous operation.

As droplet coalescence rate is very slow the column needs a larger diameter in that part where the principal interface is located. Many extractors have a larger diameter at the top as well as at the bottom to allow for any mode of dispersion.

An important question to be answered is: Which phase should be dispersed? The answer depends on a variety of conditions that are sometimes contradicting (Blaß 1992). In aqueous two-phase systems water constitutes the continuous phase in most cases since the inventory of organic liquids should be small. Often recommended is the dispersion of the larger volumetric flow because more droplets are formed and, in turn, the interfacial area becomes higher. Furthermore, the interfacial mass transfer should take place out from the continuous phase into the dispersed phase (see Sect. 6.4.2). The dispersed phase must not wet the column internals to avoid premature coalescence of the droplets.

The proper selection of the dispersed phase is an essential prerequisite of efficient extraction processes. Often, this problem can only be solved by small-scale experiments.

6.3.3 Decantation (Phase Splitting)

One of the most crucial problems of extraction processes is the splitting of the phases after completion of the interfacial mass transfer. Phase splitting is very difficult because of the small density difference between the coexisting phases (liquids) and, in particular, the very low interfacial tensions. This problem sometimes prevents the application of extraction in the process industry.

In principle, phase splitting consists of two steps:

- Approach of droplets to each other
- Coalescence of droplets

In particular, the mechanisms effective during coalescence are not known sufficiently yet. Hence, small-scale experiments are often required. As even small concentrations of contaminants have a dominant effect on drop coalescence those experiments have to be conducted with the original liquids.

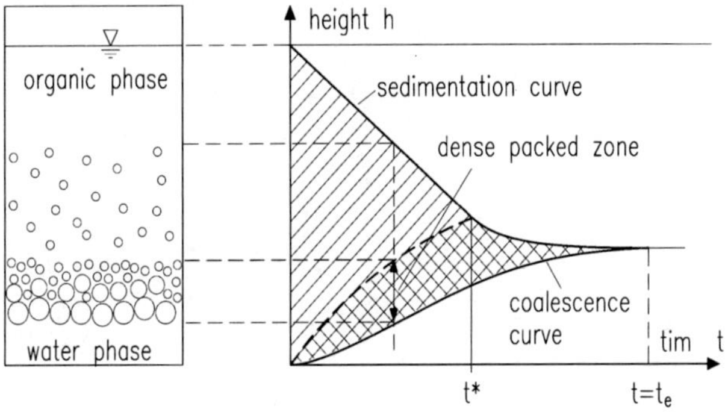

Fig. 6.3-6 Mechanisms of phase splitting (Henschke 1995)

The principal mechanism of phase splitting by decantation is shown in Fig. 6.3-6 at the example of water droplets in an organic liquid. First, a two-phase system is generated by shaking or agitating. After the shaking has been stopped the phase separation starts. Typically, clear liquid layers develop above and below the dispersion. The phase separation at the top is effected by sedimentation of droplets, which shows a linear dependence on time. Phase separation at the bottom results from coalescence. Within the dispersion two zones can be distinguished (Henschke 1995). In the upper dispersion zone sedimentation of a swarm of distant droplets is the dominant mechanism. In the lower dispersion zone (packed layer of adjacent droplets) the coalescence (droplet–droplet and droplet–clear liquid) takes place. The time-limiting mechanism of phase splitting is, in most cases, coalescence. The time t_e required for complete phase splitting can vary to a large extent depending on the system at hand and on the operation conditions.

Two experimental results of phase splitting by decantation are presented in Fig. 6.3-7. In the system butanol/water with small interfacial tension the process of

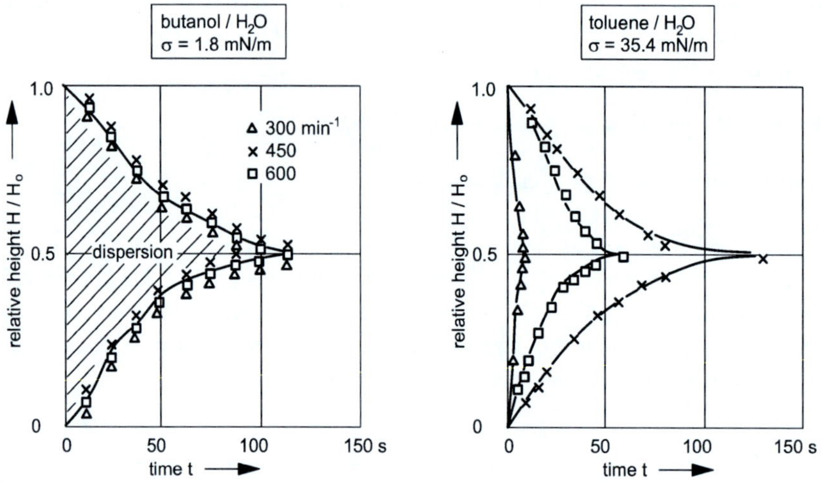

Fig. 6.3-7 Effect of agitation intensity on phase splitting (Berger 1987)

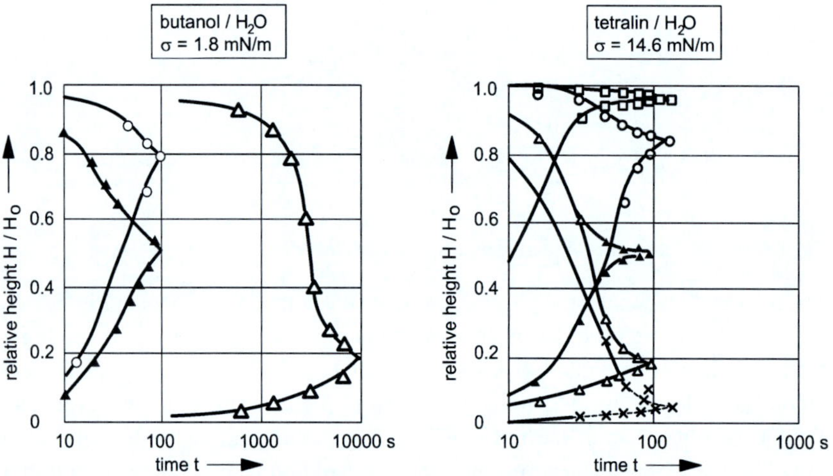

Fig. 6.3-8 Effect of phase ratio on phase splitting (Berger 1987)

phase separation does not depend on the initial agitation intensity. In a system with large interfacial tension, toluene/water, an increase of agitation intensity leads to a significant longer phase splitting time.

The influence of phase ratio on phase splitting is shown in Fig. 6.3-8 with the systems butanol/water and tetralin/water. Obviously, the phase ratio has little influence on phase splitting in the system tetralin/water (intermediate interfacial tension). In the system butanol/water, however, the time required for complete phase splitting

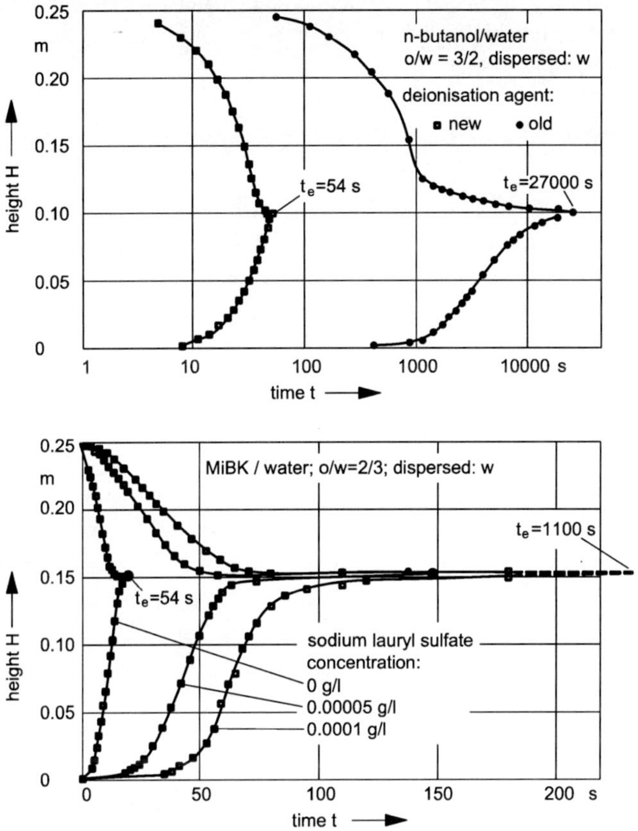

Fig. 6.3-9 Effect of traces of contaminats on phase splitting (Henschke 1995)

increases drastically, by a factor of 100, when the organic liquid is the main constituent of the two-phase system.

The influence of traces of contaminants on phase splitting is demonstrated in Fig. 6.3-9. The pretreatment of the water, either by a fresh or a used deionization agent, changes the phase splitting time by two orders of magnitude. The effect of surface active contaminants (sodium lauryl sulfate) is even larger as the experiments with MIBK/water clearly prove.

These and many other astonishing experimental results are not fully understood yet. In principle, the mechanisms of droplet sedimentation are well known and can be reliably modelled. The only problem is the prediction of drop size. The knowhow on coalescence is, however, rather poor and totally insufficient. Henschke (1995) developed a method for phase splitting modelling that is a combination of simple and smallscale experiments and a rather complex theory. At the present state of the art this is the only reliable method for decanter dimensioning.

6.4 Dimensioning of Solvent Extractors

In contrast to gas/liquid contactors the two-phase system in the mass transfer zone of solvent extractors has a well-defined structure, which is called drop regime. One phase, mostly the organic phase, is dispersed in droplets in the other (mostly aqueous) phase. Therefore, the dimensioning of solvent extractors should be less empirical than that of distillation and absorption columns.

6.4.1 Two-Phase Flow

The decisive flow mechanism in solvent extractors is the motion of swarms of droplets that is in close relation to the motion of single droplets. In liquid–liquid systems, drops reach their terminal velocity after a very short distance of free motion since the density difference between the drop and the continuous phase is very small.

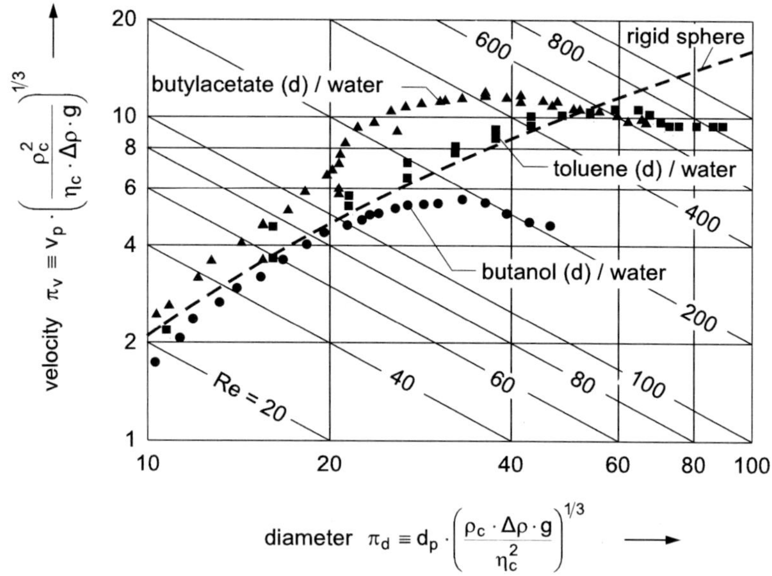

Fig. 6.4-1 Terminal velocity of organic drops in water. Experimental data taken from Haverland 1988; Qi 1992; Hoting 1996; Garthe (2006)

Present knowledge of the terminal velocity of drops in liquids is very high. Small droplets often move a little bit faster than equivalent rigid spheres due to the mobility of the drop surface. Large drops, however, move significantly slower since they lose their spherical shape. Experimental data of the terminal velocity of some organic drops in water are shown in Fig. 6.4-1. The dimensionless terminal velocity is plotted vs. the dimensionless drop diameter. For comparison, the terminal velocity of equivalent rigid spheres is also shown.

A generalized diagram of the terminal velocity of drops is shown in Fig. 3.6-3. From this diagram, the terminal velocity of drops can be predicted as a function of drop size and system properties. For large drops the velocity is nearly independent of drop size. Their velocity is (see Sect. 3.6.1)

$$v_E = 1.55 \cdot \left(\frac{\sigma \cdot \Delta\rho \cdot g}{\rho_c^2} \right)^{1/4}. \tag{6.4-1}$$

The maximum size of drops in free motion is (see Sect. 3.6.1)

$$d_E = \sqrt{\frac{9 \cdot \sigma}{\Delta\rho \cdot g}}. \tag{6.4-2}$$

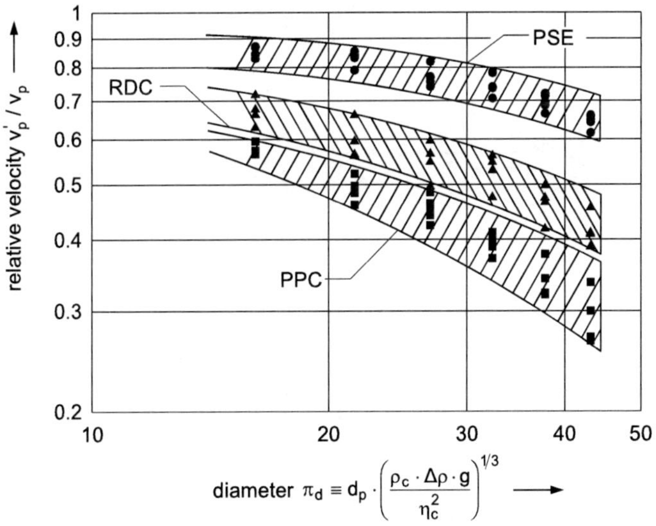

Fig. 6.4-2 Experimental results on the retardation of drops by the internals of different extractor designs (Garthe 2006). PSE pulsed sieve tray column, RDC rotating disc contactor, PPC pulsed packed column

The data shown in Fig. 6.4-1 are valid for the motion of drops in a continuous liquid phase, i.e., in a spray column. Most extractors have internals (trays, packings)

for mass transfer enhancement. These internals slow down the terminal velocity of drops as shown in Fig. 6.4-2. Drop motion can be retarded up to a factor of 2 and even more.

The motion of swarms of droplets is, generally, much slower than that of single droplets. The influence of drop concentration ε_d is accounted for by an empirical correlation (Richardson and Zaki 1954) developed for fluidized beds of rigid particles:

$$\frac{w_d}{v_p} = (1-\varepsilon_d)^m \qquad \text{with} \qquad m = f(Re) . \tag{6.4-3}$$

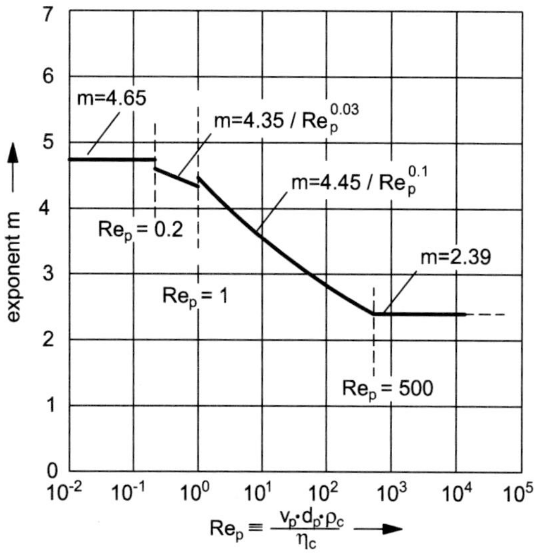

Fig. 6.4-3 Values of the exponent m vs. Reynolds number (Richardson and Zaki 1954)

The exponent m depends on the Reynolds number of the particles as shown in Fig. 6.4-3. An evaluation of (6.4-3) reveals that small drops (laminar flow) can be retarded up to a factor of 80 (Fig. 3.6-9). In turbulent flow the swarm effect is much lower, e.g., as low as a factor of 10. A simpler formulation of the swarm effect is possible when the influence of swarm concentration ε_d is accounted for in the friction factor instead as in the velocities:

$$\xi_s / \xi_p = (1-\varepsilon_d)^{-4.65} . \tag{6.4-4}$$

Hence, the swarm effect itself does not depend on Reynolds number at all. It is the same in laminar as well as in turbulent flow. Just the friction factor depends on Re. However, this dependency on Reynolds number is well known.

Prediction of the motion of single drops and swarms of drops requires knowledge of drop size that results from the flow conditions in the extractor. Some empirical correlations can be found in the literature. For RDC contactors the correlation of Fischer (1971) is recommended:

$$d_{\mathrm{p}} = 0.3 \cdot d \cdot \left(\frac{n^2 \cdot d^3 \cdot \rho_{\mathrm{c}}}{\sigma}\right)^{-0.6} \cdot \left(\frac{n^2 \cdot d}{g}\right)^{-0.27} \cdot \left(\frac{n \cdot d^2 \cdot \rho_{\mathrm{c}}}{\eta_{\mathrm{c}}}\right)^{0.14} . \tag{6.4-5}$$

A correlation of Postl and Marr (1980) is valid for agitated columns:

$$d_{\mathrm{p}} = 0.27 \cdot d \cdot \left(\frac{n^2 \cdot d^3 \cdot \rho_{\mathrm{c}}}{\sigma}\right)^{-0.6} \cdot (1 + 6.5 \cdot \varepsilon_{\mathrm{d}}) \cdot \tag{6.4-6}$$

The following correlation (Widmer 1967) has been developed for pulsed packed columns:

$$d_{\mathrm{p}} = 0.03 \cdot \frac{\psi^2 \cdot \sigma^{0.3} \cdot (\Delta\rho \cdot g)^{0.2} \cdot d_{\mathrm{F}}^{0.9}}{\rho_{\mathrm{c}}^{0.5} \cdot a \cdot f} . \tag{6.4-7}$$

For pulsed sieve tray columns, Pilhofer (1978) recommends

$$d_{\mathrm{p}} = 0.18 \cdot \frac{\sigma^{0.6}}{\rho_{\mathrm{c}}^{0.6} \cdot P^{0.4}} \quad \text{with} \quad P = \frac{\pi^2 \cdot (1 - \varphi^2) \cdot (a \cdot f)^3}{2 \cdot C_0^2 \cdot \varphi^2 \cdot H} \quad \text{with} \quad C_0 = 0.6 . \tag{6.4-8}$$

These empirical correlations have been verified for special system properties, equipment designs, and operation conditions only. Extrapolation to other conditions is risky. Typically, extractors are operated in a way that the drop size is in the range from 1 to 4 mm. It can be effectively steered by the intensity of energy input either by pulsation or by agitation.

From knowledge of drop size and drop motion (swarm) the throughput of dispersed phase can be predicted (Mersmann 1980):

$$\frac{w_{\mathrm{d}}}{v_{\mathrm{p}}} = \frac{\varepsilon_{\mathrm{d}}}{1 - \varepsilon_{\mathrm{d}}} \cdot \frac{w_{\mathrm{c}}}{v_{\mathrm{p}}} + \psi \cdot \varepsilon_{\mathrm{d}} \cdot (1 - \varepsilon_{\mathrm{d}})^{m-1} . \tag{6.4-9}$$

The symbols w_{d} and w_{c} denote the superficial velocity of the dispersed and the continuous phase, respectively. The first term in (6.4-9) stands for the motion of the continuous phase. The second term accounts for the influence of swarm concentration. The symbol ψ denotes the free cross section of the column in packed columns.

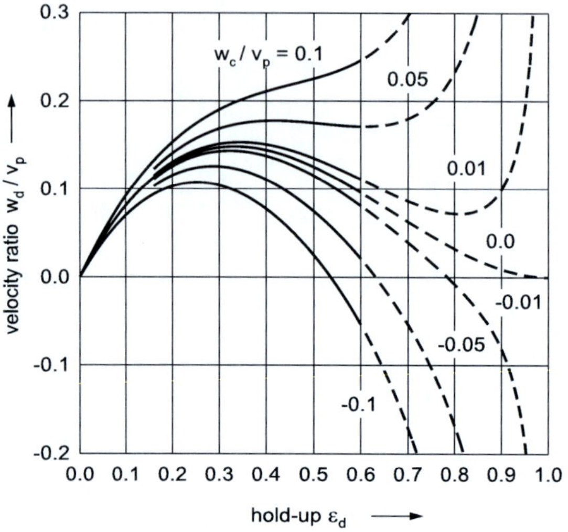

Fig. 6.4-4 Throughput of dispersed phase vs. hold-up. Realistic are values of the hold-up lower than 0.6 (dense packing of spherical particles)

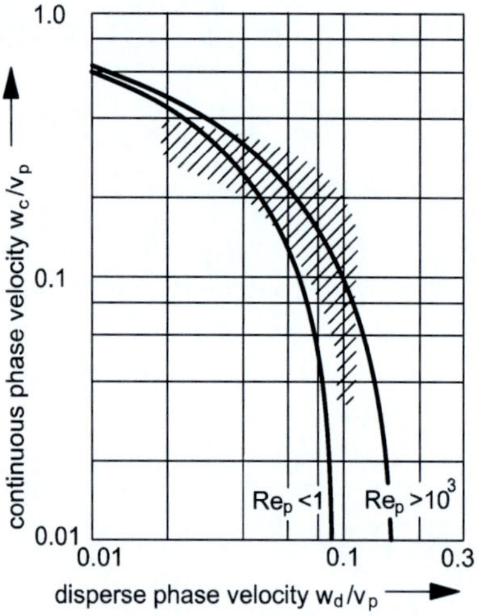

Fig. 6.4-5 Flooding in spray columns according to Mersmann (1980). The experimental data are in the *hatched zone*

The results of the evaluation of (6.4-9) with $\psi = 1$ and $m = 3$ are presented in Fig. 6.4-4. At small values of the hold-up ε_d the throughput of the dispersed phase increases since the increase of the number of drops is the dominant factor. At large values of ε_d, however, the throughput decreases drastically. Here, the swarm effect dominates the motion of the dispersed phase. At intermediate values of the hold-up the throughput curve exhibits a maximum. Equation (6.4-9) allows for the determination of the hold-up from knowledge of the loads of dispersed and continuous phases. It is valid for cocurrent as well as for countercurrent flow of the phases. In countercurrent flow there exist two different values of hold-up ε_d for a given throughput of dispersed phase. The lower value is the hold-up in the mass transfer zone, and the large one exists in the two-phase layer near the principal interphase. Good operation conditions for most extractors exist at values of the hold-up between $\varepsilon_d \approx 0.05$ and 0.15.

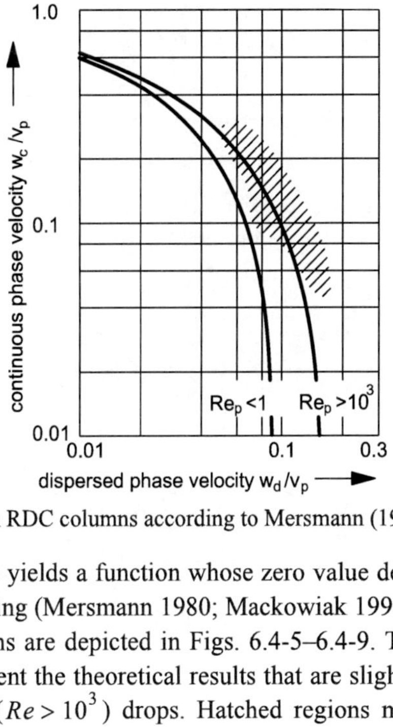

Fig. 6.4-6 Flooding in RDC columns according to Mersmann (1980)

Differentiating (6.4-9) yields a function whose zero value describes the maximum throughput, i.e., flooding (Mersmann 1980; Mackowiak 1993). The results for different extractor designs are depicted in Figs. 6.4-5–6.4-9. The parameter lines in those diagrams represent the theoretical results that are slightly different for small ($Re_p \leq 1$) and large ($Re > 10^3$) drops. Hatched regions mark the experimental data. For most extractors there exists a good agreement between theory and experiment. However, in packed columns and sieve tray columns the experimental throughput is higher than predicted by the model. One reason might be the coalescence of droplets not accounted for by the model. Another explanation for the discrepancies can be that the correlation of the swarm exponent m (Fig. 6.4-3), developed for rigid particles, is not fully correct for deformed drops (Garthe 2006) encountered in extractors.

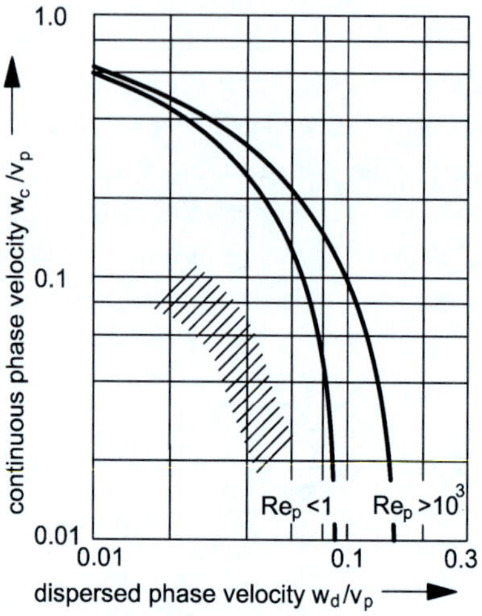

Fig. 6.4-7 Flooding in agitated columns according to Mersmann (1980). The superficial velocities w_d and w_c refer to the *open area* in the stator rings

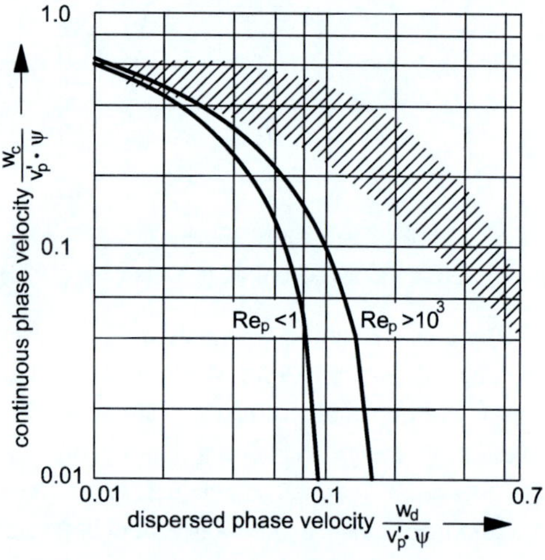

Fig. 6.4-8 Flooding in pulsed and unpulsed packed columns according to Mersmann (1980)

6.4.2 Mass Transfer

The basic relationship for interfacial mass transfer is

$$\dot{M} = \rho_d \cdot \beta_{od} \cdot A \cdot (y - y^*) . \tag{6.4-10}$$

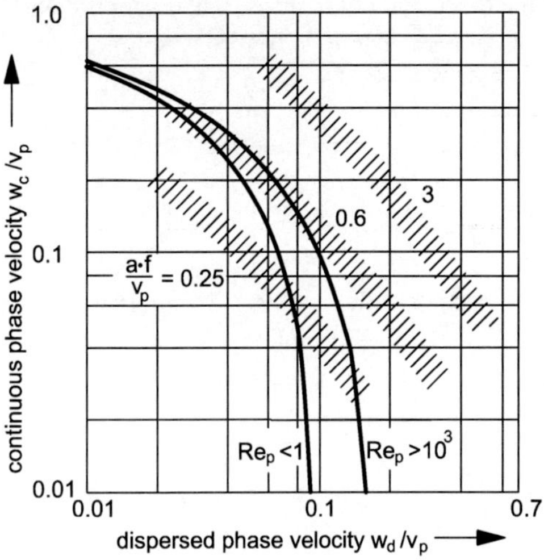

Fig. 6.4-9 Flooding in pulsed sieve tray columns according to Mersmann (1980). The experimental data show a significant dependency on the pulsation intensity $a \cdot f$

The mass transfer rate $\dot{M}$ is proportional to the product of overall mass transfer coefficient $(\rho_d \cdot \beta_{od})$, of interfacial area A, and of driving concentration difference $(y - y^*)$. All terms in this rather simple equation are very problematic.

6.4.2.1 Overall Transfer Coefficient

The overall mass transfer coefficient depends on the individual transfer coefficients of the coexisting phases:

$$\frac{1}{\rho_d \cdot \beta_{od}} = \frac{1}{\rho_d \cdot \beta_d} + \frac{k}{\rho_c \cdot \beta_c} \quad \text{with} \quad k = \mathrm{d}y^* / \mathrm{d}x. \tag{6.4-11}$$

Here, k denotes the slope of the equilibrium line. Typically, the greater part of mass transfer resistance is in the dispersed phase. The values of β_c and β_d differ by a factor 10 or even more. Wagner (1999) found values of β_{od} in the range from 10^{-5} to $10^{-4}\,\mathrm{m/s}$ for the systems toluene/acetone/water and butylacetate/acetone/water.

An empirical correlation for β_c is

$$Sh_c \equiv \frac{\beta_c \cdot d_p}{D_c} = 2 + C \cdot Re^{1/2} \cdot Sc^{1/3} \quad \text{with} \quad C = 0.6\text{–}1.1. \tag{6.4-12}$$

Well verified for rigid spheres, this correlation can also be applied to organic drops.

The prediction of the dispersed phase mass transfer coefficient β_d is more complex because it is of instationary nature. A rigorous solution developed by (Newman 1931) is valid for rigid spheres. An approximation, valid for short transfer times only, has been published by Mersmann (1986):

$$Sh_d \equiv \frac{\beta_d \cdot d_p}{D_d} = \frac{4}{\sqrt{\pi \cdot Fo_d}} \quad \text{with } Fo_d = 4 \cdot D_d \cdot t/d_p^2 \text{ for } Fo_d \leq 0.03 \,. \qquad (6.4\text{-}13)$$

In contrast to rigid particles, there exists an internal circulation in drops that enhances mass transfer rates. This internal convection is accounted for by an enhancement factor E which is combined with the diffusion coefficient $D_{eff} = E \cdot D$:

$$Sh_d \equiv \frac{\beta_d \cdot d_p}{D_d} = 4 \cdot \sqrt{\frac{E}{\pi \cdot Fo_d}} \quad \text{with } Fo_d = 4 \cdot D_d \cdot t/d_p^2 \,. \qquad (6.4\text{-}14)$$

Often, a factor of $E = 2.5$ is recommended. Several authors (e.g., Kronig and Brink 1950; Angelo et al. 1966; Olander 1966; Clift et al. 1978; Steiner 1986) developed empirical correlations for β_d that are, however, verified by few experimental data only. Brauer (1978) found in his experiments an additional influence of the term $Re_c \cdot Sc_c/(1 + \eta_d/\eta_c)$ that enhances mass transfer compared to that of rigid spheres. For longer times of contact, he recommends

$$Sh_d \equiv \frac{\beta_d \cdot d_p}{D_d} = \frac{2}{3 \cdot Fo_d} \quad \text{with } Fo_d = 4 \cdot D_d \cdot t/d_p^2 \,. \qquad (6.4\text{-}15)$$

A decisive parameter is the time of contact. In pulsed sieve tray columns the time of instationary mass transfer is equal to the residence time of the dispersed phase in the space between two trays. In agitated columns, the residence time in an agitation cell has to be taken. However, the application of the above equations to packed columns is difficult. Often recommended is the correlation of Handlos and Baron (1957) which has been developed and verified for oscillating drops:

$$\beta_d = 0.00375 \cdot v_p \cdot \frac{\eta_c}{\eta_c + \eta_d} \,. \qquad (6.4\text{-}16)$$

This equation contains neither the diffusion coefficient nor the contact time. Obviously, the internal circulation dominates the mass transfer rate within a drop.

The correlations presented above allow only a rough estimation of the mass transfer coefficients, since many relevant parameters are not accounted for, as, for instance, shape of column internals, intensity of energy input, neighboring drops, and drop size distribution (Blaß et al. 1985).

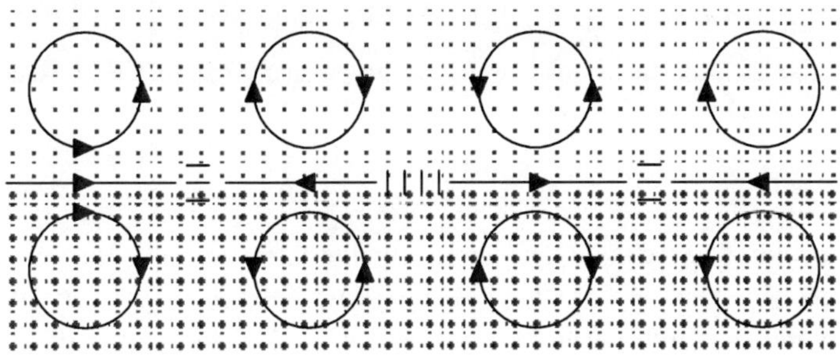

Fig. 6.4-10 Rolling cells generated by concentration differences within the interface (*horizontal lines*) regions with large interfacial tension, (*vertical lines*) regions with small interfacial tension

In real systems the overall mass transfer coefficient β_{od} is often much higher than predicted by the published correlations. One reason is the existence of Marangoni convections that originate from concentration differences caused by the mass transfer.

Local concentration differences caused by the interfacial mass transfer change the interfacial tension and, in turn, produce convections in the interface. By friction forces these convections enhance intensive motions on both sides of the interphase that can have regular (rolling cells) or irregular (eruptions) structures (Fig. 6.4-10). An example of Marangoni enhanced eruptions is shown in Fig. 6.4-11 (Wolf 1999). Such random motions have the potential to increase mass transfer rates by one order of magnitude. They are, however, not predictable with sufficient accuracy.

Fig. 6.4-11 Eruptive Marangoni convections near the interface generated by an interfacial mass transfer

6.4.2.2 Interfacial Area

The area effective for interfacial mass transfer can be easily calculated in case of drop regime in the mass transfer zone:

$$A = V \cdot a \qquad \text{with} \qquad a = 6 \cdot \varepsilon_d / d_p . \tag{6.4-17}$$

V denotes the volume of the mass transfer zone in the extractor and a the interfacial area referred to the volume. The hold-up ε_d follows from (6.4-9). A very complex task is the prediction of mean drop diameter d_p as the correlations (6.4-5)–(6.4-8) presented before are not sufficiently reliable. Typically, the values of the volumetric interfacial area are in the range 200–500 m^2/m^3.

One of the reasons that make the prediction of the interfacial area so difficult is, again, the existence of Marangoni convections. Marangoni convections can either hinder or promote the coalescence of neighboring drops as demonstrated in Fig. 6.4-12.

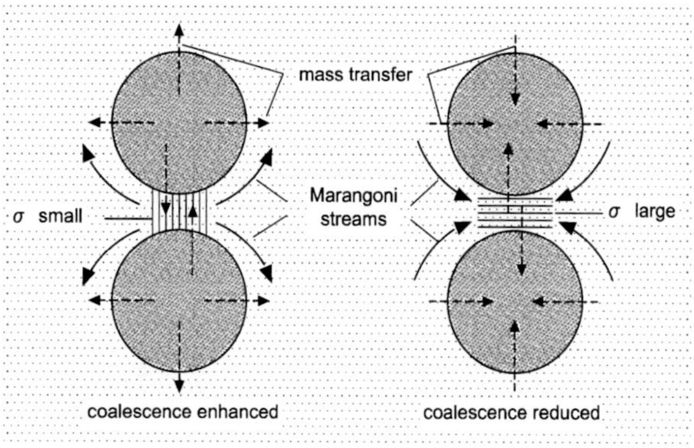

Fig. 6.4-12 Marangoni convections at different directions of the interfacial mass transfer

In case of a mass transfer directed from drops into the continuous phase, the transfer component B enriches in the space between neighboring drops and, in turn, reduces the interfacial tension there (Fig. 6.1-2).

The local difference of the interfacial tension produces Marangoni convections that are directed from regions with low interfacial tension into regions with high interfacial tension. Hence, the Marangoni convections transport some fluid out of the space between drops. The drops approach each other and, eventually, coalesce to larger drops. This mechanism reduces the size of the interfacial area and, in turn, the mass transfer rate. In case of a mass transfer that is directed from the continuous phase into the drops, the transfer component depletes in the space between drops resulting in a higher interfacial tension there. The Marangoni convections transport some liquid from the continuous phase into the space between neighboring drops and, in turn, drop–drop coalescence is hindered.

The mechanisms depicted in Fig. 6.4-12 dominate the operation of solvent extractors. Hence, solvent extractors should not be operated in a mode with mass transfer directed out of the dispersed phase (see Sect. 6.3.2) since drop size and interfacial area are small and not sufficient for a good interfacial mass transfer rate.

Marangoni convections are very important to the operation of extractors. On the one hand they can increase the mass transfer coefficients; on the other hand they can reduce the interfacial area. Which mechanism has the larger effect cannot be decided generally, it depends on the system and equipment at hand.

6.4.2.3 Driving Concentration Difference

The determination of the driving force for the interfacial mass transfer is very difficult in extraction columns since no pure plug flow exists within the column. This is one of the most important differences between gas–liquid and liquid–liquid contactors. Because of the small density differences in liquid–liquid systems there exists a relatively high rate of axial backmixing of dispersed as well as of continuous phase. Backmixing reduces the concentration differences along the column height.

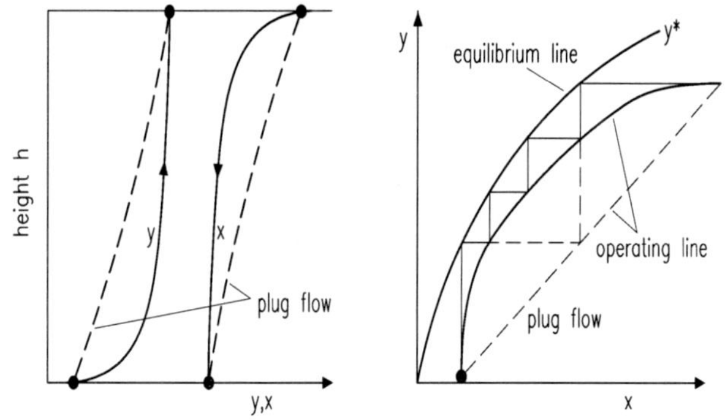

Fig. 6.4-13 Internal concentration profiles of both phases and number of equilibrium stages (*dashed line* without back mixing, *solid line* with back mixing)

The effects of axial backmixing on separation efficiency of a column are explained in Fig. 6.4-13. From experiments the concentrations of the phases at the inlet and the outlet are known. Assuming plug flow, for lack of better knowledge always done, nearly linear concentration profiles (dashed lines) are supposed to exist within the column giving a linear operating line in the y/x diagram. Obviously, the separation efficiency of the column is 2 equilibrium stages (steps between equilibrium line and dashed operating line).

However, axial backmixing drastically changes the internal concentration profiles. In particular at the feed points the concentrations are changed stepwise. The real concentration profiles within the column significantly differ from plug flow profiles. The concentration differences between the two phases are, in reality, much smaller. In consequence, the operating line is much closer by the equilibrium line and, in turn, the number of stages drawn between these two lines in the y/x diagram is much larger. Hence the separation efficiency of the column is as high as four equilibrium stages. The number of stages found by the plug flow model is obviously wrong. These stages are just pseudostages determined from a false concentration difference. However, these pseudostages are decisive for practical column dimensioning.

The mechanism of axial backmixing is described by the so-called dispersion model (e.g., Miyauchy and Vermeulen 1963). Backmixing is modelled in analogy to molecular diffusion. However, the dispersion coefficient D_{ax} is larger by several orders of magnitude than the molecular diffusion coefficient. The degree of backmixing is quantified by the Bodenstein number Bo according to the following definition:

$$Bo_c = \frac{w_c \cdot H}{(1-\varepsilon_d) \cdot D_{ax,\,c}} \quad \text{and} \quad Bo_d = \frac{w_d \cdot H}{\varepsilon_d \cdot D_{ax,\,d}} . \tag{6.4-18}$$

An open problem is the prediction of the dispersion coefficients of the continuous and the dispersed phase, $D_{ax,\,c}$ and $D_{ax,\,d}$, respectively. Godfrey et al. (1988) developed the following correlation for the continuous phase:

$$D_{ax,\,c} = 4.71 \times 10^{-3} \cdot \frac{a^{1.22} \cdot f^{0.78} \cdot H_B^{\;0.57}}{\eta_c^{\;1.19} \cdot d^{0.17} \cdot D_K^{\;0.08}} \cdot \left(1 + \frac{w_c}{2 \cdot a \cdot f}\right) . \tag{6.4-19}$$

Typically, values of $D_{ax,\,c}$ are in the range of $10^{-4} - 10^{-3}\,\mathrm{m^2/s}$. Experimental data of the dispersion coefficient $D_{ax,\,d}$ in the dispersed phase are rather rare. An approximation is

$$D_{ax,\,d} \approx (2 \div 4) \cdot D_{ax,\,c} . \tag{6.4-20}$$

The rigorous modelling of interfacial mass transfer in extractors with regard to axial backmixing is very difficult. In practice, the concept of transfer units is used, which is expanded by an additional height that accounts for the effect of axial backmixing. In case of parallel operating and equilibrium lines, often encountered, the following relation holds:

$$HDU = \frac{\varepsilon_d \cdot D_{ax,d}}{w_d} + \frac{(1-\varepsilon_d) \cdot D_{ax,c}}{w_c} . \qquad (6.4\text{-}21)$$

The column height H, required to meet the product specifications, is

$$H = (HTU + HDU) \cdot NTU . \qquad (6.4\text{-}22)$$

HTU denotes the genuine height of a transfer unit expected in a backmixing free column.

Symbols

A	m^2	Interfacial area
a	m	Amplitude
a	1/m	Interfacial area referred to volume
B		Solute
D, D_{eff}	m^2/s	Diffusion coefficient, effective diffusion coefficient
D_{ax}	m^2/s	Axial dispersion coefficient
d_p, d_h, d	m	Diameter of drops, holes, or agitator
E	kg	Extract phase
E		Enhancement factor $E \equiv D_{eff}/D$
$\dot{E}$	kg/s	Extract flow
F	kg	Feed
$\dot{F}$	kg/s	Feed flow
g	m/s^2	Acceleration of gravity
H	m	Height
h	m	Height of column
HDU	m	Height of dispersion
HTU	m	Height of a transfer unit
k		Slope of equilibrium curve
L		Solvent
M, $\dot{M}$	kg, kg/s	Mixture, mixture flow
$\dot{M}$	kg/s	Mass flow
m		Exponent in Richardson and Zaki equation for swarms
NTU		Number of transfer units
n		Number of equilibrium stages
n	1/s	Number of revolutions
R	kg	Raffinate
$\dot{R}$	kg/s	Raffinate flow
S		Solid feed component
T		Liquid feed component

t	s	Time
V	m^3	Volume
v_p, v_p'	m/s	Terminal velocity of drops
w	m/s	Superficial velocity
x		Mass fraction of solute in the raffinate
y		Mass fraction of solute in the extract

Greek Symbols

β	m/s	Mass transfer coefficient
β_{od}	m/s	Overall mass transfer coefficient
ε_d		Volume fraction of dispersed phase
η	kg/(m · s)	Viscosity
ξ		Friction factor
π		Pole
π		Circular constant
π		Dimensionless number
ρ	kg/m^3	Density
$\Delta\rho$	kg/m^3	Density difference
σ	kg/s^2	Interfacial tension
φ		Relative free area (of sieve trays)
ψ		Relative free area (of packings)

Dimensionless Numbers

Bo	Bodenstein number, $Bo = w \cdot H/D_{ax}$
Fo	Fourier number, $Fo = 4 \cdot D \cdot t/d^2$
Re	Reynolds number, $Re = v \cdot d \cdot \rho/\eta$
Sc	Schmidt number, $Sc = \eta/(\rho \cdot D)$
Sh	Sherwood number, $Sh = \beta \cdot d/D$

7 Evaporation and Condensation

Heat transfer to a liquid leads to an increase of the temperature up to the boiling temperature at which evaporation starts. The vapor pressure becomes equal to the pressure of the system. In the case of the evaporation of a liquid mixture all components or only some of them or perhaps only one component can be present in the vapor. Dealing with the evaporation of an aqueous solution of an inorganic salt with a very low vapor pressure approximately pure steam is leaving the liquid. (Note that an entrainment of small drops can take place with the result of small salt contents in the steam.) In general, all components of the liquid mixture will be present in the vapor when there are no great differences of the vapor pressure of the components.

In Fig. 7.0-1 a simple scheme of a single effect and continuously operated evaporator is depicted together with a condenser. The brine flow $\dot{L}_0$ enters the evaporator and the transfer of heat causes the production of the vapor flow $\dot{G}_1$. The vapor is condensed in the condenser. The heat flow $\dot{Q}_1$ is transferred by the heat exchanger. Later questions concerning the change of concentration of the liquid, the area of heat exchangers, and the arrangements of several evaporators will be discussed. At first some types of evaporators are depicted in the following figures.

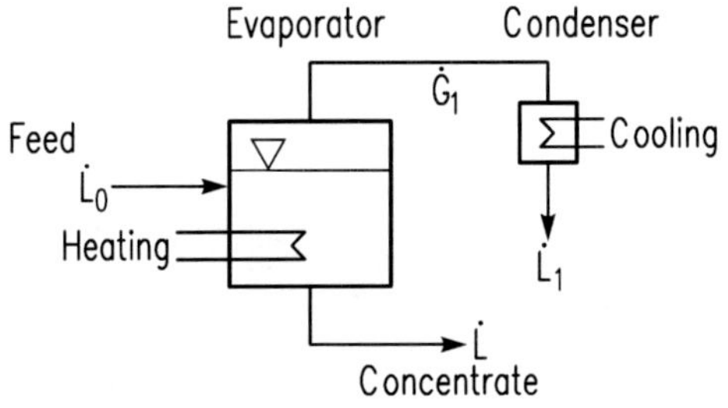

Fig. 7.0-1 Single effect continuously operated evaporator and condenser

A. Mersmann et al., *Thermal Separation Technology: Principles, Methods, Process Design*, VDI-Buch, DOI 10.1007/978-3-642-12525-6_7,
© Springer-Verlag Berlin Heidelberg 2011

7.1 Evaporators

In the case of a favorable equilibrium, evaporation is a simple unit operation; however, with respect to a big variety of the properties of liquids, a great number of evaporator types is known.

Some evaporators are depicted in Figs. 7.1-1–7.1-7. First, the heat transfer coefficient should be high to minimize the heat transfer area and the investment cost. Next, the design and the operation of an evaporator must be chosen in such a way that some undesirable side effects are avoided. These effects are the entrainment of droplets above the liquid surface, the fouling of the heat transfer areas, and the decomposition of liquids during the boiling process. Especially, organic liquids with high molar masses show a tendency to decomposition which increases with an increase of the temperature and of the residence time. Therefore, many evaporators are operated under vacuum to reduce the boiling temperature. Additionally, the residence time $\tau = V_L/\dot{V}_L$ of the liquid volume V_L should be short for a given volumetric liquid flow $\dot{V}_L$.

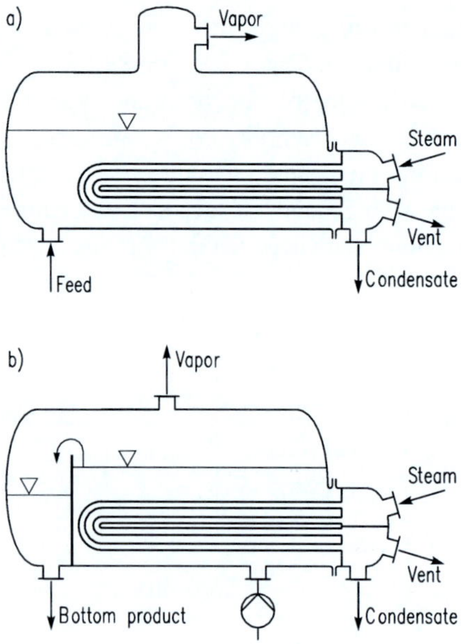

Fig. 7.1-1 Horizontal-tube evaporators: (**a**) Batch or vessel evaporator, (**b**) Reboiler

If these restrictions are not decisive, kettle-type reboilers are simple with respect to their design and investment cost, see Fig. 7.1-1. Figure 7.1-1a shows an evaporator in which the feed is completely evaporated. The apparatus depicted in Fig. 7.1-1b is continuously operated and the feed is separated into vapor and a concentrated solution. The heating medium is flowing within the tubes. As a rule, steam of ambient or elevated pressure is used and condensed in the tubular system. Organic liquids applicable up to 400 °C and melts of salts above 400 °C are applied. Sometimes evaporators are heated by hot flue gases or electrical devices. The advantages of reboilers are a wide operation range, a small liquid height in the reboiler, a large space for vapor–liquid separation, and a simple operation control. After scaling or fouling of the tubes the heating equipment can be removed for cleaning.

Reboilers according to Fig. 7.1-1b can also be operated for liquids close to the critical point. Principally, speaking, the heating media and systems described here can also be applied for other types of evaporators.

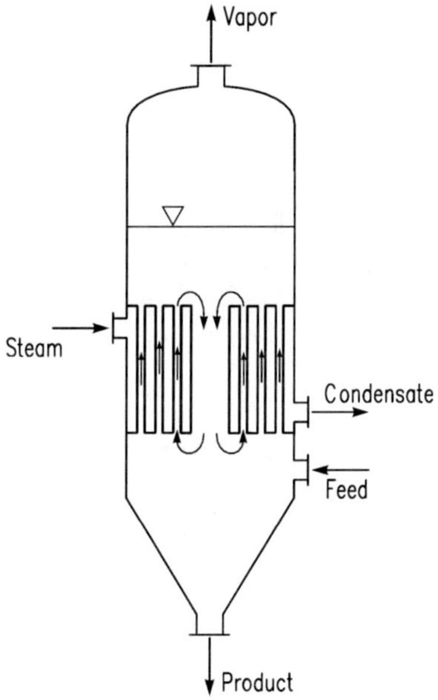

Fig. 7.1-2 Short-tube vertical evaporator with wide recirculation passage

Heat transfer is improved by an increase of the velocity of the liquid. According to natural convection and the buoyancy force of rising bubbles a circulation of the liquid is induced in a short-tube vertical evaporator, see Fig. 7.1-2. In the narrow tubes, liquid is evaporating to a certain degree and recirculated in the central return passage after separation of the vapor. This type is less recommended when a fouling of the heat transfer area by product components takes place. Then it is reasonable to separate the tube bundle and the head space in which the entrainment of liquid is reduced, see Fig. 7.1-3. The cleaning of the tubes can be easily carried out.

Dealing with evaporators operated under vacuum the static pressure of the liquid at the bottom of the bundle can be high in comparison to the pressure at the top. In such a case it is recommended to use evaporators with an inclined tube bundle, see Fig. 7.1-4. Such an arrangement leads to a reduction of the differences of the boiling temperatures at the bottom and the top of the tube bundle.

The small operation range and a minimum circulation rate are drawbacks in the case of evaporators operated due to natural convection. In addition, problems may arise for evaporators under vacuum or for liquid mixtures with a wide range of boiling temperatures.

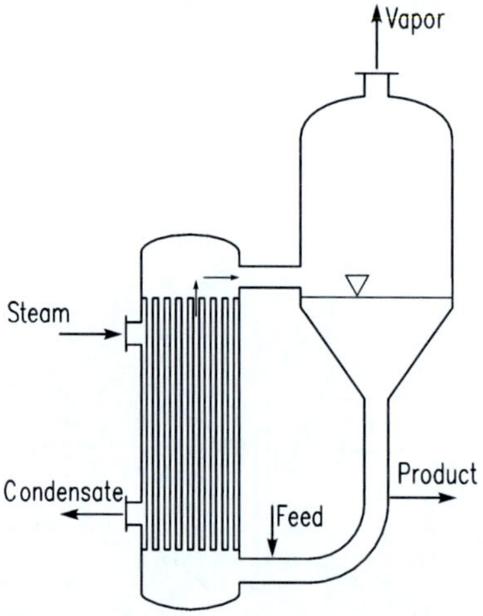

Fig. 7.1-3 Recirculation long-tube vertical evaporator with separated vapor head

The heat transfer of the product side of the heat exchanger can be improved when the natural convection is replaced by forced convection using a pump, see Fig. 7.1-5. Therefore, the heat transfer area can be reduced for a given heat flow with the result that the decomposition of heat-sensitive products is decreased. Forced circulation evaporators with or without flash depressurization are recommended for viscous liquids and suspensions. Advantages are a wide operation range and a stable process control. Falling film evaporators (see Fig. 7.1-6) are advantageous with respect to high heat transfer coefficients and the absence of differences of the hydrostatic pressure. The vertical tubes are heated from the outside of the tubes whereas the product liquid is irrigated on their inside surface. Liquid films or – sometimes – rivulets with a small layer thickness and a short residence time are running downward. Such evaporators are operated under cocurrent (see Fig. 7.1-7) or countercurrent flow of the liquid and the vapor.

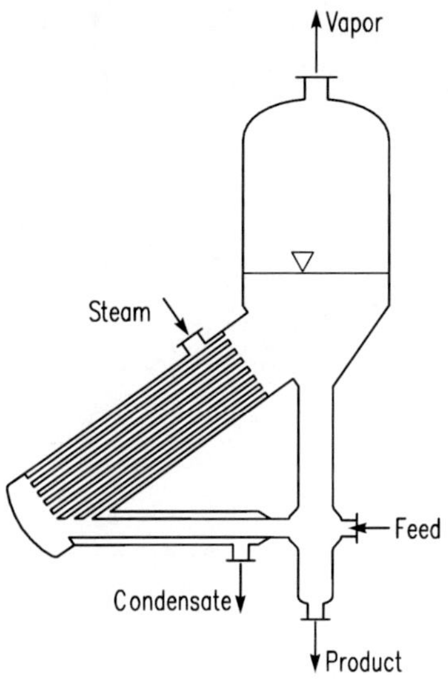

Fig. 7.1-4 Recirculating evaporator with inclined tube bundle

It is difficult to distribute the entire product liquid to achieve the same thin liquid film in all tubes and to avoid the disrupture of the liquid film. Surface tension effects (the surface tension can increase or decrease in the film downward flow) are decisive for the film stability (Zuiderweg and Harmens 1958; Ford and Missen 1968). Cocurrent once-through rising film evaporators can be used for a complete evaporation of the feed. However, fouling may occur at high evaporation rates.

The problem of film disrupture may be tempered by recirculation of the film to avoid very thin films. Another possibility is the use of a falling film evaporator (only one tube) equipped with a rotor and blades which have the duty to distribute and thin the film. Figure 7.1-7 shows an evaporator with motionless blades fixed on the rotor. The distance between the blade surface and the inside of the tube is approximately 1–2 mm.

Other evaporators are equipped with mobile blades which are pressed against the tube according to centrifugal forces. The advantages are a very good distribution of the product liquid and very thin films. Therefore, the heat transfer coefficients are high and the liquid residence time is short. With respect to a low surface temperature of the tube this expensive type can be recommended for heat-sensitive products. The wall temperature can further be decreased by operation under vacuum. The evaporation of suspensions can be carried out in falling film evaporators. The properties of the suspension decide on the question whether or not a dry solid material can be withdrawn as product without massive incrustation. An empirical inves-

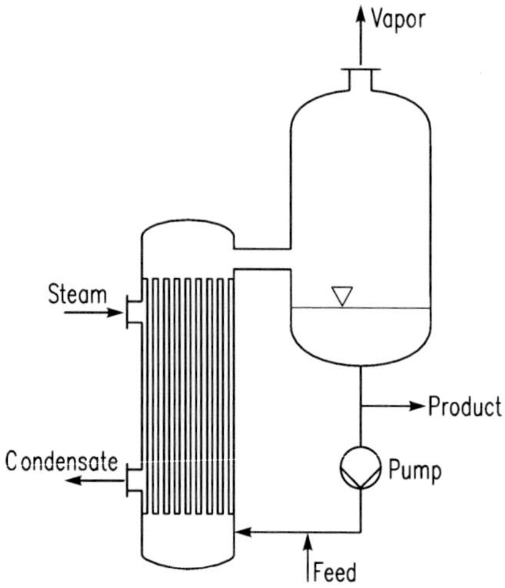

Fig. 7.1-5 Forced circulation evaporator with separated vapor head

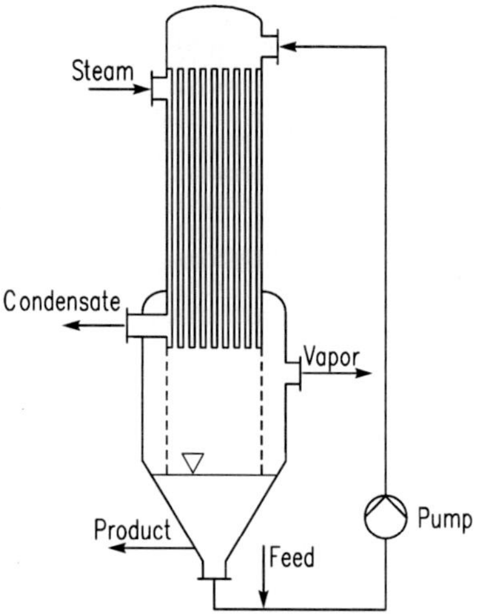

Fig. 7.1-6 Falling film evaporator with recirculation

tigation is necessary. Other possibilities of the evaporation of suspensions are flash evaporation and evaporation in stirred vessels or fluidized beds.

The advantages of falling film evaporators according to Fig. 7.1-7 are a wide operation range, a low pressure drop of the vapor, and the operation under vacuum down to 10 Pa.

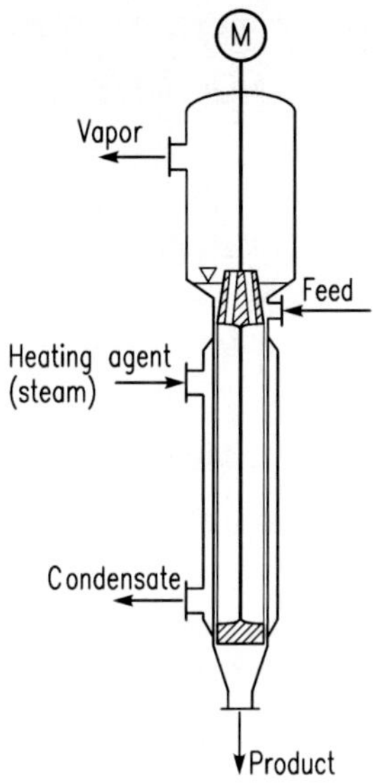

Fig. 7.1-7 Agitated thin-film evaporator equipped with a rotor with fixed blades

7.2 Multiple Effect Evaporation

Based on an energy balance, the mass flow rate of the heating steam can be calculated for the production of the vapor flow rate $\dot{G}_1$. The question arises whether the energy of this vapor can be utilized. This is achieved in a multiple effect evaporation unit. When several evaporators are arranged and connected according to Figs. 7.2-1–7.2-3 the energy cost for a mass unit vapor are lower in comparison to a single evaporator because the vapor produced in the effect k is condensed in the effect $k+1$. Consequently, most of the heat of condensation is utilized.

In the case of a unit with n effects only the vapor flow $\dot{G}_n$ leaves the unit. Therefore, the mass of heating steam is much smaller compared with a single evaporator. The smallest steam consumption can easily be calculated for an ideal process with the following conditions:

- The heating steam entering the first effect and the liquid to be evaporated are the same substance (for instance water, see later seawater desalting)
- The feed flow $\dot{L}_0$ is entering the unit with the boiling temperature
- The heat transfer areas of all effects are infinite large with the consequence that the temperature difference as the driving force for heat transfer is nearly zero

- All temperatures and pressures in the effects are the same
- The vapor flow rates between all effects are the same

The specific steam consumption D is then with $D = G_k$:

$$\frac{D}{L_0 - L_n} = \frac{D}{\sum_1^n G_k} = \frac{D}{n \cdot G_k} = \frac{1}{n} \quad \frac{\text{kg steam}}{\text{kg vapor}}. \tag{7.2-1}$$

This means that very small energy consumptions are obtained by an increase of the number of effects.

In any real multiple effect evaporation unit, the above mentioned conditions are not fulfilled; however, the reduction of the energy consumption is substantial. In industrial plants the specific steam consumptions according to Table 7.2-1 are obtained.

Table 7.2-1 Specific steam consumption

Number of effects	1	2	3	4	5
$\frac{\text{kg steam}}{\text{kg vapor}}$	1.1	0.6	0.4	0.3	0.25

This reduction of operation cost leads to an increase of investment cost with a rising number of effects. Let us assume a minimum temperature difference $(\Delta T)_{\min}$, say 10 K, as driving force for the heat transfer with respect to an economical heat transfer area. With the overall heat transfer coefficient k, the heat flow $\dot{Q}_k$ of the effect k is given by

$$\dot{Q}_k = k \cdot A_k \cdot (\Delta T)_{\min} = \dot{G}_k \cdot r_L = \frac{(\dot{L}_0 - \dot{L}_n)}{n} \cdot r_L. \tag{7.2-2}$$

Here, $r_L = q$ is the heat of evaporation plus an amount due to superheating when the vapor is leaving a solution according to an increase of the boiling temperature in comparison to that of the pure solvent. Now it is assumed that the same vapor mass flow is generated in all effects. The total heat transfer area of n effects is

$$A = n \cdot A_k = \frac{(\dot{L}_0 - \dot{L}_n) \cdot r_L}{k \cdot (\Delta T)_{\min}}. \tag{7.2-3}$$

Let us now assume that instead of a multiple effect evaporation unit only a single evaporator with the heat transfer area A' is operated with the total temperature difference $\Delta T = n \cdot (\Delta T)_{\min}$ as driving force.

This results in (7.2-4)

$$\dot{Q} = (\dot{L}_0 - \dot{L}_n) \cdot r_L = k \cdot A' \cdot n \cdot (\Delta T)_{\min}$$

or

$$A' = \frac{(\dot{L}_0 - \dot{L}_n) \cdot r_L}{k \cdot (\Delta T)_{\min} \cdot n}. \tag{7.2-5}$$

Based on a given total temperature difference as driving force for heat transfer, the total heat transfer area of a multiple effect evaporation unit is n times greater in comparison to a single evaporator. According to Fig. 7.2-4 the investment cost is rising and the energy cost is decreasing with an increasing number of effects. The most economical number is given where the total cost is passing a minimum.

A multiple effect evaporation unit can be operated in different ways. In Figs. 7.2-1, 7.2-2, and 7.2-3, respectively, the arrangements of parallel-feed operation, forward-feed (or cocurrent of vapor and liquid) operation, and backward-feed (or countercurrent of vapor and liquid) operation are shown. The effects are numbered in the direction of the liquid. The low-concentrated solution mass flow $\dot{L}_0$ enters the first effect in which the (ideally mixed) liquid has the temperature T_{L1}, the pressure p_1, and the concentration x_1. Parallel-feed operation means that the feed is distributed to the effects. Dealing with the parallel-feed and the forward-feed operations the vapor of the first effect is condensed in the heating element of the second effect and the vapor of the second effect is used to evaporate liquid in the third effect. Note that a temperature difference ΔT_k is necessary for heat transfer and that the vapor is superheated with ΔT_1^0 above the boiling temperature of the solvent according to boiling point elevation, see Chap. 2. Therefore, the temperature of the vapor decreases according to the temperature span ΔT_1^0 and afterward the vapor is condensed at temperature T_1^0. The pressure in the second effect must be lower than in the first to evaporate the liquid with the concentration x_2 at the temperature $T_{L2} = T_1^0 - T_2$ with the consequence that the liquid entering the second stage must be depressurized. Therefore, the pressure decreases from effect to effect but the concentration increases. Starting with the first effect under an elevated pressure the design can be carried out with the result of an ambient pressure in the last effect. However, when the first effect is operated at ambient pressure, vacuum must be maintained in all following effects. Dealing with parallel-feed operation, the pressures and temperatures are decreasing with a rising number of effects. In all cases the concentration increases from effect to effect.

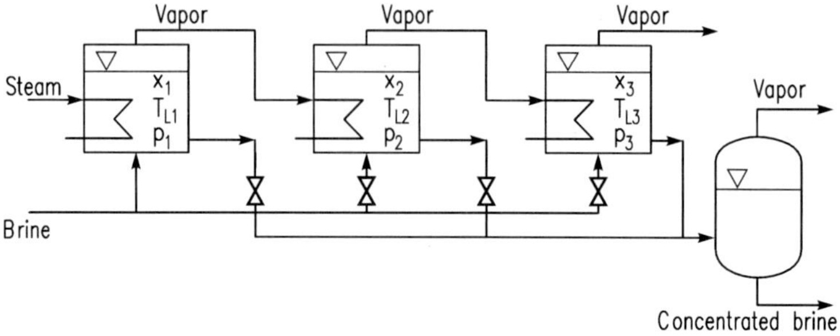

Fig. 7.2-1 Three effect evaporation unit with parallel-feed operation

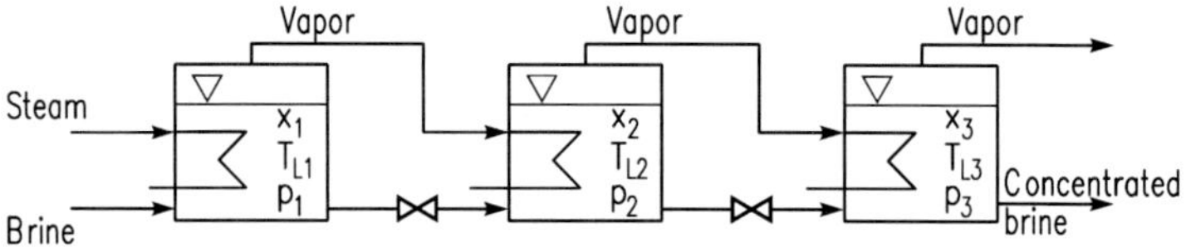

Fig. 7.2-2 Three effect evaporation unit with forward-feed operation

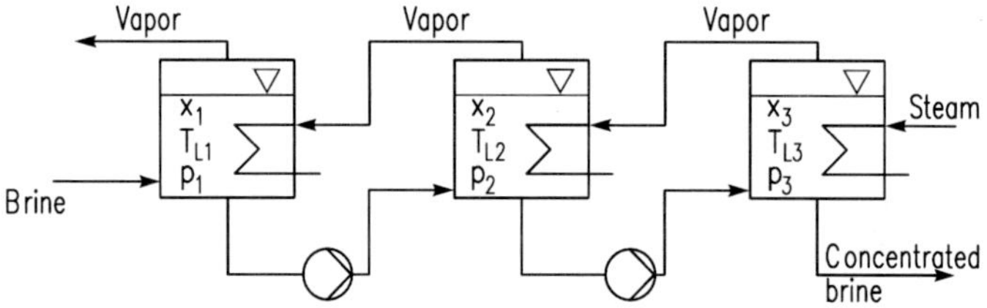

Fig. 7.2-3 Three effect evaporation unit with backward-feed operation

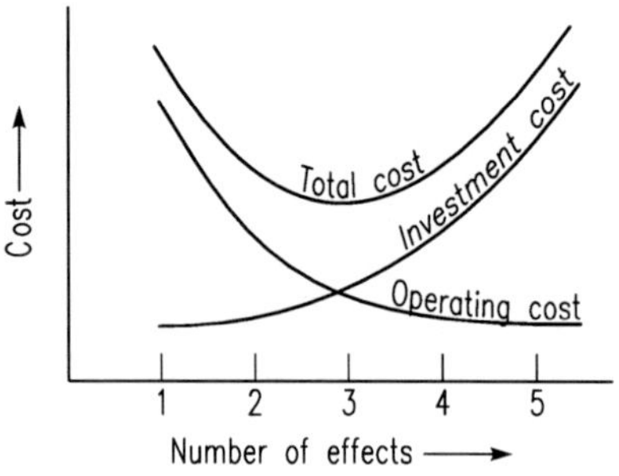

Fig. 7.2-4 Investment, operating and total costs as a function of the number of effects

Dealing with the forward-feed operation according to Fig. 7.2-2, the total liquid mass flow rate $\dot{L}_0$ is entering the first effect and the leaving liquid flow is depressurized and fed in to the second effect. The vapor of the first effect enters the heating element of the second. Liquids and vapors are moving in the same direction. The vapor of the first effect is cooled down to its condensation temperature for condensation. With respect to heat transfer the boiling temperature T_{L2} of the liquid in the second effect is ΔT_2 degrees lower than the condensation temperature T_1^0 of the vapor coming from the first effect. Depressurization of the liquid results in flash evaporation and a certain decrease of the temperature of the liquid.

Fig. 7.2-3 shows the arrangement of a countercurrent-feed operation. The vapor of the third effect is condensing in the second and that of the second in the first. This arrangement requires that the vapor of the third effect is condensing at a higher temperature than the boiling temperature of the liquid in the second effect. Therefore, the pressure of the liquid leaving the second effect must be increased by means of a pump. The liquid leaving the first effect has to be pressurized, too. Contrary to parallel-feed and forward-feed operations, temperatures and pressures are rising from effect to effect. This can be problematic for heat-sensitive substances. In the case of the other two arrangements, the consequences of a low temperature are poor heat transfer coefficients due to high viscosities and the danger of crystallization and scaling of the heat transfer element.

In multiple effect evaporation units the steam consumption $\dot{D}$ and the liquid flow rates between the various effects can be calculated by means of energy and material balances of the first effect or any other effect. The steam consumption $\dot{D}$ (here only the heat of condensation r is utilized) of a forward-feed operation unit according to Fig. 7.2-2 is given by

$$\dot{D} = \frac{\dot{L}_0 \cdot (r_{L1} + c_{L1} \cdot T_{L1} - c_{L0} \cdot T_{L0}) - \dot{L}_1 \cdot r_{L1}}{r}. \quad (7.2\text{-}6)$$

with c_{L1} as the specific heat of the liquid and r_{L1} as the heat resulting from cooling down and subsequent condensation. According to the balances already mentioned, the liquid flow leaving the effect k is

$$\dot{L}_k = \frac{\dot{L}_{k-1} \cdot (r_{L(k-1)} + r_{Lk} + c_{Lk} \cdot T_{Lk} - c_{L(k-1)} \cdot T_{L(k-1)}) - \dot{L}_{k-2} \cdot r_{L(k-1)}}{r_{Lk}}. \quad (7.2\text{-}7)$$

In the case of a countercurrent arrangement of the effects (see Fig. 7.2-3) the corresponding equations are

$$\dot{D} = \frac{\dot{L}_{k-1} \cdot (r_{Lk} + c_{Lk} \cdot T_{Lk} - c_{L(k-1)} \cdot T_{L(k-1)}) - \dot{L}_k \cdot r_{Lk}}{r} \tag{7.2-8}$$

and

$$\dot{L}_k = \frac{\dot{L}_{k+1} \cdot r_{L(k+1)} + \dot{L}_{k-1} \cdot (r_{Lk} + c_{Lk} \cdot T_{Lk} - c_{L(k-1)} \cdot T_{L(k-1)})}{r_{L(k+1)} + r_{Lk}}. \tag{7.2-9}$$

After the decision of the number of effects, the concentration spread $(x_n - x_0)$ has to be fixed to determine the flow masses of the liquid $(\dot{L}_k)$ and the vapor $(\dot{L}_{k-1} - \dot{L}_k)$ under the condition that the energy released in one effect is the same necessary for evaporation in the next. Such calculations are extensive and can be carried out in an easy way when an enthalpy–concentration diagram is available for the system under discussion.

In Fig. 7.2-5 it is shown how such calculations can be carried out in the case of a double-effect forward-feed operation. It is assumed that the vapor pressure of the solute is so low that the leaving vapor consists only of the component solvent. The enthalpy h of the pure solvent can be found at $x = 0$. In this diagram h vs. x, three curves of constant vapor pressure (p_1, p_2, p_3) are drawn. The enthalpy h at which boiling starts can be found for a given concentration x and a given pressure in the evaporation unit. The enthalpy of the solution depends on the temperature and the concentration. Therefore, isotherms (dotted lines) can be drawn in such a diagram. The deviations of the curves valid for constant temperature and for constant vapor pressure can be explained according to the effect of boiling point elevation, see Chap. 2. The dotted lines (isotherms) are a bit steeper in comparison to isobars because the boiling point elevation increases with rising concentrations.

The liquid mass flow rate $\dot{L}_0$ with the concentration x_0 is fed in to the first effect with the pressure p_1. The concentration increases due to the evaporation of the solvent. The liquid mass flow rate $\dot{L}_1$ with the concentration x_1 is throttled by means of a valve and assumes the pressure p_2. The vapor mass flow rate $\dot{G}_1 = \dot{L}_0 - \dot{L}_1$ leaving the first effect enters the heating element of the second effect and is condensed according to the progress of heat transfer. It is assumed that the entering liquid solution is undercooled and that a certain increase of its enthalpy is necessary to start boiling at pressure p_1. The heat flow rates $\dot{Q}_1$ and $\dot{Q}_2$ based on the mass flow rates $\dot{L}_0$ and $\dot{L}_1$, respectively, can be expressed by the following enthalpy differences:

$$\Delta h_1 = \frac{\dot{Q}_1}{\dot{L}_0} \text{ and } \Delta h_2 = \frac{\dot{Q}_2}{\dot{L}_1}. \tag{7.2-10}$$

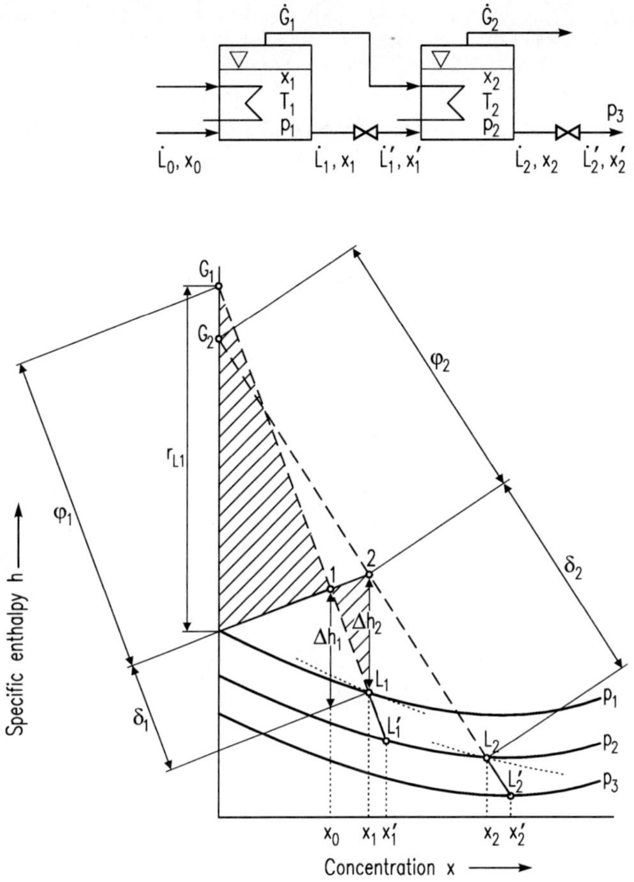

Fig. 7.2-5 Representation of a two-effect feed forward evaporation unit in an enthalpy-concentration diagram

The enthalpy of the liquid $\dot{L}_0$ is increased by the amount Δh_1 without any change of concentration. In this way point 1 can be found. This point represents the enthalpy of the two-phase system with the phases vapor at $x = 0$ and liquid L_1 with concentration x_1 and pressure p_1, compare information on balances in previous chapters. The enthalpy of the solvent vapor is the sum of the enthalpy of the liquid, the heat of evaporation, and the enthalpy according to the superheated vapor. By this way the point G_1 is fixed. According to the balances of energy and mass, the point L_1 of the liquid can be found as the intersection point of the straight line through the points G_1 and 1 and the isobar curve $p_1 = \text{const}$. Therefore, the relationship

$$\varphi_1 \cdot \dot{G}_1 = \delta_1 \cdot \dot{L}_1 \tag{7.2-11}$$

is valid.

The heat Q_2 of the vapor G_1 can be utilized in the heating element of the second effect:

$$\dot{Q}_2 = \dot{G}_1 \cdot r_{L1} . \tag{7.2-12}$$

A combination of the last two equations leads to

$$\frac{\Delta h_2}{r_{L1}} = \frac{\delta_1}{\varphi_1} . \tag{7.2-13}$$

In Fig. 7.2-5, the hatched triangles are similar to each other.

The addition of the enthalpy difference Δh_2 to the enthalpy of the liquid L_1 at $x = \text{const}$. leads to point 2 which again represents a two-phase system vapor and liquid. The enthalpy of the vapor G_2 is valid for $p_2 < p_1$ and the concentration $x = 0$. The liquid mass flow $\dot{L}_1$ is depressurized and a small mass is converted to vapor which has nearly the same enthalpy as G_2. The state of the liquid changes from L_1 with x_1 to L_1' and x_1', and this point L_1' ,x_1' is approximately the point of intersection of the straight line $\overline{G_1 L_1}$ and the isobaric curve $p_2 = \text{const}$. In the same way the point L_2' ,x_2' can be found.

In the case of a triple-effect unit, the relationship

$$\frac{\Delta h_3}{r_{L2}} = \frac{\delta_2}{\varphi_2} \tag{7.2-14}$$

is valid and for the effect k the equation

$$\frac{\Delta h_k}{r_{L(k-1)}} = \frac{\delta_{k-1}}{\varphi_{k-1}} \tag{7.2-15}$$

can be written.

It is difficult to find the appropriate number of effects for a given concentration spread $(x_n - x_0)$ because the iteration is a bit tedious and requires an appropriate computer program. Applying a diagram as shown in Fig. 7.2-5, it is recommended to start at concentration x_n.

7.3 Condensers

Vapors or mixtures of vapor and gas consisting of one or several components are cooled down with the objective to liquefy condensable components. It can happen that the liquid consists of two or more liquid phases according to heterogeneous azeotropes. This results in emulsions which have to be separated in decanters or in centrifuges (compare with the section "Extraction"). In the following, it is

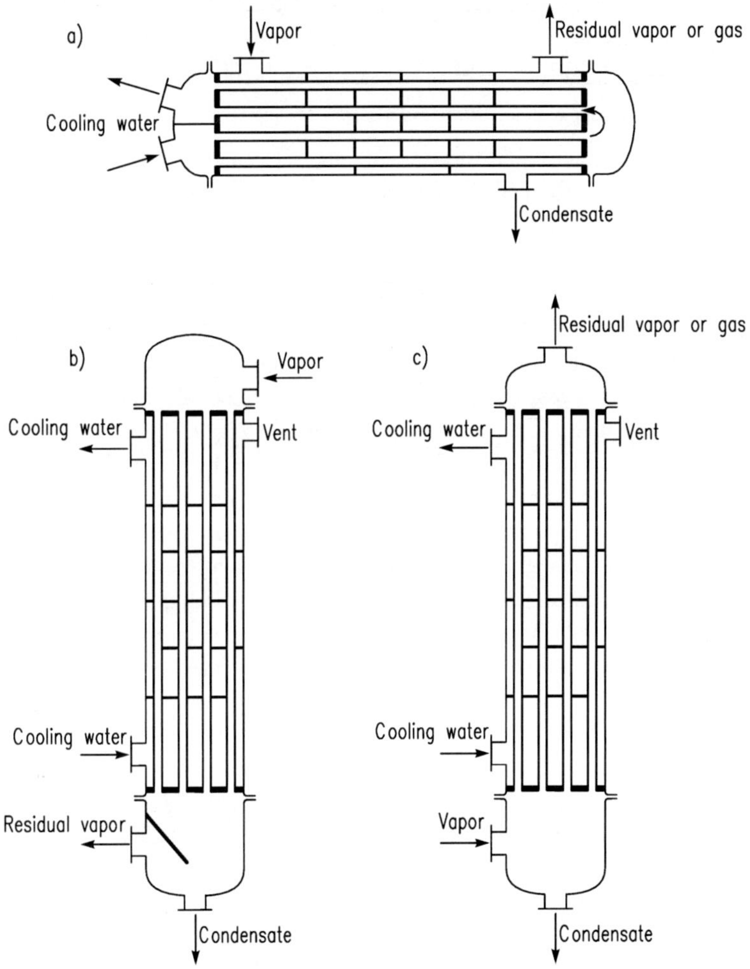

Fig. 7.3-1 Condensers operated with cooling water. (**a**) Condensation on the outside surfaces of horizontal tubes; (**b**, **c**) condensation on the inside surfaces of vertical tubes in cocurrent (**b**) or countercurrent (**c**) flow of vapor and condensate

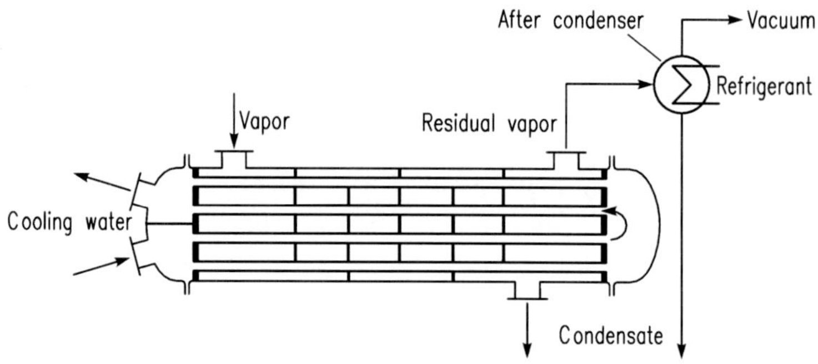

Fig. 7.3-2 Condenser with an after condenser

assumed that only one liquid phase is obtained. Besides condensers with cooled surfaces, sometimes direct or quench condensers are used. Quenching means that a cold liquid (in the rule the same liquid as the condensate) is brought into direct contact with the vapor. Direct contact condensers are simple and nonexpensive apparatus; however, only surface condensers (shell and tube) are in the position to avoid any mixture between the cooling medium and the product. Principally speaking, shell and tube apparatus already shown in the section "Evaporators" can be used as surface condensers. Vapor or steam is condensed on the inside walls of the tubes or outside and the condensate is running downward. In Fig. 7.3-1 surface condensers are shown. Vapor or vapor–gas mixtures are condensed by heat transfer to cooling water or refrigerants. Vapor is condensed on the outside surface of the tubes, see Fig. 7.3-1a. Figure 7.3-1b, c shows heat exchanges where the vapor is condensed inside of the tubes and the cooling medium is flowing outside. As a rule the medium with the greatest potential of fouling is passed through the tubes which can be cleaned more easily. It is important to introduce the cooling at the bottom to remove air or other permanent gases which can be desorbed from the cooling medium and accumulate. Dealing with the construction according to (b) condensate and residual vapor are moving downward in cocurrent flow whereas in the case (c) a countercurrent flow occurs. Note that flooding can take place when a vapor with a big amount of noncondensable gases flows upward against a downward directed liquid film. Since noncondensable gases reduce the heat transfer after accumulation they have to be removed by a nozzle which is close to the coldest spot of the heat exchanger because there is the lowest partial pressure of condensables. The exact position of this nozzle depends on the densities of the vapor and the inert gases. In most cases well water ($\approx$10°C) or cooling-tower water ($\approx$ 25°C in central Europe) is used as cooling medium. Some plants are equipped with cold water ($\approx$ 5°C) provided refrigeration. Sometimes a cooler operated with a refrigerant is installed after the main condenser to reduce the contents of very volatile components. Figure 7.3-2 shows a condenser with a withdrawal of residual vapor which is cooled down in such an after condenser. With respect to the shortness of cooling water (wells, rivers, lakes) the use of air cooling has extended. Figure 7.3-3 shows a forced-draft air-cooled heat exchanger with finned tubes. The operation of cooling towers is problematic during hot and humid days. Then either the reduction of the production rate or the operation of additional refrigerators is necessary.

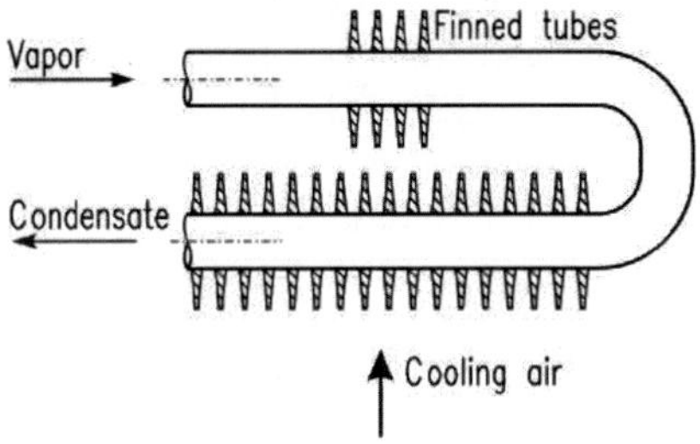

Fig. 7.3-3 Finned tube condenser operated with cooling air

7.4 Design of Evaporators and Condensers

Before starting with the calculation of the heat transfer area of such heat exchangers it is recommended to discuss temperature profiles in the apparatus because product damage at hot spots of the transfer area and product losses due to insufficient condensation can reduce the economics of the process. With respect to the reduction of the boiling temperature at low pressure, vacuum evaporators are often used for the evaporation of heat-sensitive organic products. Condensers operated at elevated pressures are advantageous because vapors can be condensed either without outside cooling or at such temperatures that water from cooling towers or ambient air can be used. In Fig. 7.4-1 the temperature profile of a vertical evaporator tube is depicted above and the corresponding profile of a vertical condenser tube is illustrated below. The lower the pressure (vacuum) the more the profile moving downward and vice versa. Note that for a given pressure at the top of tubes the boiling temperatures can differ within a tube due to different hydrostatic pressures. Therefore, the temperature profiles shown in Fig. 7.4-1 are only valid for a certain level and a given pressure of the system. The heat transfer area $A = n_r \cdot \pi \cdot d_a \cdot L$ of tubes with the outer diameter d_a and the length L is given by

$$\dot{Q} = k \cdot A \cdot \Delta T \text{ with } \frac{1}{k} = \frac{1}{\alpha_a} + R_a + \frac{s}{\lambda_s} + R_i + \frac{1}{\alpha_i} \tag{7.4-1}$$

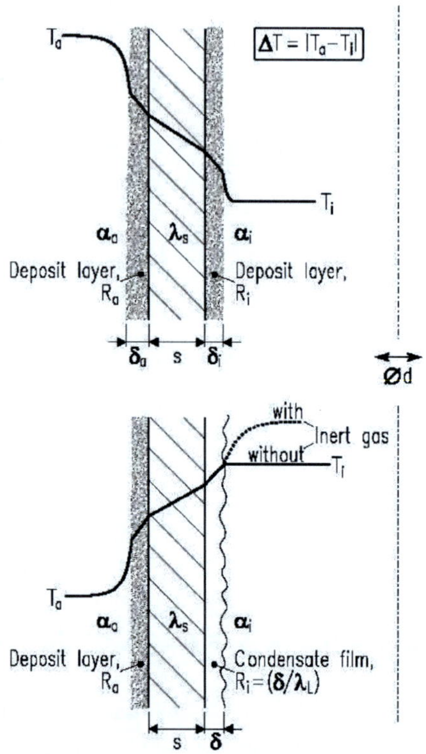

Fig. 7.4-1 Temperature profile around a tube wall: (*above*) Tube of an evaporator; (*below*) tube of a condenser

with $\Delta T = |T_a - T_i|$ as the mean difference, compare with Sect. 4.3.3.2. The heat flow $\dot{Q}$ rises with increasing overall heat transfer coefficients k and temperature differences ΔT. The heat transfer coefficients α_a and α_i depend on the properties of the fluid and its velocity. The ratio s/λ_s of the thickness s of the tube wall based on the conductivity λ_s of the solid material of construction can be dominant (glass, plastic) in comparison to the other heat transfer resistances. With respect to Fig. 7.4-1 above it is assumed that both the outside and the inside walls of the tube are covered with a fouling film of the thicknesses δ_a and δ_i, respectively. In the case that these thicknesses and the conductivities of the fouling materials are known the fouling resistances $R_a = \delta_a/\lambda_a$ and $R_i = \delta_i/\lambda_i$ can be calculated (Müller-Steinhagen 2000; Bohnet 2003). However, as a rule empirical values of these resistances are taken because a more theoretical approach is difficult. On the other side, heat transfer coefficients can be calculated by means of equations which can be found in Chap. 4 (see also the references of this chapter). However, calculation procedures are not available for every geometry of evaporators and condensers and sometimes the properties of the fluids (thermal conductivity, viscosity) are not known. Therefore, some data of orientation are presented in Table 7.4-1. As a rule heat transfer coefficients are increasing in the following order: coefficients for gases under vacuum, normal pressure, high pressure, highly viscous liquids, low viscous liquids without phase change, evaporation, film and drop condensation.

Table 7.4-1 Approximate heat transfer coefficients

		Heat transfer coefficients $(\mathrm{W/m^2 \cdot K})$
Fluids without phase change	Gases	70
	Organic liquids	1,500
	Water	2,000
Evaporation	Organic liquids	800 – 1,000
	Water	1,500 – 2,000
Condensation	Organic liquids without inert gases	1,500
	Organic liquids with inert gases	70 – 1,500
	Water	7,000

The heat transfer coefficients for water drop condensation (the drops of the condensate are separated from the cooling surface without the formation of a liquid film) can be $\alpha \approx 260000\ \mathrm{W/(m^2K)}$. These coefficients are four orders of magnitude higher than the coefficients valid for gases at ambient pressure ($\alpha \approx 20\ \mathrm{W/(m^2K)}$). The heat transfer coefficients of water are much higher than those for organic liquids because its thermal conductivity is by a factor ≈ 5 greater than many data valid for organic substances. It is important to note that the calculation of the heat flux density $\dot{q} = \dot{Q}/A = \alpha_\mathrm{i} \cdot \Delta T_\mathrm{i} = \alpha_\mathrm{a} \cdot \Delta T_\mathrm{a}$ is problematic because the temperature difference ΔT, the fluid properties, and the heat transfer coefficients as local quantities are different from point to point. Unfortunately, little is known about the changes of these parameters from the bottom to the top of a tube. This is especially true for evaporator tubes in which phase changes take place (Wettermann and Steiner 2000; VDI-Wärmeatlas 2002).

The design of natural convection evaporators is difficult because a complex interrelationship between the liquid circulation rate due to density differences and heat transfer coefficients exists. The circulation flow rate depends on the amount of evaporated liquid and is not controlled by an external device as in forced circulation evaporators. With respect to the hydrostatic pressure at the bottom of the evaporator tubes the liquid entering the heat transfer zone is undercooled. In this section, the heat transfer is poor due to small velocities of the liquid. In the upper section where evaporation takes place the fast two-phase flow favors heat transfer. On one hand, increasing circulation rates improve the heat transfer in the entire tube; however, on the other hand, the preheating zone at the bottom is extended. Therefore, the impact of the density of circulation on the heat transfer is ambiguous. An exact calculation of the heat transfer of such natural convection evaporators requires the determination of the length of the preheating and the subsequent two-phase evaporation zone and the calculation of the heat transfer coefficients in these different zones in more detail (Fair 1960; Arneth 1999; Arneth and Stichlmair 2001). As illustrated in Figs. 7.4-2 and 7.4-3 the flow in an evaporator tube is very complex and this is also true for the temperature field. In Fig. 7.4-2 it is assumed that a boiling liquid with the temperature T_s enters a tube at the bottom. The pressure decreases with increasing height and this is also true for the boiling temperature of a single component liquid. In the case of a mixture the volatile components are mainly evaporated with the result that heavy boiling components in the liquid are enriching and the boiling temperature rises in spite of a decrease of the pressure. Let us now discuss the evaporation of a mixture with boiling temperatures of the various components in a narrow range. After a nearly complete evaporation the boiling temperature at the top can be different in comparison to this temperature at the bottom because of the hydrostatic pressure. Furthermore, the vapor can be

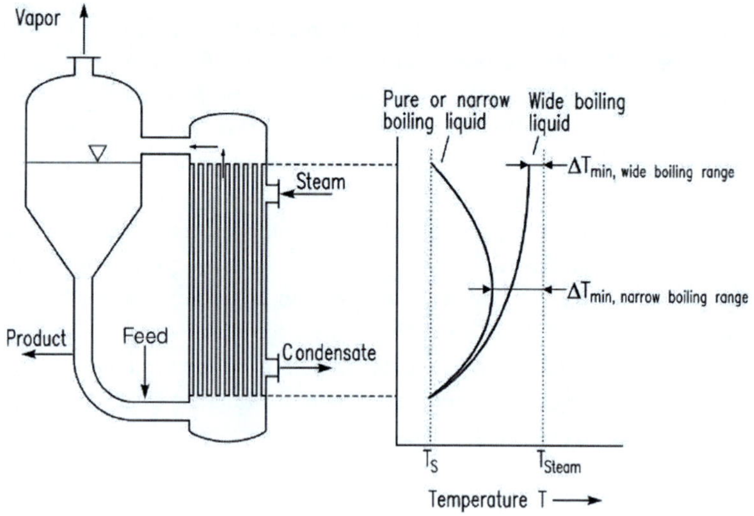

Fig. 7.4-2 Possible temperature profiles of the liquid in the tubes of a natural convection evaporator

superheated. Dealing with a mixture of components boiling in a wide temperature range the temperature in a tube is changing steadily.

According to Fig. 7.4-3 a subcooled liquid is heated in the low part of a tube. After the beginning of boiling vapor bubbles lead to a bubble flow which results in plug and slug flow with increasing vapor fraction and finally annular and spray flow can be observed. It is understandable that the heat transfer coefficients are different in these zones because of different flow patterns. Therefore, approximate coefficients are used in industrial design, see Table 7.4-1. In any case it is recommended to draw diagrams in which both the temperature of the product and the temperature of the heating or cooling agent are plotted against the heat flow $\dot{Q}$. In Fig. 7.4-4 this is shown for a heat exchanger and a condenser.

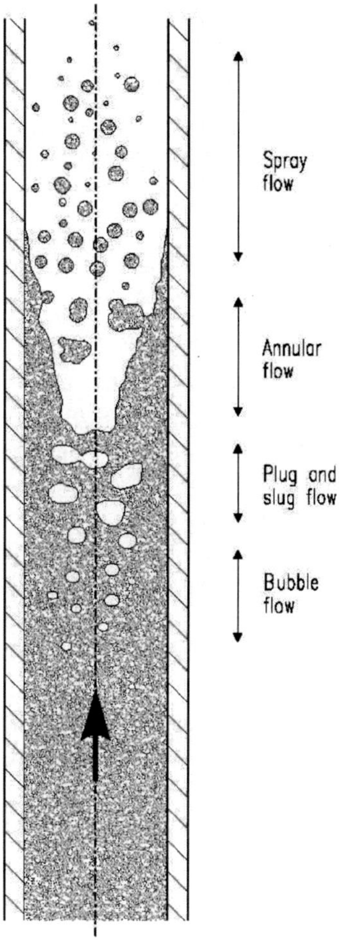

Fig. 7.4-3 Flow patterns in a vertical evaporator tube

The calculation of the heat transfer coefficient and the driving temperature difference is not difficult for a one tube pass product cooler, see Fig. 7.4-4 left which illustrates a cocurrent flow of the product flow and the flow of the cooling agent. However, things are much more complex for condensers with an entering stream of a superheated vapor. In such exchangers the heat transfer and the heat flow $\dot{Q}$ are appreciably reduced by an increasing amount of noncondensable gas in the vapor. Then it is important to remove such inert gases by means of a refrigeration cooler, of an active carbon adsorber or a vacuum pump, for instance a liquid piston type of rotary compressor filled with a high boiling liquid.

It will be shortly discussed how high-boiling and heat-sensitive substances can be separated from viscous residuals with a very high boiling temperature. Especially, falling film evaporators with or without a rotor and operated under vacuum are

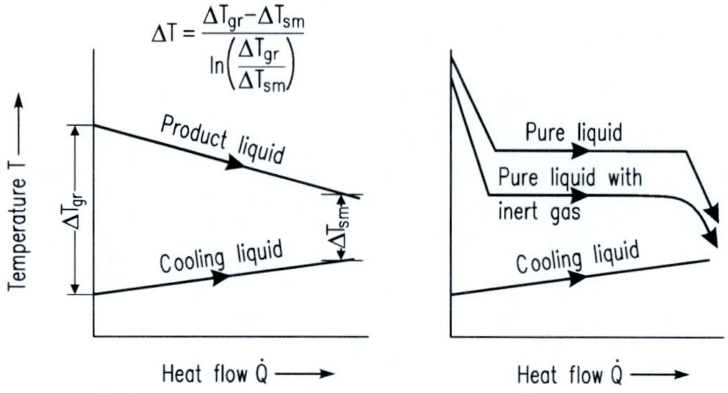

Fig. 7.4-4 Temperature–heat flow diagrams for a product cooler (*left*) and a condenser (*right*)

appropriate tools for such separations. In Fig. 7.4-5 a falling film evaporator is illustrated. The high-boiling liquid is withdrawn at the bottom and fed in to the second falling film evaporator which is equipped with a rotor system to distribute the liquid and to obtain a very thin liquid film. In this apparatus the decomposition of the liquid and the cracking of components are widely reduced with respect to low temperatures and short residence times. The feed is separated into condensates 1 and 2 and the final residual as the bottom product of the second evaporator. As a rule the vapor–gas mixture which leaves the water-cooled condensers is further cooled down by means of a refrigerant (here not illustrated). Vacuum pumps or steam ejectors are used to maintain low pressure in the first condenser and even lower pressure in the second one. The steam-jet ejector illustrated in Fig. 7.4-6 can reduce the pressure down to 100 Pa when five ejectors are connected in series or stages. The purity of the condensate can be deteriorated by the entrainment of drops. Therefore, much space is recommended for the apparatus segment where the vapor is leaving to reduce the vapor velocity. Sometimes impingement separators are installed.

7.5 Thermocompression

Besides multiple effect evaporation, thermocompression is employed to save energy. The vapor of an evaporator can be used for its own production if the leaving vapor is compressed to such a degree that the condensing temperature of the compressed vapor is higher than the boiling temperature of the solution in the evaporator. Reciprocating (small units), rotary positive displacement as well as centrifugal or axial flow compressors are used. In the case of steam thermocom-

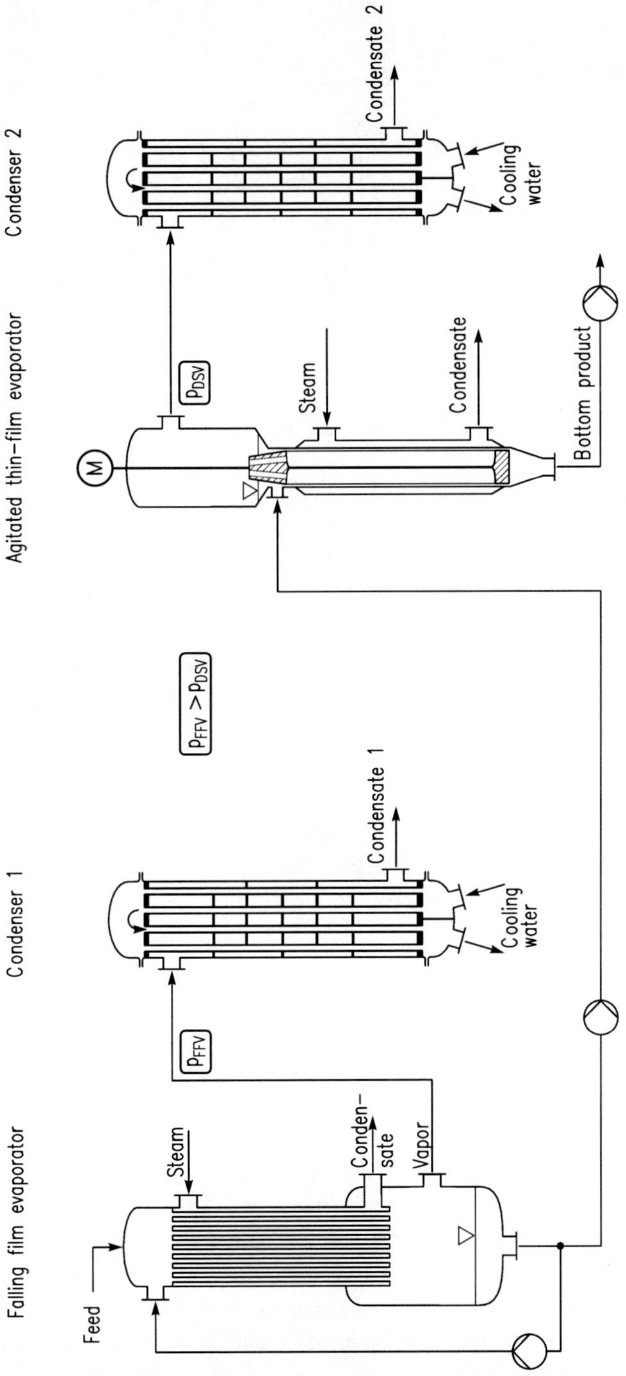

Fig. 7.4-5 Combination of a falling film evaporator and an agitated thin-film evaporator

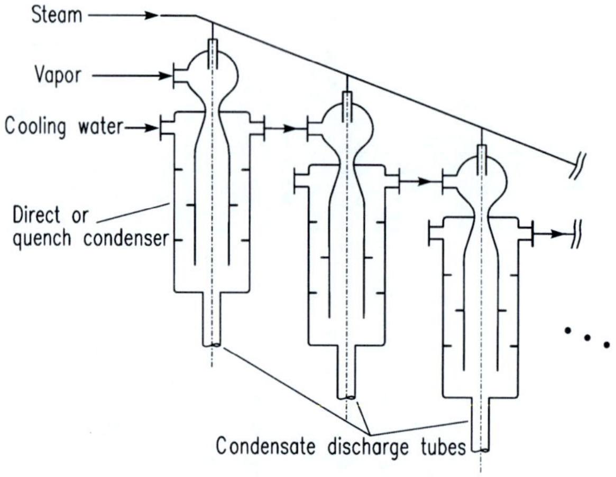

Fig. 7.4-6 Multiple stage steam ejectors with discharge tubes

pression applied for aqueous solutions both the vapor produced in the evaporator and the jet steam can be condensed; however, the compression efficiency is low. A certain problem is the fouling of mechanical compressors because the entrainment of small drops and their evaporation on the rotor blades leave solid particles behind. Additional energy is necessary if the heat provided by the vapor flow rate $\dot{G}_1$ is not sufficient for its production. In the case of aqueous solutions fresh steam is added to the vapor flow. The scheme of a thermocompression is illustrated in Fig. 7.5-1 on the left side. The temperature T is plotted against the entropy s in a T–s diagram, compare Sect. 7.1. In this diagram three isobaric curves for the pressures p'_a, $(p'_1)_{min}$ and p'_1 are drawn. The superheated vapor with the temperature $T_1^0 + \Delta T_1^0$ (ΔT_1^0 is the superheating) leaving the evaporator has the pressure p_1 which is increased to the pressure p'_1 by means of a compressor. The superheated vapor (point 4 in the T–s diagram) is cooled down. Condensation starts at point 5 and is finished at point 6. The pressure $(p'_1)_{min}$ is the minimum pressure which allows heat transfer to the boiling solution with the pressure p_1 and the temperature T_{L1}; however, in this case the heat transfer area would be infinite with respect to the driving temperature difference $\Delta T_1 = 0$. Therefore, the pressure p'_1 must be higher than $(p'_1)_{min}$ to obtain an economical heat transfer area in the evaporator. In Fig. 7.5-1 it is assumed that the compressor is operated at constant entropy or reversibility is given. (In real compression, however, this is never true with the consequence that the entropy s is increasing, see dotted line in Fig. 7.5-1). The increase of s is small in compressors with a high efficiency. The work W consumed by the compressor is equal to the difference $h_4 - h_3$ ($W = h_4 - h_3$) of the enthalpies at points 4 and 3, respectively. This difference can also be expressed by the area within the lines through the points 1, 2, 3, 4, 5, and 6. The heat transferred

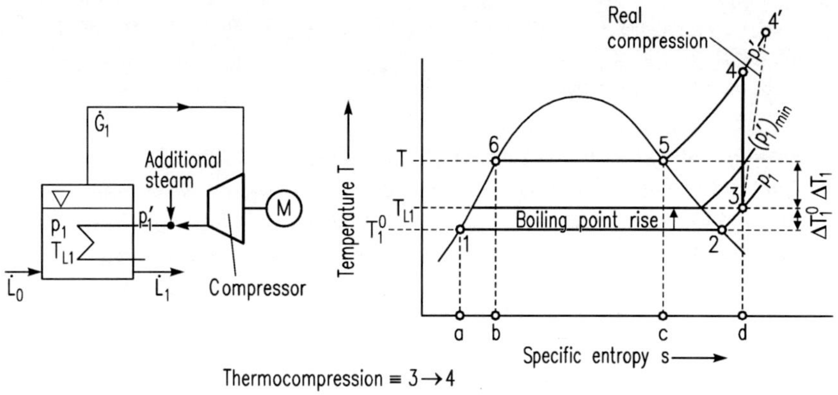

Fig. 7.5-1 Thermocompression unit (*left*) and presentation of the thermocompression in a temperature-specific entropy diagram (*right*)

in the heating element consists of two parts, e.g., the heat of superheating according to the area "c d 4 5" and the heat of condensation (area "b c 5 6").

The economics of thermocompression is dependent on the performance ratio ε which is the ratio of the utility energy or heat based on the work of compression:

$$\varepsilon = \frac{\text{Heat of condensation} + \text{Heat of superheating}}{\text{Work of compression}} \quad \text{or}$$

$$\varepsilon = \frac{\text{areas "bc56" and "cd45"}}{\text{area "123456"}}.$$

The higher this performance ratio the more economical is the process. A small ratio p_1'/p_1 leads to a small work; however, this must be balanced against a large heat transfer area as the result of a low condensation temperature close above the boiling temperature. The most economical pressure ratio can be found when the sum of the investment cost and the energy cost are plotted against this ratio, compare Fig. 7.2-4. In this case the investment cost or the depreciation cost decreases with an increasing ratio $p_1'/(p_1')_{min}$; however, the energy costs are rising. As a rule the total costs are passing through a minimum.

7.6 Evaporation Processes

Some evaporation processes are known for which the energy costs are very dominant because the liquid to be evaporated is very cheap. This is true for saline water as the basis for the production of potable water. Seawater is cheap and potable water is the only product in most desalination processes. Multiple effect evaporation is applied to reduce the energy cost. As a rule such evaporation processes are more economical for seawater with 40,000 ppm salt than the reversed osmosis or desalination by freezing. Especially, this is true when waste heat from a power station or solar energy is available.

Problems encountered in seawater evaporators are the precipitation of salt, mainly calcium carbonate and calcium bicarbonate with increasing temperature and the subsequent fouling of heat transfer areas. Fouling can be reduced by the addition of polyelectrolytes, for instance special phosphates. Seawater is deaerated to remove dissolved gases (O_2, CO_2). There are two kinds of evaporation processes, e.g., the multiple effect evaporation and the multistage flash evaporation. In Fig. 7.6-1 a multiple effect evaporation unit with falling film evaporators is illustrated. The preheated seawater is fed in to the first effect where it is heated with fresh steam. The vapor enters the second effect and its heat is transferred to seawater which flows from the first effect to the second according to a cocurrent process. The last condenser is cooled with seawater. Such multiple effect units are installed with eight up to ten effects which result in a consumption of ≈ 0.2 kg fresh steam necessary for 1 kg of potable water.

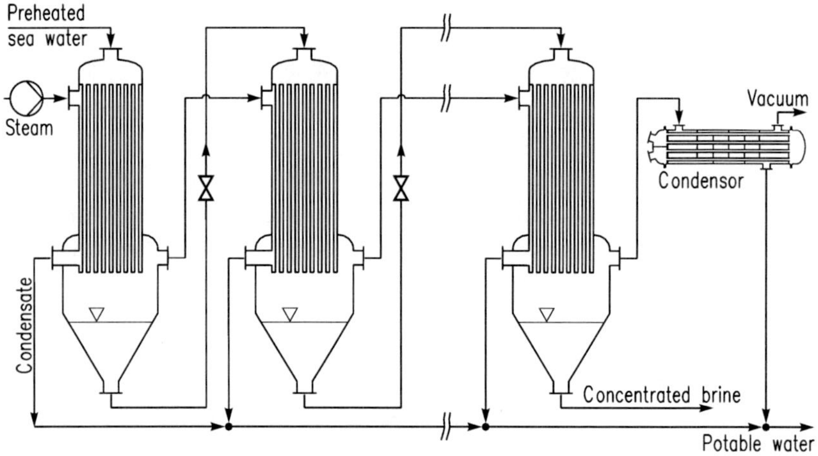

Fig. 7.6-1 Multiple effect evaporation unit with falling film evaporators for desalination of seawater

With respect to the incrustation of heat transfer areas of multiple effect plants, multistage flash evaporators can be more favorable. In Fig. 7.6-2 such a unit is illustrated. As a rule 20 stages are installed. The seawater is preheated in the various stages and is fed into the first stage after the passage through a final heater. A certain part of the seawater is flash evaporated by depressurization and is con-

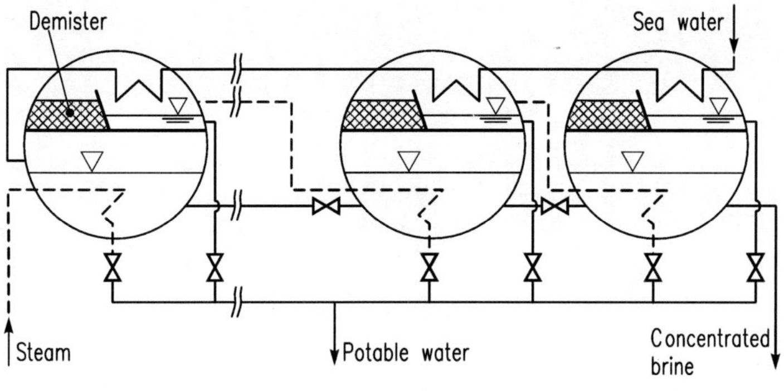

Fig. 7.6-2 Multistage flash evaporation unit for seawater (without preheating)

densed on the above installed heat transfer area. Heat is transferred to the brine passing the tubes. The seawater enters the next stage after the passage of a special overflow sluice. The condensate collected in the various stages is fed into the container for potable water. The brine with the elevated salt concentration is returned to the sea. Such multistage flash evaporation units are advantageous because scaling of heat transfer areas can be avoided to a large extent. The consumption of prime steam is approximately 0.12 kg steam based on 1 kg potable water.

Symbols

A	m^2	Heat transfer area
c	kg/m^3	Concentration
$\dot{D}$	kg/s	Steam flow
$\dot{G}$	kg/s	Vapor flow
h	kJ/kg	Specific enthalpy
k	$W/(m^2 \cdot K)$	Overall heat transfer coefficient
$\dot{L}$	kg/s	Flow of solution
n		Number of effects
n_r		Number of tubes
p	Pa	Pressure
$\dot{Q}$	W	Heat flow
$\dot{q}$	W/m^2	Heat flow density
R	$m^2 \cdot K/W$	Fouling resistance
r	kJ/kg	Heat of evaporation
r_L	kJ/kg	Heat of (evaporation + superheating)
s	m	Thickness of a wall

Greek Symbols

α	$W/(m^2 \cdot K)$	Heat transfer coefficient
δ	m	Thickness of a layer
ε		Performance ratio
λ	$W/(m \cdot K)$	Heat conductivity
τ	s	Residence time

Indices

a	Outer
i	Inner
k	kth effect
L	Liquid
n	nth effect
S	Solid
0	Entrance in the first stage
1	Outlet of the first stage

8 Crystallization

Crystallization is the transformation of one or more substances from the amorphous solid, liquid, or gaseous phase to the crystalline phase. Above all, crystallization is of great importance as a thermal separation process for the concentration or purification of substances from solutions, melts, or the vapor phase.

8.1 Fundamentals and Equilibrium

A phase, e.g. a solution has to be supersaturated so that new crystals can arise or existing crystals can grow. Supersaturation can be achieved by cooling a solution or by evaporation of the solvent. This is called cooling and evaporative crystallization. For vacuum crystallization flash evaporation is used to create supersaturation. In this case cooling and evaporation superimpose.

Sometimes a drowning-out medium is added to a solution. It reduces the solubility of the solute and hence leads to supersaturation. This is called drowning-out crystallization. The solubility of many aqueous solutions of inorganic salts can be reduced by the addition of organic solvents (e.g., acetone, methanol). In reactive crystallization two or more reactants form a product which is less soluble and therefore crystallizes. For example, reactions between an acid and a base lead to the precipitation of a solid salt. This is called precipitation crystallization. However, it should be mentioned that this term is neither clearly defined nor uniformly used.

Although there is no strict and universally valid distinction between a "solution" and a "melt," it is purposive to differentiate between crystallization from solution and from melts. The process is called crystallization from the melt, if the crystallization temperature is close to the melting temperature of the crystalline phase. In solution crystallization this is not the case. While the kinetics of crystallization from solution is often limited by mass transport, the limiting factor of crystallization from melt is often heat transport.

Sometimes solutes in a solution have to be concentrated by freezing the solvent. This is called freeze crystallization or freeze concentration.

A. Mersmann et al., *Thermal Separation Technology: Principles, Methods, Process Design*, VDI-Buch, DOI 10.1007/978-3-642-12525-6_8,
© Springer-Verlag Berlin Heidelberg 2011

To produce crystals of a certain crystal size distribution, crystal shape, and purity, the local and average supersaturation as well as the distribution and the residence time of the crystals in the supersaturated solution have to be controlled. In general, the crystals have a higher density than the solution. Therefore an upward flow in the crystallizer is required to keep the crystals suspended. This flow can be generated by a stirrer or by external circulation with a pump. It will be shown later how fluid mechanics influence the grain size of crystals.

In the References section a list of general textbooks and journal articles is provided about fundamental aspects like modeling, nucleation and growth (Garside et al. (2002), Lacmann et al. (1999), Randolph and Larson (1988)), about industrial crystallization (Hofmann (2004), Mersmann (2001), Mullin (2004)), about precipitation and colloids (Israelachvili (1995), Lyklema (1991), Söhnel and Garside (1992)), and about melt crystallization (Arkenbout (1995), Ulrich and Glade (2003)).

8.1.1 Fundamentals

Phase equilibrium of solid–liquid systems, as well as solubility and melt diagrams, are given in Chap. 2.

Crystals are solids with a three-dimensional periodical arrangement of elementary structural units (atoms, ions, molecules) in space lattices. Due to its highly regular structure the crystal differs from amorphous matter. The regular structure is caused by various bond forces. The resulting crystal types are shown in Table 8.1-1 together with typical properties and exemplary substances.

The crystal lattice of an ideal crystal consists of completely regular elementary cells. The elementary cell defines a coordinate system with axes x, y, and z and the angles α, β, and γ. Crystals of different substances vary in the elementary lengths a, b, and c and in the angles. Figure 8.1-1 shows an elementary cell of that kind. According to the spatial periodic arrangement of the structural units, seven crystalline systems (Table 8.1-2) can be differentiated.

The shape of a regularly built crystal is not entirely defined by the lattice but also by the growth rates of its faces. The different types of crystal shapes are called prismatic, acicular, dendritic, straticular, or – in the case of uniform growth in all directions – isometric.

By appropriate choice of temperature, supersaturation, solvent, and additives, it is often possible to influence crystal habit. The orientation of crystal faces with respect to the lattice is defined by the proportion h:k:l of the reciprocal values of their axis intercepts, see Fig. 8.1-1. The quantities h, k, and l are the so-called Miller indices, for which the notation (hkl) is common.

Table 8.1-1 Crystal types

Crystal type	**Structural units**	**Lattice forces**	**Properties**	**Examples (bond energy in kJ/mol)**
Metal lattice	Atomic core with free outer electrons	Metallic bond	Low-volatile, high electric and thermal conductivity	Fe (400) Na (110) Brass
Ion lattice	Ions	Ionic bond (Coulomb force)	Low-volatile, nonconductor, conductive in a melt, mostly soluble	NaCl (750) LiF (1000) CaO (3440)
Atom lattice	Atoms	Atomic bond = covalent bond (shared electron pairs)	Low-volatile, nonconductor, insoluble, very hard	Diamond (710) SiC (1190) Si, BN
Molecule lattice	Molecules	van der Waals forces (induced dipoles) consistent dipoles (e.g., hydrogen bridges)	High volatile, nonconductor	CH_4(10) I_2 $SiCl_4$

Table 8.1-2 Crystal systems

Crystal system	Elemental lengths	Shaft angles
1) Triclinic	$a \neq b \neq c$	$\alpha \neq \beta \neq \gamma$
2) Monoclinic	$a \neq b \neq c$	$\alpha = \gamma = 90° \neq \beta$
3) (ortho)rhombic	$a \neq b \neq c$	$\alpha = \beta = \gamma = 90°$
4) Tetragonal	$a = b \neq c$	$\alpha = \beta = \gamma = 90°$
5) Hexagonal	$a = b \neq c$	$\alpha = \beta = 90°; \gamma = 120°$
6) trigonal–rhombohedral	$a = b = c$	$\alpha = \beta = \gamma \neq 90°$
7) Cubic	$a = b = c$	$\alpha = \beta = \gamma = 90°$

In general, real crystals contain inhomogeneities (inclusions of gaseous, liquid, or solid impurities) and lattice defects (imperfections, dislocations, grain boundaries, and distortions). They also deviate from the ideal forms, because corners and edges are abraded due to mechanical wear in the crystallizer. The surfaces are often impure due to adhering residues of mother liquor.

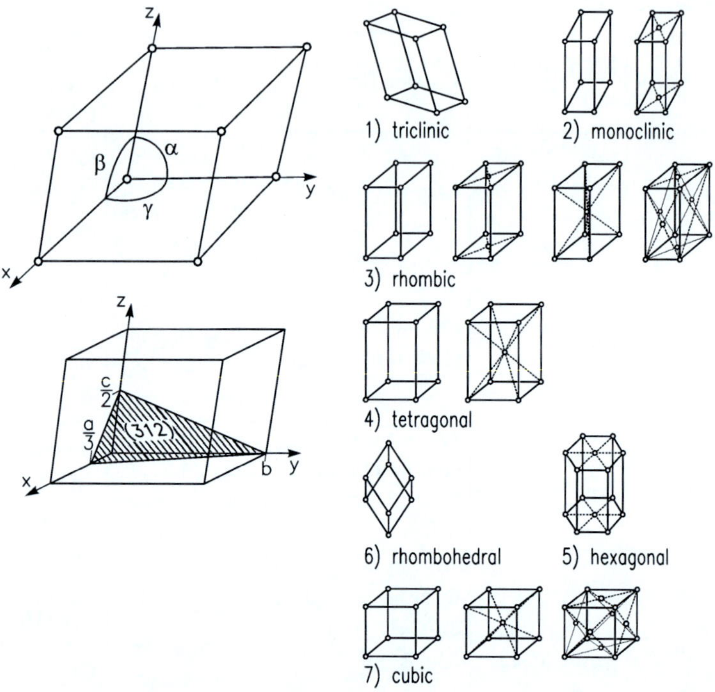

Fig. 8.1-1 Elementary cell (*upper left*); various crystal systems (*right*); explanation to Miller indices (*lower left*)

It will be shown later that for real crystals the abrasion behavior determines the quality of the product, e.g., the particle size distribution and the particle shape. The abrasion behavior depends on the following physical crystal properties:

- Elasticity modulus E
- Shear modulus μ
- Fracture resistance $(\Gamma/(K')$

The elasticity modulus and the shear modulus are linked through Poisson's number ν_c ($\nu_c \approx 1/3$)):

$$\nu_c = \frac{E}{2 \cdot \mu} - 1 . \qquad (8.1\text{-}1)$$

For a first approximation it is sufficient to determine one of the two moduli experimentally. The fracture resistance $(\Gamma/(K')$ can be estimated with the equation (Orowan 1949)

Table 8.1-3 Characteristic strength values of crystals

Elasticity modulus E	$10^{10} - 5 \times 10^{10}$ $[\mathrm{N/m^2}]$ or $[\mathrm{J/m^3}]$
Shear modulus μ	$10^{9} - 10^{10}$ $[\mathrm{N/m^2}]$ or $[\mathrm{J/m^3}]$
Fracture resistance (Γ/K')	2–20 $[\mathrm{J/m^2}]$

$$\left(\frac{\Gamma}{K'}\right) \approx 1.7 \cdot E \cdot \left(\frac{1}{n \cdot \tilde{c}_{\mathrm{c}} \cdot N_{\mathrm{A}}}\right)^{1/3}. \tag{8.1-2}$$

Herein n is the number of atoms in a molecule. In Table 8.1-3 guideline values of the quantities E, μ, and (Γ/K') are shown for inorganic and for organic crystals.

8.1.2 Equilibrium

Important fundamentals of solution equilibria have already been introduced in Chap. 2. In the following it will be discussed how solid–liquid equilibria can be determined experimentally. It is necessary to wait until the equilibrium of a solution with its precipitate is reached, before its concentration and the temperature are measured. However, it has to be considered that in the case of multicomponent precipitates with corollary components or impurities, the occurring concentration of the noncrystallizing corollary components depends on the mass of precipitate weighed in. In this case, the added amount of precipitate can affect the result. Additionally, concentrations may be difficult to measure. In binary systems concentrations can best be determined by measuring the density of the solution. In case of multicomponent equilibria it is useful to weigh the exact masses of the individual components and create an undersaturated solution. Via temperature variation (heating or cooling) or variation of the "solvent" component, the solubility curve is exceeded in one or the other direction. It should be mentioned that the unknown concentration of a solution can be determined by complete evaporation of the solvent and weighing the mass of the dry crystals.

The experimental determination of equilibria becomes much more complicated in the case of polymorphic and pseudopolymorphic multiple component systems. Among these are hydrates, solvates, and racemates. Often equilibrium is only reached after a very long period of time, because at first amorphous substances or unstable modifications are formed after the induction of supersaturation. This is mainly observed if the supersaturation is high. Measuring the concentration in a fluid and especially a solid phase is often complicated. Since phase transitions often release or consume latent heats, differential thermal analysis (DTA) and differential scanning calorimetry (DSC) are useful tools.

8.2 Crystallization Processes and Devices

Depending on the method of creation of supersaturation, it is common to distinguish four different kinds of crystallization from solution:

- Cooling crystallization
- Evaporative crystallization
- Drowning-out/salting-out crystallization
- Reactive crystallization

The so-called vacuum crystallization is a superposition of cooling and evaporative crystallization. The boundaries between crystallization by displacement media and reactive crystallization can be difficult to distinguish, depending on how and to what extent a third substance reacts chemically with components in the solution. Crystallization can occur if a strong electrolyte is added (salting out). The term "precipitation" is used when dealing with a very fast, usually hard to control crystallization during which a large number of seed crystals is formed.

8.2.1 Cooling Crystallization

The process of cooling crystallization can always be used, if the solubility of the substance, which is to be crystallized, considerably increases with temperature, see Figs. 8.2-1 and 8.2-2. Aqueous solutions of potassium, sodium, and ammonium nitrate as well as copper sulfate are typical examples. The hot, undersaturated solution is fed into the crystallizer and then cooled.

A simple method of batch operation is cooling the solution with a constant cooling rate. This is not optimal however, because at the beginning of the cooling process either no or just little surface of the seeding material is available and high supersaturation with subsequently strong nucleation occur. At the end of the cooling process the crystals have a large surface but they grow only very slowly due to low supersaturation. Therefore, it is advantageous to adjust the cooling rate such that the supersaturation is kept roughly constant during the cooling process. If the crystallizer is well mixed, the supersaturation is approximatey the same everywhere in the crystallizer.

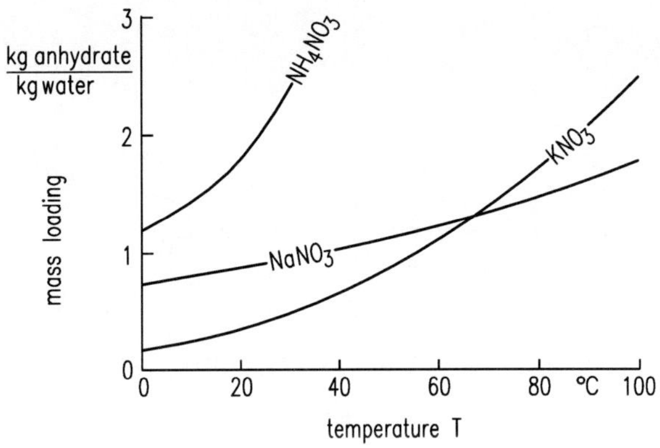

Fig. 8.2-1 Solubility of inorganic systems; steep solubility curves

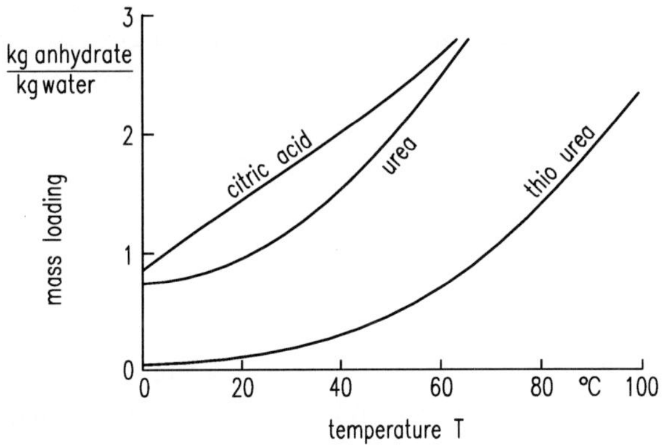

Fig. 8.2-2 Solubility of organic systems; steep solubility curves

8.2.2 Evaporative Crystallization

Evaporative crystallization is advantageous, if the solubility hardly rises with temperature, remains almost constant, or even decreases, see Fig. 8.2-3. Typical systems for this are aqueous solutions of sodium chloride, ammonium sulfate, and potassium sulfate as well as methanolic solutions of dimethyl terephthalate. The undersaturated solution is fed into the crystallizer and heated up to the boiling point, so that the solvent evaporates. Since the boiling point of the solution is a function of pressure, the boiling process preferably takes place at the liquid's surface. This can lead to high supersaturation in this region. If the crystallizer is operated continuously, the average occurring supersaturation depends on the evaporation

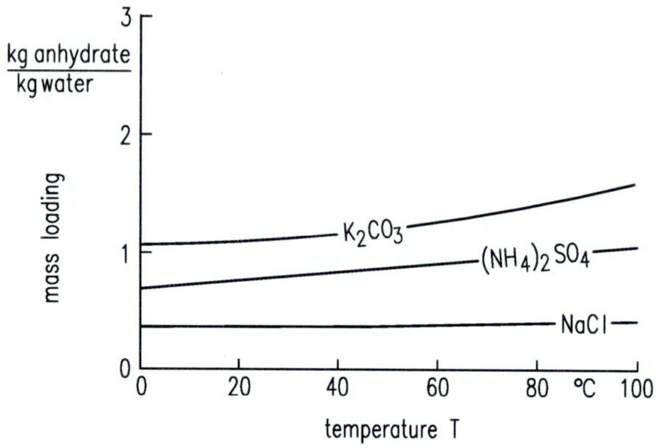

Fig. 8.2-3 Solubility of inorganic systems; flat solubility curves

rate. If the device is operated batch-wise, the same statements as for cooling crystallization remain valid: at a constant evaporation rate, unfavorably high supersaturations are observed at the beginning and uneconomically low supersaturations at the end. Here again it is advantageous to adjust the evaporation rate so that the supersaturation remains roughly constant with time.

8.2.3 Vacuum Crystallization

During vacuum crystallization the solution is simultaneously evaporated and cooled by decreasing temperature and pressure. Because enthalpy of vaporization is removed from the solution it cools down. Therefore it is possible to operate without cooling surfaces, which may be prone to incrustation. However, the steam leaving the surface of the fluid can entrain highly supersaturated drops, which spray onto the wall and lead to incrustation. As a countermeasure, it is recommended to wash the wall of the apparatus with solvent, condensate, or undersaturated solution. Furthermore, the entrainment of drops can be limited, if a certain F-factor – well known from vapor–liquid columns – is not exceeded ($F \equiv u_G \cdot \rho_G^{1/2}$ where u_G is the velocity of the steam), see Chap. 5.

8.2.4 Drowning-Out and Reactive Crystallization

In comparison with other processes the drowning-out of inorganic salts from aqueous solutions (with the help of organic substances) offers the advantage of lower energy consumption, because the enthalpy of vaporization of many displacement media is considerably lower than that of water. However, those processes compete with the multistage evaporative crystallization and with processes using compression of water vapor or combinations of such processes, which all make energy saving during crystallization possible, see Chap. 7. The drowning-out of sodium sulfate and

Table 8.2-1 Examples for reactive crystallizations

Homogeneous reaction		
$Ba(OH)_2 + H_2SO_4$	$\leftrightarrows$	$BaSO_4\downarrow + 2H_2O$
$BaCl_2 + Na_2SO_4$	$\leftrightarrows$	$BaSO_4\downarrow + 2NaCl$
$AgNO_3 + KCl$	$\leftrightarrows$	$AgCl\downarrow + KNO_3$
$NaClO_4 + KCl$	$\leftrightarrows$	$KClO_4\downarrow + NaCl$
$Ti(OC_2H_5)_4 + 4H_2O$	$\overset{C_2H_5OH}{\leftrightarrows}$	$TiO_2\downarrow + 2H_2O + 4C_2H_5OH$
$MgCl_2 + Na_2C_2O_4$	$\leftrightarrows$	$MgC_2O_4\downarrow + 2NaCl$
$Ba(NO_3)_2 + 2NH_4F$	$\leftrightarrows$	$BaF_2\downarrow + 2NH_4NO_3$
$NiSO_4 + (NH_4)_2SO_4 + 6H_2O$	$\leftrightarrows$	$NiSO_4\cdot(NH_4)_2SO_4 \cdot 6H_2O\downarrow$
Heterogeneous reaction		
$Ca(OH)_2 + CO_2(g)$	$\leftrightarrows$	$CaCO_3\downarrow + H_2O$
$Ca(OH)_2 + 2HF(g)$	$\leftrightarrows$	$CaF_2\downarrow + 2H_2O$
$K_2CO_3 + CO_2(g) + H_2O$	$\leftrightarrows$	$2KHCO_3\downarrow$
$Ca(OH)_2 + SO_2$	$\leftrightarrows$	$CaSO_3 + H_2O$

potash alum from aqueous solutions with methanol and the drowning-out of ammonium alum with ethanol have been thoroughly examined (Fleischmann and Mersmann 1984; Wirges 1986; Liszi and Liszi 1990).

During homogeneous reactive crystallization one or more reactants react with one or more components in the liquid phase. In Table 8.2-1 some examples are shown. In case of a heterogeneous reaction one reactant is often added as a gas. Aspects of reactive crystallization like macro- and micromixing as well as how and where the reactants are added are discussed later.

8.2.5 Crystallization Devices

The choice and design of crystallization devices depend on the properties of the involved phases and on the flow required for mixing and suspending.

Regarding the crystallization devices, crystallization from solution and crystallization from melts can be distinguished. The process principles of crystallization from melts can again be divided into two groups:

- Processes during which coherent crystal layers are deposited onto a cooled surface, so that the remaining melt can be separated from it without further separation operations.
- Processes during which the whole melt is continuously transformed into a suspension of crystals by cooling.

The total freezing of a melt is called solidification and is not regarded here.

8.2.5.1 Crystallization from Solution

During crytallization the suspension has to be mixed and the sedimentation of crystals has to be prevented. More or less abrasion of big crystals may be observed. Since abrasive particles can act as effective secondary seed crystals, the grain size distribution of a technical crystalline product and therefore also its average grain size are often impaired by abrasion processes. Among typical industrial crystallizers shown in Fig. 8.2-4 (Wöhlk and Hofmann 1984), the fluidized bed crystallizer is an exception. Here only small crystals (e.g., <100 μm) pass the pump. Fluidized bed crystallizers therefore can provide rather coarse crystals. In contrast to stirred vessels, forced circulation (FC) and fluidized bed crystallizers offer the advantage that with the help of external heat exchangers the heat transfer area can be adjusted to the needs.

In Figs. 8.2-5 and 8.2-6 (left) this is illustrated for fluidized bed crystallizers. These are classifying crystallizers which aim at a spatial separation of supersaturation creation and growth. In a zone with very few crystals the solution is supersaturated, while in the growth zone the supersaturation is depleted by the crystals. The growth zone is designed to form a fluidized bed.

Figure 8.2-5 shows a fluidized bed cooling crystallizer with an external heat exchanger. Despite low admissible temperature differences – mostly below 2 K – high heat fluxes can be achieved between circulated solution and coolant. The solution is supersaturated in the heat exchanger, enters the region of crystallization at the bottom of the crystallizer, and suspends the crystals.

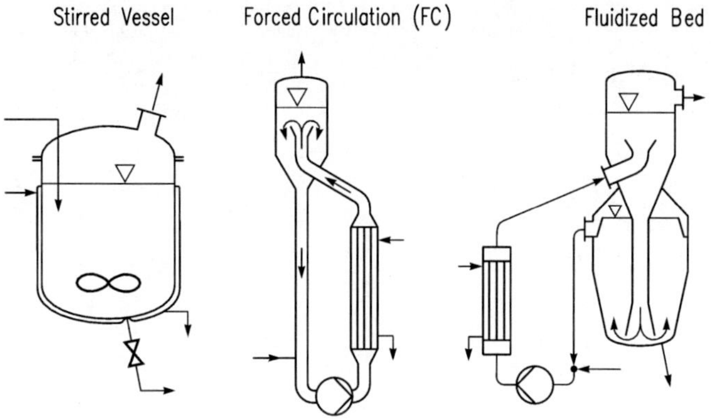

Fig. 8.2-4 Typical industrial crystallizers

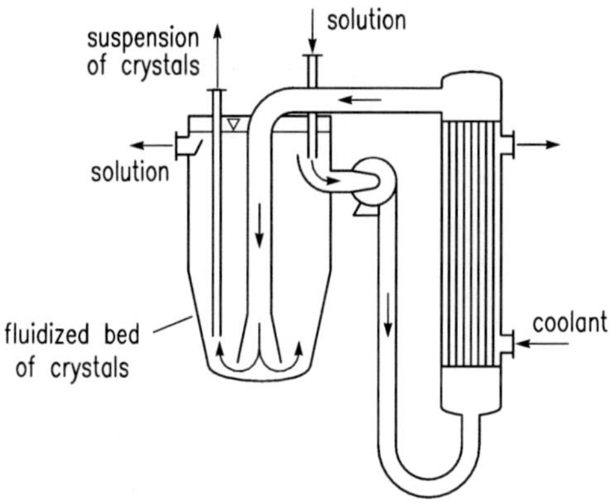

Fig. 8.2-5 Fluidized bed cooling crystallizer with circulation pump and external cooler

Figure 8.2-6 (left) shows this type of crystallizer as an evaporative crystallizer. The evaporation and crystallization parts are directly built as one. The crystallization vessel is connected to the heat exchanger. The fresh solution is fed into the circulating flow.

In case of the vacuum crystallizer given in Fig. 8.2-6 (right) the heat exchanger is missing. The pressure difference between the crystallization region and the vacuum region is compensated by the hydrostatic pressure of the liquid.

Figure 8.2-7 (left) shows a vacuum crystallizer with upward flow in the draft tube and with baffles. With this type of crystallizer the growth conditions required for the creation of coarse crystals can be realized. In Figs. 8.2-6 and 8.2-7,

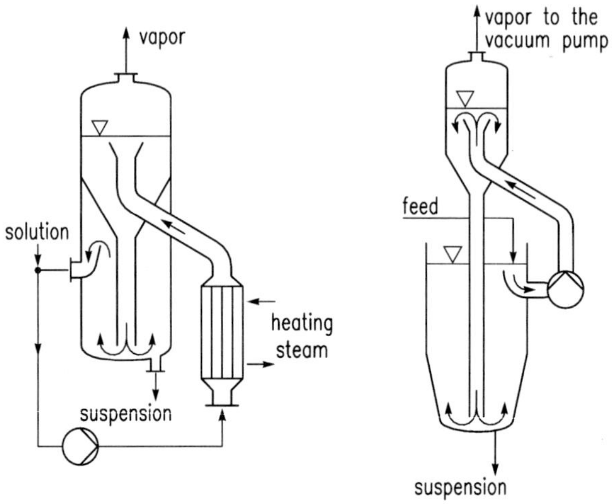

Fig. 8.2-6 Evaporative crystallizer with external heat register (*left*) and vacuum crystallizer with separate space for crystallization and evaporation (*right*)

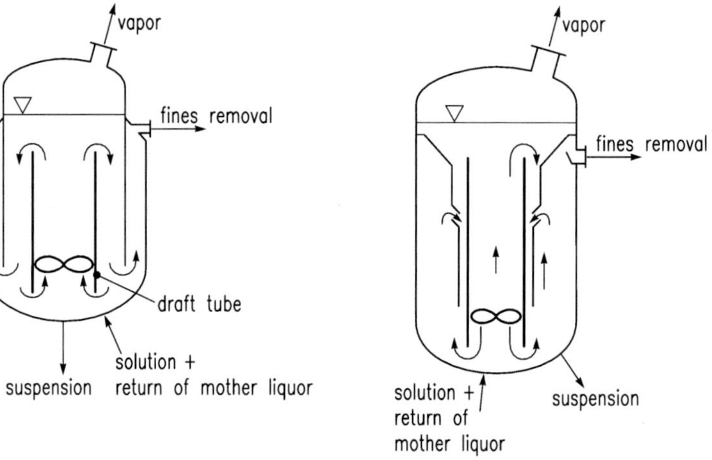

Fig. 8.2-7 Continuously operated routing tube crystallizers with fine grain dissolving. (*Left*) draft-tube-baffled (DTB); (*right*) MESSO crystallizer

instead of a circulation pump an axial pumping impeller is installed in the lower part of the draft tube. The crystals move up to the evaporation zone, where the supersaturation is at its maximum. Fines are removed by the classifying ring space and are dissolved.

The MESSO crystallizer (Fig. 8.2-7, right) has two concentric tubes – a lower draft tube with an axially pumping stirrer and an outer ejector tube with a circumferential gap – there are two closed loops of suspensions. In the inner loop – due to fast upward flow in the inner draft tube and high supersaturation in the evaporation area – mainly fine material is found. The solution flowing out through the ring space and the circumferential gap creates the outer loop with a zone of classification in the lower jacket space. In this outer jacket space a classifying fluidized bed is developed, in which preferably coarse crystals are found; finer crystals are carried out and pulled into the inner loop. The concentration of crystals can be influenced by an overflow of solution above the zone of classification. Fresh solution is directly fed into the draft tube. The product is drained from the zone of classification. The control of the crystallizer is quite flexible due to a multitude of control possibilities (e.g., impeller speed, solution overflow, ejector setting).

Figure 8.2-8 shows a horizontal multistage crystallizer, appropriate for vacuum cooling crystallization. Through several walls the steam of the different stages is separated; the suspension flows from one stage to the next. Fresh solution is added

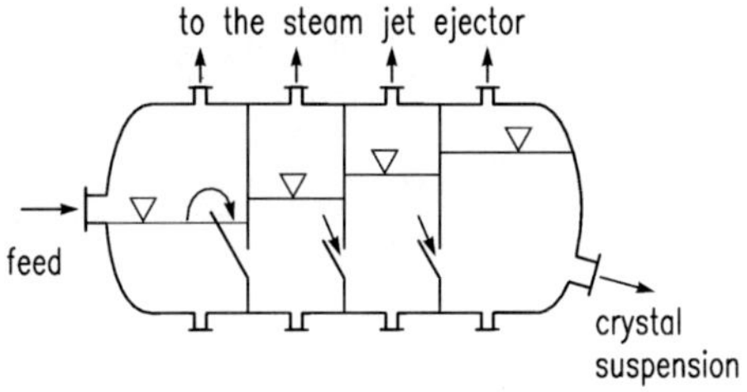

Fig. 8.2-8 Lying four-stage vacuum crystallizer

only in the first stage and it cools down from stage to stage due to pressure reduction. In the last stage, working at the lowest pressure, the product is drained. Steam jet ejectors provide the various pressures. Sometimes the movement of the liquid is generated by bubbles of gas (air) in each individual stage.

8.2.5.2 Crystallization from Melts

As mentioned in Sect. 8.1.2, the term "melt" can only be vaguely distinguished from the term "solution." In the general usage, melts are rather pure substances. Solutions are liquid mixtures, which contain considerable amounts of one or more components which are not to be crystallized. From a procedural point of view solution and melt can be distinguished in the following way: a melt can be frozen to a large extent even by a small decrease in temperature below the liquidus line; for a solution, however, considerable temperature differences or evaporation ratios are necessary to accumulate considerable amounts of crystals.

Often melts are liquid reaction products, which have gone through preceding separation operations (rectification, extraction, etc.). If the purity of the melt is sufficient for further processing, the melt only has to be cooled down to ambient temperature for transport and storage. However, if the temperature falls below the liquidus line during the cooling process, the melt solidifies. To keep the solidified melt manageable (e.g., easy to dose), the solidification is usually carried out by cooling rolls or cooling bands.

If the reaction product has not reached the required purity in the preceding separation operations (rectification, extraction, etc.), it may be purified by crystallization.

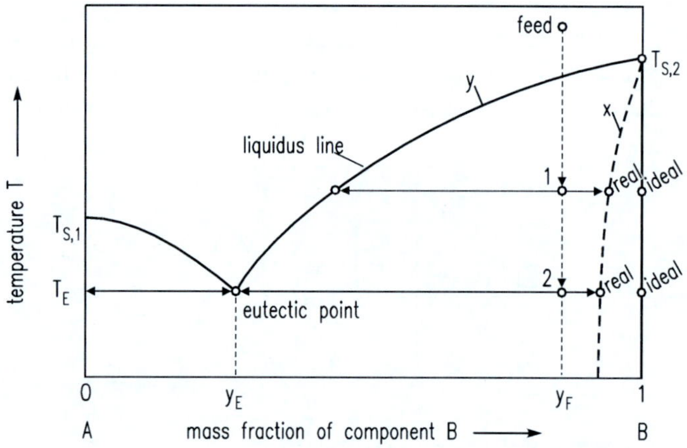

Fig. 8.2-9 Phase diagram

In Figure 8.2-9 the phase diagram for an eutectic solidifying binary mixture is shown. According to the given phase diagram (binary system without formation of mixed crystals) this solid should contain less A than the feed. In this case the thermodynamic distribution coefficient for an impurity component i is defined by

$$k_i = \frac{x_i}{y_i}. \tag{8.2-1}$$

It describes the thermodynamic equilibrium between the concentrations of impurities x_i in the solid and y_i in the solution. The thermodynamic equilibrium is only achieved at low crystal growth rates $v \to 0$. In case of systems without formation of mixed crystals this distribution coefficient should ideally be close to zero. In reality, however, the crystallized solid will not possess a distribution coefficient of zero. The reasons for this are kinetics of mass transfer during crystallization, morphology of the created crystals, and the crystallization technology (Wintermantel and Wellinghoff 2001).

In engineering two procedures for cleaning of melts by crystallization are established (Fig. 8.2-10): layer and suspension crystallization. During layer crystallization the heat is always transferred through the wall of the device and there a crystal layer is formed. Tube bundle devices, where the melt flows through the tubes, as well as plate apparatus with static melt have proved to give good results. In suspension crystallization the cold wall is usually cleaned mechanically in periodic intervals to prevent the formation of a crystal layer on its surface (scraping cooler).

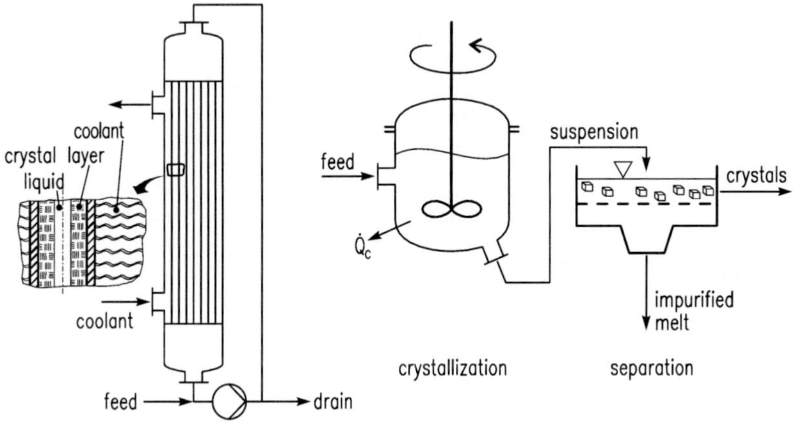

Fig. 8.2-10 (*Left*) layer crystallization; (*right*) suspension crystallization

The crystals have to be separated from the remaining melt (mother liquor) to achieve the intended purification. In case of layer crystallization, this is done by draining the remaining melt, collecting it separately, and melting down the crystal layer afterward (Fig. 8.2-11). In case of suspension crystallization the solid–liquid separation is done either by conventional filtration or by a sedimentation apparatus, with or without support of centrifugal forces. Another device repeatedly discussed in the context of solid–liquid separation is the wash column (Arkenbout 1995).

Kinetics of mass transfer prevents a growing crystal from achieving the thermodynamically possible value of the distribution coefficient. Figure 8.2-12 shows that the impurities enrich close to a growing crystal surface (phase boundary *Ph*). Due to the single-sided mass transfer, impurity *i* enriches near the growing phase boundary to $y_{i,\mathrm{Ph}}$. The definition of the thermodynamic distribution coefficient

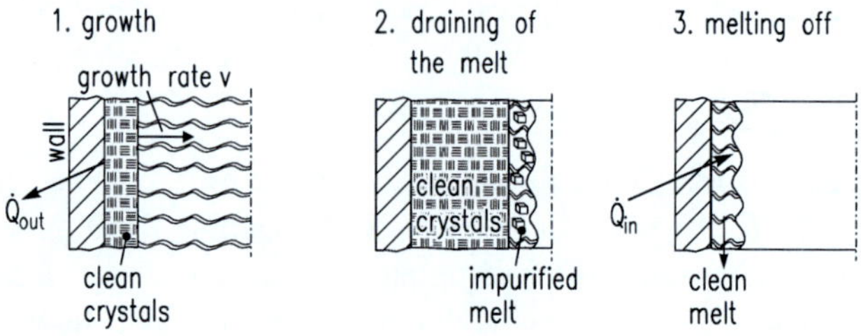

Fig. 8.2-11 Principle of purification by layer crystallization

provides $k_{i,0} = x_i/y_{i,\mathrm{Ph}}$. Because the mass fraction of the impurity in the melt at the phase boundary $y_{i,\mathrm{Ph}}$ is not a measurable quantity, an effective distribution coefficient $k_{i,\mathrm{eff}}$ is introduced. According to Burton et al. (1951) it depends on the

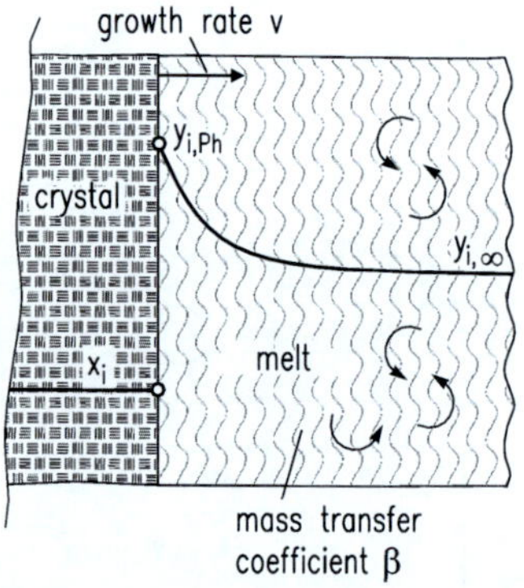

Fig. 8.2-12 Concentration profile and distribution coefficient of a growing crystal

ratio of the rate of crystal growth v to the mass transfer coefficient β, as well as on $k_{i,0}$ as follows:

$$k_{i,\mathrm{eff}} \equiv \frac{x_i}{y_{i,\infty}} = \frac{k_{i,0}}{k_{i,0} + (1 - k_{i,0}) \cdot \exp\left(-\frac{v}{\beta} \cdot \frac{\rho_\mathrm{S}}{\rho_\mathrm{L}}\right)} . \quad (8.2\text{-}2)$$

For realistic mass transfer coefficients β, Fig. 8.2-13 shows how the effective distribution coefficient depends on the rate of crystal growth. At a certain rate of crystal growth the cleaning effect of the crystallization step fully disappears. During layer crystallization high rates of crystal growth are intended, especially to reach high area specific production rates.

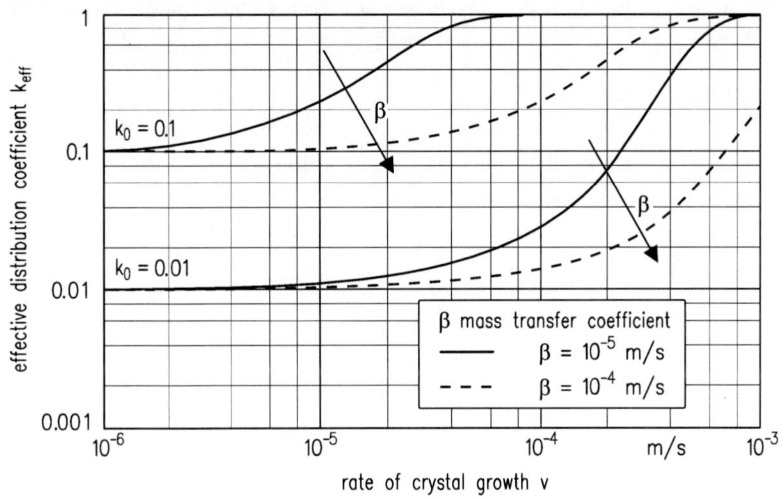

Fig. 8.2-13 Effective distribution coefficient

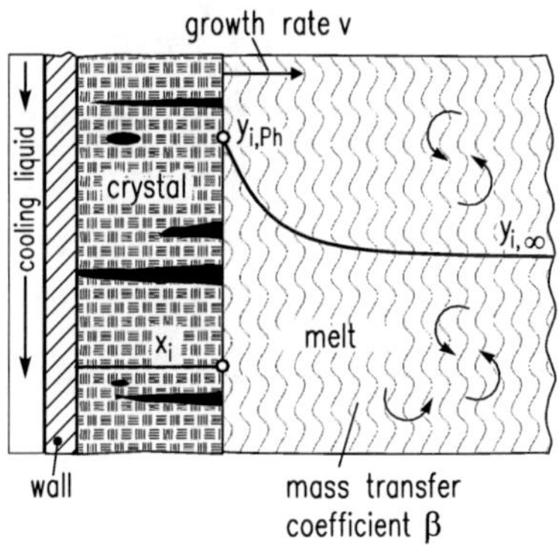

Fig. 8.2-14 Concentration profile and distribution coefficient of a crystal layer

Looking at a crystal layer with a microscope shows that it consists of crystals which are separated by grain boundaries (Fig. 8.2-14). Along these grain boundaries and also within the crystals, mother liquor can be included. Therefore, further decrease of the cleaning effect is obtained in real crystal layers compared to the thermodynamically possible. Wintermantel (1986) was able to show that this influence of the morphology of the crystals can be correlated to the dimensionless parameter

$$\frac{y_{i,\infty}}{1-y_{i,\infty}} \cdot \left(\exp\left[\frac{v}{\beta} \cdot \frac{\rho_S}{\rho_L}\right] - 1\right) . \qquad (8.2\text{-}3)$$

This parameter allows for the description of measured distribution coefficients, see Fig. 8.2-15.

Apart from mass transfer and morphology, the design of the plant with respect to its ability of completely discharging the impure melt plays an important role for the achievable purification effect. Washing and melting steps may lead to a considerable improvement of the purification effect.

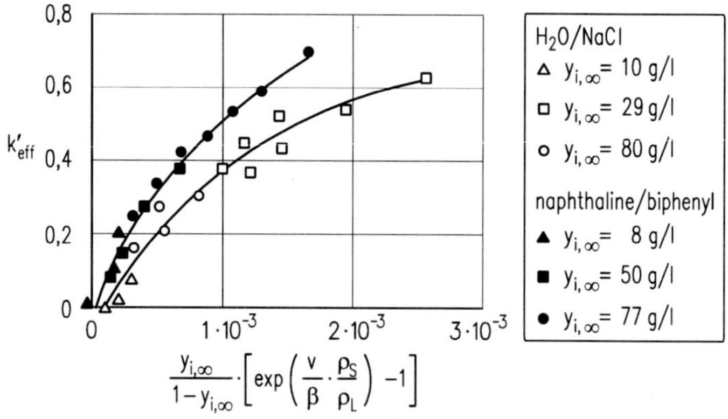

Fig. 8.2-15 NaCl/H2O and naphthaline/biphenyl: effective distribution coefficient

Because in a crystallization stage the impurity of the residual melt increases during crystallization, the impurity of a crystal layer increases with $r_f \equiv M_S/M_{L0}$. Here, M_S is the mass of the crystals and M_{L0} is the initial mass of the melt. This effect can be quantified by the following equation:

$$k_{i,\,\text{stage}} \equiv \frac{x_i}{y_{i,\,0}} = \frac{1-(1-r_f)^{k_{i,\,\text{eff}}}}{r_f}\,. \qquad (8.2\text{-}4)$$

In its graphic representation (see Fig. 8.2-16) this equation shows that the cleaning effect disappears if r_f, the fraction of crystals, attains too high values.

When designing a technical plant for cleaning melts by suspension or layer crystallization, apart from the growth rate and the mass transfer coefficient, the fraction of crystals, r_f, also has to be considered as a parameter to be optimized. Often the requested purity cannot be achieved with a single stage and the requested yield cannot be achieved because of the limited fraction of crystals, r_f, in a single stage. In these cases crystallization has to be carried out in multiple stages. The schematic diagram of such a multistage process is shown in Fig. 8.2-17. The crystals of a stage are molten and repeatedly crystallized in further stages. The impure residual melt from a stage is treated in the same way. In the lower part of Fig. 8.2-17, the

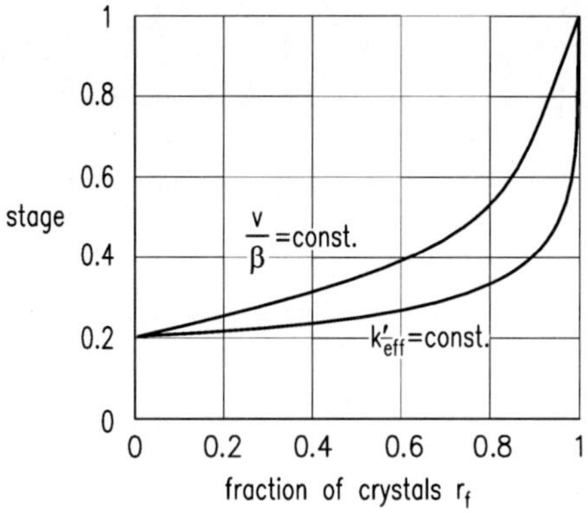

Fig. 8.2-16 Stage and stage distribution coefficient

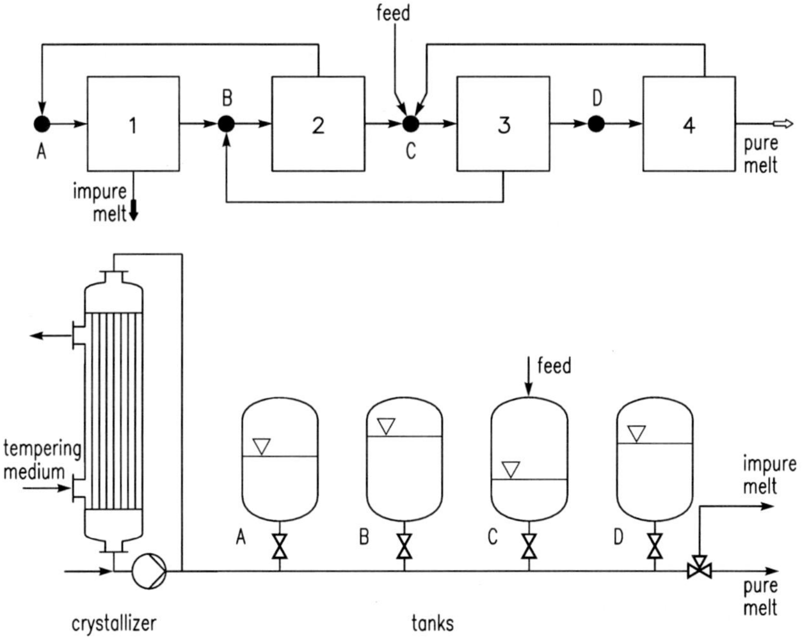

Fig. 8.2-17 Multistage process: process principle (*above*) and technical design of a layer crystallizer (*below*)

technical realization is shown for the case of layer crystallization. There is only one crystallizer and the obtained melts are stored in tanks, from which they are taken out for further crystallization. In case of multiple stage suspension crystallization, it is not sufficient to use one single crystallizer. Instead a combination of suspension crystallizer, solid–liquid separator, and melting device is necessary.

8.3 Balances

As with other unit operations the design of crystallizers is based on mass and energy balances. Here, the balances for a continuously operated crystallization apparatus at steady state are formulated for a stirred vessel crystallizer.

8.3.1 Mass Balance of the Continuously Operated Crystallizer

In crystallization technology it is common to use mass concentrations c in kg/m^3 apart from mass fractions and loadings. In Fig. 8.3-1, a continuously operated crystallizer is shown. At steady-state conditions, the mass flow $\dot{L}_0$ of the solution with the concentration c_0 entering the crystallizer is equal to the sum of the mass

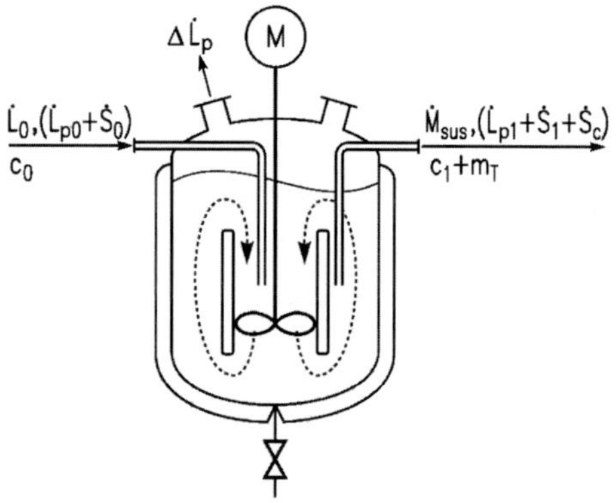

Fig. 8.3-1 Mass balance of a continuously operated crystallizer

flows of vapor $\Delta\dot{L}_r$ (index ‘p’ for pure solvent) and suspension flow $\dot{M}_{sus}$:

$$\dot{L}_0 = \Delta\dot{L}_p + \dot{M}_{sus}. \tag{8.3-1}$$

Here, it is assumed, that the exhaust vapors do not contain any dissolved substance and no droplets from the solution are entrained. The leaving suspension consists of solution with the concentration c_1 and crystals. The suspension density is m_T (in kg crystal/m^3 suspension). The balance for the dissolved substance (with φ_T being the volume fraction of the crystals in suspension) is

$$\dot{V}_0 \cdot c_0 = \dot{V}_{sus} \cdot (1-\varphi_T) \cdot c_1 + \dot{V}_{sus} \cdot \varphi_T \cdot \rho_c$$

and with $\dot{M}_{sus} = \dot{V}_{sus} \cdot \rho_{sus}$ (8.3-2)

$$\dot{L}_0 \cdot \frac{c_0}{\rho_{L,0}} = \frac{\dot{M}_{sus}}{\rho_{sus}} \cdot [(1-\varphi_T)\cdot c_1 + m_T]. \tag{8.3-3}$$

In this equation, the quantity ρ_c is the density of the compact crystals, i.e., the density of the solid. If the suspension density ($m_T < 200\ \text{kg}/\text{m}^3$ is common in industrial crystallizers) is much lower than the density of the suspension ρ_{sus} and if the densities ρ_L (solution) and ρ_{sus} (suspension) are nearly the same, the combination of (8.3-1) and (8.3-3) leads to

$$\frac{c_0}{1-\Delta\dot{L}_p/\dot{L}_0} - c_1 - m_T \approx 0. \tag{8.3-4}$$

Here, the term $\Delta\dot{L}_p/\dot{L}_0$ is the evaporation ratio which is zero for cooling-crystallizers. Then the above expression simplifies to

$$c_0 - c_1 - m_T \approx 0\,.$$

Hence in cooling crystallizers the concentration difference $c_0 - c_1$ precipitates as crystals of suspension density $m_T = (c_0 - c_1)$. In case of a batch cooling crystallizer the suspension density is $m_T = (c_\alpha - c_\omega)$ with the initial concentration c_α and the final concentration c_ω.

The term

$$\Delta c_0 = c_0 \cdot \frac{1}{1-\Delta\dot{L}_p/\dot{L}_0} - c^* \tag{8.3-5}$$

is the theoretical supersaturation, which would exist everywhere in an ideally mixed crystallizer if it were not depleted by nucleation or crystal growth. In reality, the supersaturation $\Delta c < \Delta c_0$ exists. The magnitude of Δc is mainly determined by the kinetics (nucleation and growth) and the crystal surface. In case of a continuously operated crystallizer Δc should be optimized regarding space time yield. According to the solubility curve $c^* = f(T)$, the saturation concentration c^* depends on the temperature T. The supersaturation Δc is the driving force for crystal growth. If the solubility curve $c^* = f(T)$ is known, Δc can be determined by the temperature difference ΔT between the current temperature and the saturation temperature:

$$\Delta c = \frac{\mathrm{d}c^*}{\mathrm{d}T} \cdot \Delta T \tag{8.3-6}$$

For fast growing systems the supersaturation is much lower than the suspension density m_T. In these cases, the magnitude m_T is close to the calculated initial supersaturation Δc_0:

$$m_T \approx \Delta c_0 = \frac{c_0}{[1 - \Delta \dot{L}_p / \dot{L}_0]} - c^* \ . \tag{8.3-7}$$

In case of a batch crystallizer with $c_\omega \approx c^*_\omega$, the correlation

$$m_T \approx \Delta c_0 = \frac{c_\alpha}{[1 - \Delta L_p / L_\alpha]} - c^*_\omega \tag{8.3-8}$$

is obtained, where L_α represents the initial amount of solution and ΔL_r is the amount of evaporated vapor.

The formulation of mass balances becomes a little more difficult if molecules of the solvent are integrated into the crystal lattice as solvates or hydrates. A hydrate is a crystal that includes water. If S_{hyd} is the mass of the hydrate, the anhydrate mass S_c can be calculated with the molar mass $\tilde{M}_c$ of the substance containing no water and that of the hydrate $\tilde{M}_{hyd}$:

$$S_c = S_{hyd} \cdot \frac{\tilde{M}_c}{\tilde{M}_{hyd}} \ . \tag{8.3-9}$$

The following applies:

$$\tilde{M}_{hyd} - \tilde{M}_c = \frac{\text{kg of solvent in the crystal}}{\text{kmol of solvent-free crystals}} \ . \tag{8.3-10}$$

This leads to

$$\frac{\tilde{M}_{hyd} - \tilde{M}_c}{\tilde{M}_c} = \frac{\text{kg of solvent in the crystal}}{\text{kg of solvent-free crystals}} \ . \tag{8.3-11}$$

According to Fig. 8.3-1 the mass balance of the solute is:

$$\dot{S}_0 = \dot{S}_1 + \dot{S}_c \ , \tag{8.3-12}$$

or expressed with the mass ratio Y in $\frac{\text{kg solute}}{\text{kg solvent}}$:

$$\dot{S}_c = \dot{S}_0 - \dot{S}_1 = Y_0 \cdot \dot{L}_{p0} - Y_1 \cdot \dot{L}_{p1} \ . \tag{8.3-13}$$

Index p denotes pure solvent. A balance of the solvent results in

$$\dot{L}_{p0} = \dot{L}_{p1} + \Delta \dot{L}_p + \dot{S}_c \cdot \left[\frac{\tilde{M}_{hyd} - \tilde{M}_c}{\tilde{M}_c}\right] . \tag{8.3-14}$$

Finally, this leads to the following equation, which can be formulated with mass ratios Y [kg solute/kg solvent] as well as with mass fractions y [kg solute / kg solution]:

$$\dot{S}_c = \frac{\dot{L}_{p0} \cdot \left(Y_0 - Y_1 \cdot \left[1 - \frac{\Delta \dot{L}_p}{\dot{L}_{p0}} \right] \right)}{1 - Y_1 \cdot \left[\frac{\tilde{M}_{hyd} - \tilde{M}_c}{\tilde{M}_c} \right]} = \frac{\dot{L}_0 \cdot (y_0 - y_1) + \Delta \dot{L}_p \cdot y_1}{1 - y_1 \cdot \frac{\tilde{M}_{hyd}}{\tilde{M}_c}} \quad . \tag{8.3-15}$$

The mass of solvate (i.e., hydrate for aqueous solutions) is

$$\dot{S}_{hyd} = \dot{S}_c \cdot \frac{\tilde{M}_{hyd}}{\tilde{M}_c} \quad . \tag{8.3-16}$$

The maximum crystal mass is received, if the solution leaves the crystallizer at equilibrium concentration c_1^*, at equilibrium ratio Y_1^*, or at equilibrium mass fraction y_1^*:

$$c_1 = c_1^* \text{ or } Y_1 = Y_1^* \text{ or } y_1 = y_1^* .$$

For the special case of cooling crystallization ($\Delta \dot{L}_p = 0$) and solvent-free crystals ($\tilde{M}_{hyd}/\tilde{M}_c = 1$), the mass balance of the solute simplifies to

$$\dot{S}_c = \dot{L}_{p0} \cdot (Y_0 - Y_1) = \dot{L}_0 \cdot \left[\frac{y_0 - y_1}{1 - y_1} \right] \quad . \tag{8.3-17}$$

Furthermore, if there are two substances dissolved in a solvent, the triangular coordinate diagram can be used to represent the crystallization process. The yield and composition of the crystals can be obtained from the lever rule as explained in the triangular diagram in Fig. 8.3-2. There is a region of undersaturation at the top of this triangular diagram. The two-phase regions GCD and BED contain a solution and a solid at equilibrium. The three-phase region GBD represents the presence of a solution that corresponds to point D, as well as solid crystals of both components.

If a solution corresponding to point Q exists and is evaporated, its state changes along the conjugation line through points L_r and Q. At point F, the first crystals, consisting of substance B, precipitate. If point H is reached, more crystals have precipitated and the solution is depleted of B according to the change from F to K. Finally, when the line BD is reached, the three-phase region is entered, where also crystals of substance G precipitate. The ratio of crystal types B and G and of the solution corresponding to point D can be determined for each point in the three-phase region by applying the lever rule. Point N, for example, splits into solution D and a mixture of crystals corresponding to point P. According to the lever rule this mixture segregates into the two crystal types B and G. If the lever rule is applied to the points of the conjugation line, it is possible to determine the amount of evaporated solvent.

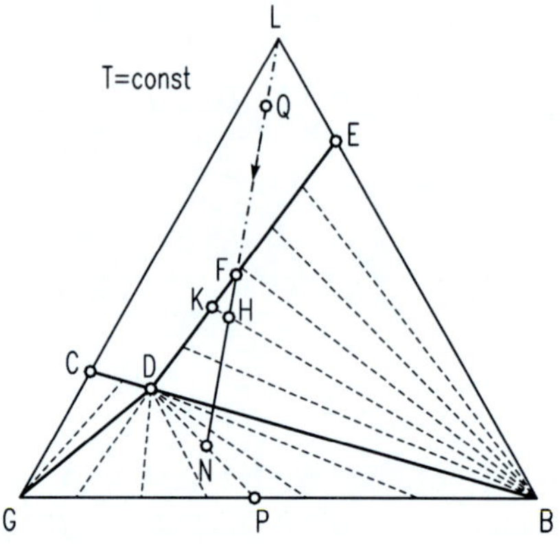

Fig. 8.3-2 Crystallization process in the triangular diagram

The average grain size of a product, generated in a continuously operated crystallizer, decisively depends on the average supersaturation, on the average residence time of the crystals in the device, and on the abrasion behavior of the crystals. Apart from this, local and momentary peaks of supersaturation can be decisive. The longer the average residence time of the suspension, the lower the average supersaturation and the rate of crystal growth and above all the nucleation rate (Mersmann 2001).

8.3.2 Mass Balance of the Batch Crystallizer

In a batch crystallizer supersaturation can be achieved by cooling, evaporation of solvent, or both combined if $(\mathrm{d}c^*/\mathrm{d}T) > 0$. If only one component crystallizes, the mass balance for this component is (suspension density $m_\mathrm{T} = \rho_\mathrm{c} \cdot \varphi_\mathrm{T}$, volume fraction $\varphi_\mathrm{T} = m_\mathrm{T}/\rho_\mathrm{c}$)

$$(1-\varphi_\mathrm{T}) \cdot V_\mathrm{sus} \cdot \left(\frac{\mathrm{d}c^*}{\mathrm{d}T}\right) \cdot \frac{\mathrm{d}T}{\mathrm{d}t} + V_\mathrm{sus} \cdot \left(1 - \frac{c^*}{\rho_\mathrm{c}}\right) \cdot \frac{\mathrm{d}m_\mathrm{T}}{\mathrm{d}t} + \cdots$$
$$+ [\varphi_\mathrm{T} \cdot \rho_\mathrm{c} + (1-\varphi_\mathrm{T}) \cdot c^*] \cdot \frac{\mathrm{d}V_\mathrm{sus}}{\mathrm{d}t} = 0 \cdot \qquad (8.3\text{-}18)$$

This balance is only valid as long as the supersaturation Δc is very small compared to the saturation concentration c^*. It is useful to seed a batch crystallizer and to cool it according to a certain temperature program, so that the rate of crystal growth G remains constant during the cooling process. This is important to prevent spontaneous primary nucleation and its harmful influence on the product quality. It is assumed that the monodisperse, volume-related seed mass m_s [kg seed material/m^3

solution] is fed into the crystallizer right at the beginning of the saturation process. With

$$m_\mathrm{s} = \frac{N_\mathrm{s} \cdot \alpha \cdot L_\mathrm{s}^3 \cdot \rho_\mathrm{c}}{V_\mathrm{sus}} \quad \text{and} \quad m_\mathrm{T} = \frac{N_\mathrm{s} \cdot \alpha \cdot L^3 \cdot \rho_\mathrm{c}}{V_\mathrm{sus}} \ , \tag{8.3-19}$$

where N_s is the total amount of seed crystals and L_s is their particle size, the terms

$$m_\mathrm{T} = m_\mathrm{s} \cdot \left(1 + \frac{G \cdot t}{L_\mathrm{s}}\right)^3 \quad \text{and} \tag{8.3-20}$$

$$\frac{\mathrm{d}m_\mathrm{T}}{\mathrm{d}t} = \frac{3 \cdot G \cdot m_\mathrm{s}}{L_\mathrm{s}} \cdot \left(1 + \frac{G \cdot t}{L_\mathrm{s}}\right)^2 \tag{8.3-21}$$

are finally obtained.

Equation (8.3-21) indicates that the suspension density m_T only increases slowly at the beginning, but faster with time, because the available crystal surface increases accordingly. If the average rate of crystal growth G and its driving force the supersaturation Δc are to remain constant during the whole batch time, the cooling rate $\dot{T} = (\mathrm{d}T/\mathrm{d}t)$ and the evaporation rate $\Delta \dot{L}_\mathrm{p}$, respectively, have to increase with time, too. Without evaporation, the cooling rate $\dot{T}$ follows the equation

$$\dot{T} = -\left(\frac{1 - \frac{c^*}{\rho_\mathrm{c}}}{1 - \left(\frac{m_\mathrm{s}}{\rho_\mathrm{c}}\right) \cdot \left(1 + \frac{G \cdot t}{L_\mathrm{s}}\right)^3} \cdot \frac{3 \cdot G \cdot m_s}{\left(\frac{\mathrm{d}c^*}{\mathrm{d}T}\right) \cdot L_\mathrm{s}} \cdot \left(1 + \frac{G \cdot t}{L_\mathrm{s}}\right)^2 \right) \ . \tag{8.3-22}$$

If the temperature remains constant during evaporative crystallization, the evaporation rate of the pure solvent (index r) is given by

$$\Delta \dot{L}_\mathrm{r} = \frac{M_{\mathrm{sus},\alpha}}{c^*} \cdot \left(1 - \frac{c^*}{\rho_\mathrm{c}}\right) \cdot \frac{3 \cdot G \cdot m_\mathrm{s}}{L_\mathrm{s}} \cdot \left(1 + \frac{G \cdot t}{L_\mathrm{s}}\right)^2 \ , \tag{8.3-23}$$

where $M_{\mathrm{sus},\alpha}$ is the mass of suspension (or solution) at the beginning of the process. During evaporative crystallization both processes superpose. In the absence of any growth dispersion, i.e., if all seeds and crystals grow at the same growth velocity $G = (\mathrm{d}L/\mathrm{d}t)$, monodisperse crystals with the grain size $L = L_\mathrm{s} + G \cdot \tau$ will be present after the batch time τ.

It should be emphasized that in a real crystallizer the assumptions of monodisperse seed material and of absence of nucleation and growth dispersion does not apply. The formation of secondary seeds by abrasion can considerably reduce the required batch time in case of a constant supersaturation but impairs the particle size distribution toward smaller sizes.

In Fig. 8.3-3 the suspension density m_T is plotted vs. time t. The curve is dependent not only on time but also on parameters like average supersaturation and flow intensity. With increasing average supersaturation Δc the curve rises faster from m_s to $m_{T,\omega}$. In other words, the batch time shortens for high values of the parameter Δc. The parameter flow intensity controls the mechanical stress caused by the circumferential speed of the rotor and its specific power. High flow intensities lead to high secondary nucleation. Therefore, just like in case of high Δc, the batch time shortens for higher flow intensities. With this shortening of the batch time the average grain size L_{50} decreases simultaneously, just as it does for higher flow intensities. The average supersaturation Δc has to be kept as constant as pos-

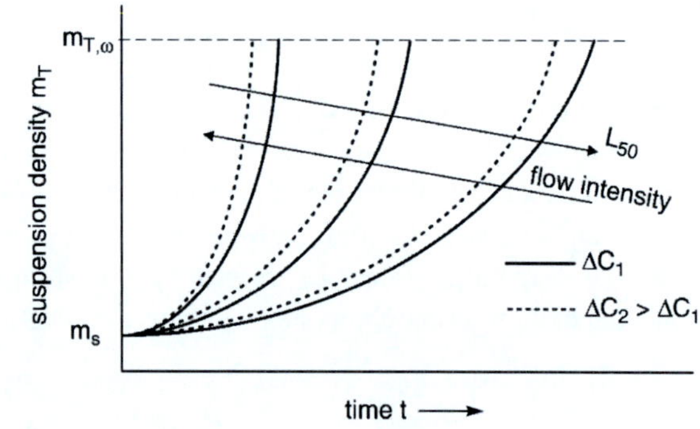

Fig. 8.3-3 Suspension density transient time for various flow intensities and supersaturations

sible which is equivalent to a constant crystal growth rate.

The operational parameters fluid flow and supersaturation have to be accurately adjusted according to the crystalline material (abrasion and growth habit) to form products of a desired quality (grain size distribution, grain shape, purity, etc.). The prediction of the product quality is very difficult if spontaneous nucleation occurs – even if only for a short period of time. For products in micro- or even nanometer scale, agglomeration can crucially influence the grain size distribution.

8.3.3 Energy Balance of the Continuously Operated Crystallizer

In Fig. 8.3-4, a schematic of an evaporative crystallization stage is depicted. During cooling crystallization the heat flow $\dot{Q}_{out}$ is removed, during evaporative crystallization the heat flow $\dot{Q}_{in}$ is supplied.

The mass flow $\Delta \dot{L}_p$ of the evaporated solvent with a corresponding enthalpy flow $\dot{H}_{\Delta L_p}$ leaves the crystallizer.

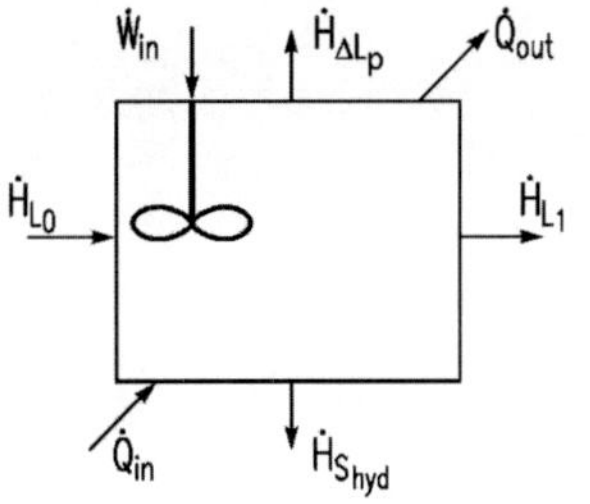

Fig. 8.3-4 Energy balance for a single stage crystallizer

In addition energy can be supplied by the circulation device and, for nonadiabatic operation, heat can be exchanged with the environment. If the crystallizer is operated at steady-state conditions, the following energy balance for the crystallizer is obtained:

$$\underbrace{\dot{Q}_{in}}_{\text{supplied heat flow}} + \underbrace{\dot{H}_{L0}}_{\text{enthalpy of the feed solution}} + \underbrace{\dot{W}_{in}}_{\text{supplied power}}$$

$$= \underbrace{\dot{Q}_{out}}_{\text{removed heat flow}} + \underbrace{\dot{H}_{L1}}_{\text{enthalpy of the product suspension}} + \underbrace{\dot{H}_{S_{hyd}}}_{\text{enthalpy of the crystals}} + \underbrace{\dot{H}_{\Delta L_p}}_{\text{enthalpy of the vapors}}.$$

The heat of crystallization is the heat that has to be supplied or removed during crystallization at constant temperature. It is equal to the negative value of the heat of solution during the dissolution of crystals in an almost saturated solution. The heat of crystallization is accounted for in the enthalpy values. Processes in crystallizers can easily be tracked, if an enthalpy concentration diagram is available for the respective system. The pure component's enthalpy is zero at reference temperature, not the enthalpy of real mixtures however. In such diagrams, the lever rule is applicable. This is shown for the system calcium chloride/water in Fig. 8.3-5, where the specific enthalpy is plotted vs. the mass fractions.

During cooling crystallization (1–2) heat is removed and the enthalpy decreases from point 1 to point 2. Point 2 lies in a two-phase region, in which solution and hexahydrate are in equilibrium. The line segments (2–2″) and (2′–2) denote the solids content. The process of evaporative crystallization in vacuum at 0.5 bar can also be seen in the diagram. During heating of the feed solution ($x = 0.45$,

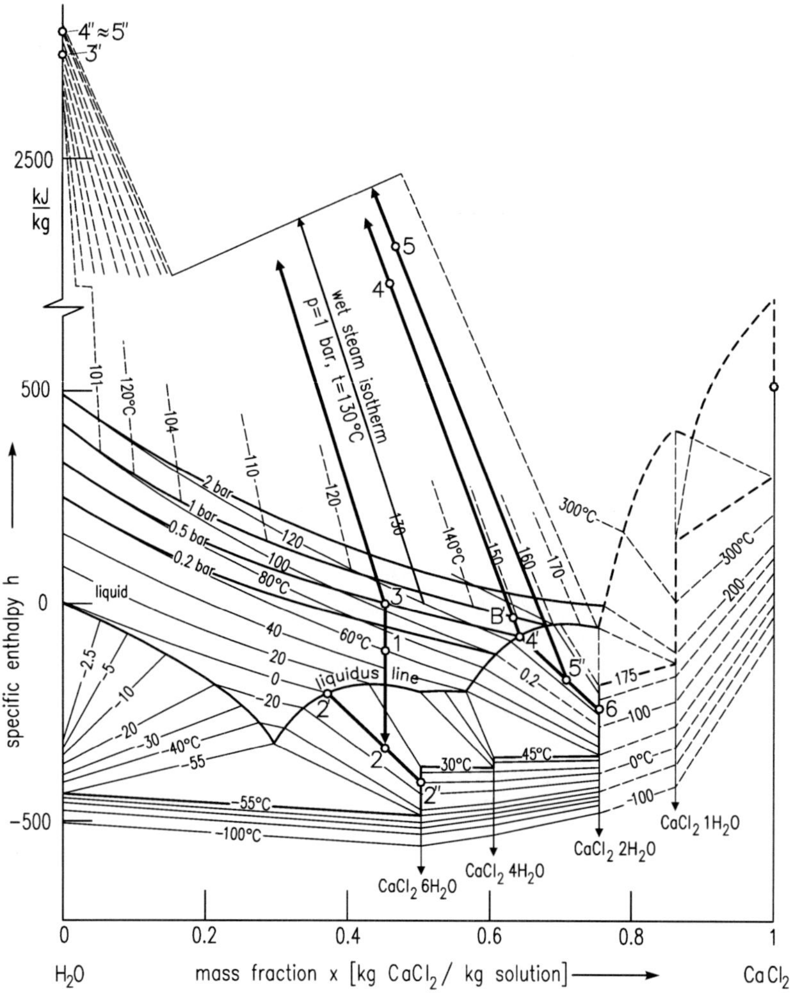

Fig. 8.3-5 Representation of a crystallization process in the enthalpy–concentration diagram for calcium chloride/water – for cooling crystallization (1–2) and evaporation-crystallization (1–3–4–5)

$T = 60\ °C$, point 1), the boiling point is reached at about 105°C (point 3). The solution is then in equilibrium with vapor which contains no salt (point $3''$, point of intersection of the wet steam isotherm with the abscissa $x = 0$). If more heat is supplied, e.g., $\Delta h = 830$ kJ/kg (point 4) the system splits into a gaseous phase (point $4'' \approx 3''$) and a liquid phase (point $4'$). The steam and the solution are at 130°C and the solution is saturated. If more heat is supplied, crystals (6), saturated solution ($4'$), and superheated steam ($5'' \approx 4'' \approx 3''$) are produced.

8.3.4 Population Balance

The mathematical description of a crystal size distribution and of its change in space and time makes use of the conservative character of the number of particles in space and state (i.e., particle size L). In the respective number balance, the particle size distribution is represented by the number density, see Hulburt and Katz (1964) and Randolph and Larson (1988),

$$n(L) = \frac{\mathrm{d}N}{\mathrm{d}L} . \tag{8.3-24}$$

Here, the quantity N is the amount of crystals per unit volume of crystal suspension and L is the crystal size. The number density $n(L)$ represents the number of crystals in a size interval ΔL per volume of suspension:

$$n(L) = \frac{\text{number of crystals}}{\text{m}^3\,\text{suspension} \cdot \text{m class width}} . \tag{8.3-25}$$

The number balance for the crystals in a size interval dL is

$$\frac{\partial n}{\partial t} + \frac{\partial(G \cdot n)}{\partial L} + n \cdot \frac{\partial V}{V \partial t} + D(L) - B(L) - \sum_i \frac{n_i \cdot \dot{V}_i}{V} = 0. \tag{8.3-26}$$

The term $\partial n/\partial t$ represents the change of the number density with time and therefore equals zero at steady state. The expression $\partial(G \cdot n)/\partial L$ describes the number of crystals growing into or out of a grain size interval dL with crystal growth rate $G = \mathrm{d}L/\mathrm{d}t$. The term $n \cdot \partial V/(V\partial t)$ changes for the crystallizer's volume. The values $D(L)$ and $B(L)$ denote "death" and "birth" rates, which may occur because of abrasion, breakage, or agglomeration of crystals. If for example two crystals coalesce by agglomeration they disappear from their size intervals and the newly formed crystal will appear in another interval.

Finally, the term $\sum n_i \cdot \dot{V}_i / V$ represents the sum of all particle flows entering and leaving the crystallizer. Solving (8.3-26) is difficult, due to the fact that until today, the birth rate $B(L)$ and the death rate $D(L)$ cannot be formulated universally for any case. Processes like breakage and abrasion of crystals are caused by mechanical and fluid dynamic stress and are not influenced by crystallization kinetics. If fragments and abraded particles are present in a supersaturated solution, however, they are able to grow. Their ability to grow and their rate of growth are then mainly influenced by the supersaturation Δc. The complex interaction of mechanical and kinetic effects leads to the difficulty of describing birth and death rates. In the laboratory, crystallization experiments can often be carried out in a manner that hardly any breakage or abrasion occurs. If in addition the supersaturation is the same everywhere in the crystallizer (due to good mixing), no crystals will dissolve and all will grow at roughly the same growth velocity. In moderately supersaturated solutions, with low suspension densities and rather coarse product, no agglomeration should occur. If all requirements mentioned here are fulfilled, the terms $B(L)$ and $D(L)$ can be neglected in the above population balance. For continuously operated cooling crystallizers at steady state, both terms $\partial n/\partial t$ and $n \cdot \partial V/(V\partial t) = n \cdot \partial(\ln V)/\partial t$ equal zero. The population balance is reduced to

$$\frac{\partial(G \cdot n)}{\partial L} + \sum_i \frac{n_i \cdot \dot{V}_i}{V} = 0. \tag{8.3-27}$$

Often, the solution fed into a continuously operated crystallizer is free from crystals and only a single volume flow $\dot{V}$ is drained continuously which is containing a particle size distribution representing the distribution in the crystallizer. In this case, the population balance can be further simplified to

$$\frac{\partial(G \cdot n)}{\partial L} + n \cdot \frac{\dot{V}}{V} = 0 \quad .$$

Since the ratio $\dot{V}/V$ of the volume flow $\dot{V}$ and the volume V is equal to the inverse of the average residence time τ of the suspension the following expression is obtained for a cooling crystallizer:

$$\frac{\partial(G \cdot n)}{\partial L} + \frac{n}{\tau} = 0 \ . \tag{8.3-28}$$

This simplification is done with the assumption that the solution and the crystals have the same average residence time in the crystallizer. Theoretically, the rate of crystal growth G can depend on the particle size. For bigger crystals, the crystal growth is often almost independent of the grain size. This is mainly because the mass transfer coefficient of particles with a grain size range of $100\ \mu m < L < 2{,}000\ \mu m$ is hardly affected by the particle size for diffusion-limited growth. In case of crystal growth limited by integration and sufficient supersaturation the rate of crystal growth also depends only weakly on the crystal size. If the value G is not a function of the grain size L then the above equation further simplifies to

$$G \cdot \frac{dn}{dL} + \frac{n}{\tau} = 0 \ . \tag{8.3-29}$$

This extremely simplified relation for the number density balance is only valid for the so-called MSMPR (mixed suspension mixed product removal) crystallizers. Integration with the integration constant n_0 as the number density at grain size $L = 0$ leads to

$$n = n_0 \cdot \exp\left(-\frac{L}{G \cdot \tau}\right) \tag{8.3-30}$$

or

$$\ln\left(\frac{n}{n_0}\right) = -\frac{L}{G \cdot \tau} \ .$$

If the logarithm of the number density n is plotted vs. the crystal size L, a straight line with the negative slope $-1/(G \cdot \tau)$ is obtained, as can be seen in Fig. 8.3-6. Since the slope of the straight line is $-1/(G \cdot \tau)$ and the residence time $\tau = V/\dot{V}$ is known, the average rate of crystal growth G of all crystals can be determined from the slope of the straight line.

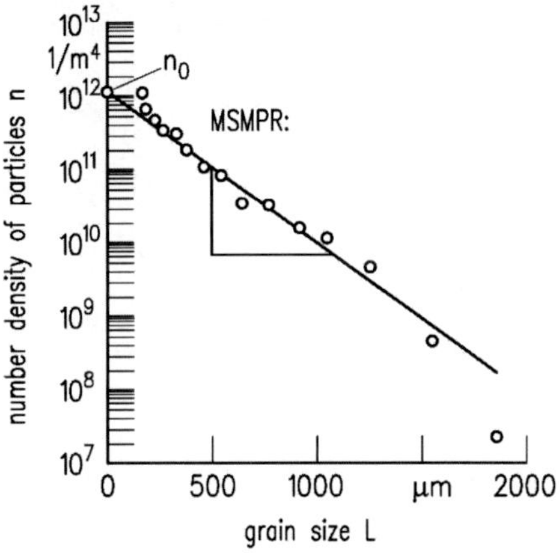

Fig. 8.3-6 Number density vs. the crystal size for the system ammonium sulfate–water with an average residence time of $\tau = 3{,}432$ s in the MSMPR crystallizer

Newly developing seeds are very small and are in the range of nanometers, i.e., in the range $L \rightarrow 0$. With the ordinate intercept n_0 for $L = 0$ the nucleation rate B_0 is given by

$$B_0 = \frac{dN_0}{dt} = \frac{dN_0}{dL} \cdot \frac{dL}{dt} = n_0 \cdot G\ . \tag{8.3-31}$$

Both kinetic parameters, the nucleation rate and the rate of crystal growth, can therefore be determined with the slope and the ordinate intercept n_0 of the straight line in the number density diagram. These quantities give according to the following equation, the median value L_{50} of the grain size distribution:

$$L_{50} = 3.67 \sqrt[4]{\frac{G}{6 \cdot \alpha \cdot (B_0/\varphi_T)}}\ . \tag{8.3-32}$$

The shape factor α results from the particle volume V_P in $\alpha = V_P/L^3$. In Fig. 8.3-7 the nucleation rate B_0 related to the volumetric crystal concentration φ_T (φ_T = volume of all crystals/volume of the suspension) is plotted against the average rate of crystal growth with the average crystal size L_{50} as a parameter. With the relation

$$L_{50} = 3.67 \cdot G \cdot \tau, \tag{8.3-33}$$

it is possible to include the average residence time τ as an additional parameter.

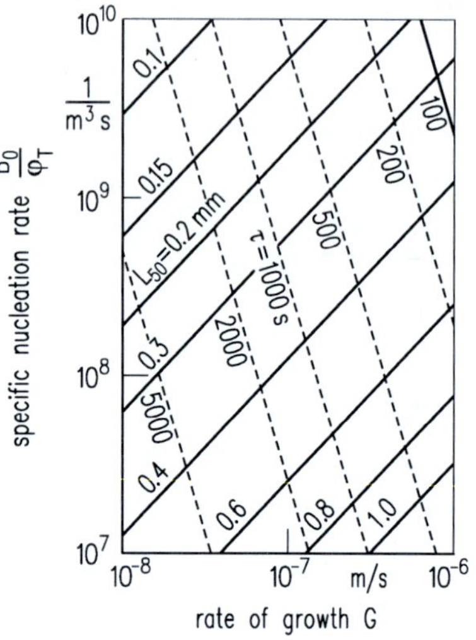

Fig. 8.3-7 Specific nucleation rate vs. the growth rate for MSMPR crystallizers

8.4 Crystallization Kinetics

In the next chapter the rates of nucleation and crystal growth are discussed. Both mechanisms are decisive for the quality of crystalline products.

8.4.1 Nucleation and Metastable Zone

The subsequent kinetic steps of the formation of nuclei and their growth lead to the production of crystals. A prerequisite for these kinetic steps is a supersaturation which can be achieved by a change of the temperature (cooling for a positive slope dc^*/dT or heating for a negative derivative dc^*/dT), the removal of a component (solvent) or the addition of an antisolvent or a drowning-out substance. As a rule the solvent is removed by evaporation; however, the removal by membranes or adsorbents may be interesting especially in the laboratory. Furthermore, supersaturation can be created by a chemical reaction after the addition of one or more reactants or a change of an equilibrium. The supersaturation as the decisive driving force leads to nucleation and crystal growth until equilibrium is reached. If neither foreign particles nor crystals are present in a liquid or melt, new crystals can only be born by homogeneous nucleation. In the presence of foreign particles, nucleation becomes easier and is called heterogeneous nucleation. Homogeneous and heterogeneous nucleation, or in general "primary nucleation," take place in the absence of crystals of the crystallizing component. A noticeable nucleation is only observed when a metastable supersaturation $\Delta c_{\mathrm{met,\,hom}}$ in the system is present.

Experiments in laboratory and industrial crystallizers have shown that nuclei are born at supersaturations $\Delta c \ll \Delta c_{\text{met, hom}}$ in the presence of crystals (either product crystals or added seed crystals). Such nuclei are called "secondary nuclei." This secondary nucleation caused by the removal of preordered species on a crystal surface and attrition fragments can take place at very small supersaturations; however, $\Delta c_{\text{met, sec}} < \Delta c_{\text{met, het}} < \Delta c_{\text{met, hom}}$ is necessary to initiate the growth of secondary nuclei. In Fig. 8.4-1 the solubility c^* and the three metastable zone widths $\Delta c_{\text{met, hom}}$, $\Delta c_{\text{met, het}}$, and $\Delta c_{\text{met, sec}}$ valid for homogeneous, heterogeneous, and secondary nucleation, respectively, are shown as a function of temperature T.

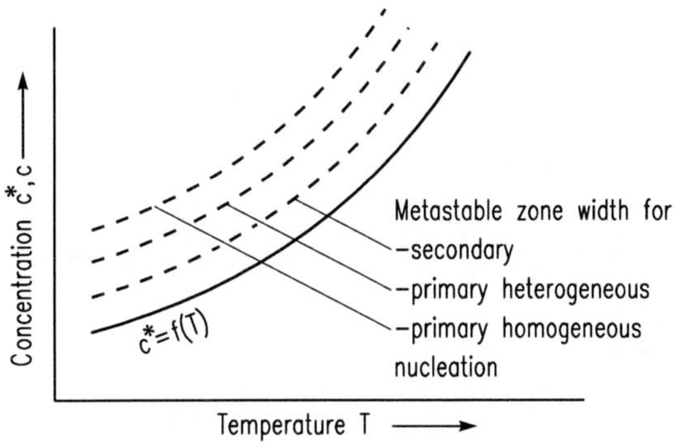

Fig. 8.4-1 Metastable supersaturation as a function of temperature for different nucleation processes

8.4.1.1 *Activated Nucleation*

According to the classical theory of nucleation nuclei are born by the successive addition of units following the formation scheme:

$$A_1 + A = A_2; \quad \cdots A_2 + A = A_3; \quad \cdots A_n + A \underset{k_Z}{\overset{k_A}{\leftrightarrows}} A_{n+1}. \tag{8.4-1}$$

Here k_A is the rate constant of the addition of units to an associate species or cluster and k_Z is the rate constant of decay. Note that the addition is a random process. If the supersaturation is high enough, more and more units can be added with the result that stable macronuclei or clusters are formed. The change of the positive free surface enthalpy ΔG_A increases with the surface tension γ_{CL} between a solid crystal and the surrounding solution and also with the surface of the nucleus. This energy has to be added to the system and is positive. On the other hand, the change in free volume enthalpy ΔG_V during solid phase formation is set free and is thus negative. The magnitude ΔG_V of this enthalpy is proportional to the volume of the nucleus and increases with increasing energy $RT \ln S$, where $S = a/a^*$ or in ideal systems $S = c/c^*$, when the concentration c of the elementary units changes to the lower equilibrium concentration $c^* = c - \Delta c$.

In Fig. 8.4-2 the free enthalpies ΔG_A and ΔG_V and also the total enthalpy $\Delta G = \Delta G_A + \Delta G_V$ are shown as a function of the size of a nucleus. This enthalpy ΔG can be written with the surface A_N and the volume A_V of a nucleus:

$$\Delta G = \Delta G_A + \Delta G_V = A_N \cdot \gamma_{CL} - V_N \cdot \frac{\rho_c}{\tilde{M}} \cdot \tilde{R} \cdot T \cdot \ln\left(\frac{c}{c^*}\right). \tag{8.4-2}$$

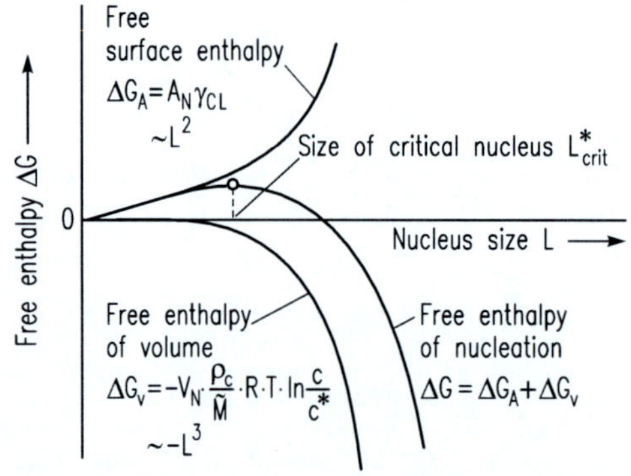

Fig. 8.4-2 Free enthalpy ΔG against the nucleus size L

The change of the total free enthalpy as a function of the size L passes through a maximum. A thermodynamically stable nucleus is given when the total free enthalpy is not changed by the addition or removal of few elementary units:

$$\frac{\partial \Delta G}{\partial L} = 0. \tag{8.4-3}$$

In this case the rate constant k_A of addition is equal to the rate constant k_Z of removal and neither dissolution nor growth of such a nucleus takes place. The last equations lead to the following result for a spherical nucleus:

$$L^*_{crit} = \frac{4 \cdot \gamma_{CL} \cdot \tilde{M}}{\tilde{R} \cdot T \cdot \rho_c \cdot \ln(c/c^*)} = \frac{4 \cdot \gamma_{CL} \cdot \tilde{M}}{\tilde{R} \cdot T \cdot \rho_c \cdot \ln(1 + \Delta c/c^*)}. \tag{8.4-4}$$

With the molecule diameter

$d_m \approx \sqrt[3]{\frac{\tilde{M}}{N_A \cdot \rho_c}}$ and the relative supersaturation $\sigma \equiv \Delta c / c^*$ and $S \equiv 1 + \sigma$, (8.4-5) the size L^*_{crit} of a critical nucleus based on the molecule diameter d_m is

$$\frac{L^*_{crit}}{d_m} = \frac{4 \cdot d_m^2 \cdot \gamma_{CL}}{k \cdot T \cdot \ln S}. \tag{8.4-6}$$

In Fig. 8.4-3 the ratio L^*_{crit}/d_m is plotted against the natural logarithm of the supersaturation S for two different surface tensions and two different molecule sizes (20°C).

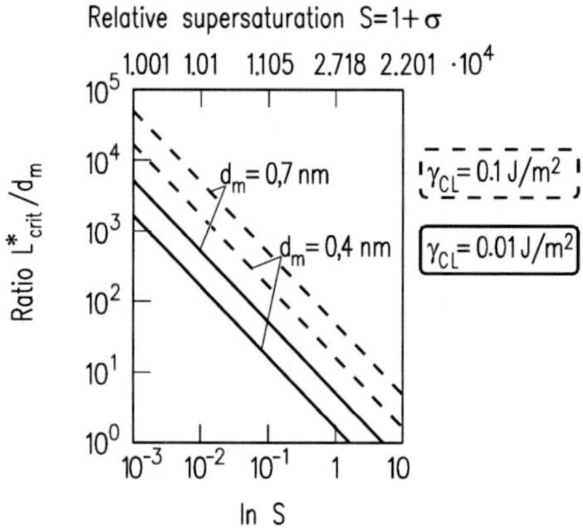

Fig. 8.4-3 Ratio L^*_{crit}/d_m as a function of the natural logarithm of the supersaturation S for different interfacial tensions and molecule diameters

Since the free enthalpy ΔG decreases for $L > L^*_{crit}$ in a system which is not in equilibrium, nuclei are growing. In the case $L < L^*_{crit}$, however, the change of the free enthalpy increases with increasing sizes. Therefore, according to $k_z > k_A$ the nucleus will dissolve.

The rate of the primary homogeneous nucleation $B_{0,hom}$ can be derived by multiplying an impact coefficient s with the total surface of all clusters n_c present in a given volume V. The impact coefficient is the number of molecules which are hitting the surface based on a unit of time and surface. The total surface of all critical clusters is given by the number n_c of clusters in a volume V and the surface A_c of a cluster. The rate of homogenous nucleation is

$$B_{0,hom} = s \cdot A_c \cdot \frac{n_c}{V} \cdot Z. \tag{8.4-7}$$

The imbalance factor Z takes into consideration that clusters which have just got the critical size are removed and are no longer in the cluster size distribution with the consequence that clusters in the different intervals of the distribution are in dynamic equilibrium to one another. According to Becker and Döring (1935) this factor is

$$Z = \sqrt{\frac{\Delta G_c}{3 \cdot \pi \cdot k \cdot T \cdot i_c^2}}. \tag{8.4-8}$$

Here ΔG_c is the free enthalpy of nucleation of a critical nucleus with i_c elementary units. This enthalpy is given (Volmer and Weber 1926) by

$$\Delta G_c = \frac{1}{3} \cdot A_c \cdot \gamma_{CL}. \tag{8.4-9}$$

Now it is assumed that the cluster distribution n_i/V is the result of random collisions of molecules and can be described by a Boltzmann distribution. According to Kind and Mersmann (1983), the distributions are

$$\frac{n_i}{V} = \frac{n_S}{V} \cdot \exp\left[-\frac{\Delta G_i}{k \cdot T}\right], \tag{8.4-10}$$

or for critical clusters

$$\frac{n_c}{V} = \frac{n_S}{V} \cdot \exp\left[-\frac{\Delta G_c}{k \cdot T}\right]. \tag{8.4-11}$$

With $\Delta G_c = 1/3 \cdot A_c \cdot \gamma_{CL}$ and the equation

$$L_c^* = \pi \cdot \sqrt{A_c} \tag{8.4-12}$$

for the critical cluster size, the distribution is

$$\frac{n_c}{V} = \frac{n_S}{V} \cdot \exp\left[-\frac{16 \cdot \pi}{3} \cdot \left(\frac{\gamma_{CL}}{k \cdot T}\right)^3 \cdot \left(\frac{\tilde{M}}{N_A \cdot \rho_c}\right)^2 \cdot \frac{1}{(\ln S)^2}\right]. \tag{8.4-13}$$

The number i_c of elementary units in a cluster with the diameter L_c is

$$i_c = \frac{\pi}{6} \cdot L_c^3 \cdot \frac{\rho_c \cdot N_A}{\tilde{M}}. \tag{8.4-14}$$

A combination of (8.4-7) and (8.4-13) leads to

$$B_{0,\text{hom}} = 2 \cdot s \cdot (\tilde{c} \cdot N_A) \cdot \sqrt{\frac{\gamma_{CL}}{k \cdot T}} \cdot \left(\frac{\tilde{M}}{\rho_c \cdot N_A}\right) \cdot \exp\left[-\frac{16 \cdot \pi}{3} \cdot \left(\frac{\gamma_{CL}}{k \cdot T}\right)^3 \cdot \left(\frac{\tilde{M}}{N_A \cdot \rho_c}\right)^2 \cdot \frac{1}{(\ln S)^2}\right]. \tag{8.4-15}$$

With the collision factor s (Kind 1990) which depends on the molecular diffusion coefficient D_{AB},

$$s = \frac{3}{4} \cdot (\tilde{c} \cdot N_A)^{4/3} \cdot D_{AB}, \tag{8.4-16}$$

we get

$$B_{0,\text{hom}} = 1.5 \cdot D_{AB} \cdot (\tilde{c} \cdot N_A)^{7/3} \cdot \sqrt{\frac{\gamma_{CL}}{k \cdot T}} \cdot \left(\frac{\tilde{M}}{\rho_c \cdot N_A}\right) \cdot \exp\left[-\frac{16 \cdot \pi}{3} \cdot \left(\frac{\gamma_{CL}}{k \cdot T}\right)^3 \cdot \left(\frac{\tilde{M}}{N_A \cdot \rho_c}\right)^2 \cdot \frac{1}{(\ln S)^2}\right] \tag{8.4-17}$$

or

$$\phi = \exp\left[-\frac{\Gamma}{(\ln S)^2}\right] \tag{8.4-18}$$

with $$\phi = \frac{B_{0,\text{hom}}}{1.5 \cdot D_{AB} \cdot (\tilde{c} \cdot N_A)^{7/3} \cdot \sqrt{\frac{\gamma_{CL}}{k \cdot T}} \cdot \left(\frac{\tilde{M}}{\rho_c \cdot N_A}\right)} \tag{8.4-19}$$

and $$\Gamma = \frac{16 \cdot \pi}{3} \cdot \left(\frac{\gamma_{CL}}{k \cdot T}\right)^3 \cdot \left(\frac{\tilde{M}}{N_A \cdot \rho_c}\right)^2. \tag{8.4-20}$$

In Fig. 8.4-4 this relationship is illustrated with an information on the operating range which is relevant for industrial crystallizers with respect to physical properties of substances and resulting nucleation rates.

A certain supersaturation $\Delta c_{\text{met, hom}}$ is necessary for a certain nucleation rate $B_{0,\text{hom}}$ for a given temperature, surface tension γ_{CL}, and molecule diameter $d_m \approx (\tilde{M}/(\rho_c \cdot N_A))^{1/3}$. This supersaturation $\Delta c_{\text{met, hom}}$ is termed the width of the metastable zone valid for homogeneous nucleation. In the range $0 < \Delta c < \Delta c_{\text{met, hom}}$, nearly no nuclei are produced; however, crystals can grow. The curve $(c^* + \Delta c_{\text{met, hom}})$ as a function of the temperature is called supersolubility curve which is dependent on thermodynamic and kinetic parameters and a given nucleation rate.

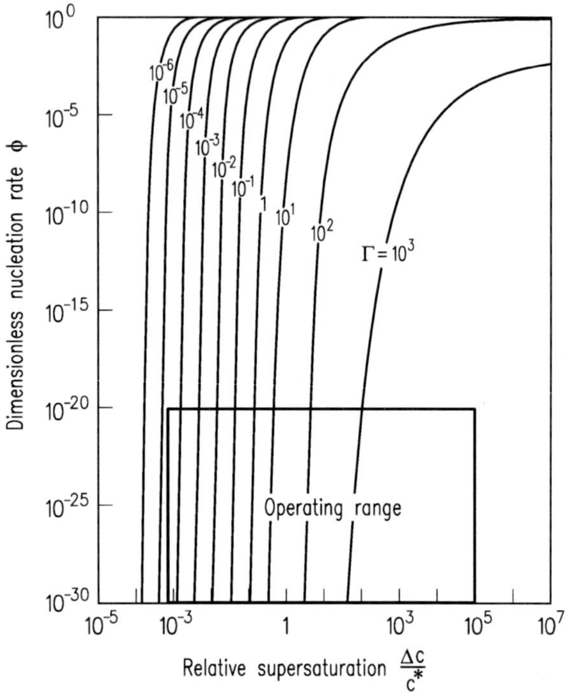

Fig. 8.4-4 Dimensionless nucleation rate ϕ as a function of the relative supersaturation $\Delta c/c^*$ for different material parameters Γ

In connection with the relationship (Mersmann 1990)

$$\gamma_{\mathrm{CL}} = K \cdot k \cdot T \cdot \left(\rho_{\mathrm{c}} \cdot \frac{N_{\mathrm{A}}}{\tilde{M}}\right)^{2/3} \cdot \ln\left(\frac{\rho_{\mathrm{c}}}{c^*}\right) \tag{8.4-21}$$

with $0.31 < K < 0.414$ (Mersmann 2001; Garside et al. 2002), the metastable supersaturation $\Delta c_{\mathrm{met,\,hom}}$ for a given rate of homogeneous nucleation can be calculated by (8.4-17). In Fig. 8.4-5 this metastable supersaturation based on the crystal density ρ_{c} is plotted against the dimensionless solubility c^*/ρ_{c} for nondissociating substances. Note that the molar crystal density $\tilde{c}_{\mathrm{c}}$ is $\tilde{c}_{\mathrm{c}} = \rho_{\mathrm{c}}/\tilde{M}$.

8.4.1.2 Heterogeneous Nucleation

Up to now it has been assumed that clusters of different sizes are the result of collisions of elementary units and that clusters greater than a critical size are able to grow. As an important prerequisite of such nucleation processes the solution must be absolutely pure without the presence of any foreign particles. Dealing with

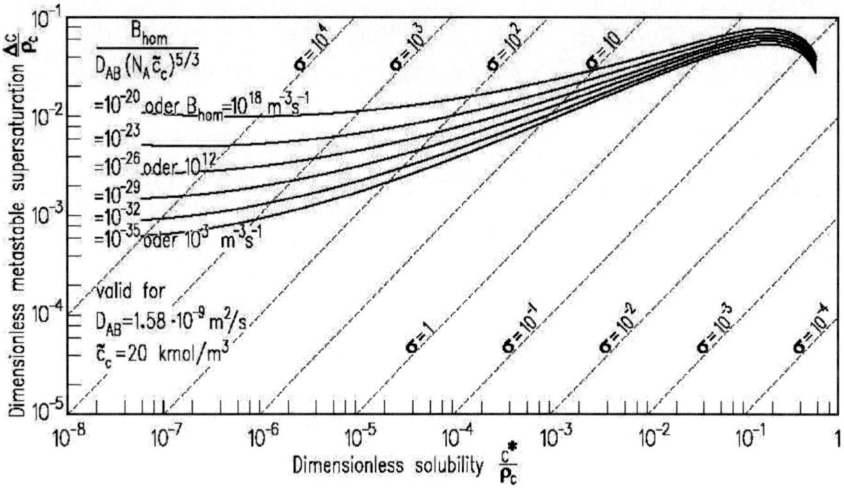

Fig. 8.4-5 Relationship between the dimensionless supersaturation $\Delta c/\rho_c$ and the dimensionless solubility c^*/ρ_c with the parameter σ (valid for $K = 0.414$)

industrial solutions this condition is never fulfilled. The question arises how nucleation works in the presence of foreign particles (sand, rust, etc.). This is explained in Fig. 8.4-6 in which a foreign particle is surrounded by a supersaturated solution.

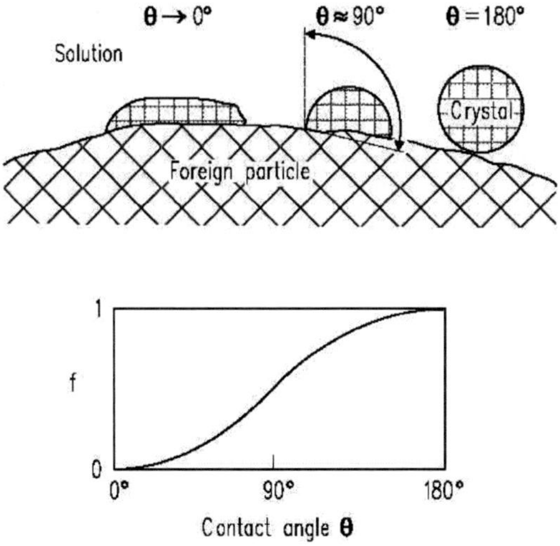

Fig. 8.4-6 Nucleation on a foreign particle for different contact angles θ (*above*); factor f as a function of the contact angle θ (below)

Elementary units are hitting the surface of the foreign particles and such impacts result in surface nuclei. The intensity of this process depends on the supersaturation of the solution, the surface energy, and the surface and lattice structure of the solid particle. In Fig. 8.4-6 three different contact angles ($\theta \to 0$; $\theta \approx 90°$; $\theta = 180°$) are depicted. According to the degree of "wetting" the contact angle θ can be in the range $0° < \theta < 180°$. In the case of $\theta = 180°$, the surface of the foreign particle is not "wetted" (the nucleus is attached on the foreign particle only at a point). This corresponds to homogeneous nucleation. For $0° < \theta < 180°$ the nucleation energy on the "wetted" surface is reduced and this reduction can be taken into account by the factor f in (8.4-22). In Fig. 8.4-6 the factor f is shown as a function of the contact angle θ according to ideas developed by Volmer:

$$\Delta G_{\mathrm{c,\,het}} = f \cdot \Delta G_{\mathrm{c}} = f \cdot \frac{A_{\mathrm{c}}}{3} \cdot \gamma_{\mathrm{CL}} \text{ and } f = \frac{(2+\cos\theta)\cdot(1-\cos\theta)^2}{4}. \tag{8.4-22}$$

In the case $\theta \to 0$ the foreign particle is completely wetted, and the nucleation energy and the supersaturation necessary for nucleation tend to zero. The supersaturation $\Delta c_{\mathrm{met,\,het}}$ which enables the growth of "wetted" foreign particles to become heterogeneous nuclei is called the metastable zone for heterogeneous nucleation with

$$\Delta c_{\mathrm{met,\,het}} < \Delta c_{\mathrm{met,\,hom}}. \tag{8.4-23}$$

The rate B_{het} of heterogeneous nucleation is (Schubert 1998)

$$B_{\mathrm{het}} = \frac{1}{2\cdot\pi}\cdot a_{\mathrm{f}}\cdot d_{\mathrm{m}}\cdot He_{\mathrm{ad}}\cdot(\tilde{c}\cdot N_{\mathrm{A}})^{7/3}\cdot\left(\frac{\gamma_{\mathrm{CL}}}{k\cdot T}\right)^{1/2}\cdot\frac{1}{\tilde{c}_{\mathrm{c}}\cdot N_{\mathrm{A}}}$$
$$\cdot\left[\frac{D_{\mathrm{s}}\cdot\sin\theta}{r_{\mathrm{c}}}\cdot(He_{\mathrm{ad}}\cdot d_{\mathrm{m}})^{3/2}\cdot(\tilde{c}\cdot N_{\mathrm{A}})^{1/6}+(1.5\cdot\pi\cdot D_{\mathrm{AB}})\cdot(1-\cos\theta)\right]$$
$$\cdot\exp\left[-f\cdot\frac{16\cdot\pi}{3}\cdot\left(\frac{\gamma_{\mathrm{CL}}}{k\cdot T}\right)^3\cdot\left(\frac{1}{\tilde{c}_{\mathrm{c}}\cdot N_{\mathrm{A}}}\right)^2\cdot\frac{1}{(\nu\cdot\ln S_{\mathrm{a}})^2}\right]. \tag{8.4-24}$$

In this equation a_{f} is the solid volume-based surface of all foreign particles and d_{m} is the molecule diameter of the solute. The Henry coefficient He_{ad} denotes the equilibrium of the solute molecules in the liquid phase and adsorbed molecules on the surface of the foreign particles. D_{s} is the surface diffusion coefficient (compare Chap. 9) and r_{c} denotes the radius of a critical nucleus. Assuming that all foreign particles have the same size as the solute molecules and that the contact angle assumes $\theta = 180°$ (or $f = 1$), (8.4-24) can be converted with $a_{\mathrm{f}} = d_{\mathrm{m}}^{-1}$ to the equation of Kind (Kind and Mersmann1990) valid for homogeneous nucleation: The rate of heterogeneous nucleation is strongly dependent on the contact angle θ (or on the factor f) whereas the dependency on the Henry coefficient He_{ad} and on the volumetric surface a_{f} of the foreign particles is less pronounced. Note that the supersaturation $S_a = 1 + (\Delta a/a^*)$ in (8.4-24) is written with activities instead of concentrations. Principally speaking, the supersaturation S_a is the real driving force for all nucleation processes. (Concentrations can be used for nondissociating ideal systems with the activity coefficient $\gamma_i = 1$.) Note the strong influence of the number of ions per solute molecule.

In industrial seeded crystallizers a large surface of crystals is available for impacts of solute molecules. In this case the process of activated nucleation (with an exponential term containing the supersaturation $\ln S_a$) requires the lowest supersaturation in comparison to the activated homogeneous and heterogeneous nucleation. With the volumetric holdup φ_T $[\mathrm{m^3 crystals/m^3 suspension}]$ and the Sauter diameter L_{32} of all crystals this surface nucleation rate B_{surf} is (Mersmann 2001)

$$B_{surf} = E \cdot \frac{6 \cdot \varphi_T}{L_{32}} \cdot \frac{D_{AB}}{d_m^4} \exp\left[-\pi \cdot \frac{[K \cdot \ln(\rho_c/c^*)]^2}{\nu \cdot \ln S_a}\right]. \tag{8.4-25}$$

Here E is a coefficient of efficiency with $0 < E < 1$. When all surface nuclei produced by dendritic growth and dendritic coarsening are separated from the crystals by partial dissolution of the nucleus base and especially by fluid dynamic forces the factor E becomes 1.

8.4.1.3 Attrition-Controlled Nucleation

Nucleation at very low supersaturation takes place when very small particles removed from the surface of a crystal or produced by attrition fragments from crystals obtain the ability of growth and become secondary nuclei. A prerequisite of secondary nucleation is the presence of crystals in a supersaturated solution. Some authors refuse the term "secondary nucleation" because the generation of attrition fragments is partly a mechanical process and no real nucleation. The larger such an attrition fragment without noticeable lattice deformation and the higher the supersaturation, the greater the chance to become a growing secondary nucleus and finally a new crystal. In the literature a huge number of equations for the description and prediction of the nucleation rate of secondary nuclei are known. Note that this is very problematic because there is a great number of parameters (physical properties of the crystals, frequency and intensity of collisions of mother crystals, grade of deformation of attrition fragments, supersaturation, etc.) which have a more or less essential influence on the rate of newborn secondary nuclei. Here only a relatively simple equation will be presented based on an attrition model developed by Gahn (Gahn and Mersmann 1999). This model takes into account the most important parameter which, however, can only be predicted in a certain range and not exactly. According to Gahn the number and the size of attrition fragments are dependent on the Vickers hardness H_V, the shear modules μ, and the fracture resistance $\Gamma/(K'$ of the mother crystals. In the case of a stirred vessel with the power number Ne and the flow number N_V (see Chap. 3) operated at a mean specific power input $\bar{\varepsilon}$ the effective rate of secondary nucleation is

$$B_{0,\mathrm{eff}} \approx 7 \cdot 10^{-4} \cdot \varphi_T \cdot \frac{H_V^5}{\mu^3} \cdot \left(\frac{K'}{\Gamma}\right)^3 \cdot \frac{\pi^2 \cdot \rho_c \cdot \bar{\varepsilon} \cdot N_V}{2 \cdot \alpha^3 \cdot Ne} \cdot \left(\frac{N_{a,\mathrm{eff}}}{N_{a,\mathrm{tot}}}\right) \cdot \eta_w^3 \cdot \eta_g . \qquad (8.4\text{-}26)$$

The ratio $N_{a,\mathrm{eff}}/N_{a,\mathrm{tot}}$ is the number of effectively growing attrition fragments based on their total number. This ratio depends on the supersaturation and is difficult to predict for any kind of crystalline fragments. According to experimental data this ratio has been found in the range $10^{-3} < N_{a,\mathrm{eff}}/N_{a,\mathrm{tot}} < 0.1$ (Mersmann 2001). The target efficiencies $\eta_w < 1$ for the collision velocity and $\eta_g < 1$ for the influence of the rotor geometry take into account deviations from a modelling based on assumptions valid for an ideal system. The letter α_v denotes the volume shape factor. Data for the solid material parameters H_V, μ, and $\Gamma/(K'$ and further information can be found in Mersmann (2001). Experimental data on the width of the metastable zone $\Delta\tilde{c}_{\mathrm{met,sec}}$ valid for the presence of mother crystals are given by Mullin (Mullin 1993) and Nyvlt (Nyvlt et al. 1970).

The effective supersaturation $\Delta a_{\mathrm{eff}} = a - a^*_{\mathrm{eff}}$ of real systems and $\Delta\tilde{c}_{\mathrm{eff}} = \tilde{c} - \tilde{c}^*_{\mathrm{eff}}$ of ideal systems are smaller than $(a - a^*)$ and $(\tilde{c} - \tilde{c}^*)$, respectively, because especially in small attrition fragments the lattice is deformed and this deformation leads to a higher chemical potential in comparison to small crystals with an uninjured lattice. It is difficult to predict this reduction of the driving force in a general way because this effect depends on a variety of different parameters (Zacher and Mersmann 1995). A progress in this area would result in a better understanding of the ratio $N_{a,\mathrm{eff}}/N_{a,\mathrm{tot}}$.

8.4.2 Crystal Growth

According to a very old growth model of Berthoud and Valeton (Berthoud 1912; Valeton 1924), elementary units (atoms, molecules, ions, dimers, etc.) are transported by diffusion and convection to a crystal surface in a supersaturated fluid and after this transport they are integrated by an integration step or a surface reaction (Mullin 1993). The properties of the fluid and solid substances, the flow conditions, and the supersaturation are the decisive parameters whether the first (diffusion) step or the second (integration) step is controlling crystal growth. This will be explained in more detail in Fig. 8.4-7 in which the concentration profile in the supersaturated solution close to the crystal surface is depicted.

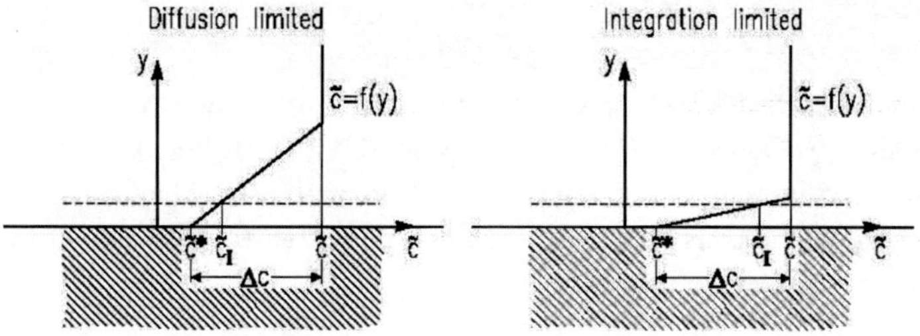

Fig. 8.4-7 Concentration profile for growth mainly limited by diffusion or surface integration

The solubility concentration is $\tilde{c}^*$ which is smaller than the bulk concentration $\tilde{c}$. The total concentration drop $\Delta\tilde{c} = \tilde{c} - \tilde{c}^*$ is split up into two contributions. The first part $(\tilde{c} - \tilde{c}_I)$ within the concentration boundary layer is the driving force for diffusion and convection whereas the second part $(\tilde{c}_I - \tilde{c}^*)$ in the very thin layer where the integration step takes place is effective for this step. The index I means Interface. In the case of growth completely controlled by diffusion and convection, $(\tilde{c}_I - \tilde{c}^*) \ll (\tilde{c} - \tilde{c}_I)$ or $(\tilde{c}_I - \tilde{c}^*)/(\tilde{c} - \tilde{c}_I) \ll 1$ is valid. Contrary to this with the ratio $(\tilde{c} - \tilde{c}_I)/(\tilde{c}_I - \tilde{c}^*) \ll 1$ crystal growth is controlled by the integration step. The molar flux density $\dot{n}$ directed toward the crystal surface is

$$\dot{n} = \beta \cdot (\tilde{c} - \tilde{c}_\mathrm{I}) = k_\mathrm{r} \cdot (\tilde{c}_\mathrm{I} - \tilde{c}^*)^r . \tag{8.4-27}$$

Here β is the mass transfer coefficient, k_r is the rate constant of the integration reaction, and r denotes the order of this reaction. As a rule the temperature dependency of the rate constant is described by an Arrhenius term:

$$k_\mathrm{r} = k_{r0} \cdot \exp\left(-\frac{\Delta E_\mathrm{r}}{R \cdot T}\right) . \tag{8.4-28}$$

Here $k_{\mathrm{r}0}$ is the rate constant and ΔE_r denotes the activation energy.

Instead of the molar flux density $\dot{n}$ the crystal growth rate is often described by the displacement rate $\bar{v}$ of a crystal face (for instance $\bar{v}_{111}$ for the 111 face). Note that every face can have different growth rates $\bar{v}$ at the same supersaturation. Let us assume spherical (poly)crystals. The growth rate is equal to the derivative of the radius with respect to time t ($\bar{v} = \mathrm{d}r/\mathrm{d}t$) or the derivative $G = \mathrm{d}L/\mathrm{d}t$ with L as the decisive length which is the diameter L for spheres. With the volume shape factor $\alpha = V_\mathrm{p}/L^3$ and the surface shape factor $\beta' = A_\mathrm{p}/L^2$, the following relationship between the mass flux density $\dot{m}$ ($\dot{m} = \dot{n} \cdot \tilde{M}$), the displacement rate $\bar{v}$ of the crystal surface, and the rate $G = 2\bar{v}$ of crystalline particles is given

$$\dot{m} = \frac{1}{A_p} \cdot \frac{dm}{dt} = \frac{6 \cdot \alpha}{\beta'} \cdot \rho_c \cdot \frac{dr}{dt} = \frac{6 \cdot \alpha}{\beta'} \cdot \rho_c \cdot \bar{v} = \frac{3 \cdot \alpha}{\beta'} \cdot \rho_c \cdot G. \quad (8.4\text{-}29)$$

The concentration profile depicted in Fig. 8.4-7 must be known for the prediction of growth rates; however, this is not the case. Before reflecting on the general case of a superposition of mass transport and integration mechanisms, two special regimes should be considered first: the control of crystal growth by diffusion/convection and by the integration reaction.

8.4.2.1 Growth Controlled by Diffusion

If the integration reaction is very fast or the rate constant $k_r \to \infty$, the growth rate is controlled by the diffusive/convective transport of elementary units. With $(\tilde{c} - \tilde{c}_I) \cong (\tilde{c} - \tilde{c}^*) = \Delta\tilde{c}$ the molar flux density valid for small rates is given by

$$\dot{n} = \beta \cdot \Delta\tilde{c} \quad (8.4\text{-}30)$$

or

$$\bar{v}_{\text{dif}} = \frac{\beta'}{6 \cdot \alpha} \cdot \beta \cdot \frac{\Delta c}{\rho_c} \quad (8.4\text{-}31)$$

or

$$G_{\text{dif}} = \frac{\beta'}{3 \cdot \alpha} \cdot \beta \cdot \frac{\Delta c}{\rho_c}. \quad (8.4\text{-}32)$$

Considering mass transfer coefficients β published in the literature the question is important whether the coefficient is valid for equimolar counterdiffusion or for a semipermeable interface or crystal surface (no transport from the solid into the fluid phase is assumed). Furthermore the transfer coefficient can be based on diffusion only or on a combined transport by diffusion and convection. The differences can become essential for high mass transfer rates occurring in systems with great solubilities.

8.4.2.2 Growth Controlled by Integration

If the mass transfer coefficient β is very high ($\beta \to \infty$ for very high velocities of low-viscous fluids with high diffusivities) the growth is controlled by the integration step or the integration reaction of the units. Besides other parameters the growth rate is dependent on the structure of the crystal surface (smooth or rough on a molecular scale) which is a function of the supersaturation and the solubility of the solute. In addition, the purity of the system can play an important role. Besides impurities, especially added additives or surfactants and their adsorption behavior on the various faces of a crystal have an effect on crystal growth.

The "birth and spread model" (B+S) describes the formation of critical nuclei on a smooth crystal surface and their subsequent growth. The so-called "nucleus above nucleus" model leads to

$$\bar{v}_{\mathrm{B+S}} = k_{\mathrm{B+S}} \cdot \left(\frac{\Delta c}{c^*}\right)^{5/6} \cdot \exp-\left(\frac{K_{\mathrm{B+S}}}{T^2} \cdot \left(\frac{c^*}{\Delta c}\right)\right). \tag{8.4-33}$$

In the case of a very small supersaturation this equation results in very small growth rates because growth is controlled by a low nucleation rate of two-dimensional nuclei.

With rising supersaturation, surface nuclei are formed (compare with the contribution on surface nucleation) and growth is favored. The growth rate valid for polynuclear (PN) growth is

$$\bar{v}_{\mathrm{PN}} = \frac{D_{\mathrm{AB}}}{3 \cdot d_{\mathrm{m}}} \cdot \left(\frac{\Delta c}{c_{\mathrm{c}}}\right)^{2/3} \cdot \exp\left(-\pi \cdot \frac{[K \cdot \ln(\rho_{\mathrm{c}}/c^*)]^2}{\nu \cdot \ln S_{\mathrm{a}}}\right). \tag{8.4-34}$$

In this equation the interfacial tension γ_{CL} has been replaced by the expression given in (8.4-21). With increasing supersaturation the crystal surface becomes more and more rough with the effect of stimulated growth.

Some authors have observed that the growth rate at very small supersaturation is greater than predicted by the nucleation models. This can be explained by the so-called BCF model (Burton et al. 1951). The authors assume that the presence of spiral dislocations which end somewhere on the crystal surface creates steps, which are thus a continuous source of favorable integration sites. The source of such screw dislocations is a lattice imperfection which prevents an ideally smooth crystal surface. The steps of these spiral dislocations are remote from the centers and considered to be parallel and the same distance apart from each other. The linear displacement rate of a face is controlled by surface diffusion. With the surface diffusion coefficient D_{s} the growth rate $\bar{v}_{\mathrm{BCF}}$ according to Burton, Cabrera, and Frank is

$$\bar{v}_{\mathrm{BCF}} = \frac{2 \cdot k \cdot T \cdot \Gamma^* \cdot D_{\mathrm{s}}}{19 \cdot x_{\mathrm{s}} \cdot \gamma_{\mathrm{CL}}} \cdot (\nu \cdot \ln S_{\mathrm{a}}) \cdot \sigma \cdot \tanh\left(\frac{19 \cdot V_{\mathrm{m}} \cdot \gamma_{\mathrm{CL}}}{2 \cdot x_{\mathrm{s}} \cdot k \cdot T \cdot \nu \cdot \ln S_{\mathrm{a}}}\right). \tag{8.4-35}$$

Here x_s is the mean displacement of adsorbed units and Γ^* denotes the equilibrium constant of adsorbed units on the crystal surface (units per area). V_m is the volume of a growth unit with $V_m \sim d_m^3$.

Both the BCF model and the B+S model predict a strong (approximately quadratic) increase of the growth rate with supersaturation but for greater values of this driving force σ the growth rate is proportional to σ. A disadvantage of these models is that some parameters in the equations are not available for any system. Therefore, the crystal growth rate is often described by the reaction kinetics equation

$$\bar{v} = k_g' \cdot \sigma^g . \tag{8.4-36}$$

As a rule the exponent g is $1 < g < 2$. The rate constant k_g' is dependent on the temperature and has to be determined experimentally. In general the mean growth rate of a face can be obtained by a superposition according to

$$\bar{v} = \left(\frac{1}{\bar{v}_{BCF} + \bar{v}_{B+S} + \bar{v}_{PN}} + \frac{1}{\bar{v}_{dif}} \right)^{-1} . \tag{8.4-37}$$

This relationship is depicted in Fig. 8.4-8.

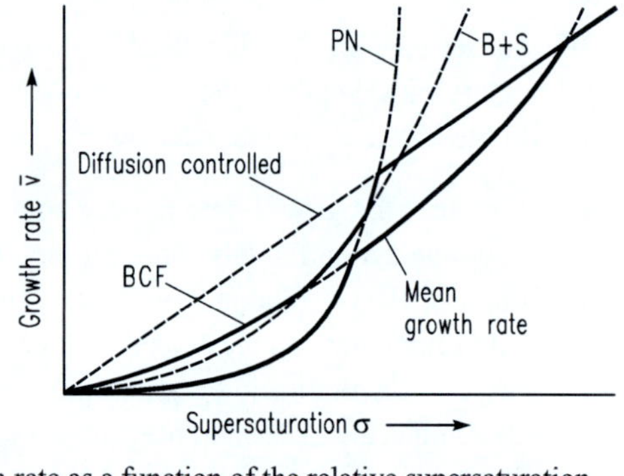

Fig. 8.4-8 Growth rate as a function of the relative supersaturation

Dealing with industrial crystallization the most important limiting steps are the growth rates according to the BCF model and to diffusion. In the following, simplified equations are presented. These relationships take into account these two controlling steps.

8.4.2.3 Growth Controlled by Diffusion and Integration

In most industrial crystallizers growth is controlled by both diffusion and the BCF integration step. As a rule the molar flux density $\dot{n}$ is described by the following equation:

$$\dot{n} = k_{\rm g} \cdot (\Delta \tilde{c})^{g} \tag{8.4-38}$$

with $1 < g < 2$ as the exponent. Taking into account the relationship

$$\dot{n}_{\rm dif} = \beta \cdot (\tilde{c} - \tilde{c}_{\rm I}) \tag{8.4-39}$$

valid for pure diffusion control and combining this equation derived for pure integration control the unknown concentration $\tilde{c}_{\rm I}$ at the interface can be eliminated (compare Fig. 8.4-7):

$$\dot{n} = k_{\rm r} \cdot \left(\Delta \tilde{c} - \frac{\dot{n}}{\beta}\right)^{r}. \tag{8.4-40}$$

This equation reads for the special cases $r = 1$ and $r = 2$:

$$r = 1: \qquad \dot{n} = \frac{\Delta \tilde{c}}{1/\beta + 1/k_{\rm r}}. \tag{8.4-41}$$

$$r = 2: \qquad \dot{n} = \beta \cdot \Delta \tilde{c} + \frac{\beta^2}{2k_{\rm r}} - \left(\frac{\beta^4}{4k_{\rm r}^2} + \frac{\beta^3 \cdot \Delta \tilde{c}}{k_{\rm r}}\right)^{1/2}. \tag{8.4-42}$$

A comparison of the reaction constants of approximately 40 different systems has shown that the mean integration-controlled growth rate $\bar{v}_{\rm int}$ valid in the temperature range between 20 and 30°C can be described by (Mersmann 1995)

$$\bar{v}_{\rm int} = 2.25 \times 10^{-3} \cdot \frac{D_{\rm AB}}{d_{\rm m}} \cdot \left(\frac{\rho_{\rm c}}{c^*}\right)^{2/3} \cdot \frac{1}{\ln(\rho_{\rm c}/c^*)} \cdot \left(\frac{\Delta c}{\rho_{\rm c}}\right)^{2} \cdot \nu^2. \tag{8.4-43}$$

Here ν is the number of ions of dissociating solutes. The introduction of this equation in (8.4-42) leads to a relationship which is depicted in Fig. 7.4-9 where the dimensionless growth rate $G/(2 \cdot \beta) = \bar{v}/\beta$ is plotted against the dimensionless supersaturation $(\Delta c)/\rho_{\rm c}$ with the crystallization parameter P^* according to

$$P^* = \frac{\beta \cdot d_{\rm m}}{D_{\rm AB}} \cdot \left(\frac{c^*}{\rho_{\rm c}}\right)^{2/3} \cdot \ln\left(\frac{\rho_{\rm c}}{c^*}\right). \tag{8.4-44}$$

The diagonal in this figure corresponds to the growth rate which is limited by diffusion. Note that this is the maximum growth rate for a given supersaturation Δc. The growth rate $\bar{v} < \bar{v}_{\rm dif}$ depends on the dimensionless supersaturation $\Delta c/\rho_{\rm c}$ and the crystallization parameter P^* and takes into account an additional growth resistance provided by the integration reaction. Data of growth rates limited only by integration can be found in the shaded area which lies one order of magnitude below the diagonal. It is important to note that Fig. 8.4-9 gives only approximate values for the temperature of 20°C and systems without an essential influence provided by impurities or additives which reduce the growth rate. Note that there is no dependency of the frequency and intensity of kinks or – in general – of the roughness of the crystal surface which plays a role in the BCF equation. Finally many experiments have shown that crystal growth is dependent on the stress and lattice deformation within a crystal (Zacher and Mersmann 1995). Hence it can be explained that a significant growth rate dispersion has been observed. This means that the growth rate differs from crystal to crystal despite the fact that macroscopic parameters such as the supersaturation, the temperature, the flow conditions, and the degree of turbulence are exactly the same for a given collection of crystals.

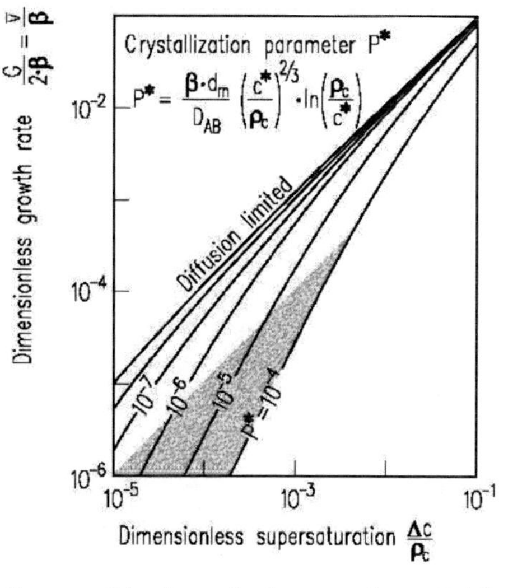

Fig. 8.4-9 General diagram of the dimensionless growth rate $\bar{v}/\beta$ vs. the dimensionless supersaturation ($\nu = 1$)

As a first approach it can be assumed that crystal growth in a solution is an isothermal process. Principally speaking, the negative heat of crystallization has to be added or the positive heat must be removed; however, the heat capacity of the solution is very high in comparison to the heats of phase change that an isothermal modelling is sufficient in low viscous aqueous solutions.

8.4.3 Aggregation and Agglomeration

The quality of crystalline products can be strongly influenced by processes such as aggregation (absence of supersaturation) and agglomeration in supersaturated solutions (Judat and Kind 2004; Schwarzer and Peukert 2004). There is a great number of collisions between particles suspended in a stirred vessel or in a fluidized bed. Such collisions are favored by attractive forces (for instance van der Waals forces) or hindered by repulsive forces (for instance electrostatic forces). In a supersaturated solution aggregated crystals can grow together and build crystalline bridges. This is called agglomeration (Israelachvili 1995). Contrary to nucleation, supersaturation is not the driving force for the first collision step introducing processes as aggregation and agglomeration. (Sometimes it is difficult to distinguish whether the birth of a detectable tiny crystal is an agglomeration, a growth or a real nucleation event (Mersmann 2002)). The combination of two or more particles requires

- A collision (which can be favored or hindered by interparticle forces)
- A cohesion (by cohesion forces in not supersaturated solutions or by crystalline bridges in supersaturated solutions)

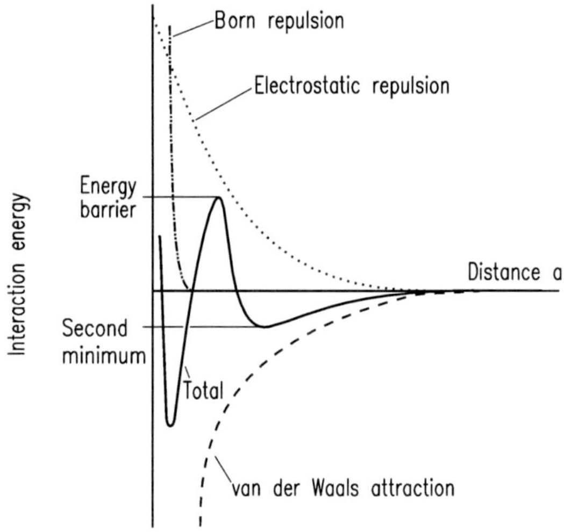

Fig. 8.4-10 Interaction energies as a function of the distance a according to the DLVO theory

In Fig. 8.4-10 (negative) attractive van der Waals forces or interaction energies and (positive) repulsive electrostatic forces or interaction energies are plotted against the distance a. Furthermore the repulsive force according to Born and the resulting total interaction energy curve with two relative maxima (which corresponds to an energy barrier) and two relative minima are depicted in the figure.

At first the simple case will be discussed that the different forces are neutralizing and the resulting interaction energy is zero. It is practical to introduce a volume-based population balance n_V according to

$$n_V \equiv \frac{\text{number of particles}}{(\text{m}^3 \text{ particle volume in a volume interval})\ (\text{m}^3 \text{ suspension})}. \tag{8.4-45}$$

This is reasonable because the volume of all particles remains constant in the absence of nucleation and crystal growth. The volume-based population balance is given by

$$\frac{\partial n_V}{\partial t} + \frac{\partial (n_V \cdot G_V)}{\partial u} + n_V \cdot \frac{\partial V}{V \cdot \partial t} = \\ B_{agg}(u) - D_{agg}(u) + B_{dis}(u) - D_{dis}(u) + B_u \cdot \delta(u - u_0) + \frac{n_{Vf} - n_{Vp}}{\tau}. \tag{8.4-46}$$

Here G_V is the volume-based growth rate in m^3/s. $B_{agg}(u) - D_{agg}(u)$ and $B_{dis}(u) - D_{dis}(u)$ are the net formation rates of birth and death events of particles with the volume u by aggregation (index agg) and disruption (index dis). The source term B_u is the birth rate of the smallest particles with the volume u_0 which can be detected by measurement or which play a role in a crystalline product. The combination of two particles with the volumina u and $(v - u)$ to a particle with the volume v is given by

$$D_{agg}(u) = n_V(v, t) \cdot \int_0^\infty \beta(u, v) \cdot n_V(u, v) \cdot du. \tag{8.4-47}$$

Here β (in m^3/s) is the rate constant of aggregation or the aggregation kernel which depends on the particle volumina u and $(v - u)$.

The perikinetic aggregation is controlled by diffusion, and the aggregation kernel of collisions in a monodisperse suspension of particles with the size L is

$$\beta_{col} = 4 \cdot \pi \cdot D_{AB} \cdot L \approx \frac{8 \cdot k \cdot T}{3 \cdot \eta_L}. \tag{8.4-48}$$

This equation is based on the assumption that the equation for the diffusion coefficient according to Einstein is valid.

The aggregation of particles with sizes $L \gg 2 \cdot k \cdot T/(3 \cdot \pi \cdot D_{AB} \cdot \eta_L)$ in a monodisperse suspension in a shear field with the shear rate $\dot{\gamma}$ is called orthokinetic aggregation and the collision kernel is

$$\beta_{col} = \frac{4}{3} \cdot \dot{\gamma} \cdot L^3. \tag{8.4-49}$$

Dealing with industrial crystallization a collection of crystals with a more or less broad crystal size distribution is suspended and not only collisions of two crystals take place. However, the probability of collisions between three or more crystals is very low. Therefore, a modelling based on collisions of only two particles comes close to reality. The collision kernel β_{col} of the collision of two crystals with the sizes L_1 and L_2 is given by

$$\beta_{col} = 4\pi \cdot D_{AB} \cdot L \cdot \left(\frac{1}{L_1} + \frac{1}{L_2}\right) \cdot (L_1 + L_2) \text{ with } L \approx (L_1 + L_2)/2 \qquad (8.4\text{-}50)$$

valid for perikinetic aggregation and

$$\beta_{col} = \frac{4}{3} \cdot \dot{\gamma} \cdot \left(\frac{L_1}{2} + \frac{L_2}{2}\right)^3 \qquad (8.4\text{-}51)$$

for orthokinetic aggregation.

In table 8.4-1 some equations for both kinds of aggregation are listed. The aggregation rate $dN(t)/dt$ is proportional to N^2 which is the square of the volume-based number of particles. This rate decreases with increasing time provided that not new crystals by nucleation or disruption are born.

Let us now consider aggregation processes in suspensions with arbitrary particle size distributions. At first the particle density of the volume-based population balance will be converted into the particle density of a population balance with the particle size as the chosen parameter. In Fig. 8.4-11 the combination of a particle

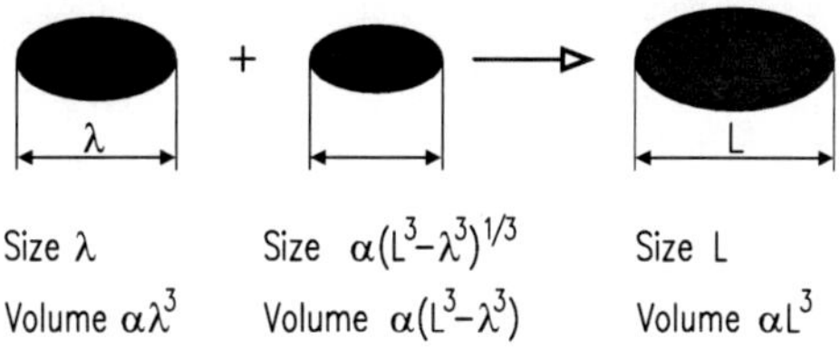

Fig. 8.4-11 Aggregation of two particles with different sizes and volumes

with diameter λ and the volume $u = \alpha \cdot \lambda^3$ and a particle with diameter $\alpha \cdot (L^3 - \lambda^3)^{1/3}$ is depicted. This unification leads to an aggregate with the size L and the volume $\alpha \cdot L^3$:

$$\begin{aligned} (u) + \quad (v) \quad &= (u + v) \quad \text{or} \\ \alpha \cdot \lambda^3 + \alpha \cdot (L^3 - \lambda^3) &= \alpha \cdot L^3. \end{aligned} \qquad (8.4\text{-}52)$$

The number of particles with diameter λ is N_λ, and $N_{L-\lambda}$ denotes the number of particles with diameter $\alpha(L^3-\lambda^3)^{1/3}$. In a given crystal size interval, the birth rate (index B) due to aggregation is proportional to the kernel β and to the product $N_\lambda \cdot N_{L-\lambda}$. Dealing with particle size distributions the contributions of various intervals have to be added up:

Table 8.4-1 Perikinetic and orthokinetic agglomeration

Perikinetic agglomeration
$\frac{dN(t)}{dt} = -4 \cdot \pi \cdot D_{AB} \cdot L \cdot N^2$ or $\frac{d(1/N)}{dt} = 4 \cdot \pi \cdot D_{AB} \cdot L$
$\frac{dZ}{dt} = \frac{d}{dt}\left(\frac{N_0}{N(t)}\right) = 4 \cdot \pi \cdot D_{AB} \cdot L \cdot N_0$
$Z \equiv \frac{N_0}{N(t)} = 1 + 4 \cdot \pi \cdot D_{AB} \cdot L \cdot N_0 \cdot t$
$N(t) = \frac{N_0}{1 + 4 \cdot \pi \cdot D_{AB} \cdot L \cdot N_0 \cdot t}$
Orthokinetic agglomeration
$\frac{dN(t)}{dt} = -\frac{2}{3} \cdot A_{agg} \cdot \dot{\gamma} \cdot L^3(t) \cdot N^2(t)$
or with the volumetric holdup $\varphi_T = \alpha \cdot L^3(t) \cdot N(t) \neq f(t)$
$\frac{dN(t)}{dt} = -\frac{2}{3 \cdot \alpha} \cdot A_{agg} \cdot \dot{\gamma} \cdot \varphi_T \cdot N(t)$
$Z \equiv \frac{N_0}{N(t)} = \exp\left(\frac{2 \cdot A_{agg} \cdot \dot{\gamma} \cdot \varphi_T \cdot t}{3 \cdot \alpha}\right)$
$N(t) = \frac{N_0}{\exp(2 \cdot A_{agg} \cdot \dot{\gamma} \cdot \varphi_T \cdot t/(3 \cdot \alpha))} = \frac{\varphi_T}{\alpha \cdot L^3(t)}$

$$\left(\frac{dN_L}{dt}\right) = \frac{1}{2} \sum_{\lambda=0}^{\alpha(L^3-\lambda^3)^{1/3}} \beta(\lambda, \alpha(L^3-\lambda^3)^{1/3}) \cdot N_\lambda \cdot N_{L-\lambda}. \tag{8.4-53}$$

At the same time, particles which have been subject to a successful collision of two or more particles disappear in the interval (death rate D). The net rate as the result of birth and death events is given by

$$\left(\frac{\mathrm{d}N_L}{\mathrm{d}t}\right)_{\mathrm{netto}} = \underbrace{\frac{1}{2}\sum_{\lambda=0}^{\alpha(L^3-\lambda^3)^{1/3}} \beta(\lambda,\alpha(L^3-\lambda^3)^{1/3})\cdot N_\lambda \cdot N_{L-\lambda}}_{\left(\frac{\mathrm{d}N_L}{\mathrm{d}t}\right)_B} - \underbrace{N_L \sum_{\lambda=0}^{\infty} \beta(L,\lambda)\cdot N_\lambda}_{\left(\frac{\mathrm{d}N_L}{\mathrm{d}t}\right)_D} \tag{8.4-54}$$

Hounslow (1990) has shown how the volume-based rates $B_{agg}(V)$ and $D_{agg}(V)$ can be converted into the rates $B_{agg}(L)$ and $D_{agg}(L)$ which are related to the crystal size L:

$$B_{agg}(L) = \frac{L^2}{2}\cdot\int_0^L \beta(\lambda,(\alpha\cdot L^3-\alpha\cdot\lambda^3)^{1/3})\cdot\frac{n(\lambda)\cdot n\cdot(\alpha\cdot L^3-\alpha\cdot\lambda^3)^{1/3}}{(\alpha\cdot L^3-\alpha\cdot\lambda^3)^{2/3}}\,\mathrm{d}\lambda \tag{8.4-55}$$

and

$$D_{agg}(L) = n(L)\cdot\int_0^\infty \beta(L,\lambda)\cdot n(\lambda)\cdot\mathrm{d}\lambda. \tag{8.4-56}$$

Thanks to these equations we are now in the position to take into account agglomeration besides processes driven by supersaturation like nucleation and growth; however, there are restrictions. With increasing size, aggregates and agglomerates are running the risk to be destroyed by forces due to fluid flow. Then the intraparticle forces are decisive whether a disruption of particles takes place. David and Villermaux (1991) have introduced a model for the volume-based disruption rate by birth (B) and death (D) with the disruption frequency $\Gamma(\alpha\cdot L^3)$ and the disruption probability $p(\alpha\cdot L^3)$ as the decisive parameter:

$$B_{dis}(u) = \int_{\alpha L^3}^{\infty} \Gamma(\alpha\cdot\lambda^3)\cdot\nu(\alpha\cdot\lambda^3)\cdot p(\alpha\cdot\lambda^3,\alpha\cdot L^3)\cdot n_v(\alpha\cdot\lambda^3)\cdot\mathrm{d}(\alpha\cdot\lambda^3) \tag{8.4-57}$$

and

$$D_{dis}(u) = \Gamma(\alpha\cdot L^3)\cdot n_v(\alpha\cdot L^3). \tag{8.4-58}$$

The parameter $\nu(\alpha \cdot \lambda^3)$ is the number of fragments which are born after the disruption of an aggregate or agglomerate and is based on the volume $(\alpha \cdot L^3)$. The stability of a particle is a function of its size L and the intraparticle forces F (or the energy $dE = F \cdot dD$ with D as the distance). In Fig. 8.4-12 the intraparticle strength in N/m^2 (or J/m^3) is shown against the size L of a particle for adhesion forces within an agglomerate or aggregate (van der Waals forces, liquid and solid bridges between primary particles, and strength of solid materials as inorganic salts and sugar). For instance the van der Waals energy E_{vdW} between two solid spheres with radii R_1 and R_2 is given by

$$E_{vdW} = \frac{-Ha_{12}}{6 \cdot a} \cdot \left(\frac{R_1 \cdot R_2}{R_1 + R_2}\right). \tag{8.4-59}$$

The Hamaker constant Ha has already been introduced in the first chapter and data are often in the range between 10^{-21} and 10^{-20} J.

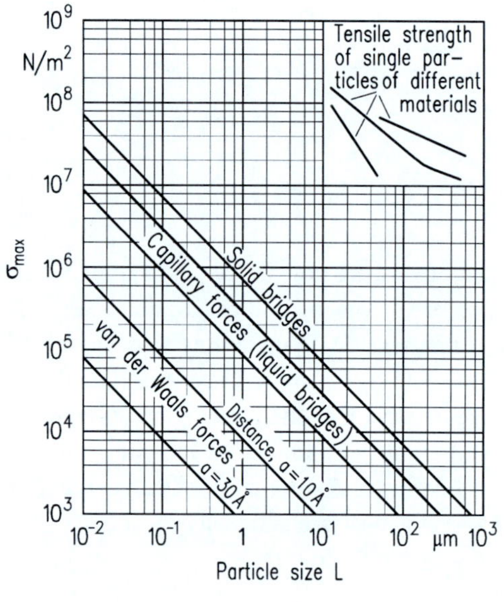

Fig. 8.4-12 Relationship between the maximum tensile strength and the size L of primary particles for different kinds of bonding

The capillary forces according to liquid bridges between two particles are approximately two orders of magnitude greater than van der Waals forces. In Fig. 8.4-13 the dimensionless adhesion force $F^* = F_{max}/(\gamma_{LG} \cdot L)$ is plotted against the dimensionless distance a/L of two particles for different adhering liquid volumes φ based on the solid volume expressed as

$$\varphi = \frac{\text{liquid volume}}{\text{solid volume}} . \qquad (8.4\text{-}60)$$

The maximum bonding force is

$$F_{max} \approx (2-2.5) \cdot \gamma_{LG} \cdot L \qquad (8.4\text{-}61)$$

for a very small distance with $a \to 0$. In the case of a solid bridge built by crystallization which is induced by supersaturation the fracture resistance (Γ/K') is the decisive material parameter which can be estimated by the Orowan equation:

$$(\Gamma/K') \approx 1.7 \cdot E \cdot \left(\frac{1}{n \cdot \tilde{c}_c \cdot N_A}\right)^{1/3} . \qquad (8.4\text{-}62)$$

Note that the fracture resistance of polycrystals is approximately two orders of magnitude greater than the surface tension γ_{LG} of liquids. ($\Gamma/K' \approx 5-15\ J/m^2$, see Mersmann 2002).

Crystals which have been grown at a low supersaturation or small growth rate show the highest tensile strength. With the modulus (Young) $E \approx 2\times10^{10}\ N/m^2$ (or $2\times10^{10}\ J/m^3$) the maximum tensile strength σ_{max} is $\sigma_{max} \approx 10^8\ N/m^2$ for polycrystals with a size of 1 mm according to $\sigma_{max} \approx 0.005 \cdot E$. Note that there are at least six orders of magnitude between van der Waals strengths and maximum tensile strengths of solid crystals for a particle size of $L = 1\ \mu m$.

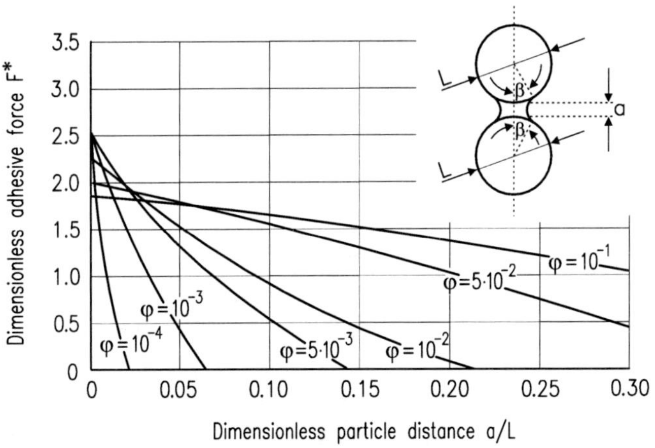

Fig. 8.4-13 Adhesion force F^* vs. the ratio a/L of the distance a based on the size L with the liquid holdup φ as parameter (Schubert 1981)

Up to now it has been assumed that there are no forces between the particles. As a rule this is not true. Section 8.4.3.1 deals with forces between particles or interparticle forces.

8.4.3.1 Forces Between Particles

Here we will discuss in more detail the so-called DLVO forces (Derjaguin and Landau 1941; Vervey and Overbeek 1948) and only mention other forces such as originating from solvation, hydration, and structure-induced. The van der Waals force has already been described. The repulsive force $F_{\mathrm{rep}} = \mathrm{d}E_{\mathrm{rep}}/\mathrm{d}D$ between two spheres with the radii r is given by (Israelachvili 1995)

$$F_{\mathrm{rep}} = 2\pi \cdot r \cdot \varepsilon_{\mathrm{r}} \cdot \varepsilon_0 \cdot \kappa \cdot \psi_0^2 \cdot \exp(-\kappa \cdot D) = 2\pi \cdot r \cdot \sigma^2 \cdot \frac{\exp(-\kappa \cdot D)}{\kappa \cdot \varepsilon_{\mathrm{r}} \cdot \varepsilon_0} . \quad (8.4\text{-}63)$$

Here $\varepsilon_0 = 8.857 \times 10^{-12}$ (As)/(Vm) is the electrical permittivity of the free space, ε_{r} is the relative permittivity and D denotes the distance between the particles. The decay length or the Debye–Hückel parameter $1/\kappa$ is dependent on the concentration N_i^0 of the ion species i with the valence z_i and the charge e of an electron ($e = 1.60 \times 10^{-19}$ C):

$$\frac{1}{\kappa} = \left(\frac{\varepsilon_{\mathrm{r}} \cdot \varepsilon_0 \cdot k \cdot T}{\sum (z_i \cdot e)^2 \cdot N_i^0} \right)^{1/2} . \quad (8.4\text{-}64)$$

In the case of a low electrolyte concentration with $|z_i \cdot e \cdot \psi| < k \cdot T$, the electrical potential ψ can be described by a simple equation:

$$\psi = \psi_0 \cdot \exp(-\kappa \cdot D) . \quad (8.4\text{-}65)$$

In Fig. 8.4-14 the surface potential ψ is shown as a function of the distance D. The highly ordered Helmholtz layer exists close to the surface for $D \to 0$ and can be distinguished from the diffuse bulk layer. The greek letter σ in (8.4-63) denotes the surface charge density.

The resulting DLVO energy $E_{\mathrm{DLVO}} = E_{\mathrm{vdW}} + E_{\mathrm{rep}}$ is given by

$$E_{\mathrm{DLVO}} = \pi r \left(-\frac{Ha_{131}}{12\pi \cdot D} + \frac{64 N_i^0 \cdot k \cdot T \cdot \Gamma_0^2 \cdot \exp(-\kappa \cdot D)}{\kappa^2} \right) . \quad (8.4\text{-}66)$$

Here r is the radius of the spheres and the dimensionless surface potential Γ_0 is

$$\Gamma_0 = \frac{\exp\left(\frac{z \cdot e \cdot \psi_0}{2k \cdot T} - 1\right)}{\exp\left(\frac{z \cdot e \cdot \psi_0}{2k \cdot T} + 1\right)}. \qquad (8.4\text{-}67)$$

The following statements can be drawn from the DLVO theory:

- High surface charges σ in diluted electrolytes lead to long-distance repulsive forces and result in poor aggregation and agglomeration rates.
- In solutions with high concentration a stable colloid can exist with not aggregated or agglomerated particles. The concentration is decisive on the tendency of aggregation.
- The aggregation is favored with decreasing surface charge σ of the particles.
- The inhibition of particle unification can be avoided in a system in which the energy of interaction becomes zero.
- The attractive van der Waals forces which favor the unification of particles become dominant in particle systems in which the surface charge σ and the electrical potential tend to zero.

Note that the processes of aggregation and agglomeration can be considerably influenced by the addition of ions present in electrolytes. Therefore, the kernels introduced earlier have to be corrected by the efficiency W_{eff} of the collisions according to

$$\beta = \beta_{\text{col}} \cdot W_{\text{eff}} \qquad (8.4\text{-}68)$$

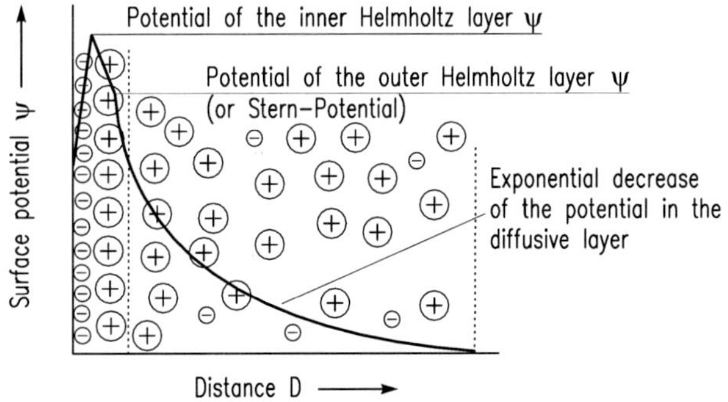

Fig. 8.4-14 Electrostatic potential ψ as a function of the distance. The potential rises from the value at $D = 0$ within the adsorbed layer to a maximum in the Helmholtz layer followed by a linear decrease to the Stern potential. The Guoy–Chapman equation describes this decrease in the double layer

with

$$W_{\text{eff}} = \frac{\text{number of successful collisions}}{\text{number of collisions in systems without forces between particles}}.$$

This efficiency mainly depends on the potential $E(r)$ of interaction and on the radius R of the particle (D is the distance):

$$W_{\text{eff}} = 2R \cdot \int_{2R}^{\infty} \frac{1}{D^2} \cdot \exp\left(\frac{E(D)}{k \cdot T}\right) \cdot \mathrm{d}D. \tag{8.4-69}$$

If the DLVO theory can be applied and no further forces between the particles in a suspension are effective, the interaction potential $E(D)$ becomes $E(D) = E_{\text{DLVO}}$.

To avoid agglomeration it is necessary to change the physicochemical conditions of the suspension in such a way that the interparticle repulsive forces are stronger than the attractive forces. As a rule, the process of aggregation can be promoted by the following measures:

- Elevated temperature
- Low viscosities
- Small particle sizes
- High particle concentration
- High diffusivities

if the physicochemical conditions allow rapid aggregation. Increasing turbulence or fluctuating velocities v_{eff}', shear rates $\dot{\gamma}$ ($\dot{\gamma} \sim n \sim \sqrt{\varepsilon/\nu_L}$), shear stress τ ($\tau = \eta_L \cdot \dot{\gamma}$ or $\tau \sim \rho_L (v_{\text{eff}}')^2$), and local specific power input ε ($\varepsilon = (v_{\text{eff}}')^3/\Lambda$) favor the collision frequency and aggregation. However, due to increasing shear stresses, the disruption of weakly bounded aggregates becomes stronger. Often the aggregation rate passes through a maximum with increasing speed.

8.4.4 Nucleation and Crystal Growth in MSMPR Crystallizers

Dealing with industrial crystallization it is necessary to limit the supersaturation and the kinetic parameter nucleation and growth to produce crystals with sizes and purities wanted by the customer. The metastable zone valid for primary homogeneous nucleation has already been discussed, see Fig. 8.4-5. Industrial crystallizers are operated at supersaturations which are one order of magnitude lower than $\Delta c_{\text{met, hom}}$ ($\Delta c \approx 0.1\ \Delta c_{\text{met, hom}}$). This means that substances with a high solubility

($\tilde{c}^* > 0.1$ mol/dm^3) are crystallized at supersaturations $\sigma < 0.1$ with the consequence that primary homogeneous nucleation can be neglected. Dealing with clean solutions the rate of heterogeneous nucleation is negligible; however; in industrial

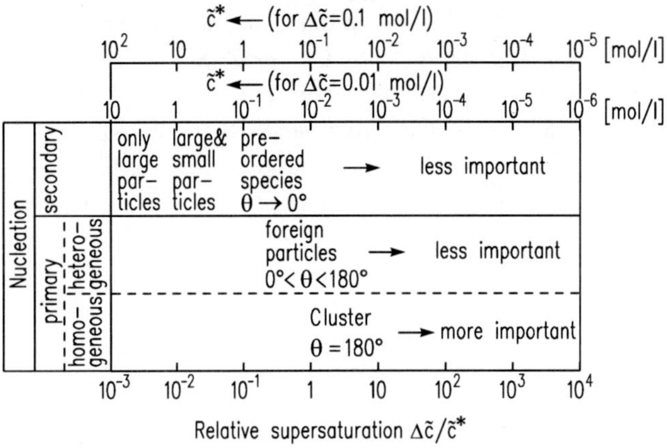

Fig. 8.4-15 Interrelationship between the solubility, the relative supersaturation, and the relevant mechanisms of nucleation

liquids a large number of foreign particles of different substances and sizes are present. Nevertheless heterogeneous nucleation plays a minor role because supersaturation in industrial crystallizers is too low. When such crystallizers are operated at high suspension densities of crystals with sizes $L_{50} > 100$ μm a huge number of attrition fragments is generated by collisions of particles with blades of stirrers or pumps, with apparatus walls or with other crystals. As a rule, attrition-induced secondary nucleation is dominant especially in systems with a high solubility and a coarse crystalline product. The smaller the solubility of a substance the higher the relative supersaturation to obtain a sufficient growth rate, compare Fig. 8.4-15. With increasing relative supersaturation the probability of the formation of heterogeneous nuclei rises and the mean crystal size decreases. When the mean crystal size is small (say $L_{50} < 100$ μm) the number of attrition fragments and secondary nuclei decreases with the consequence that primary heterogeneous nucleation becomes dominant. Primary homogeneous nucleation plays no big role because tiny particles in the nanometer range (sand, rust, etc.) are always present in any liquid. Dealing with reaction crystallization and precipitation the supersaturation can be so high in a small volume for a short time that homogeneous nuclei are born.

A general predictive calculation of nucleation rates is difficult with respect to the complex processes involved as attrition and the presence of foreign particles and substances as well as additives and inhibitors. Therefore, experiments are carried out to measure the rate of secondary nucleation. This can be done in the so-called continuously operated MSMPR crystallizers (mixed suspension mixed product removal) which are fed with a clear solution. The suspension contents are completely mixed with the consequence that the continuously removed suspension has the same properties as the suspension in the vessel. Such experiments are a bit difficult and time consuming and described in more detail by Garside et al. (2002). The crystallizer is operated with a certain feed rate which leads to a definite mean residence time τ. When a steady-state operation is reached the population density $n(L)$ is measured and plotted against the crystal size L to determine the density n_0 at $L \to 0$ and the slope of the line. This leads to effective nucleation rates $B_0 = n_0 \cdot G$ and growth rates G, compare Fig. 8.3-6, after a variation of the feed rate or the residence. A couple of B_0 and G gives a point in the diagram of Fig. 8.4-16 in which the effective nucleation rate B_0, based on the volumetric suspen-

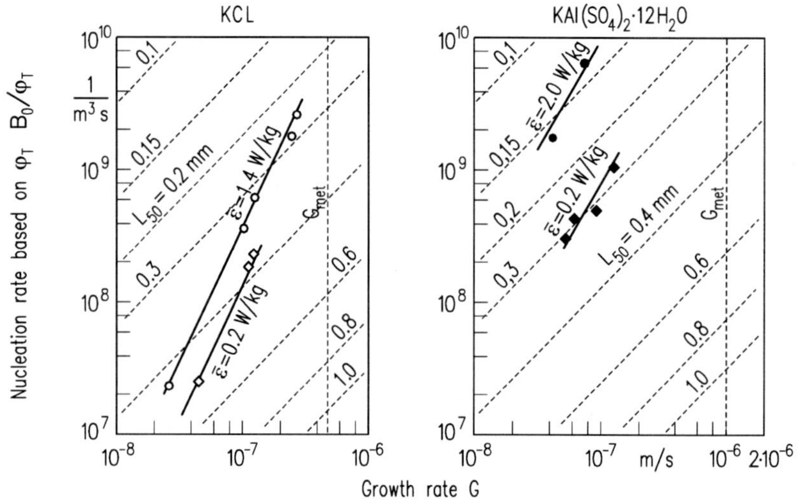

Fig. 8.4-16 Rates of secondary nucleation based on the volumetric crystal holdup as a function of the growth rate

sion holdup $\varphi_T = m_T/\rho_c$, is plotted against the mean growth rate of all crystals in the suspension. The diagrams show experimental results for potassium chloride and potash alum produced in a crystallizer which has been operated at constant mean specific power inputs $\bar{\varepsilon}$.

The results obtained for potash alum show that the effective nucleation rate increases with rising mean specific power inputs for given growth rates. This can be explained by increasing attrition rates and rising numbers of attrition fragments. The experimental determination of the rates of nucleation and growth is simple and unequivocal when the measured data points (B_0, G) can be represented by a straight line in a B_0/φ_T, G diagram. This can be expected if the mean specific power input is below $\bar{\varepsilon} = 0.5\ \mathrm{W/kg}$, the suspension density is $m_T < 50\ \mathrm{kg/m^3}$ or the volumetric suspension holdup is $\varphi_T < 0.02$, and the mean residence time is $\tau < 5{,}000\ \mathrm{s}$. Note that with increasing suspension density, specific power input, and residence time, the prerequisites of the MSMPR modelling are no longer valid because attrition becomes more and more important. More information is given in Garside et al. (2002).

8.5 Design of Crystallizers

Based on the very simple idea that the residence time τ is the ratio of the crystal size L and the growth rate G ($\tau \approx L/G$) and with the assumption $L = 10^{-3}\mathrm{m}$ and $G = 10^{-7}\mathrm{m/s}$ it can be concluded that the residence time is about several hours to obtain a coarse product. The volume V of a cooling crystallizer results from $V = \dot{V} \cdot \tau$ for a given volumetric flow of feed solution. The mean crystal size decreases with decreasing residence time. On the other hand, the supersaturation can be so low for very long residence times that the attrition rate or negative growth rate G_a is the same as the kinetic growth G due to supersaturation. Then a certain maximum crystal size cannot be exceeded. In the case of crystals prone to attrition (KNO_3) the mean crystal size L_{50} as a function of the residence time passes through a maximum after 1–2 h. As a rule the mean residence τ of a suspension in a crystallizer is chosen according to this maximum.

The heat transfer areas of cooling or evaporation crystallizers must be large enough to remove the heat (cooling) or to transfer the heat of evaporation. Let us have a look on the heat transfer area and the temperature profile in the solution/suspension very close to the solid surface because the minimum and maximum temperature is decisive for incrustation and fouling. The heat flux density $\dot{q} = \alpha \cdot \Delta T$ or the temperature difference $\Delta T = \dot{q}/\alpha$ between the suspension and the surface has to be limited.

If the concentration difference Δc according to

$$\Delta c = \frac{\mathrm{d}c^*}{\mathrm{d}T} \cdot \Delta T = \frac{\mathrm{d}c^*}{\mathrm{d}T} \cdot \dot{q}/\alpha \qquad (8.5\text{-}1)$$

exceeds the metastable zone width $\Delta c_{\mathrm{met,\,het}}$, heterogeneous nucleation can become a severe problem with the result of a fine crystalline product and incrustation of heat transfer areas. In industrial crystallization vacuum-flash evaporation is applied where the maximum supersaturation can be found below the surface of the boiling suspension. Here heat transfer areas are not necessary.

There are partially contradictory requirements for the suspension flow in general and especially for the tip speed of the rotor (stirrer, pump) and the mean specific power input: the avoidance of settling of crystals and a sufficient blending of the suspension. As a rule a mean specific power input of $\bar{\varepsilon} = 0.5\ \mathrm{W/kg}$ is enough for these requirements with the tendency of lower inputs in large vessels. A scale-up at constant mean specific power input ($\bar{\varepsilon} = \mathrm{const.}$) results approximately in the same mean crystal size; however, the degree of mixing of the suspension will become worse, compare with Chap. 3. A constant mixing time in the model vessel and in the large-scale vessel would require the same speed of the rotor for geometrically similar crystallizers. The scale-up rules of stirred vessel crystallizers for geometrical similarity and $\bar{\varepsilon} = \mathrm{const.}$ predict that

- The mean crystal size L_{50} remains constant or is a bit increased
- The homogeneity of the suspension is improved
- The blending of the solution is deteriorated with the result that the differences of the local supersaturations are increased

The suspension density m_{T} in industrial crystallizers can assume values up to $500\ \mathrm{kg/m^3}$ or to volumetric crystal holdups between $\varphi_{\mathrm{T}} = 0.2$ and 0.25.

Fundamentals of fluid dynamics of stirred vessels and fluidized beds can be found in Chap. 3. In small vessels it is difficult to suspend especially large crystals with a high density. On the other hand, blending can be insufficient in large vessels with the result that there are great differences of the supersaturation and suspension density. In the case of continuously operated crystallizers a great ratio of the residence time $\tau = V/\dot{V}$ and the macromixing time $t_{\mathrm{macro,\,turb}} \approx 5 \cdot (D^2/\bar{\varepsilon})^{1/3}$ is advantageous for the degree of mixing.

In industrial batch crystallizers the ratio of the volume V based on the internal volumetric flow rate $\dot{V}_{\mathrm{circ}}$ of circulation ($\dot{V}_{\mathrm{circ}} = N_{\mathrm{v}} \cdot n \cdot d^3$ or $V/\dot{V}_{\mathrm{circ}}$ is often between 30 and 120 s (Mersmann 2002) or, in other words, the circulation time τ_{circ} is short in comparison to the batch time τ_{Batch} ($\tau_{\mathrm{circ}}/\tau_{\mathrm{Batch}} \approx 0.01$)). In very large crystallizers ($V > 100\ \mathrm{m^3}$) it can happen that the supersaturation is remarkably reduced especially close to heat transfer areas or in the vicinity of feed points with the result that in a big volume fraction of the apparatus the supersaturation tends to zero. This can be avoided if the rotor speed is above the minimum speed n_{min} according to

$$n_{\min} \approx 28 \cdot \frac{m_{\mathrm{T}}}{\tau \cdot \Delta c} \approx 28 \cdot \frac{\varphi_{\mathrm{T}} \cdot \rho_{\mathrm{c}}}{\tau \cdot \Delta c} . \tag{8.5-2}$$

Here τ is the mean residence time of crystals in a continuously operated cooling crystallizer. According to (8.5-2) a low dimensionless supersaturation $\Delta c / \rho_{\mathrm{c}}$ leads to high speed of rotors in crystallizers.

In small stirred vessels off-bottom lifting (index BL) can be critical. The minimum mean specific power input $\bar{\varepsilon}_{\mathrm{BL}}$ is given by (compare with the chapter 3)

$$\bar{\varepsilon}_{\mathrm{BL}} \sim \frac{\eta_{\mathrm{L}} \cdot \Delta\rho \cdot g}{H \cdot \rho_{\mathrm{L}}^{2}} \cdot \left(\frac{D}{d}\right)^{5/2} . \tag{8.5-3}$$

The greater the density difference $\Delta\rho = \rho_{\mathrm{c}} - \rho_{\mathrm{L}}$ between the density ρ_{c} of the crystals and the density ρ_{L} of the liquid the higher the mean specific power input necessary for off-bottom lifting of crystals. A large stirrer diameter d in a vessel with the diameter D is advantageous for a given suspension with the properties $\Delta\rho$, ρ_{L}, and η_{L} of the two-phase system. Furthermore the height H of the vessel plays a role.

In the case of large stirred vessels the avoidance of settling (index AS) is the decisive process of crystal suspending. This requires the mean specific power input

$$\bar{\varepsilon}_{\mathrm{AS}} \sim \left[L \cdot \left(\frac{\Delta\rho \cdot g}{\rho_{\mathrm{L}}}\right)^{3}\right]^{1/2} , \tag{8.5-4}$$

which rises with the crystal size L and the density difference $\Delta\rho$. Therefore, large crystals are prone to attrition with the consequence of a great number of attrition fragments which are potential secondary nuclei. The mean and especially the maximum crystal size are controlled by attrition and abrasion. Large crystals with a high density ρ_{c}, a great hardness but a low fracture resistance are remarkably endangered by these mechanical impacts. As a rule the objective of industrial crystallization is the production of coarse crystals with a narrow size distribution to minimize the investment and operating cost for the subsequent processes of liquor removal and drying. One big crystallizer is advantageous in comparison to several small units because the mean specific power input for suspending is low. This is especially true for fluidized beds which are used for coarse products. A crystallizer should be operated everywhere in the volume and during the total time at the optimum supersaturation Δc_{opt}. In the case $\Delta c > \Delta c_{\mathrm{opt}}$ the product tends to become more fine and less pure, whereas for $\Delta c < \Delta c_{\mathrm{opt}}$ the economics of the process is reduced. Data for Δc_{opt} can be found in the literature (Mersmann and Löffelmann 1999).

In Table 8.5-1 some approximate values for the design and operation of industrial crystallizers are listed. The data are valid for systems with a medium or high solubility c^*.

The kinetics of crystallization of such systems is often controlled by secondary attrition-induced nucleation and growth rates which are valid in the transition range between diffusion-limited and surface integration-limited growth. Diffusion limited means $G \sim \Delta c/\rho_{\mathrm{c}}$.

With decreasing solubilities c^* crystallizers are operated at

- Dimensionless supersaturations $10^{-4} < \Delta c/\rho_{\mathrm{c}} < 5\times10^{-2}$
- Relative supersaturations $\sigma > 0.1$
- Activated (heterogeneous, surface) nucleation
- Diffusion limited growth

Table 8.5-1 Approximate design data of industrial crystallizers (valid for products with the solubility $c^* > 10\ \mathrm{kg/m^3}$)

Type	Forced circulation	Draft tube baffled	Fluidized bed
m_{T} [kg/m^3]	200 – 300	200 – 400	400 – 600
φ_{T}	0.10 – 0.15	0.10 – 0.15	0.10 – 0.15
τ [h]	1 – 2	3 – 4	2 – 4
$\bar{\varepsilon}$ [W/kg]	0.2 – 0.5	0.1 – 0.5	0.01 – 0.5
$\Delta c/\rho_{\mathrm{c}}$	$10^{-4} - 10^{-2}$	$10^{-4} - 10^{-2}$	$10^{-4} - 10^{-2}$
L_{50} [mm]	0.2 – 0.5	0.5 – 1.2	1 – 5(10)

With respect to economics only one apparatus is installed (instead of a combination of several units). The mean crystal size L_{50} obtained in a crystallizer can be roughly estimated and depends mainly on the relative supersaturation σ, see Fig. 8.5-1 (Mersmann 2001; Garside et al 2002; Mersmann 2007). In this figure the

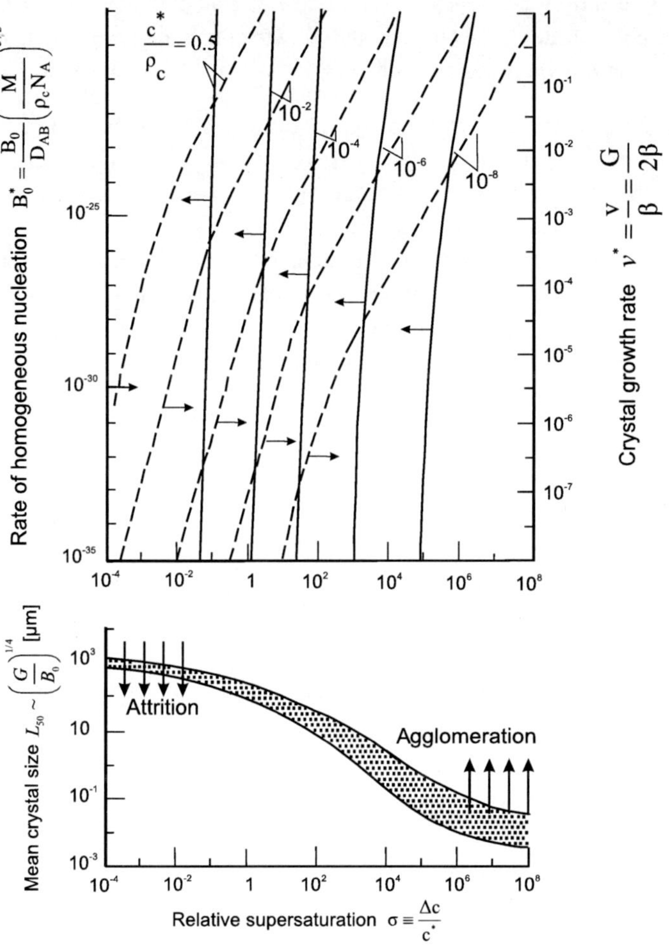

Fig. 8.5-1 Dimensionless nucleation and growth rates (*above*) and mean crystal size L_{50} (*below*) as a function of the relative supersaturation and the dimensionless solubility as parameter

dimensionless rate of primary homogeneous nucleation and the dimensionless growth rate (above) and the mean crystal size L_{50} (below) as a function of the relative supersaturation σ are given. Note that the nucleation rates are only valid for homogeneous nucleation in ideal systems (activity coefficient $\gamma = 1$, no dissociation) and $\sigma > 1$ where primary nucleation becomes dominant. The diagram above gives some information on the metastable supersaturation ($\sigma_{\text{met, hom}} = \Delta c_{\text{met, hom}}/c^*$) which occurs at the steep solid lines which depend on the ratio c^*/ρ_c. The diagram above leads to the message that the relative supersaturation is the most important parameter for the processes of nucleation and growth and consequently for the mean crystal size L_{50} (diagram below). Contrary to the diagram above (rates of nucleation and growth) the diagram below is based on a great number of experimental results.

The equation $L_{50} = 3.67 \cdot G \cdot \tau$ is only valid when all MSMPR assumptions are fulfilled (besides other no attrition and no agglomeration). However, in the case $\sigma < 0.1$ the mean crystal size L_{50} and especially the maximum crystal size with $L_{max} \approx 2 \cdot L_{50}$ is controlled by attrition, see Fig. 8.5-1 below. On the other hand, agglomeration can be the dominant process parameter for $\sigma > 10^3$. Nanocrystals can only be produced if the relative supersaturation is high with the result of high rates of activated nucleation and by the avoidance of agglomeration. Aggregates which are formed under low or zero supersaturation do not possess crystalline bridges and can be redispersed.

Symbols

A		Aggregation constant
a	m	Distance
B	$1/(m^4 \times s)$	Birth rate
B_o	$1/(m^3 \times s)$	Nucleation rate
$B_{agg}(u)$	$1/(m^6 \times s)$	Birth rate by agglomeration
$B_{dis}(u)$	$1/(m^6 \times s)$	Birth rate by disruption
D	$1/(m^4 \times s)$	Death rate
$D_{agg}(u)$	$1/(m^6 \times s)$	Death rate by agglomeration
$D_{dis}(u)$	$1/(m^6 \times s)$	Death rate by disruption
E	N/m^2	Young's modulus
ΔE_r	J/mol	Activation energy
G	m/s	Growth rate (linear)
G_V	m^3/s	Growth rate (volumetric)
H	m	Height
Ha	J	Hamaker constant
He		Henry coefficient
H_V	N/m^2	Vickers hardness
h	$J \times s$	Planck's constant ($h = 6.626 \times 10^{-34} J \times s$)
K		Constant
k		Distribution coefficient
k_r, k_{ro}	var	Rate constants
L	m	Crystal size
m	kg/m^3	Mass per volume
m_T	kg/m^3	Suspension density
N_i^o	$1/m^3$	Ion concentration
n		Number of atoms in a molecule
n	$1/m^4$	Population density
n_V	$1/m^6$	Volumetric population density
P^*		Parameter
r_f		Progress of crystallization
S		Supersaturation ($S \equiv c/c^* = 1 + \sigma$)
s	$1/(m^2 \times s)$	Collision factor
$\dot{T}$	K/s	Cooling rate
u	m^3	Particle volume
v	m^3	Particle volume
v	m/s	Growth rate
W		Collision efficiency
x_s	m	Mean displacement
Z		Imbalance factor according to Zeldovich
Z		Agglomeration ratio

Greek Symbols

α_0	$(A \times s \times m)^2/J$	Polarizability
α		Volume shape factor
β		Surface shape factor
β	m^3/s	Agglomeration kernel
$\dot{\gamma}$	$1/s$	Shear rate
γ	J/m^2	Interfacial tension
ε_0	$A \times s/(V \times m)$	Electrical field constant $\varepsilon_0 = 8.85 \times 10^{-12} A \times s/(V \times m)$
ε_r		Relative permittivity
η		Factor of efficiency
κ	$1/m$	Debye–Hückel parameter
λ	m	Particle diameter
μ	N/m^2	Shear modulus
ν		Number of ions in a molecule
ν	$1/s$	Frequency
ν_c		Poisson ratio
σ	$A \times s/m^2$	Electrical charge density
σ		Relative supersaturation
σ	N/m^2	Tensile strength
φ		Volumetric holdup
φ_T		Volumetric holdup of crystals
ψ	V	Electrical potential
Θ	grad	Contact angle
Γ	$1/s$	Disruption frequency
Γ		Material parameter
Γ/K'	J/m^2	Fracture resistance
Φ		Dimensionless nucleation rate

Indices

A	Addition, surface
a	Activity related
agg	Agglomeration
at	Atomar
BCF	Burton–Cabrera–Frank
B + S	Birth and spread
c	Crystalline, cluster
CL	Crystal / solution
col	Collision
crit	Critical
dif	Diffusion limited
dis	Disruption

eff	Effective
f	Feed
g	Growth
het	Heterogeneous
hom	Homogeneous
hyd	Hydrate
I	Interface
int	Integration limited
L	Liquid, length
m	Molecular
max	Maximum
N	Nucleus
PN	Polynucleation
p	Product
r	Reaction
R	Removal
s	Seed, semipermeable
surf	Surface
sus	Suspension
*	Equilibrium

9 Adsorption, Chromatography, Ion Exchange

Adsorption is the loading of solid surfaces with substances present in a surrounding fluid phase or, in other words, it is a surface effect between a solid and a fluid phase. Sometimes molecules of the fluid phase are not only fixed on the surface but can additionally enter the bulk of the nonporous solid phase according to a volume effect. This is called occlusion or absorption. When it is not known which of these two effects is dominant the term "sorption" is used. Adsorption means the loading of one or several components (adsorptives) on a solid material (adsorbent). The reverse process, e.g., the separation of adsorptives from the surface is called desorption. It is a question of mass transfer in a two-phase system solid/fluid:

$$\text{adsorbent} + \text{adsorptive} \quad \underset{\text{desorption}}{\overset{\text{adsorption}}{\rightleftarrows}} \quad \text{adsorbent} + \text{adsorbate}$$

Microporous adsorbents with a large inner surface possess the ability to fix adsorptives. The condensed phase called adsorbate is desorbed in most industrial processes. After this regeneration the adsorbent can again be loaded in the next step.

9.1 Industrial Adsorbents

As industrial adsorbents, silica gels, Fuller's earth, activated carbon, and molecular sieves are used. Some properties of these adsorbents are listed in Table 9.1-1. The most important characteristic is the large specific inner surface in the range between 300 and 1,800 m^2/g. This huge surface is based on a great number of narrow pores created by special production and activation processes. The pore diameter has a range between 0.3 and 10 nm. The inner surface is determined by the adsorption of nitrogen up to a monomolecular layer at 77.4 K which corresponds to the boiling temperature. The area occupied by a nitrogen molecule at this temperature is 0.162 nm^2.

A. Mersmann et al., *Thermal Separation Technology: Principles, Methods, Process Design*, VDI-Buch, DOI 10.1007/978-3-642-12525-6_9,
© Springer-Verlag Berlin Heidelberg 2011

Table 9.1-1 Properties of industrial adsorbents

Adsorbent	Silica gel (narrow pores)	Silica gel (wide pores)	Molecular sieves	Aluminium oxide	Activated carbon (narrow pores)	Activated carbon (wide pores)
Main substance	SiO_2	SiO_2	SiO_2, Al_2O_3	Al_2O_3	C	C
Particle size, mm	1–5	1–5	1–5	2–10	3–10	3–10
True density, kg/m³	2,200	2,200	2,600	3,000	2,000	2,200
Apparent density, kg/m³	1,100	1,100	1,100–1,500	1,200	800	600–700
Bulk density, kg/m³	700–800	400–800	600–900	700–850	400–500	300–500
Internal porosity	0.5	0.5	0.42–0.57	0.60	0.60	0.68–0.73
Micropore volume, cm³/g	0.35–0.45	0.30–0.45	0.25–0.30	0.40	0.25–0.40	0.20–0.30
Macropore volume, cm³/g	<0.10	0.25–0.10	0.30–0.40	0.10	0.20–0.40	0.30–0.50
Spec. surface, m²/g	600–850	250–350	500–1,000	100–400	1,000–1,800	1,000–1,500
Spec. heat capacity, kJ/(kg K)	0.92	0.92	0.92	0.88	0.76–0.84	0.76–0.84
Heat conductivity W/(m K)	0.20	0.20	0.13	0.12	0.1	0.1

Nearly the total area accessible for adsorptive molecules is provided by the micropores with diameters below 2 nm. This is true for activated carbons and molecular sieves. Macropores with diameters larger than 50 nm are decisive for the adsorption kinetics or the mass transfer, see later. Adsorbents like aluminium oxides and molecular sieves with electrical charges are hydrophilic and can be highly loaded with polar adsorptive molecules such as water, ammonia, and methanol, see Chap. 3. Adsorption isotherms of water are a good tool to characterize the capacity of these adsorbents.

Activated carbons have hydrophobic surfaces with the property that nonpolar molecules or organic compounds with a very poor solubility in water such as benzene or toluene are preferably adsorbed. However, small amounts of oxides or functional groups present in the surface can change the hydrophobic character of the adsorbent surface. The capacity of carbonic adsorbents is evaluated by means of benzene adsorption isotherms.

Industrial important adsorbents are:

- Silica gel with narrow or wide pores. This adsorbent consists of amorphous SiO_2 (97.3 %) and is commercially available as a particle material. It is chemically neutral and resistant to acids with the exception of hydrofluoric acid. Silica gel with wide pores is sometimes used for the separation of tiny droplets present in gases.
- Zeolites or molecular sieves are natural or synthetic aluminosilicates or silicates which are releasing water at high temperatures. Starting from sodium aluminosilicate the cation Na^+ is replaced by other cations (Li^+, K^+, Ca^{2+}) by means of a hydrothermal synthesis. By this process, cages within a zeolite crystal are formed. These cages have openings or windows or micropores in the range between 0.3 and 1 nm. As a consequence only molecules smaller in their effective size than the opening can pass and enter the cage. In this way a precise separation of linear and branched hydrocarbons is possible. Microcrystals of aluminosilicates in the size range between 1 and 15 μm are pelletized using a clay binder. The pellets with diameters between 1 and 5 mm are calcinated at a temperature of approximately 650°C with the result that the adsorbents possess a high adsorption capacity. The process of pelletization leads to macropores with diameters above 50 nm between the crystals.
- Activated aluminium oxides are commercially available with particle sizes between 2 mm and 10 mm and used for the drying of gases and liquids (removal of water). This adsorbent is resistant to weak alkaline solutions; however, it can be destroyed by acids.

- Carbonic adsorbents like activated carbon, activated coke, graphite, and carbon molecular sieves are very porous adsorbents with hydrophobic inner surfaces. Therefore, nonpolar organic adsorptives are preferably adsorbed. The amorphous skeleton is composed of microcrystals which form a lattice of graphite. Carbonic adsorbents are produced by the carbonization of organic materials (carbon, coke, wood, petroleum, peat, lignite) at a low temperature and the subsequent activation at elevated temperatures. Carbon molecular sieves dispose of great portions of micropores with diameters in the range between 0.3 and 0.9 nm. They are produced by the partial oxidation of anthracite and a subsequent thermal process which leads to a decomposition of polymers (polyethylene, polyvinylchloride, polyvinylidene chloride) which have been added to the anthracite. By this treatment, micropores are created. All carbonic adsorbents are oxidized and this oxidization is favored by increasing temperatures and partial pressures of oxygen. Especially the decomposition of heat-sensitive organic adsorptives can lead to hot spots with the consequence of an incineration of the entire bed.
- Organic polymer adsorbents such as polystyrene, polyacrylic ester, phenolic and phenolic amine resin without functional groups (contrary to ion exchange resins) are advantageous because properties (pore volume, pore width, specific inner surface, polarity) can be adjusted in wide ranges. By this way the selective separation of polar and nonpolar substances can be achieved. The thermal stability of these polymers is limited to 180°C.

The most important properties for the characterization of adsorbents are the porosity, the micropore and the macropore volume, the inner specific surface, the pore size distribution, and the kind of the solid substance, compare Sect. 1. The micropore volume of pores below a width of 2 nm is in the range between 0.25 and 0.40 cm^3/g and is important for the adsorption capacity. Typical pore size distributions of oxide adsorbents show a maximum in the range of micropores and a second maximum which is observed for macropores with diameters above 50 nm. In Fig. 9.1-1 the pore volume distribution is plotted against pore radius for the oxidic

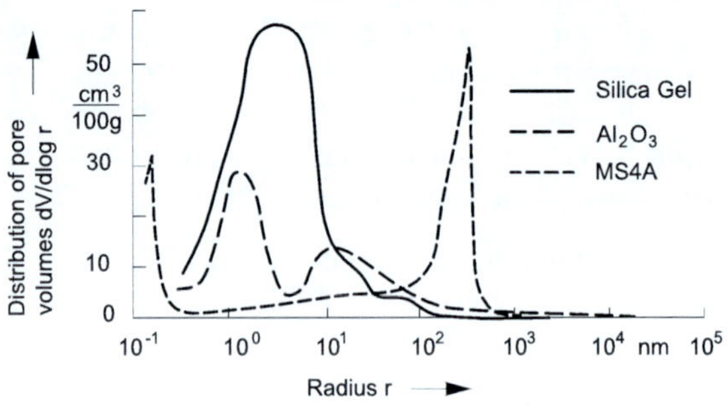

Fig. 9.1-1 Distribution of pore volumes of oxidic adsorbents

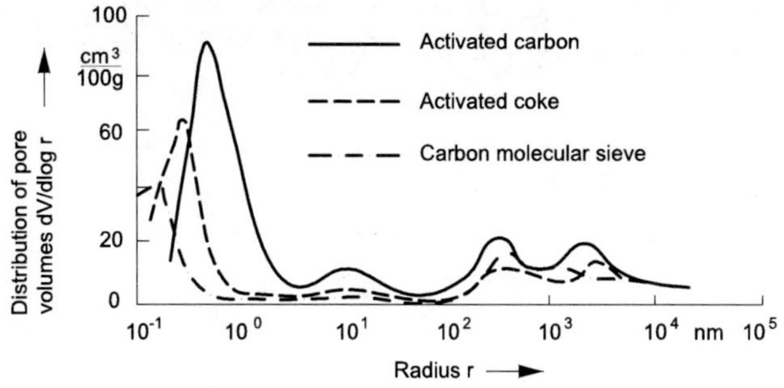

Fig. 9.1-2 Distribution of pore volumes of carbonic adsorbents

adsorbents silica gel, Al_2O_3, and the molecular sieve MS4A. The analogous diagram of Fig. 9.1-2 shows these properties for some carbonic adsorbents. The pore sizes are very decisive for the mechanisms and rates of mass transfer as will be shown later.

9.2 Adsorbers

According to the previous chapters the reader may come to the conclusion that countercurrent columns are dominant as has been shown for mass transfer equipment used in the areas of rectification, absorption, and extraction. However, this is not true because the continuous transport of solid granular material is much more difficult in comparison to a fluid. Therefore, nearly all adsorbers are fixed beds which are operated batchwise. As a rule, at least two fixed beds are installed in continuously operated industrial processes. The first bed is used for the adsorption step whereas in the second the adsorbates is removed or desorbed at the same time. The duty of the two beds is changed when the adsorption capacity is exhausted. Sometimes several beds are arranged to carry out pressurization and depressurization steps.

In Fig. 9.2-1 an adsorption unit with two fixed beds is depicted. The fluid to be treated passes downward through the bed. At the same time the second bed is regenerated by a fluid which is flowing from the bottom to the top. This can be advantageous with respect to the efficiency of the desorption step; however, the danger of fluidization exists at high velocities, compare Sect. 3.

Therefore, the bed is covered by a sieve or a grid to avoid the extension and entrainment of the granular adsorbent. In the case of temperature swing adsorption (TSA) illustrated in Fig. 9.2-1 the fixed bed is directly heated up by a hot fluid to

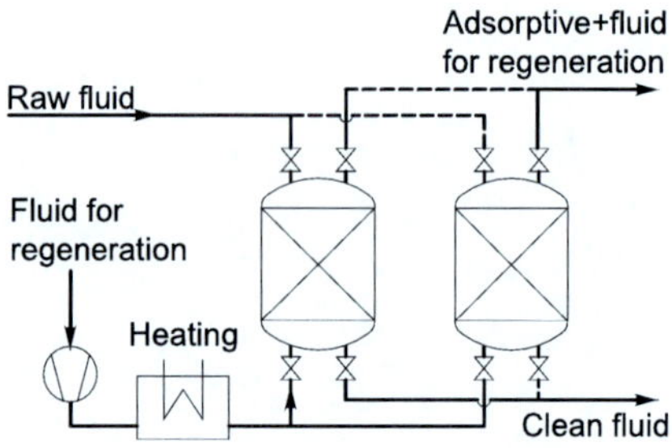

Fig. 9.2-1 Adsorption unit with two fixed beds (adsorber and desorber)

transfer the heat of desorption. This regeneration method by direct or indirect transfer of heat is usually applied when the heat of desorption is high (above 30 kJ/mol adsorptive, e.g., 50 kJ/mol water).

Pressure swing adsorption (PSA) and vacuum swing adsorption (VSA) are based on a depressurization which leads to desorption. This regeneration method is applied for adsorbates with a small heat of desorption which must be supplied by the adsorbent and the surrounding fluid, e.g., 18.7 kJ/mol N_2. In Fig. 9.2-2 an adsorption unit with four fixed beds is illustrated. The objective is the separation of the components N_2, CO, CH_4, and CO_2 from a feed gas which mainly consists of hydrogen. In the first adsorber under elevated pressure all gases but hydrogen are adsorbed and pure hydrogen is withdrawn at the top. At the same time the second bed is desorbed by depressurization.

The off-gas with the impurities (N_2, CO, CH_4, CO_2) is passed through the third bed which is further desorbed by purging with pure hydrogen. Finally pressurization is carried out with highpressure hydrogen. After the breakthrough of the impurities at the top of the first adsorber the roles of the four adsorbers are cyclically switched: The raw gas is fed into the fourth bed and the first bed is regenerated by depressurization, the second desorbed by purging, and the third is pressurized. In Fig. 9.2-2 the pressure in an adsorber is shown as a function of time for four subsequent cycles.

The special feature of a rotating adsorber is that many compact fixed beds are arranged, and in a distinct bed various process steps such as adsorption, desorption, purging, and cooling can be carried out successively. In Fig. 9.2-3, a continuously operated rotating adsorber with eight segments is illustrated. In six segments adsorption takes place whereas the bed in the seventh is regenerated by a hot gas and the adsorbent in the bed number eight is cooled by a cold gas. Such adsorption units are often used for the cleaning of waste gases or polluted air. A continuous flow of such gases can also be treated in moving and fluidized beds.

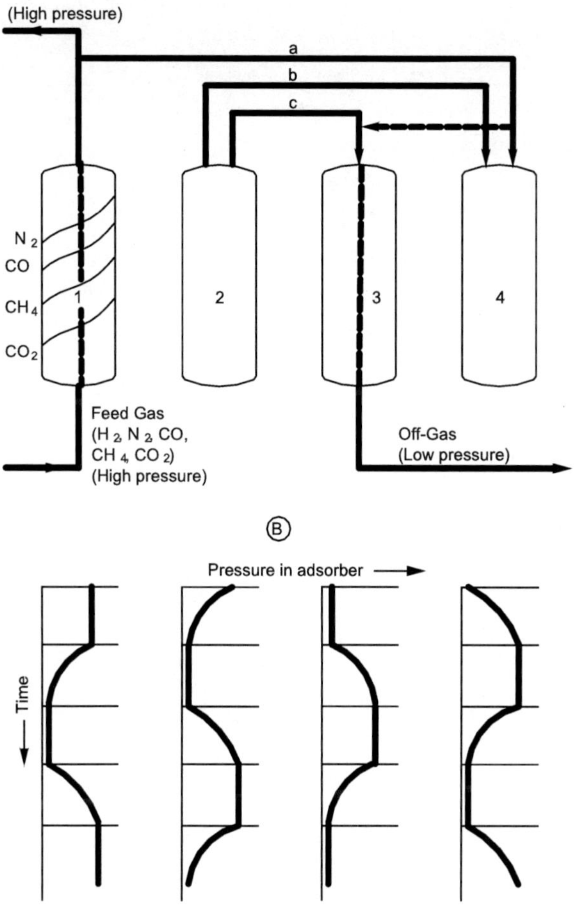

Fig. 9.2-2 Pressure swing adsorption unit with four fixed beds. (**a**) pressure rise, (**b**) depressurization, (**c**) depressurization – purging

Figure 9.2-4 shows a moving bed adsorber in which sulfur oxides and nitrogen oxides are adsorbed on downward moving activated coke. The flue gas is flowing in radial direction from the center to the rim in crossflow to the coke. The advantage of such a crossflow is that the pressure drop of a huge flue gas flow is small thanks to a short path length of the gas. This is also true for the fixed bed adsorber depicted in Fig. 9.2-5. This equipment allows the treatment of huge gas streams with the advantage that the pressure drop and also the heat losses are low.

Occasionally continuously operating adsorbers are installed. Such an adsorber is equipped with trays on which the fluidization of the granular adsorbent is maintained by the upward flowing gas. The adsorbent is transported by gravity forces

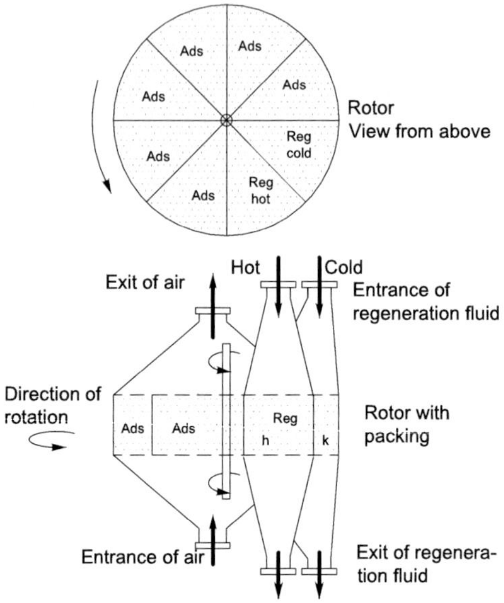

Fig. 9.2-3 Continuously operated rotating absorber for cleaning of waste gas

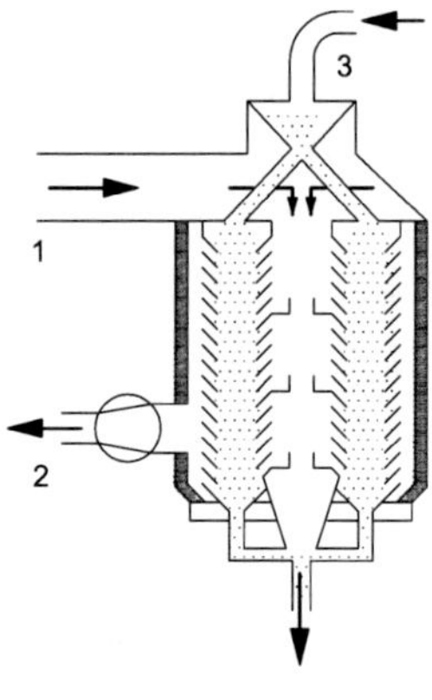

Fig. 9.2-4 Continuously operated moving bed adsorber with crossflow

from tray to tray downward. In Fig. 9.2-6 a multistage adsorber operated in countercurrent flow is illustrated. The column has two sections, an adsorption zone in the upper section and a regeneration zone below. Humid air which enters the adsorption zone at the bottom is continuously dried and dry air is leaving at the top. The adsorbent loaded with humidity is transported downward and regenerated by a hot gas which is flowing in countercurrent flow to the solid. The regenerated gran-

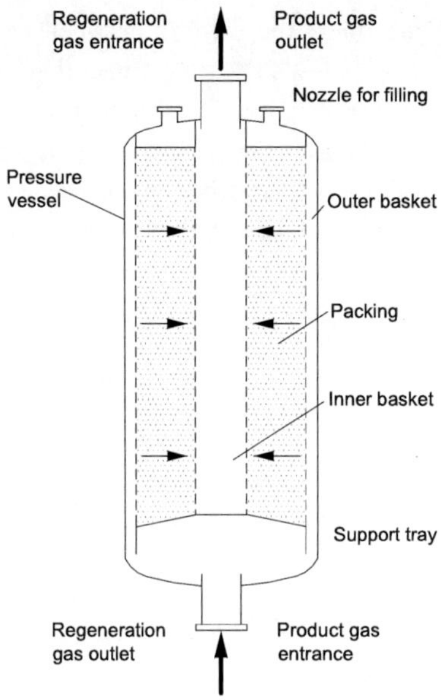

Fig. 9.2-5 Radial flow in an adsorption vessel for huge gas flows

ular adsorbent is conveyed by pneumatic transport to the top of the adsorption column.

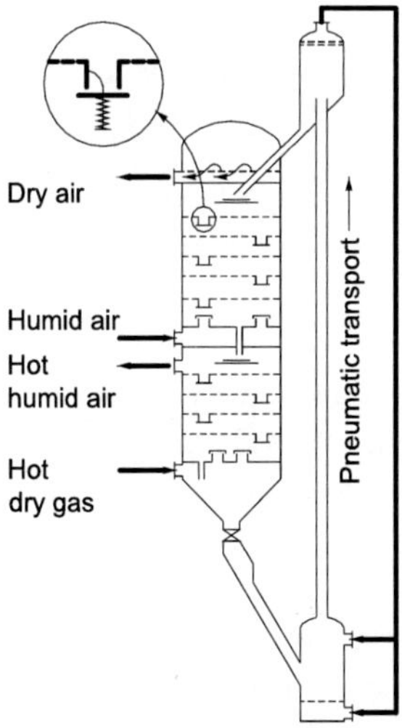

Fig. 9.2-6 Continuously operated countercurrent adsorber for drying of humid air

In a so-called hypersorber, see Fig. 9.2-7, a binary mixture is separated into the two components. The feed gas enters the column between the upper adsorption zone and the regeneration zone below. It is assumed that only one component is adsorbed by the downward moving bed of a granular adsorbent whereas the other component is withdrawn at the top. The separation efficiency depends mainly on a favorable adsorption equilibrium. The strongly adsorbed component is desorbed by stripping with steam and withdrawn as bottom product. The regenerated adsorbent is lifted to the top of the column by pneumatic transport and cooled down to increase the adsorption capacity.

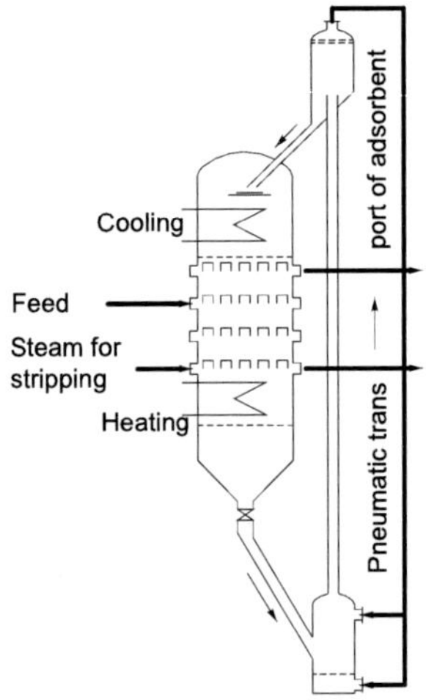

Fig. 9.2-7 Continuously operated countercurrent adsorber for the separation of a gas mixture

Liquid phase adsorption is mostly carried out in fixed beds where the liquid is flowing in axial or radial direction. Sometimes fluidized or moving beds are used. Activated carbon is applied to remove small amounts of organic substances from potable or wastewater. The regeneration of the loaded carbon is expensive because the water has to be removed from the bed and afterward out of the pores of the adsorbent. Therefore, the carbon or coke is burnt in a boiler in some cases. (Highly polluted water is treated by distillation or stripping, see other chapters) Besides fluidized or moving beds wastewater is treated in stirred vessels by contact with granular or powdery adsorbents kept in suspension by the stirrer. In Fig. 9.2-8 an installation with a stirred vessel and a filter is illustrated. The type of the agitator and its speed are chosen with the objective to keep the particles in suspension and to improve the mass transfer between the particles and the surrounding liquid, compare Chap. 3. The residence time of the liquid in the vessel to obtain approximately the equilibrium of the adsorptive depends on the mass transfer coefficient between

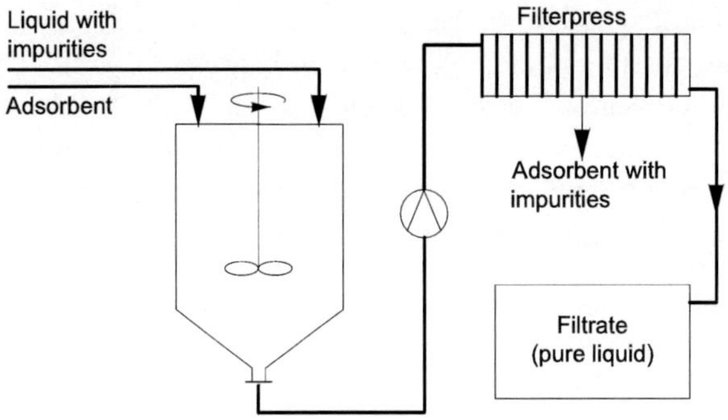

Fig. 9.2-8 Adsorption unit for the treatment of liquids

the liquid and the particle. In most cases the decisive step is the diffusion in the pore system of a particle (this will be discussed later). The suspension is separated by means of sieves, sedimenters, filters, or centrifuges. When the loaded adsorbent is regenerated by direct or indirect heat transfer (temperature swing adsorption, TSA) the mass transfer rate is at first constant (constant rate period) and decreases later as a function of time (falling rate period). This is discussed in more detail in Chap. 10.

9.3 Sorption Equilibria

In the second chapter it has been shown that adsorption equilibria are mainly dependent on the pressure, the temperature, and the physical properties of the adsorptives (molecule diameter, critical data, dipole moment, quadrupole moment, polarizability) and the adsorbents (molecule diameter, Hamaker constant, density, electrical charge). Theoretical models for the prediction of single component (type I) and multicomponent adsorption equilibria are known; however, the application is sometimes difficult (often data are missing) and time consuming. Therefore, in the following some measurement procedures will be described shortly and general problems of adsorption equilibria are discussed. It is important to note that adsorbents with the same mark can differ from producer to producer and even from charge to charge due to impurities and manufacturing though there are nearly no differences in the chemical composition, the distribution of pore sizes, and the pore volumina. The equilibrium can be changed by impurities and this fact can offer the chance to produce adsorbents with a specific selectivity. In consequence the measurement of adsorption equilibria is very important.

Microbalances are often used to measure single component adsorption isotherms. Measurements are possible in the pressure range between 1 and 10^7 Pa (Akgün 2007). The partial pressure of organic adsorptives can be adjusted by means of a bubble flask in combination with a condenser (Scholl 1991). As a rule gas concentrations are measured by gas chromatographs equipped with special calibrated analyzers. Multicomponent adsorption equilibria can be determined by

- Volumetric method (Sievers 1993; Markmann 1999)
- Fixed bed method (Schweighart 1994)
- Zero length column method (Brandani et al. 2003)

The volumetric method is described in (Mersmann and Akgün 2009). The pure components withdrawn from the feed bottles are fed into the evacuated mixing vessel with the volume V_{mix} and blended by circulation achieved by a circulation pump. After a certain circulation period adsorption is started by the flow of the mixture through the adsorber bed with the consequence that the volume occupied by the gas mixture has increased from V_{mix} to V_{ads}. Based on the law of conversation of any component and with M_{ads} as the mass of the adsorbent the loading n_i of the component i is given by (bef = before; aft = after)

$$n_i = \frac{N_{i,\text{ads}}}{M_{\text{ads}}} = \frac{1}{M_{\text{ads}}}\left[\frac{y_i^{\text{bef}} \cdot p^{\text{bef}} \cdot V_{\text{mix}}}{Z^{\text{bef}} \cdot R \cdot T_{\text{ads}}} - \frac{y_i^{\text{aft}} \cdot p^{\text{aft}} \cdot V_{\text{ads}}}{Z^{\text{aft}} \cdot R \cdot T_{\text{ads}}}\right]. \tag{9.3-1}$$

Here, p is the pressure, T_{ads} the temperature of the system kept constant by means of a thermostat, Z the real gas factor, and y_i the mole fraction of the component i in the gas mixture. Pressure and temperature are measured. Measurements have been carried out for pressures up to 10 MPa

The fixed bed method is also based on the conservation of any component in combination with its breakthrough curve (see later "Kinetics"). The equilibrium loading $n_i{*}$ of the component i is given by

$$n_i{*} = \frac{T_n \cdot \rho_n \cdot p \cdot f \cdot \dot{v}_G \cdot y_{i,\text{in}}}{p_N \cdot T_{\text{ads}} \cdot M_{\text{ads}} \cdot \tilde{M}_i} \cdot \int_0^{t*}\left[1 - \frac{y_{i,\text{out}}(t) \cdot [1 - y_{i,\text{in}}]}{y_{i,\text{in}} \cdot [1 - y_{i,\text{out}}(t)]}\right] \cdot \mathrm{d}t \tag{9.3-2}$$

with the volumetric flow rate $f \cdot \dot{v}_{\text{G}}$ (f is the cross-sectional area), the pressure p, and the temperature T_{ads}. The capital letter N means standard state ($T_N = 273.15\,\text{K}$; $p_N = 1.013 \times 10^5$ Pa). M_{ads} is the mass of the adsorbent. The mole fraction $y_{i,\text{in}}$ of the component i in the feed gas is kept constant. The outlet mole fraction $y_{i,\text{out}}$ has to be measured as a function of time until the time $t*$ is reached at which equilibrium has been established. Equilibria of CO_2, C_2H_4,

C_2H_6 and C_3H_8 adsorbed on molecular sieves have been measured starting at pressures of approximately 1 Pa (Schweighart 1994).

The zero length column (ZLC) method is based on a very short fixed bed which is treated as an ideal stirred vessel with an infinite dispersion coefficient. With the gas volume V_G in the adsorber, the carrier gas flow rate $\dot{V}_C$, the mass M_{ads} of the adsorbent, the overall concentration C, and the molar concentration y, the loading results from

$$n_i^* = \int_0^{t \to \infty} \frac{\dot{V}_C \cdot C \cdot y_{i,\,out}(t)}{M_{ads} \cdot 1 - y_{i,\,out}(t)} \cdot dt - \int_0^{t^*} \frac{\dot{V}_C \cdot C \cdot y_{i,\,out}(t)}{M_{ads} \cdot (1 - y_{i,\,out}(t))} \cdot dt - \frac{V_G \cdot c_i(t)}{M_{ads}} . \quad (9.3\text{-}3)$$

In this equation the loading n_i^* is reached after the loading time t^*. The outlet mole fraction $y_{i,\,out}(t)$ and the molar concentration $c_i(t)$ are measured as a function of time t.

Finally, the concentration pulse chromatography will be shortly discussed. A pulse of a sample is injected into a carrier gas flow which is passed through an adsorbent-packed column. The response of the column is measured as concentration $c(t)$ vs. time t. The mean retention time τ of the sample is experimentally determined. With the superficial velocity w, the bed length L, and the adsorbent density ρ_{app}, the modeling leads to the following equation for the Henry coefficient He_i:

$$He_i = \left[\frac{\int_0^\infty c_i \cdot (t - \tau_D) \cdot dt}{\int_0^\infty c_i \cdot dt} \cdot \frac{w}{L} - 1 \right] \cdot \frac{\psi}{(1 - \psi) \cdot \rho_{app} \cdot \tilde{R} \cdot T} . \quad (9.3\text{-}4)$$

Here ψ is the bed porosity. The dead time τ_D is the time required for the passage of a sample pulse through the empty volume of the connecting tube from the injection point to the detector plus the void space in the packed column. This method can be applied to determine binary adsorption isotherms (Harlick et al. 2003, 2004).

In the experimental determination of multicomponent adsorption equilibria in the liquid phase, shakers are used (Fritz 1978; Mehler 2003). The masses of the various adsorptives can be exactly measured using microbalances. Consequently, the feed concentration is known. The liquid mixture is brought into contact with a certain mass of adsorbent. After the establishment of equilibrium the concentrations of the components in the residual liquid have to be determined. This can be difficult when the equilibrium concentration is very low. Sometimes a small amount of a component in the adsorbate is extracted by an appropriate solvent and determined by gas chromatography.

9.4 Single and Multistage Adsorber

In the following it will be shown how balances of masses and components can be simply established in the case that the streams leaving a stage are in equilibrium. In industrial adsorbers the approach to equilibrium depends on mass transfer kinetics (see later) and the residence time of the fluid.

9.4.1 Single Stage

In Fig. 9.4-1 a single stage is depicted. The completely regenerated solid adsorbent is denoted by S_r (or $\dot{S}_r$ for a stream) and the pure carrier fluid by G_r (or $\dot{G}_r$ for a flow rate). In the stage the loading X_0 of the $\dot{S}_r$ stream is increased to the loading X_1 whereas the loading Y_0 of the gas flow $\dot{G}_r$ is reduced to Y_1. The balance of the substance transferred from the fluid to the solid phase can be written as follows:

$$\dot{S}_r \cdot (X_1 - X_0) = \dot{G}_r \cdot (Y_0 - Y_1) \quad \text{or} \quad -\frac{\dot{S}_r}{\dot{G}_r} = \frac{Y_0 - Y_1}{X_0 - X_1}. \tag{9.4-1}$$

The slope of the operating line corresponds to the ratio $-\dot{S}_r/\dot{G}_r$, see Fig. 9.4-1, where $Y^* = f(X)$ is the equilibrium curve. Dealing with desorption or drying, the substance is transferred from the solid adsorbent phase into the fluid phase, see Fig. 9.4-2. The loading of the solid is reduced from X_0 to X_1 whereas the loading Y_0 of the entering fluid is increased to Y_1. Note that X_1 and Y_1 are points of the equilibrium curve $Y^* = f(X)$. The starting point X_0, Y_0 is above the equilibrium curve for adsorption but below in the case of desorption. (Such diagrams and balances can be applied to any single stage operations with the leaving streams in equilibrium with each other, for instance, absorption, liquid–liquid and liquid–solid extraction.)

With respect to the slope $\dot{S}_r/\dot{G}_r$ two boundaries can be distinguished:

- $\dot{S}_r/\dot{G}_r \to 0$ means that no adsorption or desorption takes place and the loading of the fluid phase remains unchanged.
- $\dot{S}_r/\dot{G}_r \to \infty$: The loading of the solid phase at the inlet of the stage and the outlet is the same and the fluid phase is in equilibrium with the entering solid phase.

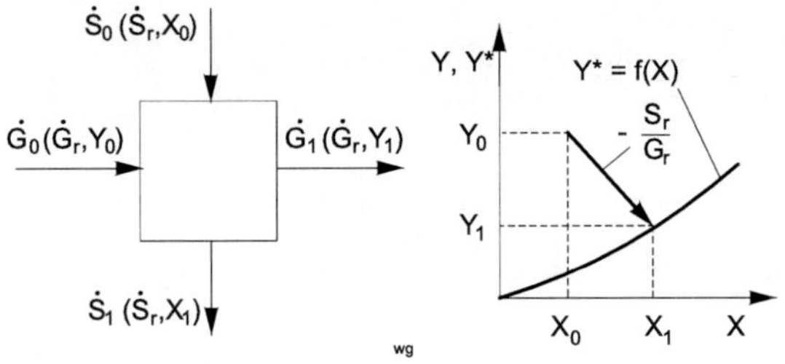

Fig. 9.4-1 Scheme of a one-stage unit and illustration in a diagram with an equilibrium curve and an operating line (adsorption)

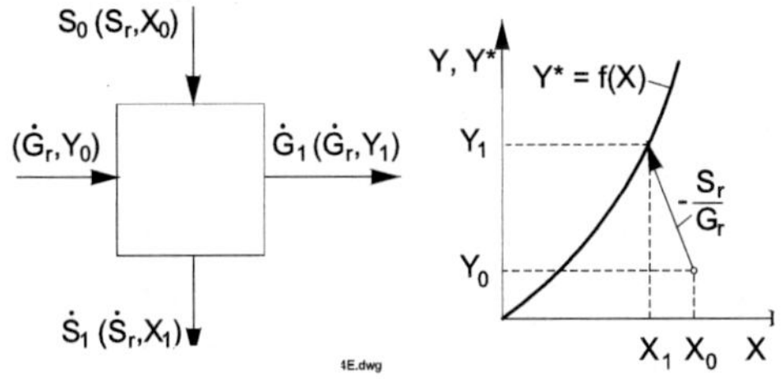

Fig. 9.4-2 Scheme of a one-stage unit and illustration in a diagram with an equilibrium curve and an operating line (desorption)

The diagrams in Figs. 9.4-1 and 9.4-2 are based on the assumption of isothermal adsorption or desorption; however, in industrial adsorbers the heat of adsorption leads to an increase of the temperature which is more pronounced at high loadings. In the case of desorption the temperature is reduced. These heat effects cause a reduction of capacity. Furthermore equilibrium would be reached after an infinite time because the driving force approaches zero. These problems will be discussed later.

9.4.2 Crossflow of Stages

In Fig. 9.4-3 an adsorption unit with three stages is illustrated. The fluid passes through the first stage, then through the second, and finally through the third whereas the three adsorbent streams fed into the three stages are moving in crossflow to the fluid. It is assumed that all adsorbent streams have the same preloading X_0 but different amounts. Such an arrangement can be recommended when the loading of the adsorptive is to be reduced in a high degree. (If it is desired to regenerate an adsorbent stream effectively the solid should be moved through the stages switched in series in crossflow with the regeneration fluid which is fed into the three stages.) A design diagram is depicted in Fig. 9.4-3. The loading Y of the fluid phase is plotted against the loading X of the solid phase with an equilibrium curve

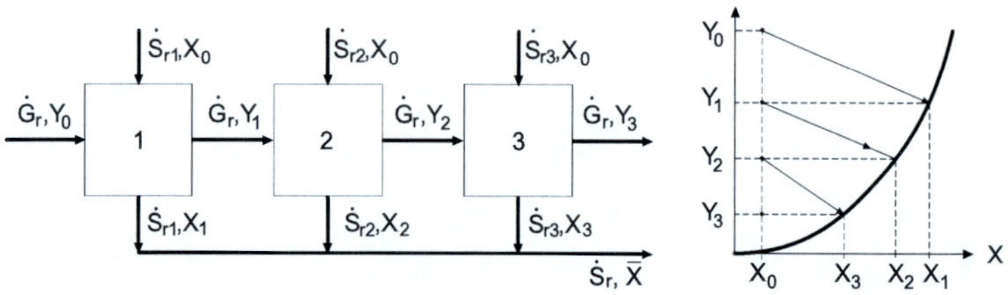

Fig. 9.4-3 Scheme of a three-stage crossflow unit; the diagram shows the equilibrium curve and three operating lines

which is favorable for adsorption. The vertical line represents the preloading of the regenerated adsorbent. The amount of adsorbent per unit time $\dot{S}_{r1}$ based on the fluid mass flow $\dot{G}_r$ or the ratio $\dot{S}_{r1}/\dot{G}_r$ is equal to the slope of the operating line of the first stage. With the assumptions of an isothermal operation and equilibrium of the streams leaving the stages the point ($Y_1^* = Y_1, X_1$) can be found on the equilibrium curve. The fluid stream with the loading Y_1 is fed into the second stage where it is brought in contact with the solid mass stream ($\dot{S}_{r2} \cdot X_0$). Again the ratio $\dot{S}_{r2}/\dot{G}_r$ decides on the slope of the operating line of the second stage. With the same assumptions already mentioned the, point ($Y_2^* = Y_2, X_2$) is a point of equilibrium curve. By this way the loadings $Y_3{*} = Y_3$ and X_3 can be found. Note that a very pure fluid stream with $Y \to 0$ can only be obtained with an unloaded adsorbent ($X_0 \to 0$) and the arrangement of a great number of stages.

In Fig. 9.4-3 the three adsorbent streams are brought together and mixed with the result that they have the loading $\bar{X}$. This loading can be found by the rule of balances or the mixing rule: This rule is at first applied on the mass streams $\dot{S}_{r1}$ and $\dot{S}_{r2}$ and then on the binary mixture ($\dot{S}_{r1} + \dot{S}_{r2}$) and $\dot{S}_{r3}$. The approach to equilibrium in a stage can be expressed by a stage efficiency, compare the sections on absorption, extraction, and rectification. This efficiency is mainly dependent on the residence time in relation to a characteristic diffusion time, see later.

Next, a special problem will be discussed: What is the minimum adsorbent mass in a two-stage unit for a given loading reduction from the loading Y_0 of the feed to the loading $Y_2{*}$ of the leaving fluid (Treybal 1968).

It is assumed that the equilibrium can be described by the equation of Freundlich:

$$Y^* = A \cdot X^B . \qquad (9.4\text{-}2)$$

and the feed adsorbent is unloaded or $X_0 = 0$. The balance of the adsorptive according to (9.4-1) can be formulated for the two stages:

First stage: $$\frac{\dot{S}_{r1}}{\dot{G}_r} = \frac{Y_0 - Y_1}{X_1} = \frac{Y_0 - Y_1^*}{(Y_1^*/A)^{1/B}}, \tag{9.4-3}$$

Second stage: $$\frac{\dot{S}_{r2}}{\dot{G}_r} = \frac{Y_1^* - Y_2^*}{(Y_2^*/A)^{1/B}}. \tag{9.4-4}$$

The total adsorbent mass flow based on the fluid stream $\dot{G}_r$ is

$$\frac{\dot{S}_{r1} + \dot{S}_{r2}}{\dot{G}_r} = (A)^{1/B} \cdot \left[\frac{Y_0 - Y_1^*}{(Y_1^*)^{1/B}} + \frac{Y_1^* - Y_2^*}{(Y_2^*)^{1/B}} \right]. \tag{9.4-5}$$

The minimum mass flow of adsorbent $(\dot{S}_{r1} + \dot{S}_{r2})$ can be found by differentiation and setting the derivative equal to zero:

$$\frac{\mathrm{d}((\dot{S}_{r1} + \dot{S}_{r2})/\dot{G}_r)}{\mathrm{d}Y_1^*} = 0. \tag{9.4-6}$$

In the case that the leaving streams of a stage are in equilibrium, the result is

$$\left(\frac{Y_1^*}{Y_2^*} \right)^{1/B} = 1 - \frac{1}{B} + \frac{1}{B} \cdot \frac{Y_0}{Y_1^*}. \tag{9.4-7}$$

This relationship is described in Fig. 9.4-4 where the ratio Y_1^*/Y_0 is a function of the ratio Y_2^*/Y_0. Parameter is the exponent B of the Freundlich equation. Note that a small value of B denotes an equilibrium favorable for adsorption. In this case the reduction of the ratio Y_1^*/Y_0 or the loading Y_1* for a given feed loading Y_0 is considerable in comparison to the second stage. Things are quite different for great exponents B which represent equilibria unfavorable for adsorption.

9.4.3 Countercurrent Flow

The balances of a component transferred in a countercurrent flow of two-phase streams have already been presented in the section on absorption. As a rule the solid adsorbent phase is moving downward in countercurrent movement to the fluid phase (moving bed).

In Fig. 9.4-5 a countercurrent unit with four stages is illustrated. On the right-hand side the loading of the fluid phase Y as a function of the loading X of the solid adsorbent phase with the equilibrium curve and the operating line are shown. The

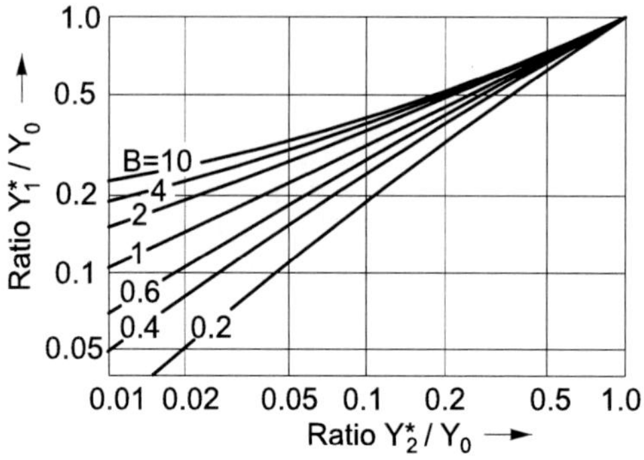

Fig. 9.4-4 Ratios of loading for a two-stage crossflow unit valid for the minimum ratio $\dot{S}/\dot{G}$

loading as concentration is used because it is assumed that the carrier fluid is not adsorbed by the adsorbent and the mass of the unloaded adsorbent remains constant (no evaporation or dissolution).

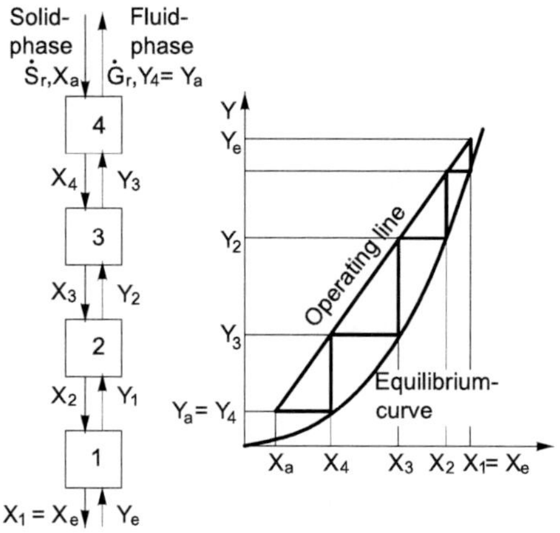

Fig. 9.4-5 Scheme of a four-stage countercurrent unit and its illustration in a diagram

The following equations can be derived from balances of the adsorptive and lead to the relationship for the operating line in Fig. 9.4-5:

$$Y = \frac{\dot{S}_r}{\dot{G}_r} \cdot X - \frac{\dot{S}_r}{\dot{G}_r} \cdot X_e + Y_e , \tag{9.4-8}$$

$$Y = \frac{\dot{S}_r}{\dot{G}_r} \cdot X - \frac{\dot{S}_r}{\dot{G}_r} \cdot X_a + Y_a . \tag{9.4-9}$$

The slope of this straight line is $\dot{S}_r / \dot{G}_r$. The number of stages can be calculated in the same way as has been shown in Chap. 5 which deals with absorption. In the case of very low loadings in the fluid phase the relationship between the loadings of the two phases is linear which leads to a straight line in the diagram of Fig. 9.4-5 according to Henry's law. Let us assume that the feed adsorbent is unloaded ($X_0 = 0$) and the equilibrium is described by the Freundlich equation (9.4-5). The fluid loading $Y_1 = Y_1^*$ in a two-stage unit is given by

$$\frac{\dot{S}_r}{\dot{G}_r} = \frac{Y_e - Y_a}{(Y_1^*/A)^{1/B}}. \qquad (9.4\text{-}10)$$

This means that the reduction of the fluid loading is low after the first stage for the given total reduction $Y_e - Y_a$ for an equilibrium favorable for adsorption (small B values).

A countercurrent unit as illustrated in Fig. 9.4-5 can also be used for desorption processes. In this case substances are transferred from the solid adsorbent phase into the fluid phase and the operating line is running below the equilibrium curve. Balances of the substance transferred around the top or the bottom of the column deliver the equation of the operating line.

9.5 Adsorption Kinetics

Adsorption takes place when the concentration c_i or the partial pressure p_i of the adsorptive i is greater than the equilibrium values c_i* or p_i* which belongs to the loading X_i of the adsorbent.

Desorption occurs as long as the values of the concentration and the partial pressure in the fluid phase are smaller than the equilibrium values for a given loading. The rates of these loading or deloading processes in the adsorbent or the adsorption kinetics can be described in a general way, e.g., it is necessary to formulate the mass transfer resistances around and within an adsorbent particle which can be in a fixed or moving or fluidized bed or suspended in a stirred vessel or in a pneumatic transport system. According to Fig. 9.5-1 these mass transfer resistances in the case of adsorption

- Transport through the fluid to the concentration boundary layer which covers the outer surface of an adsorption pellet.
- Transport through this boundary layer.

- Transport in the macro- and/or micropores in the fluid phase and the adsorbate phase of molecules attached on the pore walls. (This step is sometimes decisive for adsorbents without crystalline cages, for instance, activated carbon and silica.)
- Transport in the cages of microporous adsorbents such as zeolites.
- Attachment of an adsorptive molecule on the wall surrounding the micropores or on the adsorbate layer or capture within a cage of biporous adsorbents, as zeolites. This step leads to the liberation of the heat of adsorption which is transferred to the fluid and the solid adsorbent.

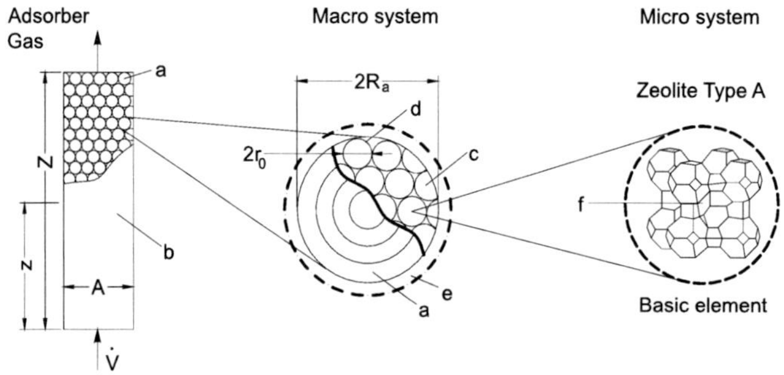

Fig. 9.5-1 Balances of a molecular sieve fixed bed (*left*: fixed bed, *center*: pellet, *right*: zeolite crystal with some cages); (**c**) zeolite crystals; (**d**) macro pore; (**e**) outer boundary layer; (**f**) micro pore

In the case of a desorption process all these resistances must be overcome in the reverse order. At first the heat of desorption to be added results in a detachment of the molecules which pass then through the micro- and macroporous system and finally through the concentration boundary layer into the bulk fluid around an adsorbent pellet. The heat of adsorption (in most cases exothermic) and the heat of desorption (endothermic as a rule) lead to the result that these processes cannot be carried out in an isothermal field. The increase of temperature of the adsorbent by adsorption and the decrease of temperature of the solid phase are the reason that the driving force is reduced and the mass transfer is retarded. It can happen that the mass transfer rates of adsorptives with great heats of adsorption result in such temperature changes that additional adsorptive can only be adsorbed after a removal of heat combined with a temperature loss. The kinetics in the adsorber is limited by heat transfer (heat transfer controlled).

In industrial adsorbers, however, often only small amounts of impurities are removed by adsorption especially in plants for the protection of the environment. The temperature changes during adsorption or desorption of gases can be so small that an isothermal modeling of an adsorber leads to results which are close to reality. As a rule this is true for adsorption in the liquid phase. In the following the modeling of fixed bed adsorbers will be described in more detail because such adsorbers are mostly used in gas and liquid phase adsorption. In this context it will be shown that the adsorption kinetics of adsorbent pellets can be formulated in a general way independent of the situation whether a pellet is suspended in a stirred vessel, a moving bed, or a fluidized bed.

In Fig. 9.5-1 a fixed bed (left side), a spherical adsorbent pellet (center), and some cages of the zeolite type A (right side) are illustrated. The radius of the pellet is denoted with R, and r is the radius of a zeolite crystal. Remember that a zeolite pellet is composed of small crystals fixed together by a bonding agent. Microporous adsorbent pellets as zeolites have radii R in the range of some millimeters and the radius r $(\mathrm{r_a} \lessapprox 1\,\mu\mathrm{m})$ of microcrystals is in the nanometer range. The diffusion coefficient in the macro- and mesopores is denoted with $D_{\mathrm{M,i}}$ (component i), and $D_{\mathrm{m,i}}$ is the diffusion coefficient in the micropores (zeolitic diffusion coefficient). The modeling of a fixed bed is based on the law of conservation of each component and the kinetics of mass transfer and diffusion within the pores. Three different systems with bounding surfaces will be distinguished (Polte 1987; Mersmann 1988):

- A thin layer of the fixed bed with the thickness $\mathrm{d}z$, the volume $f \cdot \mathrm{d}z$ (f is the cross-section area of the bed) and the surface area $\mathrm{d}A = a \cdot f \cdot \mathrm{d}z$ of the particles in the layer (a is the surface area of the pellets based on layer volume, see Sect. 3.4)
- An adsorbent pellet with the outer radius R_a
- A microporous crystal with the outer radius r_a

The balances of mass of the chemical species i and the terms for the adsorption kinetics (mass transfer, pore diffusion) are listed in Table 9.5-1 for the three systems with c_i as the concentration in the fluid phase and X_i as the mass loading of the adsorbent. β denotes the mass transfer coefficient of a pellet and ε_p is its internal porosity. The tortuosity factor μ_p will be explained later. The derivation of equations describing instationary diffusion in spheres has already been presented in Sect. 4.3.3. With respect to diffusion in macropores it is important to consider that diffusion can take place in the fluid as well as in the adsorbate phase. In Table 9.5-1 special initial and boundary conditions valid for a completely unloaded bed (adsorption) or totally loaded bed (desorption) are given. In this section only the model valid for a thin layer in a fixed bed with the thickness $\mathrm{d}z$ and the volume $f \cdot \mathrm{d}z$ will be derived, see Fig. 9.5-2.

Table 9.5-1 Material balances (fixed bed, pellet, crystal)

(1) Material balance in a fixed bed adsorber with the differential height $\mathrm{d}z$:

$$\frac{\partial c_i}{\partial t} + \frac{1-\psi}{\psi} \cdot \rho_p \cdot \frac{\partial \bar{X}_{M,i}}{\partial t} = -\frac{\dot{v}}{\psi} \cdot \frac{\partial c_i}{\partial z} + D_{ax} \cdot \frac{\partial^2 c_i}{\partial z^2}$$

$$c_i(z, t=0) = 0; \qquad c_i(z, t=0) = c_{0,i}; \quad \frac{\partial c_i}{\partial z}(z = Z, t) = 0;$$

$$c_i(z=0, t) = c_{0,1,\mathrm{ads}} + \frac{D_{ax} \cdot \psi}{\dot{v}} \cdot \frac{\partial c_i}{\partial z}; \qquad c_i(z=0, t) = c_{0,1,\mathrm{des}} + \frac{D_{ax} \cdot \psi}{\dot{v}} \cdot \frac{\partial c_i}{\partial z}$$

Adsorption — Desorption

(2) Material balance of an adsorption pellet with the outer radius R_a:

$$\frac{\partial c_{M,i}}{\partial t} + \frac{\rho_p}{\varepsilon_p} \cdot \frac{\partial \bar{X}_{M,i}}{\partial t} = \underbrace{\frac{1}{R^2} \cdot \frac{\partial}{\partial R}\left(R^2 \cdot \frac{D_{M,i}}{\mu_p} \cdot \frac{\partial c_{M,i}}{\partial R}\right)}_{\text{Fluid phase}} + \underbrace{\frac{1}{R^2} \cdot \frac{\partial}{\partial R}\left(R^2 \cdot \frac{\rho_p}{\varepsilon_p} \cdot D_{s,i} \cdot \frac{\partial X_{M,i}}{\partial R}\right)}_{\text{Sorbate phase}}$$

Transport only in the fluid phase:

$$c_{M,i}(R, t=0) = 0; \qquad c_{M,i}(R, t=0) = c_{0,i}; \qquad \frac{\partial c_{M,i}}{\partial R}(R=0, t) = 0;$$

Adsorption — Desorption

$$\beta_i \cdot (c_i - c_{M,i})\big|_{R=R_a} = \frac{\varepsilon_p}{\mu_p} \cdot D_{M,i} \cdot \frac{\partial c_{M,i}}{\partial R}\bigg|_{R=R_a}$$

(3) Material balance of a microporous crystal with the radius r_a:

$$\frac{\partial \bar{X}_{m,i}}{\partial t} = \frac{1}{r^2} \cdot \frac{\partial}{\partial r}\left(r^2 \cdot D_{m,i} \cdot \frac{\partial X_{m,i}}{\partial r}\right)$$

$$X_{m,i}(r, t=0) = 0; \qquad X_{m,i}(r, t=0) = X_i^*(c_{0,i}); \qquad \frac{\partial X_{m,i}}{\partial r}(r=0, t) = 0$$

Adsorption — Desorption

$$X_{m,i}(r = r_a, t)_R = X_i^*(c_{m,i}, t)_R$$

The balance of the adsorptive component i in a system with the bounding surfaces according to this figure as part of the fixed bed with the external void ψ is given by

$$\dot{M}_{\mathrm{con},i} + \dot{M}_{\mathrm{dis},i} = \tag{9.5-1}$$

$$= \dot{M}_{\mathrm{con},i} + \dot{M}_{\mathrm{dis},i} + \frac{\partial \dot{M}_{\mathrm{con},i}}{\partial z} \cdot \mathrm{d}z + \frac{\partial \dot{M}_{\mathrm{dis},i}}{\partial z} \cdot \mathrm{d}z + \mathrm{d}\dot{M}_{\mathrm{ads},i} + \frac{\partial}{\partial t}(c_i \cdot \psi \cdot f \cdot \mathrm{d}z).$$

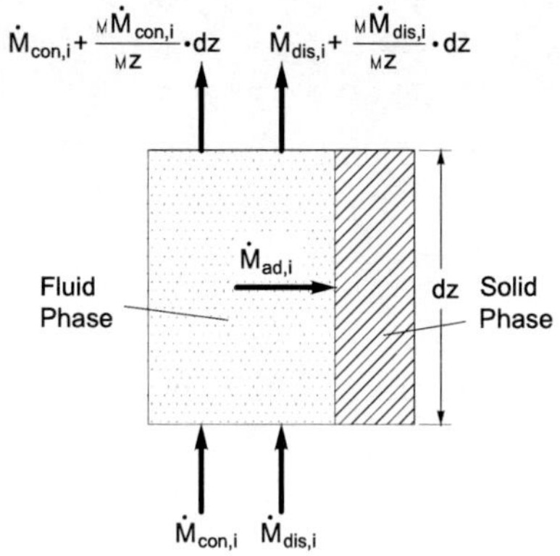

Fig. 9.5-2 Material balance of a thin fixed bed layer

The entering convection stream $\dot{M}_{\text{con},\,i}$ and the entering dispersion stream $\dot{M}_{\text{dis},\,i}$ are equal to the corresponding leaving streams plus the change of the fluid concentration c_i multiplied by the differential void volume $\psi \cdot f \cdot \mathrm{d}z$ plus the stream $\dot{M}_{\text{ads},\,i}$ taken up by the adsorbent. The stream $\dot{M}_{\text{con},\,i}$ can be expressed by the volumetric flow density or the superficial fluid velocity $\dot{v} = \dot{V}/f$:

$$\dot{M}_{\text{con},\,i} = f \cdot \dot{v} \cdot c_i\,. \tag{9.5-2}$$

The mass stream caused by axial dispersion (sometimes called axial diffusion) in the fluid phase or the distribution of residence times is given by

$$\dot{M}_{\text{dis},\,i} = -f \cdot \psi \cdot D_{\text{ax}} \cdot \frac{\partial c_i}{\partial z}\,. \tag{9.5-3}$$

After differentiation with respect to height z, we obtain finally

$$\frac{\partial \dot{M}_{\text{con},\,i}}{\partial z} \cdot \mathrm{d}z = f \cdot \dot{v} \cdot \frac{\partial c_i}{\partial z} \cdot \mathrm{d}z \tag{9.5-4}$$

and

$$\frac{\partial \dot{M}_{\text{dis},\,i}}{\partial z}\mathrm{d}z = -f \cdot \psi \cdot D_{\text{ax}} \cdot \frac{\partial^2 c_i}{\partial z^2} \cdot \mathrm{d}z\,. \tag{9.5-5}$$

The stream $\mathrm{d}\dot{M}_{\text{ads},\,i}$ which is absorbed by the solid adsorbent or pellet with the density ρ_{p} results in a differential increase $\partial \overline{X}_i$ of the mean loading $\overline{X}_i$ during the differential time ∂t:

$$\mathrm{d}\dot{M}_{\mathrm{ads},\,i} = f \cdot (1-\psi) \cdot \rho_{\mathrm{p}} \cdot \mathrm{d}z \cdot \frac{\partial \overline{X}_i}{\partial t}\,. \tag{9.5-6}$$

Note that the loading $\overline{X}_i$ is the mean loading of the pellets defined by the equation

$$\overline{X}_i = \frac{3}{R_{\mathrm{a}}^3}\int_0^{R_{\mathrm{a}}} X_i \cdot R^2 \cdot \mathrm{d}R\,. \tag{9.5-7}$$

If the volumetric flow density $\dot{v}$ is constant with respect to time a combination of these equations leads to

$$\frac{\partial c_i}{\partial t} + \frac{1-\psi}{\psi} \cdot \rho_{\mathrm{p}} \cdot \frac{\partial \overline{X}_i}{\partial t} = -\frac{\dot{v}}{\psi} \cdot \frac{\partial c_i}{\partial z} + D_{\mathrm{ax}} \cdot \frac{\partial^2 c_i}{\partial z^2}\,. \tag{9.5-8}$$

Here c_i is the concentration of the component i in the fluid phase. c_i can be greater than the concentration $c_i|_R$ at the surface of the pellet especially in the case of small superficial velocities $\dot{v}$ and high viscosities of the fluid phase. The model presented in Table 9.5-1 has been often tested experimentally without the kinetic step of molecule attachment (adsorption) or detachment (desorption). As a rule this step can be neglected if the heat of phase change is small, say <30 kJ/mol (Scholl 1991). The initial and boundary conditions given in Table 9.5-1 are valid for special cases (completely unloaded bed for adsorption and desorption of a totally loaded bed by purge gas regeneration). Of course, these conditions have to be adjusted for industrial adsorbers with the consequence that the modeling of fixed bed adsorbers becomes much more difficult and complex. In addition the modeling has to be carried out for each adsorbable component present in the feed fluid. Before the model coefficients

D_{ax} = axial dispersion coefficient

$D_{\mathrm{M},\,i}$ = diffusion coefficient in the macropores

$D_{\mathrm{m},\,i}$ = diffusion coefficient in the micropores

β = mass transfer coefficient fluid / particle

μ_{p} = tortuosity factor

D_{s} = diffusion coefficient of surface diffusion

are described in more detail, some model simplifications valid for fixed beds will be presented.

9.5.1 Simplified Models of Fixed Beds

Integration of the differential equations presented in Table 9.5-1 leads to a description of the fields of concentration in the fixed bed and in the pellets. With respect to industrial adsorption the breakthrough curve of a fixed bed adsorber is most important because this curve is decisive for the capacity of the bed.

Let us first assume that only one adsorbable component is present in a gas stream which is moving downward from the top to the bottom of a fixed bed. So potential fluidization can be avoided. The adsorptive enters the pores of adsorbent pellets in a thin layer of the fixed bed or in the mass transfer zone (mtz). When the equilibrium in this zone is established and the pellets are saturated the zone is moving downward. In such a mass transfer zone the concentration of the adsorptive decreases from the feed concentration to the small exit concentration at the bottom which tends to zero for a completely unloaded bed. After a certain operating time three zones in a fixed adsorber bed can be distinguished:

- Above a saturated equilibrium zone without mass transfer
- In the middle the mass transfer zone
- Below the mtz a second equilibrium zone without mass transfer

When the mass transfer zone arrives at the bottom and an allowable (product or environmental standard) adsorptive concentration is surpassed the raw gas stream must be switched to the second adsorber which has been regenerated. It is understandable that the mass transfer zone should be short to utilize a maximum of the adsorption capacity. Therefore, the shape of the breakthrough curve is economically very important.

9.5.1.1 The LDF model

The model presented in Table 9.5-1 is complex and calculations are time consuming. Therefore, many simplifications of the model are known and experimental breakthrough curves have been compared with results obtained from simplified models. Most known is the linear driving force (LDF) model. As a rule the following assumptions are made:

- No axial dispersion in a fixed bed (data of dispersion coefficients D_{ax} will be given later)
- No radial dispersion (this is approximately true for a ratio of the column diameter based on the pellet diameter greater than 15)

- The adsorbent pellets are spheres and the differences of the local porosities in the bed are small
- The law of ideal gases is valid

The balance of the component i in a thin layer of a fixed bed without the dispersion term ($D_{ax} = 0$) is with the pellet density $\rho_p = \rho_{app}$ (see Table 9.5-1)

$$\frac{\partial c_i}{\partial t} + \frac{(1-\psi)}{\psi} \cdot \rho_p \cdot \frac{\partial \overline{X}_i}{\partial t} = -\frac{\dot{v}}{\psi} \cdot \frac{\partial c_i}{\partial z}. \tag{9.5-9}$$

As a further simplification the transport in the adsorbate phase is neglected in the balance of the component i valid for the spherical shell with thickness dR (compare with Table 9.5-1)

$$\frac{\partial c_{M,i}}{\partial t} + \frac{\rho_p}{\varepsilon_p} \cdot \frac{\partial \overline{X}_i}{\partial t} = \frac{1}{R^2} \cdot \frac{\partial}{\partial R}\left(\frac{D_{M,i}}{\mu_p} \cdot R^2 \cdot \frac{\partial c_{M,i}}{\partial R}\right). \tag{9.5-10}$$

With the expression

$$\frac{\partial X_i}{\partial t} = \frac{\partial X_i}{\partial c_{M,i}} \cdot \frac{\partial c_{M,i}}{\partial t} \approx \frac{\partial \overline{X}_i}{\partial t}, \tag{9.5-11}$$

the last equation can be written as

$$\frac{\partial c_{M,i}}{\partial t} = \frac{(D_{M,i}/\mu_p)}{\left(1 + \frac{\rho_p}{\varepsilon_p} \cdot \frac{\partial X_i}{\partial c_{M,i}}\right)}\left[\frac{1}{R^2} \cdot \frac{\partial}{\partial R}\left(R^2 \cdot \frac{\partial c_{M,i}}{\partial R}\right)\right]. \tag{9.5-12}$$

Here it is assumed that the diffusion coefficient $D_{M,i}$ in the macropores and the tortuosity factor μ_p are not a function of the pellet radius R.

The so-called effective diffusion coefficient $D_{eff,i}$ of the component i is defined by

$$D_{eff,i} = \frac{(D_{M,i}/\mu_p)}{\left(1 + \frac{\rho_p}{\varepsilon_p} \cdot \frac{\partial X_i}{\partial c_{M,i}}\right)}. \tag{9.5-13}$$

Principially speaking, this coefficient is not a real diffusion coefficient which depends on a characteristic velocity and on a characteristic path length of the diffusing units (ions, atoms, molecules, clusters). The simplification is justified by the very simple linear driving force equation at first published by Glueckauf (Glueckauf 1955):

$$\frac{\partial \bar{X}_i}{\partial t} = \frac{15 \cdot \mathrm{D}_{\mathrm{eff,i}}}{R_\mathrm{a}^2} \cdot \left(X_i^*\big|_{R = R_\mathrm{a}} - \bar{X}_i\right). \tag{9.5-14}$$

This equation has been theoretically derived by Liaw (Liaw et al 1979) by the assumption of a parabolic profile of the loading X_i within a pellet. Note that the Glueckauf equation is only valid for a linear adsorption isotherm ($\partial X_i / \partial c_{\mathrm{M},i} = \text{const.}$).

A combination of (9.5-9) and (9.5-14) leads to

$$\frac{\partial c_i}{\partial t} = -\frac{\dot{v}}{\psi} \cdot \frac{\partial c_i}{\partial z} - \frac{15 \cdot \mathrm{D}_{\mathrm{eff,i}}}{R_\mathrm{a}^2} \cdot \frac{(1-\psi) \cdot \rho_\mathrm{p}}{\psi} \cdot \left(X_i^*\big|_{R = R_\mathrm{a}} - \bar{X}_i\right). \tag{9.5-15}$$

Note that up to now it has been assumed that the diffusion in the micropores of microporous adsorbents is not rate controlling. This assumption is also a prerequisite for the Rosen model (Rosen 1954).

9.5.1.2 Rosen model

This model leads to a compact equation for the breakthrough curve. With the expressions

$$z' = \frac{z}{\dot{v} \cdot (1-\psi)} \quad \text{and} \tag{9.5-16}$$

$$t' = t - \frac{z \cdot \psi}{\dot{v}} \tag{9.5-17}$$

the balance of the component i in a thin layer of a fixed bed can be written as follows:

$$\frac{1}{(1-\psi) \cdot \psi} \cdot \frac{\partial c_i}{\partial z'} + \frac{\partial c_i}{\partial t'} = -\frac{(1-\psi)}{\psi} \cdot \rho_\mathrm{p} \cdot \frac{\partial \bar{X}_i}{\partial t'}. \tag{9.5-18}$$

Note that in this equation the axial dispersion ($D_{\mathrm{ax}} = 0$) and also the micropore diffusion are neglected. The model takes into account only the diffusion in the macro- and mesopores and the resistance in a concentration boundary layer around a pellet.

The expression

$$\beta \cdot \left[c_i - \left(\frac{\rho_p \cdot X_i^*}{K}\right)_{R_\mathrm{a}}\right] \tag{9.5-19}$$

with the mass transfer coefficient β is the mean mass transfer rate based on the outer surface of a pellet with the radius R_a. The concentration difference in the boundary layer between the bulk concentration c_i of the adsorptive in the fluid and the concentration on the pellet surface is $[c_i - ((\rho_p \cdot X_i^*)/K)_{R_a}]$ with X_i^* as the equilibrium loading according to $(X_i^* \cdot \rho_p) = K \cdot c_i$.

Note that a linear adsorption isotherm exists only in the Henry range. With the volume-based surface a_p of a particle

$$a_p = 3/R_a, \tag{9.5-20}$$

the uptake $\partial \bar{X}_i/\partial t'$ of the component i in a particle is equal to the mass transfer rate:

$$\frac{\partial \bar{X}_i}{\partial t'} = \frac{3 \cdot \beta}{\rho_p \cdot R_a} \cdot \left[c_i - \left(\frac{\rho_p \cdot X_i^*}{K}\right)_{R_a}\right] \tag{9.5-21}$$

according to the mass transfer coefficient β.

With the expression $\partial \bar{X}_M/\partial t' = (\partial \bar{X}_M/\partial c_M) \cdot (\partial c_M/\partial t')$ the basic differential equation is obtained:

$$\frac{\partial c_{M,i}}{\partial t'} = \underbrace{\frac{\dfrac{D_{M,i}}{\mu_p}}{1 + \dfrac{\rho_p}{\varepsilon_p} \cdot \dfrac{\partial \bar{X}_i}{\partial c_{M,i}}}}_{D_{\text{eff}}} \cdot \left(\frac{2}{R} \cdot \frac{\partial c_{M,i}}{\partial R} + \frac{\partial^2 c_{M,i}}{\partial R^2}\right). \tag{9.5-22}$$

Note again that the effective diffusion coefficient $D_{\text{eff},i}$ is problematic because this coefficient is dependent on the slope of the equilibrium curve. The introduction of the Glueckauf equation and the integration of the differential equation lead finally to the Rosen model:

$$\frac{c_i(t,z)}{c_{\alpha,i}} = \frac{1}{2} \cdot \left\{1 + \text{erf}\left[\frac{\dfrac{\dot{v} \cdot (t - z \cdot \psi/\dot{v})}{K \cdot z \cdot (1-\psi)} - 1}{2\sqrt{\dfrac{1 + 5 \cdot D_{\text{eff,i}} \cdot K/(R_a \cdot \beta)}{15 \cdot D_{\text{eff,i}} \cdot K \cdot z/(R_a^2 \cdot \dot{v})}}}\right]\right\}, \tag{9.5-23}$$

Here $c_{\alpha,i}$ is the concentration of the adsorptive component i at the entrance. The model is valid if

- The fixed bed can be operated isothermally
- The dispersion can be neglected
- The adsorption isotherm is linear
- The micropore diffusion can be neglected
- The pore diffusion can be described by the Glueckauf equation

Micropore diffusion can be neglected if

$$\frac{r_a^2 \cdot D_M}{R_a^2 \cdot \mu_p \cdot D_m} \cdot \frac{\rho_p \cdot \bar{X}}{\varepsilon_p \cdot c_M} < 0.1 \qquad (9.5\text{-}24)$$

according to Polte (Polte 1987). As a rule this is true for small zeolite crystals in large adsorbent pellets. Based on the Rosen model, five breakthrough curves have been calculated, see Fig. 9.5-3.

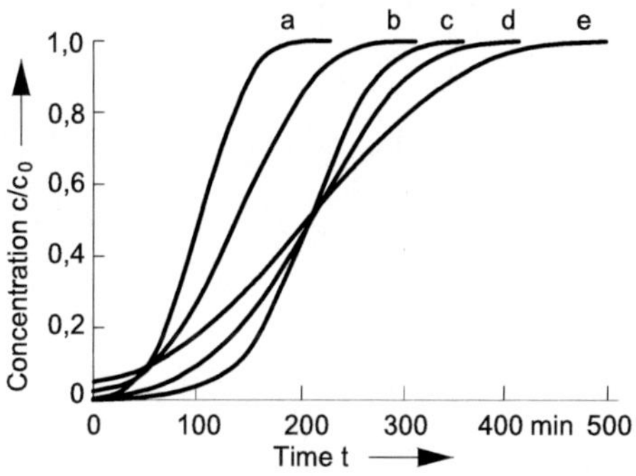

Fig. 9.5-3 Breakthrough curves calculated according to the Rosen model ($K = 2540$; $\dot{v} = 0.12\, m/s$; $D_{eff} = 1.9 \times 10^{-10} (m^2/s)$; $R_a = 4\,mm$; $Z = 1.5\,m$; $\psi = 0.4$; **(a)** $K' = K/2$; **(b)** $\dot{v}' = 1.5 \cdot \dot{v}$; **(c)** $D_{eff} = 2 \cdot D_{eff}$; **(d)** standard case; **(e)** $R'_a = 1.5 \cdot R_a$

Steep breakthrough curves which are favorable in industrial adsorption will be obtained when

- The mass transfer coefficient is high or the mass transfer resistance in the concentration boundary layer is small:
  $$\frac{R_a \cdot \beta}{3 \cdot \varepsilon_p \cdot D_M / \mu_p} \gg 1$$
- The effective diffusion coefficient D_{eff} is high or the molecular as well as the Knudsen diffusion coefficients are high, and the tortuosity factor μ_p is small (curve c)

- The radius R_a of the adsorbent pellet is small (curve e)
- The volumetric flow rate density $\dot{v}$ (equal to the mean velocity referred to the empty bed) is high but the adsorptive concentration in the fluid is small
- The slope $\partial X/\partial c$ and the equilibrium constant K are great or the adsorption isotherm is favorable for adsorption

If in a special case the mass transfer coefficient β is very high or

$$\frac{5 \cdot D_{\text{eff}} \cdot K}{R_a \cdot \beta} \ll 1$$

the Rosen model can be simplified to

$$\frac{c_i(t,z)}{c_{\alpha,i}} = \frac{1}{2} \cdot \left\{ 1 + \text{erf}\left[\frac{\frac{\dot{v} \cdot (t - z \cdot \psi/\dot{v})}{K \cdot z \cdot (1-\psi)} - 1}{2\sqrt{\frac{1}{15 \cdot D_{\text{eff},i} \cdot K \cdot z/(R_a^2 \cdot \dot{v})}}} \right] \right\}. \tag{9.5-25}$$

When all prerequisites for a steep breakthrough curve are fulfilled in a high fixed bed with the height Z the maximum breakthrough time $\tau_{\max}$ is with the assumption that the entire adsorptive in the fluid according to $f \cdot \dot{v} \cdot \tau_{\max} \cdot c_\alpha$ is taken up by the adsorbent mass $S = f \cdot Z \cdot \rho_b$ or $\tau_{\max} \approx Z \cdot \rho_b \cdot X/(\dot{v} \cdot c_\alpha)$. The higher the fixed bed and the smaller the superficial velocity $\dot{v}$ and the concentration c_α the longer is the operation time τ for the adsorption step.

9.5.1.3 General Approach

The solution of the model presented in Table 9.5-1 for a thin layer in a fixed bed, a pellet, and a microcrystal is very time consuming. Therefore, the linear driving force (LDF) model has been extended to take into account further mass transfer resistances. The following extended LDF model takes into account the resistance provided by the concentration boundary layer around a pellet with volume-based outer surface a:

$$\frac{\partial \bar{X}_i}{\partial t} = \left(\frac{R_a^2}{15 \cdot \text{D}_{\text{eff},i}} + \frac{\rho_p}{a \cdot \beta} \cdot \frac{\partial X_i}{\partial c_i} \right)^{-1} (X_i^* |_{R = R_a} - \bar{X}_i). \tag{9.5-26}$$

It is valid if the adsorption time is not too short:

$$\frac{D_{\text{eff,i}} \cdot t}{R_a^2} > 0.1 .$$

In the majority of publications the effective diffusion coefficient takes into account the pore diffusion in the fluid phase (free or molecular diffusion and Knudsen diffusion). In the case of narrow windows of zeolite cages and large adsorptive molecules, micropore diffusion can be rate controlling. Some authors have extended the LDF model in the following way:

$$\frac{\partial \bar{X}_i}{\partial t} = \left(\frac{R_a^2}{15 \cdot \mathrm{D}_{\text{eff,i}}} + \frac{r_a^2}{15 \cdot \mathrm{D}_{\text{m},i}} + \frac{\rho_p}{a \cdot \beta} \cdot \frac{\partial X_i}{\partial c_i} \right)^{-1} \cdot (X_i^* \big|_{R = R_a} - \bar{X}_i) . \qquad (9.5\text{-}27)$$

Here again the expression $\partial X_i / \partial c_i$ is approximately the slope of a linear adsorption isotherm. A typical result of breakthrough curves is the "self-sharpening" effect: These curves are moving through a fixed bed with a constant pattern if the isotherm is favorable for adsorption. When the mass transfer zone at the entrance of the column is broader or narrower than the constant pattern the breakthrough curves will assume this constant pattern after a certain introduction length. In the case of an unfavorable adsorption isotherm the pattern becomes more flat and the mass transfer zone is broader in comparison to an equilibrium which is favorable for adsorption.

With respect to multicomponent adsorption the prediction of breakthrough curves is difficult because a fluctuation of adsorption and desorption of different components in the mass transfer zone can take place due to adsorption equilibrium and adsorption kinetics. In Fig. 9.5-4 the concentration c_i at the exit of a fixed bed based on the concentration $c_{\alpha,i}$ at the entrance is plotted against the adsorption time for the binary mixture CO_2/C_3H_8 ($p_{CO_2} = p_{C_3H_8} = 3{,}000$ Pa, temperature 40°C). Data of some adsorptives are given in Table 9.5-2.

The propane front is moving faster and is travelling ahead of the CO_2 front and a roll up of both components is observed. Obviously the smaller CO_2 molecules with the higher loading for a given partial pressure and temperature in comparison to propane are preferably adsorbed in the first part of the mass transfer zone with the consequence that in the rear of this zone the propane is accumulated. Therefore, the loadings of both components in the mass transfer zone are not equal to the loadings which have established in the equilibrium zone at the entrance of the fixed bed. A pronounced roll-up of propane has also been observed for the ternary adsorptive mixture $CO_2/C_2H_4/C_3H_8$. Data of these components can be found in Table 9.5-2. In Fig. 9.5-5 the ratio $c_i/c_{\alpha,i}$ is plotted against the adsorption time for this mixture. The breakthrough curves of the binary and the ternary mixture have

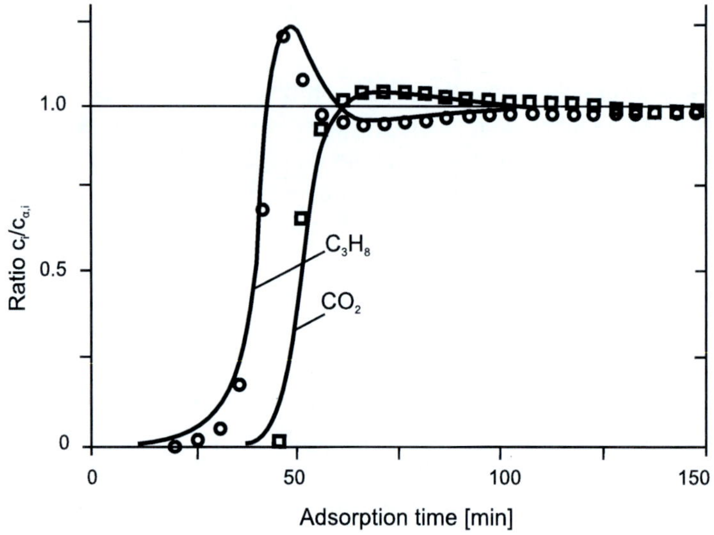

Fig. 9.5-4 Ratio $c_i/c_{\alpha,i}$ vs. the adsorption time

Table 9.5-2 Loadings of some adsorptives in mol/kg

Adsorptive	Critical molecule diameter (nm)	Loadings (mol/kg) at 40°C and	
		1,715 Pa	3,000 Pa
CO_2	0.33	≈ 1.7	≈ 2.0
C_2H_4	0.43	≈ 1.3	≈ 1.5
C_3H_8	0.43	≈ 1.0	≈ 1.2

been calculated with a modified LDF model which takes into account the micropore diffusion. The order of breakthrough curves depends on the selectivity to a certain degree (Schweighart 1994). According to a comparison of experimental and calculated results the adsorption kinetics of propane is limited by zeolitic diffusion in most cases.

9.5.2 Simplified Solution for a Single Pellet

In fixed or fluidized beds, in pneumatic or hydraulic transport systems, or in a stirred vessel, adsorbent particles are surrounded by a fluid phase. If the micropore diffusion (for instance in cages of zeolite crystals) can be neglected the adsorptive concentration in the macropores of a pellet with the radius R is given by

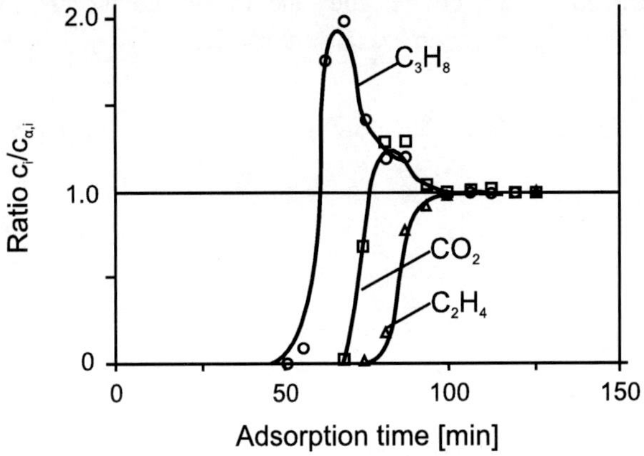

Fig. 9.5-5 Ratio $c_i/c_{\alpha,i}$ vs. the adsorption time

$$\frac{\partial c_{M,i}}{\partial t} = D_{eff,i} \cdot \left(\frac{2}{R} \cdot \frac{\partial c_{M,i}}{\partial R} + \frac{\partial^2 c_{M,i}}{\partial R^2} \right). \tag{9.5-28}$$

Despite the fact that the effective diffusion coefficient is problematic the last equation is often applied on adsorption processes because simple solutions can be found in the literature. In the case of a constant initial concentration and a constant concentration at the particle surface the mean concentration $\overline{c_{M,i}}$ according to

$$\overline{c_{M,i}} = \frac{3}{R_a^3} \cdot \int_0^{R_a} c_{M,i}(R) \cdot R^2 \cdot dR \tag{9.5-29}$$

the following equation is obtained (compare Chap. 4):

$$\frac{\overline{c_{M,i}}}{c_{M,i,\omega}} = 1 - \frac{6}{\pi^2} \cdot \sum_{n=1}^{\infty} \frac{1}{n^2} \cdot \exp\left(-\frac{n^2 \cdot \pi^2 \cdot D_{eff,i} \cdot t}{R_a^2} \right). \tag{9.5-30}$$

Here $D_{eff,i} \cdot t/R_a^2$ is the Fourier number of mass transfer. If the ratio $\overline{c_{M,i}}/c_{M,i,\omega}$ is smaller than 0.95 a further simplification of the last equation is valid:

$$\frac{\overline{c_{M,i}}}{c_{M,i,\omega}} = \frac{6}{\sqrt{\pi}} \cdot \sqrt{\frac{D_{eff,i} \cdot t}{R_a^2}} - \frac{3 \cdot D_{eff} \cdot t}{R_a^2}. \tag{9.5-31}$$

The rate of adsorption is high when the pellets are small and the diffusion is fast (gaseous adsorptives at low pressure but high temperatures). In the case that after regeneration the particle is not completely desorbed and a residual concentration $c_{\mathrm{M},\alpha,i}$ is left in the particle the last equation has to be modified:

$$\frac{\overline{c_{\mathrm{M},i}}-c_{\mathrm{M},i,\alpha}}{c_{\mathrm{M},i,\omega}-c_{\mathrm{M},i,\alpha}}=\frac{6}{\sqrt{\pi}}\cdot\sqrt{\frac{D_{\mathrm{eff},i}\cdot t}{R_{\mathrm{a}}^2}}-\frac{3\cdot D_{\mathrm{eff},i}\cdot t}{R_{\mathrm{a}}^2}. \tag{9.5-32}$$

Since the expression $D_{\mathrm{eff},i}/R_{\mathrm{a}}^2$ is smaller than $1\,\mathrm{s}^{-1}$ the last equation can be further simplified:

$$\frac{\overline{c_{\mathrm{M},i}}-c_{\mathrm{M},i,\alpha}}{c_{\mathrm{M},i,\omega}-c_{\mathrm{M},i,\alpha}}=\frac{6}{\sqrt{\pi}}\sqrt{\frac{D_{\mathrm{eff},i}\cdot t}{R_{\mathrm{a}}^2}}.$$

According to the material balance of the component i

$$\Delta X_i\cdot\rho_{\mathrm{p}}=K\cdot\Delta c_{\mathrm{M},i}$$

with K as the equilibrium constant. The change of the loading can be described by

$$\frac{\overline{X_i}-X_{i,\alpha}}{X_{i,\omega}-X_{i,\alpha}}=\frac{6}{\sqrt{\pi}}\sqrt{\frac{D_{\mathrm{eff},i}\cdot t}{R_{\mathrm{a}}^2}}.$$

According to this equation the expression of the left side is plotted against the square root of time ($\sqrt{t}$) because the effective diffusion coefficient $D_{\mathrm{eff},i}$ can be derived from the slope of the breakthrough curve, see Fig. 9.5-6.

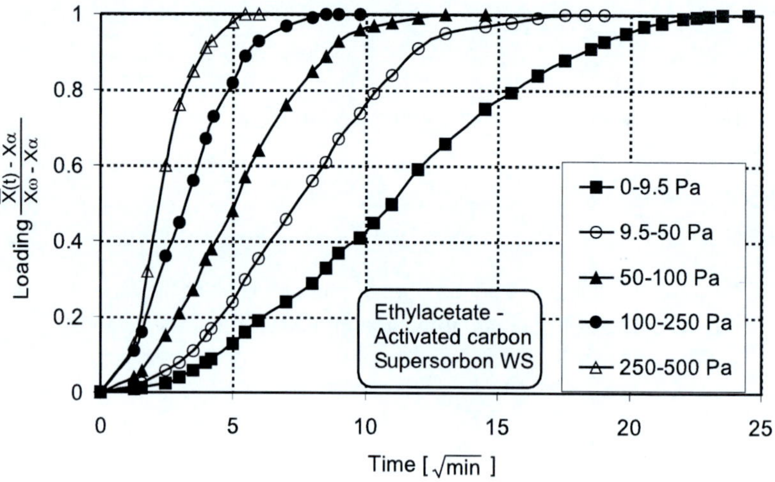

Fig. 9.5-6 Change of loading vs. time for ethylacetate on activated carbon (Kajszika 1998)

Note that these equations are simplified versions of the general model and based on an effective diffusion coefficient which is dependent on the slope of the adsorption isotherm. Principally speaking, the general model based on the material balance and all mass transfer resistances (concentration boundary layer around a pellet, macro and micropore diffusion, sometimes surface diffusion and laminar pore flow) has to be solved.

This has been done by many authors with the following results: In the case of non-linear adsorption isotherms the rates of adsorption and desorption are different. If the isotherm is favorable for adsorption (see Fig. 9.5-7 below) steep loading fronts are moving through the pellet during the adsorption period and it can happen that adsorption only takes place in a relatively thin spherical shell. Dealing with desorption the contrary is true: The changes of the loading as a function of the pellet radius are not dramatic. In the case of an adsorption isotherm unfavorable for adsorption but favorable for desorption the statements are reverse and steep loading profiles are observed for the process of desorption. The different loading patterns in an adsorbent pellet lead to the result that the sorption rates dX/dt are different for adsorption and desorption for curved adsorption isotherms.

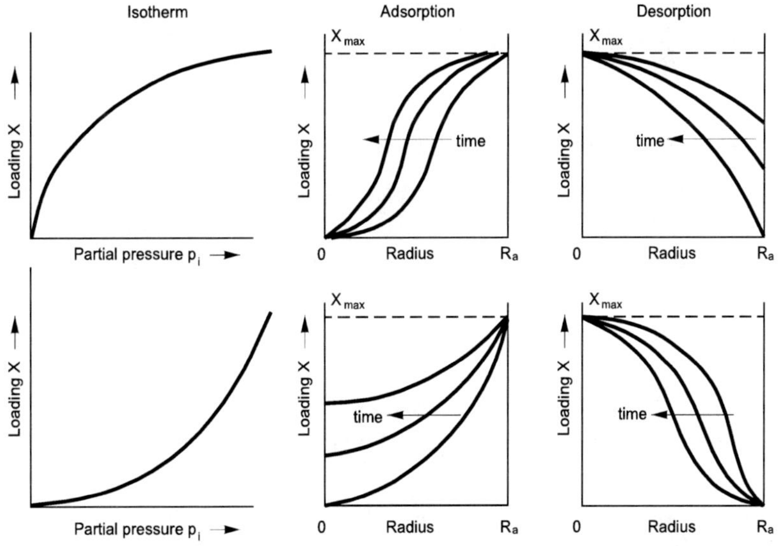

Fig. 9.5-7 Change of loading in a single particle for a concave (favorable for adsorption but unfavorable for desorption) above and a convex isotherm below

Two cases can be distinguished (Kast 1988; Mersmann 1988):

- The isotherm is favorable for adsorption. In this case the rate of adsorption is higher in comparison to the rate of desorption:

$$\left(\frac{\mathrm{d}X}{\mathrm{d}t}\right)_{\mathrm{ads}} > \left(\frac{\mathrm{d}X}{\mathrm{d}t}\right)_{\mathrm{des}} . \tag{9.5-33}$$

- The adsorption isotherm is favorable for desorption with the result that the rate of desorption is greater than the rate of adsorption:

$$\left(\frac{\mathrm{d}X}{\mathrm{d}t}\right)_{\mathrm{des}} > \left(\frac{\mathrm{d}X}{\mathrm{d}t}\right)_{\mathrm{ads}} . \tag{9.5-34}$$

These differences become more pronounced with increasing deviations of the adsorption isotherm from a linear relationship. As a rule the decision on the separation process as temperature swing adsorption (TSA), pressure swing adsorption (PSA), or vacuum swing adsorption (VSA) is a compromise between the steps of adsorption and desorption. Applying the PSA technique on approximately linear adsorption isotherms is advantageous.

9.5.3 Transport Coefficients

Here it will be shown how the coefficients of diffusion, dispersion, and mass transfer defined by the equations in the last chapters can be calculated or estimated.

9.5.3.1 Axial Dispersion Coefficient D_{ax}

In the case of fixed beds a piston or plug flow of the flow is desirable; however, this flow can only be achieved at very high superficial velocities $\dot{v}$. Deviations from the piston flow can be characterized by the dimensionless number D^* according to

$$D^* = \frac{\dot{v} \cdot Z}{\psi \cdot D_{\mathrm{ax}}} . \tag{9.5-35}$$

Here, Z is the height of the bed. Two boundary conditions can be distinguished:

- Piston flow leads to $\frac{\dot{v} \cdot Z}{\psi \cdot D_{\mathrm{ax}}} \to \infty$ or $D_{\mathrm{ax}} \to 0$
- Perfect mixing results in $\frac{\dot{v} \cdot Z}{\psi \cdot D_{\mathrm{ax}}} \to 0$ or $D_{\mathrm{ax}} \to \infty$

The axial dispersion coefficient is dependent on the Peclet number $Pe = \dot{v} \cdot d/(\psi \cdot D_{\mathrm{ax}})$ with d as the pellet diameter:

$$D_{\text{ax}} = \frac{\dot{v} \cdot d}{\psi \cdot Pe} = \frac{2 \cdot \dot{v} \cdot R_a}{\psi \cdot Pe}. \tag{9.5-36}$$

The Peclet number $Pe = \dot{v} \cdot d/(\psi \cdot D_{\text{ax}})$ is proportional to the mean velocity $\dot{v}$ and the pellet size d which is an extension cross to the flow direction (and not in flow direction as the height Z). This number is a function of the product of the Reynolds number and the Schmidt number, see Fig. 9.5-8. The Peclet number assumes small values for $Re \cdot Sc = \dot{v} \cdot d/D_{\text{AB}} < 1$. This means that dispersion becomes strong with the result that the breakthrough occurs earlier in comparison to a piston flow.

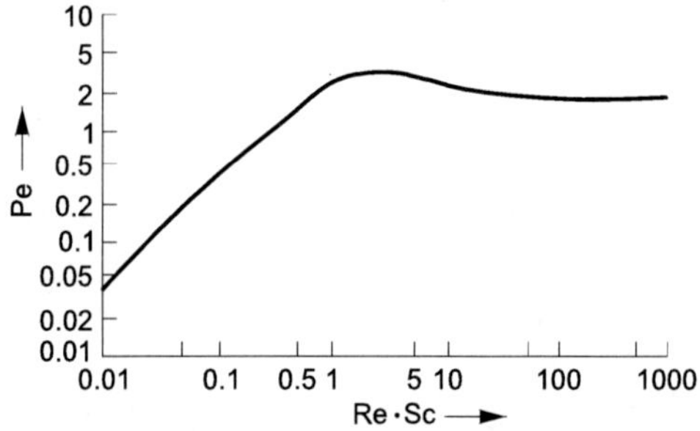

Fig. 9.5-8 Peclet number Pe as a function of the product $Re \cdot Sc$. Valid in the ranges $0.008 < Re < 400$ and $0.28 < Sc < 2.2$

As a rule the superficial velocity $\dot{v}$ is chosen according to a reasonable F-factor with $F = \dot{v}\sqrt{\rho}$. Here ρ is the density of the fluid. Then dispersion is more pronounced in fixed beds operated with liquids as fluids. In most cases the dispersion coefficient is in the order of magnitude $D_{ax} \leq 10^{-3}\ \text{m}^2/\text{s}$.

9.5.3.2 Mass Transfer Coefficient Particle / Fluid

The mass transfer coefficient between the fluid phase and a particle in a fixed bed depends on the Reynolds number $Re_\psi = \dot{v} \cdot \rho \cdot d/(\eta \cdot \psi)$ and the Schmidt number $Sc = \eta/(\rho \cdot D_{\text{AB}})$. Here ψ is the porosity of the bed. The mass transfer coefficient β can be calculated from the following equations (valid for fixed and fluidized beds):

$$Sh_{\text{fb}} \equiv \frac{\beta \cdot d_{\text{p}}}{D_{\text{AB}}} = (1 + 1.5 \cdot (1 - \psi)) \cdot Sh_{\text{b}}, \tag{9.5-37}$$

$$Sh_{\text{b}} = 2 + \sqrt{Sh_{\text{lam}}^2 + Sh_{\text{turb}}^2}, \tag{9.5-38}$$

$$Sh_{\text{lam}} = 0.664 \cdot \sqrt{Re_\psi} \cdot Sc^{1/3}, \tag{9.5-39}$$

$$Sh_{\text{turb}} = \frac{0.037 \cdot Re_{\psi}^{0.8} \cdot Sc}{1 + 2.44 \cdot Re_{\psi}^{-0.1} \cdot (Sc^{2/3} - 1)}. \tag{9.5-40}$$

The Sherwood number Sh_{fb} of the bed depends on the porosity ψ and the Sherwood number Sh_{b} of a single particle and can be mainly the result of diffusion ($Sh_{\text{b}} = 2$), laminar flow ($Sh_{\text{b}} = Sh_{\text{lam}}$), or turbulent flow ($Sh_{\text{b}} = Sh_{\text{turb}}$). The mass transfer coefficient in a fixed bed is approximately by a factor 2 higher than in a fluidized bed. This factor is between 1 and 2 with 1 valid for $\psi \rightarrow 1$. Dealing with agitated vessels the mass transfer coefficient β can be calculated from the mean specific power input $\bar{\varepsilon}$ if all particles are suspended in the liquid, compare Chap. 3:

$$Sh = \frac{\beta \cdot d_{\text{p}}}{D_{\text{AB}}} = 2 + 0.8 \left(\frac{d_{\text{p}}^{4} \cdot \bar{\varepsilon} \cdot \rho^{3}}{\eta^{3}} \right)^{1/5} \cdot Sc^{1/3}. \tag{9.5-41}$$

Note that the increase of the mass transfer coefficient with the mean specific power input $\bar{\varepsilon}$ is rather weak ($\beta \sim (\bar{\varepsilon})^{1/5}$).

9.5.3.3 Diffusion in the Macropores and Tortuosity Factor

In macropores which are wide in comparison to a molecule diameter the adsorptive can be transported by convection, molecular diffusion in the fluid phase, and surface diffusion in the adsorbate. When local differences of the total pressure are present in a pore a flow is initiated: A Poiseuille flow for $Kn < 0.01$, a slip velocity for $0.01 < Kn < 1$, or a molecular flow for $Kn > 1$. With the exception of rapid loading or deloading processes in a fixed bed (often encountered in pressure swing adsorption PSA) the mass transport by differences of the total pressure (Poiseuille flow and convection) can be neglected. Then the adsorptive is transported by the following mechanisms:

- Free or molecular diffusion in the macro- or mesopores which depends on the molecular diffusion coefficient D_{AB} and the tortuosity factor.
- Knudsen diffusion by the Knudsen diffusion coefficient D_{Kn}, when molecular collisions with the pore wall is dominant in comparison to collisions between molecules ($\Lambda > d_{\text{M}}$).
- Surface diffusion or diffusion in the adsorbate layer by the surface diffusion coefficient D_{s}. This mechanism can become important for coverages $\theta > 0.5$.

Let us consider the diffusion coefficients D_{AB} and D_{Kn}. The molecular diffusion coefficient can be calculated from

$$D_{AB} = \frac{2}{3} \cdot \sqrt{\left(\frac{k \cdot T}{\pi}\right)^3} \cdot \left[\frac{1}{2 \cdot M_A} + \frac{1}{2 \cdot M_B}\right] \cdot \frac{1}{p \cdot ((d_A + d_B)/2)^2} \quad \text{for gases} \tag{9.5-42}$$

and

$$D_{AB} = 7.4 \times 10^{-12} \cdot \frac{\sqrt{\psi_B \cdot \tilde{M}_B}}{\eta_L \cdot \tilde{v}_A^{0,6}} \cdot T \quad \text{in} \quad \left[\frac{m^2}{s}\right] \text{ for liquids.} \tag{9.5-43}$$

Here M_A and M_B are the masses of the molecules A and B, respectively, d_A and d_B are the corresponding molecule diameters, p is the total pressure, T is the absolute temperature, and η_L is the viscosity of the liquid in mPa s. $\tilde{M}_B$ denotes the molar mass of the component B in kg/kmol, and $\tilde{v}_A$ in cm^3/mol is the molar volume of the component A. The molecule association factor ψ_B is between 1 and 2 and depends on the dipole moment of the adsorptive molecule B. The molecular diffusion coefficient of gases is inversely proportional to the total pressure p and increases with the absolute temperature according to $D_{AB} \sim T^{3/2}$ whereas for liquids $D_{AB} \sim T/\eta_L$ is valid where $\eta_L(T)$ is dependent on the temperature itself. (The order of magnitude of diffusion coefficients valid for 20°C and 0.1 MPa is $10^{-5} m^2/s$ for gases and $10^{-9} m^2/s$ for low viscous liquids.)

Contrary to the molecular diffusion coefficient, the Knudsen diffusion coefficient is only a function of the absolute temperature T and the pore width d_M:

$$D_{Kn} = \frac{4}{3} \cdot \sqrt{\frac{R \cdot T}{2 \cdot \pi}} \cdot d_M . \tag{9.5-44}$$

As a rule a pore size distribution in adsorbents can be expected. Some authors apply a mean pore size width which is valid for a cumulative volume distribution 0.5 of the pores. The macropore diffusion coefficient D_M present in the balance equations is composed of the molecular diffusion coefficient D_{AB} and the Knudsen diffusion coefficient D_{Kn} according to

$$D_M = \frac{1}{\dfrac{1 - (1 + \dot{n}_B/\dot{n}_A) \cdot y_A}{D_{AB}} + \dfrac{1}{D_{Kn}}} \tag{9.5-45}$$

or

$$D_M = \frac{1}{\dfrac{1}{D_{AB}} + \dfrac{1}{D_{Kn}}} \tag{9.5-46}$$

for equimolar counterdiffusion (Münstermann 1984) for $\dot{n}_{\mathrm{A}} = -\dot{n}_{\mathrm{B}}$ as the molar fluxes of the components A and B , respectively.

High total pressures lead to $D_{\mathrm{M}} \approx D_{\mathrm{AB}}$ whereas in vacuum adsorbers $D_{\mathrm{M}} \approx D_{Kn}$ is valid. Note that the Knudsen diffusion coefficient is changing with the pore diameter and depends on the definition of a mean pore diameter.

9.5.3.4 Tortuosity Factor

The macropores are not straight cylinders but bounded with changing widths and sometimes narrow slits or dead-end pores. Therefore, a tortuosity factor is introduced in the balance equations. Principally speaking, the tortuosity factor should be dependent only on the inner geometry of the adsorbent pellet but not on operation parameter of fluid properties such as the pressure, the temperature, the density, the viscosity, or the molar fluxes of the components. However, this is not true for some publications because of the difficulty to clearly separate the contributions from the particle porosity ε_{p} and the Knudsen diffusion. Therefore, the tortuosity factors published in the literature are often greater than 3 up to 6 which can be expected due to the inner structure of the porous material.

Tortuosity factors are always problematic when Knudsen diffusion in a pore system with a wide width distribution is dominant.

Since some tortuosity factors published in the literature are "Curve-Fitting" factors based on a comparison between calculated and experimental breakthrough curves it is problematic to apply such factors on different adsorbents, adsorptives, and carrier fluids.

9.5.3.5 Surface Diffusion Coefficient

Dealing with adsorbents without microporous cages but rather flat pore walls, an additional transport of the adsorptive component in the adsorbate layer takes place when the coverage is greater than 0.5. The local loading is the driving force; however, this loading cannot be measured. Therefore, the local loading is calculated from the local concentration in the fluid phase by the assumption of equilibrium (equilibrium model). The surface diffusion is an activated process and an adsorbed molecule requires approximately half of the adsorption energy for activation to move on the surface in short jumps. The surface diffusion coefficient is the product of the self-diffusion coefficient D_{s}^{0} and an activation term according to Arrhenius:

$$D_{\mathrm{s}} = D_{\mathrm{s}}^{0} \cdot \exp\left(-\frac{a_s \cdot \Delta\tilde{h}_{\mathrm{ads}}(n)}{\tilde{R} \cdot T} \right). \tag{9.5-47}$$

Experiments carried out with the system carbon dioxide/activated carbon (Hartmann 1996) resulted in:

$$D_s^0 = 2.2 \times 10^{-6}\ \mathrm{m^2/s}$$

$$\Delta\tilde{h}_{ads} = 24.5\ \mathrm{kJ/mol}\ \text{and}\ a_s \approx 0.5\,.$$

Principally speaking, the chemical potential is the decisive driving force (and not the concentration). The Darken equation takes this effect into account. With D_s^0 as the self-diffusion coefficient the Darken diffusion coefficient D_{Dark} results in

$$D_{Dark} = D_s^0 \cdot \frac{\partial \ln p_i}{\partial \ln X_i}\,.$$

9.5.3.6 Micropore Diffusion Coefficient

The predictive calculation of micropore diffusion coefficients is difficult because the mechanisms which lead to a molecule transport are not sufficiently known. When the molecules preferably move in the fluid phase it can be assumed that the diffusion coefficient is a function of the molecule velocity and the width L_z of the (zeolitic) micropores. Molecules will be adsorbed on the pore walls and have to be accelerated to return into the fluid phase. So far the transport mechanism is an activated process according to the relationship (Schweighart 1994)

$$D_m = g \cdot \left(\frac{8 \cdot k \cdot T}{\pi \cdot M_B}\right)^{1/2} \cdot L_z \cdot \exp\left(-\frac{a_s \cdot \Delta\tilde{h}_{ads}}{\tilde{R} \cdot T}\right). \qquad (9.5\text{-}48)$$

Here M_B is the mass of the molecules of the component B. The factor a_s can be expected between 0.3 and 1. Things are quite different when the molecules have to stay in the immediate neighborhood of solid pore walls because the micropores or slits are so narrow. In this case the vibration of the solid molecules with the frequency ν is the decisive transport mechanism. The following equation is assumed to be a reasonable approach:

$$D_m = g' \cdot \nu \cdot L_z^2 \cdot \exp\left(-\frac{a_s \cdot \Delta\tilde{h}_{ads}}{\tilde{R} \cdot T}\right). \qquad (9.5\text{-}49)$$

Experimental results can deliver the constants g and g'. In Table 9.5-3 some orders of magnitude of micropore diffusion coefficients are given (Schweighart 1994).

Table 9.5-3 Orders of magnitude of micropore diffusion coefficients of some adsorptives in MS5A at 20°C

Adsorptive	Micropore diffusivity m^2/s
CO_2	10^{-11}
C_2H_4	10^{-13}
C_2H_6	10^{-13}
C_3H_8	10^{-15}

9.5.4 The Adiabatic Fixed Bed Absorber

The operation mode of fixed bed adsorbers can be isothermal (very small adsorptive concentration in the fluid and low heats of adsorption), nonisothermal, and adiabatic. The heat loss of large industrial adsorbers is often so small in comparison to the heat production by adsorption that the bed is nearly operated adiabatically. In such a case not only the mass balances but also the energy balances have to be taken into account to get information on the operating mode and the fields of concentration and temperature in a fixed bed. These balances for the adsorbent (Index S = solid) and the fluid (Index G) are

$$(1-\psi)\cdot\rho_S\cdot c_S\cdot\frac{\partial T_S}{\partial t} =$$

$$= \bar{\alpha}\cdot a\cdot(T_G - T_S) + (1-\psi)\cdot\lambda_S\cdot\frac{\partial^2 T_S}{\partial z^2} + (1-\psi)\cdot\rho_S\cdot\Delta h\cdot\frac{\partial\bar{X}}{\partial t} \quad (9.5\text{-}50)$$

and

$$\psi\cdot\rho_G\cdot c_G\cdot\frac{\partial T_G}{\partial t} = \bar{\alpha}\cdot a\cdot(T_S - T_G) - \dot{v}\cdot\rho_G\cdot c_G\cdot\frac{\partial T_G}{\partial z} + \psi\cdot\lambda_G\cdot\frac{\partial^2 T_G}{\partial z^2}. \quad (9.5\text{-}51)$$

If the Biot number $Bi \equiv \alpha\cdot d_p/(6\cdot\lambda_S) \ll 1$ the differences of the local temperature in a particle with the diameter d_p are small. Then the mean heat transfer coefficient $\bar{\alpha}$ can be replaced by the coefficient α according to the following equations:

$$Nu_{fb} \equiv \frac{\alpha\cdot d_p}{\lambda_G} = [1 + 1.5\cdot(1-\psi)]\cdot Nu_b \quad (9.5\text{-}52)$$

with the Nusselt number Nu_b of the particle

$$Nu_{\mathrm{b}} = 2 + \sqrt{Nu_{\mathrm{lam}}^2 + Nu_{\mathrm{turb}}^2}\,, \tag{9.5-53}$$

and

$$Nu_{\mathrm{lam}} = 0.664 \cdot \sqrt{Re_{\psi}} \cdot Pr^{1/3} \tag{9.5-54}$$

and

$$Nu_{\mathrm{turb}} = \frac{0.037 \cdot Re_{\psi}^{0.8} \cdot Pr}{1 + 2.44 \cdot Re_{\psi}^{-0.1} \cdot (Pr^{2/3} - 1)}\,. \tag{9.5-55}$$

Based on the adsorption equilibria and the balances of energy and mass in combination with the transport coefficients, the fields of temperature and concentration and the breakthrough curves as a function of temperature and concentration can be calculated by a simultaneous solution of the equations. Some general statements can be made. Most of the heat released by the attachment of adsorptive molecules is transported by convection and conduction in the fixed bed and results in a heat wave or temperature front moving in flow direction. In the same direction a concentration front or breakthrough curve is travelling. Both fronts are coupled with respect to the equilibrium which is strongly temperature dependent and the kinetic coefficients with their weak temperature dependence. These fronts can move through the bed with the same or different velocities.

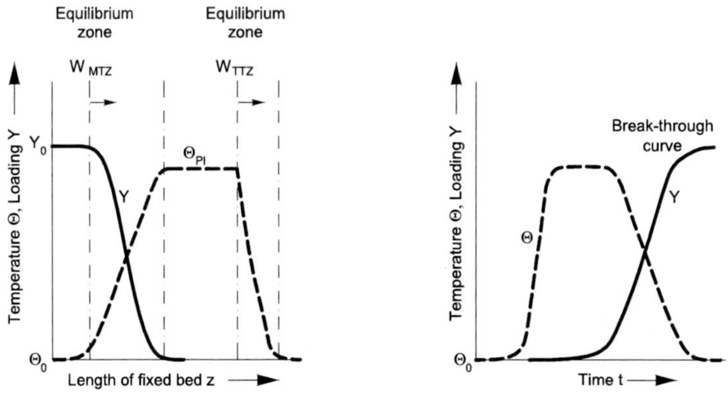

Fig. 9.5-9: Schematic illustration of the temperature and concentration break through curve for the case of a preceding temperature front

In Fig. 9.5-9 it is assumed that the temperature front (w_{TTZ} is the velocity of the "temperature transfer zone") is moving with a higher speed than the ("mass transfer Zone" or the breakthrough curve. In the left figure the courses of the fluid loading Y and the temperature rise θ in the fixed bed for a given time are depicted, whereas in the figure on the right side these operating parameters Y and θ are valid for a certain bed length, here the end of the bed.

In Fig. 9.5-10 the case is shown that both fronts are moving through the bed with the same velocity. According to Pan and Basmadjian (Pan and Basmadjian 1970) the temperature front moves faster than the breakthrough curve if

$$\frac{c_s}{c_p} < \frac{\Delta X}{\Delta Y}. \tag{9.5-56}$$

Here c_s and c_p are the specific heats of the solid adsorbent and of the fluid, respectively, and ΔX denotes the change of the loading X of the adsorbent. ΔY is the reduction of the loading in the fluid phase. Such combined fronts are existing when

$$\frac{c_s}{c_p} \geq \frac{\Delta X}{\Delta Y}. \tag{9.5-57}$$

Note that the specific heat capacity of air is $c_p \approx 1\ \mathrm{kJ/(kgK)}$ and that of silica gel and zeolites approximately $0.92\ \mathrm{kJ/(kgK)}$. With $c_s/c_p \approx 0.92$ for these systems the slope $\mathrm{d}X^*/\mathrm{d}Y$ of the adsorption isotherm is decisive for the question whether the temperature front is proceeding or the breakthrough of the temperature and concentration front occurs simultaneously. The maximum temperature rise $\Delta\theta_{max}$ in a fixed bed is the difference of the plateau temperature θ_{pl} and the temperature θ_0 of the entering fluid according to

$$\Delta\theta_{max} = \theta_{pl} - \theta_0 \tag{9.5-58}$$

and can be calculated from

$$\Delta\theta_{max} = \frac{\Delta X \cdot \Delta h / c_p}{\Delta X / \Delta Y - c_s / c_p}. \tag{9.5-59}$$

This means that strong temperature effects or deviations from an isothermal operation mode can be expected when the loading difference ΔX of an adsorptive with a great heat of adsorption Δh is high but the specific heat capacity c_p of the carrier fluid is low. In adiabatic fixed bed adsorbers high temperature rises during adsorption and very sensitive temperature drops during desorption can take place. As an example, the concentration ratio c/c_0 and the temperature difference $\Delta\theta = \theta - \theta_0$ are shown in Fig. 9.5-11 as a function of time for the system CO_2 / MS5A. When the concentration c_0 of the carbon dioxide in the feed gas is only 1 vol% the temperature front is faster than the concentration front or breakthrough curve (curve a).

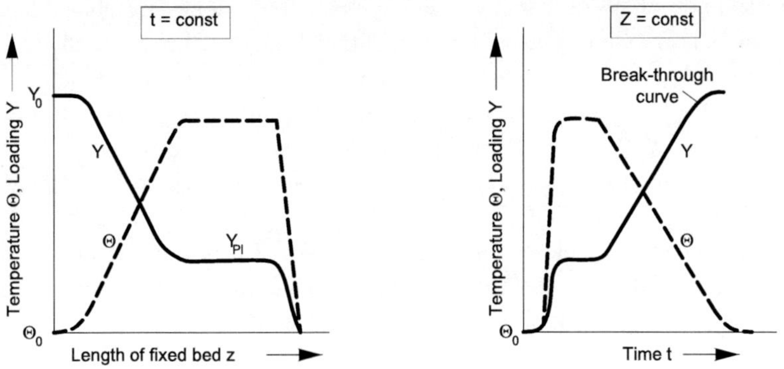

Fig. 9.5-10 Schematic illustration of the temperature and concentration breakthrough curve for the case of simultaneous breakthrough

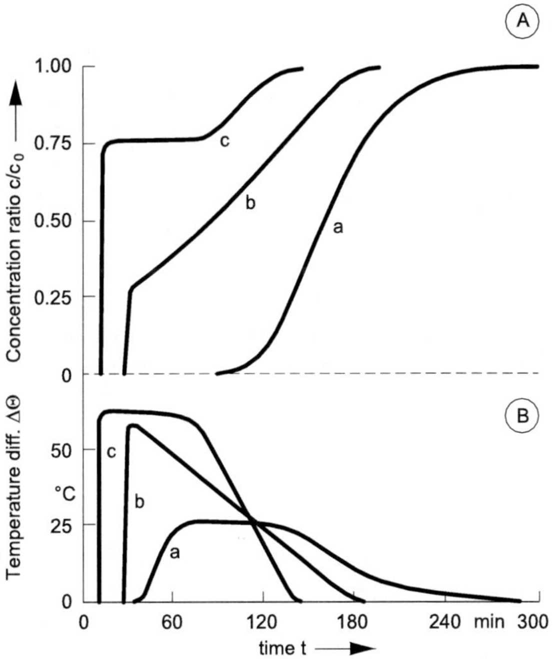

Fig. 9.5-11 Adsorption system: CO_2 molecular sieve 5A; $\Theta_0 = 25°C$; $\dot{v} = 0.2 m/s$; $d = 2mm$; $D = 100mm$; $Z = 1m$. (**a**) Effect of the feed concentration of the adsorptive on the time-dependent concentration. (**b**) Effect of the feed concentration on the time-dependent temperature (**a**) $c_0 = 1 vol\%$; (**b**) $c_0 = 3 vol\%$; (**c**) $c_0 = 10 vol\%$

Things are different for higher concentrations in the feed gas. Combined temperature and concentration fronts are moving through the fixed bed when the feed concentration is either 3 vol% (curve b) or 10 vol% (curve c). The maximum temperature rise θ_{max} is mainly dependent on the heat source ($\sim \Delta X \cdot \Delta h$) in the mass transfer zone. With increasing temperatures diffusivities are rising; however, the mass transfer in this zone decreases because the adsorbent can take up less adsorptive according to a lower equilibrium loading X^* in comparison to an isothermal bed. The intensity of these heat and temperature effects depends mainly on the shape of the equilibrium adsorption isotherm and its slope $\mathrm{d}X^*/\mathrm{d}Y$ but also on the heat of adsorption. According to the first chapter it is

$$\left|\frac{\partial \ln p_i}{\partial(1/T)}\right|_{n_i = \text{const.}} = -\frac{\Delta h_i}{R}.$$

In the case of desorption the heat of desorption is provided by the adsorbent or the purge gas or by both. When the purge gas is not preheated a drop of the temperature of the adsorbent takes place and this drop is more pronounced with increasing loading changes ΔX and heats of desorption. This is shown in Fig. 9.5-12 in which the concentration ratio c/c_0 (above) and the temperature drop $\Delta\theta$ are plotted against time, again for the system CO_2 / MS5A. This purge gas desorption results in combined temperature and concentration fronts and the drops of temperature and concentration at the end of the bed are very steep and occur after a short time. After 20–50 min – the exact time depends on the CO_2 concentration – the curves run into plateaus. The temperature in the mass transfer zone is so low for a certain period that no longer desorption takes place. At the end of desorption – in Fig. 9.5-12 after 6 h – the concentration in the leaving gas is very low and the bed has been heated up with the result that the gas temperature at the exit is only a bit lower than the temperature of the purge gas at the entrance.

Note that the regeneration with a purge gas at ambient temperature results in long desorption times and cannot be economically recommended because the recovered adsorptive component is very diluted in the off-stream. Therefore, the purging with a hot gas is used in industrial plants. This operation mode leads to plateau temperatures which are higher in comparison to regeneration with a cold gas. After a short desorption time the temperature of the gas at the bed outlet jumps up. The higher the temperature of the regeneration gas the shorter the plateau period. Figure 9.5-13 shows an example, molecular sieve MS5A loaded with CO_2 is regenerated by a gas (air) at a temperature of 200°C. The CO_2 concentrations in the outlet gas of a bed with a height of 0.5 m (curve a) or 1 m (curve c) and the outlet temperatures again for a height of 0.5 m (curve b) and 1 m (curve d) are shown as a function of time. After a certain plateau period the temperatures increase steeply whereas a rapid drop of the concentrations takes place. The combined concentration and tem-

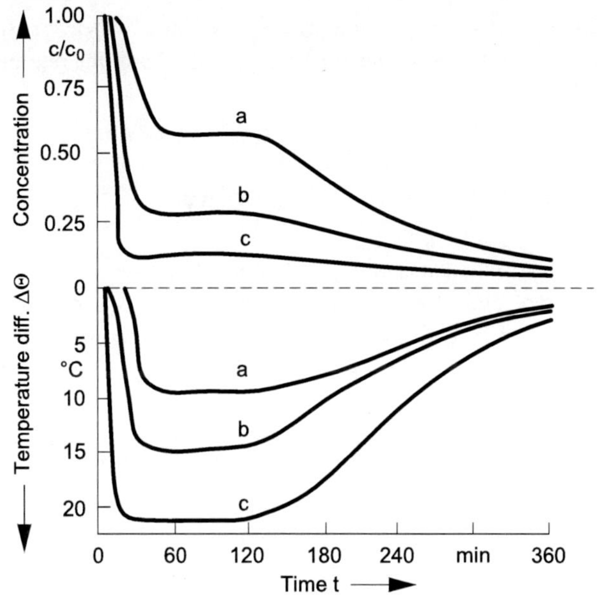

Fig. 9.5-12 Desorption by a purge gas: effect of the feed concentration of the adsorptive on the concentration and temperature. System: CO_2 molecular sieve 5A; $\Theta_0 = 25°C$; $\dot{v} = 0.2\,m/s$; $d = 2mm$; $Z = 1m$.
(**a**) $c_0 = 1vol\%$; (**b**) $c_0 = 3vol\%$; (**c**) $c_0 = 10vol\%$

perature front appear after 40 min in the short bed (0.5 m) but after approximately 80 min in the bed with a height of 1 m. Note that in the case of a preheated regeneration gas the desorption time is much shorter than for purging with a gas of ambient temperature. When the adsorbate is removed to a high degree and the heat of the regeneration gas is no longer used for the heat of desorption a sudden jump of the temperature of the gas at the outlet can be observed. It is important to note that

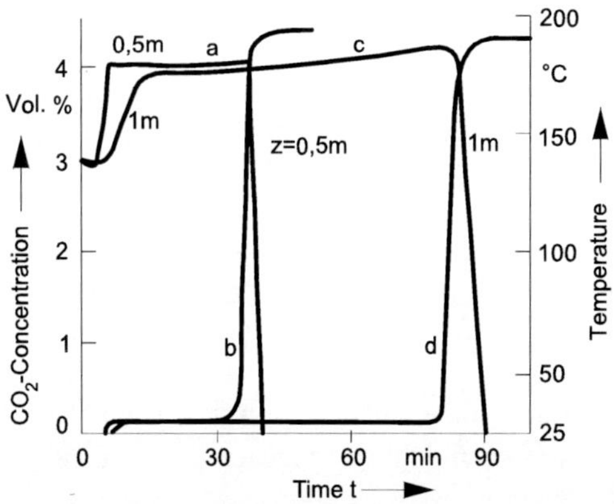

Fig. 9.5-13 Thermal desorption: temperatures and concentrations in an adiabatic fixed bed

here the kinetics of the desorption process is limited by heat transport (and not by diffusivities). The same is true for drying in the constant rate period as will be shown in the next section (heat transfer controlled).

9.6 Regeneration of Adsorbents

According to the statements on adsorption equilibria the loading X decreases with increasing temperature for a given partial pressure or concentration of the adsorptive in the fluid. In an isothermal system the loading decreases with decreasing partial pressure or concentration. As has been shown for absorption loaded adsorbents can be regenerated either by an increase of the temperature (TSA process) or by a reduction of the pressure or concentration (PSA and VSA processes). This is illus-

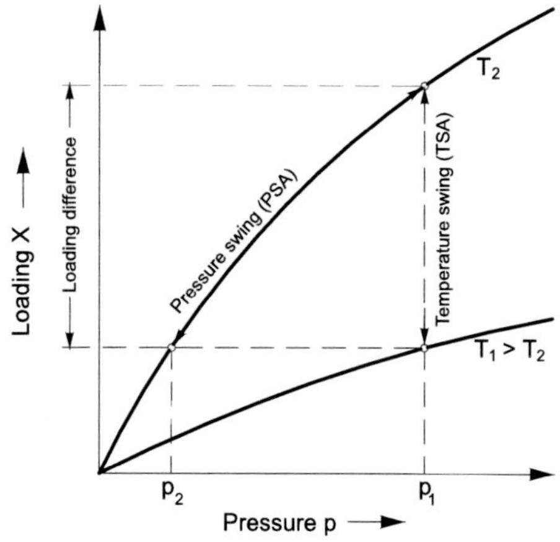

Fig. 9.6-1 Comparison of the regeneration methods TSA and PSA in a chart loading vs. pressure

trated in Fig. 9.6-1. A further regeneration process is based on the replacement of the adsorbate by another adsorptive with a greater affinity to the adsorbent (Ruhl 1971).

In Fig. 9.6-2 a unit with two fixed beds is depicted. The adsorbent is regenerated by the TSA process and deloaded by heating provided indirectly by an embedded heat transfer exchanger or directly by a hot regeneration gas or by both. The residual loading depends on the maximum temperature and a preconcentration of the perhaps recircled gas and of the kinetic approach to equilibrium. TSA is applied when the equilibrium is favorable for this regeneration method, e.g., an essential reduc-

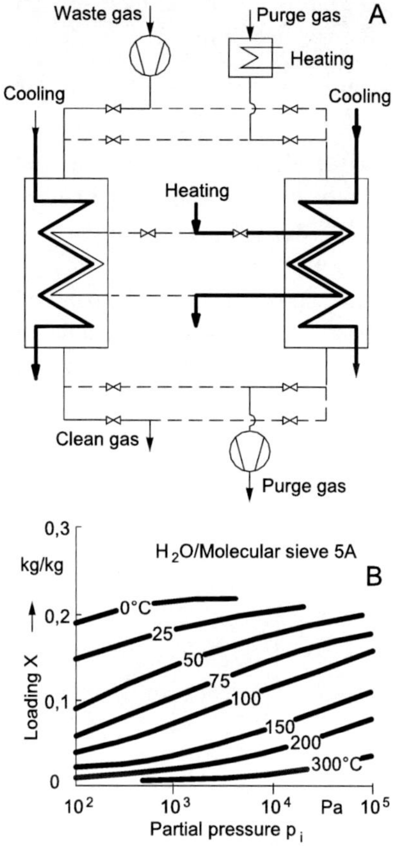

Fig. 9.6-2 (**A**) Adsorption unit with regeneration of the adsorbent by temperature swing. (**B**) Adsorption isoterms of water / MS5A

tion of the loading by a given temperature rise. As an example the adsorption of vapor on MS5A will be shortly discussed. For a given partial pressure of 100 Pa the loading is reduced from $0.15 \text{ kg H}_2\text{O / kg MS}$ to $\approx 0.02 \text{ kg / kg}$ when the temperature is increased from 25 to 200°C. Temperature swing adsorption is applied on molecular sieves which are loaded with H_2O, SO_2 and CO_2. Note that the heats of adsorption are very high ($\Delta h > 30 \text{ kJ / mol}$) for polar adsorptives adsorbed on adsorbents with electrical charges like zeolites. The TSA process is also applied on the regeneration of activated carbon, zeolites, and silica gel loaded with hydrocarbons present in natural gas when the heat of adsorption is higher than 30 kJ / mol.

Principally speaking, heat can be transferred to the loaded adsorbent

- Directly by a preheated medium as air, flue gas, nitrogen (often necessary for activated carbon), inert liquids, steam or solids (sand).

- Indirectly by heat exchangers surrounded by the adsorbent. These exchangers are heated by a hot fluid or electrically.
- By an electrical heating (according to an electrical resistance, changing electrical field, or induction).

According to the scale of the adsorber and the physical properties of the adsorbent and the adsorptive (conductivity for heat and electricity, permittivity, heat sensitivity of organic compounds), a decision has to be made on the most economical regeneration process. Note that the desorption with preheated air can lead to a decomposition of organic components and to hot spots with the result that activated carbon is burnt. When a part of the pores is plugged by polymerizates and/or coke, adsorbents are reactivated at elevated temperatures by a partial oxidation. As a rule this regeneration results in a loss of capacity. Furthermore a part of activated carbon is burnt resulting in a loss of weight. Sometimes loaded activated carbon is

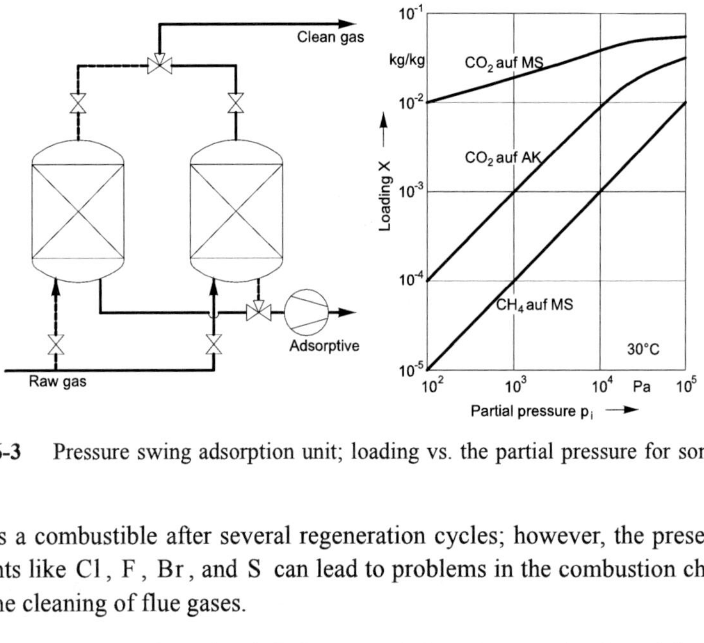

Fig. 9.6-3 Pressure swing adsorption unit; loading vs. the partial pressure for some systems

used as a combustible after several regeneration cycles; however, the presence of elements like Cl, F, Br, and S can lead to problems in the combustion chamber or in the cleaning of flue gases.

The pressure swing adsorption (PSA) process is based on a reduction of the partial pressure to desorb the adsorbate. This can be obtained either by depressurization or by vacuation or by both. In Fig. 9.6-3 an adsorption unit with two fixed beds is depicted, compare also Fig. 9.2-2. The PSA process is applied if an essential reduction of the loading is achieved by a certain pressure drop. This depends on the loading n as a function of the partial pressure at the operation temperature. In Figs. 9.6-4 and 9.6-5 the loading is plotted against the pressure for some gases and vapors according to experimental and theoretical (Maurer 1999) results. Let us compare

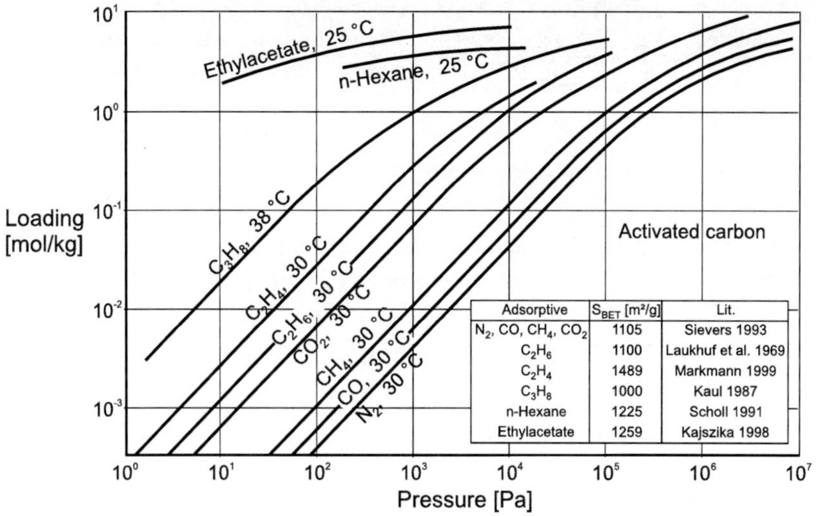

Fig. 9.6-4 Loading vs. the adsorptive pressure for different adsorptives

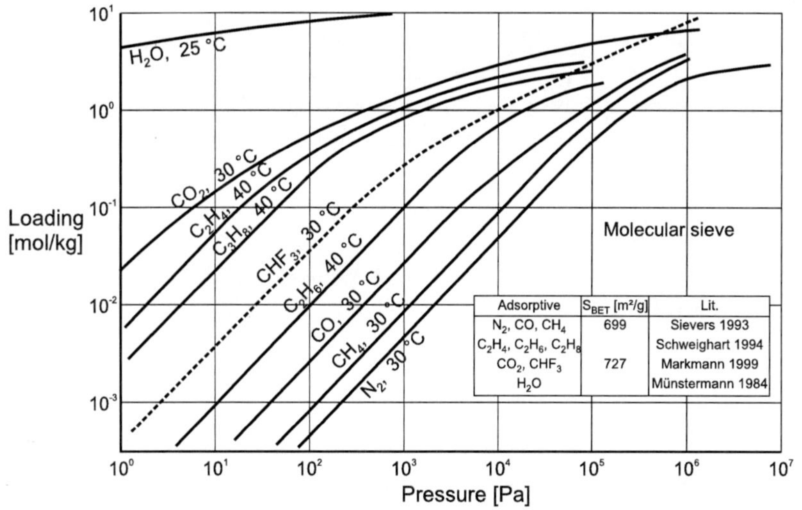

Fig. 9.6-5 Loading vs. the adsorptive pressure for different adsorptives

carbondioxide adsorbed on activated carbon and on MS5A at a temperature of 30°C. When the partial pressure is reduced from 10^5 to 10^4 Pa the loading of CO_2 on activated carbon is reduced from ≈ 2.3 mol / kg to ≈ 0.55 mol / kg but in the case of MS5A only from ≈ 5.2 mol / kg to ≈ 3.0 mol / kg. Note that the isotherms of C_3H_8 on activated carbon and MS5A are close together because propane is nonpolar. The PSA process is not favorable for both adsorbents and this adsorptive. In the case of the adsorptives *n*-hexane and ethyl acetate, an efficient desorption can only be achieved by a TSA process. This is also true for the desorption of H_2O adsorbed on MS5A, see Fig. 9.6-5 and compare with Fig. 9.6-2.

9.7 Adsorption Processes

The decisive parameters for industrial adsorption processes are the equilibrium and the desorption mode. As a rule activated carbon is used for the separation of organic solvents present in off-gases. Regeneration is carried out by passing steam through the loaded bed. The subsequent process steps are loading, steaming, drying, and cooling.

Benzene and gasoline components as main substances in coke oven gas are also isolated by adsorption; however, the pores of activated carbon can be plugged by resins or other impurities. Activation at high temperatures is a suitable process to restore the original capacity of the adsorbent to a certain degree. Liquefied gases as propane and butane can be separated by adsorption using activated carbon.

This adsorbent is also applied for the removal of hydrogen sulfide from oxygen-free off-gases. Hot flue gases are used for regeneration. Organic sulfur compounds are also separated by adsorption. Natural gas contains hydrogen sulfide and mercaptanes. Molecular sieves are excellent adsorbents for the removal of these impurities. Desulfurization of liquefied gases can be carried out with molecular sieves as adsorbents. Sulfur present in liquid combustibles is the source of sulfur dioxide in the flue gas. When this adsorptive is adsorbed on coke or activated carbon, sulfur acid comes into existence. This adsorbate is removed by a hot inert gas or hot sand at 600°C for regeneration (Noack and Knoblauch 1976; Jüntgen et al. 1972).

Adsorption processes with molecular sieves as adsorbents are a good tool to separate mixtures of *n*-paraffins, isoparaffins, olefines, and cyclic hydrocarbons according to the PSA mode.

Principally speaking, separation by adsorption can be based on a steric or a kinetic or an equilibrium effect. With respect to a big variety of adsorbents (pore width, pore size distribution, polar or nonpolar, hydrophilic or hydrophobic behavior, addition of functional groups) and wide ranges of pressure and temperature which can be chosen independently of each other, adsorption processes allow more flexibility in comparison to rectification.

Desorption can be carried out by displacement of the adsorbate with an appropriate substance.

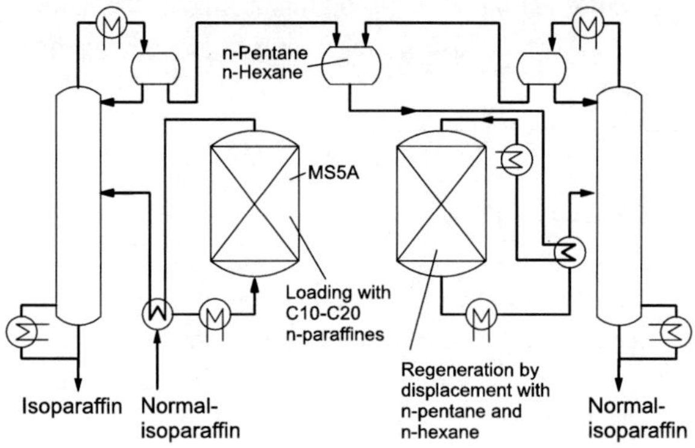

Fig. 9.7-1 Adsorption unit for the separation of a mixture of *n*-paraffins and isoparaffins

Fig. 9.7-1 shows a simplified scheme of a process by which a mixture of *n*-paraffins and isoparaffins is separated. Such a mixture is fed the adsorber on the left side. Only *n*-paraffins are adsorbed due to the steric or sieve effect. Isoparaffins with a molecule size larger than the micropores are passing a heat exchanger and enter the rectification column from which they are withdrawn as bottom product. The adsorber loaded with *n*-paraffins is regenerated by displacement with n-pentane and *n*-hexane. The mixture of short- and long-chained paraffins is fed a second rectification column and separated in to high-boiling isoparaffins and low-boiling displacement substance. Note that the mixture of *n*-pentane and *n*-hexane is circulating through the adsorber bed during desorption and the column on the right side. A rectification of a mixture of *n*-paraffins and isoparaffins is problematic because the boiling temperatures are very close together.

Adsorption is applied to a large extent in the field of fresh- and wastewater purification. Organic impurities and especially substances which are not accessible to biological treatment are adsorbed on coke or activated carbon. Some organics like aromates containing chlorine and nitrogen as well as polycyclic compounds cannot be degraded by microorganisms. Often loaded coke is burnt in a furnace instead of regeneration.

Chlorine, ozone, manganese, and iron present in freshwater are eliminated by adsorption. Activated carbon is also used for the purification of condensed steam. Not only dissolved oil constituents but also small emulsified droplets are removed. The purification of aqueous or organic solvents and solutions is often carried out by activated carbon adsorption.

Besides fixed and fluidized bed adsorbers also stirred vessels filled with an impurity loaded liquid and powdered active carbon are used. The suspension is separated by filtration or centrifugation after a certain residence time necessary for mass transfer. Decolorization of solutions occurring in the food industry can be achieved by activated carbon adsorption. Examples are the purification of glucose, sugar, wine, oil, and fats.

9.8 Chromatography

In the previous section multicomponent adsorption in a fixed bed adsorber has been discussed. As an example the breakthrough curves of the components C_2H_4, C_3H_8, and CO_2 of a gaseous ternary mixture are shown in Fig. 9.5-5. During the adsorption cycle adsorptives are continuously fed the column. Since the breakthrough curves are close together a separation of the components cannot be accomplished. Here the objective of adsorption is the cleaning of an off-gas. Contrary to this operation elution chromatography means that a mixture is fed into a chromatography column only at the beginning of a batch and then moved through the column by means of a carrier fluid. According to adsorption and desorption steps the components are travelling in mass transfer zones with different velocities with the result that the components leave the column subsequently as fractions.

By means of elution chromatography in the trace mode a mixture can be separated in more or less pure components. If the travelling velocities of the mass transfer zones or bands are sufficiently different, these zones will draw apart from each other. Bulk mode chromatography means that a feed of only few components but large amounts is separated in fractions. In the following the trace mode separation is discussed in more detail. In Fig. 9.8-1 a chromatography column (above) and component bands (below) at different times of the batch are shown. After the addition of a sorptive or solute, a carrier fluid as the mobile phase is continuously fed into the column. Bands of various components are formed and these zones are travelling to the exit. In the diagram the concentrations of the different fractions are shown as a function of time or the volume passed through the column for a given flow rate. Note that an effective separation can only be expected if there is no overlapping of the zones at the exit.

Principally speaking, the stationary phase can be

- An adsorbent (microporous solid particles)
- A liquid with an absorption capacity of the different solute components but which is immiscible with the carrier fluid

As a rule the concentration of the solute in the carrier fluid is small. Therefore, the heat effects caused by (exothermic) adsorption and (endothermic) desorption steps are small, too. Often chromatography can be considered as an isothermal process at

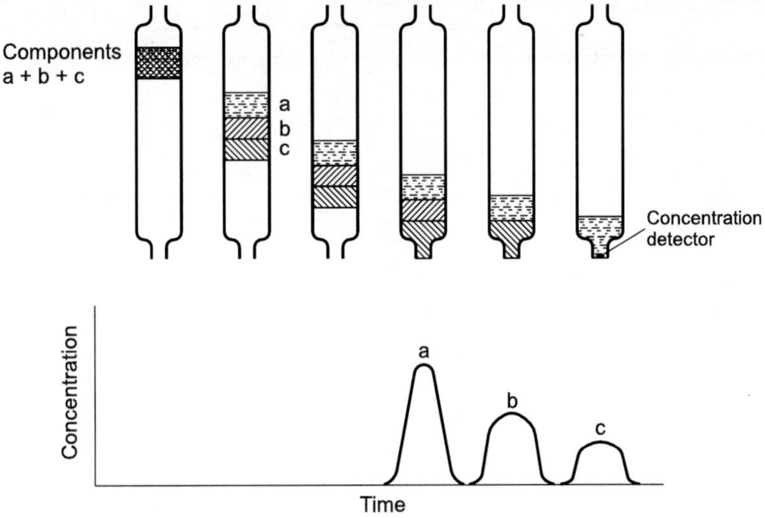

Fig. 9.8-1 Chromatography column (*above*) and component bands (*below*)

ambient temperature to a certain degree. High-molecular heat-sensitive organic products can be separated by chromatography with a liquid carrier fluid. Gas phase chromatography can be problematic because the solute must be evaporated before entering the column. In any case equilibria are the most decisive parameter for the design of chromatography columns.

9.8.1 Equilibria

The equilibrium laws of diluted systems

$$K_i^{\mathrm{id}} = \frac{p^0}{p} = \frac{\tilde{y}_i^*}{\tilde{x}_i} \quad \text{as the Raoult law (1.1-157)}$$

$$K_i = \frac{y_i^*}{x_i} \quad \text{as the Nernst law (1.2-1)}$$

$$\left.\begin{aligned} He_i &= \frac{p_i}{\tilde{x}_i} \\ He_i &= \frac{n_i}{p_i} \end{aligned}\right\} \quad \begin{aligned}&\text{as the Henry law in G/L or S/G systems, respectively}\\ &\text{(1.1-62) and (1.4-12), respectively),}\end{aligned}$$

have been introduced in previous chapters.

With the concentration q in the solid or stationary phase and c as the concentration in the fluid or mobile phase the equilibrium constants are defined here in an analogous way:

$$K_i = \frac{q_i}{c_i} \quad \text{(in binary systems } K_a = \frac{q_a}{c_a} \text{ and } K_b = \frac{q_b}{c_b}\text{).} \tag{9.8-1}$$

Dealing with ion chromatography and ion exchange it is necessary to take into account that ions (anions, cations) of the same charge can be exchanged between a mobile and a stationary phase. Let us first consider a binary system in which a species A (index a) and a species B (index b) are exchanged until equilibrium is obtained. The mass action equilibrium can often be described by the equation

$$m \cdot A + R_m \cdot B \rightleftarrows m \cdot R \cdot A + B \tag{9.8-2}$$

Here, m is an integer or the valence ratio of the species A and B. With A as the ion with the smaller charge $m > 1$ is valid. R denotes the resin. Next the

- Maximum total solid or stationary phase concentration Q
- Total ion-equivalent concentration C in the fluid or mobile phase

are introduced and the following mass fractions are defined:

$x_a \equiv q_a/Q$ is the mass fraction of species a in the solid phase

$x_b \equiv q_b/Q$ is the mass fraction of species b in the solid phase

$y_a \equiv c_a/C$ is the mass fraction of species a in the fluid phase

$y_b \equiv c_b/C$ is the mass fraction of species b in the fluid phase

Note that it is common in the literature to refer x and equilibrium constants to the solid phase which takes up a sorptive or solute. With these definitions the mass action equilibrium can be described by the equilibrium constant K_{ab} according to

$$K_{ab} = \left(\frac{q_a}{c_a}\right)^m \cdot \frac{c_b}{q_b} = \left(\frac{x_a^*}{y_a}\right)^m \cdot \frac{y_b}{x_b} \cdot \left(\frac{Q}{C}\right)^{m-1} . \tag{9.8-3}$$

Let us assume that either molecules are exchanged or a cation A is exchanged by a cation with the same charge, e.g., $m = 1$.

K_{ab}, the equilibrium of a binary system, can be described by

$$K_{ab} = \frac{x_a^* \cdot (1 - y_a)}{y_a \cdot (1 - x_a^*)}. \tag{9.8-4}$$

The last equation can be transformed:

$$x_a^* = \frac{K_{ab} \cdot y_a}{1 + (K_{ab} - 1) \cdot y_a}. \tag{9.8-5}$$

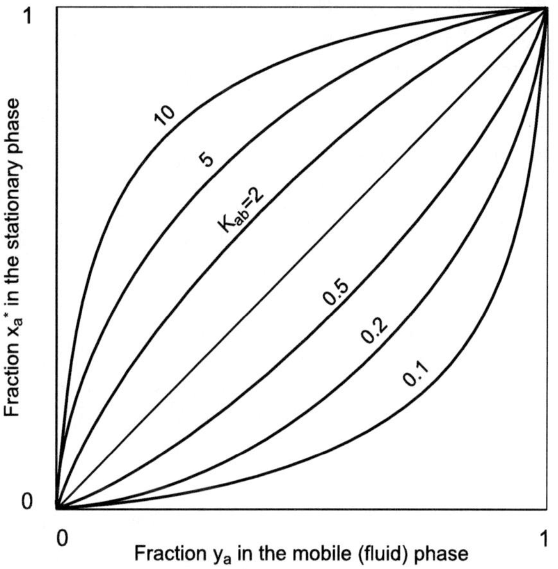

Fig. 9.8-2 Fraction x_a^* vs. the fraction y_a for some equilibrium constants

In Fig. 9.8-2 the fraction x_a^* is shown as a function of the fraction y_a of the component a in the fluid or mobile phase, compare with Fig. 2.1-16. As can be seen, the fraction x_a^* of the component in the stationary phase increases with rising equilibrium constants K_{ab} and a given fraction in the mobile phase. With the exception of bulk chromatography this separation process is applied on dilute solutions to obtain a high separation efficiency. If $(K_{ab} - 1) \cdot y_a \ll 1$ the last equation can be written as

$$K_{ab} = \frac{x_a^*}{y_a} \text{ or } x_a^* = K_{ab} \cdot y_a. \tag{9.8-6}$$

Since nearly all fluid mixtures to be treated by chromatography separation are multicomponent or multi-ion fluids or solutions, the equilibrium of an arbitrary species i will be defined by the constant

$$K_i \equiv \frac{q_i}{c_i}. \tag{9.8-7}$$

9.8.2 Theoretical Model of the Number N of Stages

In Fig. 9.8-3 a cascade of N stirred vessels all with the same volume is shown.

From reaction engineering it is known that in the case of an infinite number of vessels with a continuous feed stream, all fluid elements have the same residence time in the total system. This corresponds to a plug flow of an ideal tube reactor without any dispersion. Here the real chromatography column is replaced by a vessel cascade to quantify the separation efficiency.

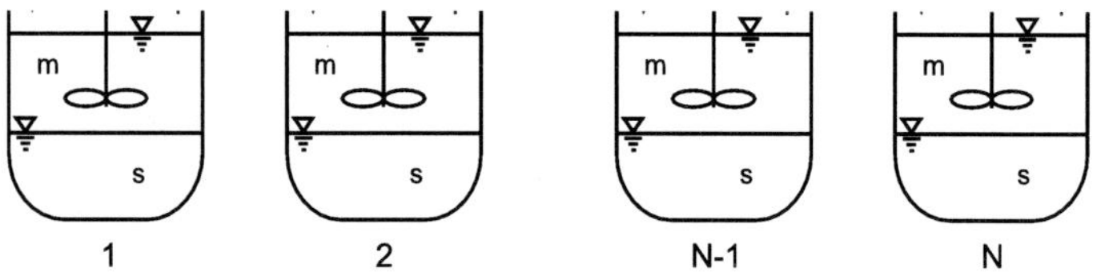

Fig. 9.8-3 Cascade of stirred vessels

Every vessel is filled with the (solid or liquid) stationary phase which takes up adsorbable or soluble components, respectively. At first, the mobile phase m consists of a carrier fluid. A small mass of a sorptive or solute to be separated is fed into the first vessel. With the assumption that mass transfer is infinitely fast (equilibrium model) and the leaving stream of every vessel is in equilibrium with the stationary phase the final concentration $c_{n,i}$ of the component i leaving the nth vessel can be calculated with the assumption that at time $t = 0$ a constant volumetric flow rate $\dot{V}$ of a carrier fluid is passed through the cascade. A balance of the component i leaving the vessel leads to

$$\frac{V \cdot \psi}{N} \cdot \frac{\mathrm{d}c_{n,i}}{\mathrm{d}t} = \dot{V} \cdot (c_{n-1,i} - c_{n,i}) - \frac{V \cdot (1-\psi)}{N} \cdot \frac{\mathrm{d}q_{n,i}}{\mathrm{d}t}. \tag{9.8-8}$$

Here, V is the total volume of all vessels and ψ is the volume fraction of the mobile phase or the voidage of the fixed bed, respectively. The last equation can be transformed to

$$\frac{V}{\dot{V}} \cdot \left[\psi \cdot \frac{\mathrm{d}c_{n,i}}{\mathrm{d}t} + (1-\psi)\underbrace{\frac{\mathrm{d}q_{n,i}}{\mathrm{d}c_{n,i}}}_{K_i} \cdot \frac{\mathrm{d}c_{n,i}}{\mathrm{d}t}\right] = N \cdot (c_{n-1,i} - c_{n,i}). \tag{9.8-9}$$

Note that the derivative $\mathrm{d}q_{n,i}/\mathrm{d}c_{n,i}$ is set equal to the equilibrium constant assuming the validity of the laws of Nernst or Henry. With the retention time of not ad- or absorbed carrier fluid

$$t_0 = \frac{\psi \cdot V}{\dot{V}}, \tag{9.8-10}$$

the last equation becomes

$$\frac{\mathrm{d}c_{n,i}}{\mathrm{d}t} \cdot \left[1 + \frac{(1-\psi)}{\psi} \cdot K_i\right] = \frac{N}{t_0} \cdot (c_{n-1,i} - c_{n,i}). \tag{9.8-11}$$

Let us assume that a certain mass M_i of the component i is fed into the first vessel. When equilibrium is obtained, a part of the substance i is present in the stationary phase. The concentrations in the mobile and the stationary phase depend on the partition coefficient K_i. At time $t = 0$ the concentration in the first vessel is

$$c_{1,i}(t=0) = \frac{M_i \cdot N}{V \cdot [\psi + (1-\psi)K_i]}. \tag{9.8-12}$$

Note that the content in every vessel is ideally mixed with the consequence that the solution leaving a vessel is the same as in this vessel. It is also assumed that equilibrium is immediately established in every vessel. The concentrations in all vessels can be calculated with the following assumptions:

- Validity of mass conservation of any substance i (no chemical reaction)
- All vessels have the same temperature, and the partition coefficient is constant

Note that the effective residence time t_R of any sorptive is longer than the residence time $t_0 = \psi \cdot V/\dot{V}$ of the carrier fluid because a part of the component i is taken up by the stationary phase:

$$t_\mathrm{R} = t_0 \cdot \left(1 + \frac{1-\psi}{\psi} \cdot K_i\right). \tag{9.8-13}$$

Only in the case of no sorption or $K_i = 0$ the two residence times would be equal $t_\mathrm{R} = t_0$.

The time-dependent concentration $c_{n,i}$ of the component i in the band leaving the last vessel n is given by a Poisson distribution:

$$c_{n,i}(t) = \frac{M_i \cdot N}{\dot{V} \cdot t_\mathrm{R}} \left(\frac{N \cdot t}{t_\mathrm{R}}\right)^{N-1} \frac{1}{(N-1)!} \exp\left[-\frac{N \cdot t}{t_\mathrm{R}}\right]. \tag{9.8-14}$$

Let us consider a column with the total volume V and the voidage ψ. If this column is operated with a given volumetric flow $\dot{V}$ of carrier fluid to separate a solute M_i with a given partition constant K_i, the concentration $c_{n,i}$ depends on the product $N \cdot t$ or the expression $N \cdot t/t_R$. In the case of a great number N of vessels (or stages) with $N > 50$ the last equation leads to a Gaussian distribution[1] (see Fig. 9.8-4):

$$c_{n,i}(t) = \frac{M_i}{\sqrt{2\pi} \cdot \sigma \cdot \dot{V}} \exp\left[-\frac{(t - t_R)^2}{2 \cdot \sigma^2}\right]. \tag{9.8-15}$$

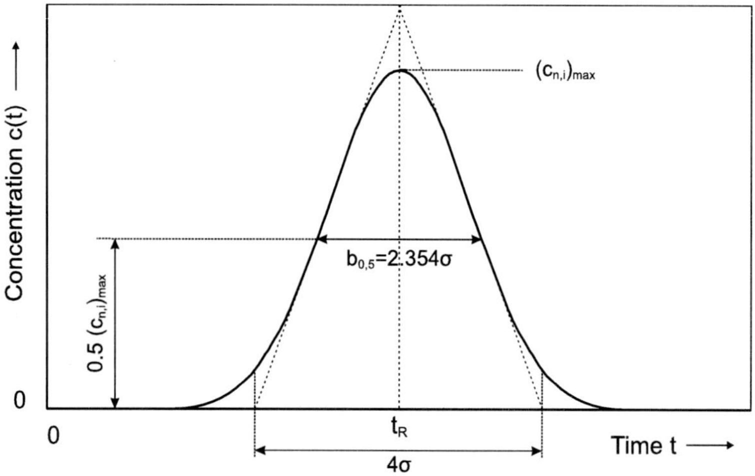

Fig. 9.8-4 Gaussian bell-shaped band: concentration vs. time

Here the time σ is the standard deviation of the Gauss bell-shaped curve, see Fig. 9.8-4. The maximum value of this curve or the maximum concentration $(c_{n,i})_{max}$ of the band leaving the last vessel is

$$(c_{n,i})_{max} = \frac{M_i}{\sqrt{2\pi} \cdot \sigma \dot{V}}. \tag{9.8-16}$$

The smaller the standard deviation σ the higher the concentration $(c_{n,i})_{max}$ for a given injected mass M_i and a given flow rate $\dot{V}$ of the carrier fluid. This concentration can be compared with the starting concentration in the first vessel at time $t = 0$:

[1] Note that $N! \approx \sqrt{2\pi N} \cdot N^N \cdot \exp(-N)$ (Stirling).

$$\frac{(c_{n,i})_{\max}}{c_{1,i}} = \frac{t_R}{\sqrt{2\pi}\cdot\sigma\cdot N} = \frac{t_0\cdot\left[1+\frac{(1-\psi)}{\psi}\cdot K_i\right]}{\sqrt{2\pi}\cdot\sigma\cdot N}. \qquad (9.8\text{-}17)$$

The maximum concentration $(c_{n,i})_{\max}$ of a band leaving a given cascade with N vessels or stages is high in comparison to the starting concentration $c_{1,i}$ in the first vessel for a great ratio t_R/σ or, in other words, for a long residence time $t_0 = V\cdot\psi/\dot{V}$ and a great equilibrium constant K_i but a small standard deviation σ. The following consideration might be helpful for an understanding of the relationship between the standard deviation σ and the number N of vessels or stages: After the addition of the amount M_i of the substance i but before starting the flow of the carrier fluid $(t = 0)$ the concentration profile in the first vessel is rectangular:

$$c_{1,i}(t=0) = \frac{M_i\cdot N}{\psi\cdot V\left[1+\frac{1-\psi}{\psi}\cdot K_i\right]}. \qquad (9.8\text{-}18)$$

As an idea let us now assume that this is replaced by the Gaussian bell-shaped profile according to (9.8-15) for $t\to 0$ or $t \ll t_R$:

$$c_{1,i}(t\to 0) = \frac{M_i\cdot t_0}{\sqrt{2\pi}\cdot\sigma\cdot\psi\cdot V}\exp\left[-\frac{1}{2}\cdot\frac{t_R^2}{\sigma^2}\right]. \qquad (9.8\text{-}19)$$

This consideration leads to the result that the square $(t_R/\sigma)^2$ is the decisive parameter for the concentration profile and the number of vessels or stages. According to the basic equation

$$\text{Number } N \text{ of stages} = \frac{\text{Column height } H \text{ (or length)}}{\text{Height equivalent to a theoretical plate (HETP)}}$$

or $N = \dfrac{H}{\text{HETP}}$ and

$$N = \left(\frac{t_R}{\sigma}\right)^2, \qquad (9.8\text{-}20)$$

defined by some authors; optimum operation is given when $HETP$ and σ are at a minimum with the result that N is at a maximum.

Every stage requires a certain column height HETP (height equivalent to a theoretical plate); however, dealing with the design of an industrial chromatography column it is necessary to take into account that

- An axial diffusion and dispersion takes place
- The mass transfer is not infinitely fast

This will be explained later.[2] Now some information is given on multicomponent solutions to be separated.

A prerequisite of an effective separation in a chromatography column is that the bands are travelling at different velocities and will draw apart from each other. The retention or capacity factor of the component i is defined as

$$k_i = \frac{t_{R,i} - t_0}{t_0}. \tag{9.8-21}$$

The retention time $t_{R,i}$ depends on the mass $M_{s,i}$ of the component i in the stationary phase (s), the mass $M_{m,i}$ of this component i in the mobile phase (m), and the mass flow rate $\dot{M}_{m,i}$ in the mobile phase:

$$t_{R,i} = \frac{M_{s,i} + M_{m,i}}{\dot{M}_{m,i}} = \frac{(1-\psi)\cdot V\cdot\bar{q}_i + \psi\cdot V\cdot\bar{c}_i}{\dot{V}\cdot\bar{c}_i} \tag{9.8-22}$$

or

$$t_{R,i} = (1-\psi)\cdot\frac{V}{\dot{V}}\cdot\underbrace{\frac{\bar{q}_i}{\bar{c}_i}}_{K_i} + \psi\cdot\frac{V}{\dot{V}} \tag{9.8-23}$$

With $t_0 = \psi V/\dot{V}$ and the definition $k_i = \dfrac{t_{R,i} - t_0}{t_0}$ the last equation reduces to

$$k_i = \frac{(1-\psi)}{\psi}\cdot K_i. \tag{9.8-24}$$

This dimensionless retention or capacity factor is the ratio of the retention time of a component adsorbed or dissolved to the retention time of a component without mass transfer and depends on the volumetric fraction ψ.

The separation factor of a sorptive or solute with the components a and b

---

[2] The terms "number of stages N" and "height equivalent to a theoretical plate HETP" are used in chromatography; however, their meanings are quite different in comparison to continuously operated countercurrent columns (absorption, extraction, rectification). Here, the column length or height must be sufficient to draw apart the bands of components or fractions. This requires favorable equilibria, certain retardation differences, and limited axial dispersion.

$$K_{\mathrm{ab}} = \frac{K_{\mathrm{a}}}{K_{\mathrm{b}}} = \frac{k_{\mathrm{a}}}{k_{\mathrm{b}}} \text{ for } \psi = \text{const}. \qquad (9.8\text{-}25)$$

is an excellent tool which quantifies the separation difficulty and the expenditure for investment and operation. The efficiency of separation or, in other words, the purity of the fractions is a function of the peak distance $(t_{\mathrm{R,\,b}} - t_{\mathrm{R,\,a}})$ based on the width $4(\sigma_a + \sigma_b)/2$ of the peaks:

$$R_{\mathrm{ba}} = \frac{t_{\mathrm{R,\,b}} - t_{\mathrm{R,\,a}}}{2 \cdot (\sigma_{\mathrm{a}} + \sigma_{\mathrm{b}})} = \frac{(1-\psi) \cdot V \cdot (K_{\mathrm{b}} - K_{\mathrm{a}})}{2 \cdot \dot{V} \cdot (\sigma_a + \sigma_b)}. \qquad (9.8\text{-}26)$$

With an increasing difference of the residence times $t_{\mathrm{R,\,b}}$ and $t_{\mathrm{R,\,a}}$ the zones are travelling in the column at different velocities. This avoids overlapping of the bands and leads to high-purity fractions. The shapes of the bands are mainly dependent on

- The equilibrium and the partition coefficient $K_i = q_i/c_i$
- The mass transfer between the mobile and the stationary phase
- The axial dispersion

As a rule a Gaussian- or bell-shaped band can only be expected if

- The partition or equilibrium constant is close to unity (special case of the laws of Henry or Nernst)
- The mass transfer of a component under discussion is very fast (small particle sizes, small tortuosity factors, low viscous fluids)
- The axial dispersion is very small (small superficial fluid velocity in a well-packed bed of small particles of the stationary phase, negligible channelling, and wall effects)

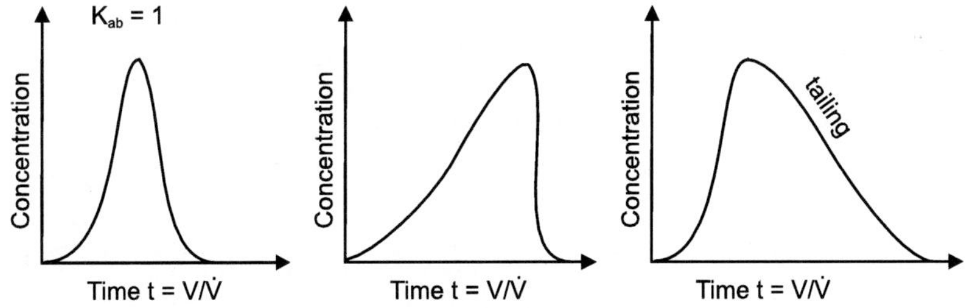

Fig. 9.8-5 Concentration bands for a linear isotherm (*left*), an isotherm favorable for desorption (*center*), and an isotherm favorable for adsorption (*right*)

In Fig. 9.8-5 the influence of the equilibrium is depicted. The concentration is shown as a function of time t or the ratio of volume V of the column based on a given volumetric flow rate $\dot{V}$ of the carrier fluid. In the front of a band or in the upstream, solute is desorbed from the stationary phase and transferred to the mobile phase because the concentration in a particle exceeds the equilibrium value. Contrary to this desorption process solute is adsorbed by the stationary phase in the downstream with the result that the concentration in the mobile phase is gradually reduced. An equilibrium constant $K_{ab} \approx 1$ results in a symmetrical and Gaussian bell-shaped curve, see Fig. 9.8-5.

The shape of chromatograms of large samples depends on the adsorption isotherms. If the equilibrium is favorable for desorption but unfavorable for adsorption the bands tend to front in the downstream. In the case of an isotherm which is favorable for adsorption, the retention time decreases with increasing concentration. This leads to the so-called tailing, see Fig. 9.8-5.

Design of Columns

In the following it is assumed that the stationary phase consists of adsorbent particles with the size d_p which can be in the range of only some micrometer to promote mass transfer. The mass balance of the component i in a thin layer with the thickness ∂z and the voidage ψ has already been described in Table 9.5-1:

$$\frac{\partial c_i}{\partial t} + \frac{(1-\psi)}{\psi} \cdot \frac{\partial \bar{q}_i}{\partial t} = -\frac{w}{\psi} \cdot \frac{\partial c_i}{\partial z} + D_{ax} \cdot \frac{\partial^2 c_i}{\partial z^2} . \tag{9.8-27}$$

The most important mass transfer resistance is pore diffusion in the adsorbent pellets. This depends on the diffusivity $D_{s,i}$ of the component i in the pores of the stationary phase (s) and the particle diameter $d_p = 2R_a$. The LDF (linear driving force) model results in the Glückauf equation (see previous section):

$$\frac{d\bar{q}_i}{dt} = \frac{15 \cdot D_{eff,i}}{R_a^2}\left(q_i^*\big|_{R=R_a} - \bar{q}_i\right) = \frac{60 \cdot D_{eff,i}}{d_p^2}\left(q_i^*\big|_{R=R_a} - \bar{q}_i\right) . \tag{9.8-28}$$

As a rule, pore diffusion is the rate-limiting step but axial diffusion and dispersion cannot be neglected. The curve in Fig. 9.5-8 can be approximated by the equation

$$\frac{1}{Pe} \approx \frac{1}{5 \cdot Re \cdot Sc} + \frac{1}{2} \quad \text{or} \tag{9.8-29}$$

$$D_{ax} \approx \frac{D_{m,i}}{5 \cdot \psi} + \frac{w \cdot d_p}{2\psi} . \tag{9.8-30}$$

Here $D_{m,i}$ is the axial molecular diffusion coefficient of the component i in the mobile phase and w denotes the mean superficial velocity in flow or z-direction. If the mean velocity w is very small, the axial dispersion coefficient D_{ax} is proportional to the diffusion coefficient $D_{m,i}$. In the case $w \cdot d_p \gg 0.4 \cdot D_{m,i}$ the axial dispersion coefficient D_{ax} increases with the velocity w and the particle size d_p.

In this section it has been shown that the integration of the general equation (9.8-27) leads to concentration profiles in adsorbent packings. In Chaps. 5 and 6, the concept of the number N of stages and the height HETP (height equivalent to a theoretical plate) was introduced. The application of this model leads to the general equation

$$\int_0^H \mathrm{d}z = H = N \cdot \mathrm{HETP}.$$

Dealing with the design of industrial chromatography columns, it is necessary to take into account

- Dispersion effects according to D_{ax}
- Diffusion effects in axial or flow direction
- Diffusion in the pore system of the adsorbent particles which is often rate controlling for the mass transfer between the two phases

The van-Deemter equation describes these effects:

$$\mathrm{HETP} = 2 \cdot a \cdot d_p + \frac{2 \cdot b \cdot D_{m,i}}{w} + \frac{2 \cdot w \cdot d_p^2}{60 \cdot D_{s,i}} \cdot \frac{k_i}{(1 + k_i)^2}. \tag{9.8-31}$$

The empirical factor a is close to unity when the packing with particles of the same size is very homogeneous with respect to porosity. The factor b takes into account the geometry of the pore system and is mainly dependent on the tortuosity factor, compare with the previous section. Note the ambiguous influence of the superficial velocity $w = \dot{V}/f$ with f as the actual cross-section area of the column, see Fig. 9.8-6. As a result the HETP value as a function of the velocity w passes through a minimum.

The concentration profiles in a chromatography column mainly depend on

- The intensity of dispersion
- The mass transfer rate

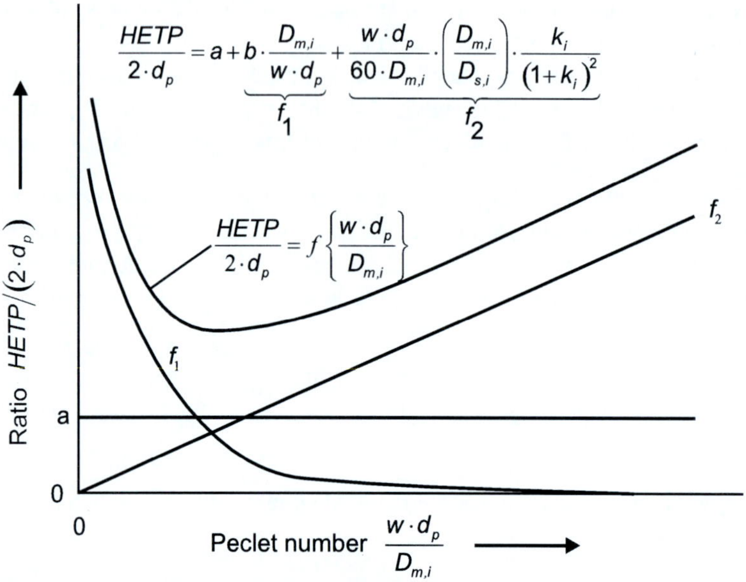

Fig. 9.8-6 Ratio $\mathrm{HETP}/(2d_\mathrm{p})$ vs. the Peclet number

In Fig. 9.8-7 some profiles are qualitatively depicted. The diagrams show that the desirable peaked symmetric shapes with high maximum concentration but small standard deviations can only be achieved in columns with weak dispersion and rapid mass transfer.

In a continuously operated production plant batchwise operation is a drawback. Principally speaking, the separation of a binary mixture with the components a and b can be carried out in a true moving bed (Seidel-Morgenstern et al. 2008). In Fig. 9.8-8 the principle of a countercurrent chromatography column is illustrated. The solid phase is moving downward whereas the mobile fluid phase is introduced at the bottom of the column. The feed of the components a and b is separated with the result that a raffinate containing a less adsorbable component a and an extract with the strong adsorbable component b are withdrawn as side streams. Therefore, the total column is subdivided into four zones:

Zone I The desorbent takes up component b and the solid adsorbent is regenerated

Zone II and zone III The feed is separated into the extract and the raffinate stream

Zone IV The desorbent takes up component a and the solid adsorbent is regenerated

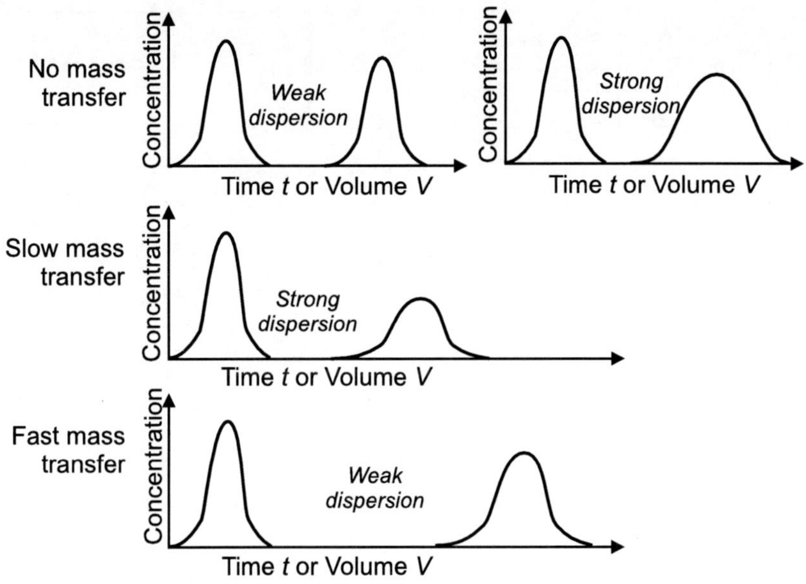

Fig. 9.8-7 Band profiles vs. the time or volume

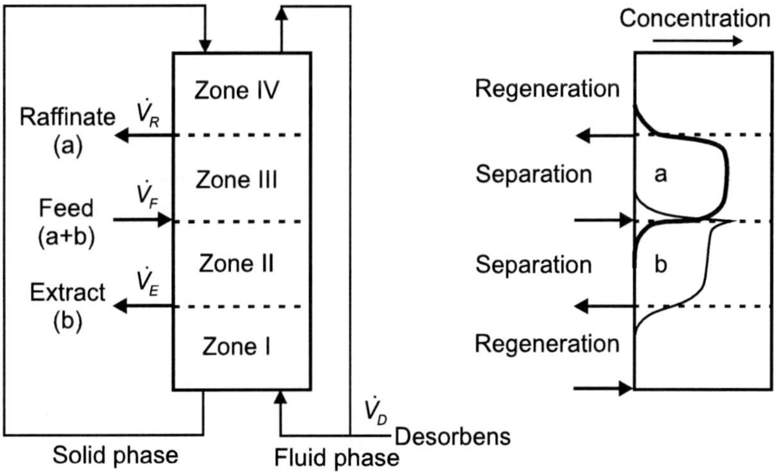

Fig. 9.8-8 True moving bed (TMB); concentration profiles (*right*)

The desorbents which are withdrawn as extract and raffinate are replaced by the addition of this substance to the fluid circle. Note that a plug flow of the moving bed without backmixing and no particle attrition is assumed; however, this operation is very difficult in industrial practice. Therefore, the true moving bed (TMB) is replaced by a simulated moving bed (SMB), see Fig. 9.8-9. The moving beds are

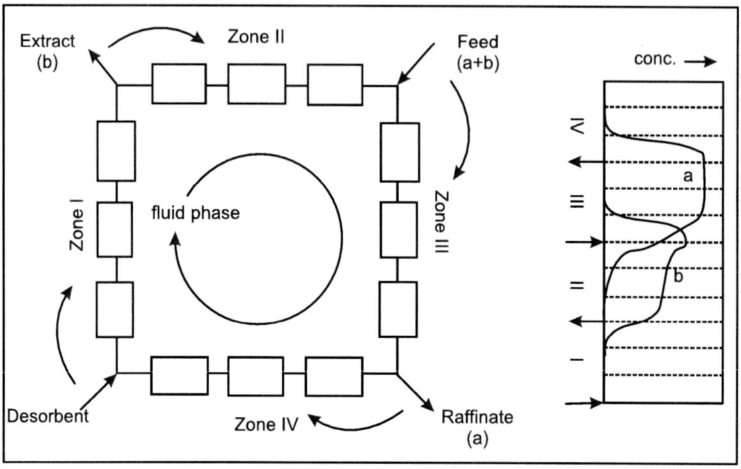

Fig. 9.8-9 Simulated moving bed (SMB); concentration profiles (*right*)

replaced by fixed beds and the countercurrent flow in a bed is obtained by a change of the flow direction of the fluid phase.

9.8.3 Chromatography Processes

There is a broad variety of chromatography processes applied in industry. A prerequisite for a good design and economical operation of any equipment is the study of the properties of the solute components and the measurement of equilibria and kinetics when no reliable data are available.

Gel permeation is based on the travelling of molecules through a gel or solvated material such as swollen polymers. Solute molecules are separated according to their size which decides on the viscous force. In addition steric effects are effective and cause an exclusion of "colloidal" material. Regeneration is carried out by a solute-free liquid. In some cases substances are very strongly bonded by the stationary phase and regeneration would require large amounts of carrier fluid. Therefore, a displacement agent is added to the mobile phase. The adsorbate is removed because this substance has an even higher affinity to the adsorbent than the adsorbate molecules. Similar to displacement adsorption this process is called displacement chromatography.

Bioaffinity chromatography means that solute components which have a very specific and selective interaction with the adsorbent are separated into fractions with a high purity. Sometimes this selective bonding is based on a steric effect (key–lock-interaction) but also an equilibrium or kinetic effect can be applied for separation. As a rule also a specific eluent is necessary for regeneration.

Instead of columns, chromatography separations can be carried out in thin layers. This method is often used for screening experiments and qualitative analysis because the laboratory equipment is less expensive but more flexible.

9.8.4 Industrial processes

Separations by chromatography are carried out in the pharmaceutical and food industry, for instance, the isolation of specific polysaccharides. As an example a mixture of glucose and fructose is separated by elution chromatography. The differences of the physical properties of isomers are often very small with the consequence that most of the known separation processes cannot be economically used. Here chromatography is an excellent tool. This is true for racemic mixtures. Enantiomers are separated by the application of chiral stationary phases. Sometimes tenside micellar solutions or mixtures of solvents are used as the mobile phase.

SMB technology is also applied in the area of enantioselective liquid chromatography. Sometimes chiral separation by ligand exchange is an appropriate process. This can also be true for enantioseparations using macrocyclic glycopeptide chiral stationary phases. Another example for separation by chromatography is the analysis of amino acids and peptides. Sometimes separations are carried out by using supercritical fluid chromatography. With respect to thermal instability of highmolecular organic substances, nearly isothermal separation processes based on any kind of chromatography are advantageous in comparison to other possibilities.

9.9 Ion Exchange

The most ion exchange columns are filled with ion exchange resin particles with a narrow particle size distribution. The mean particle size is 1 mm or less. As a rule the solution to be treated moves downward through a fixed bed with a mean superficial velocity between 10^{-3} and $10^{-2} m^3/(m^2 s)$. Ion exchange resins are polymers with electrically charged sites. Ions attached on such sites can be replaced by other ions of the same charge. Cation exchange resins R exchange cations as H^+, Na^+, K^+, Ca^{2+}, Mg^{2+}, Cu^{2+}, Fe^{3+}, NH_4^+, etc. Water softening often is obtained by an exchange process according to

$$Ca^{2+} + 2NaR \rightleftarrows CaR_2 + 2Na^+ \tag{9.9-1}$$

Anions attached on anion exchange resins can be replaced by other anions (OH^-, CO_3^{2-}, SO_4^{2-}, Cl^-, etc.)

Besides polymeric resins also zeolites Z are used for the exchange of ions. For instance free mobile Na^+ ions can be exchanged by cations Ca^{2+} according to

$$2Na^{+} + 2Z^{-} + Ca^{2+} \rightleftarrows CaZ_2 + 2Na^{+} \tag{9.9-2}$$

This exchange process is applied for the softening of water. The reaction takes place until the capacity of the ion exchange solid material is used or the total number of attached ions is replaced by ions provided by the solution according to equilibrium. Note that this capacity is constant and does not depend on the concentration of the solution – in contrary to adsorption.

In cation exchange resins an exchange of protons or metallic ions takes place. With the acid group A this reaction can be described by

$$A^{-}H^{+} + Na^{+} \rightleftarrows A^{-}Na^{+} + H^{+} \tag{9.9-3}$$

When water is at first treated by a cation exchange and in a following step by an anion exchange resin on which OH^{-} ions are attached, then protons H^{+} and OH^{-} ions are combined to water. Such an exchange unit delivers demineralized water with the same or even higher purity in comparison to distilled water.

The regeneration of the resins is carried out with a base as $NaOH$ (cation exchange resin) or HCl (anion exchange resin). In the case of water softening, $NaCl$ solutions are employed as regenerants for the deliverance of cations.

Ion exchange resins are made of cross-linked polystyrene, polyacrylate esters, polymethacrylate esters, and polymers of olefinic acids, amines, or phenols. These polymers are cross-linked with polyfunctional monomers (divinyl benzene). Cross-linking leads to insolubility in solvents and to chemical and physical stability. The higher the degree of cross-linking the lower the moisture content. Water wet ion exchange resins shrink or swell when changing from one ionic loading to another. Upon contact with nonpolar solvents shrinking of such polymers takes place. Free ions are held on resinous sorbents by bound groups of the opposite charge. Cation exchange resins contain bound sulfonic acid groups and sometimes carboxylic, phosphonic, or phosphinic groups. They can be used up to 120°C. Anion resins have quaternary ammonium groups (strongly basic) or amino groups (weakly basic). Their temperature tolerance is only 50°C. Commercial resinous sorbents are mostly granules or pellets and sometimes powders. Such materials can take up water or other liquids with the result of considerable swelling. The volume is increased by about 10–20% and more in comparison to the volume of a dry resin. The bulk wet density of the majority of resins is between 0.7 and $0.9\ kg/dm^3$. With the exception of macroreticular resins (this material contains a continuous pore phase and a continuous gel polymeric phase) there is no measurable porosity for the homogeneous gel-type resin when it is dry. Chelating resins have special functional groups with electron donors that can coordinate bonds to a single metal atom. In gel-type resins there is no permanent pore structure defined by the distance between the polymer chains and cross-links which varies with the physical properties of both phases and the operating conditions. The maximum pore size is up to a few nanometers.

9.9.1 Capacity and Equilibrium

Exchangeable ions are attached to bound groups of the opposite charge. The total capacity of an ion exchange resin is the number of bound groups or the number of attached ions either based on the mass of dry resins or the volume of wet resins. The number of ions or bound groups is expressed as the number of equivalents. According to these definitions there are two capacities:

$$\text{Capacity dry} = \frac{\text{mol ions or equivalents}}{\text{kg dry resin}}, \qquad (9.9\text{-}4)$$

$$\text{Capacity wet} = \frac{\text{mol ions or equivalents}}{\text{dm}^3 \text{ wet resin}}.$$

Table 9.9-1 Ion exchange resins (Perry 1984, 16–10)

Resin	Capacity dry equiv./kg	Capacity wet equiv./dm^3
Cation, strongly acidic polystyrene sulphonate gel resin, 4%–20% cross-linked	4–5.5	1.2–2
Cation, weakly acidic acrylic or methacrylic gel resin	8.3–10	3.3–4
Cation, weakly acidic phenol resin	6.6	3
Anion, strongly basic polystyreney based trimethylbenzyl ammonium	3.4–3.8	1.3–1.5
Anion, weakly basic aminopolystyrene gel resin	5.5	1.8

In Table 9.9-1 both the dry and the wet capacity for some ion exchange resins are given. As can be seen there are no dramatic differences of the capacities dry for strongly or weakly acidic cation resins on the one hand and strongly or weakly basic anion resins on the other hand. The exact capacity is a function of many parameters (cross-linking, homogeneous gel or macropore structure, macroreticularity) and has to be measured for any resin. As a rule a variety of ions with the concentration c_i in equiv./dm^3 of the species i is present in a solution to be treated by ion exchange, for instance, the cations Na^+, K^+, Cs^+, NH_4^+, Ca^{2+}, Mg^{2+}, Zn^{2+}, Cu^{2+}, Ni and Pb^{2+}. A given concentration c_i in the fluid phase leads to the loading q_i in the solid phase. Here the question about the equilibrium of a species i or its partition between the fluid and the solid phase arises.

The equilibrium $q_i = f(c_i)$ or the partition between an ionic component in the solid or resin phase and in the fluid phase of multicomponent systems depends on many material properties and the temperature and has to be measured. Things are easier for binary systems in which the ionic species a is exchanged. With the concentration q_a and the mass fraction x_a in the solid or resin phase and the concentration c_a and the mass fraction y_a in the liquid or solution phase the equilibrium $q_\text{a} = f(c_\text{a})$ or $x_\text{a}{}^* = f'(y_\text{a})$ can be described by the mass action equilibrium constant or selectivity coefficient K_ab (see (9.3-3)):

$$K_\text{ab} = \left(\frac{q_\text{a}}{c_\text{a}}\right)^m \cdot \frac{c_\text{b}}{q_\text{b}} = \left(\frac{x_\text{a}{}^*}{y_\text{a}}\right)^m \cdot \left(\frac{1-y_\text{a}}{1-x_\text{a}{}^*}\right) \cdot \left(\frac{Q}{C}\right)^{m-1}, \tag{9.9-5}$$

with Q in equiv./dm^3 wet resin as the total solid phase concentration and C in equiv./dm^3 solution as the total concentration in the fluid phase.

Here, the integer m is the valence ratio of the ionic species a and b. With the ion a as the ion with the smaller charge, $m > 1$ is valid. When the cation H^+ is exchanged by a cation with the same charge (for instance, Na^+, K^+, Cs^+, NH_4^+, etc.) the integer is unity and the equilibrium constant or the selectivity coefficient is

$$K_\text{ab} = \frac{q_\text{a} \cdot c_\text{b}}{q_\text{b} \cdot c_\text{a}} = \frac{x_\text{a}{}^* \cdot (1-y_\text{a})}{y_\text{a} \cdot (1-x_\text{a}{}^*)} \tag{9.9-6}$$

or in the case of diluted solutions with $y_\text{a} \ll 1$ and $x_\text{a}{}^* \ll 1$ finally

$$K_\text{ab} = \frac{x_\text{a}{}^*}{y_\text{a}}. \tag{9.9-7}$$

These equations are the same as in the section on chromatography but here with $x_\text{a}{}^*$ as the fraction of equivalents in the resin phase (instead of the stationary phase in chromatography) and y_a the fraction of equivalents in the liquid phase. For a given fraction y_a the fraction of equivalents increases with rising equilibrium constants K_ab. These constants are in the range $1.7 < K_\text{ab} < 4.90$ for the cations listed in Table 9.9-2 (Perry 1984, 16–13)

9.9.2 Kinetics and Breakthrough

In a previous section the various steps of species travelling from the bulk fluid through a sorbent particle to find a site of attachment have been described. In the case of homogeneous resins without a permanent pore system the ions have to

Table 9.9-2 K_{ab} constants of cations replacing H^+ as ion b in polystyrene sulfonate resins (25°C and 16% cross-linking)

Ion replacing H^+	Equilibrium constant, K_{ab}
Na^+	1.70
K^+	3.39
Cs^+	3.39
NH_4^+	2.14
$0.5 \cdot Mg^+$	2.82
$0.5 \cdot Ca^+$	4.90
$0.5 \cdot Zn^+$	3.39
$0.5 \cdot Cu^+$	3.47

move through a more or less viscous liquid by diffusion within a particle. After an exchange of ions a counterdiffusion of ions detached from their sites takes place. According to this model there are mainly two transport resistances:

- Film diffusion in the concentration boundary layer around a particle
- Instationary diffusion within a particle or internal diffusion

Film diffusion is usually the controlling step in dilute solutions which are moving at a low superficial velocity through the fixed bed whereas internal diffusion is limiting in high concentration solutions. Dealing with macroreticular resins with a permanent pore volume, BET surface and a defined pore size distribution, pore diffusion can be described by an approach which is based on the conservation of ions in a spherical shell and diffusion steps.

9.9.3 Operation Modes

As a rule anion and cation exchangers are two different apparatus containing the resin as a fixed bed. In the case of demineralizing water, often mixed beds are used, see Fig. 9.9-1. The containment is filled with both anion and cation exchange resins which are mixed by the introduction of air. This leads to a pneumatic mixing of the particles. Raw water is passed through the bed and demineralized. Before the regeneration step an upward backwash flow of raw water separates the resin in the anion and the cation resin which has a higher density in comparison to that of the anion resin. By this procedure the resins are mechanically separated with the result that two resin layers are obtained. An additional inert resin with a medium density forms a buffer zone between the anion resin layer above and the cation resin layer below. The upper zone is regenerated by the addition of a base which is withdrawn from the interface of the two resins. Here the acidic regenerant is introduced and withdrawn at the bottom of the apparatus. After regeneration the different resins are mixed by air as has been described. Besides this batch operation a quasistationary or continuous mode is applied. During a short stop of the solution to be treated a small part of the resin is removed and regenerated in a separate vessel. At the same time the same amount of resin taken from the regeneration unit is filled into the ion exchanger.

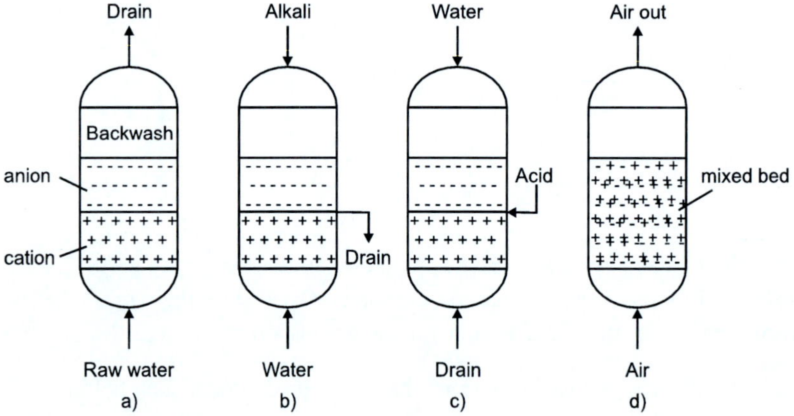

Fig. 9.9-1 Mixed bed ion exchanger (**a**) Backwash; (**b**) regeneration of anion exchange resin; (**c**) regeneration of cation exchange resin; (**d**) mixing of resins (next step: demineralization of raw water) (Perry 1984, 19–43)

9.9.4 Industrial Application

Besides the treatment of raw water there is a big variety of applications of ion exchangers. Wastewater of the metal industry is often contaminated with small amounts of metal ions. These problematic ions can be replaced by ions which are present in natural water. A subsequent treatment with anion and cation exchange resins results in an elimination of metal ions to a high degree. Rare-earth ions in water can be successfully separated by ion exchange. Organic ionic substances can also be treated by ion exchange.

Symbols

A		Constant
a	$1/\mathrm{m}$	Volumetric surface
a_s		Factor
B		Exponent
C	$\mathrm{kg/m^3}$	Max. concentration
c	$\mathrm{kg/m^3}$	Concentration
D_{AB}	$\mathrm{m^2/s}$	Molecular diffusivity
D_{ax}	$\mathrm{m^2/s}$	Axial dispersion coefficient
D_{eff}	$\mathrm{m^2/s}$	Effective diffusion coefficient
D_{Kn}	$\mathrm{m^2/s}$	Knudsen diffusion coefficient
D_M	$\mathrm{m^2/s}$	Macropore diffusion coefficient
D_m	$\mathrm{m^2/s}$	Micropore diffusion coefficient
D_s, D_s^o	$\mathrm{m^2/s}$	Surface diffusion coefficient
d	m	Molecule, sphere, particle diameter
d_p	m	Pellet diameter
f	$\mathrm{m^2}$	Cross-sectional area
$\dot{G}$	kg/s	Fluid mass flow
g, g'		Constants
H	m	Height
HETP	m	Height equivalent to a theoretical plate
Δh	kJ/kg	Specific heat of adsorption
K		Equilibrium constant
L_z	m	Pore width
M	kg	Mass
$\dot{M}$	kg/s	Mass flow
N		Number of steps
n		Variable
$\dot{n}$	$\mathrm{kmol/(m^2 \cdot s)}$	Molar flow density
Q	$\mathrm{kg/m^3}$	Max. concentration
q	$\mathrm{kg/m^3}$	Concentration
R	m	Radius in pellet
R_a	m	Outer pellet radius
r	m	Radius in a pore or microcrystal
r		Reference
r_a	m	Outer crystal radius
$\dot{S}$	kg/s	Solid mass flow
T	K	Absolute temperature
t	s	Time
t'	s	Time variable
$\tilde{v}_A$	$\mathrm{cm^3/mol}$	Molar volume
V	$\mathrm{m^3}$	Volume
$\dot{V}$	$\mathrm{m^3/s}$	Volumetric flow

$\dot{v}$	$m^3/(m^2 \cdot s)$	Volumetric flow density
w	m/s	Superficial velocity
Z	m	Height of bed
x		Mass fraction
y		Mass fraction
z	m	Length
z'	s	Modified length

Greek symbols

α		Separation factor
α	$W/(m^2 \cdot K)$	Heat transfer coefficient
β	m/s	Mass transfer coefficient
ε		Particle porosity
Θ	K	Local temperature in a fixed bed
Θ		Coverage
μ		Tortuosity factor
Λ	m	Mean free path length
λ	$W/(m \cdot K)$	Heat conductivity
ρ	kg/m^3	Density
τ	s	Breakthrough time
ψ		Porosity of a packing
ψ_B		Association factor
σ	s	Standard deviation

Indices

a	"Other" end of an apparatus
ads	Adsorption
app	Apparent
ax	Axial
b	Bead, pellet, bulk
e	End of an apparatus
con	Convection
eff	Effective
des	Desorption
dis	Dispersion
G	Gas or fluid phase
Kn	Knudsen
lam	Laminar
i	Component i
M	Macropore
m	Micropore, molecular, mobile
n	Stage n
p	Particle, pore

R	Retention
Reg	Regeneration
r	Reference
S, s	Solid
s	Surface, stationary,
$turb$	Turbulent
0	Entrance in the first stage
1	Exit of the first stage
α	Beginning (time)
ω	End (time)
*	Equilibrium

Dimensionless numbers

$Bi = \frac{\alpha \cdot d_p}{6 \cdot \lambda_s}$ Biot number

$D^* = \frac{\dot{v} \cdot Z}{\psi \cdot D_{ax}}$ Dispersion number

$Fo = \frac{D_{eff} \cdot t}{R_a^2}$ Modified Fourier number

$Kn = \frac{\Lambda}{2 \cdot R_a}$ Knudsen number

$Nu_b = \frac{\alpha \cdot d_p}{\lambda_G}$ Nusselt number of a particle

$Nu_{fb} = [1 + 1.5 \cdot (1 - \psi)] \cdot Nu_b$ Nusselt number of a packing

Nu_{lam} Nusselt number of laminar flow

Nu_{turb} Nusselt number of turbulent flow

$Pe = \frac{\dot{v} \cdot d}{\psi \cdot D_{ax}}$ Peclet number

$Re_\psi = \frac{\dot{v} \cdot \rho \cdot d}{\eta \cdot \psi}$ Reynolds number of a packing

$Sc = \frac{\eta}{\rho \cdot D_{AB}}$ Schmidt number

$Sh_b = \frac{\beta \cdot d_p}{D_{AB}}$ Sherwood number of a particle

$Sh_{fb} = [1 + 1.5 \cdot (1 - \psi)] \cdot Sh_b$ Sherwood number of a packing

Sh_{lam} Sherwood number of laminar flow

Sh_{turb} Sherwood number of turbulent flow

10 Drying

Drying is used to separate volatile components called moisture from a carrier. In many cases this implies the separation of water. The carrier may be solid, liquid, or gas. In this chapter only the drying of solid materials shall be discussed. The presented methods are also applicable to paste-like materials. Drying of fluids denotes the removal of small amount of water from gases or organic liquids. Adsorption as well as absorption and rectification processes are used for this. Thus, the drying of liquids and gases has been dealt with in other chapters.

Since the drying of solids involves the desorption of the moisture, it is necessary either to raise the temperature of the carrier to be dried or to lower the partial pressure of the moist component in the vicinity of the good. The method of pressure reduction is, for example, used in vacuum or freeze drying. Supplied heat is used to convert the liquid moisture into its vapor and is in the order of the heat of phase transition. Finally, the vapor must be led away from the drying good.

Drying processes are characterized by a simultaneous transfer of heat and mass. The driving physical processes are subject to change during the progress of drying. If in a very moist material capillary action is sufficient to keep the surface wet, surface evaporation takes place. Such behavior is denominated as the first drying section or constant rate period. In this case the kinetics of drying is governed by heat and mass transfer at the material's surface. At smaller moisture content transport phenomena of the condensed and gaseous phases inside the pores play a decisive role. It needs to be considered that the pore diameter is often in the range of nanometers. Hence the Knudsen number for vapor transport in the pores is bigger than 1 and molecular diffusion occurs, see Chap. 3.

With consideration of the large variety of drying goods and applied drying processes, there is a large variety of types of dryers. Some important ones are presented in the following.

A. Mersmann et al., *Thermal Separation Technology: Principles, Methods, Process Design*, VDI-Buch, DOI 10.1007/978-3-642-12525-6_10,
© Springer-Verlag Berlin Heidelberg 2011

10.1 Types of Dryers

Dryers are classified into contact, convective, and radiative dryers according to the mode of heat supply. A single drum dryer and a double drum dryer are shown in Fig. 10.1-1 as examples of contact dryers. The heat supply through the cylinder surface dries the film of material located on the rotating drum. In the following figures further examples of contact dryers are given. In some cases the heat is supplied through the outer jacket of a drum or a vessel. As an example a vacuum-wobble-dryer is given in Fig. 10.1-2. The vacuum-wobble-dryer is discontinously operated.

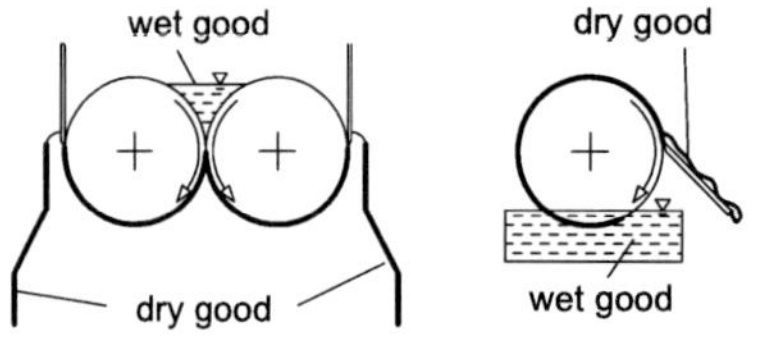

Fig. 10.1-1 Drum dryers

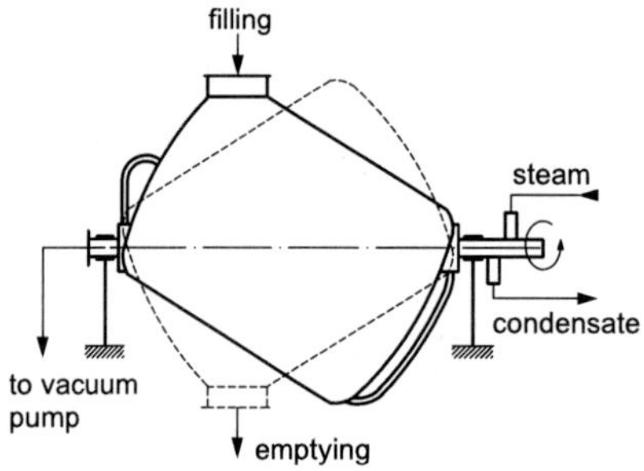

Fig. 10.1-2 Vacuum-wobble-dryer

In other cases not the vessel is rotated, but revolving rotors are placed inside it. An example is the paddle dryer shown in Fig. 10.1-3. For sticky drying goods, a self-cleaning twin screw such as in Fig. 10.1-4 can be employed, where the heat is supplied via the hollow screws as well as via the outer jacket.

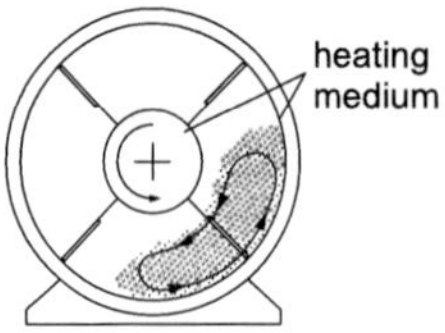

Fig. 10.1-3 Paddle dryer

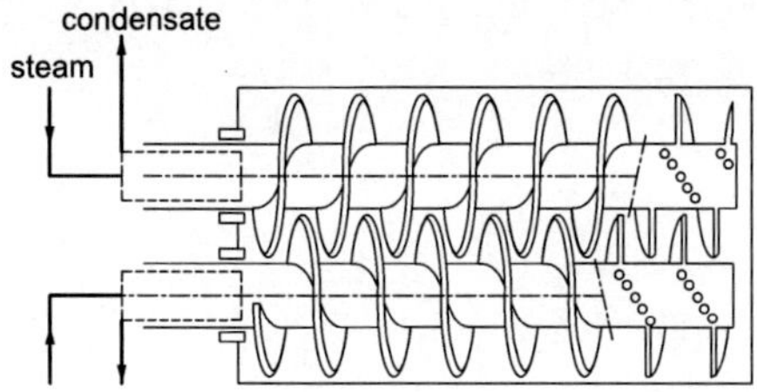

Fig. 10.1-4 Twin screw dryer

In a multistage rotary jacketed shelf dryer such as in Fig. 10.1-5, the drying good consecutively moves over various heated trays, where it is heated and dried. The drying process may be improved by passing hot gas over the drying good.

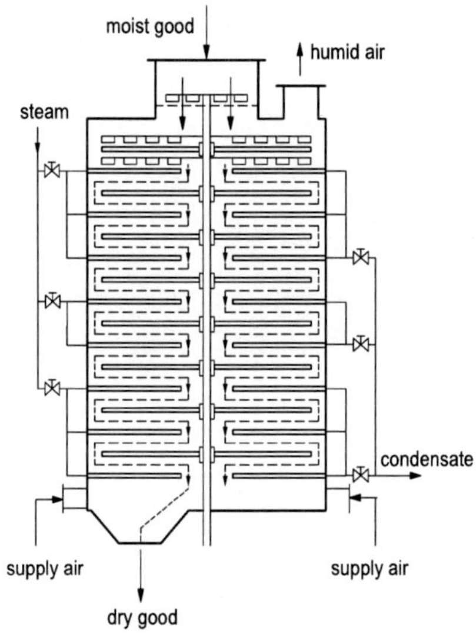

Fig. 10.1-5 Rotary jacketed tray dryer

Sometimes the necessary heat is generated electrically inside the drying good. Figure 10.1-6 on the left shows the case where the drying good itself serves as an electrical resistance and is conducting the current. On the right side of the figure the drying good is located in an alternating electrical field. Electroconductive materials may be heated by induction with alternating current fields of moderate frequency. For nonconductive drying goods dielectric heating is applied. Such drying goods are heated in high-frequency electric fields.

Convective dryers have the largest technical relevance. Here the desiccant (in general heated air or a heated inert gas) flows past the surface of the drying good or it can be transported to the surface of the drying good by nozzles and vents. In the case of coarse drying goods the desiccant often flows through. Then the drying good is either a fixed or a fluidized bed. In conveying and spray dryers, the drying good and the gas may be routed in cocurrent or in countercurrent flow.

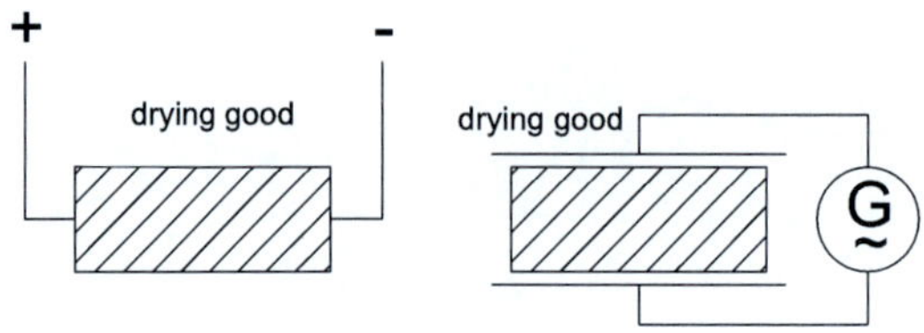

Fig. 10.1-6 Principle of resistance drying (*left*) and of high-frequency drying (*right*)

Figure 10.1-7 shows a drying cabinet where the drying good rests on trays. In this case the gas flows through the drying good. Fig. 10.1-8 shows a double-staged fluidized bed dryer. Drying air passes through the fluidized drying good. In some cases the entire bed is vibrated to prevent aggregation of the particles.

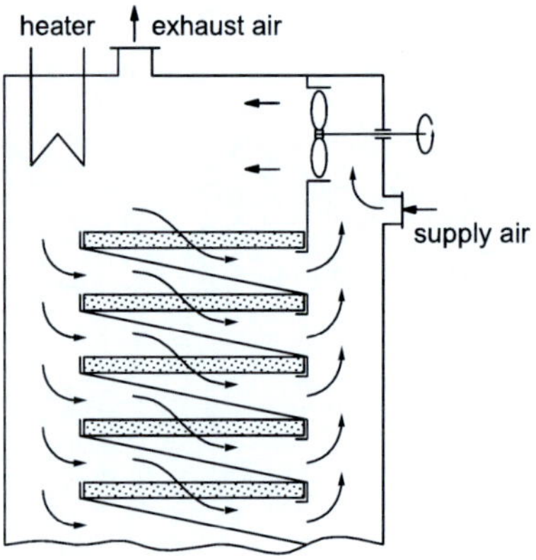

Fig. 10.1-7 Tray dryer

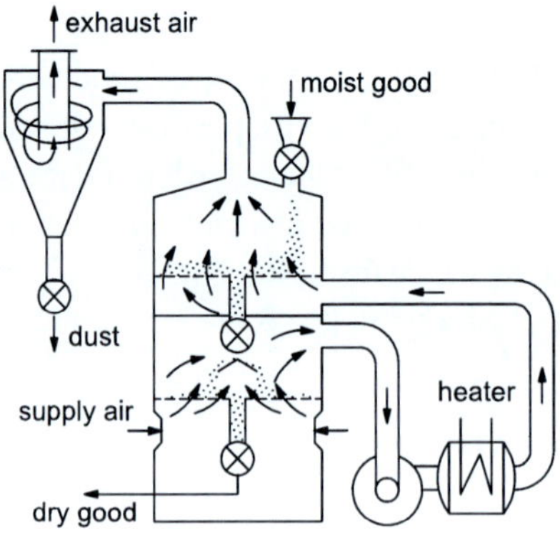

Fig. 10.1-8 Double-stage fluidized bed dryer

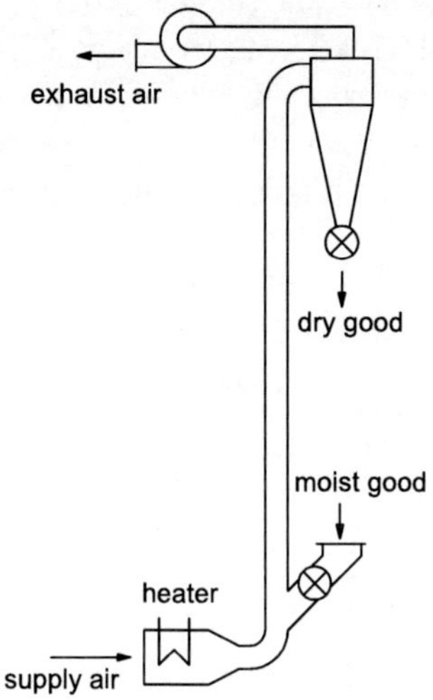

Fig. 10.1-9 Pneumatic conveyor dryer

As the flow velocity of the desiccant is increased, the drying good may be transported pneumatically. This kind of flow is found in a continuous pneumatic conveyor dryer as shown in Fig. 10.1-9. Here, the drying good is transported by the heated air. The solid as well as the gaseous phase may also be directed vertically downward in cocurrent flow. Figure 10.1-10 shows a spray dryer. The suspension to be dried is dispersed by nozzles at the top and moves downward with the heated air in cocurrent flow.

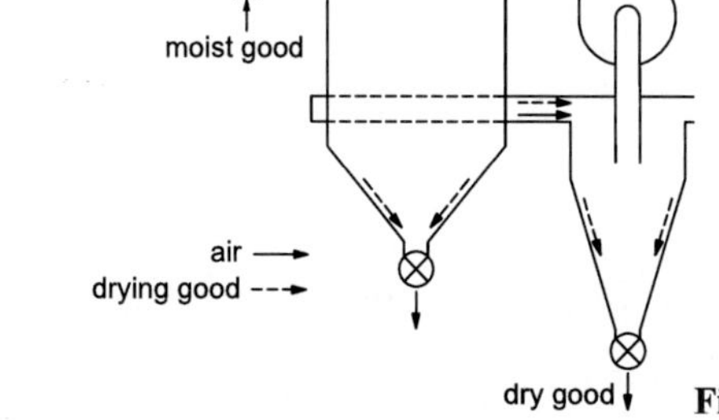

Fig. 10.1-10 Cocurrent spray dryer

In some cases the drying good is transported horizontally and the gas passes over it or through it. A five belt dryer is shown in Fig. 10.1-11, where the drying good is transported horizontally by belts and overflown with desiccant. For free-flowing bulk goods co- and counterflow rotary dryers are often used, Fig. 10.1-12. The rotating movement of the cylinder fitted with lifters or flights lifts and cascades the material through the hot air stream.

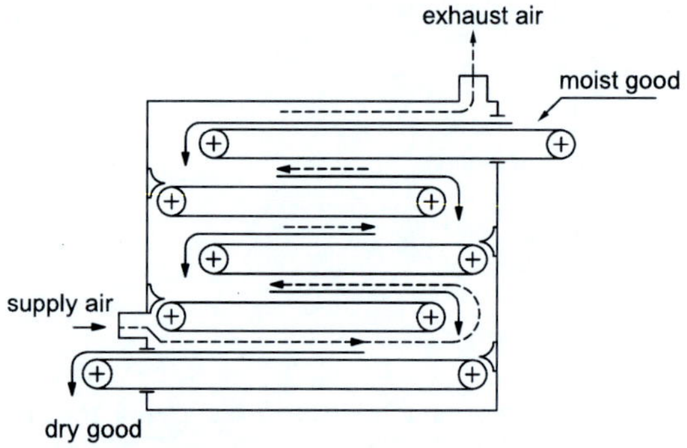

Fig. 10.1-11 Five-stage belt dryer

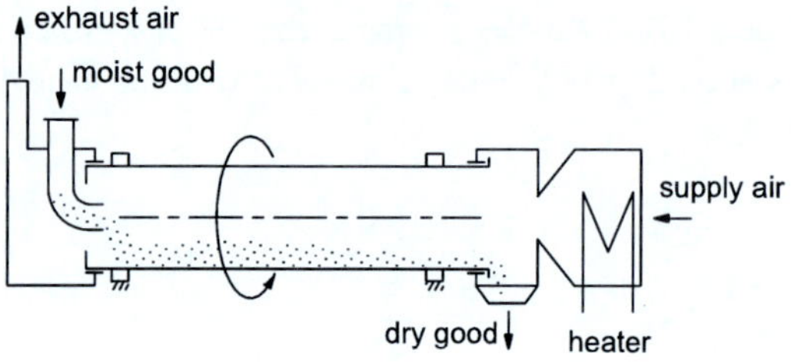

Fig. 10.1-12 Rotary dryer

Finally it should be kept in mind that, particularly, in the paper and textile industry, in food technology, and in the ceramic industry scores of specific dryers are applied, which are not discussed here.

10.2 Drying Goods and Desiccants

In many cases drying goods require desiccants with predefined characteristics depending on temperature and oxygen sensitivity.

10.2.1 Drying Goods

For some drying goods the equilibrium partial pressure depends only on the temperature and corresponds to the vapor pressure of the good's moisture. This is the case for nonhygroscopic materials. Such drying goods can be dried to moisture content $X = 0$, if the ambient relative saturation is less than 1. Nonhygroscopic drying goods are, for example, packed beds of compact materials like sand and glass beads, which have no porosity.

However, for most of the drying goods the equilibrium partial pressure depends on the temperature as well as on the moisture loading X of the material. For example, wood, paper, textiles, bricks, leather, and foods of all kinds are such hygroscopic goods. As opposed to the free or unbound moisture in nonhygroscopic materials, here the moisture is bound to the solid by various forces. Among these are capillary, sorption, and valency forces (chemical bonds). In the case of chemical bonding this is also called chemisorption. If a coherent film is formed on the outer surface of the drying good and is directly exposed to the desiccant, this is called an adhering liquid. The vapor pressure at the surface of the adhering liquid only depends on temperature.

In many porous drying goods there are numerous narrow pores and capillaries, which contain liquid. This liquid is transported to the material's surface by capillary action. For wide pores with more than about 1 μm in diameter, the vapor pressure at the liquid's surface is the equilibrium vapor pressure, which only depends on the temperature. If a drying good has pores so fine that a vapor pressure drop of $p_r^0/p^0 < 1$ occurs, it behaves as hygroscopic.

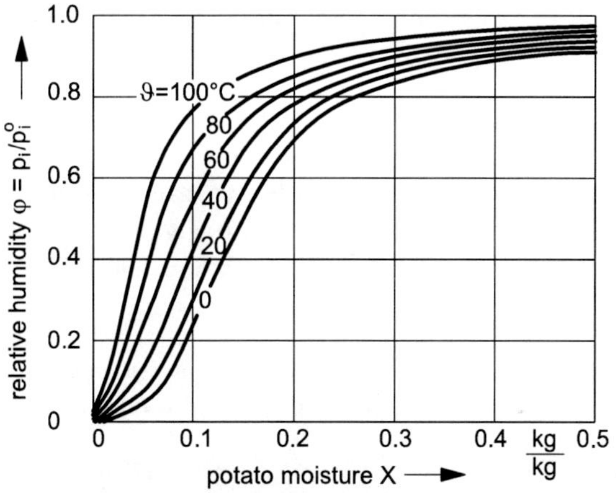

Fig. 10.2-1 Adsorption isotherms of potatoes

Colloidal materials tend to swell upon moisture uptake.

The sorption equilibrium may be described by relating the relative saturation φ to the moisture loading X of the drying good with temperature as a parameter. Figure 10.2-1 gives a few sorption isotherms for potatoes (Görling 1956). As in adsorption, the moisture of the good increases with increasing relative saturation φ and decreasing temperature. Figure 10.2-2 shows sorption isotherms for several materials after Kneule (Kneule 1975). There is, extensive literature on sorption isotherms (e.g. Kneule 1964, 1975; Krischer and Kast 1978; Tsotsas et al. 2010).

The equations of *Freundlich, Langmuir* as well as *Brunauer, Emmet, and Teller,* which were presented in Chap. 2, are suitable for the mathematical description of sorption isotherms.

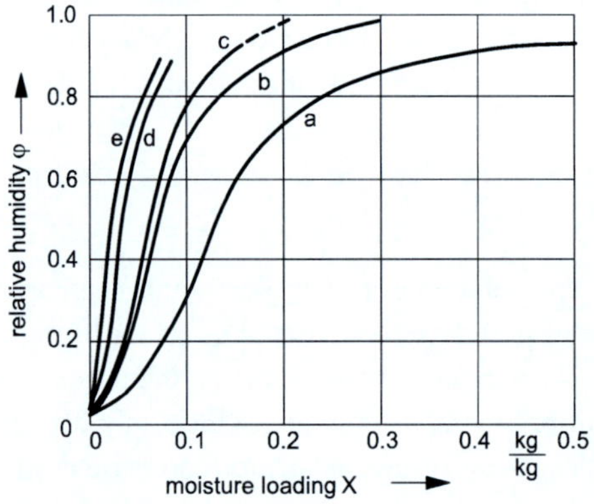

Fig. 10.2-2 Adsorption isotherms of various materials at 20°C.
(**a**) Potatoes; (**b**) copper beech wood; (**c**) sulfate paper; (**d**) 6-polyamide; (**e**) soap

The heat of phase change, q, also often called the sorption enthalpy, can be divided into the enthalpy of vaporization Δh_{LG} and the bond enthalpy Δh_B, as it was already presented in Chap. 2.

The bond enthalpy decreases with increasing relative saturation, and for this reason also with rising moisture content. In Fig. 10.2-3 the ratio $q/\Delta h_{LG}$ of the sorption enthalpy to the enthalpy of vaporization of potatoes is displayed in dependence of the moisture loading of the good (Görling 1956). Figure 10.2-3 shows that above a moisture loading of 0.2, the sorption enthalpy differs only little from the enthalpy of vaporization.

Similar to adsorption a hysteresis of the equilibrium curve may also occur in drying. In drying processes, the desorptive equilibrium is relevant.

During drying a moisture distribution appears, which results in moisture transport. The mass flux of this moisture transport can be described with the following

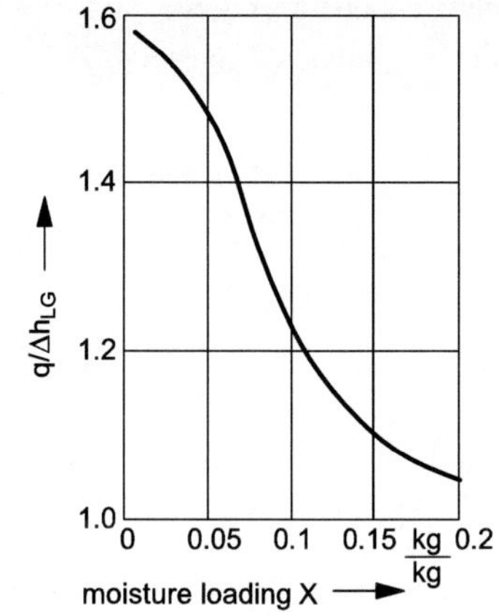

Fig. 10.2-3 Ratio of the sorption enthalpy q to the enthalpy of vaporization Δh_{LG} of potatoes dependent on the moisture content X of the drying good

approach, see Krischer and Kast (1978); see Chap. 9:

$$\dot{m} = -\kappa \cdot \rho_{\mathrm{s}} \cdot \frac{\mathrm{d}X}{\mathrm{d}s}. \tag{10.2-1}$$

Here, κ is the moisture conduction coefficient with the dimension of a diffusion coefficient. However, unlike the latter, the moisture conduction coefficient strongly depends on the moisture content of the drying good. Often, with decreasing moisture content, κ also decreases.

In addition, the moisture conduction coefficient depends on the material characteristics of the fluid, the solid as well as on the distribution of the pore radii. An idealized model including assumptions concerning the number n of capillaries of radius r (Krischer and Kast 1978) gives

$$\kappa = \frac{\sigma}{4 \cdot \eta} \cdot \frac{\int_{r_{\min}}^{r_{\max}} r^4 \cdot \frac{\mathrm{d}n}{\mathrm{d}r} \cdot \mathrm{d}r}{r^4 \cdot \left(\frac{\mathrm{d}n}{\mathrm{d}r}\right)_r} \sim \frac{\sigma \cdot r'}{\eta}, \tag{10.2-2}$$

where the radius $r_{\max}$ belongs to the widest pore that contributes to the moisture content in the corresponding cross section. In this equation the radius r' is a factor that depends on the distribution of the pore radii.

Hence, the coefficient is directly proportional to the surface tension σ and is reciprocally proportional to the dynamic viscosity η of the fluid. κ decreases with decreasing pore size. Figure 10.2-4 shows the moisture conduction coefficient for various materials as a function of the volumetric moisture content (Kneule 1975).

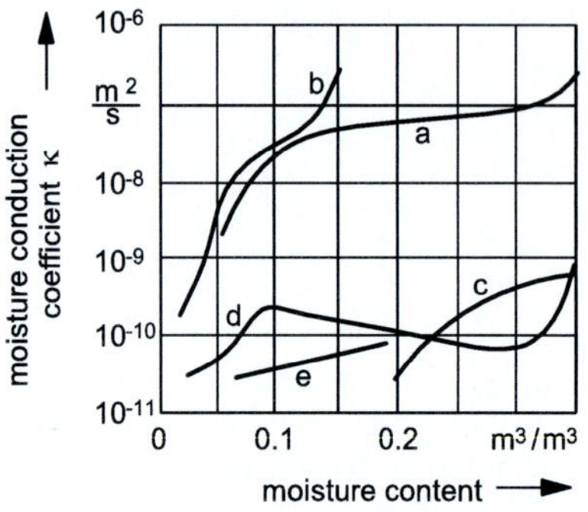

Fig. 10.2-4 Moisture conduction coefficient dependent on the volumetric moisture content for various materials. (**a**) Aerated concrete; (**b**) roof tiles; (**c**) potato disks; (**d**) beech wood in radial direction; (**e**) paste

In drying, the heat of phase transformation, q, must be delivered by conduction, convection, or radiation. Sometimes combinations of these possibilities are used. If the heat is supplied by conduction, this is called contact drying. The drying good either rests on a heated surface or is moved over it. The heat flux may be calculated with the following equation:

$$\dot{q} = -\lambda_{\text{eff}} \cdot \frac{d\vartheta}{ds}, \tag{10.2-3}$$

where λ_{eff} is the effective thermal conductivity of the moist drying good or porous material.

Thermal conductivity coefficients λ_{eff} can be found in the literature for many materials (Krischer and Kast 1978).

For a generalized description of these coefficients it is convenient to distinguish between two limiting cases, where the solid bridges and cavities either lie parallel (Fig. 10.2-5 left) or perpendicular (Fig. 10.2-5 middle) to the direction of heat flux (Krischer and Kast 1978). Porous drying goods are assumed to consist of layers, of which a fraction a is parallel to the direction of the heat flux, while a fraction $1 - a$ is perpendicular to it. The first limiting case gives the following conductivity coefficient, where ε_c is the relative pore volume of the drying good (and ε_d is the solids fraction):

$$\lambda_I = \varepsilon_c \cdot \lambda_G + \varepsilon_d \cdot \lambda_S. \tag{10.2-4}$$

The limiting case of perpendicular cavities can be described as follows:

$$\lambda_{II} = \frac{1}{\frac{\varepsilon_c}{\lambda_G} + \frac{\varepsilon_d}{\lambda_S}}. \qquad (10.2\text{-}5)$$

The effective thermal conductivity of a porous drying good with fraction a of thermally conductive bridges in the direction of the heat flux and fraction $(1\text{-}a)$ perpendicular to it (Fig. 10.2-5 on the right) is then

$$\lambda_{eff} = \frac{1}{\frac{1-a}{\lambda_I} + \frac{a}{\lambda_{II}}}. \qquad (10.2\text{-}6)$$

Relating the effective thermal conductivity to that of the gas gives

$$\frac{\lambda_{eff}}{\lambda_G} = \frac{1}{\frac{1-a}{\lambda_I/\lambda_G} + \frac{a}{\lambda_{II}/\lambda_G}}. \qquad (10.2\text{-}7)$$

These equations may also be used for a drying good with liquid-filled pores. In that case the thermal conductivity of the gas must be replaced by that of the liquid.

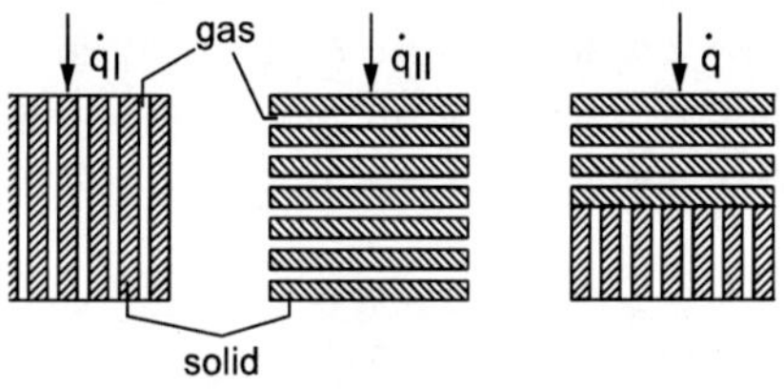

Fig. 10.2-5 Schematic model to elucidate the thermal conductivity of porous drying goods

10.2.2 Desiccants

Generally, heated air, and in special cases heated inert gases like nitrogen or carbon dioxide are used as desiccants in convection drying. If carbon monoxide, carbon dioxide, or other combustion products are not detrimental to the drying good, flue gas may be used advantageously as a desiccant. Here it is important to not cool below the dew point.

Drying with pure superheated water vapor has the advantage that no air must be heated and for this reason heat exchangers are not necessary. For water-moistened goods, the vapor pressure reaches the saturation pressure, so that the evaporated moisture does not move through the pores due to diffusion, but according to the laws of laminar flow. Therefore, drying times can be reduced. Superheated steam drying at standard pressure requires elevated temperatures and therefore temperature resistant drying goods.

10.2.3 Drying by Radiation

Because of the laws of radiative energy exchange appreciable heat flows can only be transferred by radiation if the absolute temperature of the radiation source is high, see Eq. 4.3-6. If a first body of surface A_1 and surface temperature T_1 is fully surrounded by a second body of surface A_2 and surface temperature T_2 the heat flow due to radiation is given by

$$Q_R = \frac{\sigma}{\frac{1}{\varepsilon_1}+\frac{A_1}{A_2}\cdot\left(\frac{1}{\varepsilon_2}-1\right)}\cdot A_1\cdot\left(T_1^4-T_2^4\right). \qquad (10.2\text{-}8)$$

Herein $\sigma = 5.67\cdot 10^{-8}\ W/\left(K^4 m^2\right)$ is the Stefan-Boltzmann constant, and ε_1 and ε_2 are the emissivity of the two bodies. Emissivity of a number of materials and correlations for various further geometries are tabulated in VDI-Heat Atlas (2010). Radiative drying is particulary suitable when thin layers are to be dried, e.g. paper, films, coating or varnish. Also micro-wave radiation is used in drying technology. Micro-waves penetrate into the drying good and are transformed to thermal energy, when they are absorbed by water molecules.

10.3 The Single-Stage Apparatus in the Enthalpy–Concentration Diagram for Humid Air

The enthalpy–concentration diagram is a useful tool to visualize thermodynamic aspects of drying.

Let us imagine a small water-moistened surface with air of the temperature ϑ_G passing over it. Let the surface temperature of the drying good at a certain position and time be ϑ_A. Then the transferred heat flow $\dot{Q}$ is

$$\dot{Q} = \alpha\cdot A\cdot(\vartheta_G-\vartheta_A)\,. \qquad (10.3\text{-}1)$$

First the surface temperature ϑ_A, shall be determined. Initially it is convenient to only examine a very small surface. Such a surface exists, e.g., in the wet bulb thermometer of an Assmann's psychrometer (see Fig. 10.3-1).

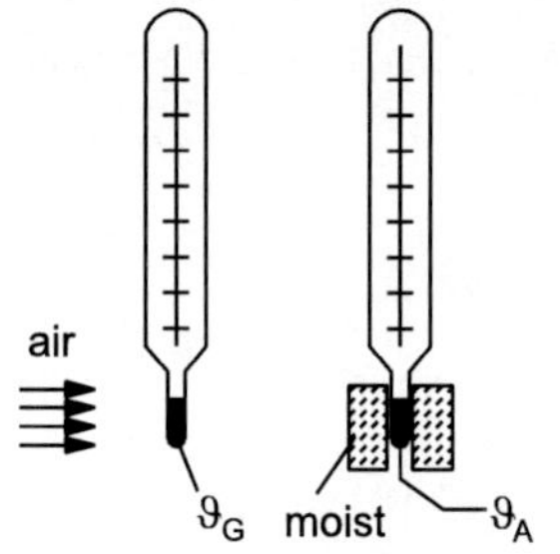

Fig. 10.3-1 Assmann's psychrometer for the measurement of the steady state temperature: air flows over a small moist surface, at steady state $\vartheta_A = \vartheta_{SS}$

The dry thermometer measures the air temperature ϑ_G. If a large amount of air flows over a small moist surface, the state of the air does not change. After a short while the moist surface of the drying good or the wet bulb thermometer of the psychrometer adopts the steady state temperature ϑ_{SS}, which can be calculated considering that the supplied heat flow $\dot{Q}$ must evaporate the mass flow $\dot{M}$ of water (Kneule 1975; Krischer and Kast 1978; Schlünder 1975). An energy balance gives

$$\dot{Q} = \dot{M} \cdot r \tag{10.3-2}$$

or with the kinetic statements for the vapor phase in Chap. 4

$$\alpha \cdot A \cdot (\vartheta_G - \vartheta_{SS}) = \dot{m} \cdot A \cdot r = \beta_h \cdot A \cdot \frac{r}{R \cdot T} \cdot (p_i^0 - p_i) \,. \tag{10.3-3}$$

The temperature difference $\vartheta_G - \vartheta_{SS}$ is

$$\vartheta_G - \vartheta_{SS} = \frac{\beta_h}{\alpha} \cdot \frac{r \cdot p_i^0}{R \cdot T} \cdot \left(1 - \frac{p_i}{p_i^0}\right) = \frac{\beta_h}{\alpha} \cdot \frac{r \cdot p_i^0}{R \cdot T} \cdot (1 - \varphi) \,. \tag{10.3-4}$$

If the analogy between heat and mass transfer is fulfilled, the ratio of the transfer coefficients β_h / α depends on the Lewis number $Le = a/D$, see Chap. 4. With this, the following relationship for the temperature difference results:

$$\vartheta_G - \vartheta_{SS} = \frac{p_i^0}{c_p \cdot \rho_G \cdot Le^{n-1}} \cdot \frac{r}{R \cdot T} \cdot (1 - \varphi) \,. \tag{10.3-5}$$

Here p_i^0 is the saturated vapor pressure at the steady state temperature ϑ_{SS}, which has to be determined by iteration. The exponent n for the Prandtl or Schmidt number of gases is 1/3 for the case of turbulent flow with a laminar boundary layer. In Fig. 10.3-2 the so-called psychrometric difference $\vartheta_G - \vartheta_{SS} = \vartheta_G - \vartheta_A$ is illustrated as a function of the air temperature ϑ_G with the relative humidity φ as a parameter. When using Assmann's psychrometer it is important to keep the wet bulb thermometer thoroughly soaked with clean water and that the air speed exceeds a certain value. This is ensured by a small fan at the head of the instrument. Assmann's psychrometer is commonly used to measure the relative humidity.

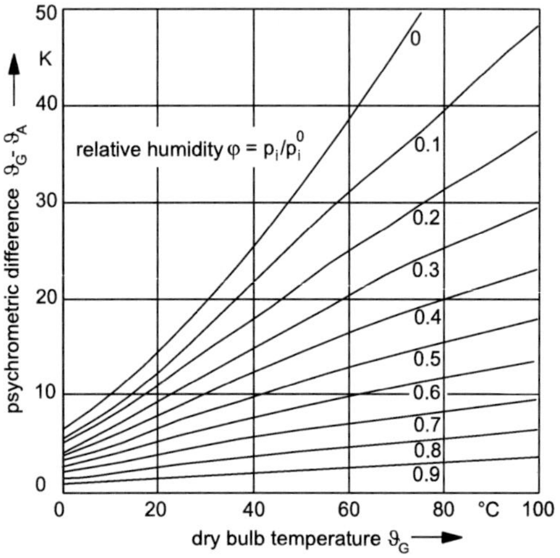

Fig. 10.3-2 Psychrometric temperature difference as a function of dry bulb temperature with relative humidity as a parameter

Now, if air flows over an extended moist surface, the surface temperature varies with position. Considering a differential length element dz the gas enters the control volume as shown in Fig. 10.3-3 with the enthalpy h and exits it with the enthalpy $h + \mathrm{d}h$. The evaporated amount of moisture $G_\mathrm{r} \cdot \mathrm{d}Y$ has the enthalpy $G_\mathrm{r} \cdot \mathrm{d}Y \cdot c_\mathrm{L} \cdot (\vartheta_\mathrm{A} + \mathrm{d}\,\vartheta_\mathrm{A}/2)$. Then, the enthalpy balance is

$$G_\mathrm{r} \cdot h + G_\mathrm{r} \cdot \mathrm{d}Y \cdot c_\mathrm{L} \cdot \left(\vartheta_\mathrm{A} + \frac{\mathrm{d}\,\vartheta_\mathrm{A}}{2}\right) = G_r \cdot (h + \mathrm{d}h) \tag{10.3-6}$$

or, when the product of two differentials is neglected,

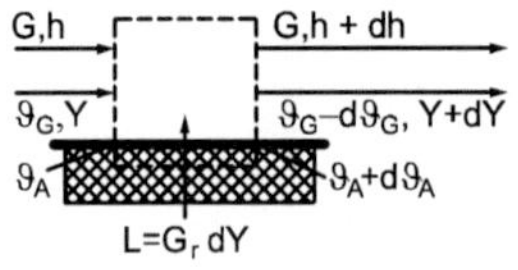

Fig. 10.3-3 Mass and energy balance of a volume element

$$\frac{\mathrm{d}h}{\mathrm{d}Y} = c_\mathrm{L} \cdot \vartheta_\mathrm{A} = h_\mathrm{L}\,. \tag{10.3-7}$$

It is thus possible to plot the change of state of air in an enthalpy–concentration diagram for humid air, see Fig. 10.3-4.

The air becomes colder and more humid while flowing over the wet surface, whereas the surface temperature ϑ_A of the moist good changes only little. Therefore, the curve of the state change of air is a curved line, which for practical calculations can be described precisely enough by a straight line, see Fig. 10.3-4. The direction of curvature depends on whether the Lewis number Le is greater or less than 1. As the length of passage approaches infinity, the gas and the good reach the same temperature, which is called the adiabatic saturation temperature ϑ_{AS}. The gas is then saturated to the core, i.e., the saturation pressure of water corresponding to the adiabatic saturation temperature is found in the entire gas. The system is at equilibrium. For practical calculations the surface temperature can be easily found by extending the corresponding saturated vapor isotherm through the point indicating the state of the air.

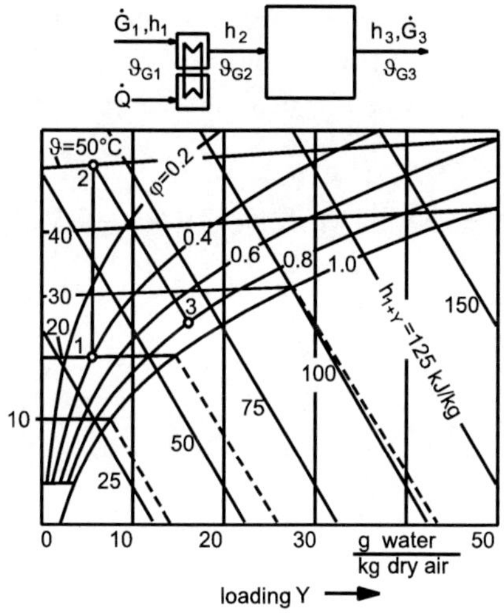

Fig. 10.3-4 Example of a single-stage drying process in the enthalpy–concentration diagram for humid air

If the surface temperature ϑ_A is very small, the approximation

$$\frac{\mathrm{d}h}{\mathrm{d}Y} \approx 0 \quad \text{or} \quad h \approx \text{const.} \tag{10.3-8}$$

may be used.

In this case the curve of the change of state of the air approximately follows a line of constant enthalpy. This has to do with the definition of the mixing enthalpy. It can be explained with the enthalpy loss of the dry air due to the cooling being approximately compensated by the enthalpy gain as a result of the increased vapor content Y according to the following equation, which was introduced in Chap. 2:

$$h = c_{pG_r} \cdot \vartheta + Y \cdot (r_0 + c_{pi} \cdot \vartheta) . \tag{10.3-9}$$

Now, these results are adapted to the single-stage dryer, depicted in Fig. 10.3-4. A heat flow $\dot{Q}$ is supplied to the incoming gas (ambient air) $\dot{G}_1$, which consists of $\dot{G}_{air}$ loaded with Y_1 kg of water vapor. Due to this heat supply, the enthalpy of 1 kg of air moistened with Y kg of water increases from h_1 to h_2, and accordingly its temperature increases from ϑ_{G1} to ϑ_{G2}. Because the humidity of the air does not change during heating, this change in state is represented by the vertical from point 1 to 2 in the enthalpy–concentration diagram for humid air. In the dryer, the air flows past or through the wet material, which is to be dried and takes up moisture which is at the adiabatic cooling temperature of the good. Hence and according to the above explanations about adiabatic saturation, the air changes its thermodynamic state from point 2 to 3 in Fig. 10.3-4. Point 3 lies on the projection of the saturated vapor isotherm, which passes through point 2 of the diagram. Because of the slope of the projection of the saturated air isotherm differs slightly from the slope of the isenthalpic line, the enthalpy of 1 kg of air moistened with Y kg of water also changes slightly from h_2 to h_3.

The enthalpy of 1 kg dry air which is moistened with Y kg water is the sum of the enthalpy of 1 kg dry air and the enthalpy of Y kg water vapor. This enthalpy remains almost constant during drying, because on the one hand, this sum is diminished by the drop of the air temperature from ϑ_{G2} to ϑ_{G3}, and on the other hand, this sum is increased due to the increasing amount of water vapor.

The difference in enthalpies $\Delta h = h_2 - h_1$ leads to the required heat input

$$\dot{Q} = \Delta h \cdot \dot{G}_{air} . \tag{10.3-10}$$

Point 1 represents the state of the incoming gas. The maximum allowable air temperature (Point 2) is determined by the possibility of thermal damage of the product. This temperature may differ in the constant rate period from the temperature in the falling rate period, see Sect. 10.6. Point 3 is set at $\varphi = 0.8$. This choice should be the result of economic considerations. In the case that this point would be set at saturation, $\varphi = 1$, an infinitely long dryer would be required, because as well as the driving temperature difference and the difference in partial pressures deplete along the dryer, and at $\varphi = 1$ they vanish.

Figure 10.3-5 depicts the case of a dryer with internal air circulation. Again, the process can be represented in the enthalpy–concentration diagram for humid air. The dryer is continuously fed with $\dot{S}$ of wet drying good, which consists of $\dot{S}_{dry}$

of dry substance initially wetted with X_α kg of moisture. The temperature of the good shall be uniform and at $\vartheta_{S\alpha}$ at the entrance and $\vartheta_{S\omega}$ at the exit of the dryer. The change of the thermodynamic state of the circulating gas flow $\dot{G}_M$ is represented by the line $\overline{M2}$, with point M being located on the projection line of the saturated vapor isotherm, which passes through point 2. Again, it is assumed that the exiting air is at relative humidity of $\varphi = 0.8$. Air at state 2 is mixed with fresh air of state 1. The state of the mixture M' can be found on the line connecting state 1 and 2 with the help of the lever rule, see Chap. 2. This mixture with humidity Y_M is heated up in the heater (line from M' to M).

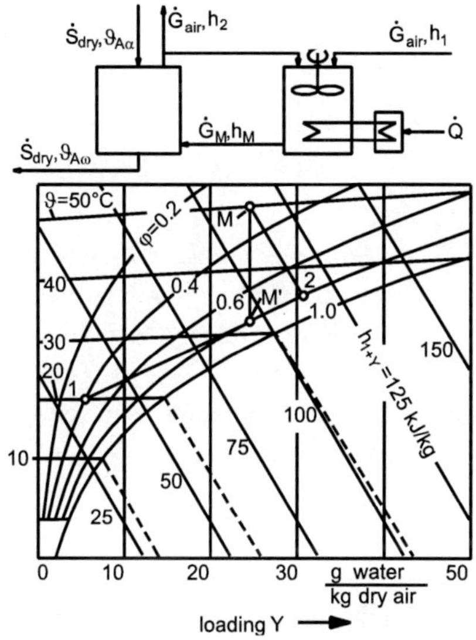

Fig. 10.3-5 Single-stage drying process with circulating air. Sketch of the process and its representation in the enthalpy–concentration diagram for humid air

The heat required to remove 1 kg of water is a measure for the energy costs of the dryer. Under the assumption of steady state, this heat can be determined by mass and enthalpy balances as follows: The amount of moisture $\dot{L}$ to be extracted from the drying good is found with the help of the water balance of the good in the dryer:

$$0 = \dot{S}_{\text{dry}} \cdot (X_\alpha - X_\omega) - \dot{L} \ . \tag{10.3-11}$$

The water balance of the air in the dryer tells how this amount of moisture increases the humidity of the air:

$$0 = \dot{L} - \dot{G}_{\text{air}} \cdot (Y_2 - Y_1). \tag{10.3-12}$$

Under the assumption of complete drying ($X_\omega = 0$) the energy balance of the dryer reads

$$0 = \dot{G}_{\text{air}}(h_1 - h_2) + \dot{S}_{\text{dry}} c_{\text{S}}(\vartheta_{\text{S}\alpha} - \vartheta_{\text{S}\omega}) + \dot{L} c_{\text{L}} \vartheta_{\text{S}\alpha} + \dot{Q}. \tag{10.3-13}$$

Rearrangement gives

$$\Delta h_{\text{L}} = \frac{\dot{Q}}{\dot{L}} = \frac{\dot{G}_{\text{air}}}{\dot{L}}(h_2 - h_1) + \frac{\dot{S}_{\text{dry}}}{\dot{L}} c_{\text{S}}(\vartheta_{\text{S}\omega} - \vartheta_{\text{S}\alpha}) - c_{\text{L}} \vartheta_{\text{S}\alpha}. \tag{10.3-14}$$

If the entrance and exit temperatures of the good do not differ significantly from the adiabatic saturation temperature

$$\vartheta_{\text{S}\alpha} \approx \vartheta_{\text{S}\omega} \approx \vartheta_{\text{AS}} \tag{10.3-15}$$

then

$$\Delta h_{\text{L}} = \frac{h_2 - h_1}{Y_2 - Y_1} - c_{\text{L}} \vartheta_{\text{AS}}. \tag{10.3-16}$$

The expenditure of energy per kilogram of removed water is favorably low, if the slope $\mathrm{d}h/\mathrm{d}Y = (h_2 - h_1)/(Y_2 - Y_1)$ of line $\overline{12}$ is small, in other words, if the exiting air is highly saturated. As mentioned before, this requirement leads to a long dryer and high investment costs. Therefore, size of the dryer and its operating parameters shall be chosen such that the total economic optimum is achieved.

10.4 Multistage Dryer

Numerous drying processes are carried out in multistage dryers, in which the drying gas is withdrawn after each stage and reheated, before it is fed into the next stage. Doing so, the required amount of heating energy per mass of removed moisture can be reduced. Figure 10.4-1 shows a three-stage dryer and its representation in the enthalpy–concentration diagram of humid air. At first, the incoming air is heated from 1 to 2, and is then fed into the dryer. Here, it cools down and acquires moist. As mentioned above, during this process-step the enthalpy of the air increases slightly.

Line $\overline{23}$ depicts the course of the state of the gas during adiabatic moisture acquisition in the first dryer. The second heater heats up the gas again, in this case as indicated by line $\overline{34}$ to the same temperature as in the first heater. This may be necessary because of the thermal sensitivity of the drying good. Then, in the second drying stage further moisture acquisition takes place, depicted by line $\overline{45}$. Finally, at the exit after the third drying stage, the humidity of the air is given by Y_7.

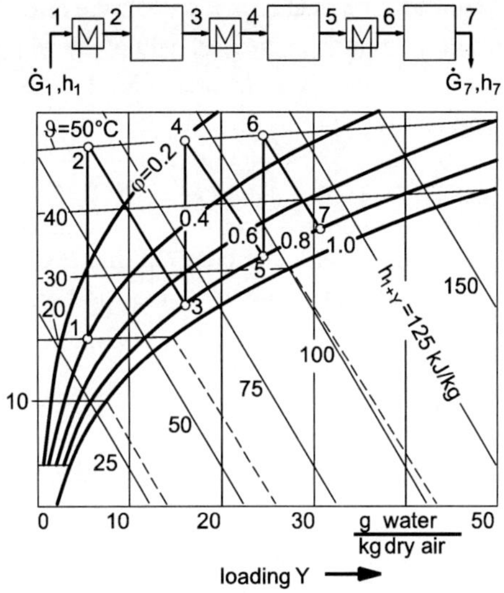

Fig. 10.4-1 Three-stage drying process. Sketch of the process and its representation in the enthalpy–concentration diagram for humid air

Furthermore, a dryer may be operated in cocurrent or countercurrent passage of the drying air and the drying good. Figure 10.4-2 shows for both modes the development of temperature and humidity in case of a belt dryer (air flows past the good) and a rotating drum dryer (air flows through the good). The good is assumed to dry in the constant rate period (see Sect. 10.6).

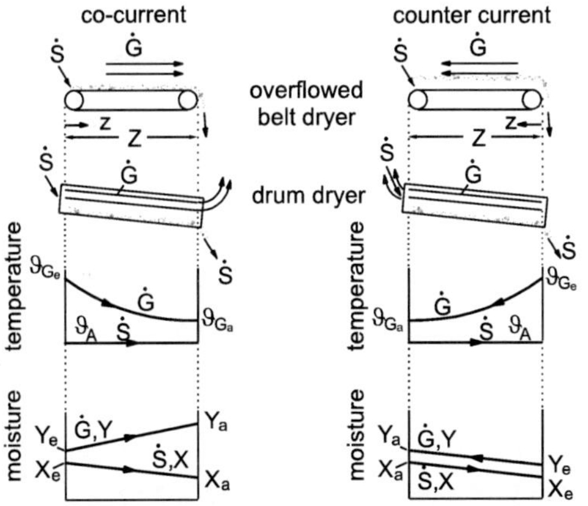

Fig. 10.4-2 Belt and rotating drum dryer operated in co- and countercurrent mode. Development of temperature and humidity in the apparatus

On the left side is shown the development of temperature and humidity in the apparatus for the cocurrent mode, whereas on the right side the countercurrent mode is shown. The spatial coordinate z is chosen to follow the direction of the gas flow. In both cases, the gas temperature decreases, while the temperature of the good remains virtually constant at the adiabatic saturation temperature. The humidity of the gas increases, and the moisture of the good decreases. The difference between both remains almost constant in the countercurrent mode, whereas in the cocurrent mode the difference decreases along the dryer.

In the case of pneumatic dryers in countercurrent mode the good falls against the upstreaming gas. In the cocurrent mode the good is transported by the air. The moisture balance for such dryers is

$$\pm\dot{S}_{\mathrm{dry}} \cdot (X_{\mathrm{out}} - X_{\mathrm{in}}) = \dot{G}_{\mathrm{air}} \cdot (Y_{\mathrm{out}} - Y_{\mathrm{in}})\,. \tag{10.4-1}$$

The plus sign holds for the cocurrent mode, and the minus sign for the countercurrent mode.

10.5 Fluid Dynamics and Heat Transfer

In the case of convective dryers, the drying medium, i.e., warm air, flows either past or in the case of a solid bed through the drying good. The air velocity is chosen according to the requirements concerning heat transfer and pressure drop.

A fixed bed exists, if the drying gas flows through a bulk material from top to bottom or from bottom to top below the minimum fluidization velocity (see Chap. 3). Pressure drop and heat transfer coefficient may be estimated using the correlations given in Chaps. 3 and 4. To achieve a fluidized bed, upward flow beyond the minimum fluidization velocity has to be chosen. Figure 3.6-7 allows estimates about the entrainment of fines. Furthermore, this diagram may in principle be used, to find favorable superficial velocities for pneumatic dryers as depicted in Fig. 10.1-9.

At first, wet bulk material, which shall be dried in a fixed bed with an upward flow of the drying gas through this bed, often dries in the constant rate period. After a certain period of time, the particles of the bottom layer of the bed begin to dry in the falling rate period. Gradually, the region of the falling rate period expands further up, until it covers the entire volume of the bed. The development of the moisture content of the material as a function of the bed height may be calculated by considering the timely development of the moisture in differential bed elements, see, i.e., Kneule (1975), Krischer and Kast (1978), Schlünder (1976), and van Meel (1958).

10.6 Drying Periods

Drying or desorption of liquid from a good requires that in the ambient the partial pressure p_i of the vapor of this liquid is lower than its partial pressure at the surface of the material $p_i < p_i^* \langle X_{i,\,surface} \rangle$. As a rule the partial pressure at the surface of the good is considered to be in thermodynamic equilibrium with the moisture of the good at its surface. To achieve this situation, the good is heated either in contact dryers through contacting hot walls or in convection dryers through a hot auxiliary gas (i.e., air), which passes by or through the material. In general, it is the task of drying to dry a solids mass S from its initial moisture content X_α to the desired final moisture X_ω by removing the corresponding amount of liquid $S_{dry} \cdot (X_\alpha - X_\omega)$. For this process the drying time τ has to be spent.

In Figure 10.6-1 the development of the moisture X is shown as a function of the elapsed drying time t. Often, the moisture at first decreases at a constant rate, and therefore linearly with time. This drying period is called either I. drying period or constant rate period. It is characterized by the fact that initially all capillaries of the good are filled with liquid and evaporation takes place from the surface of the good. This evaporation process is independent of the good and solely governed by heat and mass transfer in the gas phase. Later on in the II. drying period or the falling rate period, the moisture has fallen below a certain threshold value, and the transport and evaporation of moisture in the good becomes limiting. In addition to the above parameters, this period is determined by the ability of the good to transport as well as heat, liquid moisture, and gaseous humidity.

The mass flux of evaporated liquid from the surface of a drying good is called drying rate $\dot{m}$. In the one-dimensional case of drying a slab of thickness s and surface A, the drying rate is related to the timely change of moisture by

$$A \cdot s \cdot \rho_S \cdot \frac{dX}{dt} = -\dot{m} \cdot A \tag{10.6-1}$$

or

$$\dot{m} = -s \cdot \rho_S \cdot \frac{dX}{dt} \tag{10.6-2}$$

and the drying rate is proportional to the change of moisture X with time. On the left side of Fig. 10.6-1 $-dX/dt$ is shown as a function of the elapsed drying time, and on the right side it is shown as a function of the residual moisture.

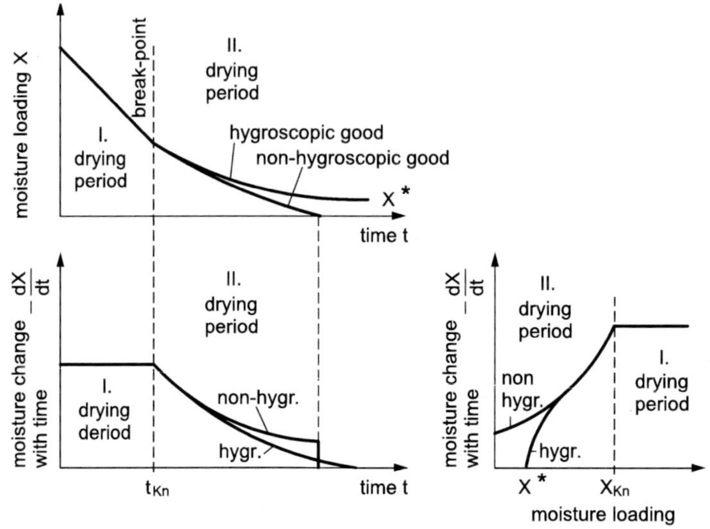

Fig. 10.6-1 Moisture vs. time (top); drying rate vs. time (bottom, left) and drying vs. moisture (bottom, right) in the constant and the falling rate periods

10.6.1 Constant Rate Period (I. Drying Period)

During the constant rate period the surface of the material is wet and thus behaves like a free liquid surface. A while after the commencement of adiabatic drying, the surface and the entire drying good attains the constant temperature ϑ_S. Because this temperature is below the temperature of the ambient, heat flows at the rate $\dot{Q}$ toward the surface. Here it is transformed into latent heat of the liquid which is evaporating at rate $\dot{M}$. Once the good has attained its steady temperature ϑ_S (adiabatic saturation temperature, wet bulb temperature) the energy balance of the drying good gives the following relation.

$$\frac{d(S \cdot h')}{dt} \left(= S \cdot \underbrace{\frac{dh'}{dt}}_{=0} + h' \cdot \frac{dS}{dt} = A \cdot \dot{m} \cdot h' \right) = A \cdot \dot{q} - A \cdot \dot{m} \cdot h'' \tag{10.6-3}$$

or

$$\dot{q} = \dot{m} \cdot \Delta h_{SL} \,. \tag{10.6-4}$$

Hence, the heat flux equals the drying rate multiplied by the heat of vaporization.

Design calculations of dryers, which operate in the constant rate period, is relatively simple, because no processes inside the product have to be taken into consideration. The temperature of the gas decreases along the flow direction z, while its humidity Y increases, see Fig. 10.4-2. The moisture X of the good decreases, while its temperature ϑ_S remains virtually constant. The differential amount of heat $\mathrm{d}\dot{Q}$ transferred in the differential volume $f \cdot \mathrm{d}z$ of the dryer is

$$0 = -\dot{G} \cdot c_{pG} \cdot \mathrm{d}\vartheta_G - \mathrm{d}\dot{Q} = -\dot{G} \cdot c_{pG} \cdot \mathrm{d}\vartheta_G - \alpha \cdot \mathrm{d}A \cdot (\vartheta_G - \vartheta_S), \tag{10.6-5}$$

or with $\mathrm{d}A = a \cdot f \cdot \mathrm{d}z$ and a being the volume specific surface it follows

$$-\frac{\mathrm{d}\vartheta_G}{\vartheta_G - \vartheta_S} = \frac{\alpha \cdot a \cdot f \cdot \mathrm{d}z}{\dot{G} \cdot c_{pG}}. \tag{10.6-6}$$

Integration yields

$$Z = \underbrace{\int_0^Z \mathrm{d}z}_{\text{length or height}} = \underbrace{-\int_{\vartheta_{G,\,in}}^{\vartheta_{G,\,out}} \frac{\mathrm{d}\vartheta_G}{\vartheta_G - \vartheta_S}}_{\text{number of transfer units } N_\vartheta} \cdot \underbrace{\frac{\dot{G} \cdot c_{pG}}{\alpha \cdot a \cdot f}}_{\text{height of a transfer unit } H_\vartheta} . \tag{10.6-7}$$

In analogy to similar problems in heat and mass transfer, the length of the apparatus is the number of transfer units N_ϑ multiplied with the height of a transfer unit H_ϑ. From this and for the case of $\vartheta_S = \mathrm{const.}$, the temperature of the gas in the dryer can be evaluated to be

$$\vartheta_G(z) - \vartheta_S = (\vartheta_{G,\,in} - \vartheta_S) \cdot \exp\left(-\frac{\alpha \cdot a \cdot z \cdot f}{\dot{G} \cdot c_{pG}}\right) . \tag{10.6-8}$$

In the entire dryer, the transferred heat $\dot{Q}$ is

$$\dot{Q} = G \cdot c_{pG} \cdot (\vartheta_{G,\,in} - \vartheta_{G,\,out}) = \alpha \cdot a \cdot f \cdot Z \cdot \frac{\vartheta_{G,\,in} - \vartheta_{G,\,out}}{\ln \dfrac{\vartheta_{G,\,in} - \vartheta_S}{\vartheta_{G,\,out} - \vartheta_S}} \tag{10.6-9}$$

$$= \alpha \cdot a \cdot f \cdot Z \cdot (\Delta\vartheta)_m$$

The local drying rate $\dot{m}(z)$ in the constant rate period is

$$\dot{m}(z) = \frac{\alpha}{\Delta h_{SL}} \cdot (\vartheta_{G,\,in} - \vartheta_S) \cdot \exp\left(-\frac{\alpha \cdot a \cdot z \cdot f}{\dot{G} \cdot c_{pG}}\right). \tag{10.6-10}$$

And finally, the mean drying rate $\bar{\dot{m}}$ in the constant rate period is

$$\bar{\dot{m}} = \frac{\alpha}{\Delta h_{SL}} \cdot (\Delta\vartheta)_m . \tag{10.6-11}$$

As shown above, the changes in the thermodynamic state of air can be read from the enthalpy-concentration diagram for humid air.

The material balance for a differential section of the dryer operated at steady state,

$$0 = \dot{G}_{air} \cdot dY + \dot{S}_{dry} \cdot dX , \tag{10.6-12}$$

tells about the relation between the change of moisture of the drying good and change of humidity of the air

$$dX = -\frac{G_{air}}{S_{dry}} dY . \tag{10.6-13}$$

Furthermore, the following relation between the drying velocity and air flow holds:

$$\dot{m} = -s \cdot \rho_S \cdot \frac{dX}{dt} = \frac{d(\dot{G}_{air} \cdot Y)}{dt} \cdot \frac{1}{A} . \tag{10.6-14}$$

The drying time τ is needed to dry a good from its initial moisture X_α to the desired residual moisture X_ω. With the above relations the drying time can be calculated as follows:

$$\tau = \int_0^\tau dt = -\int_{X_\alpha}^{X_\omega} \frac{s \cdot \rho_S}{\dot{m}} \cdot dX . \tag{10.6-15}$$

In the constant rate period the local drying rate $\dot{m}$ equals the mean drying rate $\bar{\dot{m}}$ and does not change with time. Therefore, under these conditions and for flow past a single-sided slab, the drying time τ is

$$\tau = \frac{s \cdot \rho_S}{\bar{\dot{m}}} \cdot (X_\alpha - X_\omega) = \frac{s \cdot \rho_S \cdot \Delta h_{SL} \cdot (X_\alpha - X_\omega)}{\alpha \cdot (\Delta\vartheta)_m}$$
$$= \frac{\rho_S \cdot \Delta h_{SL} \cdot (X_\alpha - X_\omega)}{\alpha \cdot a \cdot (\Delta\vartheta)_m} . \tag{10.6-16}$$

Hence, the drying time is short, if the heat transfer coefficient α and/or the mean temperature difference $(\Delta\vartheta)_m$ is high. Further on, the drying time is short, if the thickness s of the slab is small and if its volume-specific surface a is high. Figure 10.6-2 shows how thickness s and the reciprocal of the specific surface a depend on the geometry. It can be seen that small spheres are most suitable for fast drying.

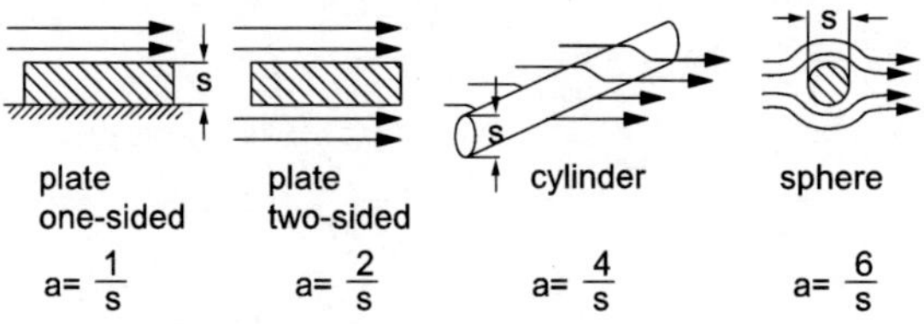

Fig. 10.6-2 Volume specific surface a of differently shaped bodies

The above can easily be transferred to continuously operated dryers, by changing from initial and final moistures X_α, X_ω to inlet and outlet moistures X_{in}, X_{out}. The drying time is then equal to the residence time of the drying good in the dryer.

10.6.2 Critical Moisture Content

The constant rate period persists only, if the surface of the drying good is sufficiently wet. During drying, the moisture of the drying good is reduced, and below a certain critical moisture content not enough capillaries of the drying good are able to transport their liquid to the surfaces. Then the surface is no longer wet and the drying rate decreases. The higher the drying rate $\dot{m}$ and the higher the thickness s of the drying good, the higher the critical moisture content X_c, see Fig. 10.6-3. But, if the term $\dot{m} \cdot s$ is plotted as a function of the moisture X, then all critical moisture contents lie on one characteristic curve, the so-called curve of critical moisture content. The larger the mean diameter of the capillaries is, the higher is the critical moisture content. Further on, dimensional analysis shows (Kneule 1975), that the critical moisture content X_c rises with the dimensionless group $(\dot{m} \cdot \eta_L)/(\rho_L \cdot \sigma)$.

Despite these fundamental insight, today it is not possible to predict the behavior of the critical moisture content without experiments.

10.6.3 Falling Rate Period (II. Drying Period)

When dealing with the falling rate period, it is necessary to distinguish between non-hygroscopic and hygroscopic drying goods. In the case of nonhygroscopic drying goods the moisture is not bound as sorptive, whereas in the hygroscopic case, the moisture is partially bound in capillaries as adsorptive on the internal surface or as swelling liquid. The amount of this moisture is in thermodynamic equilibrium with the humidity of the gas phase, and may be described by a sorption isotherm.

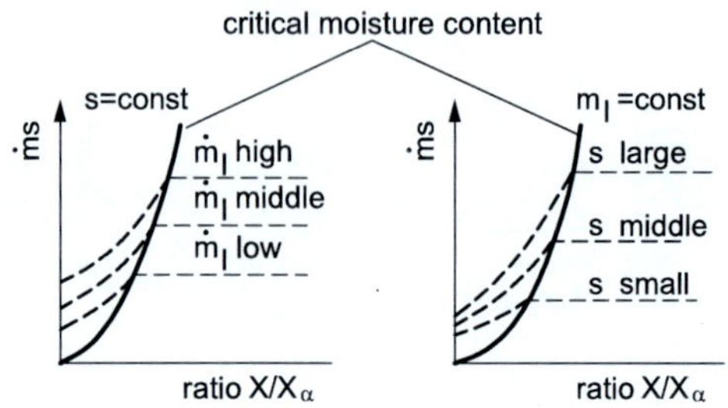

Fig. 10.6-3 Dependency of the critical moisture content as a function of the drying rate and the slab thickness s

Nonhygroscopic goods

Figure 10.6-4 shows the drying rate as a function of the moisture of the good for the case of nonhygroscopic goods. Beyond the critical moisture content, the drying rate falls until it reaches the final drying rate $\dot{m}_{\omega}$ at $X = 0$. There is no general method to predict how the drying rate falls, but some propositions can be made about the final drying rate. Following Fig. 10.6-5, the good is characterized by pores of a certain size distribution.

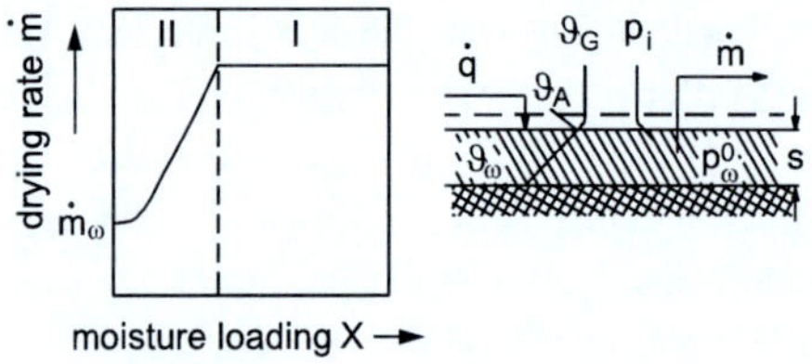

Fig. 10.6-4 Drying rate vs. moisture for a nonhygroscopic drying good

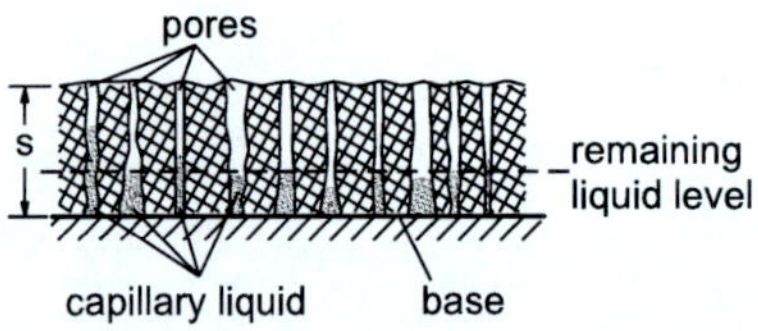

Fig. 10.6-5 Porous drying good during drying in the falling rate period and indication of the remaining liquid level in the drying good

During the falling rate period, the level of the remaining liquid in the capillaries continuously recedes into the drying good. At the end of the drying process, the drying good attains $X = 0$ and the level of the remaining liquid has receded throughout the entire thickness s of the drying good. At adiabatic drying conditions (i.e., no extra heating from the bottom or by radiation), the incoming heat flux at the liquid level effectuates a mass flux of moisture into the gas phase of the capillaries and further on out of the drying good. The heat flux has to overcome the outer heat transfer resistance in the gas phase and the resistance due to heat conduction in the drying good. If the temperature at the final liquid level is ϑ_ω, then

$$\alpha \cdot (\vartheta_G - \vartheta_{S\omega}) = \frac{\lambda}{s} \cdot (\vartheta_{S\omega} - \vartheta_\omega) \tag{10.6-17}$$

holds and the final drying rate is

$$\dot{m}_\omega = \frac{\lambda}{\Delta h_{SL} \cdot s} \cdot \left(\frac{\vartheta_G - \vartheta_\omega}{1 + \dfrac{\lambda}{\alpha \cdot s}} \right) \tag{10.6-18}$$

with λ being the heat conductivity of the dry solid. On the other hand, the mass flux is driven by the difference in partial pressure of the evaporated compound between the remaining liquid level and the bulk of the gas:

$$\dot{m}_\omega = \beta_{tot} \cdot \left(\frac{\overset{0}{p}_{i,\omega} - p_i}{R \cdot T} \right) . \tag{10.6-19}$$

The mass transfer coefficient β_{tot} is composed of the external coefficient in the gas and of the internal diffusion in the already dried porous layer of the drying good

$$\frac{1}{\beta_{tot}} = \frac{1}{\beta_{out}} + \frac{s}{D_G} \cdot \mu_p , \tag{10.6-20}$$

with D_G being the diffusion coefficient of the evaporated compound in the gas, and with μ_p being the diffusional resistance number, which scales down the diffusion in the porous medium in relation to diffusion in the gas, see Chap. 9. From the above, the final drying rate can be derived:

$$\dot{m}_\omega = \frac{1}{\dfrac{1}{\beta_{out}} + \dfrac{s}{D_G} \cdot \mu_p} \cdot \frac{\overset{0}{p}_\omega - p_i}{R \cdot T} . \tag{10.6-21}$$

Both, (10.6-18) and (10.6-21) present the unknown ϑ_ω and $\overset{0}{p}_\omega$, which can be determined by iteration. Thus, the final drying rate $\dot{m}_\omega$ is determined.

The quantity $\dot{m}_\omega$ depends on several factors as the overall drying conditions like heat and mass transfer coefficient, which depend on flow rate and flow regime and on the properties of the gas. Furthermore, the temperature and the relative humidity of the gas are important. Finally $\dot{m}_\omega$ depends on the slab thickness of the drying good, its heat conductivity, and the diffusivity. High values of the final drying rate can be achieved, if high flow rates and small slab thicknesses are chosen.

Hygroscopic goods

Drying of hygroscopic goods ends at a moisture X^* which is in equilibrium with the gas phase of a certain relative humidity, see Fig. 10.6-6. In the course of drying, the drying rate deceases with decreasing moisture of the good due to as well as the reduction of the driving force for drying and the reduction of the diffusion coefficient. Drying is determined by unsteady diffusion in the matrix of the good, and the drying process may not be influenced by the hydrodynamic conditions of the gas phase or by heat transfer inside the drying good.

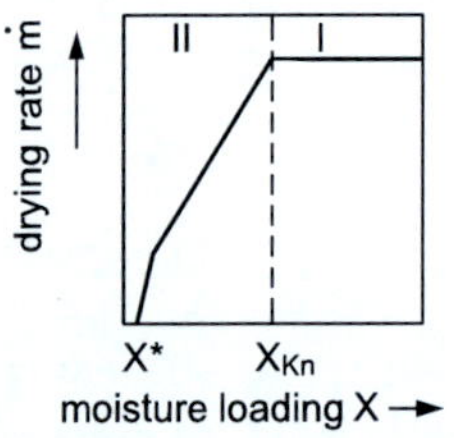

Fig. 10.6-6 Drying curve: mass flux as function of the moisture content for a hygroscopic drying good

According to Fick's law the driving force for diffusion is the moisture gradient inside the drying good. With certain simplifications the one-dimensional moisture balance for a differential slab of the drying good leads to the following partial differential equation (see Chap. 9)

$$\frac{\partial X}{\partial t} = D \cdot \frac{\partial^2 X}{\partial y^2}, \tag{10.6-22}$$

with D being the diffusion coefficient of moisture in the drying good. This diffusion coefficient may depend strongly on the moisture content itself and, as a rule, must be determined experimentally.

Solutions to this partial differential equation are provided in Chap. 9. In the case of a constant diffusion coefficient and for a homogeneous initial moisture distribution X_α and in case of drying from both sides the long-term solution of the above equation yields the mean moisture $\bar{X}$ of the slab as a function of the drying time τ:

$$\frac{\bar{X} - X^*}{X_\alpha - X^*} = \frac{8}{\pi^2} \cdot \exp\left(-D \cdot \tau \cdot \left(\frac{\pi}{s}\right)^2\right) . \tag{10.6-23}$$

Thus, the required time for drying a product from its initial moisture X_α to its final moisture X_ω is

$$\tau \approx \left(\frac{s}{\pi}\right)^2 \cdot \frac{1}{D} \cdot \ln\frac{X_\alpha - X^*}{X_\omega - X^*} , \tag{10.6-24}$$

and by differentiation of (10.6-23) with respect to time it can be shown that the mass flow density is

$$\dot{m} \equiv -\rho_S \cdot s \cdot \frac{dX}{dt} = \pi^2 \cdot \rho_S \cdot \frac{D}{s} \cdot (X - X^*) \ . \tag{10.6-25}$$

Thus the drying rate is proportional to the difference in moisture loading and is inversely proportional to the thickness s of the drying good. Also in this case it is beneficial for drying, if the drying good is rather thin.

The drying time of hygroscopic drying goods in the falling rate period can be estimated from the so-called drying curve. As a rule, the drying curve has to be determined experimentally. It tells about the course of the drying rate as a function of the moisture content of the drying good. Often the product $\dot{m} \cdot s$ of the drying rate and the thickness of the drying good do not vary as much as the drying rate itself. Therefore, in many cases this product $\dot{m} \cdot s$ is plotted vs. the difference $X - X^*$, see Fig. 10.6-7. Thus the drying time is obtained from the following integral:

$$\tau = -s^2 \cdot \rho_S \cdot \int_{X_\alpha - X^*}^{X_\omega - X^*} \frac{d(X - X^*)}{\dot{m} \cdot s} \ . \tag{10.6-26}$$

According to Fig. 10.6-7 the drying time is proportional to the area below the curve representing the inverse of $\dot{m} \cdot s$ as a function of the moisture content of the drying good $X - X^*$.

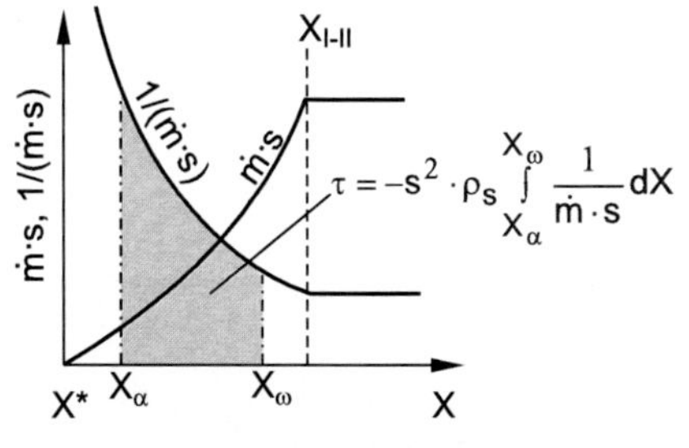

Fig. 10.6-7 Inverse of $\dot{m}s$ vs. moisture content in the falling rate period

It is common to represent the drying curve in normalized coordinates. To this end the drying rate is related to its value in the constant rate period $\dot{m}/\dot{m}_I$, and the moisture of the good $X - X^*$ is related to its value $X_{I-II} - X^*$ at the inflection point where the constant rate period ends:

$$\eta = \frac{X - X^*}{X_{I-II} - X^*}. \tag{10.6-27}$$

Figures 10.6-8 (Kamei 1934) and 10.6-9 (Kamei 1934) show normalized drying curves for paper and Kibushi-aluminum oxide. For the experiments with paper its thickness has been varied. The results for aluminum oxide show that a variation of air temperature and humidity does not severely influence the results.

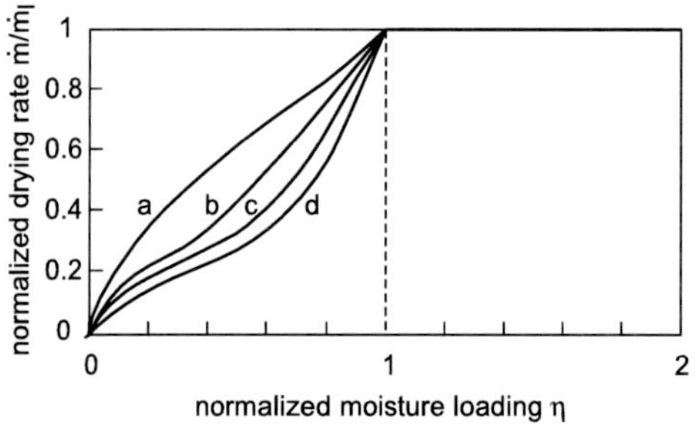

Fig. 10.6-8 Normalized drying curve for paper of varying thickness (Kamei 1934) (**a**) 1,0 cm, (**b**) 1,5 cm, (**c**) 2,0 cm, (**d**) 3,0 cm

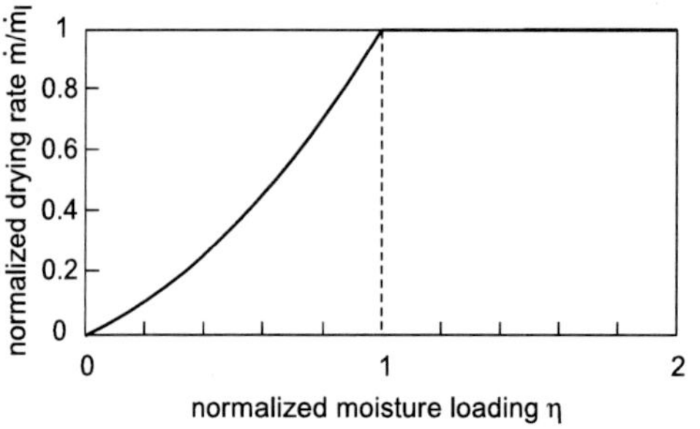

Fig. 10.6-9 Normalized drying curve for Kibushi-aluminum oxide at various air temperatures and humidities (Kamei 1934)

10.7 Some Further Drying Processes

In the past, numerous convective and contact drying processes have been developed and are successfully used to dry all kinds of products such as particulate bulk material and powder, sheets and fluids. They are developed to cope with a wide variety of physicochemical and handling properties. Another reason for the manifold of processes is that most of these processes produce different product qualities and thus can meet a wide variety of consumer needs. In principle, these processes can be discriminated according to the handling behavior of the product (i.e., free flowing, cohesive, chunky), to the residence time needed for drying (seconds, minutes), and to the mode of operation (batchwise, continuous). Drying processes for bulk materials are presented in Sect. 10.1. Here further drying processes are presented.

Impinging jet dryers are employed in the case of sheet-like materials such as cardboard, folio, polymer film, paper, textile, and veneer. Hot gas impinges from an array of nozzles onto the good. This process creates high drying rates. Sometimes the good is transported on a substrate and heated from below as well. It may also be transported on air cushions.

Agricultural goods like crop, silage and malt are dried in kiln and rotary kiln dryers, which lead the drying medium through or over the product. Often the drying medium is a mix of air and exhaust gases.

Solutions, suspensions, or pastes are often dried in spray dryers and thus formulated to powders. The liquid is dispersed into droplets by nozzles or turning wheels. Gas and good may move in co- or countercurrent manner. Droplet size and hence particle size are in the range of some 10–100 μm depending on the specific design and operational parameters. A wide variety in particle morphology can be formulated. Spray dryers may be combined with agglomerating and granulating moving or fluidized bed dryers. The drying time is in the order of seconds due to the large surface area of the drying good. Therefore this process is especially suitable for thermo- and oxidation sensitive goods like, i.e., milk products. Also dyes, detergents, and numerous chemicals are spray-dried.

Viscous solutions, polymers, and pastes are preferably dried by contact drying on the surface of hot drums and belts. Also these processes offer a certain variety in possible product forms like flakes, chips, and pastilles. Also paddle dryers and kneaders are used in this case.

Drying processes may be intensified by introduction of radiation energy. Infrared radiation is most common. In this case the applied heat is absorbed on the surface of the goods. Recently, microwave radiation is applied for certain products. Microwaves are absorbed by the water molecules of the drying good and thus penetrate into the drying good.

Freeze drying is a well-established drying technology to meet certain pharmaceutical requirements (lyophilization). It is also used for some food products like berries, which need to be quickly reconstituted (moistened) during meal preparation. A freeze drying process is composed of two process steps. First, the wet product has to be frozen in a manner that on the one hand creates a continuous network of the ice phase, and that on the other hand preserves the structure of the product. After that the ice has to be desublimated without melting. Melting would create capillary forces during drying, which destroy the structure of the product. Fig. 10.7-1 shows the principal parts of a freeze drying apparatus. Often, it consists of a chamber, which can be evacuated below the triple point of the moisture to be dried off. The heat required for desublimation must be supplied either by contact or radiation heating. Furthermore, a condenser is needed to condensate the evaporated moisture. Among other influences, the mass flux depends on the flow regime of the vapor in the chamber. If the flow regime is Knudsen-like, then the maximum flow is given by

$$\dot{m}_{i,\,\max} = \frac{p_i^*}{\sqrt{2 \cdot \pi \cdot R \cdot T}} \ . \qquad (10.7\text{-}1)$$

Thus, the maximum mass flux should only depend on vapor pressure and temperature. In practice, only a fraction of this flux can be achieved.

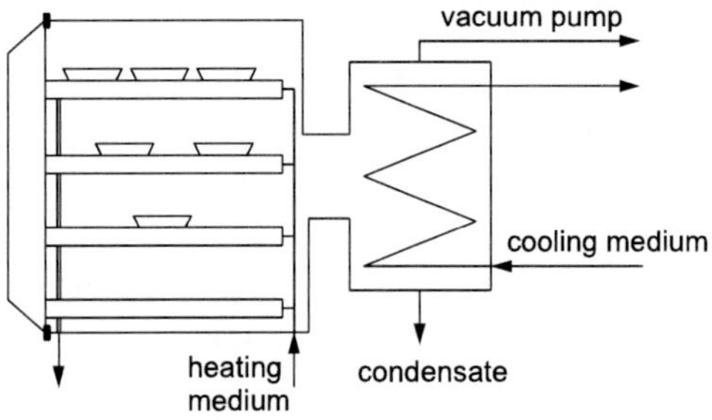

Fig. 10.7-1 Freeze drying equipment

Symbols

A	m^2	Exchange surface
a		Fraction
D	m^2/s	Diffusion coefficient
D_{eff}	m^2/s	Effective diffusion coefficient
f	m^2	Sectional area
$\dot{G}$	kg/s	Gas mass flow rate
H_ϑ	m	Height of a transfer unit
h	kJ/kg	Specific enthalpy
L	kg	Liquid mass

$\dot{L}$	kg/s	Liquid mass flow rate
$\dot{m}$	$kg/(m^2 \times s)$	Drying rate
N_ϑ		Number of transfer units
n		Number of capillaries
p^0	Pa	Vapor pressure
p_i	Pa	Partial pressure of the component i
$\dot{Q}$	W	Heat flow
$\dot{q}$	W/m^2	Heat flux
q	kJ/kg	Specific heat of adsorption
r	m	Radius
r	kJ/kg	Specific enthalpy of vaporization
s	m	Layer thickness
t	s	Time
Z	m	Apparatus length × "no dimension" moisture
z	m	Coordinate

Greek symbols

α	$W/(m^2 \times K)$	Heat transfer coefficient
β	m/s	Mass transfer coefficient
ε		Volume fraction, emissivity
η	Pa×s	Dynamic viscosity
η		Ratio of moisture content
ϑ	°C	Celsius temperature
κ	m^2/s	Moisture conduction coefficient
λ	W/(m×K)	Thermal conductivity
ρ	kg/m^3	Density
σ	N/m	Surface tension
τ	s	Drying time
φ		Relative humidity of the gas

Indices

out	Outlet
B	Bonding
c	Continuous phase
d	Dispersed phase
in	Inlet
eff	Effective
G	Gas
tot	Total
semi	Semipermeable
AS	Adiabatic saturation
L	Liquid

LG	Liquid/gas
m	Midpoint
max	Maximal
min	Minimal
Ph	Surface
R	Radiation
p	Reference (pure) substance
r	Curved surface
S, s	Solid matter
SS	Steady state
α	Start (temporal)
ω	End (temporal)
0	At 0°C, beginning
I, II	First, respectively second drying period
*	equilibrium

11 Conceptual Process Design

Industrial separation processes typically consist of various distillative and alternative separation steps that are coupled by material and energy streams. Such processes often have very complex structures caused by the properties of the systems at hand and by the constraints set by cost and energy savings. In most cases, a rather empirical approach is used for process design. Novel developments concern a conceptual process design (e.g., Douglas 1988; Smith 1995; Blass 1997; Stichlmair and Fair 1998; Seider et al. 1999; Doherty and Malone 2001; Mersmann et al. 2005), which is based on the thermodynamic properties of the mixture at hand.

Separation processes typically consist of hundreds of elements (pipes, vessels, heat exchangers, columns, pumps, compressors, engines, measuring and control devices, etc.) that are too complex for exact graphical depiction. Therefore, the process structures are depicted in a very abstract and geometrically nonsimilar manner in the form of flow sheets. Flow sheets show the process structure with symbols for the essential process elements according to international standards (e.g., DIN, ISO).

The flow sheet of a process for the production of absolute alcohol by pressure swing distillation is shown in Fig. 11.0-1. The process consists, in essence, of two distillation columns, several heat exchangers, several vessels, and a complex network of pipelines (see Sect. 11.3.2). Also shown are measuring and control devices. According to international standards (e.g., DIN 19227, ISO 3511) these devices are illustrated by thin circles with a letter code indicating the function of the device, for instance, first letters: F = flow, L = liquid level, P = pressure, T = temperature; subsequent letters: A = alarm, C = control, F = fraction, etc.

Separation processes are subject to many constraints that significantly influence process design and operation. In distillation processes, for instance, the species can be exposed to mild temperatures only to avoid thermal degradation, polymerization, or crystallization. Furthermore, the temperature of heat supply (i.e., pressure of steam) often sets limiting conditions to process design and operation. The temperature of the cooling system (river water, cooling tower water, air, etc.) is also of great importance. The same holds for maximum and minimum (vacuum) feasible operating pressures. Even the feasible size of equipment (maximum height and diameter of columns) sets severe constraints to process design (Strigle 1987).

A. Mersmann et al., *Thermal Separation Technology: Principles, Methods, Process Design*, VDI-Buch, DOI 10.1007/978-3-642-12525-6_11,
© Springer-Verlag Berlin Heidelberg 2011

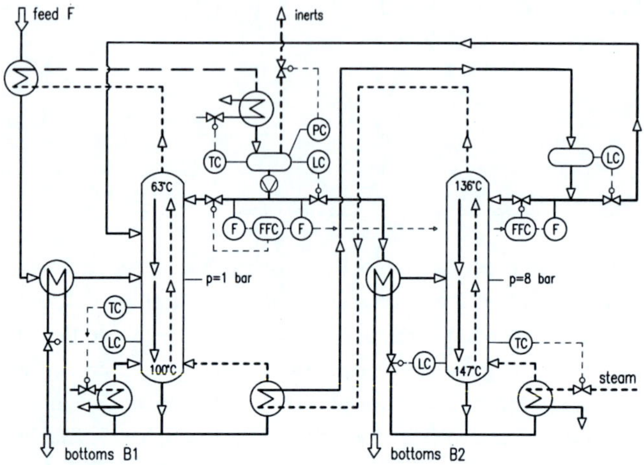

Fig. 11.0-1 Pipe and instrument flow sheet of a process for the production of absolute alcohol

11.1 Processes for Separating Binary Mixtures

The process examples presented in this section are selected to demonstrate the significance of process constraints to process design.

11.1.1 Concentration of Sulfuric Acid

Sulfuric acid is a very important chemical that is diluted and polluted in most of its industrial applications. The purification and reconcentration of sulfuric acid is an important task of process engineering.

The basis of any process design is a knowledge of vapor–liquid equilibrium. It is depicted in Fig. 11.1-1 in the form of vapor pressure lines with constant H_2SO_4 content in the liquid. Up to liquid concentrations of 75 wt% H_2SO_4, the coexistent vapor consists of pure water. Therefore, the liquid can be concentrated up by single stage distillation (evaporation). The feasible acid concentration is limited by the temperature level of the steam available at the site. Typically, the acid can be heated up to approximately 170°C, that is, at a total operating pressure of 1 bar, equivalent to an acid concentration of 72 wt%.

At higher product specifications a second evaporation step, which is operated under vacuum, has to be performed. The feasible operating pressure follows from the condition that the vapor (pure water) has to be condensed by water cooling. This constraint sets a minimum temperature of the condensate of approximately

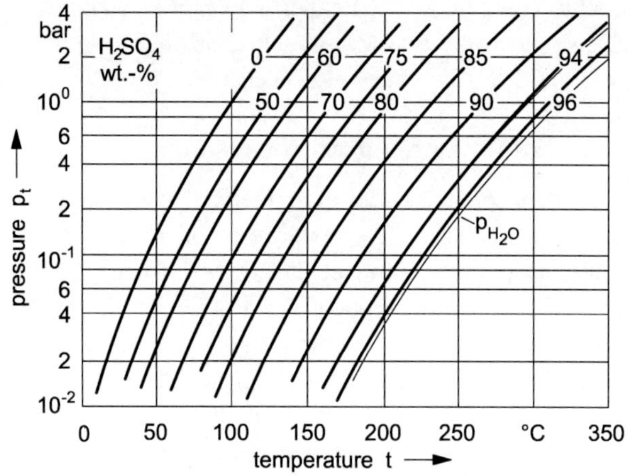

Fig. 11.1-1 Vapor–liquid equilibrium of aqueous sulfuric acid

50°C that is equivalent to a pressure of 0.12 bar (according to the vapor pressure curve of water). Any other cooling agent would be much more expensive. At a maximum bottom temperature of 170°C (set by the available steam) the highest feasible acid concentration is approximately 86 wt%. In this concentration range the vapor contains some SO_3. Therefore, the second evaporation step has to be performed in a countercurrent column.

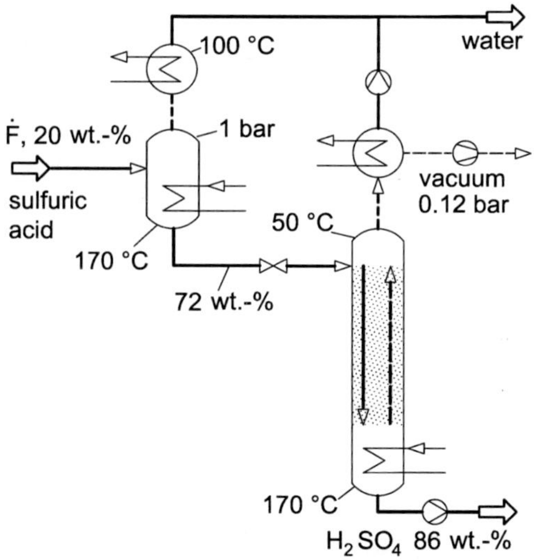

Fig. 11.1-2 Process for concentrating diluted sulfuric acid

The process depicted in Fig. 11.1-2 clearly demonstrates that the feasible product quality is determined by the temperature levels of the heating and cooling agents. A higher acid quality could be reached either by a higher heating temperature (i.e., higher steam pressure) or by a lower cooling temperature. However, such extensions of temperature levels are extremely expensive.

11.1.2 Removal of Ammonia from Wastewater

Wastewater very often contains traces of ammonia that have to be removed before biological wastewater treatment. The mixture ammonia/water is a wide boiling one that does not exhibit any azeotropes (see Fig. 11.1-3). Hence, distillation is a feasible process for ammonia removal from water. At a standard operating pressure of 1 bar the temperature at the top of the column is as low as –33.4°C (boiling temperature of ammonia). Therefore, water cooling of the condenser is not possible. Water cooling requires an operating pressure of about 20 bar. However, at 20 bar the bottom temperature (boiling temperature of water) would be higher than 200°C, which is too high for steam heating.

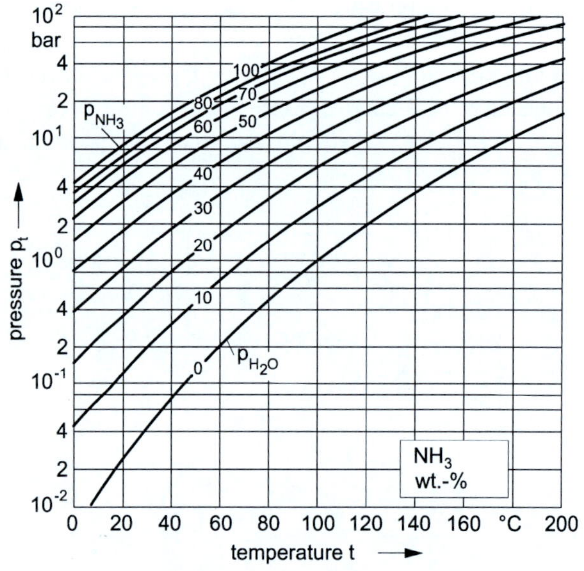

Fig. 11.1-3 Vapor/liquid equilibrium of the system ammonia/water

The solution of the problem is a process with two different operating pressures as shown in Fig. 11.1-4 (Wunder 1990). In the first column (p_1 = 1 bar) pure water is recovered as bottoms. The lowest feasible condensation temperature at the top is approximately 45°C, which is equivalent to an ammonia concentration of 20 wt%. This overhead fraction is compressed (in liquid state) to 20 bar and fed into column C-2 for further fractionation. The overhead fraction of C-2 is pure ammonia (20 bar, 45°C). The bottom fraction, which is limited by the available steam (maximum

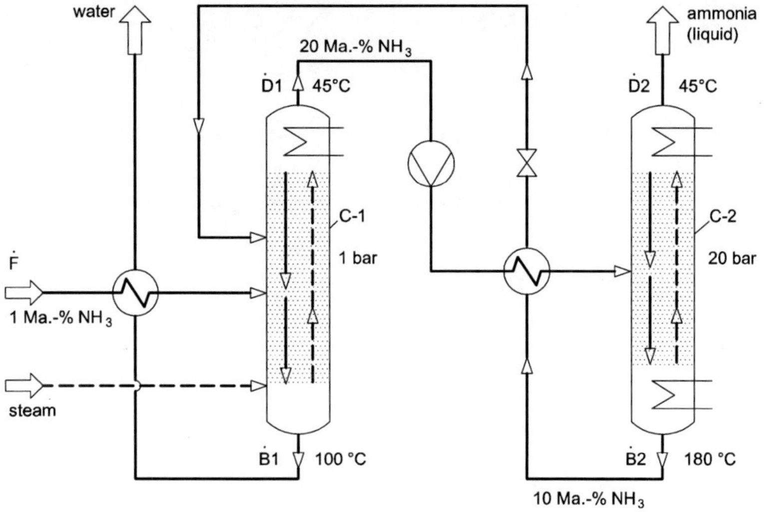

Fig. 11.1-4 Process for ammonia recovery from diluted aqueous solutions (Wunder 1990)

temperature 180°C), has an ammonia concentration of 10 wt%. This fraction is recycled to column C-1 for water recovery as bottoms $\dot{B}1$.

However, large recycle streams within processes are disadvantageous since they increase operating and investment costs. For economic reasons recycle streams have to be as small as possible. In this process, the concentration of the bottom fraction of column C-2 has to be much lower than that of the overhead fraction of column C-2. The following mass balance holds:

$$\dot{D}1 \cdot x_{D1\mathrm{NH3}} - \dot{B}2 \cdot x_{B2\mathrm{NH3}} = \dot{F} \cdot x_{F\mathrm{NH3}} \,. \tag{11.1-1}$$

Equation (11.1-1) makes it clear that the concentration x_{B2} has to be significantly lower than the concentration x_{D1}. This condition is met when the bottom temperature of column C-2 is higher than 155°C. Otherwise the process of Fig. 11.1-4 would fail.

11.1.3 Removal of Hydrogen Chloride from Inert Gases

Hydrogen chloride is a toxic and corrosive compound that must not pollute the environment. A process for air purification and hydrogen chloride recovery is shown in Fig. 11.1-5. The boiling point of HCl is too low (−85°C) for simple distillation and even for partial condensation. The process depicted in Fig. 11.1-5 uses the fact that HCl is extremely well soluble in water. Hence, absorption of HCl by water is a promising process.

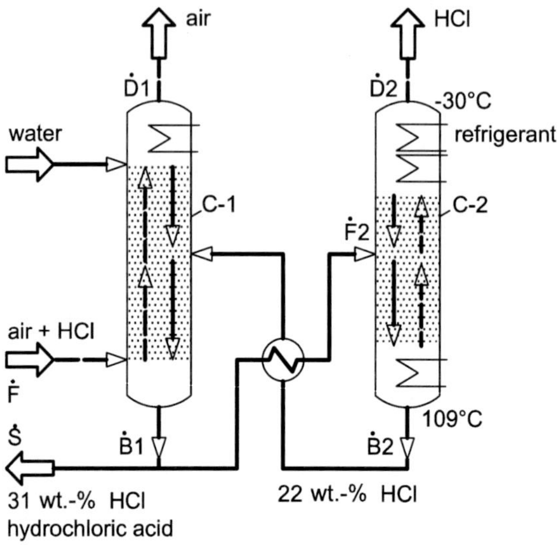

Fig. 11.1-5 Process for HCl removal from inert gases

Dissolution of the HCl in water sets free a large amount of heat of absorption (approximately as high as the latent heat of evaporation of water). Hence, the liquid in the absorber reaches boiling temperature and, in turn, some water is evaporated by the absorption of HCl. Hence, the conditions in the absorber are quite similar to those in a distillation column. The equilibrium curve of a boiling HCl/H_2O liquid is shown in the McCabe–Thiele diagram of Fig. 11.1-6. The presence of the inert gases only reduces the effective pressure of the HCl/H_2O system.

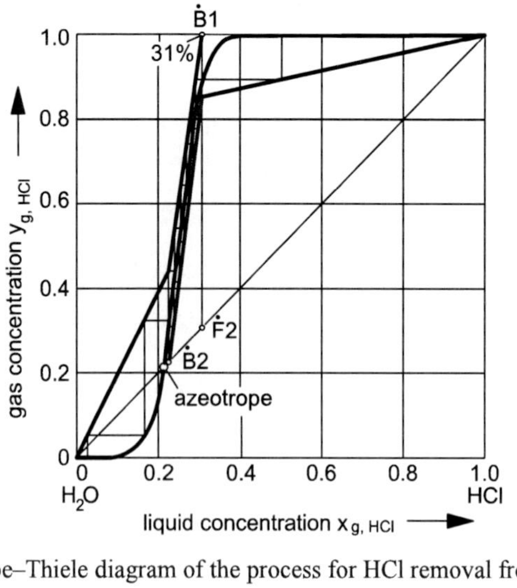

Fig. 11.1-6 McCabe–Thiele diagram of the process for HCl removal from inert gases

At the top of the absorption column the water vapor is condensed and recycled into the column. To increase the liquid reflux in the column some external water is fed into the column serving as washing agent. The operating line of column C-1 lies above the equilibrium line which is characteristic of absorbers. The azeotrope (maximum azeotrope) does not hinder the absorption process.

The concentration of the bottom fraction $\dot{B}1$ is as high as 31 wt%. Most of $\dot{B}1$ is fed via a heat exchange into distillation column C-2 to be separated into pure HCl (overhead fraction) and an azeotropic mixture (bottom fraction $\dot{B}2$). Fraction $\dot{B}2$ is recycled into the absorption column C-1. Hence, the operating line of the absorber has a sharp bend at a concentration of 22 wt%.

A mass balance around the whole process reveals that there is no exit for the water fed into the top of the absorber. Hence, some of the bottoms $\dot{B}1$ has to be withdrawn as side product $\dot{S}$. This fraction meets the specification of commercial hydrochloric acid and can be utilized elsewhere. From the McCabe–Thiele diagram in Fig. 11.1-6 follows the slope of the operating line at the top of the column:

$$(\dot{L}/\dot{G})_{top} = 2 \qquad \text{or} \qquad \dot{L}_{H2O} = \dot{G}_{HCl}\,. \tag{11.1-2}$$

Hence, the amount of external water $\dot{L}_{H_2O}$ is approximately as high as the amount of HCl in the raw gas. A HCl balance delivers

$$\dot{S}\cdot(1-0.31) = \dot{G}_{HCl} \qquad \text{or} \qquad \dot{S} = 1.45\cdot\dot{G}_{HCl}\,. \tag{11.1-3}$$

Thus, about 45% of the HCl in the raw gas is removed in the side product making the yield of pure HCl as low as 55%.

11.1.4 Air Separation

The fractionation of air into nitrogen and oxygen is a classical process with a great industrial importance, even after a history of over 100 years (Baldus et al. 1983). The special feature of the process is the very low temperature level required for the operation of the distillation column, approximately –180 to –190°C. Hence, it is not possible to heat and cool the columns by steam and water, respectively.

The system nitrogen/oxygen is a nearly ideal and very wide boiling mixture (relative volatility $\alpha \approx 4$) that can, in principle, very easily be fractionated in a single distillation column. The industrial process, however, uses two distillation columns, see Fig. 11.1-7.

In column C-1 the feed mixture (20% oxygen) is prefractionated into pure nitrogen (overhead fraction) and an oxygen-rich mixture with approximately 40% oxygen (bottom fraction). As the feed is a two-phase mixture with high vapor content, no

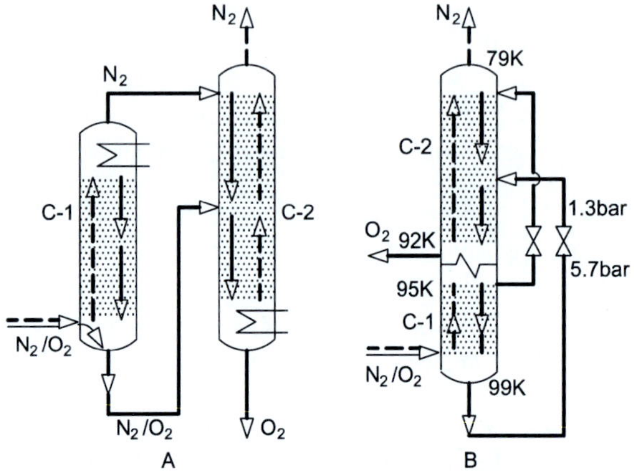

Fig. 11.1-7 Scheme of a process for air fractionation by distillation. (**A**) Basic process with two columns. (**B**) Improved process with Linde double column

reboiler is required for operating column C-1. The bottoms of C-1 is fed into column C-2 to be fractionated into the pure products nitrogen (distillate) and oxygen (bottoms). The overhead fraction of column C-1 is used as reflux in column C-2. Therefore, no condenser is required at the top of column C-2. Hence, the two columns of the basic process (Fig. 11.1-7 left) have only one condenser (C-1) and one reboiler (C-2). This enables a process simplification by thermal coupling of columns.

In thermal coupling of columns, the waste heat from any column is used for covering the heat demand of another column (see Sect. 11.2.2). The prerequisites for thermal coupling of columns are proper amounts and temperature levels of the heats. To adjust the temperature levels, column C-1 has to be operated with a higher pressure than column C-2. The famous Linde Double Column combines condenser and reboiler in single heat exchanger by arranging column C-2 directly above column C-1 (Fig. 11.1-7 right). The whole process of air separation is depicted in Fig. 11.2-18.

11.2 Processes for Separating Zeotropic Multicomponent Mixtures

Fractionation of multicomponent zeotropic mixtures (i.e., mixtures without azeotropes) into all constituents can be performed by a sequence of distillation columns. Generally, different sequences of columns are feasible to achieve the same products. However, investment costs and energy demand of the feasible processes may differ significantly. Thus, optimal column sequencing is an important task of process design. The basic principles will be demonstrated at the example of ternary mixtures.

11.2.1 Basic Processes for Fractionating Ternary Mixtures

There exist three basic column sequences for completely fractionating ternary mixtures by distillation.

11.2.1.1 a-Path

Often the low boiler *a* is removed first as overhead fraction of column C-1. The binary bottom fraction is fed into column C-2 to be fractionated into the pure intermediate boiler *b* (overhead) and the high boiler *c* (bottoms). The process depicted in Fig. 11.2-1 is called *a-path*.

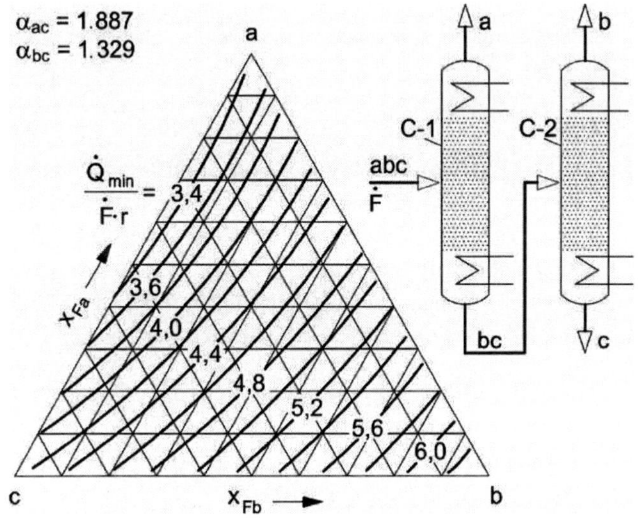

Fig. 11.2-1 Flow sheet and energy demand of the *a-path* for fractionating a zeotropic ternary mixture

11.2.1.2 c-Path

In the *c-path* (Fig. 11.2-2), the high boiler *c* is separated first as bottoms of column C-1. The binary overhead fraction is fed (in vaporous state) into column C-2 to be split into the pure low boiler *a* (overhead fraction) and the pure intermediate boiler *b* (bottoms).

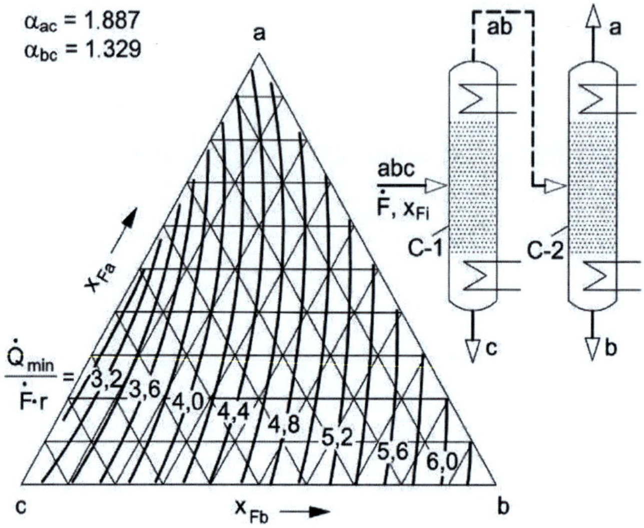

Fig. 11.2-2 Flow sheet and energy demand of the *c-path* for fractionating a zeotropic ternary mixture

11.2.1.3 a/c-Path

In the *a/c-path* (Fig. 11.2-3), the ternary feed is fractionated in column C-1 into an overhead fraction, which is free of high boiler c, and a bottom fraction, which is free of low boiler a. Thus, the first separation is performed between the low and high boiling components of the mixture a and c. The intermediate boiler b is present in both fractions produced in column C-1. The two binary fractions of C-1 are fractionated in the subsequent columns C-2 and C-3, respectively.

The energy demand of distillation processes is a very important quantity since it is the dominant factor for operating costs. Furthermore, a high energy demand also stands for high investment costs since the amount of vapor generated in the reboiler determines the size (in particular the diameter) of the columns (see Chap. 5).

The minimum energy demand of distillation processes can be determined without rigorous column calculation (see Chap. 5). A comparison of the minimum energy demand $\dot{Q}_{\min}/(\dot{F} \cdot r)$ of all three basic processes is plotted in Fig. 11.2-4. Depending on feed composition x_{Fi}, each of the three paths can have the lowest energy demand. The a-path is the best choice if the concentration of the low boiler a in the feed is very high. The c-path has the lowest energy demand if the concentration of the high boiler c is very high. However, the a/c-path is by far the better choice when the intermediate boiler b is the main constituent in the feed. It should be

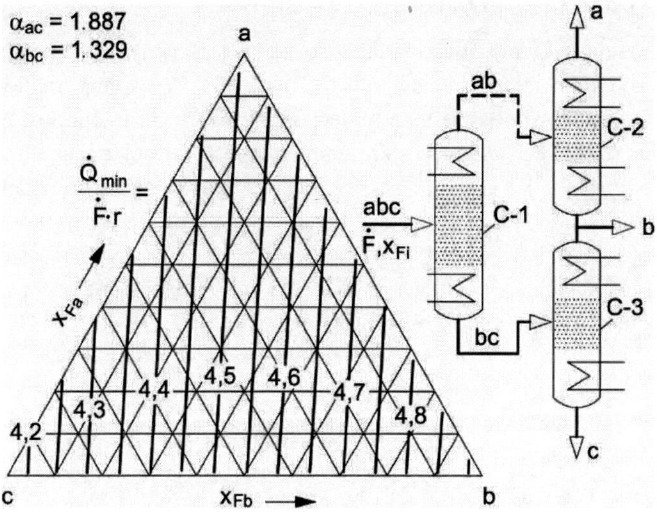

Fig. 11.2-3 Flow sheet and energy demand of the *a/c-path* for fractionating a zeotropic ternary mixture

noted that the energy demand of the a/c-path is a linear function of feed composition.

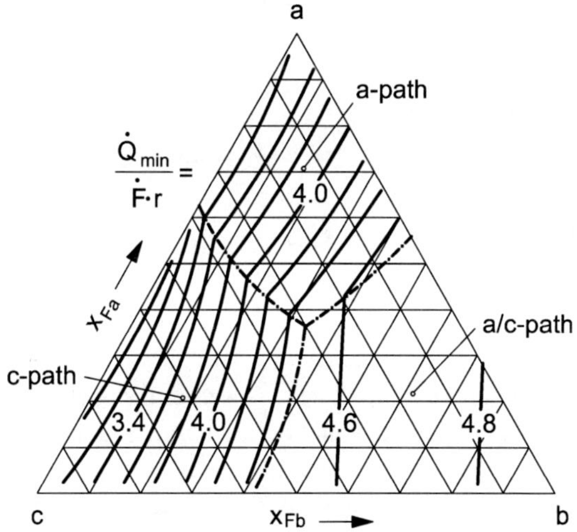

Fig. 11.2-4 Comparison of the energy demand of the three basic processes for fractionating a zeotropic ternary mixture. Valid for the relative volatilities $\alpha_{ac} = 1.887$ and $\alpha_{bc} = 1.329$

11.2.1.4 Direct Column Coupling

The a/c-path can be modified by direct (i.e., material) coupling of columns C-2 and C-3. The bottom fraction of column C-2 has, in a first approximation, the same concentration (pure intermediate boiler) as the overhead fraction of column C-3. Therefore, these two columns can be combined into one column and, in turn, a condenser and a reboiler can be discarded. The intermediate boiling component *b* is withdrawn as side product from the combined column. The energy demand of the combined column is either that one of column C-2 or that one of column C-3 depending on which is the larger one. Thus, the energy demand of the process with direct column coupling is significantly lower (Fig. 11.2-5).

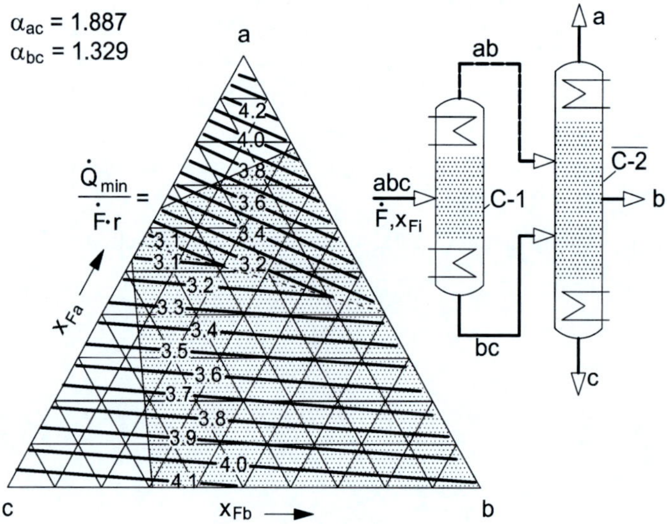

Fig. 11.2-5 Process sheet and energy demand of the a/c-path with material coupling of columns C-2 and C-3. In the *shaded area*, the coupled process is superior to all basic processes

Direct column coupling is a very effective measure for process simplification as it has the potential to reduce the investment as well as the operating costs. This technique for process simplification should be used whenever possible. The only prerequisite is the existence of gas and liquid streams with approximately same concentration in different parts of the process. Such situations can often be detected by careful process design.

11.2.2 Processes with Side Columns

The processes presented in Figs. 11.2-1 and 11.2-2 are often used in the process industry. However, they have the drawback that nearly identical separations are performed in different sections of subsequent columns.

11.2.2.1 a-Path with Side Column

In the a-path, for instance, the same mixture of components *b* and *c* is fractionated near the bottom of column C-1 and immediately above the feed point of column C-2, see concentration profiles in Fig. 11.2-6 (Mersmann et al. 2005). This energy-consuming double fractionation can be avoided by using a side column, see Fig. 11.2-7. In this modification of the basic process the stripping section of the second column (shaded section) is arranged at the bottom of the first column. The reboiler of column C-1 is discarded. The remainder of the second column is the so-called side column (rectifying side column). The internal concentration profiles of the process with side column are shown in Fig. 11.2-7. The feed is fractionated in the modified column C-1 into pure low boiler *a* (overhead fraction) and pure high boiler *c* (bottom fraction). A *b*-rich vapor fraction is taken from the main column (below the feed point) and fed into the side column. The intermediate boiling component *b* is recovered as overhead fraction of the side column. No twofold fractionation exists and, in turn, the energy demand of the process with side column is lower than that of the basic a-path. The process with side column is especially advantageous for mixtures with low content of the intermediate boiling compound *b*.

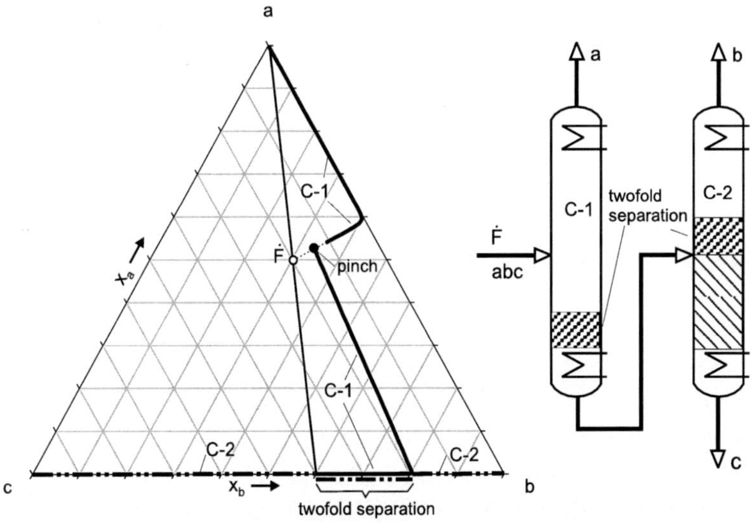

Fig. 11.2-6 Flow sheet and internal concentration profile of the a-path

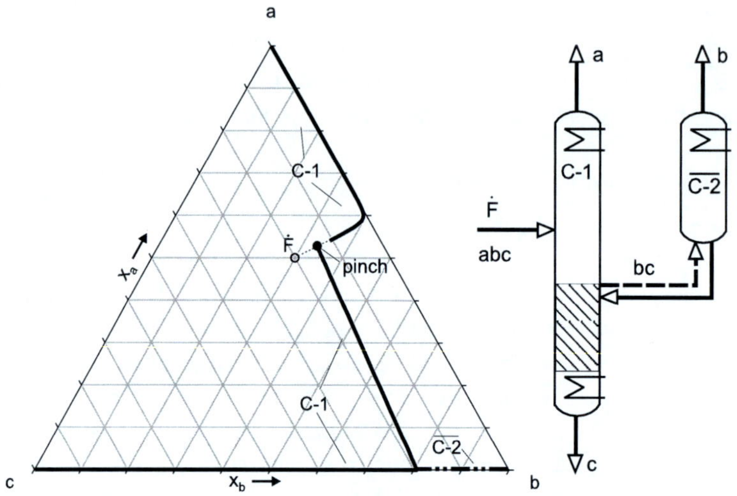

Fig. 11.2-7 Flow sheet and internal concentration profile of the a-path with rectifying side column

11.2.2.2 c-Path with Side Column

Side columns can also be applied to the c-path. Here, the rectifying section of column C-2 is attached to the top of column C-1 by discarding the condenser of C-1, see Fig. 11.2-8 right. Some liquid is withdrawn from the rectifying section (above the feed point) of the main column to be fractionated in the stripping side column. The bottom product of the side column is pure intermediate boiler *b*. Surprisingly, the energy demand of the a-path with rectifying side column is the same as that of the c-path with stripping side column (Fig. 11.2-8). These two processes are in particular advantageous when the concentration of the intermediate boiler *b* in the feed is low (shaded area in Fig. 11.2-8).

11.2.2.3 a/c-Path with Side Column

In the basic a/c-path there exist two sections with twofold separations, see Fig. 11.2-9 (Brusis 2003). These twofold separations can be avoided by arranging the rectifying section of column C-2 at the top of column C-1 and by arranging the stripping section of C-2 at the bottom of column C-1. Two heat exchangers of the basic process (Fig. 11.2-5) can be discarded. The internal concentration profiles of the process with side column are depicted in Fig. 11.2-10. No double fractionation exists in the modified process. The feeds into the side column are taken at the concentration maxima of the intermediate boiling compound *b*.

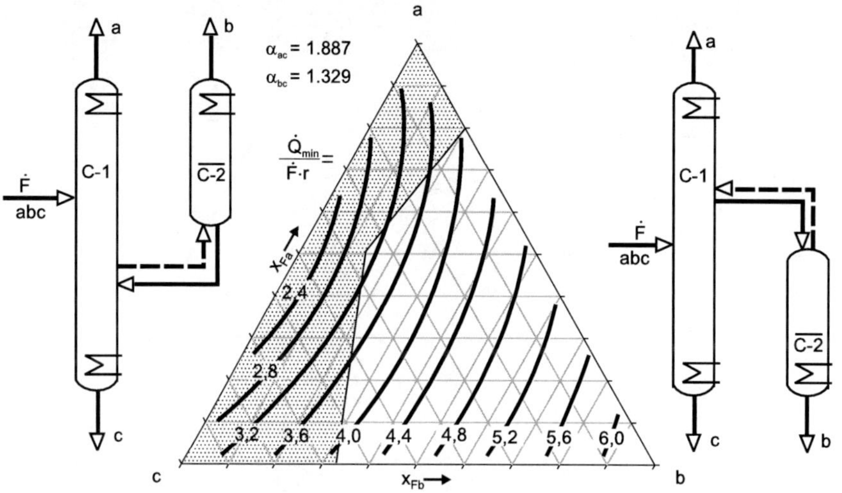

Fig. 11.2-8 Flow sheet of the a-path (*left*) and the c-path (*right*) with side columns. The energy demand of both processes is the same. In the *shaded concentration range* this process is superior to the a/c-path with direct column coupling

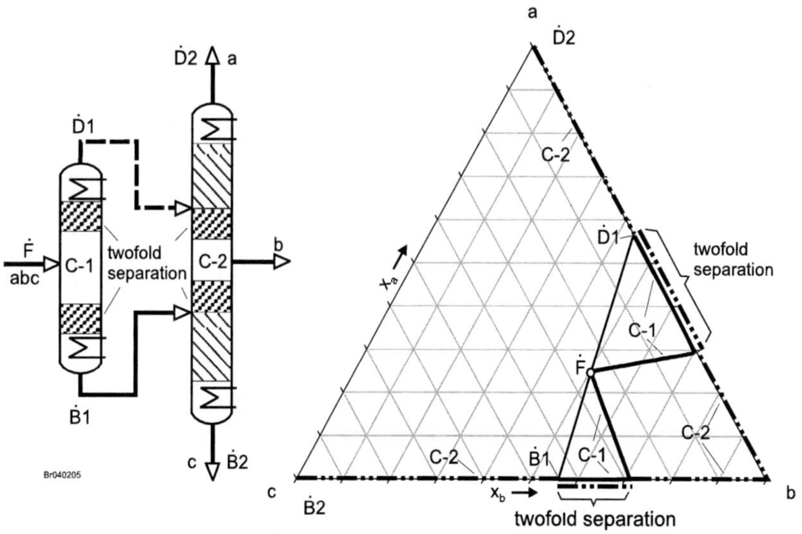

Fig. 11.2-9 Flow sheet and internal concentration profile of the a/c-path

The energy demand of the a/c-path with side column is shown in Fig. 11.2-11. The parameter lines of constant energy demand $\dot{Q}_{\min}/(\dot{F} \cdot r)$ are a strong function of the feed concentration x_{Fi}. At high contents of the low boiler *a* in the feed the energy demand is identical with the energy demand for separating a pure low boiler

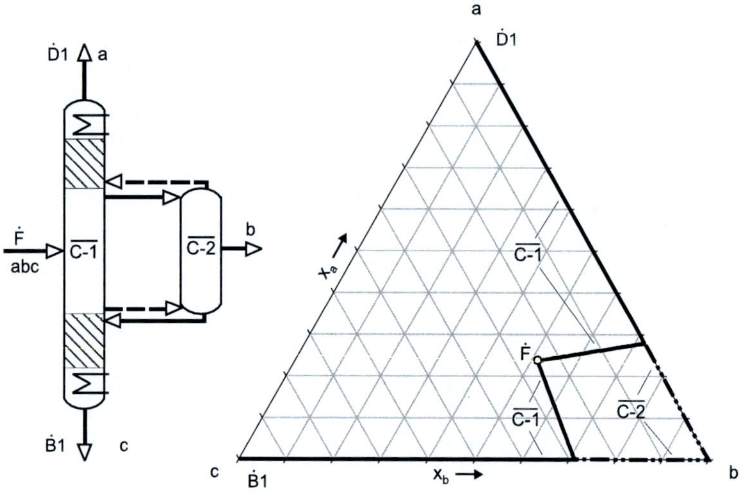

Fig. 11.2-10 Flow sheet and internal concentration profile of the a/c-path with side column

from a ternary mixture, see Fig. 5.2-31 in Chap. 5. Therefore, no energy is required for the fractionation of the resulting binary mixture *b–c*.

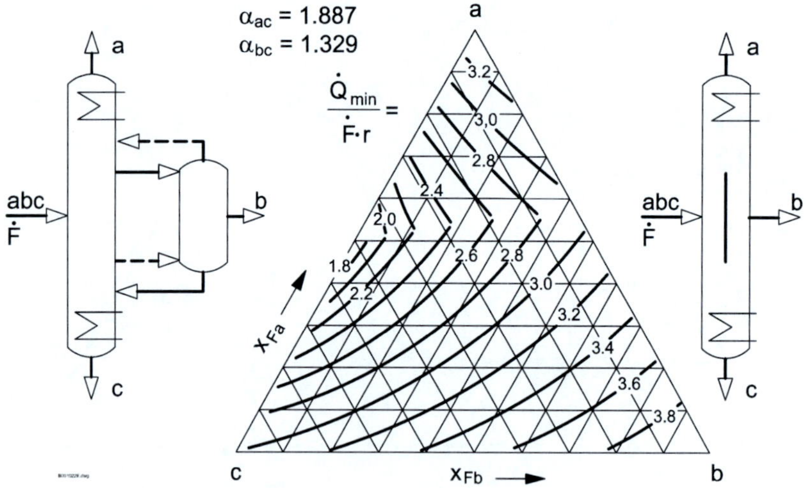

Fig. 11.2-11 Flow sheet and energy demand of the a/c-path with side column

The same holds for low concentrations of the low boiler in the feed. Here, the energy demand of the a/c-path with side column is identical with the energy demand for removing the high boiler *c* alone from a ternary mixture, see Fig. 5.2-32 in Chap. 5. This is a very import fact that demonstrates the effectiveness of the a/c-path with side column. In particular at high concentrations of the intermediate boiler *b* this process is by far better than all the other processes discussed so far.

Side columns are used, for instance, in the most important distillation processes worldwide, the fractionation of air (see Fig. 11.2-18) and the distillation of crude oil (Meyers 1996). The atmospheric tower of oil refineries consists of a main column and four stripping side columns (Fig. 11.2-12). In this tower the crude oil is split into six fractions which are processed further in several subsequent columns. Oil refineries also have some other interesting features. Steam is fed into the bottom of the main column and most of the side columns. This causes a stripping effect and reduces the temperatures in the columns (steam distillation). The overhead fractions of all side columns are fed into the main column thus increasing the vapor flow there. So-called pump arounds effect a partial condensation of the vapor in the main column and, in turn, a reduction of the vapor flow rates in the upper sections.

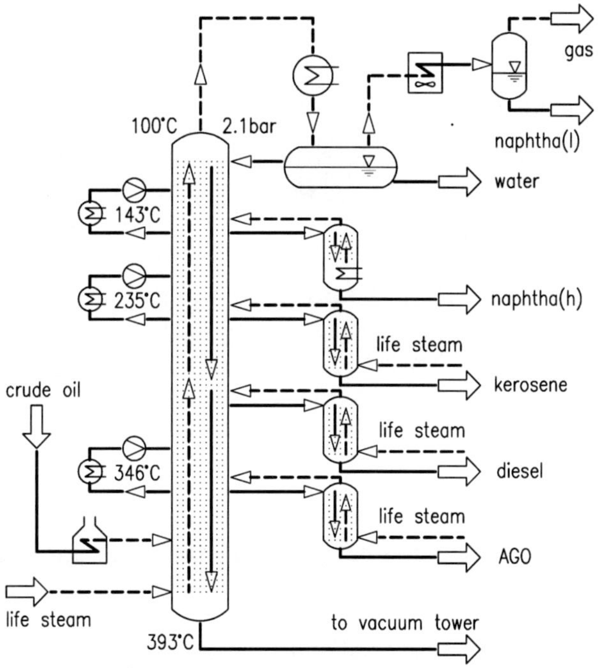

Fig. 11.2-12 Atmospheric tower of crude oil refining

11.2.2.4 Divided Wall Columns

In processes with side columns the gas load is reduced in that section of the main column where the side column is operated in parallel. Since column diameter is primarily determined by the gas load (see Chap. 5) the main column should have a smaller diameter in the sections with reduced gas load. A better choice, however, is the integration of the side column into the main column (Kaibel 1987).

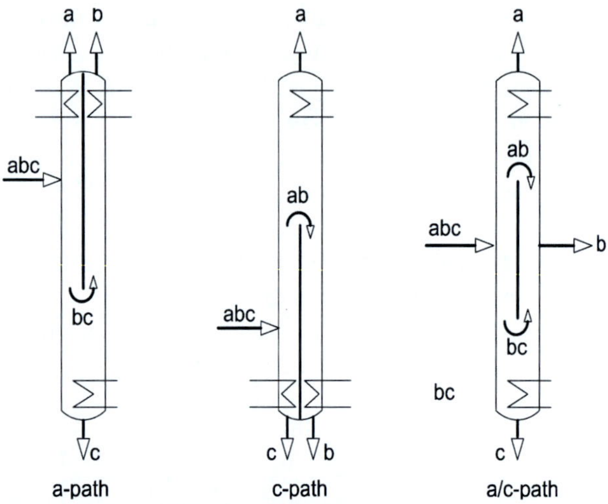

Fig. 11.2-13 Three examples of divided wall columns

This is the principle idea behind the divided wall columns shown in Fig. 11.2-13. From a process point of view, divided wall columns are identical with a combination of a main and a side column. Divided wall columns have the same energy demand as the corresponding processes depicted in Figs. 11.2-8 and 11.2-11, respectively. They are just an investment cost-saving modification of the equipment. The biggest advantage offers the realization of the a/c-path in a divided wall column. Here, the energy demand is lowest in the full range of feed concentration and only two heat exchangers (one reboiler, one condenser) are required. Divided wall columns have been increasingly used in recent years in the process industry (Kaibel et al. 2003).

11.2.3 Processes with Indirect (Thermal) Column Coupling

A further measure for process intensification is the indirect (or thermal) coupling of columns. In principle, waste heat of the process is internally used to cover the heat demand elsewhere. The prerequisites of this heat matching are appropriate amount and temperature level of the heats. In evaporators and condensers, the temperature level can be adjusted by variations of operating pressures. However, as the vapor pressure is, in a first approximation, an exponential function of temperature, rather large pressure changes are necessary for adjusting temperature levels.

11.2.3.1 Multistage Flash Process

A good example for thermal coupling, i.e., internal heat matching, is the MSF (multiple stage flash) process for seawater desalination (Fig. 11.2-14) . This process typically consists of 18 single distillation stages (Greig 1987). At each stage, a flash evaporation is effected by a stepwise pressure reduction from 0.6 to 0.07 bar. The heat of evaporation is covered by down cooling of the mixture itself. In each stage, the vapor is condensed against fresh seawater that is heated up to temperatures as high as 83°C. By this internal heat exchange most of the required heat is covered. Only 1/7 of the energy demand has to be provided by external heat. This makes the process very cost-effective. More than 13,000 desalination units are installed worldwide with very large capacities of up to 300,000 m^3 potable water per day (El Saie and El Kafrawi 1989).

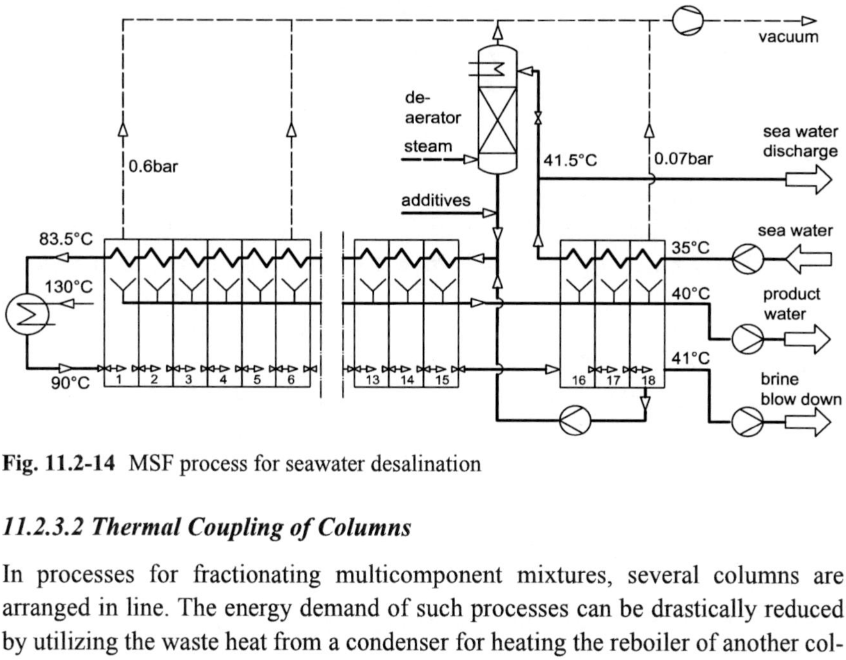

Fig. 11.2-14 MSF process for seawater desalination

11.2.3.2 Thermal Coupling of Columns

In processes for fractionating multicomponent mixtures, several columns are arranged in line. The energy demand of such processes can be drastically reduced by utilizing the waste heat from a condenser for heating the reboiler of another column.

Figure 11.2-15 illustrates the flow sheet and the energy demand of the a-path with thermal column coupling. The heat from the condenser of column C-2 is transferred to the reboiler of column C-1. Therefore, the external heat supplied to column C-1 is utilized twice in the process. For adjusting the temperature level, column C-2 has to be operated at a higher pressure than column C-1. The parameter lines of constant energy demand $\dot{Q}_{\min}/(\dot{F} \cdot r)$ presented in the triangular diagram

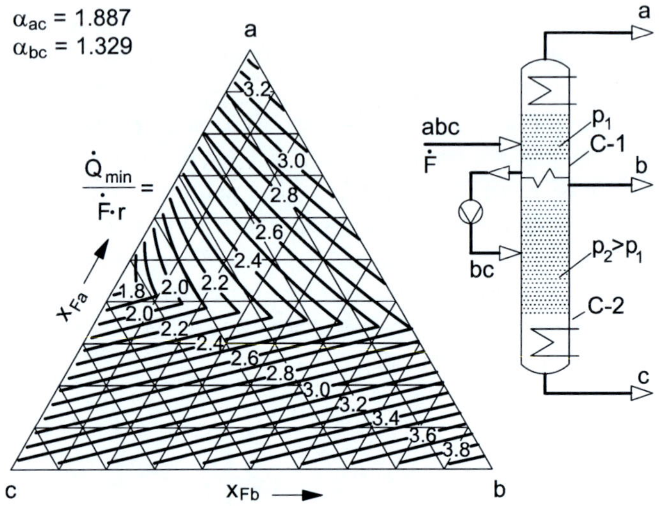

Fig. 11.2-15 Flow sheet and energy demand of the a-path with thermal coupling

of Fig. 11.2-15 show a significantly different shape and orientation at low and high concentrations of the low boiler a in the feed, respectively. At high contents of low boiler the separation of the low boiler from the mixture is decisive for the energy demand of the whole process (see Fig. 5.2-31). At low values of $x_{F\mathrm{a}}$ the energy demand of the integrated process is governed by fractionating the binary mixture in column C-2, which is a linear function of feed concentration. Thus, column C-1 is operated with a surplus of energy thus reducing the number of equilibrium stages required for the separation.

The process structure and energy demand of the thermally coupled c-path are illustrated in Fig. 11.2-16. Here, the column C-1 is operated at a higher pressure and the waste heat of this column is supplied to the reboiler of column C-2. As can be seen from the parameter lines in the triangular diagram, the fractionation of the ternary mixture (characterized by nonlinear parameter lines, see Chap. 5) in column C-1 is decisive for the energy demand of the integrated process at low values of $x_{F\mathrm{a}}$ (see Fig. 5.2-32).

Figure 11.2-17 shows the flow sheet and the energy demand of the thermally integrated a/c-path. This path has the lowest energy demand of all processes for fractionating ternary mixtures. It is important to note that the energy demand is lowest at intermediate concentrations of the low boiler in the feed. Thus, it is easier to fractionate a mixture with 50% than with 90% low boiler in the feed. This is explained by the fact that in the intermediate concentration range the energy

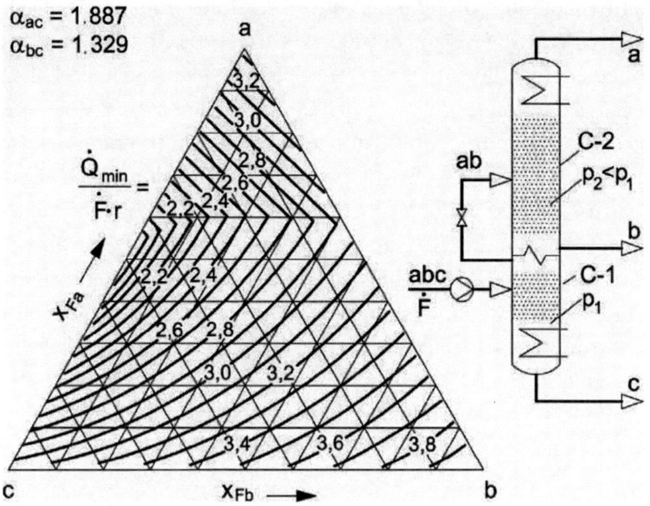

Fig. 11.2-16 Flow sheet and energy demand of the c-path with thermal coupling

demand of each column is nearly the same and, in turn, the energy saving by column coupling is the highest (up to 50%).

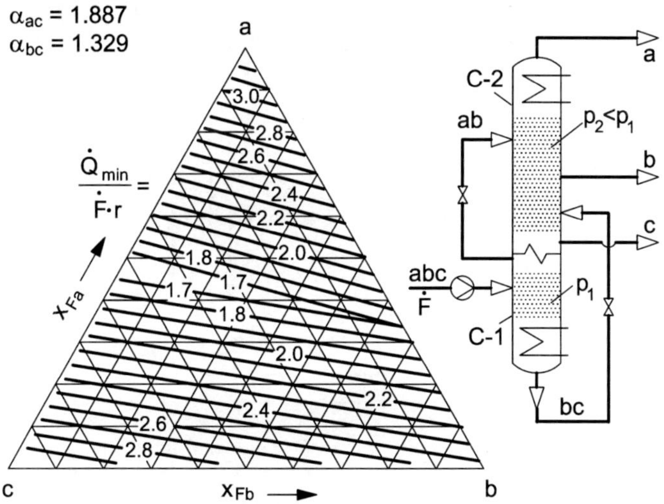

Fig. 11.2-17 Flow sheet and energy demand of the a/c-path with thermal coupling

11.2.3.3 Pinch Technology

Thermal coupling of columns is often used in industrial processes since it effectively reduces the energy demand by multiple use of the external heat supplied. An interesting example is the Linde Process for air separation (Fig. 11.2-18). In this process thermal coupling of columns as well as a side column are used. Since the process requires very low temperatures, the fresh air is first cooled down and partially liquefied against the cold products in the heat exchangers E-1, E-2, and E-3. In column C-1 the reflux (pure nitrogen) for the main column C-2 is provided. The fractionation of air into nitrogen (distillate) and oxygen (bottoms) is performed in the main column C-2 which is thermally coupled with column C-1. From the stripping section of the main column an argon-rich (up to 8% argon) vapor stream is withdrawn and fed into a rectifying side column for argon recovery. The system argon/oxygen is a very close boiling one. Thus, up to 150 equilibrium stages are necessary to get pure argon. The cooling of the condenser is established by the flashed bottom fraction of column C-1. No external heat is supplied to the process. The energy required for the fractionation is covered by the pressure difference of feed and products (Joule–Thomson effect). Modern units use an expansion machine which is more efficient since it draws some energy off the system (Baldus et al. 1983).

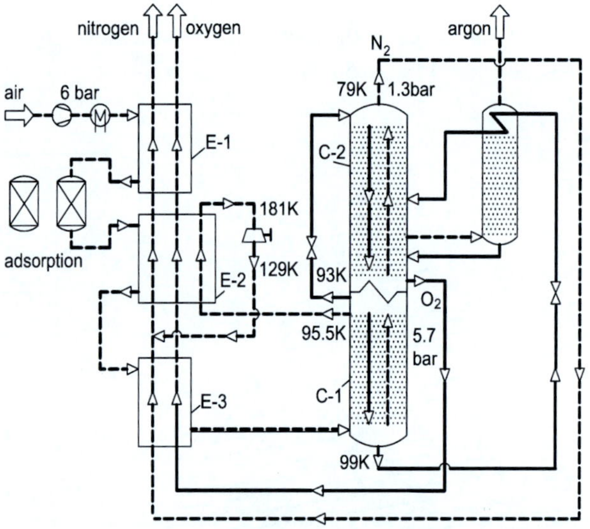

Fig. 11.2-18 Simplified Linde process for air separation and argon recovery

The design of processes with thermal coupling is best performed with the help of effective methods for heat matching. Well known and often applied is the so-called Pinch Technology originally developed for the design of cryogenic separation processes like air separation. Its application to hot processes has been greatly promoted by Linnhof and coworkers (e.g., Linnhoff 1983; Linnhoff and Dhole 1983; Linnhoff and Sahdev 1988).

11.3 Processes for Separating Azeotropic Mixtures

Azeotropes form a barrier to distillation as the concentrations of vapor and liquid are the same at the azeotropic point. Thus, no driving force for interfacial mass transfer exists at the azeotrope. Fractionation of azeotropic mixtures is only possible if either the concentration of the azeotrope is changed by any means, or alternative separation processes, which can overcome azeotropes, are combined with distillation.

11.3.1 Fractionation of Mixtures with Heteroazeotropes

Mixtures with miscibility gaps in the liquid phase very often – but not always – form azeotropes within the miscibility gap. Since miscibility gaps decrease the boiling point of mixtures (see Chap. 5), such heteroazeotropes are always low boiling azeotropes (minimum azeotropes).

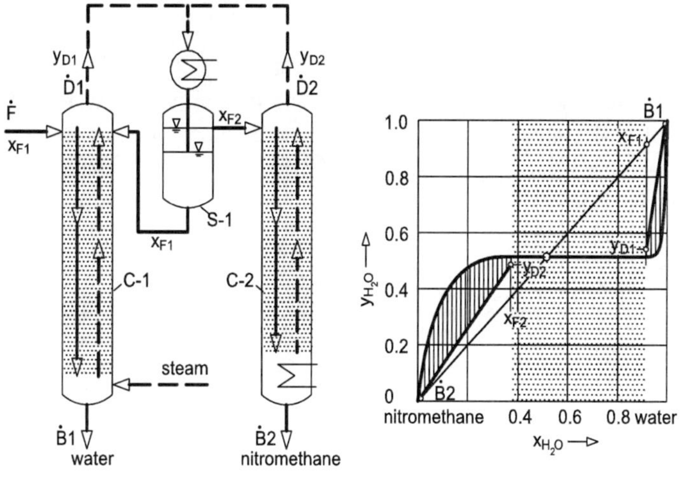

Fig. 11.3-1 Process for fractionating binary mixtures with heteroazeotrope

Processes for completely fractionating binary mixtures with heteroazeotropes consist of two distillation columns and one decanter (Fig. 11.3-1). As the azeotrope lies within the miscibility gap of the liquid the azeotrope can be broken by decantation. The two fractions from the decanter are at different sides of the azeotropic point. Purification of these two rather impure fractions is performed by distillation. The pure products are recovered as bottoms from the distillation columns C-1 and C-2.

The process shown in Fig. 11.3-1 utilizes the specific advantages of distillation and decantation. The advantage of distillation is the ability to produce pure fractions. However, distillation cannot break azeotropes. On the other hand, decantation can break heteroazeotropes but cannot produce pure products. Thus, the combination of distillation and decantation is a very effective process for fractionating mixtures with heteroazeotropes. Such processes are extensively used in industry. Table 11.3-1 lists some important binary mixtures fractionated by the process shown in Fig. 11.3-1.

Table 11.3-1 Systems processed by distillation and decantation

Toluene/water	Ethyl propyl ether/water
Benzene/water	Heptane/water
Chloroform/water	Butanol/water
Dichloromethane/water	Ethyl acetate/water
Butyl alcohol/water	Nitromethane/water

The combination of distillation and decantation can also be applied to multicomponent mixtures. A process for the separation of the ternary mixture acetone/water/1-butanol is depicted in Fig. 11.3-2. One organic compound (acetone) is miscible with water, the other one (1-butanol) is inmiscible. The mixture water/1-butanol exhibits a heteroazeotrope. A boundary distillation line runs from the heteroazeotrope to the low boiler acetone.

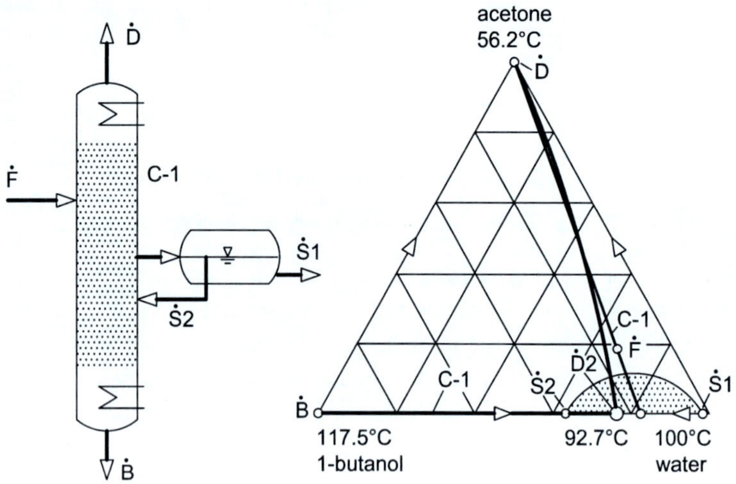

Fig. 11.3-2 Process for fractionating mixtures of water and two organic compounds (miscible and inmiscible)

From the ternary feed pure acetone is recovered as overhead fraction. Below the feed point the liquid within the column enriches in 1-butanol and, in turn, splits into two liquid phases. The aqueous phase with very low 1-butanol content is removed in a decanter, which is externally arranged. The organic phase is recycled into the column to recover pure 1-butanol as bottoms.

The process shown in Fig. 11.3-2 can always be applied when several organic compounds (some miscible, some immiscible) have to be removed from water.

11.3.2 Pressure Swing Distillation

A well-proven technique for separating azeotropic mixtures is the so-called pressure swing distillation. This process is applied to systems where the concentration of the homogeneous azeotrope strongly depends on pressure. The whole process consists of two distillation columns having different operating pressures (Fig. 11.3-3). In binary systems with a minimum azeotrope, the pure products are recovered as bottom fractions. The two overhead fractions having azeotropic concentrations are recycled into the other column. At a higher (or lower) pressure, the recycles are no longer at the azeotropic point and can, in turn, be further processed there.

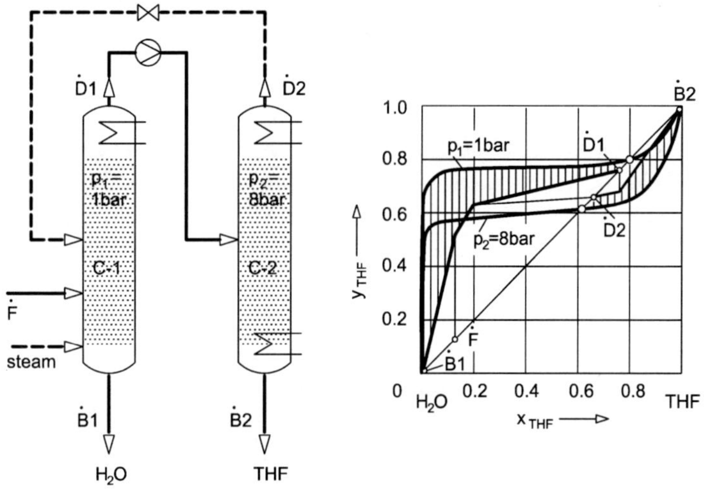

Fig. 11.3-3 Pressure swing distillation of the azeotropic system water/tetrahydrofuran

Table 11.3-2 lists some azeotropic systems that are separated by pressure swing distillation on industrial scale. It is important to note that this process can be applied to systems with minimum as well as maximum azeotropes. However, systems with minimum azeotropes are more sensitive to variations of the operating pressure. Therefore, pressure swing distillation is preferentially applied to systems with minimum azeotropes. In processes with maximum azeotropes (e.g., hydrogen chloride/water) there is a risk of accumulation of high boiling impurities in the internal recycle streams that makes the operation of the process more difficult.

Table 11.3-2 Systems processed by pressure swing distillation

Methyl ethyl ketone/water (1 bar/7 bar)	Water/2-butanone
Methanol/acetone (1 bar/0.25 bar)	Water/ isobutyl alcohol
Methanol/methyl ethyl ketone (1 bar/7.6 bar)	Ethanol/2-pentanone
Tetrahydrofuran/water (1 bar/7.6 bar)	Methanol/2-butanone
Ethanol/water	Hydrogen chloride/water (0.1 bar/6 bar)

11.3.3 Processes with Entrainer

The addition of an external substance (called *entrainer* e) is a very effective means for fractionating azeotropic mixtures by distillation. In the multicomponent mixture generated by the addition of the entrainer the azeotrope is circumvented by distillation. This principle is explained in Fig. 11.3-4 at the example of a binary mixture *a–b* having a minimum azeotrope.

If the entrainer e has a lower boiling point than the minimum azeotrope there exists a boundary distillation line running from the minimum azeotrope to the entrainer e. The boundary distillation line divides the ternary mixture into two distillation fields having different origins of distillation lines and, in turn, different feasible products (see Chap. 5). In Fig. 11.3-4 the low boiler rich feed $\dot{F}$ lies in that distillation field where the low boiling constituent a can be recovered as bottoms of distillation column C-1. The overhead fraction $\dot{D}1$ with (nearly) azeotropic concentration cannot be separated by simple distillation. Therefore, fraction $\dot{D}1$ is mixed with the entrainer e (or with an entrainer-rich fraction) yielding a mixture $\dot{M}2$ that lies in the other distillation field. The mixture $\dot{M}2$ is fractionated into the high boiler b (bottoms $\dot{B}2$) and the overhead fraction $\dot{D}2$, which is close by the boundary distillation line. If the boundary distillation line is curved in a sufficient degree, fraction $\dot{D}2$ can be separated in column C-3 into an entrainer-rich overhead fraction $\dot{D}3$ and a bottom fraction $\dot{B}3$ that lies in the same distillation field as the feed. Thus, the fraction $\dot{B}3$ can be recycled into column C-1.

The process shown in Fig. 11.3-4 consists of three distillation columns and two recycles. A material balance around columns C-2 and C-3 reveals that the fraction $\dot{D}1$ is split into the fraction $\dot{B}2$ (pure high boiler b) and the recycle fraction $\dot{B}3$.

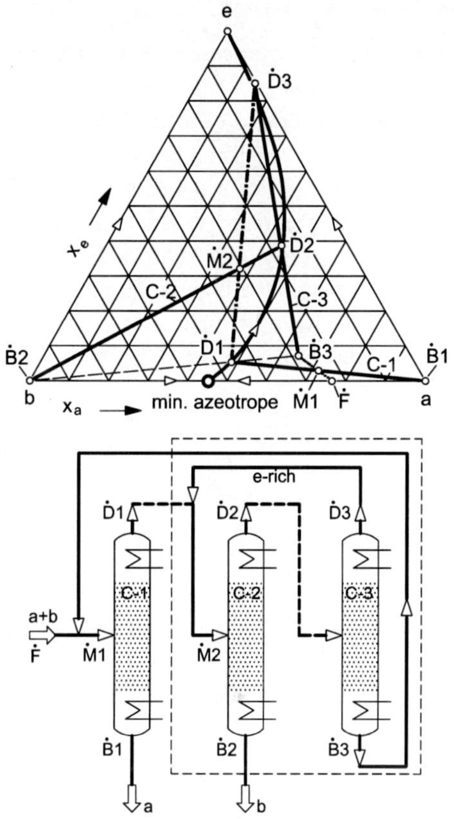

Fig. 11.3-4 Process for fractionating a binary mixture with minimum azeotrope using a low boiling entrainer *e* (Stichlmair and Fair 1998)

Thus, the states of fractions $\dot{D}1$, $\dot{B}2$, and $\dot{B}3$ have to lie on a straight line (dashed line in the concentration diagram of Fig. 11.3-4).

An important prerequisite of the process shown in Fig. 11.3-4 is that the amount of fraction $\dot{D}2$, which has to be fractionated further in column C-3 into the recycle fractions $\dot{D}3$ and $\dot{B}3$, is small. This condition requires that the distance between points $\dot{D}2$ and $\dot{M}2$ in the concentration space is as large as possible. Hence, the boundary distillation line has to be strongly curved and the position of mixing point $\dot{M}2$ has to be in the middle of the concentration space (Stichlmair and Fair 1998).

The process of Fig. 11.3-4 can also be applied to mixtures with maximum azeotropes. Here, a top side down version of the flow sheet has to be used.

11.3.3.1 Criteria for Entrainer Selection

Decisive for the effectiveness of the process shown in Fig. 11.3-4 is a proper selection of the entrainer e. In principle, the entrainer has to be chosen so that the substances a and b are origins or termini of distillation lines in the ternary mixture (Chap. 5). Six different situations are illustrated in Fig. 11.3-5 (Stichlmair and Fair 1998). According to these graphs, the separation of a mixture with a minimum azeotrope requires a low boiling entrainer or an entrainer that forms new low boiling azeotropes with the mixture (see also Table 11.3-3). Separation of a mixture with a maximum azeotrope requires a high boiling entrainer or an entrainer that forms high boiling azeotropes with the mixture at hand. Thus, only azeotropes of the same type can exist in such processes (Stichlmair and Fair 1998).

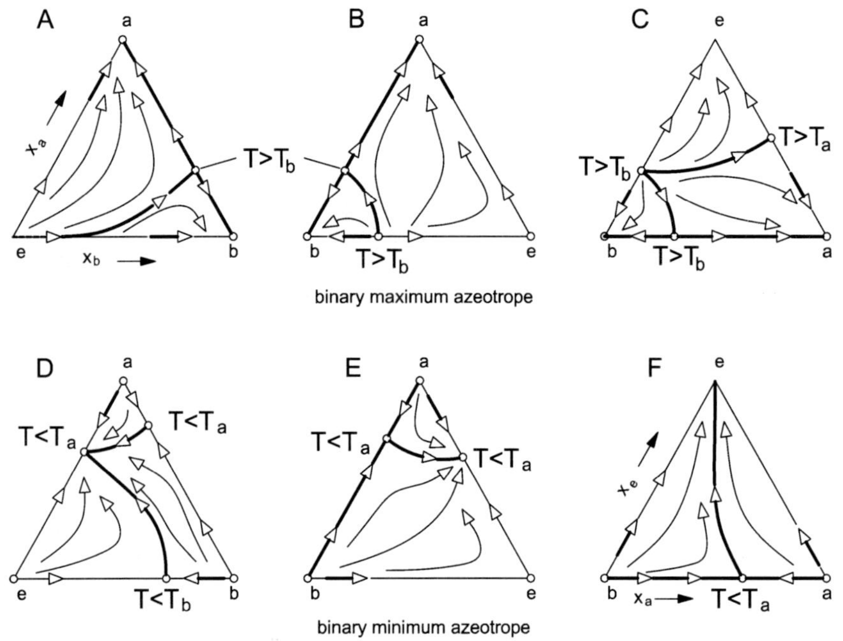

Fig. 11.3-5 Criteria for entrainer selection (Stichlmair and Fair 1998)

11.3.3.2 Process Simplification

In special cases, the general process consisting of three distillation columns can be simplified to a process with two columns only. An example of such a process is the fractionation of hydrochloric acid with the entrainer sulfuric acid (Stichlmair and Fair 1998). The boundary distillation line in Fig. 11.3-6 runs along the binary side of the triangular diagram at high concentration of sulfuric acid. Therefore, the

Table 11.3-3 Criteria for entrainer selection

Type of azeorope	Required properties of the entrainer
Maximum azeo-trope	High boiler Intermediate boiler that forms a maximum azeotrope with the high boiling component of the feed Low boiler that forms maximum azeotropes with the low boiling and the high boiling components of the feed mixture
Minimum azeo-trope	Low boiler Intermediate boiler that forms a minimum azeotrope with the low boiling component of the feed High boiler that forms minimum azeotropes with the low boiling and the high boiling components of the feed mixture

entrainer-rich bottom fraction of column C-1 is a binary and not a ternary mixture. In those situations column C-3 of the general process can be discarded.

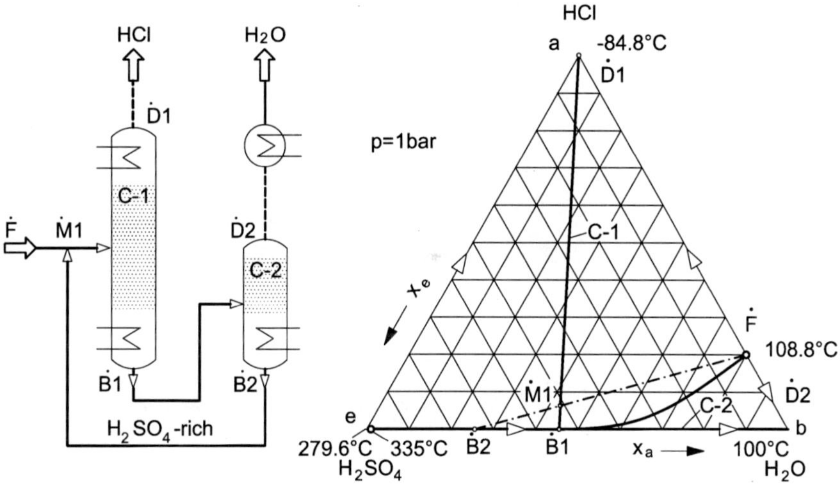

Fig. 11.3-6 Process for fractionating hydrogen chloride with the entrainer sulfuric acid

11.4 Hybrid Processes

Hybrid processes are combinations of distillation with alternative separation processes. These processes make use of the advantages of distillation (i.e., fractionation into pure substances) as well as of alternative processes (i.e., breaking azeotropes). In principle, all alternative separation techniques can be combined with distillation (Schweitzer 1997). Especially well suited are decantation, absorption, extraction, stripping, adsorption, and membrane permeation. In most cases hybrid processes consist of two distillation columns and one alternative separation unit.

11.4.1 Azeotropic Distillation

The combination of distillation and decantation is used in the well-known azeotropic distillation process shown in Fig. 11.4-1. The separation of the homogeneous azeotropic feed (e.g., ethanol/water) is achieved by the admixture of an entrainer (e.g., toluene) which forms a large mixing gap with one of the feed components (here water). In column C-1, water is recovered as bottoms $\dot{B}1$. The azeotropic overhead fraction $\dot{D}1$ is mixed with the fraction $\dot{S}2$ (toluene-rich) from the decanter. In column C-2 pure ethanol is recovered as bottoms $\dot{B}2$. The overhead fraction $\dot{D}2$ is condensed, subcooled, and split into the fractions $\dot{S}1$ (water-rich) and $\dot{S}2$ (toluene-rich) in the decanter. Both fractions are recycled within the pro-

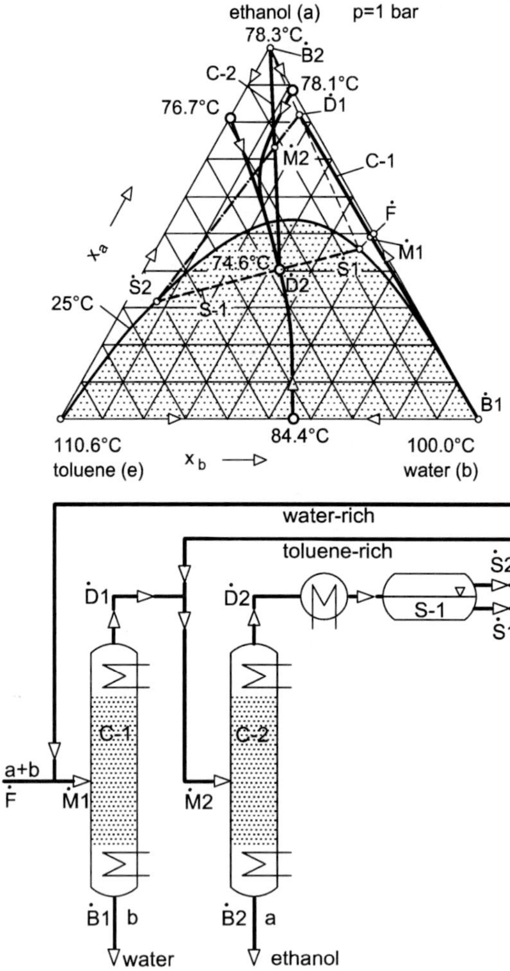

Fig. 11.4-1 Azeotropic distillation

cess. Azeotropic distillation is a very important process often used in industry. Some example systems are listed in Table 11.4-1. However, this process can only be applied to systems with a minimum azeotrope.

Table 11.4-1 Systems processed by azeotropic distillation

Mixture	Entrainer	Mixture	Entrainer
Water/ethanol	Benzene	Acetic acid/formic acid	Chloroform
	Toluene	Water/pyridine	Benzene
	Pentane		Toluene
	Trichloroethylene	Water/acetic acid	Butyl acetate
	Cyclohexane		Propyl acetate
	Ethy lacetate	Water/propanol	Benzene
	Ethy lether	Benzene/cyclohexane	Acetone

11.4.2 Extractive Distillation

A process with comparable industrial importance is extractive distillation. Here, two distillation columns are combined with an absorption column. This process needs an entrainer (absorbent) that selectively absorbs one of the two feed components. In aqueous systems, for instance, the entrainer should be a hygroscopic liquid, e.g., ethylene glycol (Fig. 11.4-2).

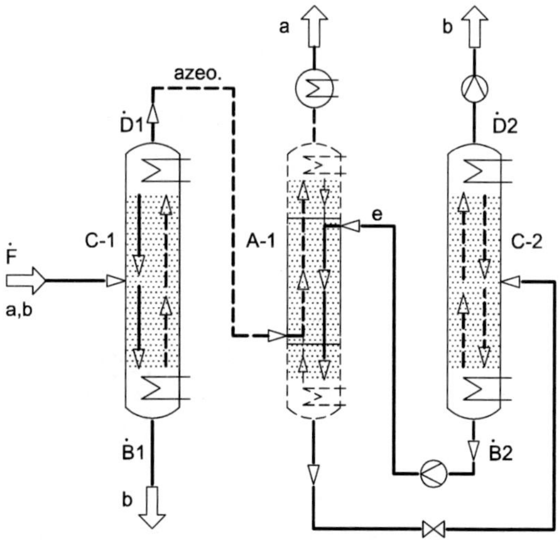

Fig. 11.4-2 Extractive distillation

In a first distillation column C-1 the high boiler (i.e., water) is recovered as bottoms $\dot{B}1$. The azeotropic overhead fraction $\dot{D}1$ is fed (in the vapor state) into the absorber A-1 for removal of the residue of the high boiler (water). The pure low boiler (ethanol) is recovered at the top of the absorber. The loaded absorbent is regenerated in distillation column C-2 for recycling.

To remove traces of the absorbent always present in the overhead product of the absorber, some of the ethanol is recycled as reflux at the top of the absorber A-1. Analogously, a small part of the loaded absorbent is boiled up at the bottom of A-1 to remove most of the coabsorbed ethanol. However, reboil and reflux rates are rather small. Thus, the countercurrent flow within the column A-1 is dominated by the vapor feed (low boiling fraction) at the bottom and the liquid feed (high boiling fraction) at the top of the column what is characteristic of absorption columns. Thus, the process should better be called absorptive distillation instead of extractive distillation.

Table 11.4-2 lists some important systems often processed by extractive distillation. This process is only suited for separating mixtures with minimum azeotropes. However, it is often also used for separating close boiling zeotropic mixtures.

Table 11.4-2 Systems processed by extractive distillation

Mixture	Entrainer	Mixture	Entrainer
Butane/butadiene	Furfural	Butene/isoprene	Dimethyl formamide
	Acetonitrile	Acetone/methanol	Water
	Dimethyl acetamid	Chloroform/methanol	Water
	n-Methyl pyrroliedone	Tetrahydrofuran/H_2O	Dimethyle formamide
Benzene/cyclohexane	Aniline	Ethanol/water	Ethylene glycol
	n-Methyl pyrrolidone		Aqueous salt solutions
	n-Formyl morpholine	Propylene/propane	Acrylonitrile

11.4.3 Processes Combining Distillation and Extraction

A process, which combines distillation with solvent extraction, is presented in Fig. 11.4-3. Such processes are used in the process industry for the regeneration of solvents (e.g., tetrahydrofuran, THF) diluted by water (Schoenmakers 1984). The system tetrahydrofuran/water forms a minimum azeotrope at approximately 80 mol% THF. Most of the water is removed as bottoms in distillation column C-1. The overhead fraction $\dot{D}1$ is fed into the extractor for removal of the residual water by solvent extraction with concentrated $NaOH/H_2O$. The diluted sodium fraction is regenerated in the single stage distillation unit D-1. If the specified concentration of tetrahydrofuran product is very high a further purification step has to be performed in column C-2 since not all water can be removed by extraction with aqueous sodium (Schoenmakers 1984).

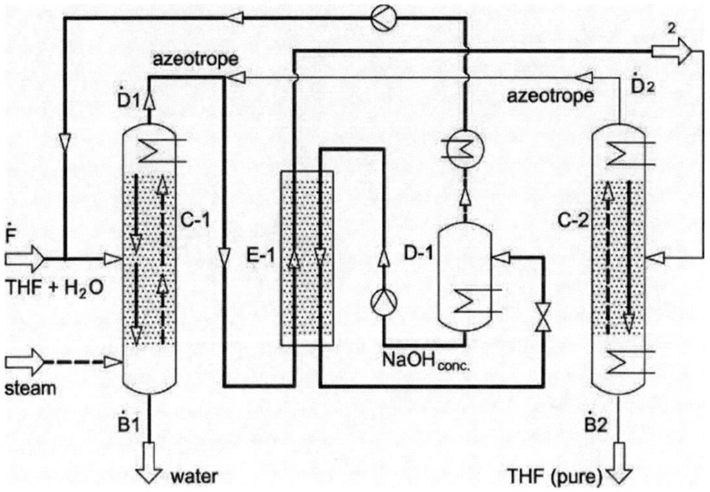

Fig. 11.4-3 Process for dewatering of tetrahydrofuran (THF) by distillation and solvent extraction

11.4.4 Processes Combining Distillation with Desorption

Distillation can also be combined with desorption (or stripping). An example of such a process is the fractionation of nitric acid with the entrainer sulfuric acid (Geriche 1973). Nitric acid has a maximum azeotrope at a concentration of 37 mol% HNO_3. Thus, a high boiling entrainer has to be used.

Water is removed as overhead fraction in column C-1 (Fig. 11.4-4). The bottom fraction $\dot{B}1$ with azeotropic concentration is fed into column C-2 to be mixed with the entrainer and further fractionation. Pure HNO_3 is recovered as overhead product $\dot{D}2$. The bottom fraction lying close by the boundary distillation line is stripped with steam to remove residual HNO_3. Thus, the boundary distillation line is crossed by the stripping step. The diluted sulfuric acid $\dot{B}2$ is fed into a single stage distillation unit for removal of most of the water. Nitric acid, when boiled, always forms some NO_x that colors the liquid. Therefore, NO_x is removed by further stripping with air to get a clear liquid.

11.4.5 Processes Combining Distillation with Adsorption

In some processes distillation is combined with adsorption (Westphal 1987). Such a hybrid process can be applied to the recovery of organics from aqueous mixtures. Most of the water is recovered as bottoms in column C-1 (Fig. 11.4-5). The over-

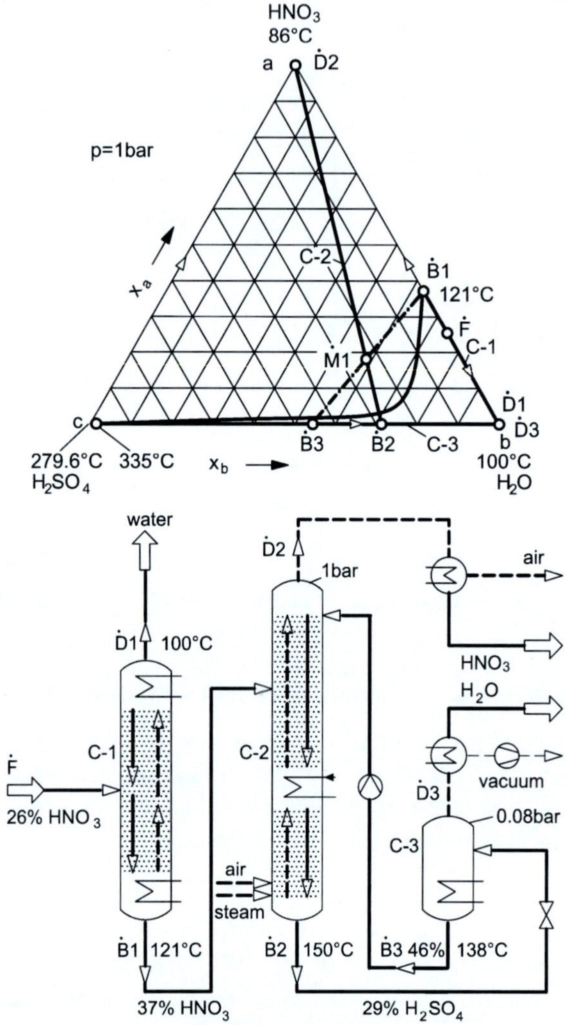

Fig. 11.4-4 Process for concentrating diluted nitric acid with the entrainer sulfuric acid

head fraction with azeotropic concentration is fed, in vapor state, into the adsorber for removal of the residual water. Pure organics are withdrawn from the adsorber.

Adsorption in fixed beds is normally performed batchwise. Therefore, three adsorbers have to be installed to enable a quasicontinuous operation. The adsorbers are cyclically operated in the modes adsorption, regeneration, and pressure build up. Regeneration is best performed by stripping with pure organics at reduced pressure. The organics/water mixture generated in the regeneration mode is fed into column C-2 for further processing. The overhead fraction $\dot{D}2$ is used as reflux in

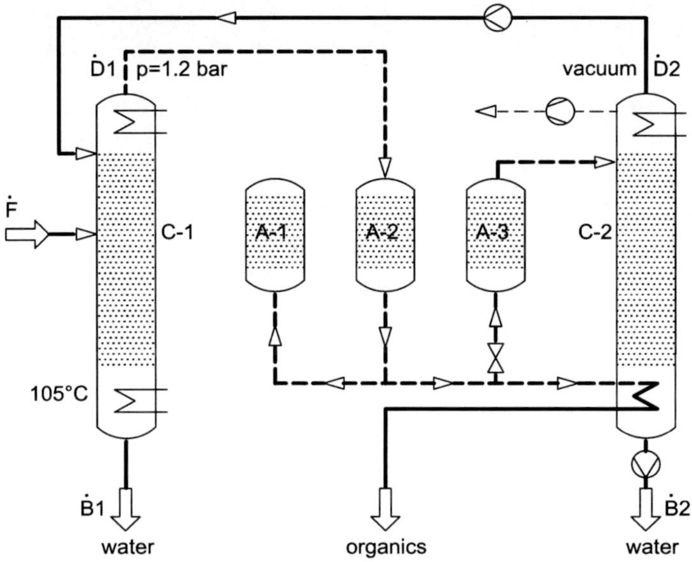

Fig. 11.4-5 Process for dewatering of organic compounds by distillation and adsorption

column C-1. In some cases, the latent heat of the vaporous organic product can be used for heating column C-2.

11.4.6 Processes Combining Distillation with Permeation

A modern process is the combination of distillation with membrane permeation, often called pervaporation. This technique is often applied to dewatering of organic compounds in industry (Rautenbach and Albrecht 1989).

Most of the water is removed as bottoms in column C-1 (Fig. 11.4-6). The liquid azeotropic overhead fraction is pressurized to approximately 4 bar and fed into a membrane stack (5–7 stages). Water preferably penetrates the membrane. Due to the high pressure difference the water-rich permeate is flashed in the membrane. After condensation it is recycled into column C-1. The retentate is very rich in organic compounds and, in turn, often meets the product specification. If not, the retentate is fed into column C-2 for further purification. Decisive for the process is the availability of efficient membranes with both a high capacity and a high selectivity.

The combination of distillation and membrane separation can, for instance, be applied to processes for the production of bioethanol. In the process presented in Fig. 11.4-7 (Weyd et al. 2010) the azeotropic overhead fraction of column C-1 is

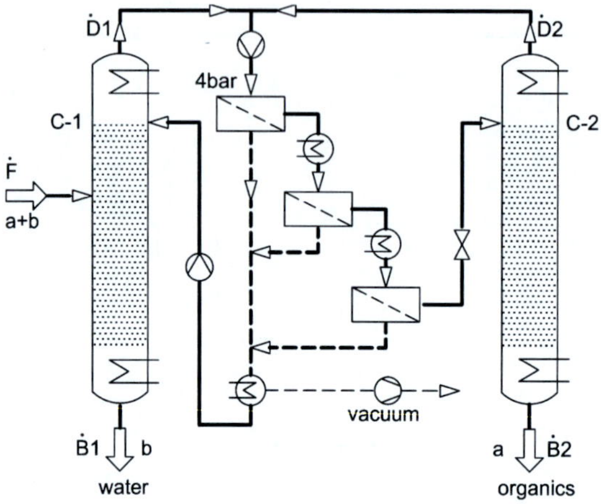

Fig. 11.4-6 Combination of distillation and membrane permeation

fed in vaporous state to the membrane stack. Favorable is a high pressure difference at the membrane. Therefore, the overhead fraction $\dot{D}1$ is compressed to approximately 8 bar what increases the dew point of the mixture to about 140°C. The pressure at the permeate side should be as low as possible. A value of 0.1 bar is a good choice since it allows the condensation of the permeate (water) by cooling water. Under these operating conditions the capacity of NaA-zeolite membranes is in the range of 10–12 kg/(m^2 h) (Weyd et al. 2010).

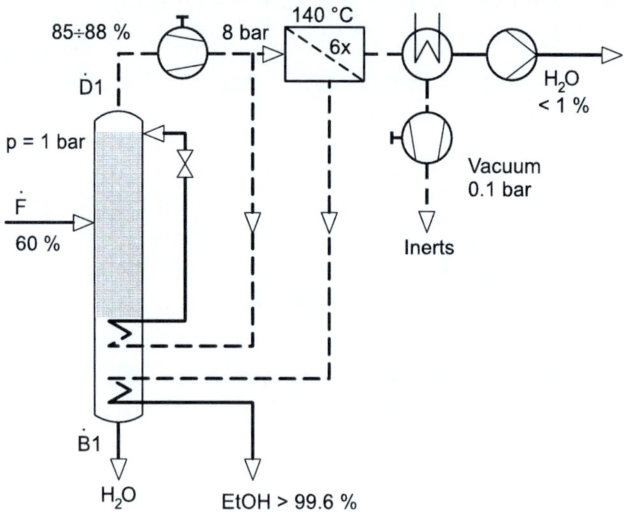

Fig. 11.4-7 Process for the production of bioethanol by distillation and vapor permeation (Weyd et al. 2010). Concentrations are wt% of ethanol

The high pressure of the overhead fraction $\dot{D}1$ offers the possibility of heating the column by condensation of most of the the overhead fraction (integrated heat pump). Just the permeate is condensed by cooling water. The energy required for the evaporation of this fraction of the feed is provided by the compressor engine. The energy demand of the process in Fig. 11.4-7 is much lower than the energy demand of all alternative processes discussed so far (azeotropic distillation, extractive distillation, distillation combined with adsorption, pervaporation, etc.). However, since a compressor driven by electric energy is required, the cost advantages are rather small.

The processes presented in this section very clearly demonstrate the fundamental principle of hybrid processes. The azeotrope is broken by an alternative separation process. The pure products, however, are recovered from distillation columns in most cases.

11.5 Reactive Distillation

Reactive distillation combines the unit operations reaction and distillation in a single apparatus. This fairly new technology has reached some importance in special fields of the process industry (Sundmacher and Kienle 2003).

Advantages of reactive distillation are:

- Total conversion even in reversible reactions
- Simplification of downstream processing
- Suppression of unwanted side reactions
- Energy reduction of the whole process

Disadvantages of reactive distillation are the lack of effective catalytic column internals, the risk of formation of reactive azeotropes, and the limitation to low operating pressures.

An important industrial application of reactive distillation is the production of methyl acetate from methanol and acetic acid. This system exhibits several azeotropes, which make the downstream processing of the products very difficult, see Fig. 11.5-1 (Siirola 1996). The conventional process consists of eight distillation columns, one extractor, and one decanter.

In this case, it is possible to replace a conventional process by a single reactive distillation column (Fig. 11.5-2) (Siirola 1996; Frey 2001) and a single stage distillation. Within the column there exists a rectifying section, a stripping section, an

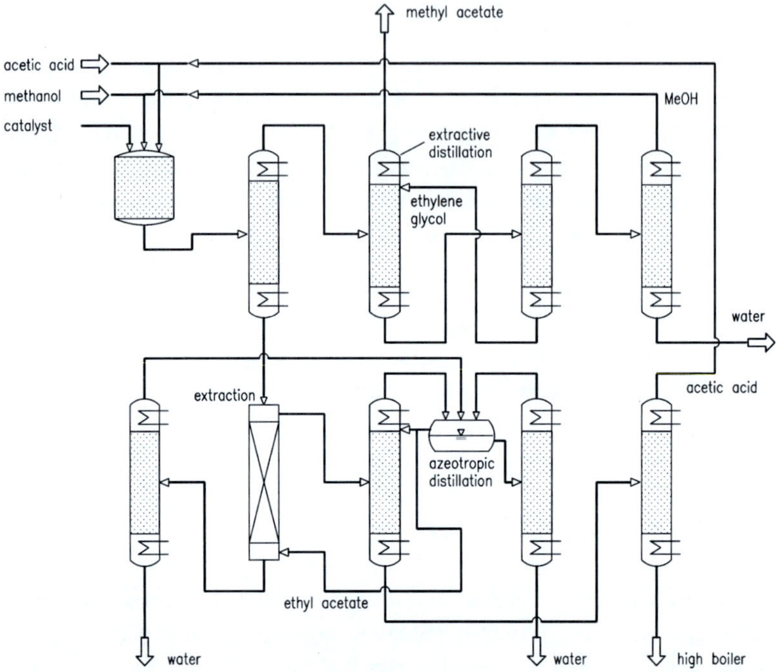

Fig. 11.5-1 Conventional process for the production of methyl acetate from methanol and acetic acid via reactive distillation

extraction section, and a reactive distillation section. Essential for the effectiveness of the process is that the reactants acetic acid and methanol are fed into the column at separate points. The high boiler acetic acid is fed above the reactive section, the low boiler methanol below that section. The process in the reactive section is, in essence, a chemical absorption with superimposed distillation of the reaction products. The single stage distillation effects the recovery of the homogeneous catalyst (sulfuric acid). This process step can be discarded when a heterogeneous catalyst (e.g., acidic ion exchanger) is used.

A comparison of the processes shown in Figs. 11.5-1 and 11.5-2 demonstrates the high potential of reactive distillation for process simplification. This type of processes is generally applicable to systems with reversible chemical reactions, e.g., to esterification and etherification of alcohols, to alkylations, to dimerization of olefins, and to hydrogenation of aromatics (Sundmacher and Kienle 2003).

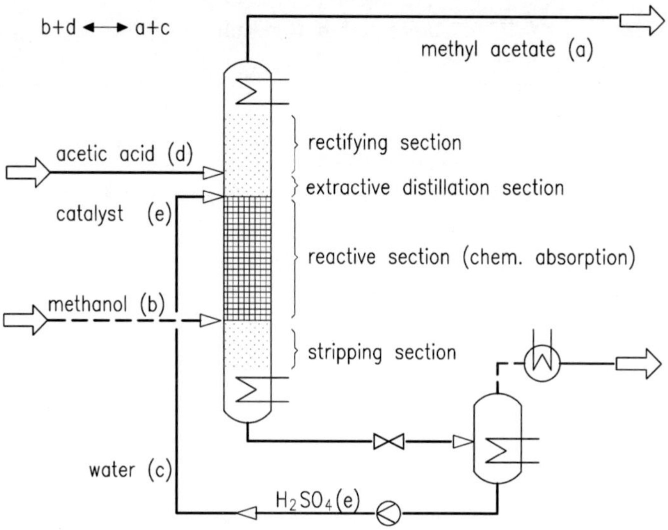

Fig. 11.5-2 Process for the production of methyl acetate from methanol and acetic acid via reactive distillation (Frey 2001)

Symbols

$A-1$		Absorber, adsorber
a		Compound (low boiling)
$\dot{B}$	kmol/s	Bottoms
b		Compound (intermediate boiling)
$C-1$		Column
c		Compound (high boiling)
$\dot{D}$	kmol/s	Distillate
$E-1$		Heat exchanger, extractor
e		Entrainer
$\dot{F}$	kmol/s	Feed flow
$\dot{G}$	kmol/s	Gas, vapor flow
$\dot{L}$	kmol/s	Liquid flow
$\dot{M}$	kmol/s	Mixture flow
$\dot{Q}$	kJ/s	Heat flow
p	bar	Pressure
p^0	bar	Vapor pressure
r	kJ/kmol	Heat of evaporation
$\dot{S}$		Fraction from decanter
$S-1$		Decanter
t	°C	Temperature
x		Mole fraction of liquid phase
x_g		Mass fraction of liquid phase

y	Mole fraction of gas phase
y_g	Mass fraction of gas phase

Greek Symbol

α	Relative volatility, $\alpha_{ac} \equiv p_a^{\ 0}/p_c^{\ 0}$

References

Chapter 1

Romm, J. J.: Der Wasserstoff Boom, Wiley-VCH, Weinheim 2006

Kreysa, G.: Methan – Chance für eine klimaverträgliche Energieversorgung, Chem.-Ing.-Techn. 80 (2008) 7, 901-908

Chapter 2

General

Baehr, H. D.: Thermodynamik, 7. Auflage, Springer (1989)

CRC Handbook of Chemistry and Physics 2010 – 2011; ed. by William M. Haynes, 91st ed. Taylor & Francis (2010)

D'Ans-Lax: Taschenbuch für Chemiker und Physiker, Bd 1: Makroskopische physikalisch-chemische Eigenschaften, 3. Auflage, Springer (1967)

DECHEMA Chemistry Data Series, Dechema Frankfurt, Bände 1-15

Gmehling, J.; Kolbe, B.: Thermodynamik, 2. Auflage, VCH Weinheim (1992)

Green, W. D.; Perry, R. H.: Perry's Chemical Engineer's. Handbook, 8th ed. New York McGraw-Hill (2008)

International Critical Tables Vol. IV, Mc-Graw-Hill, New York (1933)

Landolt-Börnstein: Zahlenwerte und Funktionen aus Physik, Chemie, Astronomie, Geophysik und Technik, 6. Auflage, Springer

Laskowski, J. S.; Ralston, J. (editors): Colloid Chemistry in Mineral Processing, Vol. 12, Elsevier (1992)

Linke, W. F.: Solubilities: Inorganic and metal-organic compounds; a compilation of solubility data from the periodical literature; a revision and continuation, origin. by Seidell, A. and Linke, W.F., Am.Chem.Soc. Washington (1958) Band 1 and (1965) Band 2

A. Mersmann et al., *Thermal Separation Technology: Principles, Methods, Process Design*, VDI-Buch, DOI 10.1007/978-3-642-12525-6,
© Springer-Verlag Berlin Heidelberg 2011

Mayinger, F.; Stephan, K.: Thermodynamik - Grundlagen und technische Anwendungen, Bd. 1 & 2, Springer (1992)

Poling, B. E.; Prausnitz, J. M.; O'Connell, J. P.: The properties of gases and liquids, 5. ed., McGraw-Hill (2007)

Prausnitz, J.; Lichtenthaler, R.; Gomes, E.: Molecular Thermodynamics of fluid-phase equilibria, PTR Prentice-Hall (1986)

Springer Materials – The Landolt-Börnstein Database: electronic resource, www.springermaterials.com

Stephan, P.; Schaber, K.; Stephan, K., Mayinger F.: Thermodynamik Vol. 1 (2009) and Vol. 2 (2010), Springer

VDI-Wärmeatlas: / Hrsg.: Verein Deutscher Ingenieure; VDI-Gesellschaft Verfahrenstechnik und Chemieingenieurwesen (GVC), 10. Auflage, Springer (2006) and VDI-Heat Atlas, Ed. VDI-Gesellschaft Verfahrenstechnik und Chemieingenieurwesen (GVC), 2. Edition (2010)

Walas, S. M.: Phase Equilibria in Chemical Engineering, Butterworth (1985)

Wagner, W.: Cryogenics, August 1973, 470-482

Special

Akgün, U.: Prediction of Adsorption Equilibria of Gases, PhD Thesis, Technical University Munich, Germany (2007)

Abrams, D. S.; Prausnitz, J. M.: Statistical Thermodynamics of Liquid Mixtures: A New Expression for the Excess Gibbs Energy of Partly or Completely Miscible Systems, AIChE J. 21 (1975), p. 116/128

Bosnjakovic, F.: Technische Thermodynamik Tl. 2, 4. Auflage, Steinkopff, Dresden (1965), p. 114

Brunauer, S.; Deming L. S.; Deming, E.; Teller, E.: On a Theory of the van der Waals Adsorption of Gases, J. Am. Chem. Soc. 62 (1940), p. 1723/1732

Cerofolini, G. F.: Adsorption and Surface Heterogeneity, Surface Science 24 (1971), p. 391/403

Cerofolini, G. F.: On the Choice of the Condensation Pressure in the Condensation Approximation, Surface Science 47 (1975), p. 469/476

Eiden, U.: Über die Einzel- und Mehrstoffadsorption organischer Dämpfe auf Aktivkohle, Ph.D. Thesis Universität Karlsruhe (1989)

Förg, W.; Stichlmair, J.: Die Gewinnung von Rein-Wasserstoff aus Gemischen mit leichten aliphatischen Kohlenwasserstoffen mit Hilfe tiefer Temperaturen, Kältetechnik-Klimatisierung, 21 (1969) 4, p. 104/110

Fredenslund, A.; Gmehling, J.; Rasmussen, P.: Vapor-liquid equilibria using UNIFAC, Elsevier, Amsterdam (1977)

Funakubo, E.; Nakada, S.: Jpn. Coal Tar (1950), p. 201

Hasselblatt, M.; Z. Phy. Chem. 83 (1913), p. 1

Hecht, G. et al.: Berechnung thermodynamischer Stoffwerte von Gasen und Flüssigkeiten, VEB Dt. Verl. f. Grundstoffind Leipzig (1966), p. 32 and p. 68.

Hougen, O. A.; Watson, K. M.; Ragatz, R. A.: Chemical process principles, 2nd ed., Vol. 2: Thermodynamics, Wiley, New York (1959)

Kaul, B. K.: Modern Vision of Volumetric Apparatus for Measuring Gas-Solid Equilibrium Data, Ind. Eng. Chem. Res. 26 (1987), p. 928/933

Margules, M.: Sitz.-Ber. Akad. Wiss. (Wien), math. naturwiss. Kl. 104 (1895) p. 1243, cit in (Schmidt et al. 1977), p. 144

Markmann, B.: Zur Vorhersagbarkeit Adsorptionsgleichgewichten mehrkomponentiger Gasgemische, Ph.D. Thesis, TU München (1999)

Matschke, D. E.; Thodos, G.: Vapor-Liquid Equilibria for the Ethane-Propane System, J. Chem. Eng. Data 7 (1962), No. 2, p. 232/234

Maurer, S.: Prediction of Single-Component Adsorption Equilibria, Ph.D. Thesis, TU München (2000) (Herbert Utz, Wissenschaft, München, 2000)

Mersmann, A.; Fill, B.; Maurer, S.: Single and Multicomponent Adsorption Equilibria of Gases, VDI Fortschrittsbericht Reihe 3 No. 735, VDI Verlag, Düsseldorf (2002)

Mersmann, A.; Akgün, U.: A Priori Prediction of Adsorption Equilibria of Arbitrary Gases on Microporous Adsorbents, VDI-Fortschrittsberichte Reihe 3, No. 906, VDI Verlag, Düsseldorf (2009)

Moon, H.; Tien, C.: Further Work on Multicomponent Adsorption Equilibria Calculations based on the Ideal Adsorbed Solution Theory, Ind. Chem. Res. 26 (1987), p. 2042

Münstermann, U.: Adsorption von CO_2 aus Reingas und aus Luft am Molekularsieb-Einzelkorn und im Festbett, Berücksichtigum von Würmeeffekten und H_2O-Prä adsorption, Ph.D. Thesis TU München 1984

Myers, A. L.; Prausnitz, J. M.: Thermodynamics of Mixed-Gas Adsorption, AIChE J. 11 (1965), p. 121/127

O'Brian, J.; Myers, A. L.: Physical Adsorption of Gases on Heterogenous Surfaces, J. Chem. Soc., Faraday Trans. 1 (1984) 80, p. 1467/1477

O'Brian, J.; Myers, A. L.: Rapid Calculations of Multicomponent Adsorption Equilibria from Pure Isotherm data, Ind. Eng. Chem. Process Des. Dev. 24 (1985), p. 1188/1191

Polaczkowa, W. et al.: Preparatyka organiczna, Panstwowe Wydawnictwo Techniczne, Warschau (1954)

Quesel, U.: Untersuchungen zum Ein- und Mehrkomponentengleichgewichten bei der Adsorption von Lösungsmitteldämpfen und Gasen an Aktivkohle und Zeolithen, Ph.D. Thesis, TH Darmstadt (1995)

Reid, R. C. et al.: The Properties of Gases and Liquids, Singapore; McGraw-Hill Book Co. (1988)

Reinhold, H.; Kircheisen, M.: J. Prakt. Chem. 113 (1926), p. 203 and p. 351

Renon, H.; Prausnitz, J. M.: Local compositions in thermodynamic excess functions for liquid mixtures, J. Am. Inst. Chem. Eng 14 (1968), p. 135/144

Rittner; Steiner: Die Schmelzkristallisation von organischen Stoffen und ihre großtechnische Anwendung, Chem.-Ing.-Techn. 57 (1985) 2, p. 91/102

Roseboom, H. W. B.: Z. Phys. Chem, Leipzig, 10 (1982), p. 145

Rudzinsky, W.; Everett, D. H.: Adsorption of Gases on Heterogenous Surfaces, Academic Press, London (1992)

Sakuth, M.: Messung und Modellierung binärer Adsorptionsgleichgewichte, PhD Thesis, Universität Oldenburg (1993)

Schildknecht, H.: Zonenschmelzen, Verlag Chemie, Weinheim (1964)

Schweighhart, P. J.: Adsorption mehrerer komponenten in biporösen Adsorbentien, Ph.D. Thesis, TU München (1994)

Stephan, P.; Schaber, K.; Stephan, K.; Mayinger, F.; Thermodynamik - Grundlagen und technische Anwendungen Vol. 2: Mehrstoffsysteme und chemische Reaktionen, 15th Ed., Springer-Verlag Berlin Heidelberg , 2010.

Sievers, W.: Über das Gleichgewicht der Adsorption in Anlagen zur Wasserstoffgewinnung, Ph.D. Thesis, TU München (1993)

Szepesy, L.; Illes, V.: Adsorption of Gases and Gas Mixtures, II. Measurement of the Adsorption Isotherms of Gases on Active Carbon under pressures of 1 to 7 atm, Acta Chim. Hung. 35 (1963), p. 53/60

Thiesen: Z. phys. Chem. 107 (1923), p. 65/73, cit. in Große, L.: Arbeitsmappe für Mineralölingenieure, J2, VDI-Verlag Düsseldorf

van Laar, J. J.: Die Thermodynamik einheitlicher und binärer Gemische, Groningen (1935), cit. in (Schmidt et al. 1977), p. 144

Wilson, G. M.: J. Amer. Soc. Chem. 86 (1964), p. 127, cit. in (Schmidt et al. 1977), p. 144

Watson: Ind. End. Chem. 35 (1943), p. 398, cit. in (Perry und Chilton 1973), p. 239

Chapter 3

General

Anderson, B.: Pressure drop in ideal fluidization, Chem. Eng. Sci. 15 (1961), p. 276/297

Batchelor, G. K.: The Theory of Homogeneous Turbulence, Cambridge University Press 1953

Brodkey, R. S.: Turbulence in Mixing Operations. Theory and Application to Mixing Reaction, Academic Press, New York 1975

Corrsin, S.: The Isotropic Turbulent Mixer: Part II. Arbitrary Schmidt Number, AIChE J. 10 (1964) No. 6, p. 870/877

Hinze, J. O.: Turbulence, McGraw Hill, New York 1975

Kolmogoroff, A. N.: Sammelband zur statistischen Theorie der Turbulenz, Akademie-Verlag, Berlin 1958

Special

Brauer, H.: Strömung und Wärmeübergang bei Rieselfilmen, VDI-Forsch.-Heft 457, VDI-Verlag, Düsseldorf 1956

Feind, K.: Strömungsuntersuchungen bei Gegenstrom von Rieselfilmen und Gas in lotrechten Rohren, VDI-Forsch.-Heft 481.VDI-Verlag Düsseldorf 1960

Geldart, T.: Types of Gas Fluidization, Powder Technol. 7 (1973), p. 285/292

Haas, U.; Schmidt-Traub, H.; Brauer, H.: Umströmung kugelförmiger Blasen mit innerer Zirkulation, Chem.-Ing.-Tech. 44 (1972), p. 1060/1067

Hadamard, J. M.: Mouvement permanent lent d'une liquide et visqueuse dans un liquide visqueux, Compt. Rend. Acad. Sci. Paris 152 (1911), p. 1735 /1738

Mersmann, A.: Auslegung und Maßstabsvergrößerung von Blasen- und Tropfensäulen, Chem.-Ing. Techn. 49 (1977), p. 679/691

Mersmann, A.; Einenkel, W.-D.; Käppel, M.: Auslegung und Maßstabvergrößerung von Rührapparaten, Chem.-Ing.-Techn. 47 (1975), p. 953/964

Mersmann, A.: Volumenanteil der dispersen Phase in Sprühkolonnen sowie Blasen- und Tropfensäulen, Verf. Techn. 12 (1978), p. 426/429

Mersmann, A.; Grossmann, H.: Dispergieren im flüssigen Zweiphasensystem. Chem.-Ing.-Techn. 52 (1980), p. 621

Mersmann, A.: Stoffübertragung, Springer Verlag, Berlin 1986

Mersmann, A.; Werner, F.; Maurer, S.; Bartosch, K.: Theoretical prediction of the minimum stirrer speed in mechanically agitated suspensions, Chem. Eng. Process. 37 (1998), p. 503/510

Molerus, O.: Interpretation of Geldart's Type A, B, C and D Powders by Taking into Account Interparticle Cohesion Forces, Powder Technol. 33 (1982), p. 81/87

Nusselt, W.: Die Oberflächenkondensation des Wasserdampfes. VDI-Z. 60 (1916), p. 541/569, cit. in: VDI-Wärmeatlas, 2. Aufl. p. Ja 1, VDI-Verlag, Düsseldorf 1974

Reh, L.: Das Wirbeln von körnigem Gut im schlanken Diffusor als Grenzzustand zwischen Wirbelschicht und pneumatischer Förderung, Ph.D. Thesis TH Karlsruhe 1961

Richardson, J. F., Zaki, W. N.: Sedimentation and Fluidization, Part I, Trans. Inst. Chem. Eng. 32 (1954), p. 35/5

Ruszkowski, S.: A Rational Method for Measuring Blending Performance and Comparison of Different Impeller Types, Proc. Eighth Europ. Conf. on Mixing, Ichem E Symp. Series No. 136, p. 283/291 (1994)

Rybczynski, W.: Über die fortschreitende Bewegung einer flüssigen Kugel in einem zähen Medium, Bull. Int. Acad. Sci. Cracovic A (1911), p. 40/46

Schäfer, M.: Charakterisierung, Auslegung und Verbesserung des Makro- und Mikromischens in gerührten Behältern, Ph.D. Thesis Univ. Erlangen-Nürnberg 2001

Voit, H.; Zeppenfeld, R.; Mersmann, A.: Calculation of Primary Bubble Volume in Gravitational and Centrifugal Fields, Chem. Eng. Technol. 10 (1987), p. 99/103

Werner, F. C.: Über die Turbulenz in gerührten newtonschen und nicht-newtonschen Fluiden, Ph.D. Thesis TU München 1997

Zehner, P.; Benfer, R.: Modelling fluid dynamics in multiphase reactors, Chem. Engng. Sci. 51 (1996) 6, pp.1735/1744

Chapter 4

General

Baehr, H. D.; Stephan, K.: Wärme- und Stoffübertragung, Springer (2010)

Bird, R. B.; Stewart, W. E.; Lightfoot, E. N.: Transport Phenomena, 2nd ed., Wiley (2002)

Cussler, E.: Diffusion -Mass transfer in fluid systems, Cambridge University Press (1997)

Gröber, H.; Erk, S.; Grigull, U.: Die Grundgesetze der Wärmeübertragung, 3. Aufl., Springer (1981)

Hausen, H.: Wärmeübertragung im Gegenstrom, Gleichstrom und Kreuzstrom, Springer (1976)

Mersmann, A.: Stoffübertragung, Springer (1986)

Schlünder, E. U.; Martin, H.: Einführung in die Wärmeübertragung, 8. Aufl., Vieweg (1995)

Schlünder, E. U.: Einführung in die Stoffübertragung, 2. Aufl., Vieweg (1996)

Taylor, R.; Krishna, R.: Multicomponent mass transfer, Wiley series in chemical engineering, Wiley (1993)

VDI-Heat Atlas, Ed.: Verein Deutscher Ingenieure: VDI-Gesellschaft Verfahrenstechnik und Chemieingenieurwesen (GVC), 2nd ed. (2010)

Wesselingh, J. A.; Krishna, R.: Mass transfer, 1. publ, Horwood New-York, (1990)

Special

Bandel, J.; Schlünder, E. U.: Pressure drop and heat transfer by vaporization of boiling refrigerants in a horizontal pipe, Int. Symp. Recent Developments in Heat Exchangers, Trogir (1972)

Chawla, J. M.: Wärmeübergang und Druckabfall in waagrechten Rohren bei der Strömung von verdampfenden Kältemitteln, VDI-Fortschr.-Heft No. 523 (1967)

Chawla, J. M.: Impuls- und Wärmeübertragung bei der Strömung von Flüssigkeits/Dampf-Gemischen, Chem. Ing. Tech. 44 (1972), p. 118/120

Haffner, H.: Wärmeübergang an Kältemittel bei Phasenverdampfung, Filmverdampfung und überkritischem Zustand des Fluids, Bundesminist. f. Bildg. & Wiss. Bonn (1970), p. 70/24, cit. in VDI-Wärmeatlas Ha 1

Hausen, H.: Neue Gleichung für die Wärmeübertragung bei freier oder erzwungener Strömung, Allg. Wärmetech. 9 (1959), p. 75/79

Kast, W.; Krischer, O.; Reinicke, H.; Wintermantel, K.: Konvektive Wärme- und Stoffübertragung, Springer (1974)

Krischer, O.; Kast, W.: Die wissenschaftlichen Grundlagen der Trocknungstechnik, 3. Aufl. Springer (1978)

Kutateladse, S. S.: Kritische Wärmestromdichte bei einer unterkühlten Flüssigkeitsströmung (russ). Energetika 7 (1959) p. 229/239, cit. in VDI-Wärmeatlas Ha 2

Mersmann, A.; Wunder, R.: Heat transfer between a dispersed system and a vertical heating or cooling surface, 6. Int. Heat Transfer Conf. Toronto, Nat. Res. Counc. Canada Vol. 4 (1978), p. 31/36

Nusselt, W.: Die Oberflächenkondensation des Wasserdampfes, VDI-Z. 60 (1916), p. 541/546 und p. 569/575, cit. in VDI-Wärmeatlas A 32

Pohlhausen, E.: Der Wärmeaustausch zwischen festen Körpern und Flüssigkeiten mit kleiner Reibung und kleiner Wärmeleitung, Z. ang. Math. Mech. 1 (1921), p. 115/121

Saunders, O. H.: The effect of pressure upon natural convection in air, Proc. Roy. Soc. London (A) 157 (1936), p. 278/291

Saunders, O. H.: Natural convection in liquids, Proc. Roy. Soc. London (A) 172 (1939), p. 55/71

Schmidt, E.; Beckmann, E.: Das Temperatur- und Geschwindigkeitsfeld vor einer wärmeabgebenden senkrechten Platte bei natürlicher Konvektion, Tech. Mech. Thermodyn. 1 (1930), p. 341/349

Schlünder, E. U.: Über eine zusammenfassende Darstellung der Grundgesetze des konvektiven Wärmeüberganges, Verf.-Tech. (1970), p. 11/60

Schlünder, E. U.: Einführung in die Wärme- und Stoffübertragung, Vieweg (1972)

Schlünder, E. U.: Einführung in die Stoffübertragung, Vieweg (1996)

Stephan, K.: Beitrag zur Thermodynamik des Wärmeüberganges beim Sieden, Abh. deutsche Kältetech. Verein, No. 18, Karlsruhe, Müller (1964); see also Chem. Ing. Tech. 35 (1963), p. 775/784, cit. in VDI-Wärmeatlas Ha 1

Stephan, K.: Wärmeübergang und Druckabfall bei nicht ausgebildeter Laminarströmung in Rohren und ebenen Spalten, Chem.-Ing.-Tech. 31 (1959), p. 773/778

Wesselingh , J. A.; Bollen, A. M.; Single Particles , Bubbles and Drops: Their Velocities and Mass Transfer Coefficients, Trans J Chem E, Vol 77, Part A, March (1999) 89-96

Chapter 5

Billet, R.; Mackowiak. J.: How to use absorption data for design and scale-up of packed columns, Fette, Seifen, Anstrichmittel 86 (1984) No.9, p. 349/358

Billet, R.; Schultes, M.: Fluid Dynamics and Mass Transfer in the Total Capacity Range of Packed Columns up to the Flood Point, Chem. Eng. Technol. 18 (1995), p. 371 /379

Billet, R.; Schultes, M.: Prediction of Mass Transfer Columns with Dumped and Arranged Packings, Trans IChemE 77 (1999), Part A, p. 498/504

Bravo, J. L.; Fair, J. R.: Generalized Correlation for Mass Transfer in Packed Distillation Columns, Ind. Eng. Chem. Process Des. Dev. 21 (1982) No. 1, p. 162/170

Engel, V.: Fluiddynamik in Packungskolonnen für Gas-Flüssig-Systeme, Fortschritt-Berichte VDI, Reihe 3, Verfahrenstechnik No. 605, VDI-Verlag, Düsseldorf 1999

Fair, J. R.: How to predict sieve tray entrainment and flooding, Petro/Chem. Eng. 33 (1961), No. 10, p. 45/52

Fair, J. R.: Distillation: King in Separations, Chem. Processing (1990) No. 9, p. 23/31

Fair, J. R.: Gas Absorption and Gas-Liquid System Design, in: Perry, R. H; Green, D. W.: Maloney, J. D.: Perry's Chemical Engineers' Handbook, McGraw-Hill, New York 1997, p. 14-1/14-98

Frey, T.; Nierlich, F.; Pöpken, T.; Reusch, D.; Stichlmair, J.; Tuchlenski, A.: Application of Reactive Distillation and Strategies in Process Design, in: Sundmacher, K.; Kienle, A.: Reactive Distillation, Wiley-VCH, Weinheim 2003

Frey, Th.; Stichlmair, J.: Thermodynamische Grundlagen der Reaktivdestillation, Chem.-Ing.-Tech. 70 (1998) No. 11, p. 1373/1381

Gmehling, J.; Onken, U.: Vapor-Liquid Equilibrium Data Collection, Chemistry Data Series, Vol. I ff, Dechema, Frankfurt 1977 ff.

Gmehling, J.; Menke, J.; Krafczyk, J.; Fischer, K.: Azeotropic Data I and II, Wiley-VCH Weinheim 2004

Hoek, P. J.; Wesselingh, J. A.; Zuiderweg, F. J.: Small Scale and Large Scale Liquid Maldistribution in Packed Columns, Chem. Eng. Res. Des. 64 (1986) p. 431/449

Kammermaier, F.: Neuartige Einbauten zur Unterdrückung der Maldistribution in Packungskolonnen, VDI Verlag Reihe 3, No. 892 (2008)

Kirschbaum, E.: Destillier and Rektifiziertechnik, 4. Aufl. Berlin, Heidelberg, New York, Springer 1969

Kister, H. Z.: Distillation Operation, McGraw-Hill, New York 1989

Kister, H. Z.: Distillation Design, McGraw-Hill, New York 1992

Klöcker, M.; Kenig, E. Y.; Hoffmann, A.; Kreis, P.; Gorak, A.: Rate-based modeling and simulation of reactive separations in gas/vapor-liquid systems, Chemical Engineering and Processing 44 (2005) p. 617/629

Kolev, N.: Packed Bed Columns. For absorption, desorption, rectification and direct heat transfer, Elsevier Amsterdam 2006

Krishna, R.: Hardware Selection and Design Aspects for Reactive Distillation Columns, in: Sundmacher, K.; Kienle, A.: Reactive Distillation, Wiley-VCH, Weinheim 2003

Landolt-Börnstein, IV, 24c, Springer Verlag, Berlin 1980

Lewis, W. K.: Rectification of binary mixtures, Ind. Eng. Chem. 28 (1936), No. 4, p. 399/402

Lockett, M.-J.: Distillation Tray Fundamentals, Cambridge University Press, Cambridge UK 1986

Mersmann, A.: Wann werden alle Löcher einer Siebbodenplatte durchströmt? Chem.-Ing.-Tech. 35 (1963) No.2,p. 103/107

Mersmann, A.: Zur Berechnung des Flutpunktes in Füllkörperschüttungen, Chem.-Ing.-Tech. 37 (1965), p. 218/226

Mersmann, A.: Thermische Verfahrenstechnik, Springer-Verlag, Berlin 1980

Mersmann, A.; Bornhütter, K.: Druckverlust und Flutpunkt in berieselten Packungen, VDI wärmeatlas. VDI-Verlag Düsseldorf 2002

Mersmann, A.; Kind, M.; stichlmair, J.: Thermische Verfahrenstechnik, Grundlagen und Methoden, Springer-Verlag Berlin Heidelberg 2005

Naphtali, L. M.; Sandholm, D. P.: Multicomponent Separation Calculations by Linearization, AIChE J., 17 (1971), p. 148/153

Onda, K.; Takeuchi, H.; Okumoto, Y.: Mass Transfer Coefficients Between Gas and Liquid Phases in Packed Columns, Journal of Chemical Engineering of Japan 1 (1968), p. 56/62

Prausnitz, J. M.; Lichtenthaler, R. N.; deAzevdeo, E. G.: Molecular Thermodynamics of Fluid Phase Equilibria, Prentice-Hall Englewood Cliffs 1998

Perry, R. H; Green, D. W.; Maloney, J. D.: Perry's Chemical Engineers' Handbook, McGraw-Hill, New York 1997

Perry, R. H.; Green, D. W.: Perry's Chemical Engineer's Handbook, McGraw-Hill, New York 2008

Ruff, K.; Pilhofer, Th.; Mersmann, A.: Vollständige Durchströmung von Lochböden bei der Fluid-Dispergierung, Chem.-Ing.-Tech. 48 (1976) No. 9, p. 759/764

Schmidt, R.: The Lower Capacity Limit of Packed Columns, Instit. Chem. Eng., Symp. Ser. 56 (1979) No. 2, p. 3.1/1-14

Souders, M.; Brown, G. G.: Design of Fractionating Columns, Ind. Eng. Chem. 26 (1934) No. 1, p. 98/1

Spiegel, L.: Packungen in der Prozessindustrie, Process 4 (1994) p. 39/44

Stichlmair, J.: Grundlagen der Dimensionierung des Gas/Flüssigkeit-Kontaktapparates Bodenkolonne, Verlag Chemie, Weinheim (1978)

Stichlmair, J.; Bravo, L. J.; Fair, J. R.: General Model for Prediction of Pressure Drops and Capacity of Countercurrent Gas/Liquid Packed Collumns, Gas Separation & Purification 3 (1989) No. 3, p. 19/28

Stichlmair, J. G.; Fair, J. R.: Distillation - Principles and Practice, Wiley-VCH New York 1998

Stichlmair, J.; Frey, Th.: Prozesse der Reaktivdestillation, Chem.-Ing.-Tech. 70 (1998) No. 12, p. 1507/1516

Strigle, R. F.: Random Packings and Packed Towers, Design and Applications, Gulf Publishing Company, Houston 1987

Sulzer: Prospekt No. d/22.13.06-v.82-45, Winterthur, Schweiz 1982

Sundmacher, K.; Kienle, A.: Reactive Distillation, Wiley-VCH, Weinheim 2003

Taylor, R.; Krishna, R.: Mulitcomponent Mass Transfer, John Wiley and Sons, New York 1993

Taylor, R.; Kooijman, H. A.; Hung, J.-S.: A Second Generation Non-equilibrium Model for Computer Simulation of Multi-component Separation Processes, Comp Chem. Eng. 18 (1994) p. 205/217

Chapter 6

Angelo, J. B.; Lightfoot, E. N.; Howard, D. W.: Generalization of the penetration theory for surface stretch: application to forming and oscillating drops, AIChE J. 12 (1966), No. 4, p. 751/760

Berger, R.: Einflüsse auf das Verhalten bei der Phasentrennung von nichtlöslichen Systemen organisch-wässrig, Lecture at RWTH Aachen 1987

Blaß, E.; Goldmann, G.; Hirschmann, K.; Milhailowitsch, P.; Pietzsch, W.: Fortschritte auf dem Gebiet der Flüssig/Flüssig-Extraktion, Chem.-Ing.-Tech. 57 (1985), No. 7, p. 565/581

Blaß, E.: Engineering Design and Calculation of Extractors for Liquid-Liquid Systems, in: Rydberg, J.; Musikas, J.; Choppin, G. R.: Principles and Practices of Solvent Extraction, Marcel Dekker, New York 1992, p. 267/313

Blaß, E.; Göttert, W.; Hampe, M. J.: Selection of Extractors and Solvents, in: Godfrey, J. C.; Slater, M. J.: Liquid-Liquid Extraction Equipment, John Wiley & Sons, New York 1994, p. 737/767

Brauer, H.: Unsteady state mass transfer through the interface of spherical particles, Int. J. Heat Mass Transfer 21 (1978) p. 445/453, 455/465

Clift, R.; Grace, J. R.; Weber, M. E.: Bubbles, drops and particles, Academic Press, New York 1978

Fischer, A.: Die Tropfengröße in Flüssig-Flüssig-Rührextraktionskolonnen, Verfahrenstechnik 5 (1971), No. 9, p. 360/365

Garthe, D.: Fluiddynamics and Mass Transfer of Single Particles and Swarms of Particles in Extraction Columns, Ph.D. Thesis TU München 2006

Godfrey, J. C.; Houlton, D. A.; Marley, S. T.; Marrochelli, A.; Slater, M. J.: Continuous phase axial mixing in pulsed sieve plate liquid-liquid extraction columns, Chem. Eng. Res. Des. 66 (1988), p. 445/457

Godfrey, J. C.; Slater, M. J.: Liquid-Liquid Extraction Equipment, John Wiley & Sons, New York 1994

Hampe, M.: Flüssig/flüssig-Extraktion: Einsatzgebiete und Lösungsmittelauswahl, Chem.-Ing.-Tech. 50 (1978), p. 647/655

Handlos, A. E.; Baron, T.: Mass and heat transfer from drops in liquid-liquid extraction, AIChE J. 3 (1957), p. 127/136

Hansen, C. M.: The Universality of the Solubility Parameter, Ind. Chem. Prod. Res. Dev. 8 (1969), p. 2/11

Haverland, H.: Untersuchungen zur Tropfendispergierung in flüssigkeitspulsierten Siebboden-Extraktionskolonnen, Ph.D. Thesis TU Clausthal 1988

Henschke, M.: Dimensionierung liegender Flüssig-flüssig-Abscheider anhand diskontinuierlicher Absetzversuche, Fortschritt-Berichte VDI, Reihe 3, Verfahrenstechnik, No. 379, VDI-Verlag, Düsseldorf 1995

Hildebrand, J. H.; Scott, R. L.: Regular solutions, Englewood Cliffs, Prentice Hall 1962

Hoting, B.: Untersuchungen zur Fluiddynamik und Stoffübertragung in Extraktionskolonnen mit strukturierten Packungen, Fortschritt-Berichte VDI, Reihe 3: Verfahrenstechnik, No. 439, VDI-Verlag, Düsseldorf 1996

Kronig, R.; Brink, J.: On the theory of extraction from falling droplets, Appl. Sci. Res. A2 (1950), p. 142/154

Mackowiak, J.: Grenzbelastung von unpulsierten Füllkörperkolonnen bei der Flüssig/Flüssig-Extraktion, Chem.-Ing.-Tech. 65 (1993) No. 4, p. 423/429

Mersmann, A.: Zum Flutpunkt in Flüssig/Flüssig-Gegenstromkolonnen, Chem.-Ing.-Tech. 52 (1980) No. 12, p. 933/942

Mersmann, A.: Stoffübertragung, Springer-Verlag, Berlin 1986, p.151/154

Miyauchi, T.; Vermeulen, T.: Longitudinal Dispersion in Two-Phase Continuous-Flow Operations, Ind. Eng. Fund. 2 (1963), p. 133

Newman, A. B.: The drying of porous solids: diffusion and surface emission equations, Trans. AIChE J. 27 (1931), p. 203/220

Olander, D. R.: The Handlos-Baron drop extraction model, AIChE J., 12 (1966) 5, p. 1018/1019

Pilhofer, Th.: Habilitationsschrift, TU München 1978

Postl, J.; Marr, R.: Lecture at GVC-Fachausschuss „Thermische Zerlegung von Gas- und Flüssigkeitsgemischen" on 8./9.5.1980 in Mersburg

Qi, M.: Untersuchungen zum Stoffaustausch am Einzeltropfen in flüssigkeitspulsierten Siebboden-Extraktionskolonnen, Ph.D. Thesis TU Clausthal 1992

Richardson, J. F; Zaki, W. N.: Sedimentation and Fluidization, Part I, Trans. Inst., Chem. Eng. 32 (1954), p. 35/53

Rydberg, J.; Musikas, J.; Choppin, G.R.: Principles and Practices of Solvent Extraction, Marcel Dekker, New York 1992

Sorensen, J. M.; Arlt, W.: Liquid-Liquid Equilibrium Data Collection, Chemistry Data Series Vol. 5, Parts 1-4, Dechema, Frankfurt 1980ff

Steiner, L.: Mass-transfer rates from single drops and drop swarms, Chem. Ing. Sci. 41 (1986), p. 1979/1986

Stichlmair, J.: Leistungs- und Kostenvergleich verschiedener Apparatebauarten für die Flüssig/Flüssig-Extraktion, Chem.-Ing.-Tech. 52 (1980) No. 3, p. 253/255

Stichlmair, J.; Steude, H. E.: Abtrennung und Rückgewinnung von Stoffen aus Abwasser und Abfallflüssigkeiten, in: Stofftrennverfahren in der Umwelttechnik, GVC.VDI-Gesellschaft Verfahrenstechnik und Chemieingenieurwesen, Düsseldorf 1990, p. 175/205

Wagner, I.: Der Einfluss der Viskosität auf den Stoffübergang in Flüssig-flüssig-Extraktionskolonnen, Ph.D. Thesis TU München 1999

Widmer, F.: Tropfengröße, Tropfenverhalten und Stoffaustausch in pulsierten Füllkörper-Extraktionskolonnen, Chem.-Ing.-Tech. 39 (1967), No. 15, p. 900/906

Wolf, S.: Phasengenzkonvektionen beim Stoffübergang in Flüssig-flüssig-Systemen, Fortschritt-Berichte VDI, Reihe 3: Verfahrenstechnik, No. 584, VDI-Verlag, Düsseldorf 1999

Chapter 7

General

Baehr, H. D.; Stephan K.: Wärme- und Stoffübertragung, Springer (2004)

Billet, R.: Evaporation technology -principles applications economics, VCH Weinheim (1989)

Bosnjakovic, F.: Technische Thermodynamik, Teil 2, 4. Aufl., Steinkopff Dresden (1965)

Hsu, Y. Y.; Graham, R. W.: Transport processes in boiling and two-phase systems: including near-critical fluids, Hemisphere Publ. Cor. Washington (1976)

Müller-Steinhagen, H. (editor): Handbook heat exchanger fouling: mitigation and cleaning technologies, Publico Publ. Essen (2000)

Rant, Z.: Verdampfen in Theorie und Praxis, Aarau Sauerländer (1977)

Schlünder, E. U.; Martin, H.: Einführung in die Wärmeübertragung -für Maschinenbauer, Verfahrenstechniker, Chemie-Ingenieure, Physiker, Biologen und Chemiker ab dem 4. Semester, 8. Aufl., Vieweg (1995)

Ullmann's encyclopedia of industrial chemistry (ed. advisory board: Bohnet, M.), completely revised ed., Wiley-VCH Weinheim (2002)

VDI-Wärmeatlas: Berechnungsblätter für den Wärmeübergang / Hrsg.: Verein Deutscher Ingenieure; VDI-Gesellschaft Verfahrenstechnik und Chemieingenieurwesen (GVC), 9. Aufl., Springer (2002)

Special

Arneth, S.: Dimensionierung und Betriebsverhalten von Naturumlaufverdampfern, Hieronymus Verlag, München, 1999

Arneth, S.; Stichlmair, J.: Characteristics of thermosiphon reboilers, International Journal of Thermal Sciences (2001) Vol. 40, p. 392/399

Bohnet, M.: Influence of the Transport Properties of the Crystal/Heat Transfer Surface Interfacial on Fouling Behavior, Chemical Engineering and Technology, Band 26, Heft 10 (2003), p. 1055/1060

Fair, J. R.: What You Need to Design Thermosiphon Reboilers, Petroleum Refiner, Vol. 39, No. 2, p. 105/123

Ford, J. D.; Missen, R. W.: On the conditions for stability of falling films subject to surface tension disturbance, the condensation of binary vapors, Canad. J. Chem. Eng. 46 (1968), p. 308/312

Wettermann, M.; Steiner, D.: Flow boiling heat transfer characteristics of wide-boiling mixtures, International Journal of Thermal Sciences, Band 39 (2000) 2, p. 225/235

Zuiderweg, F. F.; Harmens, H.: The influence of surface phenomena in the performance of destillation colums, Chem. Eng. Sci. 9 (1958), p. 89/103

Chapter 8

General

Arkenbout, G.: Melt crystallization technology, Lancaster Techn. publ. (1995)

Garside, J.; Mersmann, A.; Nyvlt, J.: Measurement of Crystal Growth and Nucleation Rates, 2nd ed., I ChemE (Institution of Chemical Engineers) Rugby (2002)

Hofmann, G.: Kristallisation in der industriellen Praxis, Wiley-VCH Weinheim (2004)

Israelachvili, J. N.: Intermolecular and surface forces, Academic press (1995)

Lacmann, R.; Herden, A.; Meyer, C.: Review -Kinetics of nucleation and crystal growth, Chem. Eng. Technol. 22 (1999) 4, p. 279/289

Lyklema, J.: Fundamentals of Interface an Colloid Science, Vol. I & II, Academic Press (1991)

Mersmann, A.: Crystallization Technology Handbook, 2nd ed., Marcel Dekker, New York (2001)

Mullin, J. W.: Crystallization, 4th ed. Oxford, Butterworth (2004)

Myerson, A. S. (ed): Handbook of Industrial Crystallization, Butterworth-Heinemann Boston (1993)

Randolph, A. D.; Larson, M. A.: Theory of Particulate Processes, 2nd ed., Academic Press, San Diego (1988)

Söhnel, O.; Garside, J.: Precipitation -basic principles and industrial applications, : Butterworth-Heinemann Oxford (1992)

Tadros, T. F. (ed): Solid/Liquid Dispersions, Academic Press (1987)

Ulrich, J.; Glade, H.: Melt Crystallization -Fundamentals Equipment and Applications, Shaker Aachen (2003)

Wöhlk, W.; Hofmann, G.: Types of crystallizers, Int. Chem. Eng. 24 (1984), p. 419/431

Special

Becker, R.; Döring, W.: Kinetische Behandlung der Keimbildung in übersättigten Dämpfen, Ann. Phys. 24 (5) (1935), p. 719

Berthoud, A.: Théorie de la formation des fases d'un cristal, J. Chim. Phys. 10 (1912), p. 624

Burton, J. A.; Prim, R. C.; Slichter, W. P.: Distribution of solute in crystals grown from the melt, J. Chem. Phys. 21 (1953), p. 1987

Burton, W. K.; Cabrera, N.; Frank, F. C.: The growth of crystals and the equilibrium structure of their surface, Phil. Trans. Roy. Soc. London, 243 (1951), p. 299

David, R.; Villermaux, J.: Crystallization and precipitation engineering III. A discrete formulation of the agglomeration rate of crystals in a crystallization process, Chem. Eng. Sci. 46 (1) (1991), p. 205/213

Derjaguin, B. V.; Landau, L.: Theory of the stability of strongly charged lyophobic solid and of the adhesion of strongly charged particles in solutions of electrolytes, Acta Physiochim. URRS 14 (1941), p. 633/622

Gahn, C.; Mersmann, A.: Brittle fracture in crystallization processes, Part A, Attrition and abrasion of brittle solids, Chem. Eng. Sci. 54 (1999), p. 1273

Fleischmann, W.; Mersmann, A.: Drowning out crystallization of sodium sulphate using methanol, in Proc. 9th Symp. on Industrial Crystallization (S. J. Janic and E. J. de Jong. eds.) Elsevier, Amsterdam (1984)

Hulburt, H. M.; Katz, S.: Some problems in particle technology, Chem. Eng. Sci. 19 (1964), p. 555

Hounslow, M. J.: Nucleation, growth and aggregation rates from steady-state experimental data, AICHE J. 36 (1990), p. 1748

Judat, B.; Kind, M.: Morphology and internal structure of barium sulfate-derivation of a new growth mechanism, Journal of Colloid and Interface Science, Band 269 (2004) Heft 2, p. 341/353

Kind, M.; Mersmann, A.: Methoden zur Berechnung der homogenen Keimbildung aus wässrigen Lösungen, Chem. Ing. Techn. 55(1983), p. 270

Kind, M.; Mersmann, A.: On Supersaturation during mass crystallization from solution, Chem. Eng. Technol. 13 (1990), p. 50

Liszi, I.; Liszi, J.: Salting out crystallization in KCl-water-methanol system, in Proc. 11th Symp. on Industrial Crystallization (A. Mersmann, ed) (1990), p. 515/520

Mersmann, A.: Calculation of interfacial tensions, J. Cryst. Growth, 102 (1990), p. 841

Mersmann, A.: General prediction of statistically mean growth rates of a crystal collective, J. Cryst. Growth 147 (1995), p. 181

Mersmann, A.; Löffelmann, M.: Crystallization and Precipitation: The optimal supersaturation, Chem. Ing. Techn. 71 (1999), p. 1240/1244

Nyvlt, J.; Rychly, R.; Gottfried, J.; Wurzelova, J.: Metastable zone width of some aqueous solutions, J. Cryst. Growth 6 (1970), p. 151

Orowan, E.: Fracture and strength of solids, Rep. Progr. Physics 12 (1949), p. 185/ 232

Schubert, H.: Principles of agglomeration, Int. Chem. Eng. 21 (1981), p. 336/377

Schubert, H.: Keimbildung bei der Kristallisation schwerlöslicher Stoffsysteme, Ph.D. Thesis Technische Universität München (1998)

Schwarzer, H. C.; Peukert, W.: Tailoring particle size through nanoparticle precipitation, Chemical Engineering Communications, Band 191 (2004) Heft 4, p. 580

Valeton, J. J. P.: Wachstum und Auflösung der Kristalle, Z. Kristallogr. 59, 135, 335 (1923); 60: p. 1 (1924)

Vervey, E. J. W.; Overbeek, J. T. G.: Theory of stability of lyophobic colloids, Elsevier, Amsterdam (1948)

Volmer, M.; Weber, A.: Keimbildung in übersättigten Lösungen, Z. Phys. Chem. 119 (1926), p. 277

Wintermantel, K.: Die effektive Trennwirkung beim Ausfrieren von Kristallschichten aus Schmelzen und Lösungen: Eine einheitliche Darstellung, Chem.-Ing.-Tech. 58 (1986), p. 498/499

Wintermantel, K.; Wellinghoff, G.: Layer Crystallization and Melt Solidification in Industrial Crystallization, ed. A. Mersmann, Marcel Dekker (2001)

Wirges, H. P.: Beeinflussung der Korngrößenverteilung bei Fällungskristallisationen, Chem. Ing. Techn. 58(7) (1986), p.586/587

Zacher, U.; Mersmann, A.: The influence of internal crystal perfection on growth rate dispersion in a continuous suspension cristallizer, J. Crystal Growth 147 (1995), p. 172

Chapter 9

General

Do, D. D.: Adsorption Analysis: Equilibria and Kinetics, Imperial College Press, London 1998

Guiochon, G.; Felinger, A.; Shirazi, D.G.; Katti, A.M.: Fundamentals of Preparative and Nonlinear Chromatography, 2nd ed Elsevier, Amsterdam 2006

Holland, C. D.; Liapis, A. I.: Computer Methods for Solving Dynamic Separation Problems, McGraw-Hill, New York 1983

Kast, W.: Adsorption aus der Gasphase, VCH Verlagsgesellschaft, Weinheim 1988

Mersmann, A.: Adsorption, in: Ullmann's Encyclopedia of Industrial Chemistry Vol. B3, VCH Verlagsgesellschaft, Weinheim 1988, p. 9-1/9-52

Mersmann, A.; Fill, B.; Maurer, S.: Single and Multicomponent Adsorption Equilibria of Gases, Fortschritt-Bericht VDI, Reihe 3, No. 735, VDI Verlag, Düsseldorf 2002

Mersmann, A.; Akgün, U.: A Priori Prediction of Adsorption Equilibria of Arbitrary Gases on Microporous Adsorbents, Fortschritt-Bericht VDI, Reihe 3, No. 906, VDI Verlag Düsseldorf 2009

Perry, R. H.; Green, W. D.: Chemical Engineers Handbook, 6th ed., McGraw Hill New York 1984

Rudzinski, W.; Everett, D. H.: Adsorption of Gases on Heterogeneous Surfaces, Academic Press, London 1992

Ruthven, D. M.: Principles of Adsorption and Adsorption Processes, John Wiley and Sons, New York 1984

Seidel-Morgenstern, A.; Kessler, L. C.; Kaspereit, M.: Neue Entwicklungen auf dem Gebiet der simulierten Gegenstromchromatographie, Chem.-Ing.- Techn. 80 (2008) No. 6, p. 726/740

Valenzuela, D. P.; Myers, A. L.: Adsorption Equilibrium Data Handbook, Prentice Hall, Englewood Cliffs, USA 1989

Special

Akgün, U.: Prediction of Adsorption Equilibria of Gases, Ph.D. Thesis TU München 2007

Brandani, F. et al.: Measurement of Adsorption Equilibria by the Zero Length Column (ZLC) Technique, Part A: Single-Component Systems. Industrial & engineering chemistry research 42(7), pp 1451/1461 (2003)

Fritz, W.: Konkurrierende Adsorption von zwei organischen Wasserinhaltsstoffen auf Aktivkohlekörnern, Ph.D. Thesis TH Karlsruhe 1978

Glueckauf, E.: Theory of Chromatography, Trans Faraday Soc. 51 (1955), p. 1540/1551

Hartmann, R.: Untersuchungen zum Stofftransport in porösen Adsorbentien, Ph.D. Thesis TU München 1996

Harlick, P. J. E.; Tezel, F. H.: Use of Concentration Pulse Chromatography for Determining binary Isotherms: comparison with Statistically Determined Binary Isotherms, Adsorption 9 (2003), p. 275/286

Harlick, P. J .E.; Tezel, F. H.: An Experimental Adsorption Screening Study for CO2 Removal from N2, Microporous and Mesoporous Materials 76 (2004) p. 71/79

Jüntgen, H.; Knoblauch, K.; Mayer, F.: Die Entschwefelung von Abgasen nach dem Verfahren der Bergbau-Forschung, gwf-gas/erdgas 113 (1972), p. 127/131

Kajszika, H. W.: Adsorptive Abluftreinigung und Lösungsmittelrückgewinnung durch Inertgasregenerierung, Ph.D. Thesis TU München 1998

Kaul, B. K.: Modern Version of Volumetric Apparatus for Measuring Gas-Solid Equilibrium Data, Ind. Eng. Res. 26 (1987), p. 928/933

Laukhuf, W. L. S.; Plan, K. C. A.: Adsorption of Carbon Dioxide, Acetylene, Ethane and Propylene on Charcoal Near Room Temperatures, J. Chem. Eng. Data 14 (1969), p. 48/51

Liaw, C. H. et al.: Kinetics of Fixed bed adsorption: a new solution, AICHEJ. 25, p. 376 (1979)

Markmann, B.: Zur Vorhersagbarkeit von Adsorptionsgleichgewichten mehrkomponentiger Gasgemische, Ph.D. Thesis TU München 1999

Maurer, S.: Prediction of Single Component Equilibria, Ph.D. Thesis TU München 1999

Mehler, C.: Charakterisierung heterogener Feststoffoberflächen durch Erweiterung eines dielektrischen Kontinuumsmodells, Ph.D. Thesis TU München 2003

Münstermann, U.: Adsorption von CO_2 aus Reingas und aus Luft am Molekularsieb-Einzelkorn und im Festbett, Berücksichtigung von Wärmeeffekten und H_2O-Präadsorption, Ph.D. Thesis TU München 1984

Noack, R.; Knoblauch, K.: Betriebserfahrungen mit dem Babcock-B.-F.-Verfahren zur Rauchgasentschwefelung, VDI-Bericht No. 267, VDI-Verlag, Düsseldorf 1976

Pan, C. J.; Basmadjian, D.: An Analysis of Adiabatic Sorption of Single Solutes in Fixed Beds: Pure Thermal Wave Formation and it's Practical Implications, Chem. Eng. Sci 25 (1970), p. 1653/1664

Polte, W. M.: Zur Ad- und Desorptionskinetik in bidispersen Adsorbentien, Ph.D. Thesis TU München 1987

Rosen, J. B.: General Numerical Solution of Solid Diffusion in Fixed Beds, Ind. Eng. Chem. 46 (1954) 8, p. 1590/1594

Ruhl, E.: Temperaturwechsel-, Druckwechsel- und Verdrängungssorptions-Verfahren, Chem.-Ing.-Techn. 43 (1971) 15, p. 870/876

Scholl, S.: Zur Sorptionskinetik physisorbierter Stoffe an festen Adsorbentien, Ph.D. Thesis TU München 1991

Schweighart, P. J.: Adsorption mehrerer Komponenten in biporösen Adsorbentien, Ph.D. Thesis TU München 1994

Sievers, W.: Über das Gleichgewicht der Adsorption in Anlagen zur Wasserstoffgewinnung, Ph.D. Thesis TU München 1993

Treybal, R. E.: Mass Transfer Operations, Mc Graw Hill, New York 1968

Chapter 10

General

Keey, R. B.: Drying Principles and Practice, Pergamon, New York 1972

Kneule, F.: Das Trocknen, 3. Aufl. Aarau: Sauerländer 1975

Krischer, O.; Kast, W.: Die wissenschaftlichen Grundlagen der Trocknungstechnik, 3. Aufl. Berlin, Heidelberg, New York: Springer 1978

Kröll, K.: Trockner und Trocknungsverfahren, 2. Aufl. Berlin, Heidelberg, New York: Springer 1978

Mujumdar A. S.: Handbook of Industrial Drying, Marcel Dekker, Inc., New York 1995

Perry, R. H.; Green D. (eds.) Perry's Chemical Engineer's Handbook, Mc Graw Hill (2008)

Tsotsas, E.; Metzger, Th.; Gnielinski, V.; Schlünder, E.-U.: Drying of Solids, in Ullmann's Encyclopedia of Industrial Chemistry, Wiley-VCH, 2010

Tsotsas, E.; Mujumdar A.; Modern Drying Technology, Vol. 1 - Computational Methods (2007), Vol. 2 - Experimental Methods (2009) and Vol. 3 - Product Quality and Formulation (2011) Wiley-VCH

Special

Görling, P.: Untersuchungen zur Aufklärung des Trocknungsverhaltens pflanzlicher Stoffe, VDI-Forsch. 458, Düsseldorf: VDI-Verlag 1956

Kamei, S.: Untersuchung über die Trocknung fester Stoffe. MEm. Coll. Eng. Kyoto Imp. Univ., VIII (1934) p. 42/64

Krischer, O.; Sommer, E.: Kapillare Flüssigkeitsleitung in porigen Stoffen bei Trocknungs- und Befeuchtungsvorgängen. Chem.-Ing.-Tech. 43 (1971) p. 967/974

Van Meel, D. A.: Adiabatic convection batch drying with recirculation of air. Chem. Eng. Sci. 9 (1958) p. 36/44

Chapter 11

Baldus, H.; Baumgärtner, K.; Knapp. H.; Streich, M.: Verflüssigung und Trennung von Gasen, in: Winnacker Küchler: Chemische Technologie, Vol. 3, Carl Hauser Verlag, München 1983, p. 567/650

Blass, E.: Entwicklung verfahrenstechnischer Prozesse, Springer Verlag, Berlin 1997

Brusis, D.: Synthesis and Optimisation of Thermal Separation Processes with MINLP Methods, Ph.D. Thesis TU München 2003

Doherty, M. F.; Malone, M. F.: Conceptual Design of Distillation Systems, McGraw-Hill, New York 2001

Douglas, J. M.: The Conceptual Design of Chemical Processes, McGraw-Hill, New York 1988

El Saie, A.; El Kafrawi, K.: Study of Operation Conditions for Three Large MSF Desalination Units, Desalination 73 (1989), p. 207/230

Frey, T.: Synthese und Optimierung von Reaktivrektifikationsprozessen, Ph.D. Thesis TU München 2001

Geriche, D.: Konzentrieren von Salpetersäure mit Schwefelsäure, Schott-Informtion, Schott GmbH, Mainz 1973

Greig, H. W.: An Ecomonomic Comparison of Two 2000 m³ per Day Desalination Plants, Desalination 64 (1987), p. 17/50

Kaibel, G.: Distillation Columns with Vertical Partitions, Chem. Eng. Technol. 10 (1987), p. 92/98

Kaibel, G.; Miller, C.; Watzdorf, R. von; Stroezel, M.; Jansen, H.: Industrieller Einsatz von Trennwandkolonnen und thermisch gekoppelten Destillationskolonnen, Chem.-Ing.-Tech. 75 (2003), p. 1165/1166

Linnhoff, B.: New Concepts in Thermodynamic for Better Chemical Process Design, Chem. Eng. Res. Dev., Vol. 61 (1983), p. 207/223

Linnhoff, B.; Sahdev, V.: Pinch Technology, in: Ullmanns Encyclopedia of Industrial Chemistry, Vol. B3, p. 13-1/13-6, VCH Verlagsgesellschaft, Weinheim 1988

Linnhoff, B.; Dhole, R. V.: Distillation Column Targets, Computers and Chemical Engineering Journal, Vol. 17 (1983), p. 549/560

Mersmann, A.; Kind, M.; Stichlmair, J.: Thermische Verfahrenstechnik - Grundlagen und Methoden, Springer, Berlin 2005

Meyers, R. A.: Handbook of Petroleum Refining Processes, McGraw-Hill, New York 1996

Rautenbach, R.; Albrecht, R.: Membrane Processes, Wiley, New York 1989

Schoenmakers, H.: Alternativen zur Aufarbeitung desorbierter Lösungsmittel, Chem.-Ing.-Tech. 56 (1984), No. 3, p. 250/251

Schweitzer, P. A.: Handbook of Separation Techniques for Chemical Engineers, McGraw-Hill, New York 1997

Seider, W. D.; Seader, J. D.; Levin, D. R.: Process Design Principles, John Wiley&Sons, Inc., New York 1999

Siirola, J. J.: Industrial Applications of Chemical Process Synthesis, Adv. Chem. Engng, 23 (1996), p. 1/62

Smith, R.: Chemical Process Design, McGraw-Hill, New York 1995

Stichlmair, J. G.; Fair, J. R.: Distillation - Principles and Practice, Wiley-VCH, New York 1998

Strigle, R. F.: Random Packings and Packed Towers, Design and Applications, Gulf Publishing Comp., Houston 1987

Sundmacher, K.; Kienle, A.: Reactive Distillation - Status and Further Directions, Wiley-VCH, Weinheim 2003

Weyd, M.; Richter, H.; Kühnert, J.-T.; Voigt, I.; Tusel, E.; Brüschke, H.: Effiziente Entwässerung von Ethanol durch Zeolithmembranen in Vierkanalgeometrie, Chemie Ingenieur Technik 82 (2010), No. 8, pp 1257/1260

Westphal, G.: Kombiniertes Adsorptions-Rektifikationsverfahren zur Trennung eines Flüssigkeitsgemisches, Deutsches Patent DE 3712291, 1987

Wunder, R.: Abtrennung und Rückgewinnung von anorganischen Stoffen durch Absorption, in: Stofftrennverfahren in der Umwelttechnik, GVC-Gesellschaft Verfahrenstechnik und Chemieingenieurwesen, Düsseldorf 1990

Index